ELECTRORHEOLOGICAL FLUIDS

ELECTRORHEOLOGICAL FLUIDS

MECHANISMS, PROPERTIES, TECHNOLOGY, AND APPLICATIONS

Editors

R. Tao
Southern Illinois University, Carbondale, Illinois

G. D. Roy
Office of Naval Research, Arlington, Virginia

World Scientific
Singapore • New Jersey • London • Hong Kong

Published by

World Scientific Publishing Co. Pte. Ltd.

P O Box 128, Farrer Road, Singapore 9128

USA office: Suite 1B, 1060 Main Street, River Edge, NJ 07661

UK office: 73 Lynton Mead, Totteridge, London N20 8DH

Proceedings of the Fourth International Conference on Electrorheological Fluids, held July 20 – 23, 1993, in Feldkirch, Austria

ISBN 981-02-1639-4

Printed in Singapore.

FOREWORD

The Office of Naval Research was pleased to cosponsor the Fourth International Conference on Electrorheological Fluids. The field of electrorheological fluids has come a long way since Willis Winslow's seminal experiments with mixtures of corn starch and mineral oil in 1947. The pace of innovation has been particularly rapid in the last five or six years. The Conference provided a timely assessment of the current state of the art. A variety of novel materials was reported that promises to expand the limits of performance of these fascinating fluids. Significant results were reported in the areas of properties and mechanisms. The agenda included an excellent balance between basic results and potential application. In that regard, it is a pleasure to note the significant level of industrial representation. It is also gratifying to note the breadth of the international participation with eighty participants from fifteen countries. All of these factors led to presentations and lively discussion that more than met our expectations for the conference. The Office of Naval Research wishes to express its gratitude to our cosponsors, Technikum Voralberg and Southern Illinois University at Carbondale. Special thanks are due to our host, the Music School Stella Matutina, Feldkirch, Austria, for providing superb support and a most inspiring environment.

Dr. Bruce B. Robinson
Director
Science Directorate
Office of Naval Research
Arlington, Virginia
November 22, 1993

PREFACE

The Fourth International Conference on Electrorheological (ER) Fluids took place at the right time when both the US Department of Energy and the Government of Japan released reports, assessing the great potential of ER fluids for future industrial and technological applications. The thrust areas in which research is needed were also identified.

The conference was held at the ancient city of Feldkirch, Austria, from July 20 to 23, 1993. A total of 58 papers were presented, which described the cutting edge of research in ER fluids, and covered materials technology, physical mechanisms, properties, and applications of ER fluids. Highly productive and scientifically stimulating discussions followed the presentation, and the state-of-art of technology was disseminated. Since the Third International conference on ER Fluids in 1991, significant progress has been made in both R&D and applications.

The conference was sponsored by the Office of Naval Research (USA), Technikum Vorarlberg (Austria), and Southern Illinois University at Carbondale (USA).

In spite of the current economic stagnation in Europe and Japan, about 80 people from 15 countries participated in the conference. The following countries participated: Austria, Belarus (Former Soviet Union), Canada, China, France, Germany, Greece, Italy, Japan, Hong Kong, Slovakia, South Korea, South Africa, United Kingdom, and the United States of America. In addition to the universities and national laboratories, major industries from all over the world were also represented. These included: Ford Motor Company, Exxon, Lord Corporation, Bridgestone, Asahi Chemical, Bayer AG, BASF AG, and Hoechst AG.

Our special thanks are due to

Dipl.-Ing. Markus Linhart, Technikum Vorarlberg,

Dr. Hubert Regner, Vorstand der Abteilung Wissenschaft und Weiterbildung, and

Dr. Maurice A. Wright, Director of Materials Technology Center, Southern Illinois University at Carbondale,

whose support and help were essential ingredients for the success of the Conference. We express our special appreciation to Mrs. Karen Palmer for her valuable assistance in coordinating the Conference.

Gabriel D. Roy
Office of Naval Research

Rongjia Tao
Southern Illinois University
November 1993

CONTENTS

Foreword . . . v

Preface . . . vii

I. MATERIALS TECHNOLOGY

The Polarization, Structuring and Rheology of ER Fluids . . . 3
K. M. Blackwood, H. Block, P. Rattray, G. Tsangaris, and D. N. Vorobiev

First Experiments on Magnetoelectrorheological Fluids (MERFs) . . . 22
W. I. Kordonsky, S. R. Gorodkin, and E. V. Medvedeva

Cryogenic Electrorheological Fluids . . . 37
R. N. Zitter, X. Zhang, T. J. Chen, and R. Tao

Controllable Fluids: The Temperature Dependence of Post-Yield Properties . . 43
K. D. Weiss and T. G. Duclos

High Dielectric Constant Particulate Materials For Electrorheological Fluids . 60
C. A. Randall, D. E. McCauley, C. P. Bowen, T. R. Shrout, and G. L. Messing

Electrorheological Fluids Based on Polyurethane Dispersions . . . 67
R. Bloodworth

Surfactant-activated Electrorheological Suspensions . . . 84
Y. D. Kim and D. J. Klingenberg

Yield Process of Electrorheological Fluid with Polyaniline Particle . . . 100
K. Koyama, K. Minagawa, T. Yoshida, N. Kuramoto, and K. Tanaka

II. PHYSICAL MECHANISM

Light Scattering Studies of a Model Electrorheological Fluid . . . 115
T. C. Halsey and J. E. Martin

Simulation of Solid Structure Formation in an Electrorheological Fluid . . . 129
R. Tao and Q. Jiang

A Conduction Model of Electrorheological Effect . . . 139
N. Felici, J. N. Foulc, and P. Atten

Theoretical Analysis of Field Induced Structures in ER and MR Fluids . . . 153
G. Bossis, H. Clercx, Y. Grasselli, and E. Lemaire

Field-Induced Structure of Confined Ferrofluid Emulsion 172
J. Liu, E. M. Lawrence, M. L. Ivey, G. A. Flores, J. Bibette, and J. Richard

The Evolution of Field-Induced Structure of Confined Ferrofluid Emulsions . 190
J. Liu, T. Mou, G. A. Flores, J. Bibette, and J. Richard

The Role of Suspension Structure in the Dynamic Response of Electrorheological Suspensions 202
M. Parthasarathy, K. H. Ahn, B. M. Belongia, and D. J. Klingenberg

Electric-Field-Induced Phase Separation in Electrorheological Fluids 223
X. L. Tang, K.-Q. Zhu, E. Guan, and X.-P. Wu

Some New Evidence on Electro-Rheological Mechanisms 233
R. Liang and Y. Xu

Effects of Electrode Morphology on the Electrorheological Response 251
P. V. Katsikopoulos and C. F. Zukoski

III. PROPERTIES

Viscoelasticity of Electrorheological Fluids: Role of Electrostatic Interactions 267
J. M. Ginder and L. C. Davis

Fluid Flow and Falling Ball Experiments in ER Fluids 283
R. N. Zitter, T. J. Chen, X. Zhang, and R. Tao

Influence of Particle Size on the Dynamic Strength of Electrorheological Fluids 294
Y.-H. Shih and H. Conrad

Static Rheological Properties of Electrorheological Fluids 314
G. L. Gulley and R. Tao

Dynamics of Structure Deformation and Rheology of Electrorheological Fluids 328
E. Lemaire, G. Bossis, Y. Grasselli, and A. Meunier

Determination of Rheological and Electrical Parameters of ER Fluids Using Rotational Viscometers 344
H. Janocha and B. Rech

Experimental Study of Yield Stresses in Electrorheological Fluids 358
J. N. Foulc and P. Atten

The Effect of Solid Fraction Concentration on the Time Domain Performance of an ER Fluid in the Shear Mode 372
J. Makin, W. A. Bullough, R. Firoozian, A. R. Johnson, and A. Hosseini-Sianaki

Electrostatic Interactions for Particle Arrays in Electrorheological Fluids: I. Calculations 393
Y.-H. Shih, A. F. Sprecher, and H. Conrad

Electrostatic Interactions for Particle Arrays in Electrorheological Fluids: II. Measurements 412
Y. Chen and H. Conrad

Pressure Coupling in the Electrical Response of Electro-Rheological Valves . 421
M. Whittle, R. Firoozian, W. A. Bullough, and D. J. Peel

Properties of Electrorheological (ER) Fluids under Periodic Deformation 440
W. I. Kordonsky, A. D. Matsepuro, S. A. Demchuk, and Z. A. Novikova

Influence of the Electric Field Frequency on the Electrorheological Fluids Properties 453
C. Boissy, J. N. Foulc, and P. Atten

Preliminary Optical Study on ER Fluids 463
L. W. Zhou, J. F. Ye, R. B. Tao, Y. Tang, J. F. Peng, Z. Gao, L. Y. Liu, S. H. Ma, and W. C. Wang

Pressure Responses of ER Fluid in a Piston Cylinder-ER Valve System . . . 477
M. Nakano and T. Yonekawa

IV. APPLICATIONS

Electro-Rheological Catch/Clutch: Inertial Simulations 493
A. R. Johnson, J. Makin, and W. A. Bullough

Dielectrophoretic Assembly: A Novel Concept in Advanced Composite Fabrication 516
C. A. Randall, C. P. Bowen, T. R. Shrout, G. L. Messing, and R. E. Newnham

Two Dimensional Bingham Plastic Flow in a Cylindrical Pressurised Clutch . 526
R. J. Atkin, T. J. Corden, T. G. Kum, and W. A. Bullough

Bingham Plastic Analysis of ER Valve Flow 538
D. J. Peel and W. A. Bullough

Simulation and Experimental Study of a Semi-Active Suspension with an Electrorheological Damper 568
X. M. Wu, J. Y. Wong, M. Sturk, and D. L. Russell

Applications of Electrorheological Fluid in Shock Absorbers 587
Wei Chenguan and Fu Zhao

On-Off Excitation Switch for ER Devices 597
R. C. Tozer, C. T. Orrell, and W. A. Bullough

Hydrodynamic Pressure Generation with an Electro-Rheological Fluid Part I - Unexcited Fluid 608
T. H. Leek, S. Lingard, W. A. Bullough, and R. J. Atkin

Hydrodynamic Pressure Generation with an Electro-Rheological Fluid Part II - Excited Fluid 625
T. H. Leek, S. Lingard, W. A. Bullough, and R. J. Atkin

Selection of Commercial Electro-Rheological Devices 643
D. A. Brooks

Heat Transfer Modelling of a Cylindrical ER Catch 657
R. Smyth, K. H. Tan, and W. A. Boullough

Index 676

I. MATERIALS TECHNOLOGY

THE POLARIZATION, STRUCTURING AND RHEOLOGY OF ER FLUIDS.

K. M. BLACKWOOD[1], H. BLOCK[1], P. RATTRAY[1], G. TSANGARIS[2] and D. N. VOROBIEV[3]

[1]*Centre for Molecular Electronics, School of Industrial and Manufacturing Science, Cranfield Institute of Technology, Cranfield, Bedford, UK*

[2]*Department of Chemical Engineering, Section III Materials Science and Engineering, National Technical University, Zografou, Athens, Greece*

[3]*Institute of Physico-Organic Chemistry, BSSR Academy of Sciences, Minsk, Belorus*

ABSTRACT

Recent dielectric and rheological measurements of ER fluids based upon poly(anthracene quinone radical) under continuous and pulsed dc fields are described. It is shown that given sufficient time, large fields induce interfacial polarizations which act to structure the fluid in quiescent conditions, leading to enhancement of relative permittivity and shifts in the dielectric relaxation frequencies. These dielectric experiments and with microscopic observations confirm that fibrillation and column formation occur, which are very time persistent for undisturbed fluids, even after the fields are removed. Alternative methods of applying the field, either in a single step or as a sequence of steps to gain the final level, result in differing outcomes dielectrically and cause differing structures as observed optically.

It is shown that the dynamics of structuring can be probed by applying single rectangular pulses of field for differing periods (10^{-4} - 10^{2} s) and following these by measurement of the fluids dielectric spectrum. From such data, characteristic times descriptive of the polarization abilities for a range of fluids was derived and these were successfully correlated to the field strength (E), base fluid viscosity (η) and volume fraction (Φ) using a simple theoretically based relation.

The rheology of such fluids under a field of rectangular unidirectional pulses of varying frequency have been measured and the results are presented and related to the dielectric measurements. In combination these provide the basis for a discussion on how and when the rates of polarization may influence the switching possibilities of ER fluids in use.

1. Introduction

Electrorheological (ER) activity is believed to involve the polarization of the particles and subsequent interactions[1-4]. In conditions where there is no flow these cause extensive structuring by the formation of fibrils which are the likely cause of the mechanical strength of the solid state[4,5]. The physics behind such structuring has been discussed and to some extent modelled[6,7] as has the dynamics of the interacting processes in flow[8]. Amongst several difficulties in correlating observations with theoretical prediction is the problem of visualizing the structures formed at depth, because of the turbidity of ER fluids, particularly for the more concentrated and practically significant ER dispersions. Looking at dilute fluids or thin layers of fluids under the microscope may well not give representative data but by extending the wave length of the probing radiation many of these problems disappear. This is an unusual but instructive view of the role of dielectric spectroscopy involving the frequency dependence of relative permittivity (ε') and dielectric loss (ε'') under radio frequency radiation. Since such spectroscopy can be done under dc organizing electric fields and in flow, the method is particularly apposite for ER fluids. What is valuable in such an approach is that particle interactions and consequent full or partial structuring markedly changes the fluid's dielectric spectrum.

We have previously reported preliminary observations of large field induced enhancements of ε' and ε'' of ER fluids, particularly if the solidified form develops[2]. These parameters reflect the polarization and, in terms of their dependence on electrical field frequency f, the dynamics of the polarization, which in the case of many if not all ER fluids involves interfacial processes. When this is the case the magnitude of the polarization can be strongly affected by the shape of the polarizing entities[9-11]. As described below, the field induced permittivity changes of ER fluids provide a means of investigating structuring in opaque situations whilst aspects of the dynamics of such structuring can be studied by using pulsed field techniques.

The rheology of ER fluids under pulsed fields can provide insight into the dynamics of particle-particle interactions and we have already reported some preliminary work in this area[2]. Below we present further observations involving pulsed field effects on the rheology of ER fluids.

2. Experimental

The ER fluids used were based upon dispersions of poly(anthracene quinone radical) (PAnQR)[3,12]. Thus they belong to the anhydrous class of ER fluids and are particularly effective in that they retain field induced stress enhancements up to considerable shear rates (*circa* $10^4 s^{-1}$). To make these fluids, synthesized PAnQR was ground and sieved to exclude particles ≥ 38 μm and dispersed at known volume fractions (Φ) in dried silicone oils of various viscosities (η).

Dielectric measurements under both still and flowing situations and with or without fields present were made using the equipment previously described[2]. This included a pulse generating system capable of delivering high voltage rectangular pulses at set frequencies from dc to 10^4Hz with rapid rise and fall times ($\approx 10^{-5}$s). For the purposes of some of the dielectric experiments, single pulses of selected period were electronically isolated and applied to the ER fluid.

The rheological properties of ER fluids under pulsed fields were measured using a modified[2] Carri-Med controlled stress rheometer with a Couette cell. Here the applied field (E) was in the form of a continuous train of rectangular pulses and applied to the rheometer cell when driven under a constant stress, so that any effect of a time varying field was reflected in real time by the variation of the angular velocity ω.

Optical microscopic studies of fibrillation were undertaken using Leitz Laborlux microscopic set at 16x magnification. The ER fluids were placed in 1 mm wide channels formed by fixing stainless steel electrodes on the surface of microscope slides with the ends and upper surface of the channels unsealed. Because of the high turbidity of the ER fluids, optical visualization was only found possible for dilute dispersions ($\Phi \approx 5\%$) even with the shallow layers of fluids used.

3. Results and Discussion

3.1. Dielectric changes of quiescent ER fluids caused by the application of electric fields.

We have already reported[2] that an electric field has the effect of enhancing ε' of an ER fluid and also causes a shift in its dielectric relaxation frequency f_R. Further investigations have shown that these changes, which were ascribed to structure formation under field, are very persistent when the field is removed and depend upon the way the field is applied. The dielectric spectra shown in Fig. 1 illustrates the long term stability of such ER fluids which makes possible post-field dielectric measurements on quiescent ER fluids, thereby removing the need for the protecting circuitry[2] between dc source and impedance bridge in such cases.

The persistence of this effect in the absence of mechanical agitation is not surprising considering the very small Brownian diffusion path that particles of this size undergo at ambient temperatures. Much less expected was that the method of applying the field affected the dielectric outcome as shown in Fig. 1 and also illustrated in Fig. 2 where Cole-Cole[13] representations are given. Applying a dc field in one step induces a much larger increase in permittivity than doing so in smaller incremental stages. That this effect was not due to any insufficiency of time for the changes to occur was established by allowing the system to come to dielectric equilibrium as shown by the constancy of its

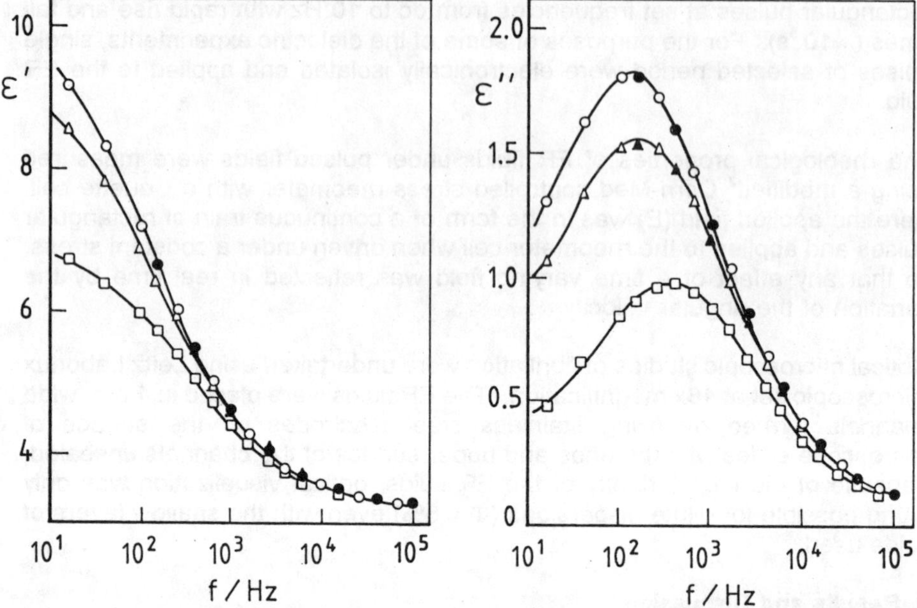

Fig. 1. The relative permittivities (ε') and dielectric losses (ε") as a function of frequency (f) of a 20% PAnQR/silicone ER fluid. □ shows the behaviour before the application of a field, O the result of a 1kVmm^{-1} field applied in one step and ▵ the effect of building up that field in staged 100Vmm^{-1} increments. Filled points refer to measurements made with the final field present and open points with the field removed but the fluid left mechanically undisturbed. The viscosity of the silicone oil was 98 mPas.

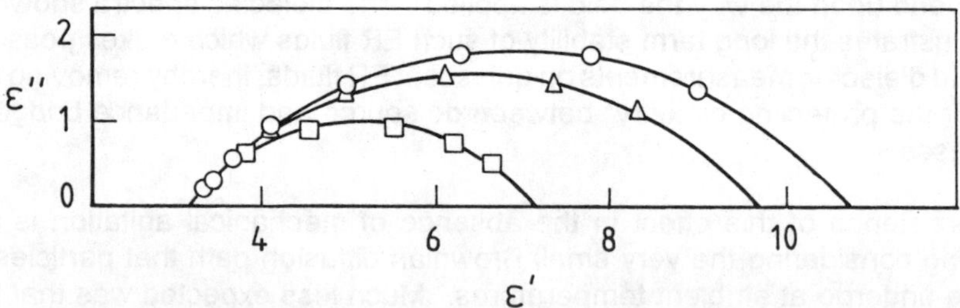

Fig. 2. Cole-Cole plots for a 20% PAnQR/silicone ER fluid with no field (□) and after 1 kVmm^{-1} applied as a single step (O) and in multiple 100 Vmm^{-1} steps (▵). The base viscosity of the silicone oil was 98 mPas. Cole-Cole parameters α were 0.412, 0.428 and 0.432 respectively. The dielectric increment, Δε' is the difference in ε' between the low and high frequency values for the dielectric process under consideration and corresponds to the length of the Cole-Cole arc.

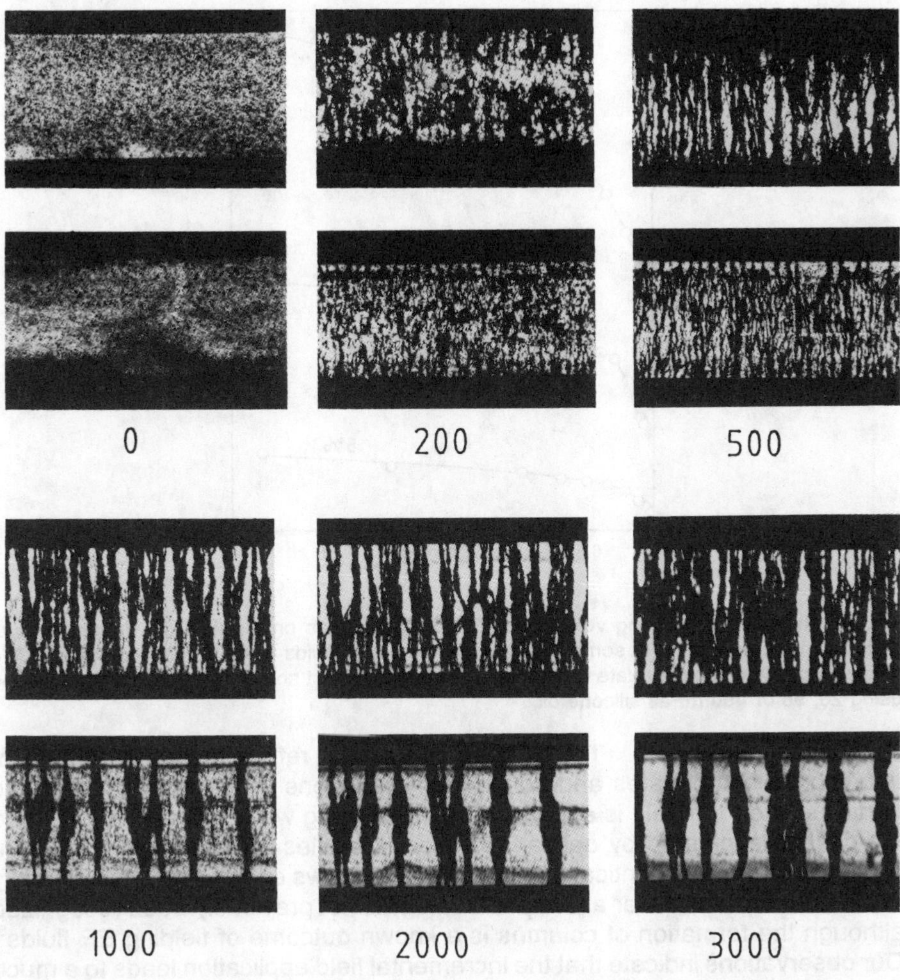

Fig. 3. The development of columnar structures under electric fields (voltage over a 1mm gap quoted) for a ≈ 5% PAnQR/silicone ER fluid. Observed microscopically when the field had been applied "all at once" (upper plates) or in steps of 0.2, 0.3, 0.5, 1 and a further 1 kV (lower plates) with equilibrium having been individually achieved in the single step and each of the stages of the multi-step alternative methods. The viscosity of the silicone fluid was 98 mPas.

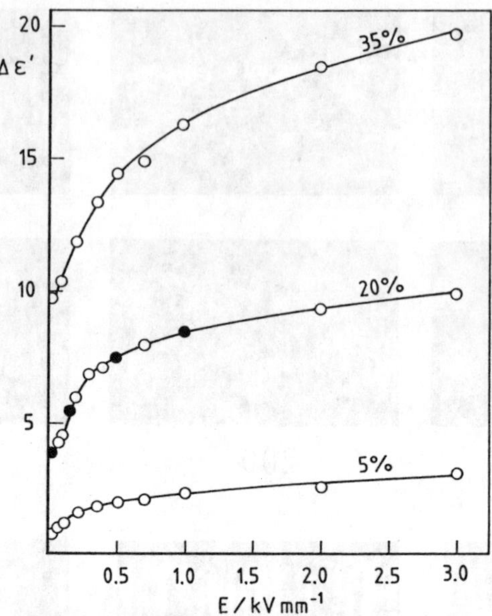

Fig. 4. The effect of altering volume fraction of particulates on the field (E) dependence of dielectric increment (Δε') for some PAnQR/silicone based fluids when the field is applied as a single step. Open points relate to 98 mPas silicone oil whilst solid points show data obtained using 20, 98 or 490 mPas silicone oil.

impedance under field. The phenomenon must reflect the development of differing dielectric states and consequent variations in the structuring of the particles when the field is applied in these differing ways. That this is indeed the case is confirmed by optical microscopy studies on thin smears of more dilute but otherwise identical ER fluids. Fig. 3 shows examples of this diversity of structures which as far as we are aware has not previously been recognized although the formation of columns is a known outcome of fielding ER fluids[6]. Our observations indicate that the incremental field application leads to a much coarser columnar structure and that building the columns "all at once" causes more of these to develop but that they are thinner. Since the dynamics of the growth process must involve a self diffusion of particles from an initially uniformly distributed source, the explanation for these differences in behaviour could be the consequence of differences in the effective capture distance and force fields involved in these differing fielding regimes. A full analysis of this would require modelling the nucleation and growth of columns under these

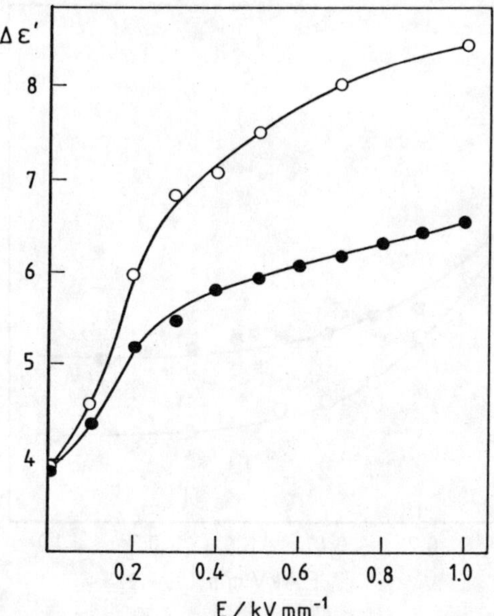

Fig. 5. The result of varying the method of fielding on Δε′. Field (E) applied as a single step (O) or in 100 Vmm^{-1} (●) steps with 20% PAnQR in 98 mPas silicone fluid.

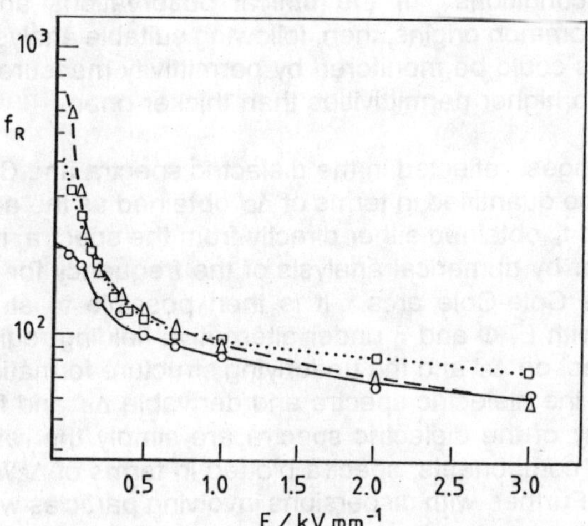

Fig. 6. The relaxation frequencies (f_R) for single step fielded PAnQR/silicone fluids. Φ = 5 (□), 20 (O) and 35% (△) dispersed in 98 mPas silicone fluid.

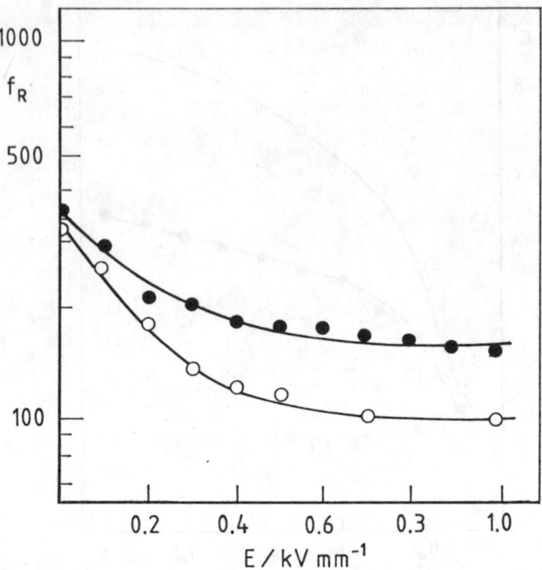

Fig. 7. The effect of varying the method of field (E) application on the relaxation frequency (f_R). Open points indicate single step and closed points that the field was incremented at 100 Vmm^{-1} steps.

differing fielding conditions. If the optical observations and thedielectric differences have common origins, then, following suitable analysis, the build up of these structures could be monitored by permittivity measurements: thinner columns leading to higher permittivities than thicker ones.

The dielectric changes, reflected in the dielectric spectra and Cole-Cole plots, can conveniently be quantified in terms of $\Delta\varepsilon'$ obtained as the arc length of the Cole-Cole plot and f_R obtained either directly from the spectra, if the maximum in loss is visible, or by numerical analysis of the frequency for maximum loss obtained from the Cole-Cole arcs. It is then possible to study how these parameters vary with E, Φ and η under alternative fielding regimes (Figs 4 to 7). η has little effect on $\Delta\varepsilon'$ and the underlying structure formation, but Φ does have an effect on the dielectric spectra and derivable $\Delta\varepsilon'$ and f_R. In situations where the intensity of the dielectric spectra are simply the result of additive contributions from components, spectra plotted in terms of $\Delta\varepsilon'/\Phi$ would follow a common locus. Further, with dispersions involving particles whose sizes are much larger than molecular dimensions, one would expect all polarizations involving the constituent molecular dipoles to occur at the same relaxation frequencies regardless of Φ. As Figs 6 and 8 show this is far from the case with fluids which have not been structured under fields but is approached when fields are applied to enhance $\Delta\varepsilon'/\Phi$ and reduce f_R.

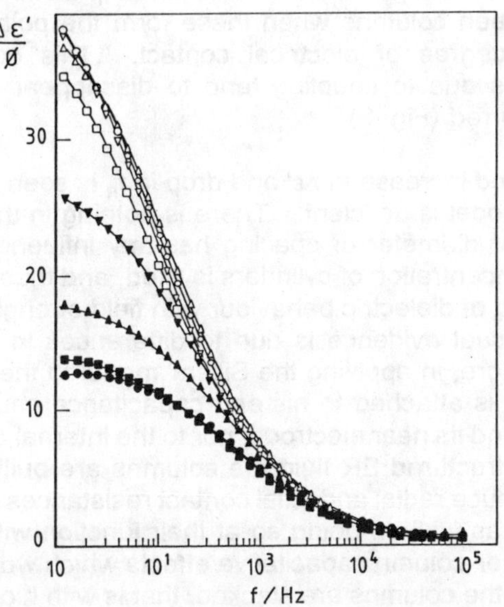

Fig. 8. The variation of $\Delta\varepsilon'/\Phi$ with frequency (f) for PAnQR/silicone at various Φ [5 (○), 10 (□), 20 (△) and 35% (▽) when no field is applied (solid points) and 1 $kVmm^{-1}$ had been applied as a single step (open points). The viscosity of the silicone oil was 98 mPas.

With the PAnQR (conductance $\sigma \approx 10^{-6}$ Sm^{-1}) based fluids studied here and probably in many if not all cases of effective ER fluids, the polarization mechanism is likely to be interfacial with electronic charge migration to the particle-fluid boundary being the probable polarizing mechanism. The dielectric outcome for such interfacial polarization processes have been extensively modelled with the simplest and most apposite being that due to Wagner[10], dealing with a dilute dispersion of spheres and that of Sillars[11], particularly in his treatment of conducting cylinders embedded in an insulator and aligned along the field direction. The former could reflect the unstructured state before fielding since the PAnQR particles, although by no means regular spheres, are not very anisometric. If columnar structures form under field then the Sillars' model has some similarities to the structured state which however is based on more irregular structures than simply uniform cylinders in array. These models indicate that for the same Φ, $\Delta\varepsilon'$ is much larger with normally aligned cylinders rather than with a dispersion of spheres which fits in well with observed behaviour. However for dilute conditions to which both analyses were restricted, the models predict that $\Delta\varepsilon'$ is proportional to Φ and f_R is independent of Φ. Concentration dependant effects on $\Delta\varepsilon'$ or f_R are often associated with interactions between polarizing regions which depend on their separation and thus concentration. Relating this to ER fluids we note that in the absence of field the interfacial process is limited to isolated particles in much nearer proximity than between columns when these form the polarizing entities via particles in some degree of electrical contact. This could explain why concentration effects due to coupling tend to dissappear under field when structuring has occurred (Fig. 8).

Although the expected increase in $\Delta\varepsilon'$ and drop in f_R is seen, in other respects the simple Sillars' model is deficient. There is nothing in that analysis which predicts that cylinder diameter or spacing has any influence on the dielctric spectra once the concentration of cylinders is fixed, and thus the model fails to address the variation of dielectric behaviour with field strength and application history, which on visual evidence is due to differences in column girth and spacing. Furthermore, in applying the Sillars' model to the present situation no precise meaning is attached to his end capacitance caused by the "gap" between a cylinder and its near electrode, nor to the internal capacitance of the "cylinder". With a structured ER fluid the columns are built up of contacting particles which introduce radial and axial contact resistances and capacitances within the columns, as well as doing so at their junction with the electrodes. There will also be inter-column capacitative effects which would become more important the closer the columns are stacked, that is with higher concentration or with thinner structures located closer to each other. Both internal contact and inter-columnar impedances would change with the level and mode of application of the field if the visual observations on thin layers of dilute fluids are any guide to the structuring which occurs. The effects within the columns would dominate over inter-column effects, which however should not be neglected as they are involved in the different dielectric outcomes of the alternative fielding methods which gives rise to the structure variation. To quantify the effects on such a basis requires numerical modelling of the network impedances of contacting lattices of particles for the columns and the inter-column impedances for arrays of these which finally approximates to the structured fluid as a whole.

3.2. Dielectric measurements following the application of pulsed fields.

In order to study some of the dynamics of the structuring, fields of a single pulse were applied for a predetermined time ($10^{-4} \leq \tau \leq 10^{2}$ s) and the dielectric properties measured afterwards, this being possible because of the long term stability of such structured ER fluids when left undisturbed . Figs. 9, 10 and 11 respectively show how τ influences $\Delta\varepsilon'$ under varying E, Φ and η. They show that for very short pulses structuring does not occur and that full structuring can only be achieved if the field is applied for long enough. For intermediate τ, an

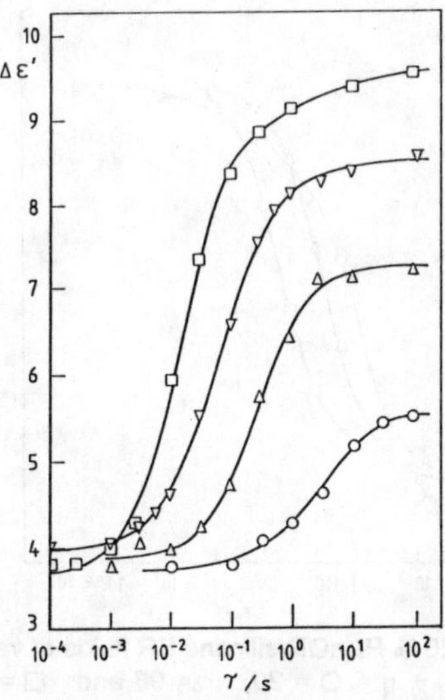

Fig. 9 The variation of Δε' for a 20% PAnQR/silicone ER fluid with pulse duration (τ) for selected E: 0.2 kVmm^{-1}(○), 0.5 kVmm^{-1}(▵), 1 kVmm^{-1}(▿) and 2 kVmm^{-1}(□). Dispersions in 98 mPas silicone oil.

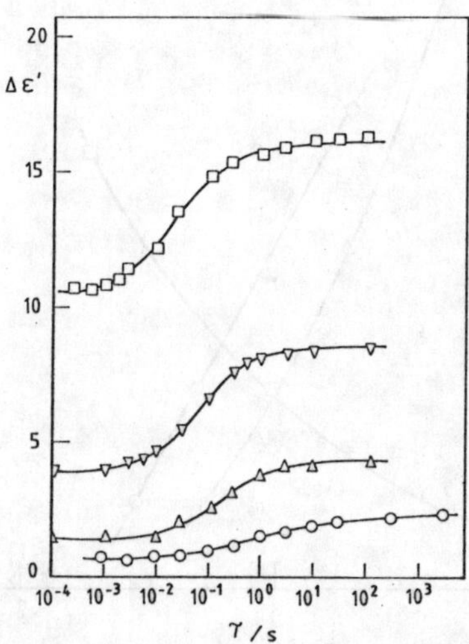

Fig. 10 The variation of Δε' for PAnQR/silicone ER fluids at various Φ with 1 kVmm^{-1} pulses of duration τ. Φ: ○ = 5, ▵ = 10, ▿ = 20 and □ = 35 %. Dispersions in 98 mPas silicone oil.

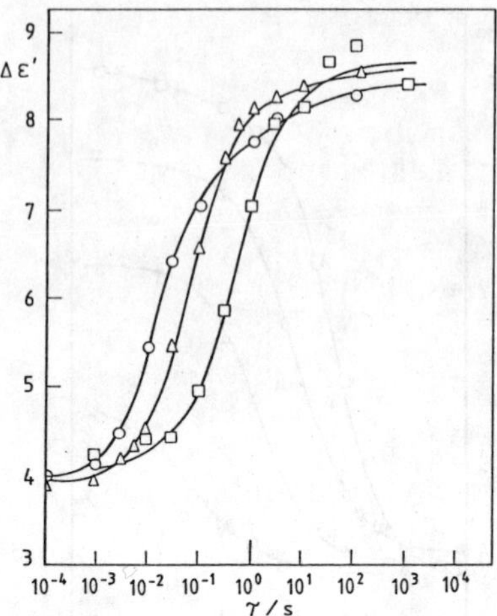

Fig. 11 The variation of Δε' for 20 % PAnQR/silicone ER fluids at various silicone oil viscosities with 1 $kVmm^{-1}$ pulses of period τ. η: O = 20, ▵= 98 and □ = 490 mPas.

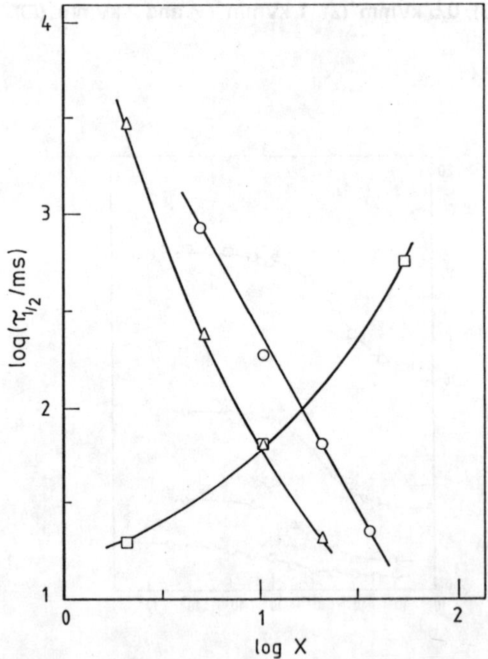

Fig 12. The variation of $\tau_{½}$ with E (O), Φ (▵) and η (□) for the various ER fluid. Abscissa scaling X = 10E/$kVmm^{-1}$ or 10^2Φ or 10^{-1}η/mPas.

incomplete attainment in $\Delta\varepsilon'$ occurs and this reflects the incomplete structuring. Thus the levels of $\Delta\varepsilon'(\tau)$ as a function of pulse duration provides a method for monitoring the dynamics of the structuring, with increasing E or Φ or decreasing η speeding up the process. In order to aid quantification of these effects we start by defining a time scale in terms of $\tau_{1/2}$, the period which would establish 50% of the equilibrium $\Delta\varepsilon'$ level. Fig. 12 shows how $\tau_{1/2}$ varies with E, Φ and η for silicone based PAnQR fluids. Forced diffusion towards structuring would appear to be the underlying mechanism as confirmed by the following simple analysis.

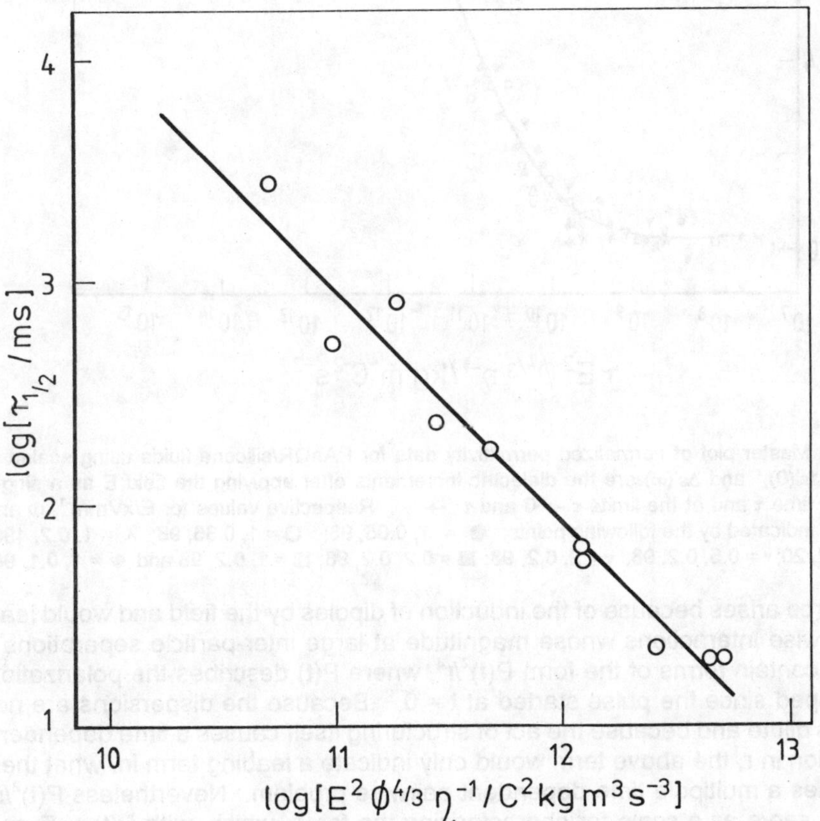

Fig 13. The data of Fig. 12 as a function of $\eta\Phi^{-4/3}E^{-2}$. Least mean square line shown solid whilst best fit with unit slope shown broken.

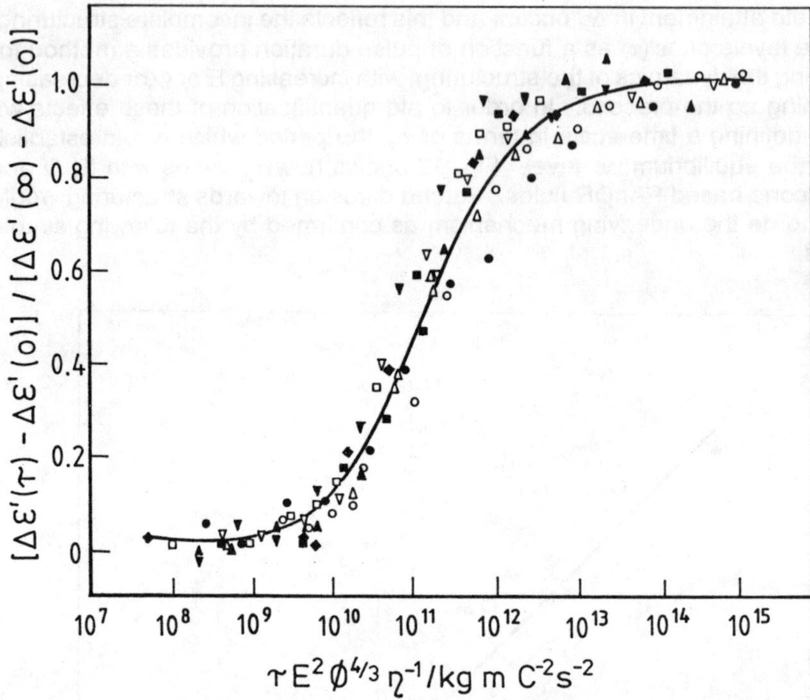

Fig 14. Master plot of normalized permittivity data for PAnQR/silicone fluids using scaled τ. $\Delta\varepsilon'(\tau)$, $\Delta\varepsilon'(0)$, and $\Delta\varepsilon'(\infty)$ are the dielectric increments after applying the field E as a single step for time τ and at the limits $\tau \rightarrow 0$ and $\tau \rightarrow \infty$. Respective values for E/kVmm^{-1}, Φ and η/mPas indicated by the following points: ● = 1, 0.05, 98; ○ = 1, 0.35, 98; ▲ = 1, 0.2, 490; △=1, 0.2, 20; ▼ = 0.5, 0.2, 98; ▽ = 2, 0.2, 98; ■ = 0.2, 0.2, 98; □ = 1, 0.2, 98 and ◆ = 1, 0.1, 98.

The force arises because of the induction of dipoles by the field and would lead to pairwise interactions whose magnitude at large inter-particle separations r would contain terms of the form $P(t)^2/r^4$, where P(t) describes the polarization developed since the pulse started at t = 0. Because the dispersions are not always dilute and because the act of structuring itself causes a time dependent reduction in r, the above term would only indicate a leading term in, what then becomes a multipole time dependent pairwise problem. Nevertheless $P(t)^2/r^4$ should serve as a scale for characterizing the force, which, with $P(t) \propto E$ and the mean separation of particle centres scaling as $\Phi^{-1/3}$ at the start of the process, gives a force scaling as $E^2\Phi^{4/3}$. In this we have ignored the time dependence of polarization which could be allowed for by time averaging using the frequency dependence of ε', but in the light of the other approximations and the fact that $\tau_{½} < (2\pi f_R)^{-1}$ this is needlessly complicated, particularly as the E^2 dependence would remain. The inter-particle force causes particle motion whose velocity $\propto$ force/η, so that any characteristic period to cover a selected distance becomes a function of $\eta\Phi^{-4/3}E^{-2}$ and $\tau_{½}$ should scale with this. Fig. 13 shows that this is indeed the case with the experimentally determined best straight line shown having a slope of -1.02 ± 0.10 which at a 90% confidence level encompasses the expected unit negative slope. Scaling individual pulse data to $E^2\Phi^{4/3}\eta^{-1}$ and the extent of structuring in terms of the achievable polarization levels measured in terms of the step in $\Delta\varepsilon'$ between very short and very long pulses has then enabled us to construct a master plot for all the data. This is shown in Fig. 14 in which explicit values for E, Φ and η are subsumed.

3.3. Rheology under pulsed fields.

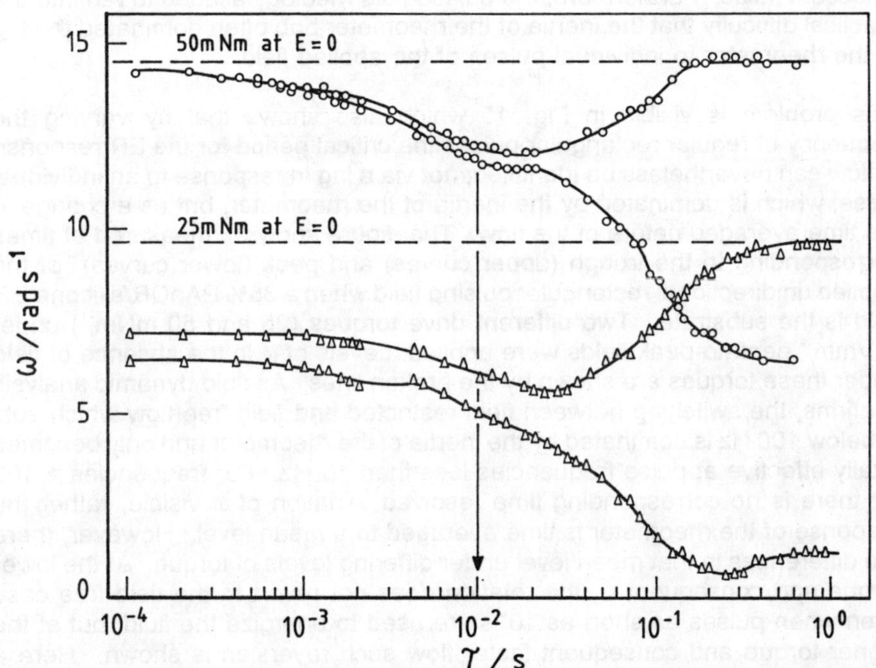

Fig 15. The response in terms of the angular velocity (ω) of a rheometer driven at 25 or 50 mNm set torques and containing a 35% PAnQR dispersion in 98 mPas silicone oil. The ER fluid is being energized by rectangular pulsed fields of period τ with each upper curve corresponding to the response to the trough (0) and lower curve to the peak (2 kVmm^{-1}) field. The dashed lines show the levels of ω in the total absence of field and the position of the arrow indicates the dielectric relaxation time for the system in that case.

In flow, fibrillation is partial and dynamic and even the polarization of particles can become shear rate (γ) as well as field frequency dependent, thus effecting the interacting dipolar forces. However, provided the vorticity of the flow field is not so high as to interfere with the ability of the particles to polarize, a level of induced dipolar interactions persist, which by resisting the action of the flow cause the phenomenon of ER, even though extensive fibrillation is absent. A fluid with useful ER behaviour is one in which these forces are rapidly switched in flow and persist at high γ. ER fluids vary in these respects and it is important that dynamic shear data and electrical response times be examined if a fluid is to find practical application.

Investigations have already shown that our PAnQR based ER fluids have a rapid response to pulsed fields[2], much faster than the time scale for the full columnar structuring discussed above, and this reflects the shorter time scale for forces to switch compared to the time scale needed to build columns in quiescent fluids. Furthermore, the pulsed field rheology alluded to ran into the practical difficulty that the inertia of the rheometer bob often dominated the lag of the rheometer to individual pulses of the applied field.

This problem is visible in Fig. 15 which also shows that by varying the frequency of regular rectangular pulses the critical period for the ER response in flow can nevertheless be identified; not via a lag in response to an individual pulse, which is dominated by the inertia of the rheometer, but as a change in the time averaged nature of the flow. That figure shows ω measured at times corresponding to the trough (upper curves) and peak (lower curves) of the applied unidirectional rectangular pulsing field when a 35% PAnQR/silicone ER fluid is the substrate. Two different drive torques (25 and 50 mNm^{-1}) under $2kVmm^{-1}$ peak-to-peak fields were applied. Levels of ω in the absence of field under these torques are shown by the broken lines. As fluid dynamic analysis confirms, the switching between field restricted and field free flow which sets in below 100 Hz is dominated by the inertia of the rheometer and only becomes totally effective at pulse frequencies less than 10 Hz. For frequencies > 100 Hz there is no corresponding time resolved variation of ω visible, rather the response of the rheometer is time averaged to a mean level. However, there are differences in that mean level under differing levels of torque. At the lower torque and consequent ω, the rotation does not revert to the field free case even when pulses as short as 10^{-4}s are used to energize the fluid, but at the higher torque and consequent faster flow such reversion is shown. Here a more pronounced drop in ER effect starts at field "on" times which correlate with the dielectric relaxation time under that field (marked by the arrow in Fig. 15) and progresses until at 10^{-4} s pulse duration ω regains the field free value within experimental error. This is evidence that the faster flow in combination with the limited time for polarization destroys the polarization and consequent interactions. This data also shows that the fluid under test is also an effective one with an ability to respond rapidly at high shear rates.

4. Conclusions

We have shown that dielectric measurements on ER fluids can provide an insight into the field induced structuring in the bulk state and the dynamics of this process. The time scales involved for such large scale ordering are longer than the dynamic response of the ER fluid in flow so that one can conclude that there are at least two characteristic time scales involved in ER behaviour: a long time structuring process relevant in the main to the Bingham state and a dynamic response which reflect the polarization and consequent interactive forces between the dipersed particles.

The types of study described provide methods by which the physics of the phenomena may be probed and correlated with chemical and physico-chemical attributes of the materials which go into ER fluid formulations. That is an important aspect of the described investigations. However, another phenomenon has come to light in the observation of the substantial enhancements of the permittivities of still ER fluid when fields are applied. This together with their stability may lead to non rheological application for ER fluids or composites derived from them.

Finally there remains the question of the types of structure which build up under various fielding regimes. We have shown that dielectric measurements may well provide a method of probing these, provided satisfactory models relating structuring parameters to permittivity can be developed. Success in such an analysis would provide data on the structuring of ER fluids in still conditions and might well lead further to include shorter range ordering under conditions of flow. The dielectric "visualization" which could result thereby, using measured dielectric data in flow, would provide much useful data for the dynamic modelling of ER.

5. Acknowledgements

The authors wish to thank the SERC for supporting this work.

6. References

1. Yu. F. Deinega and G. V. Vinogradov, *Rheol. Acta.* **23** (1984) 636 Z. P. Shul'man, R. G Gorodkin, E. V. Korobko and V. K. Gleb, *J Non-Newtonian Fluid Mechanics* **8** (1981) 29; H. Block and J. P. Kelly, *J. Phys. D Appl. Phys.* **21** (1988) 1661; P. A. Gast and C. F. Zukoski, *Adv. Coll. Interf. Sci.* **30** (1989) 153; M. Whittle, *Physics World* **2** (August 1989) 39; H. Block and J. P. Kelly in *First Internat. Symp. on ER Fluids*, ed. H. Conrad, J. D. Carlson and A. F. Sprecher (North Carolina State

Univ. Engineering Pub. Raleigh N. C. 1989), p. 1; T. C. Jordan and M. T. Shaw, *MRS Bulletin* (August 1991) 38; H. Block, J. P. Kelly and T. Watson in *High Value Polymers*, ed. A. H. Fawcett (RSC Special Pub. **87** London, 1991), p. 151; F. E. Filisko, *Chem and Ind.* (1992) 370; H. Block in *Actuator 92 Conference Proceedings* ed. K. Lenz and H. Borgmann (VDI/VDU-Technologiezentrum Informationstechnik Gmbh, Berlin 1992), p. 105; H. Block; *Chemtech* (1992) 368.

2. H. Block, P. Rattray and T. Watson in *Proc. Int. Conf. Electrorheological Fluids* ed. R. Tao (World Sci. Press, Singapore, New Jersey, London, Hong Kong, 1992), p. 93.

3. H. Block, J. P. Kelly, A. Qin and T. Watson, *Langmuir*, **6** (1990) 6.

4. J. C. Jordan and M. T. Shaw *IEEE Trans. Elec. Inst.* **24** (1989) 849; N. Webb, *Chem. Brit.* (1990) 338; M. Whittle and W. A. Bullough, *Nature* **358** (1992) 373; W. A. Bullough, *Advanced Materials Optics Electronics* **1** (1992) 159; K. M. Blackwood and H. Block, *Trends in Polymer Science*, **1** (1993) 98.

5. W. M. Winslow, *US Patent* (1947), 2417850; *J. Appl. Phys.* **20** (1949) 1137; D. J. Klingenberg and C. F. Zukoski, *Langmuir* **6** (1990) 15.

6. R. Tao and J. M. Sun, *Phys. Rev. Lett.* **67** (1991) 398; *Phys. Rev. Lett. A* **44** (1991) R6181; R. Tao, in *Proc. Int. Conf. Electrorheological Fluids* ed. R. Tao (World Sci. Press, Singapore, New Jersey, London, Hong Kong, 1992), p. 3.

7. T-J. Chen, R. N. Zitter and R. Tao in *Proc. Int. Conf. Electrorheological Fluids* ed. R. Tao (World Sci. Press, Singapore, New Jersey, London, Hong Kong, 1992), p. 17.

8. P. M. Adriani and A. P. Gast, *Phys. Fluids* **31** (1988) 2757; P. Bailey, D. G. Gillies, D. M. Heyes and L. H. Sutcliffe, *Mol. Sim.* **4** (1989) 137; M. Whittle, *J. Non-Newtonian Fluid Mechanics* **37** (1990) 233; T. C. Halsey and W. Toor, *Phys. Rev. Lett.* **65** (1990) 2820; D. M. Heyes and J. R. Melrose, *Mol. Sim.* **5** (1990) 293; J. R. Melrose, *Phys. Rev. A* **44** (1991) R4789; T. C. B. McLeish, T. Jordan and M. T. Shaw, *J. Rheol.* **35** (1991) 427; J. I. Takimoto in *Proc. Int. Conf. Electrorheological Fluids* ed. R. Tao (World Sci. Press, Singapore, New Jersey, London, Hong Kong, 1992), p. 5; T. C. Halsey in *Proc. Int. Conf. Electrorheological Fluids* ed. R. Tao (World Sci. Press, Singapore, New Jersey, London, Hong Kong, 1992), p. 37.

9. J. C. Maxwell *Electricity and Magnetism* **1**, (Clarendon Press, Oxford, 1892), p. 452; A. R. von Hippel *Dielectrics and Waves*, John Wiley, 1954), p. 228.

10. K. W. Wagner, *Archiv Elektrotech.* **2** (1914) 371.

11. R. W. Sillars, *JIEE* **80** (1937) 378.

12. H. Block and J. P. Kelly, *US Patent* (1987) 4687589.

13. K. S. Cole and R. H. Cole, *J. Chem. Phys.*, **9**, (1941) 341.

FIRST EXPERIMENTS ON MAGNETOELECTRORHEOLOGICAL FLUIDs (MERFs)

W.I.Kordonsky, S.R.Gorodkin and E.V.Medvedeva

Luikov Heat and Mass Transfer Institute, Academy of Science, Minsk, Belarus

1. Introduction

Phenomenologically, the ER [1] and MR [2] effects are close in character but fundamentally different in the physical nature of the mechanisms responsible for structure formation. Each of these mechanisms has characteristic advantages and disadvantages. However, common to both effects is their ability to saturation. And what is more, it would be beneficial to extend the possibilities for structure control using two independent physical channels.

2. Description of MER Fluids and Experimental Setup

The main problem on preparing a dual fluid consists in specifying the procedures of manufacturing a dispersed solid phase sensitive both to electric and magnetic fields. One of the procedures was based on coating a ferromagnetic particle surface with an electrorheologically active layer. Non-colloidal γ-iron oxide (Fe_2O_3) particles (0.6-1 μm in mean size) were selected as a ferromagnetic dispersed phase (Fig.1). An amine-containing activator was applied over a particle surface to provide the sensitivity to an electric field.

Transformer oil served as a carrier fluid. Sedimentation and aggregation stability of a fluid was gained by using a stabilizing addition (high-dispersed activator-treated colloid) forming within a volume of a matrix that prevented the sedimentation of heavy ferroparticles and their coalescence. The matrix did not hinder the approach and interaction of particles being formed their structure, was easily deformed under a shear, and was reconstructed at a stop.

To conduct rheological measurements when a MER fluid is under separate and combined electric and magnetic fields a special co-cylindrical viscometric bell-type cell (Fig.2) serving as an attachment to torque meter 1 (TM) of the viscometer RV-12 manufactured by Firm "HAAKE" has been designed. The cell contains metal fixed cup 2 and rotating bell 3 located in the cup cavity. The cup, with its outer surface coated with electrical insulating layer 4, is slipped over the core of magnetic field inductor 5 and is connected to high voltage supply (HVS)6. The bell is earthed. Milliampermeter 7 in the earthed circuit is meant for determining fluid electrical conductivity in the cell. Inductor coil 8 is connected to DC supply (DCS)9. The MER fluid is placed into the cup cavity and TM is connected with the bell shaft. When TM rotates in a gap between the bell and the cup the Couette flow is achieved. The magnetic and electric fields are applied when DCS and HVS are switched on separately or simultaneously. In this case, the force lines of the fields are normal to the shear. In experiments, measurements have been made of shear stress, τ, at varying: magnetic field intensity H=(0-80) kA/m, a.c. (50 Hz) field strength E=(0-1) kV/mm, shear rate $\dot{\gamma}$=(6-445) s^{-1}, and volume concentration, φ=(2, 6, 8, 10)%, of activated γ-Fe_2O_3 particles.

3. Measurement Results

At the first stage, an optimal amount of the activating addition in the ferromagnetic powder has been determined. From experiment it has been found that a relative viscous stress increment $(\tau-\tau_o)/\tau_o$ (τ_o's are the viscous stresses under no fields) vs weight concentration of the activator at an electric field and at magnetic and electric fields has a maximum around 13%. Available extremal concentration is associated with the fact that with small additions the activator volume is insufficient to completely coat the particles and to create optimal conditions for charge transfer and buildup; in our case, large additions have reduced the MER fluid quality, thus causing particle coalescence and coarse conglomerate formation.

Further, the MER fluid containing an optimal amount of the activator on the particles has been used.

Rheological measurements have shown that under no fields the MER fluid behaves as a linear viscoplastic medium with a small yield point, τ_o ,occurring due to the available stabilizer (Fig.3). The magnetic field gives rise to τ_o owing to structure formation in the MER fluid and results in the nonlinear behaviour of $\tau_H = f(\dot{\gamma})$ at small and moderate shear rates characteristic of MR fluids. At an electric field the MER fluid flow curve is similar to the traditional ER fluid one.

Under two combined fields, viscous stresses in ER fluid flow much grow. As this takes place, it should be especially noted that first the flow curve is almost parallel to the γ-axis and this is very convenient for control of such fluid flow; second, values of τ_{H+E} exceed a sum of τ_H and τ_E especially at small shear rates. Growing the dispersed phase concentration yields an expected linear increase of the shear stress increment, $\Delta\tau$, under separate fields, and under combined fields the relation $\Delta\tau = f(\varphi)$ is quadratic in nature (Fig.4).

4. Visualization Procedure

As mentioned above, the MER fluid refers to a class of dispersed fluids being formed their structure under external physical fields. The rheological behaviour of such a system is determined through the characteristic of a microstructure formed by a dispersed phase particle field. In doing so, of dominant role are the strength, packing density, shape, and mutual arrangement of structural elements, as well as the dynamics of their break and formation in the shear flow. The essential difference in the flow curves, concentration dependences, and the availability of the supertotal effect point to qualitative differences in microstructures formed under electric, magnetic and combined fields.

For visualization of the processes to occur in MER fluid flow the setup (Fig. 5) has been designed to visualize on the monitor and to make a video film of a 200-fold magnification structure formation picture in creep flow of low-concentrated (transparent) fluids in a narrow rectangular (0.5x2.0)mm channel between high-volt electrodes disposed in a magnetic field, allowing the mutual orientation of the force lines of two field to be changed.

Visualizations have shown that the electric field (Fig.6a) induces relatively thin and fine structural chains that "intergrow" through the entire gap, continuously break, and appear with a frequency correlating with the shear rate.

Under a magnetic field (Fig.6b) separate densely packed elongated practically non-interacting aggregates are formed, having a very small (close to point) contact area with the wall.

Under combined fields (Fig.6c) there exists a system of densely packed aggregates that intergrow through the gap. Their contact with the wall occurs through a set of "branches".

Under crossed fields oriented along the magnetic field (and channel axis) separate aggregates unite in a fine cellular structure connected with the electrodes.

Thus, combining two mechanisms, namely: magnetodipole interaction of a particle ferromagnetic base and formation of chains necessary for charge transfer and buildup on the activated particle surface gives rise to a qualitatively new microstructure responsible for the above MER fluid properties.

5. Concluding Remarks

The complex of the rheological MER fluid properties to some extent hinders the analysis and estimation of the obtained results. Owing to this, an attempt has been made to present all numerous experimental data in a unique generalized form. Phenomenologically, a relative viscous stress increment $(\tau-\tau_o)/\tau_o$ or, what amounts to the same thing - a characteristic viscosity $[\eta]=(\eta-\eta_o)/\eta_o$, with the MER fluid flowing under a field (or fields) depends on a strength, geometric dimensions, and orientation of elements (particle aggregates) of a microstructure, as well as on the magnitude of an electric charge of particles (or aggregates), charge transfer velocity, and a collision frequency of separate elements of the microstructure during charge transfer under a shear.

Logically, it may be assumed that under all other things being equal, the above physical parameters depend on the ratio of electric and magnetic field energy per unit fluid volume to the hydrodynamic energy of shear flow per unit volume. Such an energy ratio can be presented in the form of the modified Mason number as

$$S = \frac{\mu_o IH + \varepsilon\varepsilon_o E^2}{\eta_o \dot{\gamma}},$$

where $\mu_o = 1.256\cdot10^{-8}$ Gn/m, and $\varepsilon_o=8.85\cdot10^{-1}$ $Kl^2/N\cdot m^2$ are the magnetic and dielectric permittivities, respectively; I is the MER fluid saturation magnetization, ε is the relative dielectric permittivity of the MER fluid.

Note that the complex S contains all assigned parameters of shear MER fluid flow under magnetic and electric fields. To determine S, the known procedures have been adopted to measure real values of I and ε of MER fluid samples. Values of the relative dielectric permittivity for an appropriate concentration have been taken at electric field frequencies close to the ones used in rheological experiments.

Figure 7a plots the characteristic concentration - normalized viscosity vs complex S for a small sampling experimental data. Points on the plot are broken down into three lines according to three variants of the external action: electric field alone; magnetic field alone; two combined fields. To our mind, the reason for such a scatter may be as follows. In the expression for electric field energy the complex S is based on the classical polarization mechanism. However, for example, in the experiments [3] it is shown that the strength of a model ER fluid chain (and this means the ER effect) is determined not by the dipole-dipole interaction of polarized particles but by the conditions for charge transfer and buildup. The measured strength of the dielectric particle chain by an order of magnitude exceeds the one measured by the polarization model.

In this connection, we have included some dimensionless coefficient ζ allowing for the above disagreement into the complex S. A new writing of the complex will be as

$$S_* = \frac{\mu_o I H + \zeta \varepsilon \varepsilon E^2}{\eta_o \dot{\gamma}}$$

In a general case, the coefficient ζ must be a complex function of a number of the parameters which characterize electrophysical processes in the MER fluid, physicochemical properties of components, etc. but having regard to cited experiments and our estimates the mean value of ζ has been taken equal to 20.

As seen from Fig.7b, on the lg $[\eta]/\varphi$–lgS_* coordinates the experimental points fall on one line with a sufficient accuracy.

Thus, the generalization of all experimental results by a unique relation supports the reasonable use of the complex S_* as a determining parameter and points to the correctness of our ideas of the mechanism of microprocesses in the MER fluid under fields, at least, on a qualitative level. In addition, such a universal relation allows the mechanical properties of a fluid to be predicted with reasonable simplicity in a specific process.

References

1.Klass, D.L. and Martinek. 1967. "Electroviscous Fluids. 1-Rheological Properties. 11-Electrical Properties",Journal of Applied Physics, 38 (1): 67-75.

2.Shulman, Z.P., W.I.Kordonsky, E.A.Zaltsgendler, I.V.Prokhorov, B.M.Khusid and S.A.Demchuk. 1986. "Structure, physical properties and dynamics of magnetorheological suspensions", Int.J.Multiphase Flow, Vol.12,No. 6, pp. 935-955.

3.Sprecher, A.F., Y.Chen and H.Conrad. 1989. "Measurement of Forces Between Particles in a Model ER Fluid", Proc. 2nd Int. Conf. on ER Fluids, pp. 82-89

List of Figures

Figure N1 Microphoto of ferromagnetic Fe_2O_3 particles

Figure N2 Schematic of viscometric setup

Figure N3 Effect of electric and magnetic fields on shear stresses

Figure N4 Solid phase concentration effect on MER fluid rheological properties

Figure N5 Visualization setup

Figure N6a Microstructure in electric field (E=1 kV/mm)

Figure N6b Microstructure in magnetic field (H=160 kA/m)

Figure N6c Microstructure in combined field (E=1 kV/mm, H=160 kA/m)

Figure N7a,b Characteristic viscosity vs modified Mason number

Figure N1 Microphoto of ferromagnetic Fe_2O_3 particles

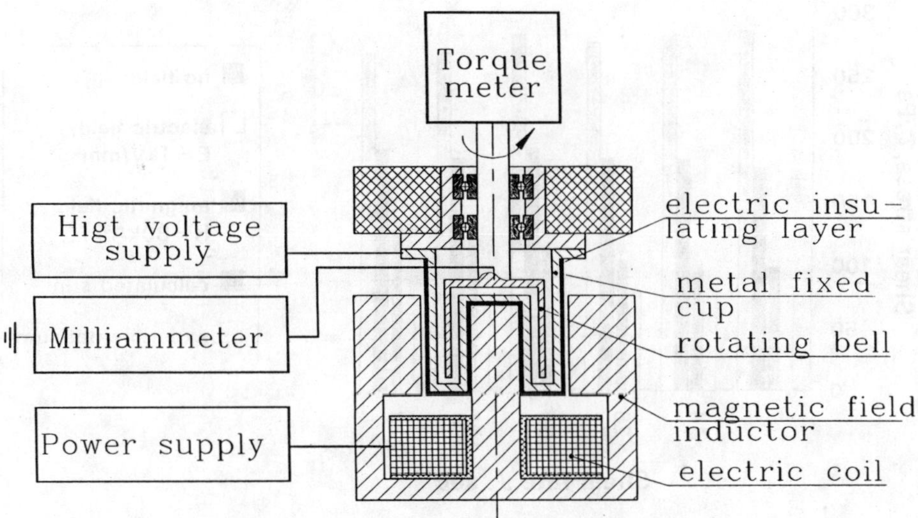

Figure N2 Schematic of viscometric setup

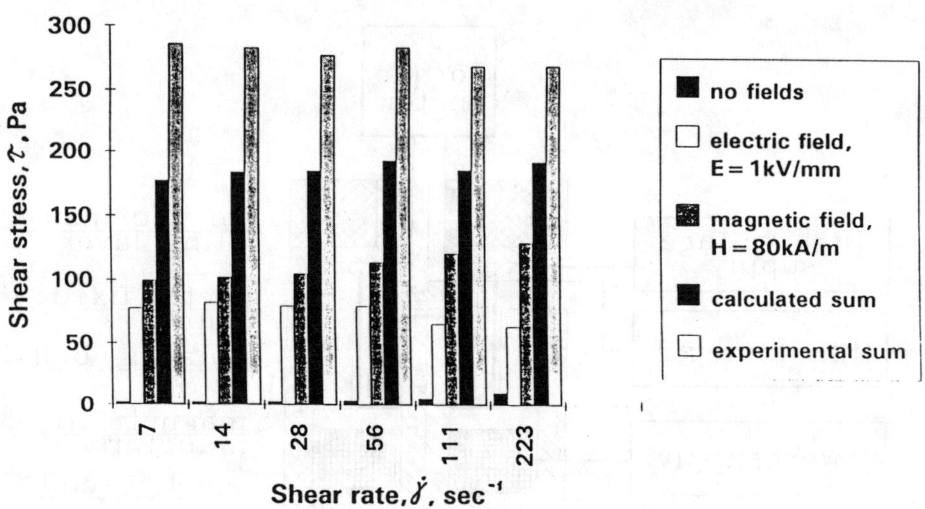

Figure N3 Effect of electric and magnetic fields on shear stresses

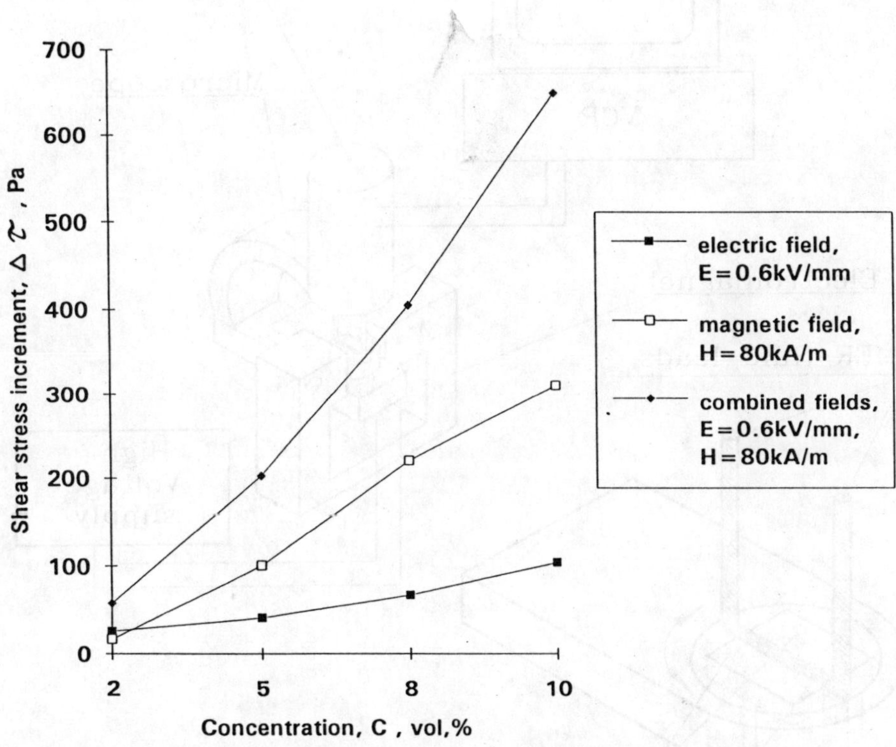

Figure N4 Solid phase concentration effect on MER fluid rheological properties

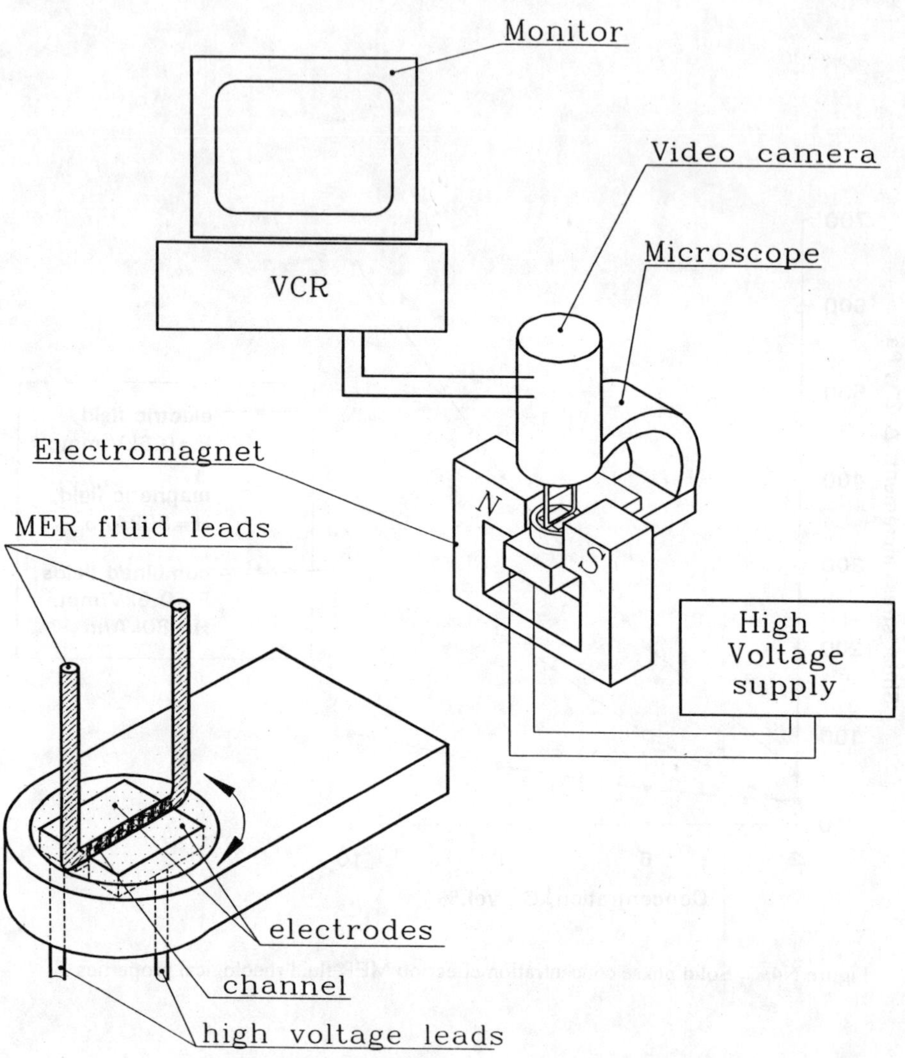

Figure N5 Visualization setup

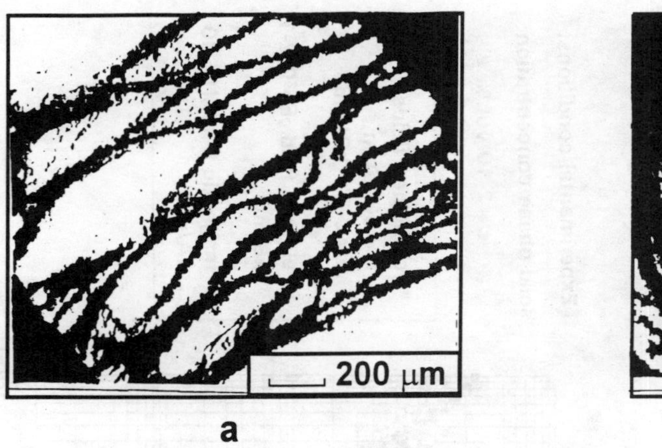

a

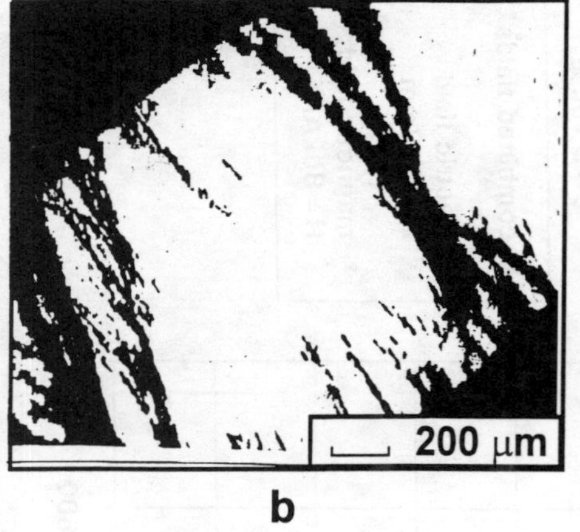

b

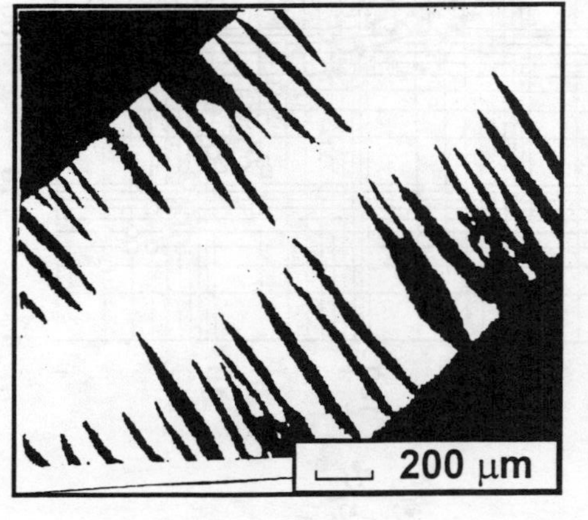

c

Figure N6a Microstructure in electric field (E=1 kV/mm)

6b Microstructure in magnetic field (H=160 kA/m)

6c Microstructure in combined field (E=1 kV/mm, H=160 kA/m)

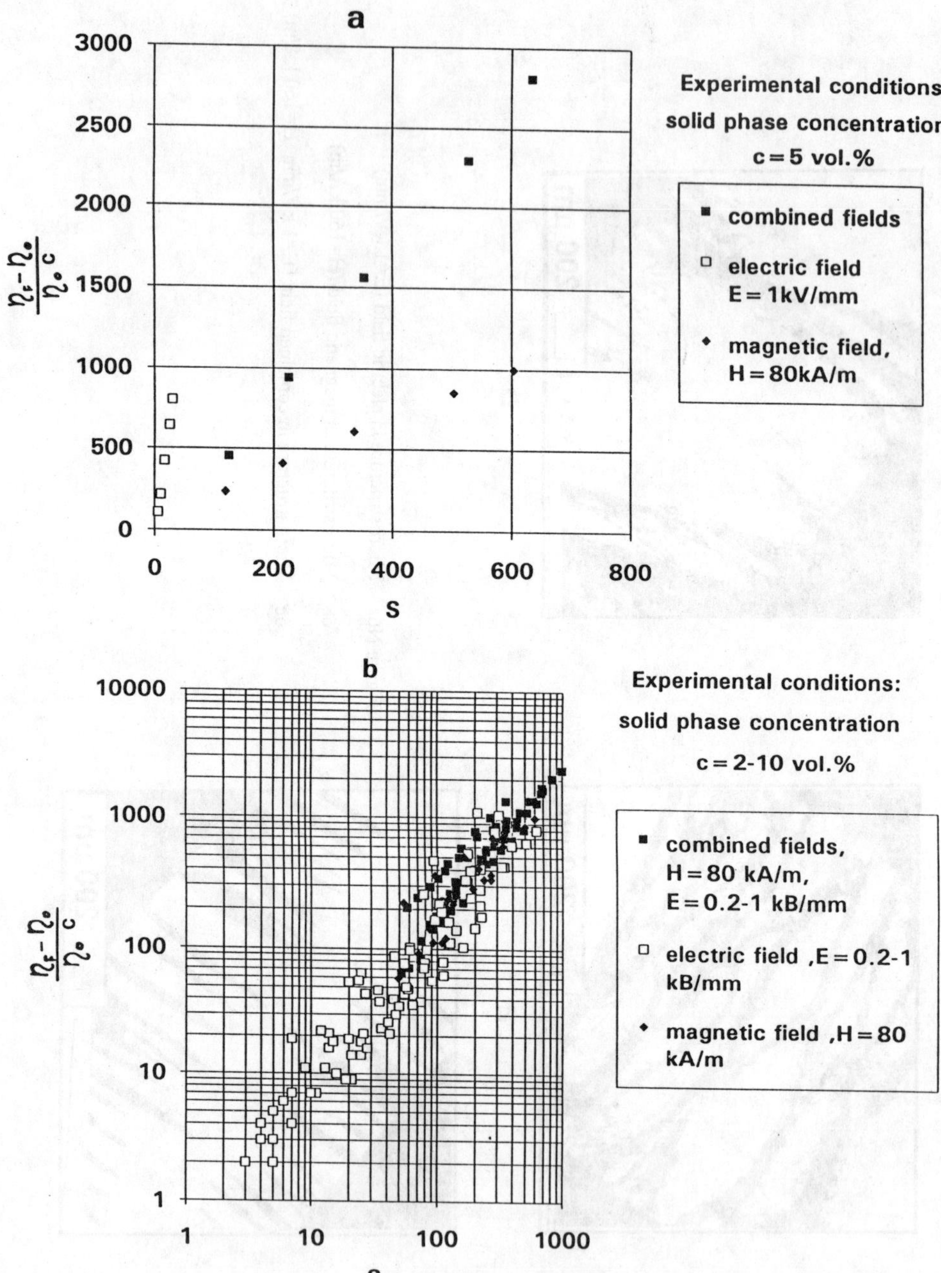

Figure N7a,b Characteristic viscosity vs modified Mason number

Cryogenic Electrorheological Fluids

R. N. Zitter, X. Zhang, T. J. Chen, and R. Tao
Department of Physics, Southern Illinois University
Carbondale, Illinois 62901, USA

ABSTRACT

Preliminary experiments on slurries of aluminum powders in liquid nitrogen find significant electrorheological (ER) effect. Though the aluminum particles have an insulate oxide surface, they still have a small conductivity which makes an ac field much more effective than a dc field in producing the ER effect. Applications of cryogenic ER fluids to rocket fuels are also discussed.

I. Introduction

Electrorheological (ER) effects at cryogenic temperatures can have valuable applications as well as offering new avenues for scientific study.

The base liquids in all of the ER fluids used or studied in the past[1–3] will freeze at temperature not too far below 300 K. In those wet ER materials where the dielectric particles contain water, the ER effect generally disappears below 273 K. Though the dry ER materials have the ER effect in a wider temperature range than the wet ER materials, no experiments have ever been reported about the ER effect far below 273 K.[1–3] The truly cryogenic ER fluids should operate at temperature $\sim$ 100 K or less and, therefore, would use a liquified gas (N_2, O_2, H_2, He, etc.) as the base fluid.

From a scientific standpoint, it would be very interesting to see if the ER effect can exist at cryogenic temperature. There are novel and unique aspects to investigations of cryogenic ER fluids: compared to room temperature, the conductivity of the particles and fluids will typically be orders of magnitude smaller, the recurring question of the effects of dissolved or adsorbed water is eliminated by freeze-out, and as a result of the extremely low viscosity of liquid nitrogen, liquid oxygen, or liquid helium, dynamic ER effects should operate on a different time scale. In addition, recently there are claims and some evidences that the Peierls-Landau thermal vibrations of single chains of particles in ER fluids help the formation of thick columns.[4–7] At cryogenic temperature, the thermal vibration would be much weaker than that at room temperature. Study of cryogenic ER fluids provides crucial tests of the validity of the above claims.

Cryogenic ER fluids may also have very important applications. For example, an important rocket fuel is a slurry fuel made of aluminum powder in liquid oxygen. Although aluminum is a conductor, its particles naturally form an insulating oxide coating so that an applied electric field is not short-circuited. While the metal itself has an extremely large dielectric constant, liquid oxygen has a very low dielectric constant, and that difference in dielectric constant should produce a strong ER effect. Thus an applied electric field can rapidly and greatly change the effective viscosity of the rocket slurry fuel, which, in turn, can change the flow velocity and control the combustion speed. An agile, precise computer-controlled ER flow valve which eliminates the problems related to mechanically moving parts and bearings is particularly attractive for cryogenic applications. Another intriguing application is that with the fast reversible ER phase transition in an electric field, slurry fuels can be stored and transported as solid fuels and combusted as liquid fuels. Since solid fuels are convenient for transportation and storage and liquid fuels are excellent for combustion, cryogenic ER fuels may have both advantages.

In this paper, we will report our preliminary experiments on slurries of aluminum powders in liquid nitrogen which, for the first time, find the significant ER effect at cryogenic temperature. Though the aluminum particles have an insulate oxide surface, they still have a small conductivity which makes an ac field much more effective than a dc field in producing the ER effect. Applications of cryogenic ER fluids to rocket fuels will also be discussed.

II. Preliminary Experiments

Since liquid nitrogen and liquid oxygen have comparable dielectric constant, density, and boiling point, our preliminary experiments are on cryogenic slurries, aluminum powders in liquid nitrogen. The results from liquid nitrogen should also be valid for liquid oxygen. Substitution of nitrogen for oxygen enables us to avoid problems of accidental combustion in the laboratory.

The cryogenic ER fluids were produced with aluminum powder of diameters $\sim 10\ \mu$m at a volume concentration $\sim 20\%$. Some details concerning the aluminum oxide layer are in order. As Inoue found in his experiments of aluminum powders in oils, if the oxide skin is not thick enough, electric breakdown may occur in ER experiments.[8] We have found that aluminum powders exposed to air for long time are good in cryogenic ER experiments. However, fresh aluminum powders just exposed to air do not have strong insulating layers. Accordingly, we followed the recipe outlined by Inoue: soaking of aluminum particles in an aqueous solution of sodium carbonate, followed by drying and then heating under nitrogen to 400^oC for hours.[8] Presumably, the initial reaction forms aluminum hydroxide which then decomposes to the oxide under heating. Inoue reports skin thicknesses approaching 1 μm depending on the soak time, with good resistance to electrical breakdown.

In the first experiment, an electric field was applied between two electrodes 21×8 mm^2 separated by 3 mm. We observed the structure formation and obtained a semi-quantitative estimate of the strength of the ER effect from the value of field required to keep material between the electrodes from falling under gravity when the electrodes were lifted from the bath. The slurries indeed acted as ER fluids. The effects of dc and 60 Hz ac fields were first compared and the ac fields were significantly more effective.

When the ac field of 60 Hz was below 6 kV/cm, the ER effect was weak. At 6 kV/cm, short broken chains began to form along the field direction. When the field reached 8.3 kV/cm, complete chains were formed to cross the two electrodes, but they were weak and easy to break. When the ac field reached 10 kV/cm or above, chains were strong enough to be lifted from the bath and sustained the structure under the gravity-induced deformation. As the field reached 10.7 kV/cm, thick columns of particles were formed between the two electrodes, characterizing a strong ER effect. When the field was reduced, column structure persisted until the field reached 3.3 kV/cm, below which the columns finally dispersed. The reason for the hysteresis is under investigation.

The difference between ac field and dc field was also observed by Inoue in his study of oxide-coated aluminum particles in oil at room temperature. He also found that a change in field from dc to 50 Hz ac produced a significantly higher shear strength. The reason for this difference lies in the fact that the aluminum particles have a small but non-zero conductivity σ although they have an oxide coating. Generally the dielectric constant (SI units) is given by[9,10]

$$\epsilon(\omega) = \epsilon_\infty + i\sigma/\omega. \tag{2.1}$$

In the dc field, $\omega = 0$, the second term in Eq.(2.1) becomes dominate even if σ is very small. The conductivity introduces constant currents which seem to be negative for polarization and for formation of solid ER structures.

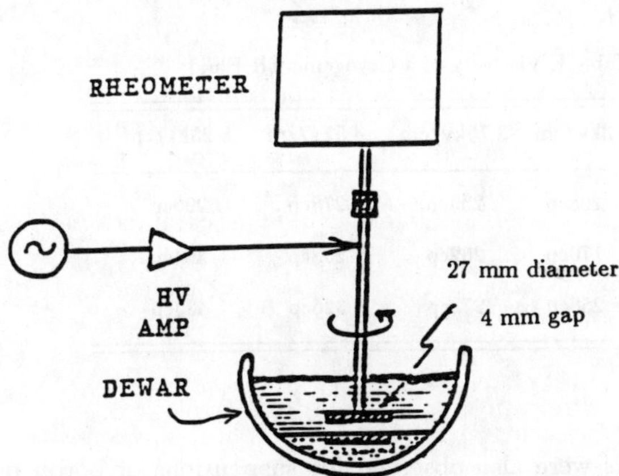

Fig.1. Experiments to determine the rheology of cryogenic ER fluids.

To suppress the negative effect of the conductivity, we need an ac field with frequency $\omega > \sigma$. It seems that the conductivity term in the dielectric constant Eq.(2.1) becomes negligible at moderate frequencies $\sim$ 60 Hz in our cryogenic ER fluids.

From Eq.(2.1), the higher the frequency ω, the smaller the negative effect from the conductance. On the other hand, since the ER effect is related to electric polarization of particles, too high frequency ω would weaken the polarization. Therefore, there must be an optimal frequency which maximizes the ER effect. To have a quantitative study, we applied a modified DV-III Rheometer to measure the viscosity of the suspension at fields of various frequency (see Fig.1).

Presently we have two ac high power supplies in our laboratory. One has a wide voltage range, but limits the output at 60 Hz frequency. The other covers the frequency up to 1000 Hz, but the voltage is limited to produce a maximum field up to $\sim$ 5 kV/cm. As noted early in the situation of 60 Hz, the ER effect is not very strong when the field is below 10 kV/cm. However, the viscosity of the suspension significantly increases with the field. When there is no field, liquid nitrogen has a negligible viscosity. At a dc field around 1 kV/cm, the viscosity is about 155 cp. However, a further increase of the dc field results little change of the viscosity. As shown in Table I, the ac field presents a different picture: The viscosity increases with the field quickly. It is clear that 100 Hz is better than other available frequencies. This implies that the optimal frequency may be above 1000 Hz. From the results at 60 Hz, we expect that the ER effect will be much stronger when the field gets above 10 kV/cm.

Accuracy and reproductibility in these measurements are far from satisfactory because the experiments were performed in a shallow dewar exposed to air, so that the ER fluid was continually being contaminated by frozen particles of water and carbon dioxide.

Table I. Viscosity of a Cryogenic ER Fluid

Freq.	3kv/cm	3.75kv/cm	4.5kv/cm	5.25kv/cm
100Hz	208cp	253cp	278cp	290cp
500Hz	170cp	252cp	258cp	304cp
1000Hz	256cp	273cp	326cp	433cp

Significant ER effects were also observed for suspensions of boron powder in liquid nitrogen. Further experiments on these and other systems will be carried out in quantitative detail, but the present preliminary results are encouraging. It clearly shows that the strong ER effect exists at cryogenic temperature. New experiments will be made in a dry box now under construction and conducted with a new ac high voltage power supply which should cover a wide range of voltages and frequencies.

III. Rocket ER Fuels

The conversion of a cryogenic fuel to an ER fluid is accomplished by the introduction of suitable particulate matter, which may or may not be removed prior to combustion. In the latter case, of course, it is desirable to use "energetic" particles which participate in and contribute energy to the combustion process. Simple inexpensive candidates are aluminum (for oxygen) and boron (for hydrogen). Aluminum, because of its conductance, must have an insulating coating. The dielectric constants of liquid oxygen, nitrogen and hydrogen are rather small and most materials will provide sufficient dielectric mismatch to obtain reasonably strong ER effects. However, the temperature involved opens the door to an entirely new realm of possible particles, namely, frozen particles of energetic substances and fuels that are normally liquid at 300K. Such particles could be fabricated with an atomizing spray in a cold chamber.

Table II. Properties of Cryogenic Liquids and Some Substances of Interest for Particulate

Substance	Density (gm/cm^3)	ϵ_∞
Liquid N_2	0.81	1.45
Liquid O_2	1.14	1.51
Liquid H_2	0.07	1.23
Al	2.70	∞
Al_2I_3	3.97	10
LiH	0.82	14
$LiAlH_4$	0.92	?
$B_{10}H_{14}$	0.94	?
$NaBH_4$	1.07	?

The liquids of oxygen, hydrogen and nitrogen unfortunately have a low density. This makes the problem of particulate setting acute. Suitable stirring and mixing procedures may therefore by a necessity in most situations. Table II lists the characteristics of cryogenic fluids and some substances of interest for particles. It may be noted that there is a fairly good density match between liquid nitrogen and lithium hydride, making this a possibly attractive pair for scientific studies.

Acknowledgements

This work was supported by U.S. Office of Naval Research Grant No. N00014-90-J-4041.

References

1. For example, see, *Electrorheological Fluids*, edited by R. Tao (World Scientific Publishing Comp., Singapore, 1992).
2. W. M. Winslow, J. Appl. Phys. **20**, 1137 (1949).
3. H. Block and J. P. Kelly, US Patent 4,687,589 (1987); F. E. Filisko and W. E. Armstrong, US patent No. 4 744 914 (1988).
4. T. C. Halsey and W. Toor, Phys. Rev. Lett. **65**, 2820(1990).
5. T. J. Chen, R. N. Zitter, and R. Tao, Phys. Rev. Lett. **68**, 2555 (1992).
6. J. E. Martin, J. Odinek, and T. C. Halsey, Phys. Rev. Lett. **69**, 1524 (1992).
7. R. Tao and Q. Jiang in this volume.
8. A. Inoue in *Electrorheological Fluids* edited by J.D. Carlson, A.F. Sprecher, and H. Conrad (Technomic Publishing Co., Lancaster, PA, 1990) p. 176.
9. L.C. Davis, J. Appl. Phys. **72**, 1334 (1992).
10. L.C. Davis, J. Appl. Phys. **73**, 680 (1993).

CONTROLLABLE FLUIDS: THE TEMPERATURE DEPENDENCE OF POST-YIELD PROPERTIES

KEITH D. WEISS and THEODORE G. DUCLOS

Advanced Technologies Group, Lord Corporation, 405 Gregson Drive, Cary, North Carolina 27511, USA

ABSTRACT

This paper represents the first detailed description of the affect of temperature on the properties exhibited by state-of-the-art electrorheological (ER) and magnetorheological (MR) fluids. In particular, shear stress versus shear strain rate curves, dynamic and static yield stress values, zero-field viscosity data, and current density measurements are discussed. Specific comments concerning the stability of both mechanical and electrical properties over broad temperature ranges are provided. Finally, insight into the advantages associated with using electrheological and magnetorheological fluids in a controllable device is provided.

1. Introduction

An engineer must be familiar with the properties exhibited by a controllable fluid, in order to evaluate that particular fluid for use in a specific device design. These properties include dynamic yield stress ($\tau_{y,d}$), static yield stress ($\tau_{y,s}$), plastic viscosity (η_p), and current density (J). The dynamic yield stress and the plastic viscosity exhibited by a controllable fluid are known from basic design equations to affect both the size and shape of a device.[1,2] Current density, on the other hand, impacts the power consumed by an electrorheological fluid device. The relationship between the properties exhibited by a controllable fluid and the performance expected for a simple sliding plate device utilizing this fluid is highlighted in the basic design equations shown in Equations [1-3].

$$V = \frac{9}{8}\left[\frac{\eta_p}{\tau_{y,s}{}^2}\right]\left[\frac{T_{CF}}{T_\eta}\right] P_M \qquad [1]$$

$$\frac{g}{R} = \frac{3}{2}\pi f\left[\frac{\eta_p}{\tau_{y,s}}\right]\left[\frac{T_{CF}}{T_\eta}\right] \qquad [2]$$

$$P_{ER} = V\,J\,E \qquad [3]$$

Equation [1] provides an indication of the overall size of the device by defining the minimum volume (V) of fluid needed in the gap of the device in order to exhibit the desired mechanical performance. Variables related to the mechanical performance of the device are present in the form of mechanical power (P_M) and a specified control ratio (T_{CF}/T_η). Within this control ratio T_{CF} and T_η represent the torque generated in the device due to the applied field induced yield strength and the viscous drag of the controllable fluid, respectively. Equation [2] provides an indication of the shape of the device by defining the minimum gap (g) to plate radius (R) ratio of the device in terms of the specified control ratio and rotational speed (f). For an electrorheological fluid device the volume of fluid in the gap of the device can be multiplied by the measured current density (J) of the fluid and the magnitude of the applied electric field (E) as shown in Equation [3] to provide an estimate of electrical power (P_{ER}) required for the operation of the device. The power required to operate a magnetorheological fluid device does not rely on the properties exhibited by the fluid. Rather in this case, the power consumption of the device is predominately dependent upon the amount of current applied to a wire coil needed to generate the magnetic field.

The key properties exhibited by a controllable fluid that influence the shape and size of a device are shown in Equations [1&2] to be $\eta_p/\tau_{y,d}$ and $\eta_p/\tau_{y,d}^2$, respectively. In order to make a device smaller or more compact the minimization of these two viscosity to dynamic yield stress ratios is desirable. This can be accomplished by either decreasing the viscosity or increasing the dynamic yield stress exhibited by the controllable fluid. A controllable fluid must exhibit a minimum dynamic yield stress and maximum plastic viscosity level of 3.0 kPa and 700 mPa-sec, respectively, in order to be utilized in a variety of devices and applications.[3]

Many scientists have addressed decreasing the viscosity and current density, as well as increasing the yield stress exhibited by controllable fluids. Typically, the data reported in the literature is related to the performance of controllable fluids at an ambient temperature, i.e., 25°C. However, in most "real world" applications a controllable fluid must be capable of either operating over a broad temperature range or surviving exposure to extreme variations in temperature. For instance, the normal temperatures encountered for devices used in automotive applications range from -40°C to 150°C. Any variation in the properties exhibited by a controllable fluid with respect to a change in temperature will inherently impact both the mechanical and electrical performance of a device.

Unfortunately very little research describing the temperature dependence of mechanical and electrical properties exhibited by controllable fluids has been reported. A few articles exist that evaluate ER fluids containing alumino-silicate particles at high temperatures (>100°C).[4,5] Although these materials operate in a substantially "anhydrous" manner at high temperatures, the ability of these fluids to operate at low temperatures (<0°C) has not been substantiated. This paper provides the first in-depth comparison between the performance of state-of-the-art electrorheological (ER) and magnetorheological (MR) fluids with respect to temperature. In particular, the dynamic yield stress, viscosity and current density for different controllable fluids have been measured over their recommended temperature ranges. The following discussion will provide an engineer with the information and data necessary to determine the feasibility of using a controllable fluid device in a variety of "real world" applications.

2. Definitions and Experimental Design

The lack of any uniform test methodology in the published literature makes a comparison of the properties exhibited by different controllable fluid formulations a difficult task. In addition, the differences that exist between research laboratories in defining the controllable fluid response adds a tremendous amount of confusion to comparing the properties exhibited by different controllable fluids. The adoption of standard definitions and test methodology by the scientific community is necessary to aid engineers in the design of controllable fluid devices. Several recent publications provide a thorough discussion of the observed behavior and properties exhibited by controllable fluids that can be used as a basis for developing an uniform understanding of this technology.[3,6]

2.1 *Description of Electrorheological and Magnetorheological Phenomenon*

The observation of a large rheological effect induced by the application of an electric field was first reported by W. Winslow in 1947.[7] His initial publication describes the electrically induced fibrillation of small dielectric particles suspended in low viscosity oils. The formation of particle chains between electrodes alters the rheological properties exhibited by an electrorheological (ER) fluid. Thus the electrorheological phenomenon observed for these fluids corresponds to the work that is required in breaking and reforming these 3-dimensional chains. The magnitude of this required work is more commonly expressed as the force necessary to initiate flow within the fluid. In other words, the force necessary to overcome the yield stress exhibited by the electrorheological fluid.

Magnetorheological (MR) fluids are the true magnetic analogs of ER fluids. These fluids exhibit particle chaining upon the application of a magnetic field. The preparation of the first MR fluids must be credited to both W. Winslow and J. Rabinow.[6,8] Magnetorheological fluids should not be confused with ferrofluids or magnetic liquids. Due to the effect of Brownian motion on the colloids present in ferrofluids, these magnetic liquids do not exhibit the ability to form particle chains.

2.2 *Definition of Properties Exhibited by Controllable Fluids*

A Bingham plastic model as described by Equation [4] typically provides an accurate description of the rheological properties exhibited by both electrorheological[3,6] and magnetorheological[6] fluids. In fact, the previously described device design equations[1,2] assume that the controllable fluid exhibits Bingham plastic behavior. According to this model, the total shear stress (τ_y) is equal to the sum of a field induced yield stress ($\tau_{y(field)}$) and a viscous component dependent upon the plastic viscosity (η_p) of the fluid and the shear strain rate ($\dot{\gamma}$). The property of a controllable fluid that is observed to change upon an increase in applied field is not the viscosity of the fluid, but rather the yield stress defining the flow regime.

$$\tau_y = \tau_{y(field)} + \eta_p \dot{\gamma} \qquad [4]$$

Two different yield stress values, the dynamic yield stress ($\tau_{y,d}$) and the static yield stress ($\tau_{y,s}$), have been commonly reported in the literature. The dynamic yield stress of a controllable fluid is typically defined as the zero-rate intercept of a linear regression curve fit to measured flow data, while the static yield stress corresponds to the stress necessary to initiate flow irregardless of whether or not the fluid behaves as a Bingham plastic. The plastic viscosity of a controllable fluid corresponds to the slope of the linear regression curve fit used in analyzing the flow data.

In order to simplify the design of a device and guarantee consistent device performance, the dynamic and static yield stress values measured for a controllable fluid must be similar in magnitude. Unfortunately, many controllable fluids due to improper formulation exhibit static yield stress values that are significantly greater than the corresponding dynamic yield stress values. This phenomenon, which is known as stiction, is a structural anomaly dependent upon the size and shape of the particles, as well as on the prior applied field and flow history of the controllable fluid. Properly formulated controllable fluids, such as Lord Corporation's VersaFlo™ Fluids, eliminate the potential of any stiction by ensuring that the static yield stress exhibited by the fluid is less than or equal to the corresponding dynamic yield stress value.

A major difference between electrorheological and magnetorheological technology is the method used to generate the corresponding field. Generation of the magnetic field used to control MR fluids is accomplished with low voltage, high current techniques, such as applying an electrical current to a coil of wire.[6] As previously mentioned the power consumed by a MR fluid device directly correlates with the electrical current needed to generate the magnetic field. In this case, the properties exhibited by a controllable fluid have no effect on the power consumption of the device. On the other hand, the generation of an electric field used to control ER fluids is accomplished with low current, high voltage power supplies. Since an ER fluid acts as a leaky capacitor, the measurement of an electric current through the fluid is possible. In this case, the charge flow through the ER fluid, which is typically represented by an average current density (J), directly affects the power consumption of the device.

A distinction between ER fluids activated with D.C. and A.C. electric fields must be made at this time. The conductivity measured for an ER fluid controlled with a D.C. electric field represents energy that must be removed from the system as heat. On the other hand, the conductivity measured for an ER fluid controlled with an A.C. electric field does not necessarily represent the energy that must be dissipated. In this system, the measured current is predominately a displacement current whose magnitude depends on the capacitance of the device.

2.3 *Experimental Conditions*

The rheological data for all ER fluids were obtained using a concentric cylinder test geometry (1 mm gap) in conjunction with a shear stress controlled rheometer. The electric field was generated by applying voltage to the inner cylinder. The rheological data measured for all MR fluids were obtained using parallel plate rheometry (1mm gap) in conjunction with a shear rate controlled rheometer. The magnetic field was applied to the cell perpendicular to the parallel plates. Both systems utilized a temperature chamber to insure that the cell and the surrounding environment was within ±2°C of the desired temperature. The dynamic current density for ER fluids was measured during the course of obtaining the flow data. The dielectric spectra obtained for all ER fluids used a test method previously described in the literature.[9]

The MR fluid damper described in this paper utilized a 46 mm outer cylinder. A current of 1 amp was supplied to a wire coil in the device in order to generate the required magnetic field. The damper was operated with a stroke of ±20mm and a stroke rate of 1 Hz. The damper was completely enclosed in a temperature chamber, which was maintained at ±2°C of the desired operating temperature. The force output of the damper reported in this paper represents the data obtained on the initial stroke of the damper after temperature equilibrium was established.

3. Controllable Fluid Performance at 25°C

The controllable fluids utilized in all evaluations are obtainable from Lord Corporation under the trade name VersaFlo™ Controllable Fluids. VersaFlo™ ER-100, ER-200 and ER-201 Fluids are electrorheological fluids, while VersaFlo™ MR-100 Fluid is a magnetorheological fluid. Measurement of the properties exhibited by these controllable fluids at an ambient temperature represents a basis to which properties obtained at various temperatures can be compared.

3.1 VersaFlo™ ER-100 Fluid

VersaFlo™ ER-100 Fluid is an optimized version of a conventional electrorheological fluid that has been previously represented in published literature as ERX02 and/or ER-II. This ER fluid, which is controlled through the use of a D.C. electric field, typically exhibits a current density that is less than 20 μA/cm² at a field of 4.0 kV/mm. The dynamic yield stress and viscosity measured for this ER fluid is approximately 1.0 kPa at 4.0 kV/mm with a zero-field viscosity between 200-300 mPa-sec as shown in Figure I. VersaFlo™ ER-100 Fluid represents a controllable fluid exhibiting design ratios, $\eta_p/\tau_{y,d}$ and $\eta_p/\tau_{y,d}^2$, on the order of 10^{-4} sec and 10^{-7} sec/Pa, respectively.

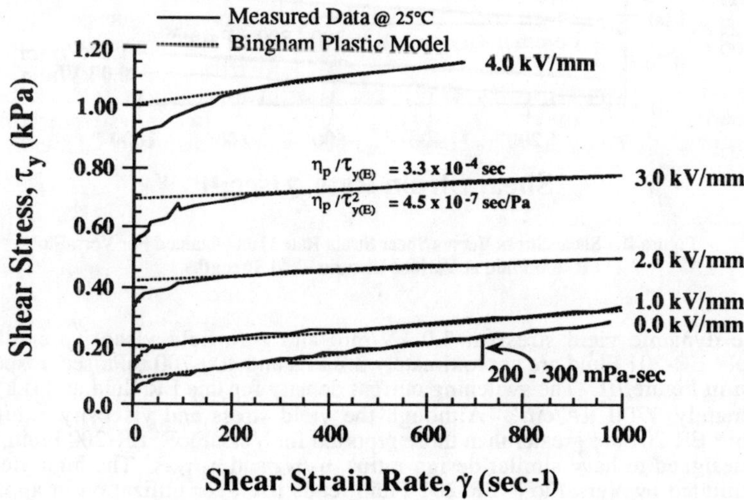

Figure I. Shear Stress Versus Shear Strain Rate Data Obtained For VersaFlo™ ER-100 Fluid at Various Electric Field Strengths.

3.2 *VersaFlo™ ER-200 and ER-201 Fluids*

VersaFlo™ ER-200 and ER-201 Fluids are examples of electrorheological fluids that exhibit both high strength and an extended temperature range of operation due to the utilization of a unique non-ionic polarization mechanism by the particle component. Due to the anhydrous nature of these fluids the application of an A.C. electric field is necessary to prevent particle electrophoresis. Optimum performance of these ER fluids is obtained using a bipolar square wave with a frequency greater than or equal to 500 Hz.

The dynamic yield stress at 3.0 kV/mm and zero-field viscosity exhibited by VersaFlo™ ER-200 Fluid are approximately 1.8 kPa and 200-300 mPa-sec, respectively, as shown in Figure II. The switching current density for this ER fluid at 3.0 kV/mm is approximately 1400 $\mu A/cm^2$. The design ratios, $\eta_p/\tau_{y,d}$ and $\eta_p/\tau_{y,d}^2$, for VersaFlo™ ER-200 Fluid are similar to those previously described for VersaFlo™ ER-100 Fluid.

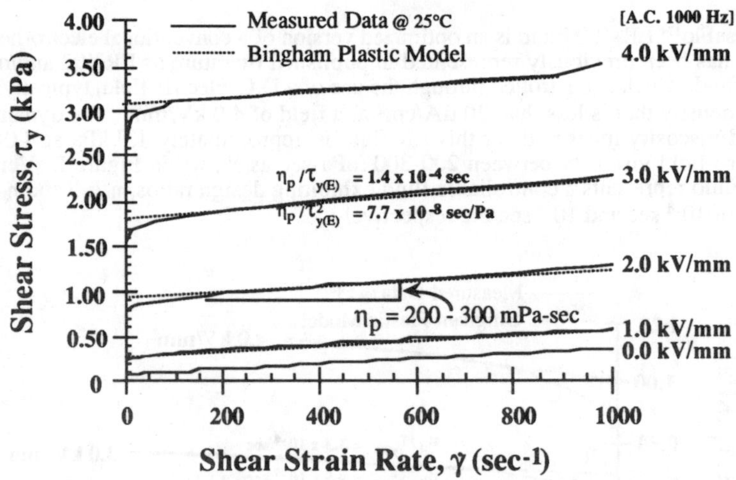

Figure II. Shear Stress Versus Shear Strain Rate Data Obtained For VersaFlo™ ER-200 Fluid at Various Electric Field Strengths.

The dynamic yield stress at 3.0 kV/mm and zero-field viscosity exhibited by VersaFlo™ ER-201 Fluid are approximately 3.2 kPa and 400-700 mPa-sec, respectively, as shown in Figure III. The switching current density for this ER fluid at 3.0 kV/mm is approximately 1700 $\mu A/cm^2$. Although the yield stress and viscosity exhibited by VersaFlo™ ER-201 are greater than those reported for VersaFlo™ ER-200 Fluid, this ER fluid is designed to have similar design ratios, $\eta_p/\tau_{y,d}$ and $\eta_p/\tau_{y,d}^2$. The high yield stress value exhibited by VersaFlo™ ER-201 Fluid lends itself for utilization in applications where the ultimate force output of a device is the critical factor. For applications, that are more concerned with very low force output in the absence of any electric field (off-state) the utilization of VersaFlo™ ER-200 Fluid is recommended.

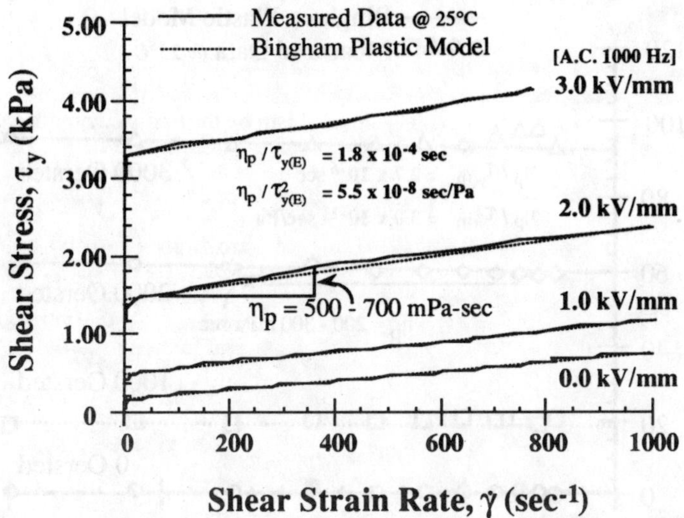

Figure III. Shear Stress Versus Shear Strain Rate Data Obtained For VersaFlo™ ER-201 Fluid at Various Electric Field Strengths.

3.3 VersaFlo™ MR-100 Fluid

VersaFlo™ MR-100 Fluid typically exhibits a zero-field viscosity (200-300 mPa-sec) comparable to most electrorheological fluids and a dynamic yield stress value (>80 kPa at 3000 Oersted) that is more than an order of magnitude greater than the best electrorheological fluids. The flow data obtained for VersaFlo™ MR-100 Fluid is shown in Figure IV. The high yield strength associated with this MR fluid provides for a several order of magnitude increase in the design ratios, $\eta_p/\tau_{y,d}$ (~10^{-6} sec) and $\eta_p/\tau_{y,d}^2$ (~10^{-11} sec/Pa), as compared to those exhibited by ER fluids. This result tends to lead one to believe that a MR fluid device could be made smaller than an ER fluid device. However, this result does not take into consideration that the generation of a magnetic field through the use of a coil of wire requires a certain amount of steel in the device to complete the magnetic circuit and drop the field across the MR fluid filled gap. The size or amount of this additional steel must be taken into account when designing a MR fluid device.[6] Magnetorheological fluid devices are in reality approximately the same size as ER fluid devices that utilize a high yield strength ER fluid, such as VersaFlo™ ER-200 Fluid.

4. Affect of Temperature on Controllable Fluids' Properties

Many of the properties exhibited by a controllable fluid are in some way affected by a change in operating temperature. For instance, the viscosity exhibited by the carrier oil present in the controllable fluid decreases as the temperature is elevated. Lower carrier oil viscosity ultimately correlates with a decrease in the viscous component of the shear stress measured for a Bingham plastic fluid.

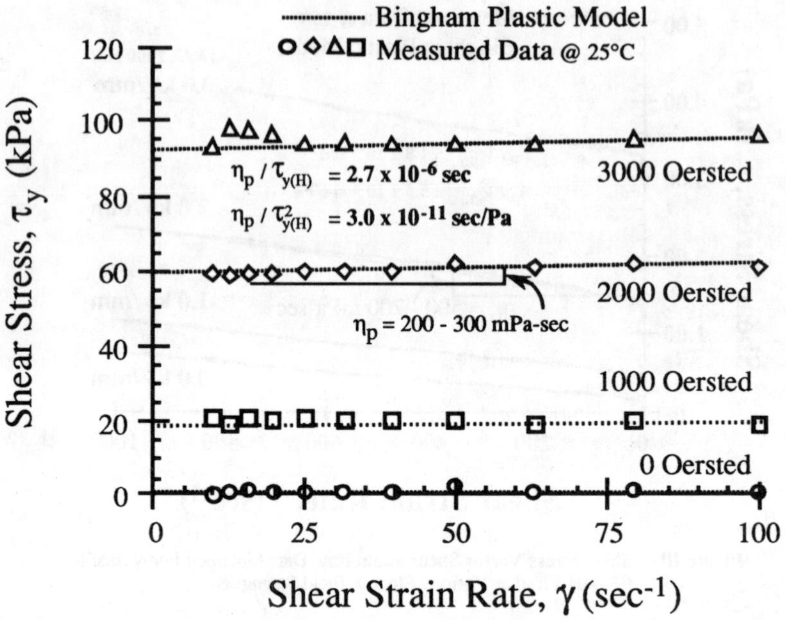

Figure IV. Shear Stress Versus Shear Strain Rate Data Obtained For VersaFlo™ MR-100 Fluid at Various Magnetic Field Strengths.

The dynamic yield strength exhibited by controllable fluids also is expected to decrease as the operating temperature is elevated. Primarily, the expansion of the carrier oil in the controllable fluid alters the volume that this oil occupies. The particle volume fraction (ϕ), which is known from quantitative modeling of ER fluids to be directly proportional to yield stress[10,11], decreases as the operating temperature of the controllable fluid is increased. Considering the typical values for the thermal expansion coefficient of conventional oils (i.e., silicone and mineral oils), a change in yield stress of about 10% over the temperature range of -40 to 150°C is expected to result from the thermal expansion/contraction of the carrier oil.

In addition to thermal expansion/contraction considerations, the yield stress exhibited by an electrorheological fluid is influenced by the ability of the particles to polarize. Particle polarization only occurs when there is a difference in the permittivity of the particle (ε_2) and carrier oil (ε_1) present in an electrorheological fluid. In most quantitative models the polarizability of the particle component is described through the use of a complex dipolar coefficient, ß, as shown in Equation [5].[10,11] This complex dipolar coefficient inherently contains contributions from the conductivity and relative permittivity (dielectric constant) associated with both the carrier oil and particles in an ER fluid. Any variation in these parameters will influence the polarizability of the particle component and alter the magnitude of the observed ER effect.

$$\beta = \left[\frac{\varepsilon 1 - \varepsilon 2}{\varepsilon 2 + 2\varepsilon 1}\right] \qquad [5]$$

The conductivity of a controllable fluid is expected to increase as the operating temperature is elevated due to the ability of ions to more freely migrate. In addition to the potential changes that may occur in the complex dipolar coefficient as previously described, an increase in the current density exhibited by an ER fluid at elevated temperatures will result in excessive power consumption and heat generated by the device.

4.1 VersaFlo™ ER-100 Fluid

Since VersaFlo™ ER-100 Fluid employs an ionic mechanism for particle polarization, the operating temperature range for this fluid is limited (10 to 90°C). A decrease in the plastic viscosity and dynamic yield stress (@ 3.0 kV/mm) of 73% and 53%, respectively, is observed for this ER fluid over the operating temperature range as shown in Figure V. The measured decrease in the dynamic yield stress exhibited by this ER fluid is considerably greater than that anticipated to occur as a result of the thermal expansion/contraction associated with the carrier oil. Thus the sensitivity of the complex dipolar coefficient to variation in operating temperature impacts the dynamic yield stress exhibited by this ER fluid. In particular, a change in the conductivity exhibited by the particle component in this ER fluid affects the ability of these particles to polarize at elevated temperatures.

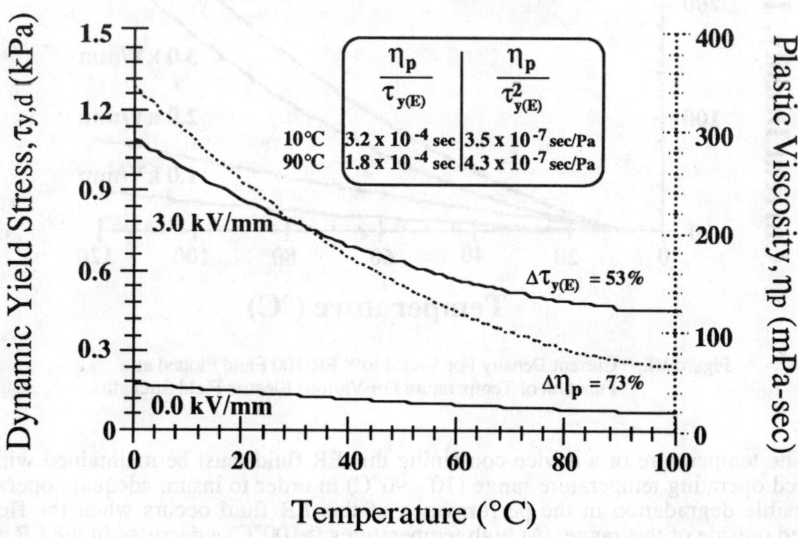

Figure V. Dynamic Yield Stress (Solid Lines) and Plastic Viscosity (Dotted Line) For VersaFlo™ ER-100 Fluid Plotted as a Function of Temperature.

Interestingly, the overall affect of temperature on the design ratios of viscosity to yield stress, $\eta_p/\tau_{y,d}$ and $\eta_p/\tau_{y,d}^2$, is negligible. These two ratios remain approximately constant over the entire temperature range due to the changes observed in both the plastic viscosity and dynamic yield stress exhibited by the ER fluid. Although these design ratios are not altered by a variation in temperature, the changes observed in yield stress and viscosity exhibited by this ER fluid with respect to temperature will impact the performance of a device. A discussion concerning the affect of temperature on device performance is presented in Section 5.

A major concern in designing a device to use an ER fluid whose particles polarize through an ionic mediated mechanism relates to both the magnitude and stability of the power consumed by the device. A large increase in the current density measured for VersaFlo™ ER-100 Fluid is observed upon elevating the operating temperature as shown in Figure VI. For instance, at a D.C. electric field of 4.0 kV/mm the current density exhibited by this ER fluid increases from 10-15 μA/cm² at 25°C to ~200 μA/cm² at 90°C. Thus the power consumed by a device containing an ER fluid that utilizes an ionic mediated polarization mechanism will dramatically increase with an elevation in operating temperature.

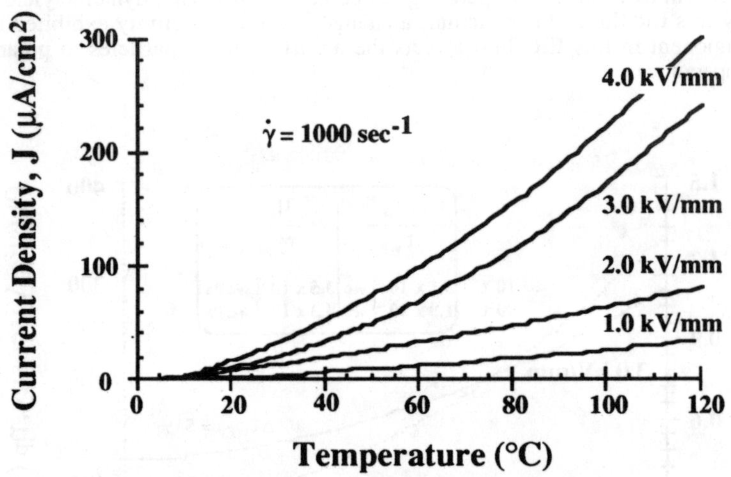

Figure VI. Current Density For VersaFlo™ ER-100 Fluid Plotted as a Function of Temperature For Various Electric Field Strengths.

The temperature of a device containing this ER fluid must be maintained within a specified operating temperature range (10 - 90°C) in order to insure adequate operation. Irreversible degradation in the performance of this ER fluid occurs when the fluid is operated outside of this range. At high temperatures (>100°C) a decrease in the ER effect occurs, while at low temperatures (<0°C) particle electrophoresis to one of the electrodes is observed. Although improvements in the performance of this ER fluid at high temperatures is possible, dramatic increases in current density will still occur. An increase in current density at high temperatures and irreversible particle electrophoresis at low temperatures are two problematic areas that must be addressed for any device incorporating an ER fluid that utilizes an ionic polarization mechanism.

4.2 VersaFlo™ ER-200/201 Fluid

Due to the non-ionic nature of the polarization mechanism utilized in VersaFlo™ ER-200/201 Fluids, the operating temperature may range from -25°C to 125°C. As shown in Figure VII, a 70% decrease in the dynamic yield stress and a 95% decrease in plastic viscosity is observed for VersaFlo™ ER-200 Fluid over this operating temperature range. Similar behavior is observed for VersaFlo™ ER-201 Fluid. The overall affect of temperature on the design ratios, $\eta_p/\tau_{y,d}$ and $\eta_p/\tau_{y,d}^2$, is negligible due to the changes observed in both the plastic viscosity and dynamic yield stress of the ER fluid. Although these two ratios are not significantly changed by a variation in temperature, the affect of temperature on the yield stress and viscosity for this ER fluid will impact the performance of a device (see Section 5).

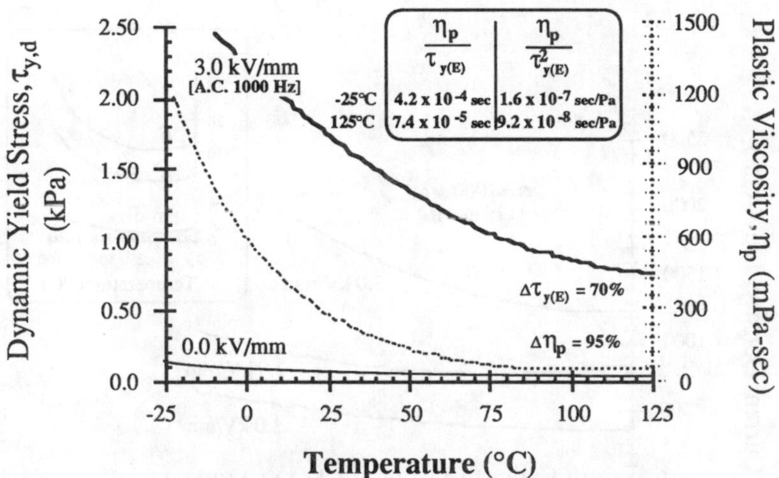

Figure VII. Dynamic Yield Stress (Solid Lines) and Plastic Viscosity (Dotted Line) For VersaFlo™ ER-200 Fluid Plotted as a Function of Temperature.

The large observed change in yield stress with respect to temperature for VersaFlo™ ER-200 and ER-201 Fluids provides evidence that variation in the complex dipolar coefficient occurs. The expansion/contraction of the carrier oil in these ER fluids cannot explain the 70% change in dynamic yield stress that is observed over the operating temperature range. Since these ER fluids utilize a non-ionic mechanism for particle polarization, the variation that occurs in the complex dipolar coefficient is related to temperature induced changes in the relative permittivities associated with the particle and oil components that make up the fluid.

The current density measured for VersaFlo™ ER-200 Fluid is relatively constant over the operating temperature range as shown in Figure VIII. Similar behavior is observed for VersaFlo™ ER-210 Fluid with respect to operating temperature. Since an A.C. electric field (bipolar square wave) is used to control this ER fluid, the measured current density does not represent energy that must be dissipated from the system. This current density, which depends on the capacitance of the device, actually represents the magnitude of a displacement current that must be switched between electrodes. The dielectric spectra for VersaFlo™ ER-200 Fluid shown in the insert in Figure VIII plots the storage and loss factors measured at 1000 Hz as a function of temperature. The real (storage factor) and imaginary (loss factor) components of the complex permittivity associated with the fluid are observed to increase upon elevating the temperature of the system. However, the loss tangent, which is the ratio between the loss and storage factors, is observed to be relatively constant and significantly less than one over the entire temperature range. This data provides credibility to the previous conclusion that variation in the relative permittivity of the carrier oil and particle component in the ER fluid affects the magnitude of the complex dipolar coefficient and ultimately the yield stress exhibited by the ER fluid.

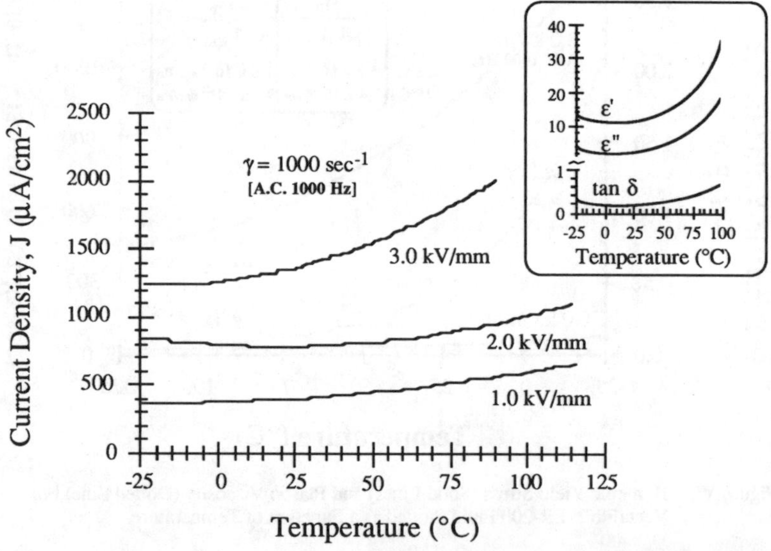

Figure VIII. Current Density Measured For VersaFlo™ ER-200 Fluid Plotted as a Function of Temperature. [Dielectric Data Obtained For ER Fluid at 1000 Hz is Provided in Insert.]

As important as the variation in the properties exhibited by an ER fluid over its operating temperature range, is the ability of the ER fluid to survive exposure to and operation at temperatures outside the recommended range. VersaFlo™ ER-200/201 Fluids exhibit this quality. These fluids can be operated at temperatures below -25°C or above 125°C for extended periods of time. Upon returning these ER fluids to within their recommended temperature range, the properties exhibited by these ER fluids are unchanged from those previously described.

4.3 VersaFlo™ MR-100 Fluid

A 10% decrease in the dynamic yield stress for VersaFlo™ MR-100 Fluid is observed over the temperature range of -40 to 150°C as shown in Figure IX. The polarization mechanism utilized in a magnetorheological fluid is not affected by variation in the relative permittivity or conductivity of the carrier oil and particle component as is the case in ER fluids. Thus the variation in dynamic yield stress observed for a MR fluid is due entirely to a change in particle volume fraction caused by the expansion at elevated temperatures and contraction at low temperatures of the carrier oil.

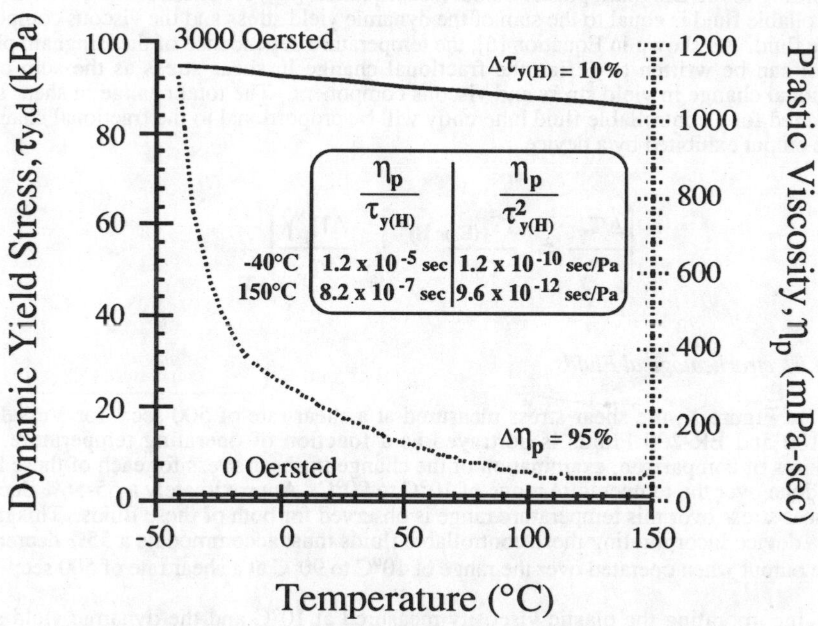

Figure IX. Dynamic Yield Stress (Solid Lines) and Plastic Viscosity (Dotted Line) For VersaFlo™ MR-100 Fluid Plotted as a Function of Temperature.

The plastic viscosity of VersaFlo™ MR-100 Fluid decreases by 95% over the temperature range of -40 to 150°C. The overall affect of this large decrease in viscosity and limited change in dynamic yield stress is a several order of magnitude decrease in the design ratios, $\eta_p/\tau_{y,d}$ and $\eta_p/\tau_{y,d}^2$, as the temperature is increased. Since minimization of these two ratios is desired, the observed decrease in these ratios represents an improvement over electrorheological fluid technology.

5. Affect of Temperature on Device Performance

In order to insure that the size and shape of a device is appropriate for use over a specified temperature range, an engineer usually must design the device to accommodate the dynamic yield stress exhibited by the controllable fluid at the highest operating temperature (T_{high}) and the plastic viscosity exhibited by the controllable fluid at the lowest operating temperature (T_{low}). Although this situation will not occur during the normal use of a device, utilization of the plastic viscosity at T_{low} and the dynamic yield stress at T_{high} in the design ratios, $\eta_{p[low]}/\tau_{y,d[high]}$ and $\eta_{p[low]}/\tau_{y,d}{}^2{}_{[high]}$, provides a first approximation in determining the feasibility of using a particular controllable fluid in a specific application.

The affect of temperature on the mechanical performance of a controllable fluid device is manifested in the Bingham plastic behavior exhibited by the controllable fluid. According to the Bingham plastic model (see Equation [5]), the shear stress observed for a controllable fluid is equal to the sum of the dynamic yield stress and the viscous component of the fluid. As shown in Equation [6], the temperature dependence of the Bingham plastic model can be written to define the fractional change in shear stress as the sum of the fractional change in yield stress and viscous component. The total change in shear stress measured for a controllable fluid inherently will be proportional to the fractional change in force output exhibited by a device.

$$\left(\frac{\Delta\tau_y}{\tau_y} = \frac{\Delta\tau_{y(E\text{ or }H)}}{\tau_y} + \frac{\Delta\eta_p\dot{\gamma}}{\tau_y}\right)_{\Delta T} \qquad [6]$$

5.1 *Electrorheological Fluids*

In Figure X, the shear stress measured at a shear rate of 500 sec^{-1} for VersaFlo™ ER-100 and ER-200 Fluids is portrayed as a function of operating temperature. For purposes of comparison, examination of the change in shear stress for each of these fluids was done over the temperature range of 10°C to 90°C. Approximately a 55-56% decrease in shear stress over this temperature range is observed for both of these fluids. This means that a device incorporating these controllable fluids must accommodate a 55% decrease in force output when operated over the range of 10°C to 90°C at a shear rate of 500 sec^{-1} .

Incorporating the plastic viscosity measured at 10°C and the dynamic yield stress measured at 90°C for the VersaFlo™ ER-100 and ER-200 Fluids into Equation [6] provides a check on the validity of the Bingham plastic model in predicting the behavior exhibited by controllable fluids. Over this temperature range a decrease in the fractional yield stress of 44% and 45% is calculated for VersaFlo™ ER-100 and ER-200 Fluids, respectively. In addition, a decrease in the viscous component for VersaFlo™ ER-100 and ER-200 Fluids of 10% and 7%, respectively, is calculated using a shear rate of 500 sec^{-1}. Thus the total expected change in shear stress of 52-54% as calculated from Equation [6] is in good agreement with the actual measured values of 55-56%.

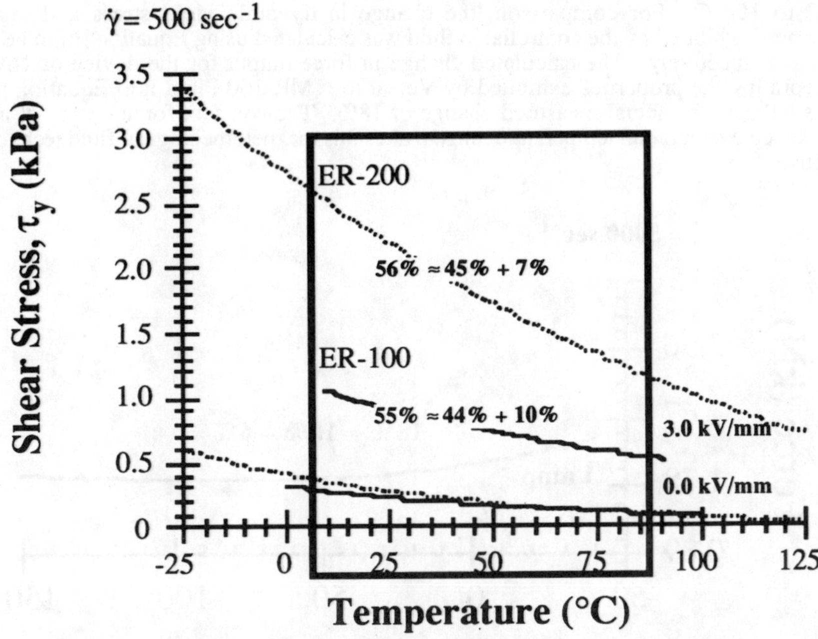

Figure X. Shear Stress Measured at 500 sec^{-1} For VersaFlo™ ER-100 [Solid Lines] and ER-200 [Dotted Lines] Fluids Plotted as a Function of Temperature.

In order to provide stability to the force output of a device over a specified temperature range, variation in the viscous component of the shear stress exhibited by a controllable fluid must be limited. In the previous comparison, the change in the viscous component of the shear stress was relatively small (7-10%). However, these calculations assumed that the shear rate of the device was 500 sec^{-1}. If a device were designed to operate at a higher shear rate (i.e., 5000 sec^{-1}), the variation in the viscous component with respect to temperature would have a more pronounced affect on the total force output of the device. In this particular example, the change in the viscous component of the shear stress would be between 70 to 100% over the temperature range of 10 - 90°C. This increased variation in the viscous component of the shear stress exhibited by the ER fluid would cause the overall force output of the device to decrease by as much as 115-144% over this temperature range.

5.2 *Magnetorheological Fluids*

In Figure XI, the force output of a controllable fluid damper operated at a shear rate of 5000 sec^{-1} using VersaFlo™ MR-100 Fluid is plotted as a function of temperature. The total force output of this MR fluid damper was measured to decrease by 18% in going from

-40°C to 150°C. For comparison, the change in dynamic yield stress and viscous component exhibited by the controllable fluid was calculated using Equation [6] to be 10% and 6%, respectively. The calculated change in force output for the device of 16% by incorporating the properties exhibited by VersaFlo™ MR-100 Fluid into Equation [6] is very similar to the actual measured change of 18%. The constant force output of a MR fluid device over a broad temperature range makes this magnetorheological fluid technology attractive.

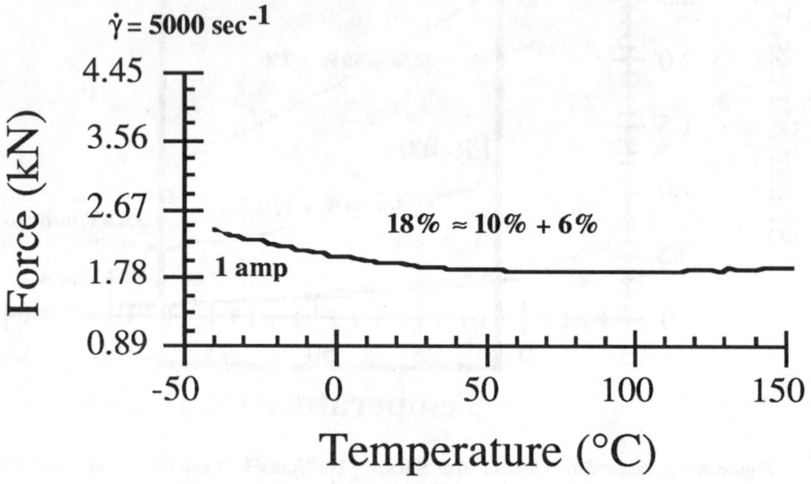

Figure XI. Force Output of MR Fluid Damper (46 mm diameter outer cylinder, ±20 mm stroke, 1 Hz stroke rate, 5000 sec^{-1} shear rate) Containing VersaFlo MR-100 Fluid Plotted as a Function of Temperature.

6. Conclusion

Extreme variation in current density, viscosity and dynamic yield stress has been reported for electrorheological fluids that utilize an ionic mediated polarization mechanism. Irreversible particle electrophoresis to one electrode is observed to occur for these fluids when operated below 0°C. At elevated temperatures, large increases in the observed current density of these ER fluids correlates with a dramatic increase in device power consumption. The variation in dynamic yield stress values with respect to a change in temperature is suggested to be predominately due to a change in the conductivity associated with the particle and oil components of the ER fluid. This change in conductivity affects the ability of the particles to polarize as described by the complex dipolar coefficient reported in quantitative models for the ER effect.

Increasing the operating temperature range of an electrorheological fluid is possible by using particles that undergo a non-ionic mediated polarization mechanism. However, these state-of-the-art ER fluids still experience a change in both viscosity and dynamic yield stress upon variation in the operating temperature. The variation in dynamic yield stress is suggested to be predominately due to a change in the relative permittivity associated with the particle and oil components of the ER fluid. This change in relative permittivity also affects the ability of the particles to polarize as described by the complex dipolar coefficient.

Although magnetorheological fluids exhibit variations in viscosity similar to ER fluids, these magnetically activated fluids exhibit only limited change in the dynamic yield stress value over an extremely broad temperature range. This small amount of change in dynamic yield stress is due entirely to the change in particle volume fraction that occurs over the temperature range. The encountered variation in particle volume fraction is caused by the expansion/contraction of the carrier oil component over this temperature range.

The fractional change in the force output of a controllable fluid device over a specified temperature range has been found to be predictable using an extension of the Bingham plastic model. Actual measured data for electrorheological and magnetorheological fluids were found to be in good agreement with the calculated values obtained from this model.

Controllable fluids currently exist that exhibit dynamic yield stress values in excess of 3 kPa and plastic viscosities less than 700 mPa-sec. The properties exhibited by these ER and MR fluids make the use of controllable fluid devices in a variety of applications feasible. In general, these controllable fluids exhibit no mechanical stiction and low current density. VersaFlo™ ER-100 Fluid is an example of a standard ER fluid exhibiting stable performance over a limited temperature range (10 - 90°C). VersaFlo™ ER-200/201 Fluids represent state-of-the-art ER fluids offering high yield stress values and stability over a broader temperature range (-25 to 125°C). VersaFlo™ MR-100 fluid is an example of a magnetorheological fluid offering extremely high yield stress values and stability over a very broad temperature range (-40 to 150°C).

7. Acknowledgements

The authors wish to acknowledge the many hours of work required by Donald A. Nixon to measure the properties of the controllable fluids described in this publication. In addition, the authors would like to thank J. David Carlson for his intellectual contributions in developing the formulations and test methodology described in this paper.

8. References

1 Duclos, T. *SAE Technical Paper #881134*, Society for Automotive Engineers, Warrendale, PA, (1988), 1-5.

2 Carlson, J. D.; Duclos, T. G. *Proceedings of the 2nd International Conference on ER Fluids*, eds., Carlson, J.; Sprecher, A.; Conrad, H. Technomic Publishing, Lancaster PA (1990), 353-367.

3 Weiss, K. D.; Coulter, J. P.; Carlson, J. D. J. Intell. Mat. Syst. & Struct. (1993), 4(1), 13-34.

4 (a) Filisko, F.; Armstrong, W. US Patent #4,744,914 (1988); (b) Filisko, F.; Armstrong, W. US Patent #4,879,056 (1989).

5 Conrad, H.; Sprecher, A.; Choi, Y.; Chen, Y. J. Rheology (1991), 35(7), 1393-1410.

6 Weiss, K.; Duclos, T.; Carlson, D.; Chrzan, M.; Margida, A. *SAE Technical Paper #932451*, Society for Automotive Engineers, Warrendale, PA, (1993), 1-6.

7 Winslow, W. US Patent 2,417,850 (1947).

8 Rabinow, J. US Patent 2,575,360 (1951).

9 Weiss, K.; Carlson, D. Int. J. Mod. Physics B (1992), 6(15-16), 2609-2623.

10 Jordan, T.; Shaw, M. IEEE Trans. Elect. Insulation (1989), 24(5), 849-878.

11 Gast, A.; Zukoski, C. Adv. Col. Interface Sci. (1989), 30, 153-202.

HIGH DIELECTRIC CONSTANT PARTICULATE MATERIALS FOR ELECTRORHEOLOGICAL FLUIDS

C.A. RANDALL, D.E. McCAULEY, C.P. BOWEN,
T.R. SHROUT and G.L. MESSING
Particulate Materials Center and
Center for Dielectric Studies
The Pennsylvania State University
University Park, PA 16802

ABSTRACT

The aim of this investigation was to establish the relative role of high dielectric constant particulates on the yield stress in silicone oil-based electrorheological fluids. Electrorheological behavior was observed in fluids containing high dielectric particulates under alternating and direct electric fields. At low frequencies (5-50 Hz) optimum field conditions were established, which indicate an advantage in the use of high dielectric constant particles for electrorheological fluids. Other factors, in addition to the intrinsic bulk dielectric properties, which influence the electrorheological properties include particle morphology and surface crystallinity .

Introduction

Electrorheological fluids (ERFs) have potential for electrically controllable shock absorbers, clutches, and engine mounts in automotive applications.[1,2,3] However, there exist a number of limitations in present electrorheological fluids including insufficient yield stress, poor temperature stability, and high power consumption.[4] Davis' (1992[5]) recent model for ERF performance suggests that high dielectric constant particulates could enhance the yield stress and yield moduli in ERFs.

The objective of this work was to explore the use of high dielectric constant particulates, based on well-known ferroelectric and related dielectric particulates, for ERFs. In the course of this study, other particulate properties relevant to the ERF performance were discovered to be important; these include morphology and surface crystallinity.

Experimental Procedure

Powders from a range of sources were obtained to test the relative role of material dielectric constant on the yield stress of an ERF. Table I outlines a variety of materials investigated, the associated bulk properties and sources. Powders with a median particle size of ~ 0.5 to 1.0 μm permitted physical consistency between ERFs and also allowed suspensions to be stable beyond the time of experimentation. The silicone oil carrier fluid had a kinematic viscosity of 50 centistokes. The powders were characterized using an Electroscan ESO-3 scanning electron microscope and BET gas adsorption using a Quantachrome MS-12.

Table I. Properties of Selected Dielectric Powders

Material Name	Chemical Formula	Density g/cm^3	Bulk Dielectric Constant - 1 KHz Room Temperature	B.E.T. Equivalent Spherical Diameter (μm)	Material Supplier
Silica	SiO_2	2.25	4	8.1 μm	Morton Thiokol Inc.
Alumina	Al_2O_3	3.97	10	0.2 μm	Apache Chemical Inc.
Rutile	TiO_2	4.26	90	0.6 μm	TAM
Bismuth Titanate	$Bi_4Ti_3O_{12}$	8.04	150	---	Molten Flux Growth*
Strontium Titanate	$SrTiO_3$	5.12	300	0.7 μm	Transelco
Barium Titanate	$BaTiO_3$	6.04	1000	0.8 μm	Transelco
Lead Magnesium Niobate	$Pb(Mg_{1/3}Nb_{2/3})O_3$	8.02	20,000	0.3 μm	Solid-State Synthesis*

*prepared at Pennsylvania State University

Particle dispersions of 20 volume % solids were examined. The powders were dispersed into silicone oil with a high shear mixing technique to ensure good dispersion and breakup of soft agglomerates.

Electrorheological measurements were made by determining the Bingham fluid response on a Brookfield-rotational viscometer DV-1. Modifications were made which permitted the application of both direct and alternating fields through a TREK620A amplifier/power supply. Various frequencies (1 Hz - 500 Hz) were obtained using a Global Specialties V-A waveform generator. The root-mean square voltage was measured using an HP-digital multimeter model 3478A.

Owing to the strong influence of electrophoresis, alternating field measurements preceded the direct field measurements. This ensured no effective loss of particles from the suspension to the electrodes until the last run of the experiment. Principally, two a.c. frequencies were used. The first gave the maximum ERF response and generally ranged between 5 to 50 Hz; this was determined by monitoring the maximum shear stress at a given shear rate while scanning through the applied frequency for a constant field strength. The second a.c. frequency was fixed at 300 Hz. The ERFs were all measured at room temperature and within a few hours of mixing to prevent any effects of suspension instability.

Observations of the electrically induced fibril structures due to particle alignment were performed in thermoset-based suspensions using silicone elastomer (Sylgard-184 Dow Corning) Alignment was induced under electric fields in the unpolymerized state and cross-linked at 60°C for approximately 40 minutes. Sections from the polymerized silicone elastomer were studied using SEM to investigate fibril structures.

Results and Discussion

Using dielectric particles of similar size and volume fraction, various ERFs were mixed such that the intrinsic dielectric constant of the materials was the major difference between the various suspensions examined. The ERFs all demonstrated typical Bingham behavior, especially under the alternating fields, as seen in Figures 1a,b.

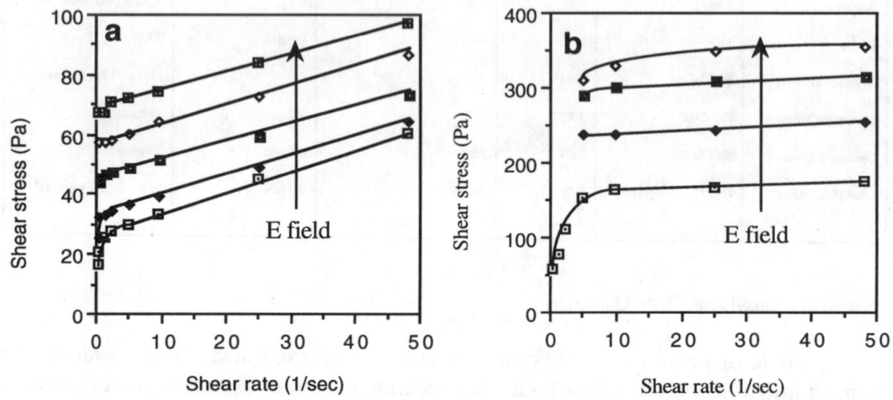

Figure 1. Examples of the Bingham fluid response for the ERFs studied (A) $BaTiO_3$-20 volume% at 300 Hz for fields between 0.2 to 0.8 kV/mm and (B) $Bi_4Ti_3O_{12}$-10 volume% for fields between 0.2 to 0.4 kV/mm.

From the rheology plots, yield stresses could be extrapolated from the dynamic Bingham regime. Also by measuring the yield stresses for various electric field conditions, the quadratic dependence of the ERF response as a function of the field strength could be evaluated. The yield stress-field square dependences for various ERFs are given in Figure 2(a-g).

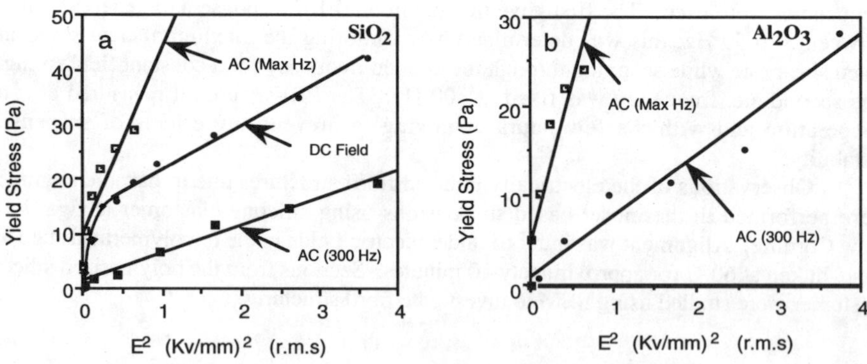

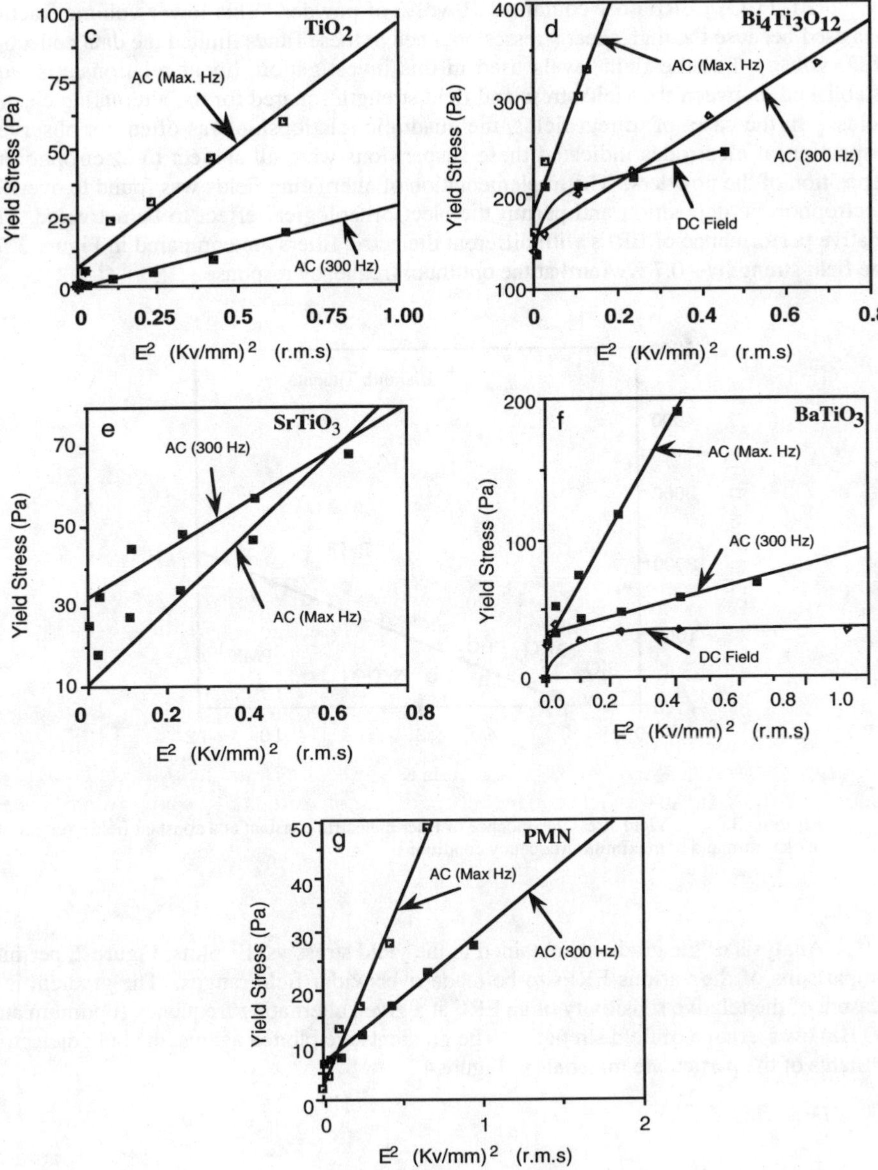

Figure 2(a-g) Series of yield stress - electric field plots for the various ERFs.

$Bi_4Ti_3O_{12}$ ERF only contained 10 vol% of powder. This lower volume fraction was used because the high shear stresses obtained in these fluids limited the data collection at 20 vol%. For the field levels used in this investigation, linear relationships were established between the yield stress and field strength squared for the alternating electric fields. In the case of direct fields, the quadratic relationship was often not observed. Inspection of electrodes indicated these suspensions were all subject to electrophoretic deposition of the powders. The implementation of alternating fields was found to override electrophoretic deposition and permit the electrorheological effect to be activated. The relative performance of ERFs with different dielectric fillers are compared in Figure 3 for one field strength (~ 0.7 Kv/mm) at the optimum frequency response.

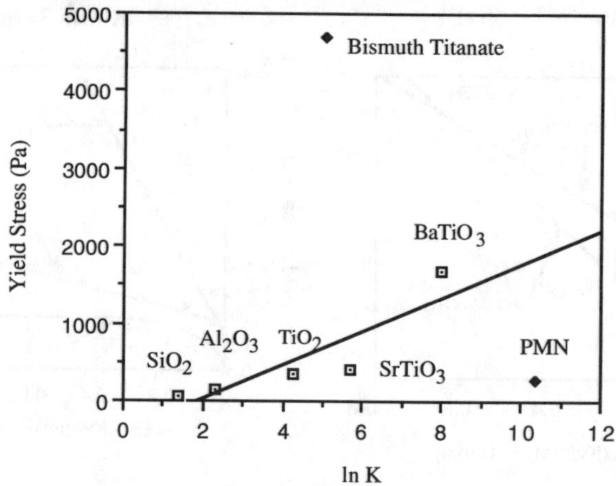

Figure 3. Yield stress dependence of filler dielectric constant at a constant field strength of 0.7 kV/mm and at maximum frequency conditions.

Analysis of the gradients obtained in the yield stress vs. E^2 plots, Figure 2, permits comparisons of the various ERFs to be made over wider field ranges. The gradient is a measure of the relative sensitivity of an ERF at a given alternating frequency (optimum and 300 Hz) over a range of field strengths. The gradients are plotted against the bulk dielectric constants of the particulate materials in Figure 4.

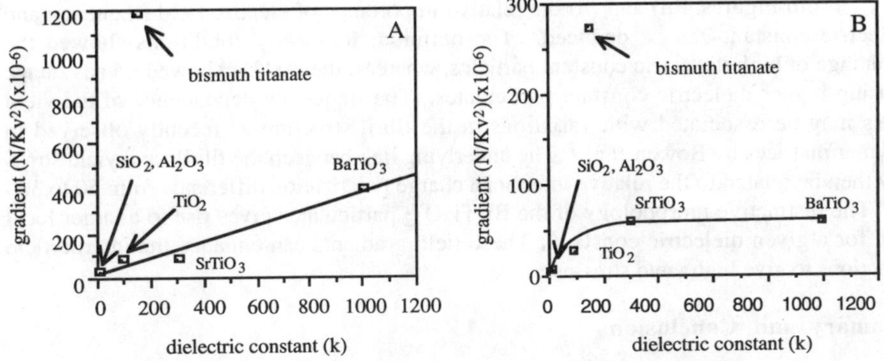

Figure 4. Gradient plots determining the relative dependence of ERF - response as a function of filler dielectric constant at (A) maximum frequency and (B) 300 Hz.

From Figures 3 and 4a,b a general increase of yield stress with bulk dielectric constant for ERFs under optimum frequency conditions (5-50 Hz) was found. This trend corresponds closely to the dodecane-based ERFs studied by Garino et al.[6] From Figure 3 there exist two fluids which did not follow the general trend. These exceptional cases contained 10 vol% $Bi_4Ti_3O_{12}$ and 20 vol% $Pb(Mg_{1/3}Nb_{2/3})O_3$. From a scanning electron microscope investigation, the $Pb(Mg_{1/3}Nb_{2/3})O_3$ particles showed an amorphous coating, perhaps resulting from the excess PbO added during powder synthesis to compensate its high volatility during calcination. This amorphous coating would effectively lower the dielectric constant of the particle to levels far below the expected bulk properties. The $Bi_4Ti_3O_{12}$ ERFs showed properties which exceeded their expected diclectric constant performance. Figure 5 is a scanning electron micrograph of a cross-linked sample showing the platelet morphology of the $Bi_4Ti_3O_{12}$ aligned corner to corner along the fibril axis.

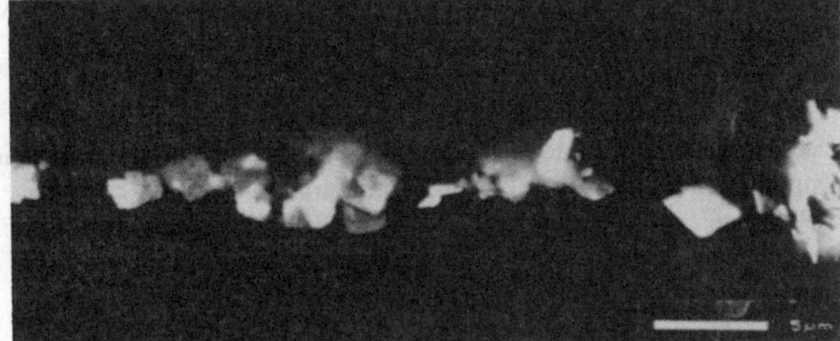

Figure 5. Scanning electron micrograph of aligned platelet $Bi_4Ti_3O_{12}$ particles in a cross-linked silicone elastomer matrix.

From Figures 4(a) and (b) the relative importance of electric field frequency and dielectric constant can be deduced. The optimum frequency conditions showed the advantage of high dielectric constant particles, whereas, the 300 Hz showed no advantage in using higher dielectric constant particulates. The frequency dependence of the yield stress may be associated with variations in the fibril structure as recently observed in polymer matrices by Bowen et al.[7] The underlying link between the fibrils and yield stress may then be related to the relative interfacial charge polarization difference from 50 to 300 Hz. The distinctive morphology of the $Bi_4Ti_3O_{12}$ particulates gives rise to a larger local field for a given dielectric constant. These field gradients can enhance the interparticle attractions to give high yield stresses.

Summary and Conclusion

High dielectric constant particulates suspended in silicone oil are suitable for ERF components at low frequencies ~5 - 50 Hz. Particle morphology was found to strongly enhance the yield stress as observed with $Bi_4Ti_3O_{12}$ platelets. Based on these observations, there exists great potential in designing ERFs with modern powder synthesis techniques and using suitable high dielectric materials. Using technologies established in the capacitor and piezoelectric ceramics-based industries, further manipulation of other ERF properties, including temperature stability and power consumption may be possible through material choice and dopant control. The tailoring of dielectric properties and particle morphology makes high dielectric particulates very attractive for ERF applications.

References

1. W.M. Winslow, J. Appl. Phy., 20(92) 1137 (1949)
2. T.G. Dulos, D.N. Acker, J.D. Carlson, Machine Design, 42-46 (1988)
3. P.J. Mullins, Automotive Industries 48-49 (1989)
4. J.E. Barnes and J.D. Carlson, Proceedings of Second International Conference on Electrorheological Fluids. Edited by J.D. Carlson, A.F. Sprecher, M. Conrad 93-114 (1989)
5. L.C. Davis, Appl. Phys. Letts., 60(3) 319-321 (1992)
6. T. Garino, D. Adolf and B. Hance, Proceedings of the International Conference of Electrorheological Fluids., Edited by R. Tao, 167,(1991) World Scientific, Singapore, New Jersey, London, Hong Kong
7. C.P. Bowen, C.A. Randall, A.S. Bhalla, R.E. Newnham Submitted to J. Mat. Res. (1993)

ELECTRORHEOLOGICAL FLUIDS BASED ON POLYURETHANE DISPERSIONS

ROBERT BLOODWORTH

Central Research, Bayer AG, 511368 Leverkusen, Germany

ABSTRACT

This paper describes a new class of electrorheological fluids based on non-aqueous polyurethane dispersions. The fluids are water-free and were developed as an alternative to first generation silica gel-based ERF's. The dispersed phase is a specially developed polyurethane which stabilizes and solvates metal ions. The fluids exhibit an attractive combination of properties: low viscosity, high ER-effect, and low conductivity. The fluids resist settling and are readily redispersible. The ER-properties have been correlated with dielectric loss measurements and can be directly related to the rate of polarization and shear rate. Using the concept of time-temperature equivalence, a master curve describing the behavior of a polyurethane ER-fluid at any temperature or shear rate was constructed from couette viscometer data obtained at different temperatures.

This paper is concerned with a new class of electrorheological fluids which have been developed at Bayer during the past several years. These materials are the result of an intensive effort to develop stable, reproducible ER-fluids for practical applications without the problems associated with water based systems.

1. History of ER-Fluid Development at Bayer AG

Work on ER-Fluids at Bayer started in 1981. The aim was to find ER-fluids for positively controlled engine mounts and other technical applications. Practical ER-fluids must fulfill stringent requirements to meet this target. In automotive applications, for example, the ER-construction must remain effective over a wide temperature range from -30°C (-22°F) to more than 110°C (230°F). In addition, the ER-fluid must be resistant to settling and aging, and stable in storage. Perhaps most important, the fluids should not attack, dissolve or cause swelling of materials coming into contact with them, in particular elastomers.

At the time of this work, ER-fluids which satisfied the above requirements were neither commercially available, nor did appropriate information exist in the literature describing such fluids. Therefore both the ER-fluids and the appropriate test methods were developed at Bayer AG during a research project lasting several years.

1.1. ER-Fluids Based on Zeolite and Silica Gel Dispersions

The results of these research and development efforts, which included the examination of a large number of different systems, were dispersions of silica gel or silicates in silicone oil or hydrocarbons. The fluids were ER-active by virtue of adsorbed water on the silica gel (1-5 Wt.%).

An ER-fluid based on a silica gel dispersions had already been described by Winslow[1] and others. However, Oppermann and Gossens[2-4] discovered that the addition of special functional polysiloxanes and/or polymers as dispersing agents was of critical importance in improving the resistance to settling and the electrorheological properties of the dispersions. These special dispersing agents made it possible to increase the volume fraction of solids and to decrease the quantity of adsorbed water. Consequently, the electroviscous effect of these fluids is quite strong even at temperatures up to and exceeding 110 °C (230°F). More recent work by Filisko[5] has shown that such zeolite based systems can also function without the presence of water.

Despite the improvement that these "new" silica gel fluids could be activated by very small quantities of adsorbed water, they still had a number of serious problems which prevented their use in practical applications.

Table 1: Advantages and disadvantages of silica gel and zeolite based ER-fluids

Advantages	Disadvantages
high shear stresses	high zero field viscosity
low minimum field strength	relatively high conductivity
	abrasive
	sedimentation; stability

Table 1 lists the advantages and disadvantages of silica gel based fluids. It is apparent that although the fluids are capable of high shear stresses, this property alone is in and of itself insufficient for practical applications. Especially troubling is the high conductivity (a direct result of the adsorbed water), abrasivness and perhaps most importantly, the sedimentation and stability of these fluids. The major problem with water based fluids is the strong dependence of the ER-effect upon the concentration of water, however, an additional problem of silica based fluids is the self-condensation of activated silica gels in the presence of water. When such a fluid sediments and is left standing over along period of time, not only is it not redispersible, it has often solidified into something resembling cement. Although it is possible to compensate these problems to some extent through surface modification of the particles (e.g. surfactants), the goal of the present work was to avoid water-based systems, and if possible, to choose a dispersed phase that would be non-abrasive and inherently less prone to sedimentation.

1.2. Considerations in Developing Improved ER-Fluids

Table 2 lists the design criteria for a new generation of ER-fluids. Contrary to the generally followed approach of maximizing the ER-effect at all cost, additional design parameters of practical importance were given priority. The varied requirements for an ER-Fluid can be separated into "necessary" and "advantageous" properties, guiding the choice of the dispersed and continuous phases.

Table 2: Goals for the development of new generation ER-fluids.

Primary Goals	Secondary Goals
stability	low zero-field viscosity
acceptable, reproducible ER-effect	low conductivity
water-free	wide effective temperature range
non-abrasive dispersed phase	
low sedimentation; re-dispersible	

2. ER-Fluids Based on Ionic Conductors

The relatively low densities of suitable non polar liquids for the continuous phase limits the choice of dispersed phases to materials with low densities. One interesting class of materials is the organic conductors. An examination of the literature concerning ERF dispersed phases based on organic conductors reveals that as a result of the methods used in their preparation, most contain ionizable groups (admittedly sometimes inadvertently), and are in fact ionic conductors.

Table 3 depicts a number of known dispersed phases whose polarization could be explained by an ion-conduction mechanism. Among the inorganic phases, the aluminosilicates with the highest known effects contain lattice defects which allow ion mobility, or are "doped" with additional metal atoms. Metal hydroxides have been of reported to show good ER-activity when activated with water or a polar solvent which facilitates the mobility of the ions. A number of interesting systems based on anhydrous inorganic conductors such as zirconium doped aluminosilicates have been reported in the recent patent literature.[6-8]

Table 3: Examples of ERF dispersed phases based on ionic conduction mechanisms.

Inorganics	Important Mechanisms
- zeolites - aluminosilicates - metal hydroxides	1) Polarization of the ions is assisted by the presence of water. 2) Dopant cations are conductive. 3) Stabilization via an ionic matrix of counterions.
Organics - conducting polymers (doped) - polyelectrolytes such as: poly(lithium methacrylate) poly(sodium styrene sulfonate)	O O⁻Li⁺ O O⁻Li⁺ / O O⁻Li⁺ O O⁻Li⁺ / COO⁻ Li⁺ COO⁻ Li⁺ Li⁺ COO⁻ Li⁺

On the polymer side, the well known polyelectrolyte based systems (including functionalized ion-exchange resins) are usually activated with water to increase the ER-response. However, a number of reports suggest that anhydrous polar solvents (even aprotic ones like propylene carbonate) can also be used to activate these systems.[9-12] Depicted in Table 3 is the structure of poly(lithium methacrylate), a well known dispersed phase in ER-fluids. As a dry powder, the lithium ions in the particles are strongly bound by the carboxylate groups. The addition of water or another substance which can solvate the lithium increases the ionic mobility and allows for a surface polarization of the particle in an electric field. We propose that the special role of water for the ER-effect lies in its solvating ability, rather than in its inherent polarizability in an electric field.

The conducting polymers hare listed here as a small caveat to those interested in using such materials as dispersed phases. Such polymers are able to function as electron conductors, and should have advantages over ionic systems when one considers the inherently low temperature coefficient of electronic conduction. However, many of these polymers must be doped in order to achieve appreciable conductivities and most are polymerized via initiation with metal ion complexes. Because most of these initiators and doping agents are oxidants or strong Lewis acids which can generate charged byproducts, the polymer must be carefully purified of free ions from the initiating system or the dopant. Otherwise, the measured ER-effects are likely to result from ionic polarization mechanisms.

A careful review of the available literature reveals that the only known truly strong ER-fluids are based on systems with ionic polarization mechanisms. As previously mentioned, a great deal of work has been done using ionic polymers activated with water or solvents, however, the use of liquid activators of any type is fraught with the problems associated with the evaporation or leaching of the activator.

2. 1 Polyurethane-based Ionic Conductors as ERF Dispersed Phases

Polymers based on functional polyethers (as well as aliphatic polyesters and polycarbonates) can be cross linked with isocyanates to form polyurethane elastomers which are solid solvents for metal salts. Especially interesting are systems containing sequences of polyethylene oxide. The high solvating power of the PEO segments for metal atoms stems from the interaction of the electron pairs on oxygen. The powerful cryptand 18-crown-6, an excellent complexing agent for lithium and sodium ions is based on an advantageous circular arrangement of the oxygen atoms coordinating the metal cation. Similar structures, in which a central metal ion is enwrapped by a helical PEO-chain have been determined for crystalline complexes of ethylene glycol oligomers and alkali metal salts.[13]

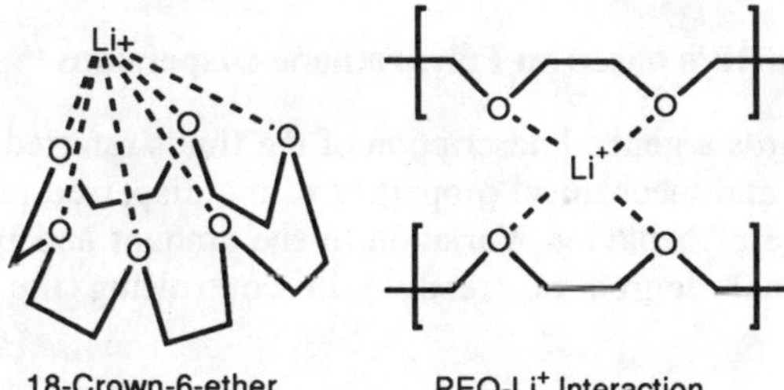

Figure 1: Schematic representation of the Polyether:Li^+ interaction.

The functional polyether sequences can, for example, can be conveniently linked with isocyanates to form polymers with the dimensional stability required for an ERF dispersed phase. Using expertise gained from many years of research into polyurethane structure and properties, we can optimize the chemistry of the polymer network to control the ion mobility, thus yielding a new degree of control over the polarization process in the dispersed phase of these ER-fluids.

Polyether or Ester Soft Segment	Urethane Crosslinking
low Tg permits ion mobility	enhances temperature stability
dissolves and binds charge carriers	furnishes structural integrity

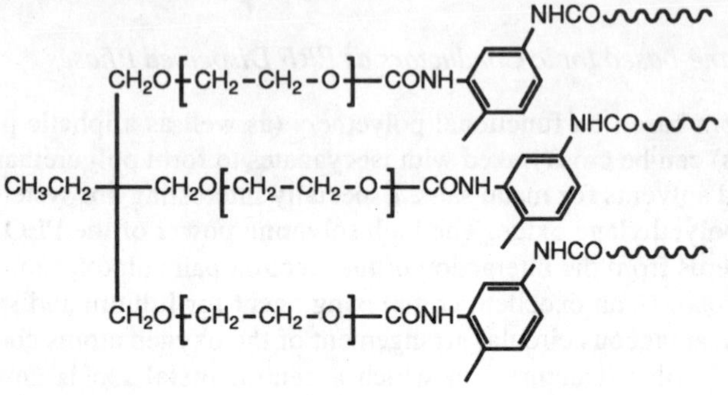

Figure 2: Example of structure and properties of a simple polyurethane ERF dispersed phase

3. Composition of ERF's based on Polyurethane Dispersions

Table 3 affords a general description of the fluids referred to throughout this talk. The electrical and mechanical properties of the dispersed phase can be varied within a wide range of behavior. Variation of the amount and type of added salts allows an additional degree of freedom in controlling the susceptibility to polarization.

Table 4: General description of an ER-Fluid with a polyurethane dispersed phase.

Dispersed Phase	Continuous Phase
crosslinked polyurethane dissolved salts up to ca. 60% solids very fine particles (app. 3-5 μm) density app. 1.09	polydimethylsiloxane oil silicone dispersing agents stabilizers density app. 0.917

Silicones remain a top choice for the continuous phase as a result of their favorable combination of properties, chiefly their high stability and low temperature coefficient of viscosity. Special dispersing agents and stabilizers are used to improve the stability against sedimentation. The small particle size allows for fluids with a high volume fraction of solids while maintaining a low zero-field viscosity. This also slows the inevitable sedimentation of the dispersed phase with time and improves the stability of the dispersion. While density matching of the two phases is possible in theory, (at one temperature!), the fluorinated and chlorinated oils with the necessary densities are expensive and raise a number of toxicological and ecological concerns.

The polyurethane dispersions are made using a proprietary in-situ polycondensation reaction which allows a good degree of control over the polymer properties and the particle size distribution. The importance of the size distribution of particles lies in obtaining stabile dispersions with low zero field viscosities.

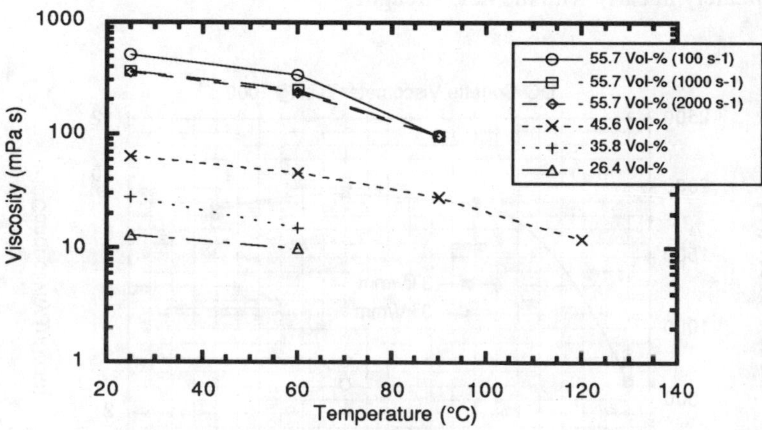

Figure 3: Effect of temperature and volume fraction solids on the zero field viscosity.

Figure 3 depicts the temperature dependence of the zero field viscosity while varying the volume fraction of dispersed phase. In general, the viscosities are quite low (< 500 mPa s), considering the high volume fraction of solids present. The temperature dependence of the viscosity increases with increasing volume fraction of polyurethane. The viscosity appears virtually shear rate independent up to a volume fraction of 45.6%. The 55.7% sample shows some shear thinning upon going from 1000 - $2000 s^{-1}$. We have found that volume fractions between 40-50% of the dispersed phase lead to a good all-around combination of properties, but that lower or higher volume fractions can be useful for special applications.

4. Electrorheological Properties of the PU-Based Fluids

Figure 4 depicts the electrorheological response of a typical PU-ERF optimized for a maximum ER-effect in the 60-100°C temperature range (high temperature formulation). Characteristic of the PU systems is their very low zero field viscosity and their relatively low conductivity. A low zero shear viscosity combined with a relatively high ER-response yields a large switchable range for the fluid. The so-called switching factor (Tau @ 3kV / Tau @ 0 kV) ranges from about 10 at 25° to approximately 200 at 90°. However, despite the generally low conductivity of the system as a whole (and as compared to other known systems), the values climb exponentially above 100-120°C thus preventing application at extremely high temperatures. The ER-response occurs at field strengths above 1kV/mm and increases approximately linearly with the field strength.

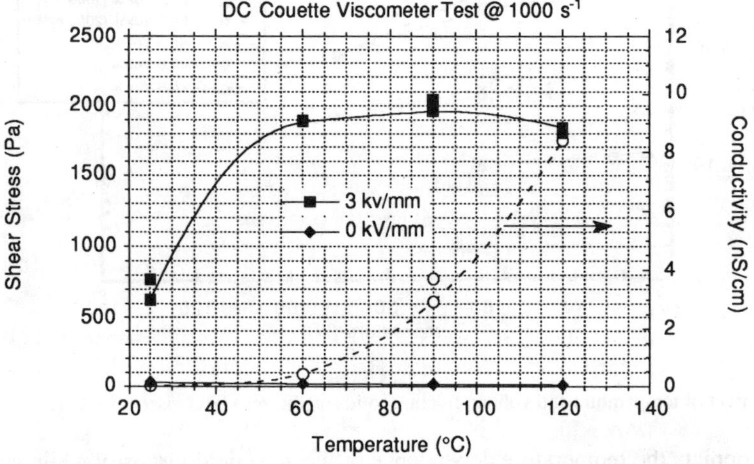

Figure 4: ER-Behavior of a polyurethane based ERF for high temperature applications.

Figure 5 depicts the ER-response of a PU-ERF formulated for applications at ambient temperatures (low temperature formulation). Here the maximum shear stress is achieved at room temperature. The switching factor and conductivity of this sample are comparable to those of the previous fluid. The response has been shifted approximately 35°to lower temperatures. Despite the decrease in excess shear stress below 25°, a switching factor of 7-8 is achieved at -10°C. This would be adequate for many applications not requiring an extremely strong response at low temperatures.

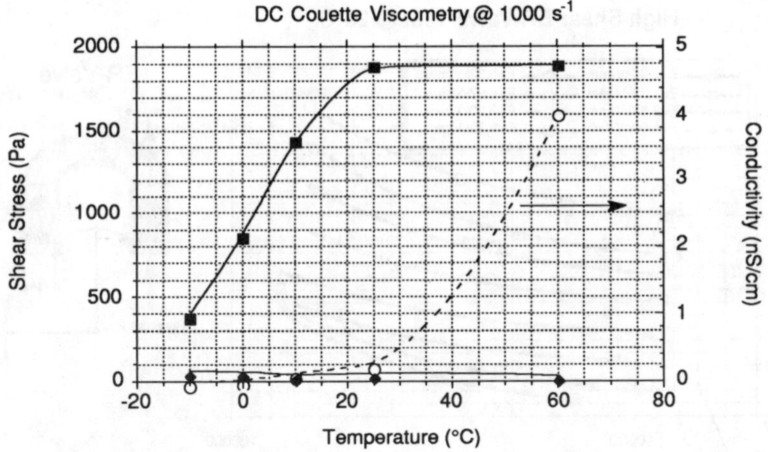

Figure 5: ER-Behavior of a polyurethane based ERF for high temperature applications.

One of the key questions concerning the usefulness of an ER-fluid is its behavior under dynamic conditions i.e. at high shear rates. The ability of an ER-fluid to respond to an electric field under high shear is a measure of how fast the fluid switches between the "on" and "off" modes and is therefore important for many of the control applications envisioned by engineers. An adequate ER-response at high shear rates is also of utmost importance for any eventual hydraulics applications of ER. The polyurethane ER-fluids are tested under high shear conditions using an ER-valve consisting of an electrified cylindrical annulus, through which the fluid is forced via a pneumatically driven piston and cylinder arrangement. Sustained performance tests have also been performed using a specially adapted hydraulic apparatus through which the fluid is mechanically pumped.

Figure 6 depicts data taken using the laboratory ER-Valve apparatus. The shear stress is plotted as a function of shear rate and field strength between 8,000 and 80,000 s^{-1}. The fluid tested was the LT-formulation depicted in Figure 5. The very high shear stresses are in part due to the increase in the zero field shear stress. However, the fluid still demonstrates a remarkable ER-activity at extremely high shear rates.

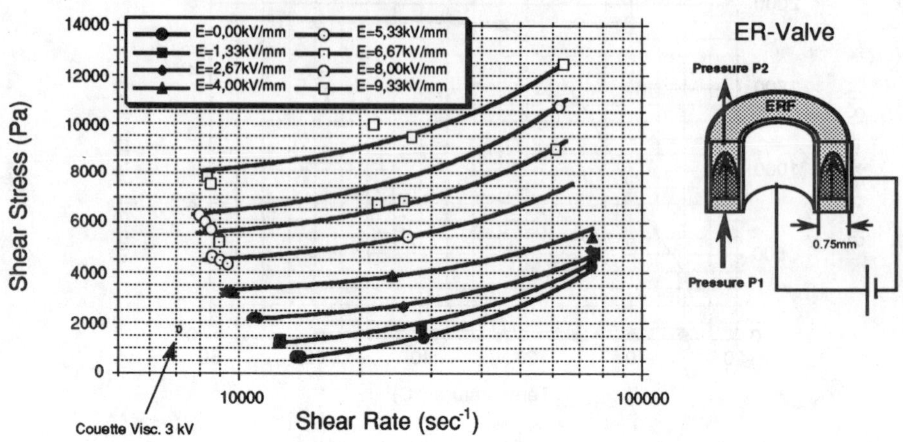

Figure 6: ER-Behavior of a polyurethane based ER-fluid in valve flow.

Although the excess shear stress decreases with increasing shear rate at fields below 4kV/mm, shear stress can still be increased 2-fold by a field strength of only 2.67 kV/mm at a shear rate of 20,000 sec^{-1}. At higher field strengths, respectable switching ranges of 6-8 are achievable at 80,000 sec^{-1} and switching factors of greater than 40 are even possible at shear rates as high as 10,000 sec^{-1}. Taking into account the length of the ER-valve as well as the volumetric flow rate, we calculate a residence time of less than 7 msec for the fluid flowing through the valve at 80,000 sec^{-1}. The actual fluid response time would have to be much faster than this in order to achieve measurable ER-effects at such high shear rates.

The question remains as to how the ER-valve data correlates with the viscosity data taken using a Couette viscometer. If one takes the shear stress measured at 1000-2000 sec^{-1} and plots it on the graph in Fig. 6, it becomes apparent that at the same field strength, the shear stress levels measured in the Couette viscometer lie somewhat lower than those measured in the ER-valve. It appears that the magnitude of the ER-effect in these (and other) fluids depends to some extent upon the geometry and mode of flow. Brunn and Vorwerk[14] of the University of Erlangen-Nürnberg recently presented measurements of the velocity profile of an ER-fluid flowing in a channel. The measured velocity profile of an ER-fluid in an electric field changed from that of a Newtonian fluid at zero field to that of an ideal Bingham fluid.

Figure 7 shows the idealized Bingham velocity profiles calculated from measurements on a PU-ERF in an hydraulic valve. The parabolic profile becomes flatter and approaches plug flow at high field strengths. Thinking in terms of the chain-like structure of an ER-fluid in an electrical field, it becomes clear that in plug flow the chains remain essentially intact in the center of the channel. This in turn leads to higher apparent ER-effect in flow mode devices. In a Couette viscometer the chains are constantly sheared within the entire active volume of the fluid, leading to somewhat lower shear stresses. The question remains if this is a universal attribute of ER-Fluids, or if some fluids (perhaps those with weak particle-particle surface interactions) are more susceptible than others.

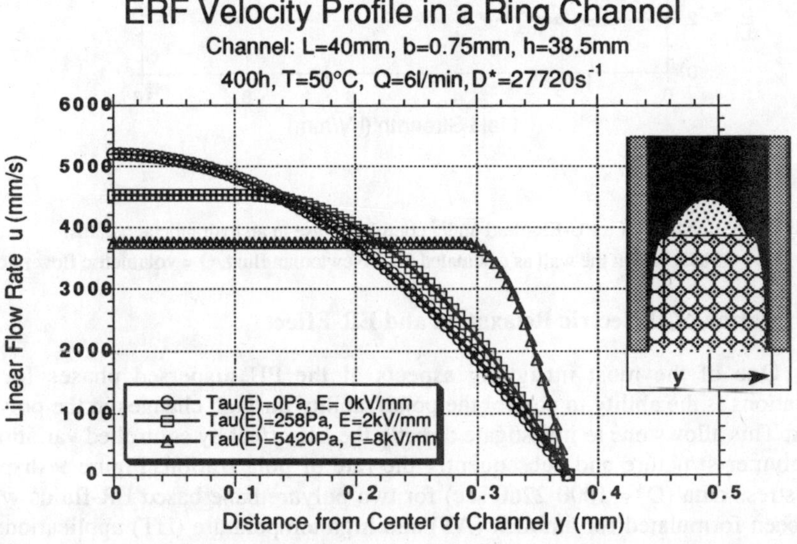

Figure 7: Calculated velocity profiles of an idealized PU-ER-fluid (Bingham behavior) in valve flow. D^* = shear rate at the wall as calculated for a Newtonian fluid, Q = volumetric flow rate

One of the questions which frequently arises concerning ER fluids is their long term stability in use. Figure 8 depicts the results of endurance tests using a modified hydraulic apparatus. The polyurethane fluid withstood shear rates of 45,000 sec^{-1} and field strengths up to 9 kV/mm for several thousand hours without significant changes in the fluid properties.

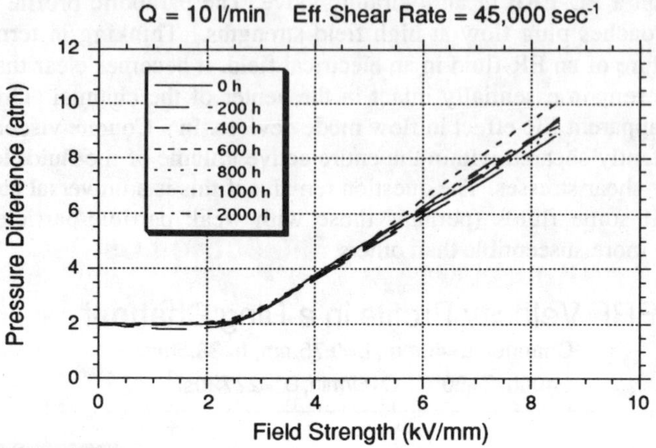

Figure 8: Behavior of a PU-ER-fluid under continuous use in an hydraulic apparatus. Shear rate at the wall as calculated for a Newtonian fluid, Q = volumetric flow rate.

5. Correlation of Dielectric Relaxation and ER-Effect

One of the most intriguing aspects of the PU dispersed phases for ER applications is the ability to control the polarizability through changes in the polymer matrix. This allows one to investigate directly the ER-effect by controlled variation of the polymer structure and subsequently the rate of polarization. Figure 9 displays shear stress data (D*= 1000-2260 sec) for two polyurethane based ER-fluids which have been formulated for ambient (AT) and high temperature (HT) applications. In this diagram, the shear stress data is plotted as a function of the frequency of maximum dielectric loss, as measured via dielectric spectroscopy. The location of the frequency maximum in the dielectric spectrum is a measure of the rate of polarization of the polymer at that temperature, and in ion-containing polymers this corresponds to the ionic polarization.

Several things are remarkable about the data taken on these samples.

1) The LT-fluid has its maximum shear stress in a fairly narrow 35° temperature range from 25 - 60°C.

2) The HT-fluid, on the other hand, has a rather broad 60° range for its maximum shear stress between 60° and 120°C.

3) Both shear stress maxima lie within a 2 decade range of frequencies centered around 1-2 kHz.

4) With a few exceptions (possibly due to surface effects) the shear stress and dielectric loss data of good polyurethane ER-fluids lie on or near this curve.

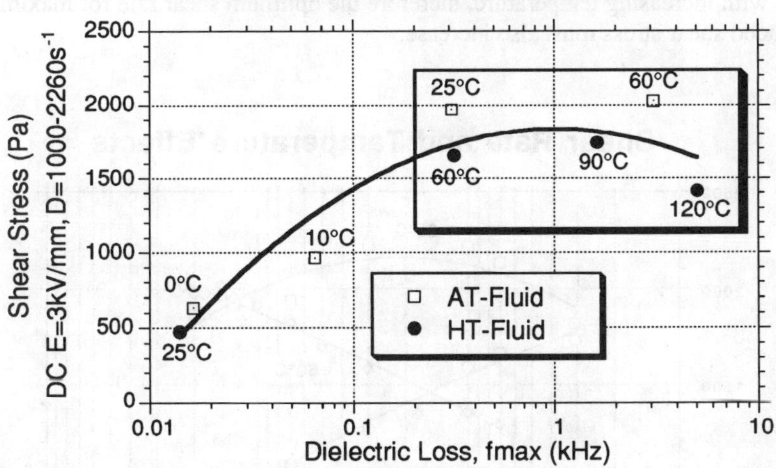

Figure 9: Correlation of shear stress and dielectric loss measurements of a polyurethane based ERF.

These results suggest a window of optimal frequencies of polarization for the polyurethane dispersed phase between 100 and 10000 Hz. Block and Kelly[15] first described the relationship between the rate of polarization, the shear rate, and the maximum achievable shear stress. The PU-ERF's also appear to follow Blocks model for the dynamics of the ER-effect under shear. Block's theory states that for the maximum ER-effect, the rate of polarization should optimally lie in the same frequency range as the shear rate. Too slow a polarization and pearl chains never have time to form, too fast, and the chains are destroyed upon shearing.

Figure 10 shows the behavior of the AT fluid depicted in Fig. 9. The excess shear stress is plotted as a function of shear rate at three different temperatures.

1) The 25°C data indicates a maximum in the shear stress at approximately 200 sec, almost exactly the dielectric loss frequency at that temperature.

2) The 60°C data increases towards a maximum with increasing shear rate.

3) The 10°C data decreases from its maximum shear stress with increasing shear rate.

The maximum of the shear stress curve appears to be shifted to higher shear rates with increasing temperature. This behavior correlates directly with Blocks theory matching the rates of polarization and shear. The rate of polarization increases naturally with increasing temperature, therefore the optimum shear rate for maximum field induced shear stress must also increase.

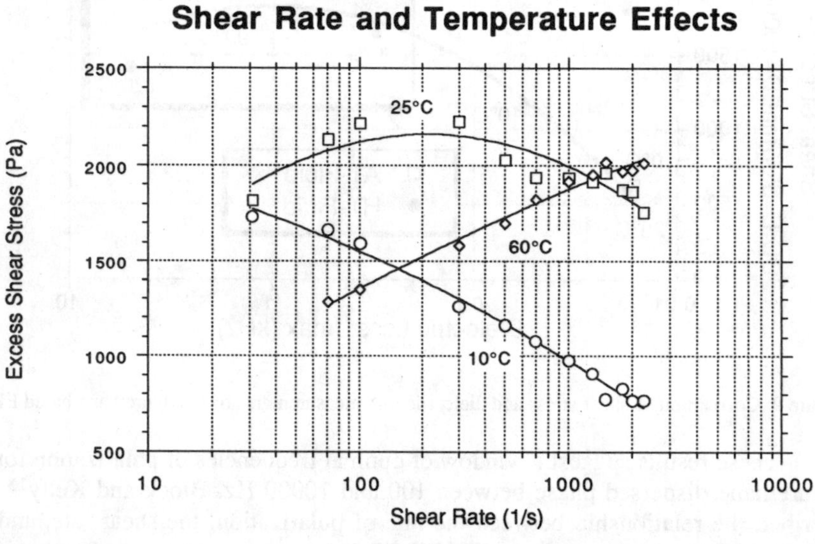

Shear Rate and Temperature Effects

Excess Shear Stress (Pa)

Shear Rate (1/s)

25°C

60°C

10°C

Figure 10: Excess shear stress (Tau_{3kV} - Tau_{0kV}) as a function of shear rate at 10°C, 25°C and 60°C.

This type of frequency-temperature correlation is well known in polymeric systems. The time-temperature superposition principle states that it is possible in such cases to construct a master curve describing the behavior of the material at any temperature simply by using reduced variables and shifting the response curves along the frequency axis relative to a reference temperature.

In Figure 11, the curves from Fig. 10 have been shifted along the frequency axis by a factor a_T to yield a so-called Master curve describing the fluid at any temperature. The values of the shift factor, a_T, were fitted using the an empirical form of the well known WLF-equation which is commonly used to fit relaxation data.[16] The constants C_1 and C_2, derived from relaxation measurements on elastomeric polyurethanes, were taken from the literature[17] and used without modification. An interpretation of the significance of $T_0 = 57°C$ is somewhat suspect, considering that only 2 decades of shear measurements at only three temperatures were used for the fit. In polymer systems, T_0, often lies about 50° above the glass transition temperature for the polymer, the temperature below which segmental motions stop. Could the concepts of free volume which played such a large role in developing the theories that describe polymer relaxation behavior be of interest in unraveling the ER-effect in such polymer-based systems?

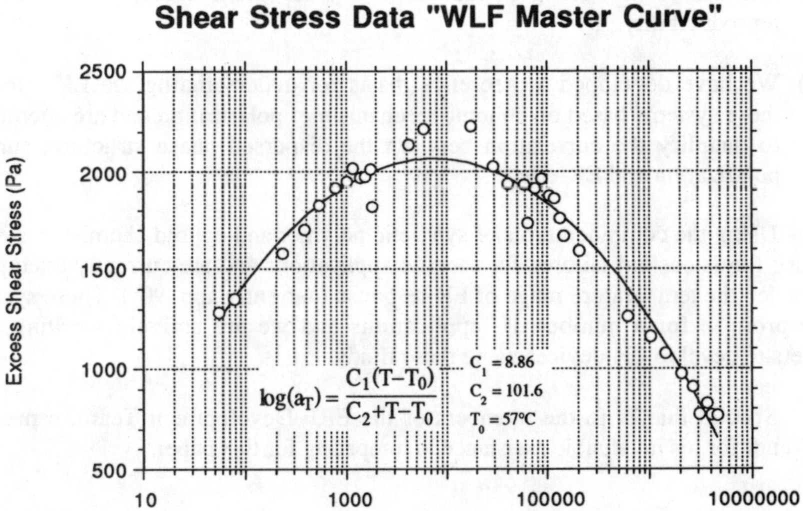

Figure 11: Master Curve for AT polyurethane ER-fluid fitted according to a WLF-type equation.

It is well known that the primary chain relaxations responsible for the mechanical properties of polymers can be related to dielectric relaxation phenomena, and we suggest that similar relaxations also play an important role in determining the polarizability of polymeric dispersed phases in ER-fluids. We propose that the correlation of shear stress response curves with temperature dependent dielectric loss measurements interpreted within the framework of known relaxation phenomena in polymers can lead us to a better understanding of the principle processes which govern interfacial polarization and the ER-effect in these systems.

6. Summary

In summary, we would like to emphasize the following achievements of this work.

1) As part of an ongoing effort to develop ER-fluids for practical applications, we have created a new generation of stable, water-free fluids based on polyurethane dispersed phases.

2) The new fluids exhibit low viscosities, good ER-effects, and low conductivity.

3) The fluids are being produced in a pilot plant scale, with excellent reproducibility.

4) We have developed a theoretical basis for understanding the ER-effect in these systems based on an ionic mechanism of polarization and are attempting to quantify the correlation between the dispersed phase structure, surface polarization and ER-effect.

Using the combined tools of synthetic polymer and colloid chemistry, we can produce fluids custom tailored for specific applications and are currently attempting to broaden the temperature range of ER-response (currently app. 90°). These systems show promise for a number of applications and we are actively working with partners to develop ER-devices using these fluids.

Special thanks to the members of the ERF Development Team, especially Dr. Wendt for his invaluable assistance in preparing for this paper.

---

1 Winslow, W. M. *J. Appl. Phys.* 20,1137(1949)

2 EP 170939

3 DE 3517281

4 DE 3536934

5 US 4,744,914
6 US 5,139,690
7 US 5,139,691
8 US 5,139,692
9 JP 1253110
10 EP 374525
11 JP 3074494
12 EP 509571
13 S. Okamura and Y. Chatini, *Polymer Preprints Japan*, **34** (1985) 2089.
14 P.O. Brunn, J. Vorwerk, Univ. of Erlangen-Nürnberg
Conference of the German Rhological Society (1993).
15 H. Block, J.P. Kelly, *J. Phys. D: Appl. Phys.*. **21(12)** (1988) 1661.
16 J.D. Ferry, *Viscoelastic properties of Polymers* (John Wiley and Sons, New York, 1980), p. 274.
17 R.F. Landel, *J. Colloid Science*, **12** (1957) 308.

Surfactant-activated Electrorheological Suspensions

Y. D. Kim and D. J. Klingenberg

Department of Chemical Engineering and Rheology Research Center
University of Wisconsin, Madison, WI 53706

Abstract

One way to improve the electrorheological activity of many suspensions is to add a small amount of surfactant. We report on preliminary experiments investigating the activation of alumina suspensions using three different types of nonionic surfactants. In most cases, adding surfactant produces a significant increase in the dynamic yield stress, with the yield stress passing through a maximum as the surfactant concentration is increased. We also find that the influence of surfactants on the response is sensitive to the amount of water in the suspension.

1 Introduction

Using electric fields to control suspension viscosity has many applications in torque and stress transfer devices [1, 2, 3, 4, 5]. Applications currently being developed include engine and truck cab mounts, shock absorbers, and robotic actuators; automotive clutches and brakes are envisioned, but conventional electrorheological (ER) fluids cannot provide the required torque transfer. Other limitations that often inhibit device development include excessive power consumption, corrosion, attrition, and limited temperature range of operation. In order to improve existing applications and exploit the full potential of ER technology, better ER fluids are needed.

The ER response can be enhanced by the addition of activators, typically small amounts of water or surfactants [6, 7, 8, 9, 10, 11, 12, 13, 14]. However, adding water to activate the suspensions also increases suspension conductivity and power consumption, promotes corrosion, and restricts the temperature range of operation. We have initiated an investigation of ER suspensions activated by surfactants. Given the large number of surfactants, it is reasonable to expect that systems can be formulated to provide a significantly enhanced ER response, without possessing all of the limitations associated with water-activated suspensions.

The goals of our research are to determine the mechanism or mechanisms by which surfactants activate the ER response, and to determine the properties that control ER activity. With this information, it will be possible to determine that extent to which the ER response can be enhanced, to identify any possible limitations, and ultimately, to design and optimize ER suspensions and devices for specific applications.

In this report, we present preliminary experimental results on the surfactant-activated response of various alumina suspensions, using three different types of non-ionic surfactants. The enhancement is characterized by the increase in the dynamic yield stress, which is measured with a modified controlled-strain rheometer. The similarities and differences among the various formulations are illustrated, and directions for future work are discussed.

2 Materials

Suspensions consisted of (Al_2O_3) particles dispersed in silicone oil (SF96, General Electric, η_c = 0.0968 Pa·s, ρ_c = 968 kg/m^3). Three different types of activated alumina particles were employed: acidic, neutral and basic (Aldrich, ρ_p=3970 kg/m^3, average pore diameter = 58 Angstroms). The alumina particles were approximately spherical and sieved to obtain diameters in the range of 63 - 90 μm. Nonionic surfactants investigated were glycerol monooleate (GMO, Chemical Service), glycerol trioleate (GTO, Chemical Service) and Brij 30 ($C_{12}H_{25}(OCH_2CH_2)_4OH$, Aldrich).

3 Suspension Preparation

Suspensions were prepared by first adding the surfactant to the pure silicone oil. The particles were either used as received ("nondried suspensions"), or were dried for 4 hours under vacuum (-10 psig) at 50°C to remove free water ("dried suspensions"). This drying procedure removed water in the amount of approximately 0.2 wt% of the dried particles. The particles were then added to the surfactant solution and stored in a desiccator to minimize contact with air. Rheological experiments were performed at least one or two days after sample preparation.

4 Rheological Measurements

Rheological experiments were performed on a Bohlin VOR rheometer fitted with parallel plates, and modified for the application of large electric fields (a schematic diagram of the modified rheometer is presented in Fig. 1). Potential differences were supplied by a function generator (Stanford Research Systems, model DS345) and amplified with a Trek model 10/10 amplifier. Experiments were conducted with an electric field frequency of 500 Hz (except for the frequency sweep experiments).

Suspensions were placed between the parallel plates, and sheared for one minute at a large shear rate (> 40 s^{-1}) and zero field strength to insure a uniform distribution of particles. The desired electric field was applied for one minute prior to the rheological

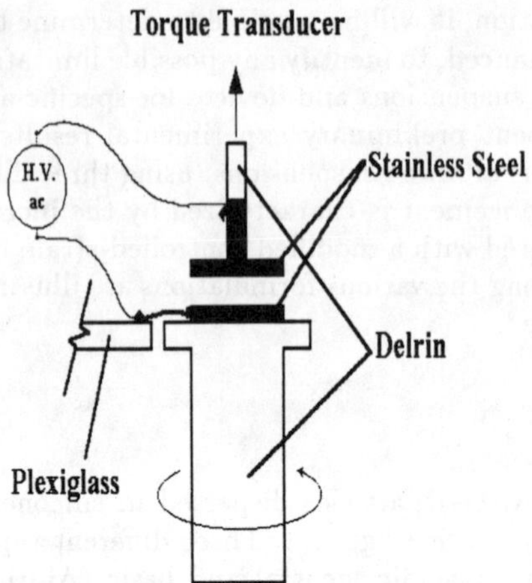

Figure 1: Schematic diagram of the rheometer modified for the application of large electric fields.

measurements. Rheological measurements were performed by shearing the suspension at constant shear rate under the applied electric field, and recording the shear stress transmitted by the suspension. Experiments were performed with decreasing and then increasing shear rates, to obtain plots of shear stress as a function of shear rate. Values for the dynamic yield stress, τ_0, were determined by extrapolating the shear stress-shear rate data to zero shear rate, as illustrated in Fig. 2.

5 Results and Discussion

Dried Particles:

Yield stress data for dried alumina suspensions (20 wt%) without surfactant are plotted as a function of the field strength squared in Fig. 3. This figure illustrates that the response is proportional to the field strength squared.

The dependence of the yield stress on Brij 30 concentration is presented in Figs. 4, 5 and 6 for acidic, neutral and basic alumina suspensions, respectively. For each alumina suspension, the yield stress initially increases with surfactant concentration and then passes through a maximum; the maximum occurs at a surfactant concentration of approximately 2-3 wt%. The maxima at electric field strengths of 500 V/mm are shifted to larger surfactant concentrations than observed for larger field strengths for both neutral and basic alumina suspensions. For these Brij 30-activated suspensions, the maximum enhancement in the yield stress varies from approximately 300

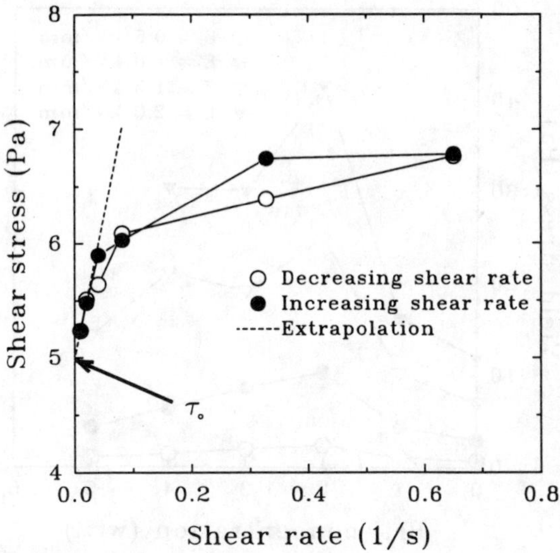

Figure 2: Extrapolation of shear stress-shear rate data to obtain a value for the dynamic yield stress, τ_0.

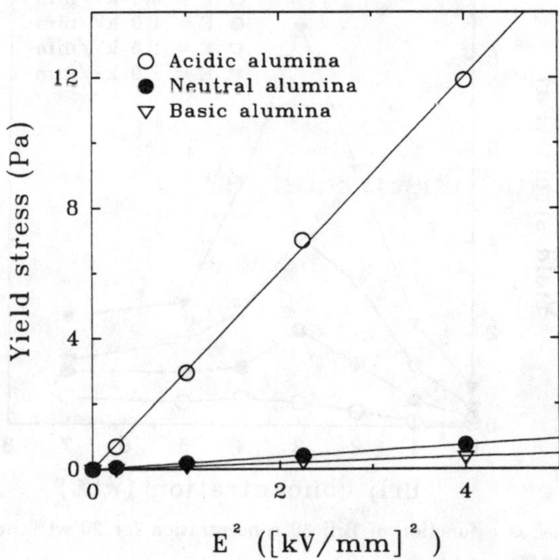

Figure 3: Yield stress as a function of the field strength squared for 20 wt% dried alumina suspensions in silicone oil.

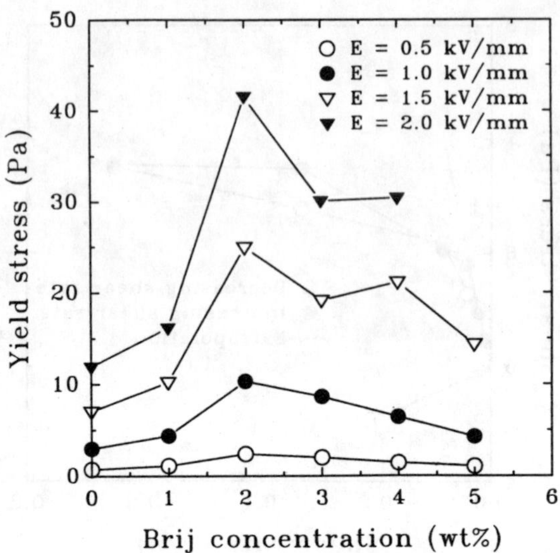

Figure 4: Yield stress as a function of Brij 30 concentration for 20 wt% dried acidic alumina suspensions in silicone oil.

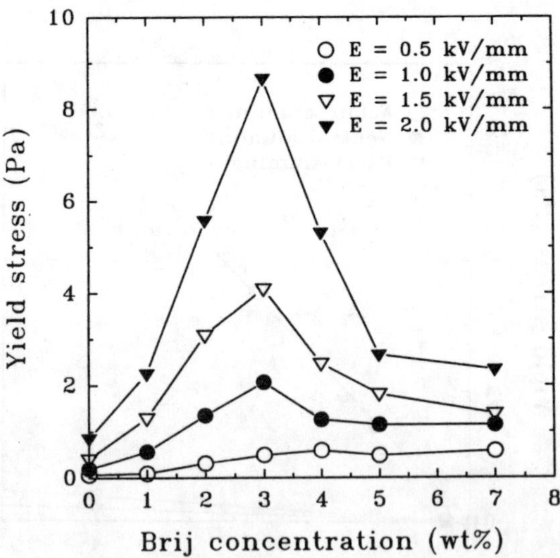

Figure 5: Yield stress as a function of Brij 30 concentration for 20 wt% dried neutral alumina suspensions in silicone oil.

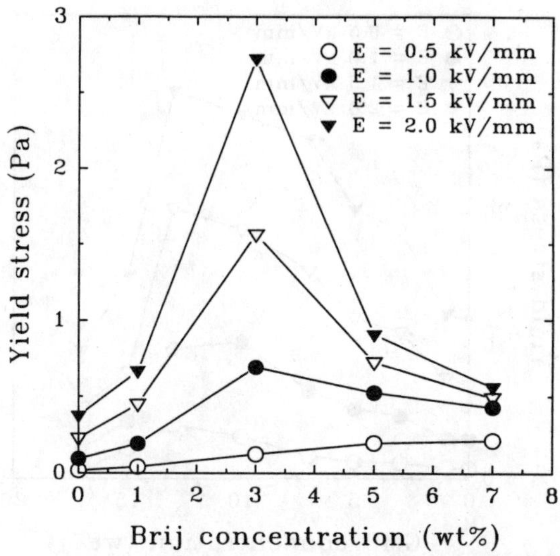

Figure 6: Yield stress as a function of Brij 30 concentration for 20 wt% dried basic alumina suspensions in silicone oil.

to 1000%; although enhancement is observed for all three types of alumina particles, the extent does depend on the particle type.

The dependence of the yield stress on GMO concentration is presented in Figs. 7, 8 and 9 for the three types of alumina suspensions. The behavior is similar to that observed for the suspensions containing Brij 30, with maximum enhancements in the range of 250 – 1000%, depending on the particle type. Maximum yield stresses are observed for GMO concentrations of approximately 13 wt% for both acidic and neutral alumina suspensions, and at approximately 7 wt% for basic alumina. The maxima shift to larger GMO concentrations as the applied electric field decreases.

The dependence of the yield stress on GTO concentration is presented in Figs. 10, 11 and 12 for the three types of alumina suspensions. The yield stress enhancement is not as significant for these suspensions, and in fact a reduction in yield stress is often observed, but yield stress maxima do exist for the acidic and neutral alumina suspensions.

For experiments performed at surfactant concentrations above that corresponding to the maximum yield stress, the samples drew a considerable amount of current, possibly indicating an increase in the number of mobile charge carriers or their mobility. In these situations, the field strength dependence of the yield stress was found to deviate from the squared dependence observed at smaller surfactant concentrations. Since the surfactant concentration corresponding to the maximum yield stress was found to be field dependent, it follows that the sample conductance is likely to depend on both the electric field strength and the surfactant type and concentration.

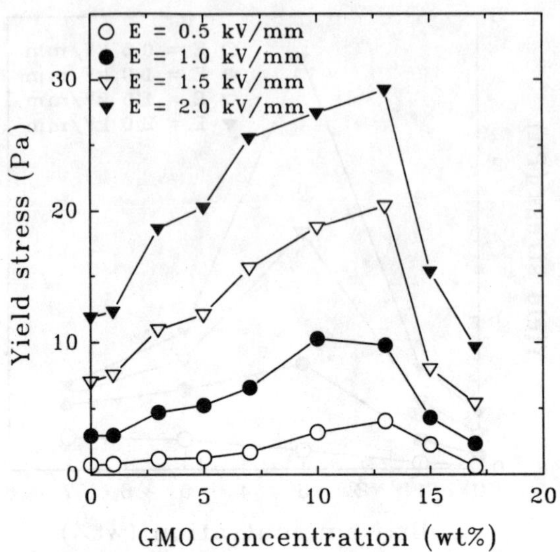

Figure 7: Yield stress as a function of GMO concentration for 20 wt% dried acidic alumina suspensions in silicone oil.

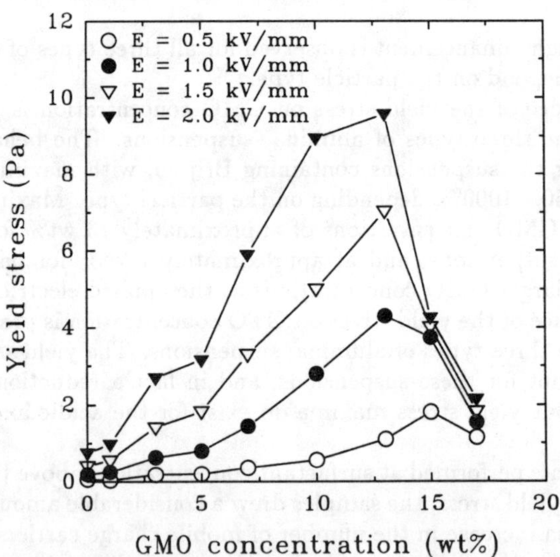

Figure 8: Yield stress as a function of GMO concentration for 20 wt% dried neutral alumina suspensions in silicone oil.

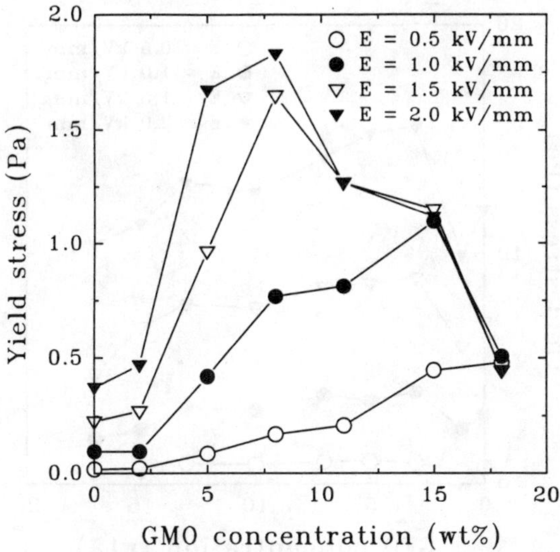

Figure 9: Yield stress as a function of GMO concentration for 20 wt% dried basic alumina suspensions in silicone oil.

GTO is not effective at enhancing the ER response, while both Brij 30 and GMO are. Possible explanations for this observation may reside in the fact that both Brij 30 and GMO contain hydroxyl functional groups, whereas GTO does not. As a result, Brij 30 and GMO should adsorb more strongly to the hydrophilic alumina surface, as well as be more effective at solvating water and ions than GTO. The significance of adsorption is also suggested by the consistent trend of basic alumina suspensions showing not only small yield stresses, but smaller enhancements in the presence of surfactants. Preliminary experiments on titanium dioxide (anatase) suspensions also show that GTO is ineffective at enhancing the ER response, while GMO provides an enhancement of approximately 700% at a concentration of 1 wt%. The precise reason for the differences among the responses for different surfactants is unclear at this time.

Nondried Particles:

The dependence of the yield stress on Brij 30 concentration is presented in Figs. 13 and 14 for nondried acidic and neutral alumina suspensions, respectively. Again, the yield stress passes through a maximum with increasing surfactant concentration; the maxima occur at a Brij 30 concentration of approximately 3 wt% for both the nondried acidic and neutral alumina suspensions. Note that the yield stress magnitudes of the nondried suspensions are often greater than those obtained for dried suspensions at equivalent Brij 30 concentrations (compare Fig. 13 and 14 with Figs. 4 and 5). For the

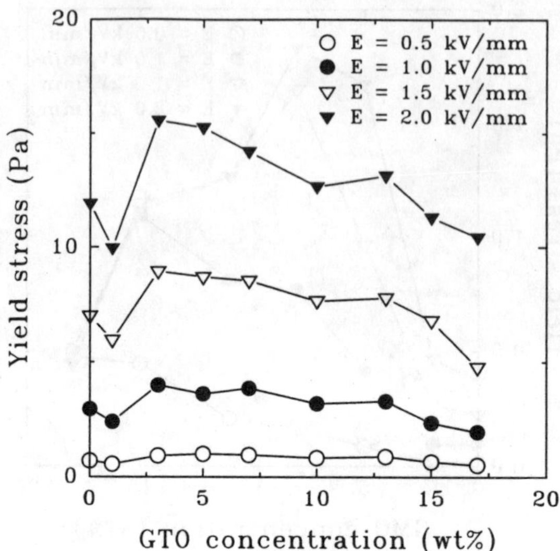

Figure 10: Yield stress as a function of GTO concentration for 20 wt% dried acidic alumina suspensions in silicone oil.

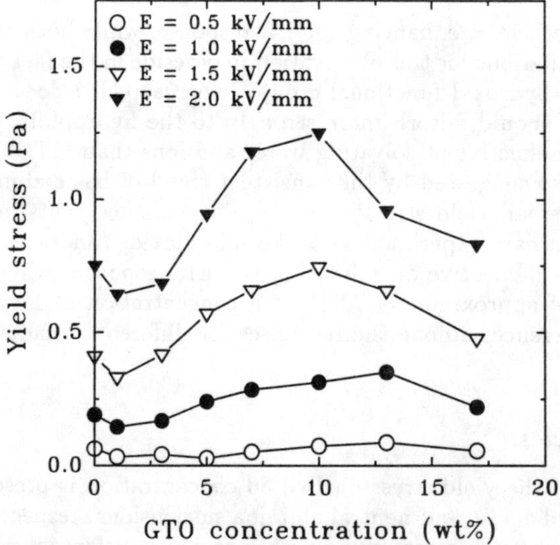

Figure 11: Yield stress as a function of GTO concentration for 20 wt% dried neutral alumina suspensions in silicone oil.

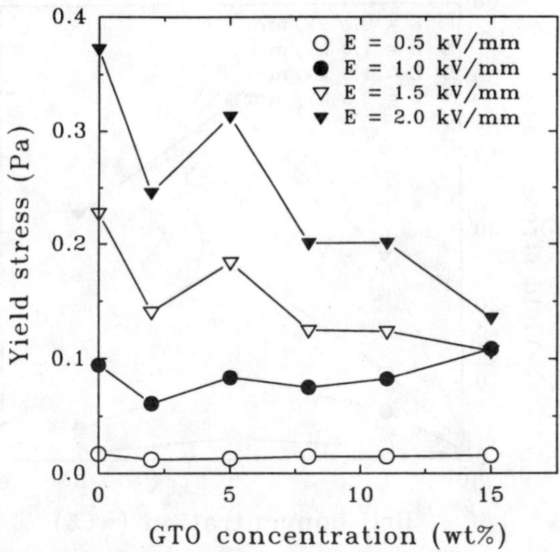

Figure 12: Yield stress as a function of GTO concentration for 20 wt% dried basic alumina suspensions in silicone oil.

nondried neutral suspensions, there appears to be a threshold surfactant concentration for yield stress enhancement.

The dependence of the yield stress on GMO concentration is presented in Figs. 15 and 16, for nondried acidic and neutral alumina suspensions, respectively, and the dependence of the yield stress on GTO concentration is presented in Figs. 17 and 18. Although maxima in the yield stress for GMO- and GTO-activated acidic alumina suspensions are not observed for the concentration ranges presented in Figs. 15 and 17, maxima may exist at larger surfactant concentrations. Such experiments were attempted, but the sample conductances were too large for the power supply to deliver the desired voltages. Figs. 15–17 also illustrate an apparent threshold concentration for surfactant activation, and again, the yield stress values are considerably larger than those observed for the corresponding dried suspensions.

Comparing Figs. 17 and 18 with Figs. 10 and 11, it is apparent that GTO can activate the ER response provided there is a sufficient amount of water in the suspension, and that the extent of activation depends on the GTO concentration. Hence, there appears to be some type of cooperative phenomenon occurring between the water and the GTO such that only together can they activate the response. Such a phenomenon may involve solvation of water by GTO molecules, but clearly more work needs to be done before any definitive statements can be made.

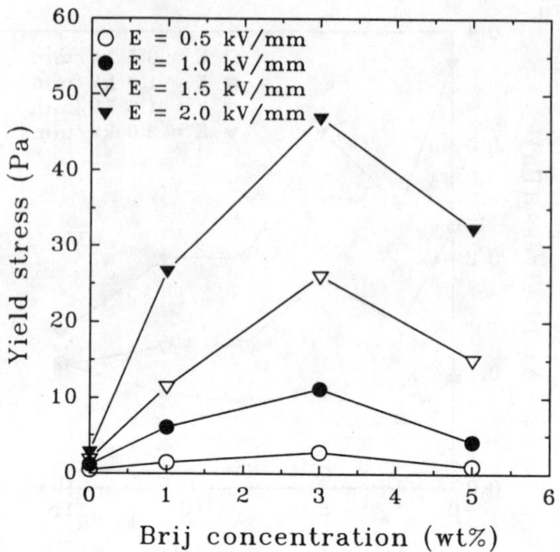

Figure 13: Yield stress as a function of Brij 30 concentration for 20 wt% nondried acidic alumina suspensions in silicone oil.

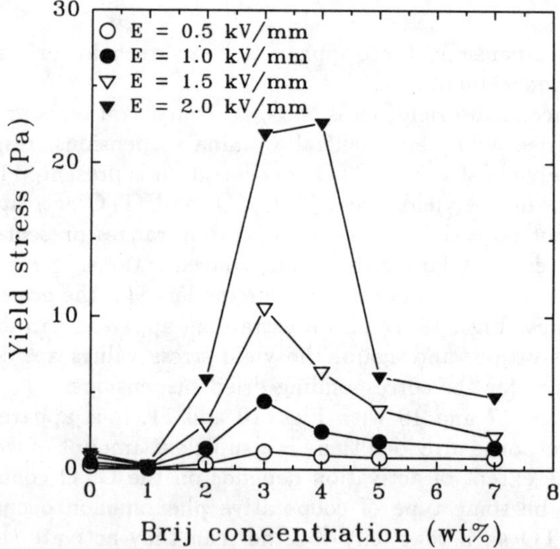

Figure 14: Yield stress as a function of Brij 30 concentration for 20 wt% nondried neutral alumina suspensions in silicone oil.

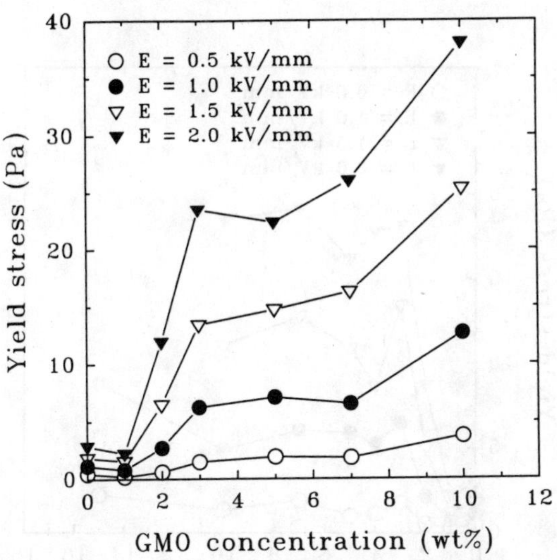

Figure 15: Yield stress as a function of GMO concentration for 20 wt% nondried acidic alumina suspensions in silicone oil.

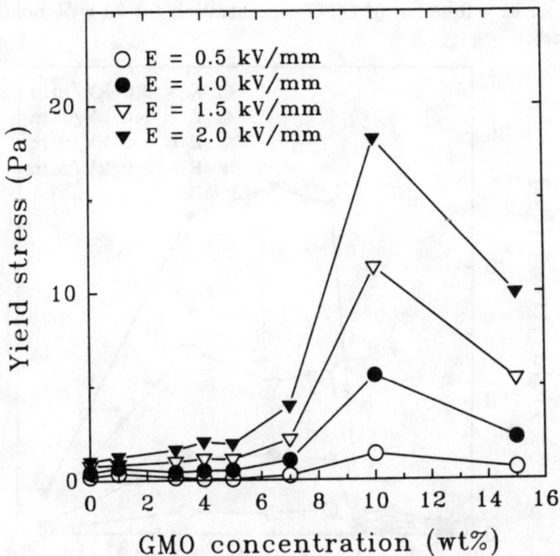

Figure 16: Yield stress as a function of GMO concentration for 20 wt% nondried neutral alumina suspensions in silicone oil.

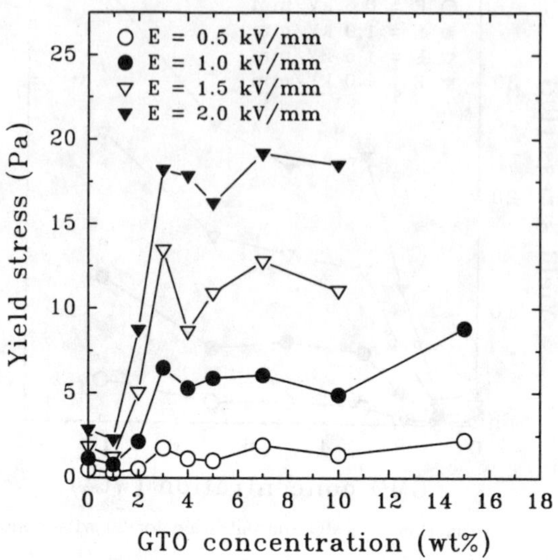

Figure 17: Yield stress as a function of GTO concentration for 20 wt% nondried acidic alumina suspensions in silicone oil.

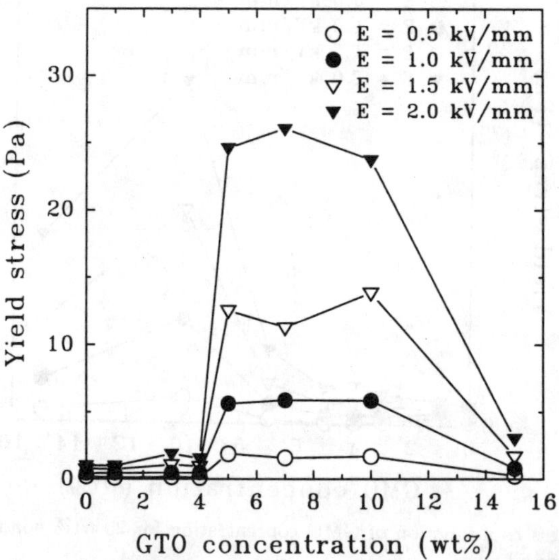

Figure 18: Yield stress as a function of GTO concentration for 20 wt% nondried neutral alumina suspensions in silicone oil.

Frequency Dependence:

The dependence of yield stress on the applied electric field frequency is presented in Figs. 19 and 20 for 20 wt% dried neutral alumina suspensions, with 0 and 3 wt% Brij 30, respectively. The two suspensions show similar trends; there appears to be a maximum yield stress at a small frequency, and the yield stress decreases as the frequency is increased. The suspension containing 3 wt% surfactant shows a decrease of more than one order of magnitude as the frequency is varied from 50 to 10^4 Hz. The surfactant-free suspension, on the other hand, decreases by a factor of approximately 5. We note also that the frequency at which the yield stress is a maximum is larger for the surfactant-activated suspension, and that this "critical" frequency varies with the electric field strength.

It is common for the frequency dependence of the ER response to generally mirror the frequency dependence of the suspension dielectric constant; hence we expect that these suspension will display a low frequency dispersion that is indicative of ionic conduction dominating at low frequencies [8]. It is possible that the role of surfactants, in combination with any water that may be present, is to increase the number and/or the mobility of charge carriers. This statement is speculative at this point, but provides direction for future work.

6 Conclusion

From the conditions investigated, we find that surfactants tend to activate the ER response; this activation is often quite significant, with few examples of a decreased response of such an extent. Plausible mechanisms for the enhancement may be related to the increase in the number or mobility of charge carriers, which can be influenced by the surfactant and water concentrations, as well as the nature of the particle. Future work will continue to probe the mechanisms responsible for surfactant activation of ER suspensions.

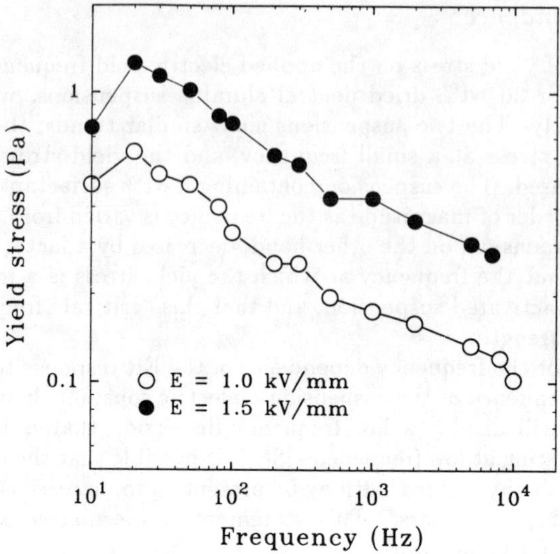

Figure 19: Yield stress as a function of electric field frequency for 20 wt% dried neutral alumina suspensions in silicone oil, without surfactant.

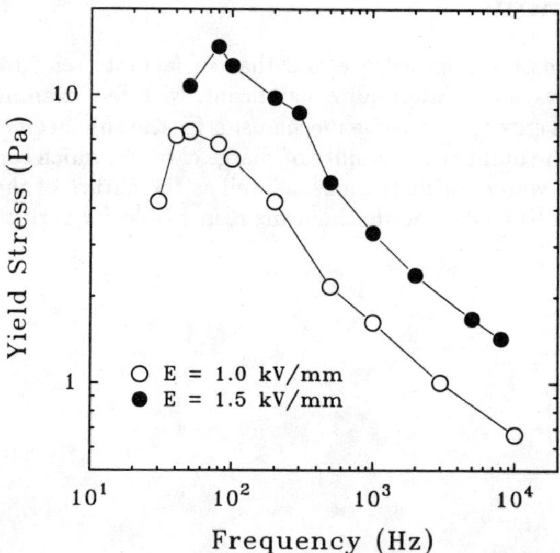

Figure 20: Yield stress as a function of electric field frequency for 20 wt% dried neutral alumina suspensions in silicone oil, with 3 wt% Brij 30.

References

[1] Z. P. Shulman, R. G. Gorodkin, E. V. Korobko, and V. K. Gleb, J. Non-Newt. Fluid Mech., **8**, 29 (1981).

[2] D. Scott, Pop. Sci., 82 (April, 1984).

[3] D. Brooks, Physics World, **2**, 35 (1989).

[4] J. E. Stangroom, Phys. Technol., **14**, 290 (1983).

[5] D. L. Hartsock, R. F. Novak, and G. J. Chaundry, J. Rheol., **35**, 1305 (1991).

[6] W. M. Winslow, J. Appl. Phys., **20**, 1137 (1949).

[7] Y. F. Deinega and G. V. Vinogradov, Rheol. Acta, **23**, 636 (1984).

[8] H. Block and J. P. Kelly, J. Phys. D: Appl. Phys., **21**, 1661 (1988).

[9] A. P. Gast and C. F. Zukoski, Adv. Coll. Int. Sci., **30**, 153 (1989).

[10] T. C. Jordan and M. T. Shaw, IEEE Trans. Elect. Insul., **24**, 849 (1989).

[11] A. A. Trapeznikov, G. G. Petrzhik, and O. A. Chertkova, Koll. Zh., **43**, 83 (1981).

[12] O. A. Chertkova, G. G. Petrzhik, and A. A. Trapeznikov, Koll. Zh., **44**, 83 (1982).

[13] H. Uejima, Jap. J. Appl. Phys., **11**, 319 (1972).

[14] N. Sugimoto, Bull. JSME, **20**, 1476 (1977).

YIELD PROCESS OF ELECTRORHEOLOGICAL FLUID WITH POLYANILINE PARTICLE

KIYOHITO KOYAMA*, KEIJI MINAGAWA, TAMOTSU YOSHIDA, NORIYUKI KURAMOTO

Department of Materials Science and Engineering, Yamagata University, Yonezawa 992, Japan

and

KATSUFUMI TANAKA

Department of Polymer Science and Engineering, Kyoto Institute of Technology, Kyoto 606, Japan

ABSTRACT

Electrorheological behaviors of polyaniline/silicone oil suspension was observed by using a modified Couette type rheometer with high resolution for shear stress. The yield behaviors were examined over a wide range of shear strain. The storage modulus and loss tangent were determined under a constant DC electric field. It was clarified that the polyaniline-based ER fluid yields at two different strain amplitudes, i.e., about 1% and 50%. The stress-strain curves obtained from shear flow experiments also suggested the existence of two-steps yield process. The yield process was found to be dependent on the electric field strength and the particle concentration in different manners. The yield behavior observed is discussed in relation to the structure of particle clusters which causes the ER effect.

*To whom correspondence should be addressed.

1. Introduction

The mechanism of electrorheological (ER) effect has been studied by many researchers. In particular, suspension systems in which dielectric particles are dispersed in an insulating solvent have been investigated in detail because of the drastic change in the rheological properties arising from an external electric field[1,2]. It was clarified from direct observations that suspended particles in ER fluid form clusters under an electric field[3]. Theoretical models have provided useful information for the understanding of the ER effect. Marshall et al. succeeded in correlating the viscosity, field strength, and shear rate with the Bingham constitutive equation[4]. The results indicate that the yield process of the fluid is important to clarify the mechanism of the ER effect.

The yield stress under shear strain for ER suspension systems were reported elsewhere[5]. However, measurements with conventional rheometers do not give sufficiently reliable data because such rheometers generally have poor resolution for the application of shear distortion and detection of the stress. ER fluids exhibit remarkable change in the apparent viscosity under external field, and thus, high resolution is required for the rheometers in order to obtain precise data. It is desirable to analyze the behavior upon cluster yield in further detail for the elucidation of the mechanism of ER effect. In the present paper, we report the measurements of the dynamic viscoelastic properties and shear flow experiments of ER suspensions by use of a viscometer which has high resolution for shear distortion.

2. Materials

Polyaniline particles with the average diameter of 0.4 μm to 7 μm were dried under vacuum prior to use. A 30cSt silicone oil was used as a solvent, which was purified carefully as reported previously[6]. The particles were dispersed in the solvent and the resulting mixture was sonicated for homogenous dispersion, and then used as the sample fluids. The particle fraction was changed from 2 to 10 wt% in the dispersion solution.

3. Apparatus

It is important to use a high-resolution rheometer for detailed examination of the response of ER fluids. We examined the difference between data obtained with a normal rheometer (Rheology Co. Ltd., MR300), which employs wire-type torque sensor, and a modified Couette type rheometer, Rheology Co. Ltd., MR300-V2. The modified rheometer is constructed with two concentric cylinders and a torque sensor which employs a torque meter instead of a torsion wire. The outer cylinder is the driving part of the shear deformation and the inner cylinder is the detecting part of the stress. The diameter of the outer and inner cylinders are 35 mm and 33 mm, respectively. The height of the inner cylinder is 14 mm. The shear stress is detected as the distortion of the center shaft of the cylinders. Torque is evaluated from input voltage V_{in} and output voltage V_{out} of crossed gages according to the following equation:

$$V_{out} = \frac{8\, V_{in}\, K\, T}{\pi d^3\, G} \tag{1}$$

where T, d, and G are the torque, diameter of the shaft, and distortion elasticity of the shaft, respectively. The two cylinders are used as electrodes for the measurements under an electric field. The cylinder electrodes were completely insulated from the main body of the apparatus.

The strain and stress were estimated with the following equations, in which the applied strain is precisely evaluated by subtracting the rotation angle of inner cylinder from that of outer cylinder;

$$\gamma = \frac{2\, R_c^2\, \theta}{R_c^2 - R_b^2}, \quad \theta = \theta_c - \theta_b \tag{2}$$

$$\sigma = \frac{T}{2\pi L R_b^2} \tag{3}$$

where γ is the shear strain, θ_c and θ_b are the rotation angle of the outer and inner cylinder, R_c and R_b are radius of outer cylinder and of inner cylinder, σ, T, and L are the shear stress, torque, and the length of the immersed fluid, respectively.

Figs. 1 and 2 show the apparent strain, applied strain, and stress response measured with the rheometers MR300 for an ER fluid with cation exchange resin at small and large strain regions, respectively. At the small strain region shown in Fig. 1, the actual applied strain (b) was quite smaller than apparent strain (a), although the stress response (c) was sufficiently detected. At larger strain region, applied strain was closer to that of apparent strain. On the other hand, the modified rheometer MR300-V2 gave more improved resolution for both the applied strain and stress detection. Figs. 3 and 4 show the applied strain and stress response under the conditions corresponding to Figs. 1 and 2, respectively. The sample fluid for these measurements was polyaniline/silicone oil dispersion. It is obvious from Fig. 3 and 4 that the strain application and stress detection over a wide range of strain amplitude were achieved with much higher resolution than those in Fig. 1 and 2.

These results indicate that the rigidity and sensitivity of the stress sensors used in rheometers are quite important for the measurements under a variety of conditions. Most appropriate apparatus should be used in order to obtain precise data according to the sample conditions. For the dynamic measurements of ER fluids, torque sensor of rheometer should have sufficiently high modulus of elasticity in torsion and enough resolution. Using the present torque sensor equipment, we have succeeded in measuring the modulus and loss tangent values with high precision even under extremely small shear distortion. All measurements in the following sections were carried out by use of the modified rheometer.

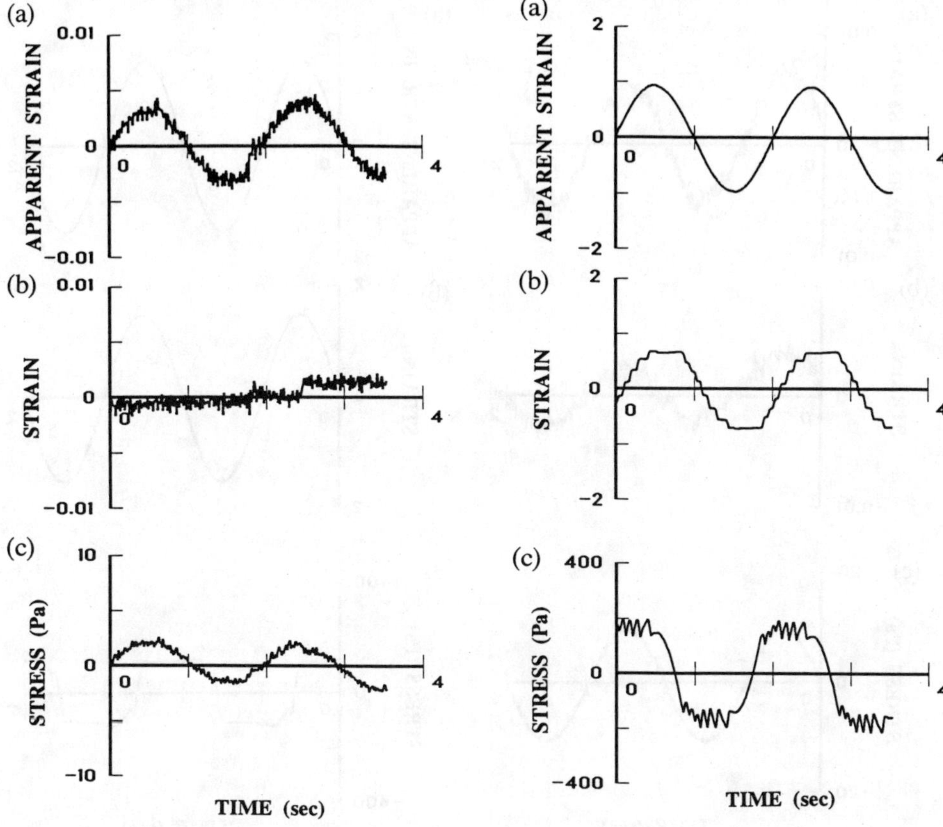

Fig. 1 Strain stimulus and stress response of cation exchange resin dispersion under small strain measured with the normal rheometer MR-300. (a) apparent strain, (b) applied strain, (c) stress response, E= 2kV/mm.

Fig. 2 Strain stimulus and stress response of cation exchange resin dispersion under large strain measured with the normal rheometer MR-300. (a) apparent strain, (b) applied strain, (c) stress response, E= 2kV/mm.

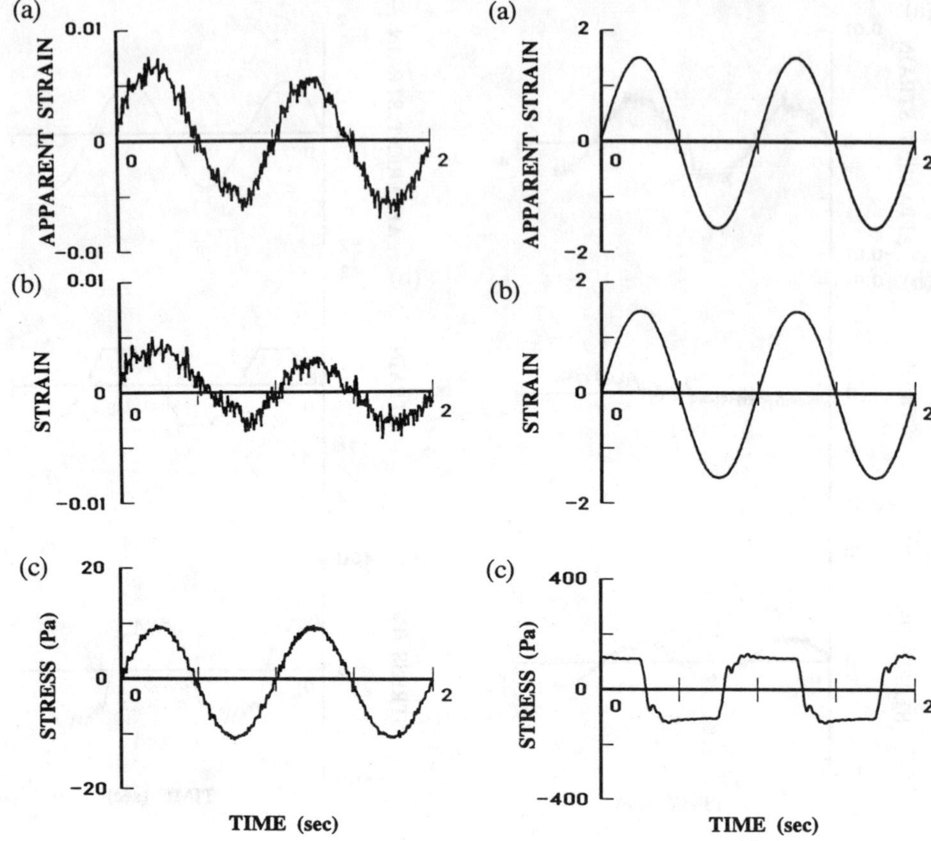

Fig. 3 Strain stimulus and stress response of polyaniline dispersion under small strain measured with the modified rheometer MR-300V2. (a) apparent strain, (b) applied strain, (c) stress response, E= 2kV/mm.

Fig. 4 Strain stimulus and stress response of polyaniline dispersion under large strain measured with the modified rheometer MR-300V2. (a) apparent strain, (b) applied strain, (c) stress response, E= 2kV/mm.

4. Dynamic Viscoelasticity

Stress response of the polyaniline dispersed in silicone oil was measured under a sinusoidal strain and a constant DC voltage. The storage modulus G' of the ER fluids was estimated with the following equation:

$$G' = \frac{|\sigma^*|}{|\gamma^*|} \cos \delta \tag{4}$$

where $|\gamma^*|$, $|\sigma^*|$, and δ are amplitude of the strain, amplitude of the stress response, and phase difference between stress and strain, respectively. The measurements were carried out under the strain frequency of 1Hz at room temperature. The results measured under 2kV/mm field are shown in Fig. 5.

As seen in Fig. 5 the measurement with the modified rheometer gave the following features for the behavior of the ER fluid: (i) Both log G' and tan δ are almost constant at the amplitude range below 0.03. In this region, the mechanical properties of the fluid is regarded as pseudo-elastic one. (ii) At the strain amplitude region between 0.03 and 0.4, the log G' decreases and tan d increases with the increase in the strain, although the changes of both values are not so drastic. (iii) At the amplitude region above 0.4, drastic changes in the log G' and tan d are observed; the former decreases and the latter increases abruptly with the increase in the strain. This phenomenon indicates the existence of two different yield processes under a constant DC voltage. Such a behavior would correlate with the change of the cluster structure.

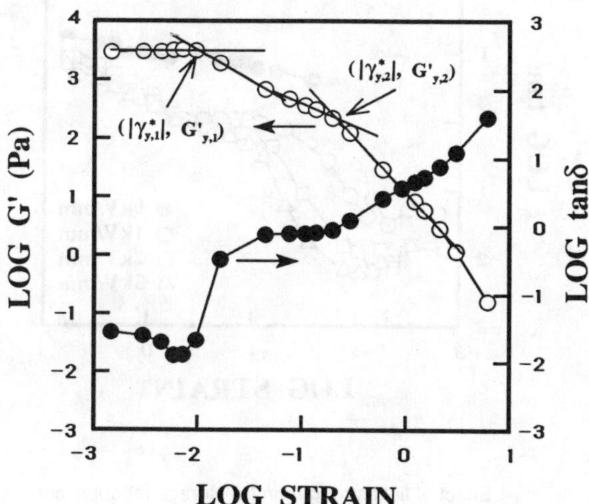

Fig. 5 Storage modulus and loss tangent for polyaniline dispersion, E=2kV/mm, C= 10wt%.

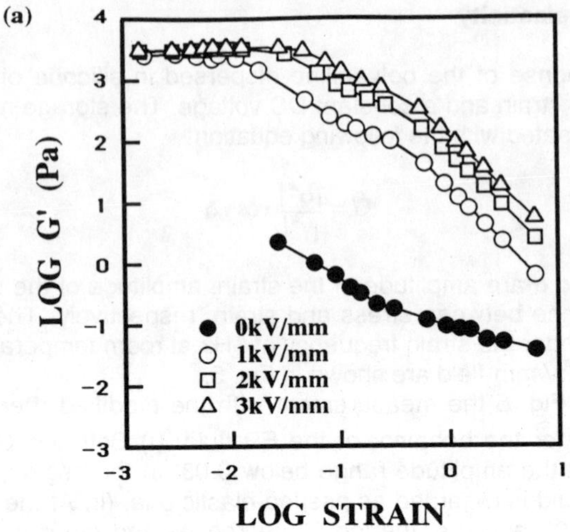

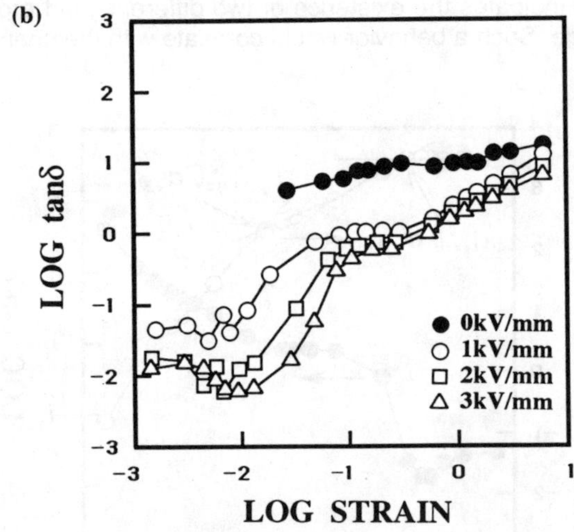

Fig. 6 Effect of field strength on (a) storage modulus and (b) loss tangent for 10wt% polyaniline dispersion.

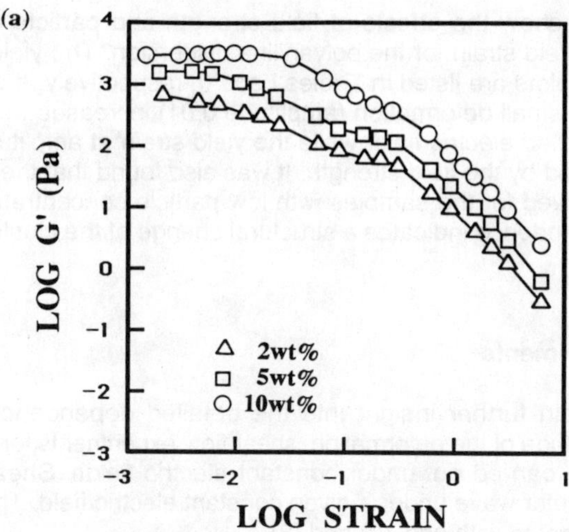

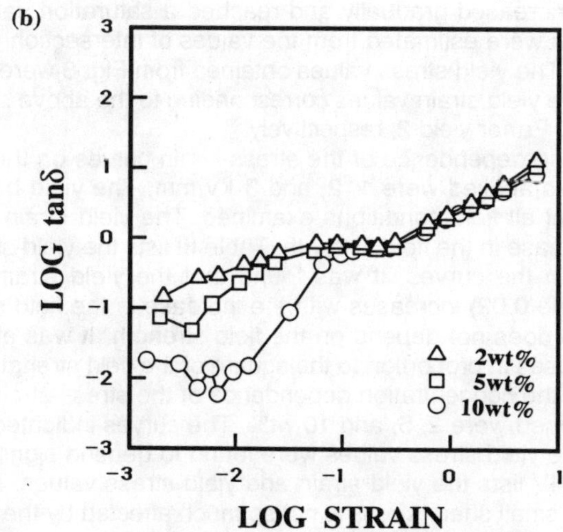

Fig. 7 Effect of polyaniline concentration on (a) storage modulus and (b) loss tangent, field strength: 2kV/mm.

Figs. 6 and 7 show the effects of field strength and particle concentration, respectively, on the yield strain for the polyaniline dispersion. The yield strain values determined from the plots are listed in Tables I and II, respectively. It was found that the yield strain under small deformation (amplitude 0.01) increases with the increase in the strength of applied electric field, while the yield strain at amplitude around 0.4 is not so much affected by the field strength. It was also found that the two-step yield behavior is not observed for the samples with low particle concentration. The result of concentration dependence indicates a structural change of the cluster bridging the electrodes.

5. Shear Flow Experiments

In order to gain further insight into the detailed dependency of the yield behavior on the amplitude of the deformation, shear flow experiments for the polyaniline dispersion fluid were carried out under constant electric fields. Shear deformation was applied as triangular wave under a given constant electric field. The shear strain and stress were estimated with eqs. (2) and (3).

Fig. 8 exemplifies the shear stress-strain curve for the polyaniline-based ER fluid obtained at a constant electric field of 2kV/mm under the strain amplitude of 1.5. Drastic increase in the stress was observed immediately after the shear strain was applied. The stress increased gradually and reached a saturation value. The yield stress and yield strain were estimated from the values of intersection of the straight lines drawn in Fig. 8. The yield stress values obtained from Fig. 8 were 0.02 (yield 1) and 0.62 (yield 2). The yield strain values corresponding to the above stress were 80 Pa for yield 1 and 160 Pa for yield 2, respectively.

Fig. 9 shows the dependence of the stress-strain curves on the electric field strength. The voltages applied were 1, 2, and 3 kV/mm. The yield behaviors were similar, as a whole, at all field conditions examined. The yield strain was found to increase with the increase in the field strength. Table III lists the yield strain and yield stress evaluated from the curves. It was found that the yield strain under small deformation (amplitude 0.02) increases with the increase in the field strength while the strain around 0.6 does not depend on the field strength. It was also found that the yield stress increases in proportion to the square of the field strength.

Fig. 10 shows the concentration dependence of the stress-strain curves. The concentrations examined were 2, 5, and 10 wt%. The curves indicated the two-step yield behavior and the yield stress values were found to depend significantly on the concentration. Table IV lists the yield strain and yield stress values. It is clear that the yield strain under small shear stress is not so much affected by the concentration of the particles, while the second yield strain increases with the increase in the concentration.

Table I Effect of field strength on yield strain, C=10wt%.

Field strength (kV/mm)	$\gamma^*_{y,1}$	$\gamma^*_{y,2}$
1	0.01	0.5
2	0.03	0.4
3	0.04	0.4

Table II Effect of particle concentration on yield strain, E=2kV/mm

Concentration (wt%)	$\gamma^*_{y,1}$	$\gamma^*_{y,2}$
2	—	0.3
5	0.007	0.3
10	0.03	0.4

Table III Effect of field strength on yield strain and yield stress, C=10wt%.

Field strength	yield 1		yield 2	
(kV/mm)	$\gamma^*_{y,1}$	$\sigma^*_{y,1}$	$\gamma^*_{y,2}$	$\sigma^*_{y,2}$
1	0.01	30	0.63	50
2	0.02	80	0.62	160
3	0.04	140	0.64	280

Table IV Effect of particle concentration on yield strain and yield stress, E=2kV/mm.

Concentration	yield 1		yield 2	
(wt%)	$\gamma^*_{y,1}$	$\sigma^*_{y,1}$	$\gamma^*_{y,2}$	$\sigma^*_{y,2}$
2	0.005	3	0.26	10
5	0.015	10	0.45	50
10	0.02	80	0.59	160

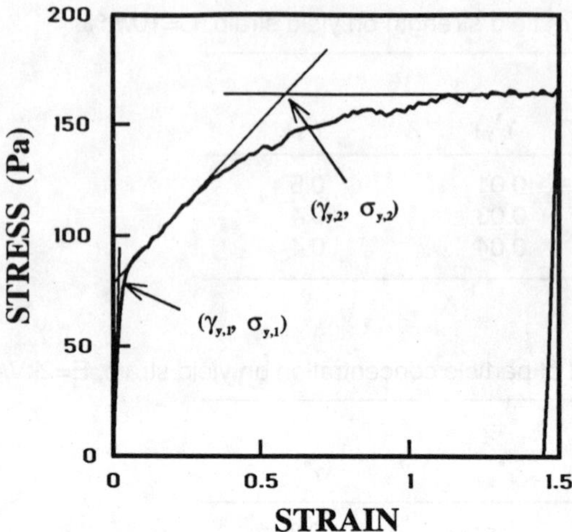

Fig. 8 Shear stress-strain curve for polyaniline based ER fluid. E=2kV/mm, C=10wt%.

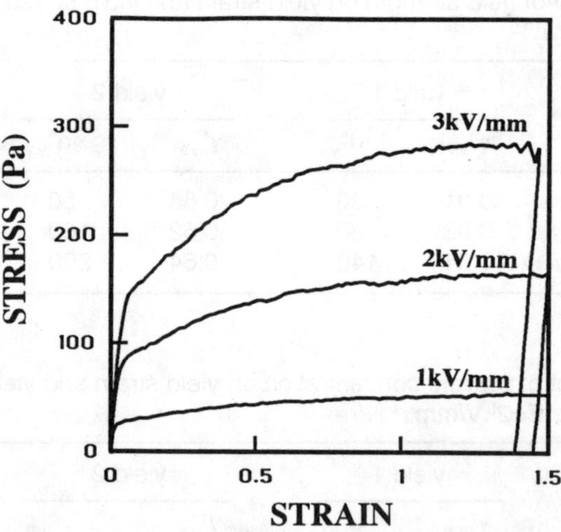

Fig. 9 Shear stress-strain curve for polyaniline based ER fluid at various field strength. C=10wt%.

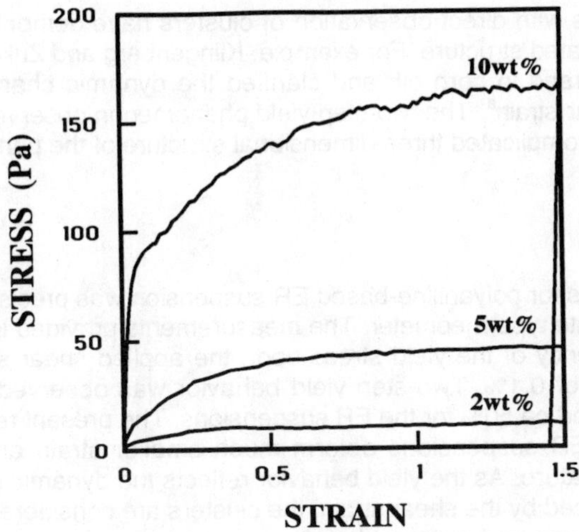

Fig. 10 Shear stress-strain curve for polyaniline based ER fluid at various particle concentrations. E=2kV/mm.

6. Discussion

Yield behaviors of ER suspensions indicate dynamical structure change of particle clusters. The yield behaviors of various ER suspensions were reported in literature[7], based on both experimental and theoretical studies, and the yield strain values were generally believed to be about 50%. In our present experiments, smaller yield strain at ca.1% was obtained in addition to the larger value of ca.50% which was in agreement with reported values. The existence of two different yield points suggests the following yield process upon the application of the shear strain to the suspension. The cluster partly deforms at the strain of ca.1%, at which the first yield behavior is observed, and the resulting structure stands up to the strain of 50%. When large strain above 50% is applied, the cluster completely breaks down and then the suspension begins to flow.

We previously reported[6] that three rheological regions were found by measuring dynamic viscoelastic properties of cation exchange resin dispersed in silicone oil, with various strain amplitudes. The existence of three regions were observed with a normal rheometer. However, it is clear from the comparison of the normal and modified rheometers that the modified rheometer used in the present study has much higher resolution for shear stress-strain measurements, and thus, the present results should be more reliable.

Recent studies with direct observation of clusters have demonstrated three-dimensionally aggregated structure. For example, Klingenberg and Zukoski observed silica particles dispersed in corn oil, and clarified the dynamic change of cluster structures under shear strain[8]. The two-step yield phenomenon observed here would be related to such a complicated three-dimensional structure of the particles.

7. Conclusion

The yield stress for polyaniline-based ER suspension was precisely estimated with a modified Couette type rheometer. The measurements provided information on the detailed dependency of the yield stress upon the applied shear strain with the amplitude resolution of 0.1%. Two-step yield behavior was observed at the strain amplitude of ca.1% and ca.50% for the ER suspensions. The present results indicate that the clusters in ER suspensions deform much smaller strain amplitude than those reported in literature. As the yield behavior reflects the dynamic change in the cluster structure induced by the shear strain, the clusters are considered to have two drastic structural changes under shear strain.

Acknowledgement

The authors wish to thank Mr. Y. Seino for the help in some experiments.

References

1. W. M. Winslow, *J. Appl. Phys.*, **20**, 1137 (1949)
2. N. Sugimoto, *Bull. of JSME,* **20**, 1476 (1977)
3. A. F. Sprecher, J. D. Karlson, H. Conard, *Mat. Sci. Eng.*, **95**, 187 (1987)
4. L. Marshall, J. W. Goodwin, and C. F. Zukoski, *J. Chem. Soc. Faraday Trans. I*, **85**, 2785 (1989)
5. D. L. Klass and T. W. Martinek, *J. Appl. Phys.*, **38**, 75 (1967)
6. K. Tanaka, T. Yoshida and K. Koyama, Proc. 3rd Int. Conf. ER Fluids, ed., R. Tao, World Scientific, pp. 289-299 (1991)
7. D. R. Gamota and F. E. Filisco, *J. Rheol.*, **35**, 399 (1991)
8. D. J. Klingenberg and C. F. Zukoski, *Langmuir,* **6**, 15 (1990)

II. PHYSICAL MECHANISM

LIGHT SCATTERING STUDIES OF A MODEL ELECTRORHEOLOGICAL FLUID

THOMAS C. HALSEY
The James Franck Institute and Department of Physics
The University of Chicago
5640 South Ellis Avenue
Chicago, Illinois 60637

and

JAMES E. MARTIN
Advanced Materials Physics Division
Sandia National Laboratories
Albuquerque, New Mexico 87185

ABSTRACT

Electrorheological suspensions typically contain particles of approximately one μm in diameter. Thus light-scattering offers a natural method of probing the microstructure of these suspensions. We report the development of an index matched single-scattering fluid, as well as light-scattering studies of this fluid in both a quiescent and sheared regime. In the first case, the results are in agreement with a phenomenological theory of coarsening based on thermal fluctuations. In the second case, they agree with an "independent droplet" model of the suspension structure under shear.

1. Introduction

In unravelling the microstructure of electrorheological (ER) suspensions, light scattering is potentially a very useful tool. This is due to the happy accident that the typical size of the suspended particles in such a fluid is comparable to the wavelength of light.[1] Thus single-scattering of light from an ER suspension will yield direct and easily interpretable information about its structure, under any flow conditions.

Unfortunately, most commercial and experimental ER fluids have particles with dielectric constants at optical frequencies significantly different from the surrounding liquid. Thus these fluids are in a multiple-scattering regime for light. This property has been used by Ginder and Elie[2] to extract structural information, but the potential of multiple-scattering techniques is limited by difficulties of interpretation.

For this reason, we developed a model ER fluid with particles whose index of refraction was closely matched to that of the surrounding liquid.[3] While probably not useful for applications, this fluid is an excellent laboratory for the study of ER fluid microstructure.

So far we have addressed two physical questions with this fluid. The first is the nature of the solidification transition of an ER suspension when a field is applied. Since the original observations of Winslow, many observers have noted the appearance of fibrous structures, often one particle thick, parallel to the applied field.[1,4] However, recent theories predict that the ultimate high-field state of an ER suspension will be a phase separated solid, rather than a set of widely dispersed fibers or particle chains.[5,6] Our results bridge this apparent contradiction by showing that, while it is true that the original state of the ER fluid after the application of a field consists of thin chains of particles aligned with the field, these fibers slowly drift together in the direction transverse to the field, forming columns of larger and larger diameter. The end product of this process will clearly be the phase-separated colloidal solid beloved of theorists. A phenomenological theory based upon the idea that the thermal fluctuations of the chains catalyze this "coarsening" process accounts well for the quantitative details of this process.[3,7].

In a shear flow, there are two competing models of the microstructure. Klingenberg and Zukoski accounted for the Bingham plastic response of the suspensions by postulating (and observing) that boundary layers of ER ordering formed near the electrodes, which were melted some way into the fluid by the shear flow.[8] Halsey, Martin, and Adolf have proposed a different mechanism, in which the electrodes play no special role, but "droplets" of condensed particles form throughout the fluid.[9] The size of these droplets is then fixed by a balance between electrical and hydrodynamical forces. We have studied the microstructure in the interior of a sheared ER fluid, and have found evidence for the existence of such droplets, and of such a balance. Of course, this does not exclude the possibility that Klingenberg-Zukoski boundary layers may form closer to the electrodes.

In this chapter, we will first report details of the synthesis of our model fluid. We will then report our experimental results in the two areas mentioned above, and briefly compare them with theory.

2. Sample Preparation

The colloids in our model fluid are synthesized by the base-catalyzed nucleation and growth of monodisperse silica spheres from tetraethoxysilicon. To reduce the Keesom interactions that lead to aggregation, this synthesis was conducted in mixed organic solvents that index-match the growing spheres. Scanning electron microscopy and elastic and quasielastic light scattering measurements indicate that 0.7 μm diameter silica spheres are easily formed at high silica concentrations under mild hydrolysis with 0.5M NH_4OH. The elastic light scattering data are consistent with a Gaussian sphere radius r_d distribution having σ_{r_d}/r_d of 10.5%. The hydrophilic silica spheres were then coated with the organophilic silane coupling agent 3-'trimethoxysilyl' propyl methacrylate via a condensation reaction.[10] After a 24-hour vacuum distillation of water and ammonia, the spheres were centrifuged at low acceleration, the supernatant decanted, and the soft colloidal solid was redispersed in 4-methylcyclohexanol, again chosen to closely index match the spheres. The two samples used in this study measure 20 wt % and 34 wt % by thermal gravimetric analysis, although the high concentration sample was diluted to 11 wt % for the kinetics studies. The refractive index increment of $dn/dc = 0.0017$ ml/g is small enough to insure single scattering from concentrated dispersions; indeed, depolarization of the scattered light was negligible.

To measure the surface charge of the colloids, electrophoresis measurements were made with a Pen Kem Laser Zee^{TM} apparatus. Since we were unable to observe any electrophoresis with this apparatus we simply applied a 1 kV/mm electric field to the particles and observed their behavior through a Nikon Microphot-FXA^{TM} optical microscope. Even at these high electric fields we were unable to observe electrophoresis of these particles, although at high applied frequencies field-induced particle chaining was observed and was found to be reversible by Brownian motion alone.

3. Quiescent Fluid Studies

3.1 Experimental Results

Representative scattering data taken after an electric field quench, shown in Figure 1, clearly indicate an unstable concentration fluctuation orthogonal to the electric field lines in the fluid. This corresponds to the formation of chains of particles parallel to the field. This scattering pattern dissipates immediately when the field is turned off, indicating that structure formation is perfectly reversible in this fluid.

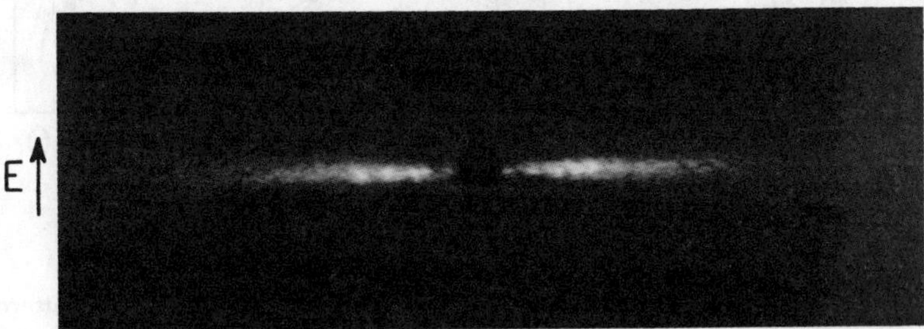

Figure 1. The scattered intensity from an electrorheological fluid shortly after the application of a strong electric field. The lobes appear at scattering wavevectors orthogonal to the electric field, indicating the appearance of chains of particles parallel to the electric field.

The evolution of the scattering intensity, or structure factor, in the direction perpendicular to the field is shown in the inset to Figure 2. This corresponds to a slice through the lobe shown in Figure 1. It is clear that as time progresses, the peak increases in intensity and moves to smaller values of wave-vector $q_\perp$ in the direction perpendicular to the field. This corresponds to the slow agglomeration of chains into columns. By dividing the intensity by the peak intensity I_{max}, and dividing the scattering wave vector by the peak wave vector $q_{\perp,max}$, we are able to demonstrate that the evolving structures scale, as evinced by the data collapse in Figure 2.

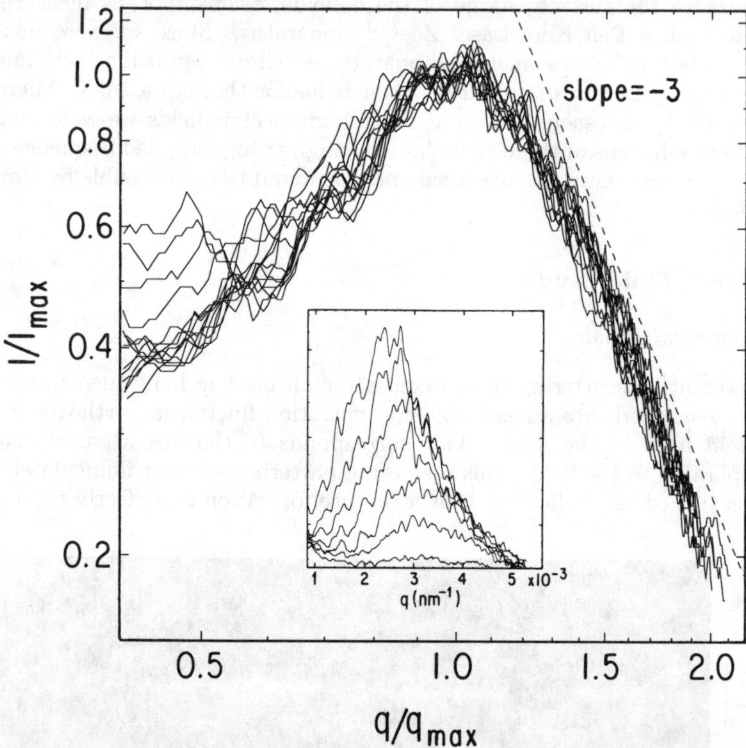

Figure 2. The inset shows intensity slices through one lobe of the scattering pattern in Figure 1 for various times. As time increases the peak intensity increases and moves to larger length scales (smaller q), corresponding to the agglomeration of chains of particles into columns. When the intensity data are plotted on dimensionless axes a master curve results, indicating scaling. On the high-q side of the peak the data fall off as q^{-3}, which indicates that the column surfaces are sharp.

These scattering data have a high-q shoulder that decays as q^{-3}. This is Porod's law in two dimensions, and indicates that the structures have sharp, non-fractal interfaces in the direction perpendicular to the field.[11] This is consistent with the picture that the agglomerating chains form compact columns. Finally, a $\log-\log$ plot shows that $I_{max} \propto q_{\perp,max}^{-2.25}$, which suggests that within experimental error the scattering intensity $I(q_\perp, t)$ obeys

$$I(q_\perp, t) \approx q_{\perp,max}(t)^{-d} f(q/q_{\perp,max}(t)) \qquad f(x) \propto x^{-(d+1)} \tag{1}$$

with $d = 2$, corresponding to the essentially two-dimensional nature of the coarsening process.

It remains to discuss the behavior of $q_{\perp,max}(t)$. In order to reduce the noise in the signal, we used the moments

$$I_k(q_l, q_m) = \int_{q_l}^{q_m} q^k I(q) dq \tag{2}$$

In practice, the lower integration limit $q_l = 0.33 \times 10^{-3}\text{nm}^{-1}$ is determined by the size of the beam stop, while the upper limit $q_m = 0.529 \times 10^{-2}\text{nm}^{-1}$ is set by the camera position. The wave vector of peak intensity can now be obtained from the ratio of I_1 to I_0.

The time dependence of the characteristic length scale (or inter-column spacing) $\tilde{R}(t) = 2\pi/q_{\perp,max}(t)$ is shown in Figure 3 for peak to peak voltages of 0.56 kV/mm, 1.25 kV/mm, and 2.6 kV/mm. At the earliest times, the characteristic length is $\tilde{R}(0) \approx 1.9\mu$m. At later times, $\tilde{R}(t)$ is observed to increase approximately as $t^{0.42}$. In the spinodal decomposition of systems with a "conserved order parameter", such as binary alloys, there is typically an intermediate time regime ("Lifshitz-Slyozov ripening") in which a characteristic length scale of the structure increases as $t^{1/3}$.[12] Thus the ER fluid coarsening is anomalously fast by comparison with ordinary spinodal decomposition; presumably this is a reflection, in some way, of the importance of long-ranged forces in this system.

However, we cannot account for this coarsening by postulating some simple mechanism of dipolar interaction between adjacent chains or columns. Any such mechanism will involve forces $\propto E^2$ acting on particles which feel a viscous drag $\propto \mu_0$, where μ_0 is the dispersing fluid viscosity. This latter assumes (reasonably) that the hydrodynamics of particle motion is in the low-Reynolds number limit. Thus we expect that the characteristic time scale t_c for electrostatically-driven coarsening would be $t_c \propto \mu_0/E^2$. However, it is clear from Figure 3 that in our experiments, t_c depends much more weakly upon E; our actual result for the three electric fields

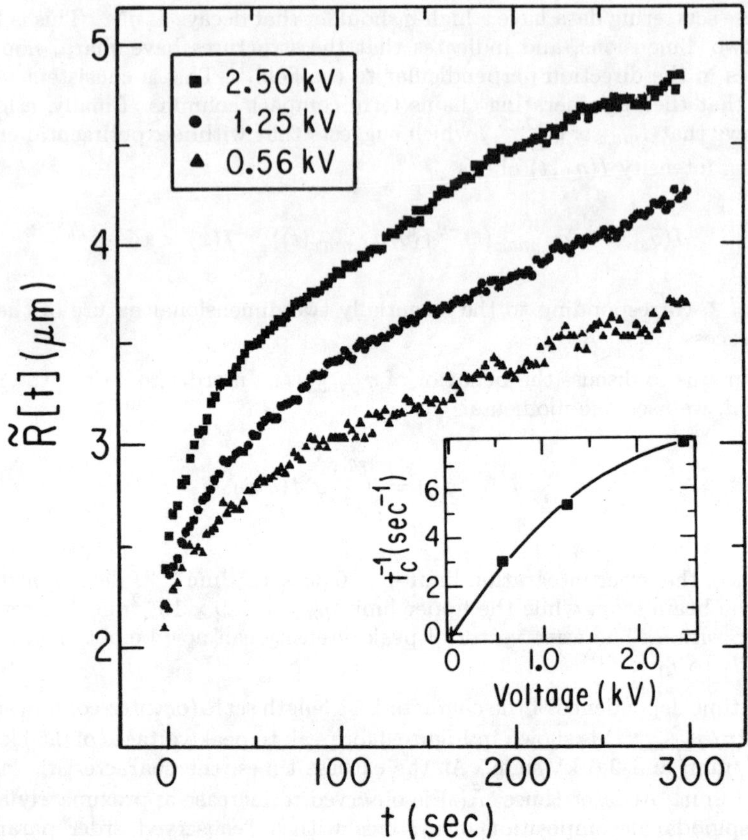

Figure 3. The distance between columns in a quiescent fluid increases as $\tilde{R}(t) = \tilde{R}(0)[1+(t/t_c)^{0.42}]$. The inset shows the increase of the growth rate t_c^{-1} with electric field. The fact that t_c^{-1} is not $\propto E^2$ implies that the coarsening is not driven simply by dipolar forces.

used is $t_c \propto E^{-0.6}$. Thus, from dimensional analysis alone, we conclude that some other energy scale must be involved in determining the coarsening process.

Since the initial structure formation parallel to the electric field does occur on a time scale $t_a \propto \mu_0/E^2$, as demonstrated, e.g., by Ginder and Elie,[2] this implies that the structure formation parallel and perpendicular to the field are driven by different mechanisms on different time scales, the latter being a much slower process.

This two-step process of structure formation was first predicted by Halsey and

Toor.[5] On a time scale determined by t_a, chains or columns of particles, whose diameter at high fields will be at most a few particles, will form parallel to the field. At high fields, this initial aggregation phase should be dominated by the ballistic motion of particles into these chain-like aggregates; the aggregates will remain one-dimensional due to their tendency to aggregate predominantly at their ends. The one-dimensional structures obtained are not the true ground state of the suspension. However, the further relaxation of these chains by motion perpendicular to the field into columnar structures is not driven by the simple electrostatic attraction of parallel chains, as this attraction is quite short-ranged.

3.2 Theoretical Results

We have proposed that the mechanism by which the chains move together is related to the thermal fluctuations of the internal configuration of a chain.[13] The importance of thermal, or Brownian effects, in ER fluids is generally expressed by a dimensionless group λ, which is defined as[14]

$$\lambda = \frac{\pi \epsilon_0 \epsilon_s (\beta E)^2 r_d^3}{k_B T}, \tag{3}$$

where ϵ_s is the dispersing fluid dielectric constant, β is the particle polarizability, k_B is Boltzmann's constant, and T is the temperature. Typically, experiments on ER effects, including ours, have been conducted at quite high values of $\lambda \sim 10^{4-5}$. Thus the proposition that thermal effects might be important, in this regime where polarization energies are very much larger than $k_B T$, is quite counter-intuitive. However, because the interactions between perfcct, zero-temperature chains are negligible if they are separated by distances greater than the intra-chain distance between particles, any long-ranged force arising from thermal effects, even if quite weak, might be significant in the long-time behavior of these suspensions.

In calculating the effect of the thermal fluctuations on the positions of particles in the chains, two possible assumptions are possible regarding the relevant time scales. The time scale over which thermal fluctuations relax defines t_f, which may be larger or smaller than the typical transverse coarsening (column formation) time t_c. If $t_f \ll t_c$, then the fluctuations of parallel chains give rise to an interaction of Van der Waals form,

$$F(\rho) = -A \frac{k_B T r_d^4 L}{\rho^5} \tag{4}$$

where $F(\rho)$ is the free energy of interaction of two parallel chains of length L, separated by a distance ρ, and A is a (large) numerical constant. For typical ER fluids, this attractive interaction would lead to coarsening with a time scale $t_c \sim 100$ sec.

On the other hand, if $t_f \gg t_c$, then a much stronger interaction between chains results. A simple scaling estimate yields

$$t_c \sim t_a \lambda^{3/5} \left(\frac{\tilde{R}(t)}{r_d} \right)^{9/5} \tag{5}$$

which determines how the coarsening time depends both on the developing distance between chains (or columns) $\tilde{R}(t)$, as well as its dependence on the electric field through λ. The coarsening predicted by Eq. (5) is in fairly good agreement with our experimental results, as it predicts that $\tilde{R}(t) \propto t^{5/9}$ and $t_c \propto E^{-4/5}$, compared to experimental exponents of 0.42 and -0.6, respectively. However, in the absence of a convincing microscopic derivation of t_f, the fluctuation approach must be regarded as only phenomenological.

This approach accounts at least qualitatively for the dependence of column formation on colloid volume fraction ϕ_c. We observed that the initial length scale $\tilde{R}(0)$ decreased with increasing concentration, but the growth rate increased. This is reasonable because we expect that $\tilde{R}(0) \propto \phi_c^{-1/2}$ in an initial array of parallel chains, and closer chains will feel a greater fluctuation-induced attraction, and thus coarsen into columns more quickly.

4. Steady Shear Studies

Our interest in shear stems from some unusual results we reported for the shear thinning viscosity of our model ER fluid.[9] In particular, we found that when the shear rate $\dot{\gamma}$ was increased, the suspension viscosity decreased as $\mu_s \propto \dot{\gamma}^{-\Delta}$, with $\Delta \approx 2/3$, at least at lower applied fields. Although qualitatively similar to the much observed "Bingham plastic" behavior, this behavior is quite different in detail from that of a Bingham plastic.

To see this, consider the shear stress τ_{xy}. In the Bingham plastic model, $\tau_{xy} = \tau_0 + \mu_\infty \dot{\gamma}$, where τ_0 is the yield stress and μ_∞ is the infinite shear rate viscosity. Thus the Bingham plastic has a constant differential viscosity $d\tau_{xy}/d\dot{\gamma} = \mu_\infty$. However, our results showed a differential viscosity $d\tau_{xy}/d\dot{\gamma} \propto \dot{\gamma}^{1-\Delta}$ which went to zero at low shear rates.

In collaboration with D. Adolf, we proposed a model for this behavior, the "independent droplet" model, in Ref. 9. This model is based upon an analysis of the response of an individual condensed droplet of dielectric particles to the combined effect of an electric field and a hydrodynamic flow. This droplet should be imagined to consist of a large number of individual colloidal particles, condensed into some ordered structure.

Suppose that the vorticity of the hydrodynamical flow is perpendicular to the electric field. Then a droplet of particles elongated in the field direction will feel a torque, which will tend to rotate it in the same sense as the hydrodynamic vorticity. However, once the long axis of such a droplet is rotated away from the electric field direction, it feels a restoring torque due to the electric field. If one restricts oneself to ellipsoidal droplets, both the electrical and the hydrodynamical torques are known exactly. For a prolate spheroidal droplet of minor radius b and major radius c, the hydrodynamical torque L obeys[15]

$$L \sim \mu_0 V \dot{\gamma} \frac{c^2}{b^2} \tag{6}$$

where V is the droplet volume. The electrical torque is given by a standard formula, which depends upon the depolarization factors of the ellipsoid.[16]

Note that we implicitly assume that the polarization of the individual particles is always parallel to the local electric field, so that the electrical torque arises from the interaction among the different particles in the droplet. This is in contrast to the work of Hemp,[17] who studied individual particles whose polarization lagged the local electric field in time by some phase; these particles can contribute an electrorheological effect without the need for significant inter-particle interactions.

For any particular droplet, the angle θ that the droplet axis makes with the electric field direction can be calculated as a function of the Mason number $\mathrm{Mn} = (6\mu_0\dot{\gamma}/\epsilon_0\epsilon_s(\beta E)^2)$. In addition, θ is a function of the aspect ratio of the droplet, the ratio of the large radius c to the small radius b. The result is

$$\theta \sim \mathrm{Mn}(c/b)^2 \tag{7}$$

so that longer droplets are rotated by a larger angle away from the field direction. In so doing, they lose much of their favorable depolarization energy. Thus a hydrodynamical flow field will tend to reduce the size of structures in an ER suspension.

An actual suspension will, of course, consist of many such droplets, which will collide with one another as well as break up under the influence of the flow. One might thus expect that the typical size of droplets in a flow will be set by the maximum stable size of a droplet, because droplets below this size which collide will tend to aggregate, while droplets above this size will break up. There are two sources of polarization energy for a droplet. The first is the depolarization energy at the equilibrium angle, as determined by the two competing torques mentioned above. The second effect is the surface energy of a droplet.[18] Balancing these two effects, one finds that

$$\begin{aligned} c &\propto \mathrm{Mn}^{-1} \\ b &\propto \mathrm{Mn}^{-2/3} \end{aligned} \tag{8}$$

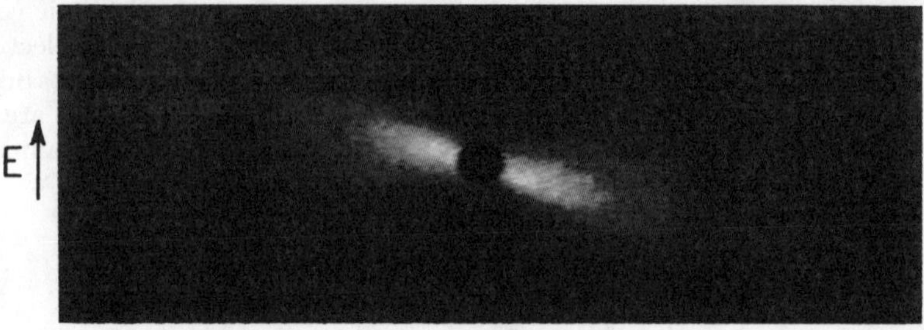

Figure 4. The scattering pattern of a sheared fluid is rotated with respect to that of a quiescent fluid (Figure 1). In this case, the shear rate $\dot{\gamma} = 1.04\mathrm{s}^{-1}$. There is also a much stronger intensity near the origin, where the scattering wave-vector is small, than for the quiescent fluid.

which sets the length and width of droplets. This implies that θ is given by

$$\theta \propto \mathrm{Mn}^{1/3} \tag{9}$$

and that the macroscopic suspension viscosity scales as

$$\mu_s \propto \mathrm{Mn}^{-2/3} \tag{10}$$

Since Mn $\propto \dot{\gamma}$, this result is in agreement with the low-field experimental results mentioned above.

In our light-scattering studies, we first applied an electric field to a quiescent fluid. When a steady shear was then applied to the sample, the scattering pattern rotated in the direction of the vorticity and the maximum intensity of the scattering lobes moved to $q = 0$, as shown in Figure 4. After a brief period, the scattering intensity no longer changed with time.

These results can be interpreted in terms of the existence of some sort of rotated structures in the fluid, be they droplets or un-fragmented columns. To proceed further, we analyze the angular distribution of the scattering pattern, and in particular the angle of its maximum $\tilde{\theta}_{max}$, as a function of $\dot{\gamma}$. It is natural to identify this angular maximum position with θ from Eq. (7), the angle a droplet makes to the electric field.

To determine $\tilde{\theta}_{max}$, we first divided a time-averaged scattering image into 360 wedges, each subtending 1° of arc. We then integrated the total intensity in each wedge, and plotted the result versus angle. The maximum of such a plot determines

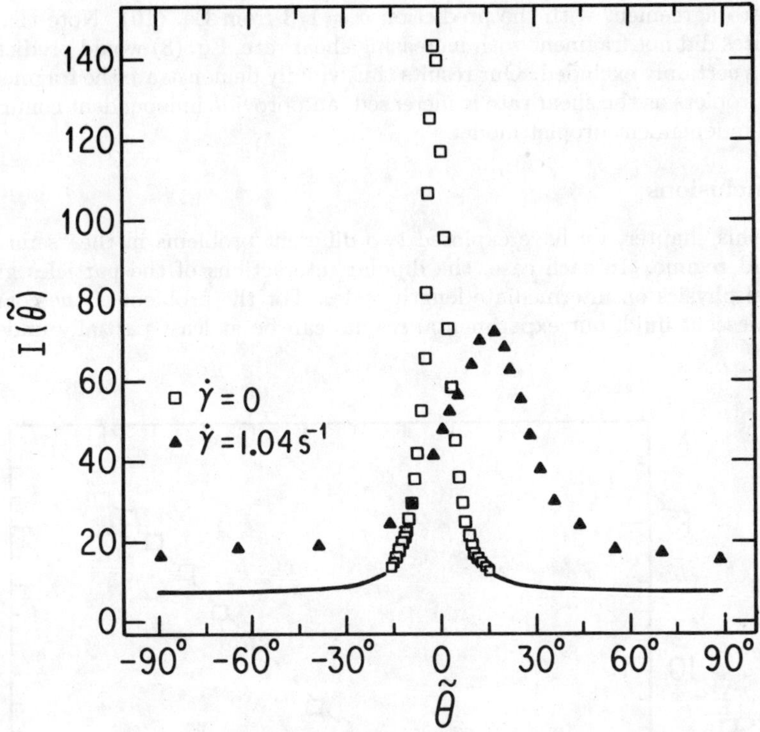

Figure 5. "Pie-o-metric" analysis indicates the angular distribution of the scattered light in quiescent and sheared fluids. Lobes perpendicular to the field, as in Figure 1, give an angular distribution sharply peaked about the direction perpendicular to the field ($\tilde{\theta} = 0$). In a sheared fluid with $\dot{\gamma} = 1.04\text{s}^{-1}$, the lobes are broader and are also rotated.

$\tilde{\theta}_{max}$. We have coined the term "pie-o-metric analysis" to describe this procedure, because it lends this somewhat arbitrary method a certain cachet otherwise lacking. Figure 5 compares the result of this procedure for a quiescent fluid with that for a fluid sheared at a rate of $\dot{\gamma} = 1.04\text{s}^{-1}$. The sheared result is considerably broader, but nevertheless has a clearly identifiable peak.

The dependence of $\tilde{\theta}_{max}$ on $\dot{\gamma}$ is shown in Figure 6. The least squares fit to the exponent gives

$$\theta_{max} \propto \dot{\gamma}^{\delta} \qquad \delta = 0.326 \tag{11}$$

in superb agreement with the prediction $\delta = 1/3$ from Eq. (10). Note that if the structures did not fragment with increasing shear rate, Eq. (8) would predict $\delta = 1$, which is certainly excluded. Our results thus vividly demonstrate the fragmentation of the droplets as the shear rate is increased, and provide independent confirmation for the independent droplet model.

5. Conclusions

In this chapter, we have explored two different problems in the "semi-dilute" ER fluid regime. In each case, the dipolar interactions of the particles give rise to novel physics on intermediate length scales. For the problem of the coarsening of a quiescent fluid, our experimental results can be at least partially understood

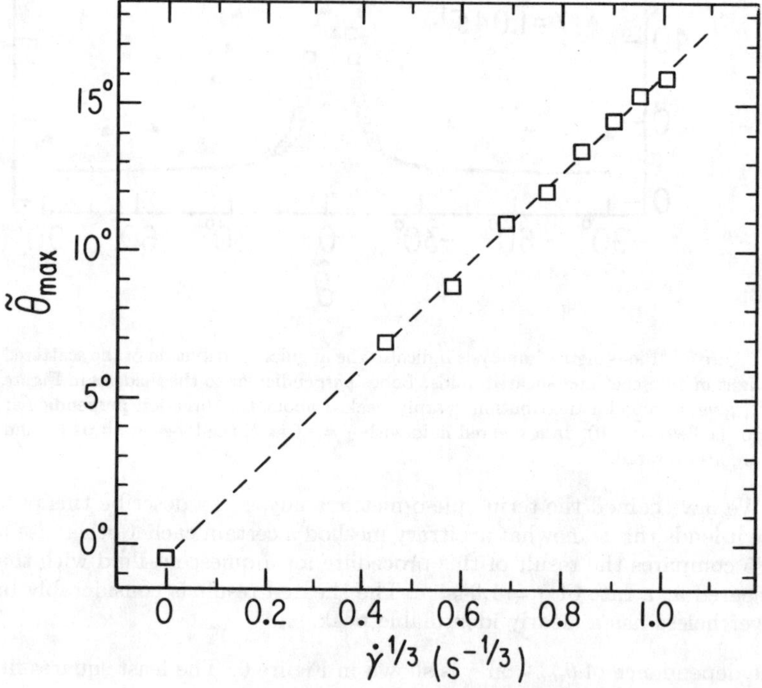

Figure 6. The maximum of the angular pattern from Figure 5, $\tilde{\theta}_{max}$, graphed versus the shear rate $\dot{\gamma}^{1/3}$. The linearity of this plot is strong evidence for the "independent droplet" model, which claims that droplets are rotated by an angle $\theta \propto \dot{\gamma}^{1/3}$ from the electric field direction.

as arising from a thermal fluctuation mechanism. Our understanding in the sheared fluid seems, at this stage, to be more complete. A model of fragmenting and recombining droplets, which was originally devised to account for the macroscopic rheological respone of the fluid, is also in excellent agreement with light-scattering observations of the fluid microstructure.

6. Acknowledgements

The work discussed in this chapter was performed in collaboration with D. Adolf, J. Odinek, and W. Toor. T.C.H. is grateful to the National Science Foundation for the support of this research through the Presidential Young Investigators program, grant DMR-9057156. Acknowledgement is made to the Donors of The Petroleum Research Fund, administered by the American Chemical Society, for the partial support of this research. The work of J.E.M. was performed at Sandia National Laboratories and was supported by the U.S. Department of Energy under Contract No. ED-AC04-76DP00789.

References

1. A.P. Gast and C.F. Zukoski, *Adv. Colloid Interface Sci.* **30** (1989) 153.
2. J.M. Ginder and L.D. Elie, in *Proceedings of the Conference on Electrorheological Fluids* ed. R. Tao (World Scientific, Singapore, 1992).
3. J.E. Martin, J. Odinek, T.C. Halsey, *Phys. Rev. Lett.* **69** (1992) 1524.
4. W. Winslow, *J. Appl. Phys.* **20** (1949) 1137.
5. T.C. Halsey and W. Toor, *Phys. Rev. Lett.* **65** (1990) 2820.
6. R. Tao and J.M. Sun, *Phys. Rev. Lett.* **67** (1991) 398; *Phys. Rev. A* **A44** (1991) R6181 . For an elegant experimental confirmation, see T.-j. Chen, R.N. Zitter, R. Tao, *Phys. Rev. Lett.* **68** (1992) 2555.
7. T.C. Halsey and W.R. Toor, *J. Stat. Phys.* **61** (1990) 1257; W.R. Toor, *J. Coll. Interface Sci.* **153** (1993) 335 .
8. D.J. Klingenberg and C.F. Zukoski, *Langmuir* **6** (1990) 15.
9. T.C. Halsey, J.E. Martin, D. Adolf, *Phys. Rev. Lett.* **68** (1992) 1519; and unpublished. A similar model was introduced for "magneto-rheological" fluids by Z.P. Shulman, V.I. Kordonsky, E.A. Zaltsgendler, I.V. Prokhorov, B.M. Khusid, and S.A. Demchuk, *Int. J. Multiphase Flow* **12** (1986) 935.
10. A.P. Philipse and A. Vrij, *J. Coll. Interface Sci.* **128** (1987) 121.
11. G. Porod, *Kolloid-Z.* **124** (1951) 83; **125** (1952) 51, 109.
12. E.M. Lifshitz and L.P. Pitaevski, *Physical Kinetics* (Pergamon Press, New York, 1981) p. 432-8.

13. For a review, see T. C. Halsey, *Science* **258** (1992) 713.

14. P.M. Adriani and A.P. Gast, *Phys. Fluids* **31** (1988) 2757.

15. G.B. Jeffery, *Proc. Roy. Soc. (London)* **A102** (1922) 161.

16. L.D. Landau, E.M. Lifshitz, and L.P. Pitaevski, *Electrodynamics of Continuous Media, 2nd ed.* (Pergamon Press, New York, 1984) p. 42.

17. J. Hemp, *Proc. R. Soc. Lond.* **A434** (1991) 297.

18. W.R. Toor and T.C. Halsey, *Phys. Rev.* **A45** (1992) 8617.

Simulation of Solid Structure Formation in an Electrorheological Fluid

R. Tao and Qi Jiang
Department of Physics, Southern Illinois University
Carbondale, Illinois 62901, USA

ABSTRACT

The temporal evolution of three-dimensional structure in an electrorheological (ER) fluid is examined by a computer simulation. A parameter B characterizing the ratio of the Brownian force to the dipolar force is introduced. For a wide range of B, the ER fluid has a rapid chain formation followed by aggregation of chains to form thick columns, which has a body-centered tetragonal (bct) lattice structure. The Peierls-Landau instability of single chains helps formation of thick columns. If B is very small, the ER system will be trapped in some local energy-minimum state.

1. Introduction

Electrorheological (ER) fluids, often referred to as smart fluids, have a wide variety of applications in industries and technologies. A typical ER fluid consists of a suspensions of fine dielectric particles in a liquid of low dielectric constant.[1–3] Its effective viscosity increases dramatically if an electric field is applied, and when the field exceeds a critical value, the ER fluid turns into a solid whose shear stress continues to increase as the field is further strengthened. These phenomena occur in milliseconds and are reversible.

The structure is fundamental in understanding the physical mechanism and properties of ER fluids. Experiments indicate that upon application of electric field, dielectric particles in ER fluids rapidly form chains which then aggregate to form thick columns between two electrodes.[4–7] A theoretical prediction[8] of a body-centered tetragonal (bct) lattice as the ground state of the thick columns has recently been verified by experiments[4].

The issue of dynamics of structure formation in ER fluids is under extensive investigation. The interest in the issue has further been enhanced, since Kamien and Nelson recently pointed out that the phase transition in ER fluids is related to the physics of directed polymer melts and quantum mechanics of bosons in 2+1 dimensions.[9–11]

Two-dimensional (2-D) computer simulations were first employed to investigate the issue. The results of 2-D simulations can be summarized as follows. If thermal fluctuations in ER fluids are ignored, 2-D simulations found separated single-chain structures.[12] If the thermal fluctuations are included in the simulations, triangular lattice, the only close-packed structure in two dimensions, is indeed found to be the ER structure (see Fig. 1).[13] The extrapolation of results in two dimensions to three-dimensional (3-D) systems, however, is hindered not only by the obvious difficulty, but also by the singular nature of the Coulomb potential.

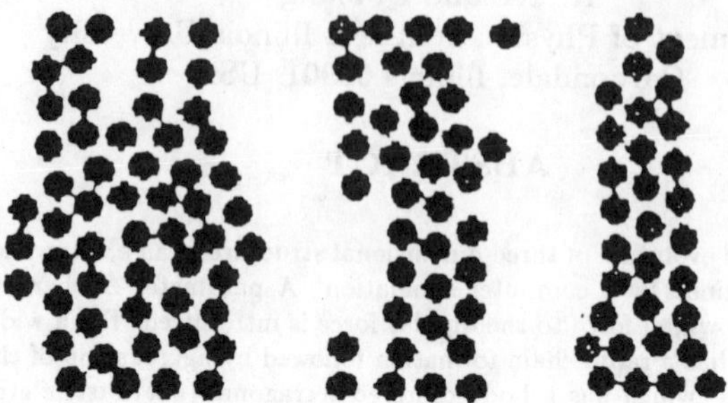

Fig.1 The close-packed triangular lattice is the ER structure found in 2-D simulations if the thermal fluctuations are included.

There have been a couple of 3-D simulations of ER fluids reported. Whittle[14] and Molrone[15] only examined the initial aggregation process in their 3-D simulations peripherally. The simulations by Bonnecaze and Brady[16] include the full hydrodynamics and electrostatics, but are performed at zero temperature, thus do not give information on the structure formed. Recently, Hass reported that a regular lattice was not formed in his 3-D simulation.[17] However, his simulation ignores the thermal fluctuations and the effect of two electrodes which have been both proved to be important in formation of thick columns.

To clarify the issue, we consider a monodisperse suspension of spherical dielectric particles in a nonconducting liquid. The particles have diameter σ and dielectric constant ϵ_p. The liquid has dielectric constant ϵ_f and viscosity η. The system is confined between two parallel electrodes which are denoted as planes $z = 0$ and $z = L$. When there is no voltage applied, the particles are randomly distributed throughout the fluid. In an electric field, each particle obtains an induced dipole moment, $\mathbf{p} = \alpha\epsilon_f(\sigma/2)^3\mathbf{E}_{loc}$ where $\alpha = (\epsilon_p - \epsilon_f)/(\epsilon_p + 2\epsilon_f)$ and $\mathbf{E}_{loc}$ is the local field. As did in other simulations[12–15,17], we take the dipolar approximation. The role of higher order multipoles will be discussed at the end of the paper. The ER solid structure is determined by the dipolar interactions, viscous drag forces, and Brownian motions.

We have found that the thermal motion plays a very important role in the structure formation of ER fluids. Though the ratio of the dipolar energy to the thermal energy is very high for a typical ER fluid, $(p^2/\epsilon_f\sigma^3)/k_BT \sim 10^6$, the thermal energy cannot be ignored. Because of the Peierls-Landau instability of one-dimensional solid[7], there is another important quantity, a ratio of the thermal energy to the lowest phonon excitation energy of a single chain, $k_BT/(hc/2L) \sim 10^6$ where c is the sound velocity in the ER fluid. Therefore, the thermal energy is more than sufficient in creating thermal vibrations of the single chains which accelerate the aggregation process. We will introduce a parameter, B, characterizing the ratio of the dipolar force to the Brownian force. Our simulation finds that for a wide range of B, with the help of the Brownian force, the ER fluid evolves into thick columns which has the bct lattice structure. When B is very small, the ER system is unable to get out from some trapped local energy minimum state to approach the ground state.[17] The structure at very small B may be related to directed polymers.[9–11] When B is too big, thermal vibrations prevent the system from formation of stable solid structure.

2. Computer Simulations

The motion of the i'th particle is described by a Langevin Equation

$$m\frac{d^2\mathbf{r}_i}{dt^2} = \mathbf{F}_i - 3\pi\sigma\eta\frac{d\mathbf{r}_i}{dt} + \mathbf{R}_i(t) \tag{1}$$

where $\mathbf{F}_i$ is the electric force acting on the particle and $-3\pi\sigma\eta\mathbf{v}_i$ is the Stokes' drag force. In Eq.(1), we include a random Brownian force $\mathbf{R}_i(t)$, representing the net effect of collisions of solvent molecules on the particle. In the previous simulation[17], $\mathbf{R}_i(t)$ was ignored.

The dipolar force acting the particle at $\mathbf{r}_i$ by a particle at $\mathbf{r}_j$ is given by

$$\mathbf{f}_{ij} = \frac{3\mathbf{p}^2}{\epsilon_f r_{ij}^4}[\mathbf{e}_r(1 - 3\cos^2\theta_{ij}) - \mathbf{e}_\theta \sin^2\theta_{ij}] \tag{2}$$

where $\mathbf{r}_{ij} = \mathbf{r}_i - \mathbf{r}_j$ and $0 \leq \theta_{ij} \leq \frac{\pi}{2}$ is the angle between the z direction and the joint line of the two dipoles. We use $\mathbf{e}_r$ as a unit vector parallel to $\mathbf{r}_{ij}$ and $\mathbf{e}_\theta$ as a unit vector parallel to $\mathbf{e}_r \times (\mathbf{e}_r \times \mathbf{E}_0)$.

A dipole $\mathbf{p}$ inside the capacitor at $\mathbf{r}_i = (x_i, y_i, z_i)$ produces an infinite number of images at $(x_i, y_i, -z_i)$ and $(x_i, y_i, 2Lk\pm z_i)$ for $k = \pm1, \pm2, \cdots$. The interaction force between a dipole and an image has the same form as Eq.(2). The jth particle and its infinite images produce an electric force on the ith particle,

$$\begin{aligned}
f_{ij,x} &= \frac{p^2}{\epsilon_f L^4}\sum_{s=1}^{\infty}\frac{4s^3\pi^3(x_i - x_j)}{\rho_{ij}}K_1(\frac{s\pi\rho_{ij}}{L})\cos(\frac{s\pi z_i}{L})\cos(\frac{s\pi z_j}{L}) \\
f_{ij,y} &= \frac{p^2}{\epsilon_f L^4}\sum_{s=1}^{\infty}\frac{4s^3\pi^3(y_i - y_j)}{\rho_{ij}}K_1(\frac{s\pi\rho_{ij}}{L})\cos(\frac{s\pi z_i}{L})\cos(\frac{s\pi z_j}{L}) \\
f_{ij,z} &= \frac{p^2}{\epsilon_f L^4}\sum_{s=1}^{\infty}4s^3\pi^3 K_0(\frac{s\pi\rho_{ij}}{L})\sin(\frac{s\pi z_i}{L})\cos(\frac{s\pi z_j}{L})
\end{aligned} \tag{3}$$

where $\rho_{ij} = \sqrt{(x_i - x_j)^2 + (y_i - y_j)^2}$ and K_0 and K_1 are modified Bessel functions. The force on the ith particle by its own images is in the z direction and denoted as

$\mathbf{f}_i^{self}$,

$$f_{i,z}^{self} = \frac{3p^2}{8\epsilon_f}\left(-\frac{1}{z_i^4} + \sum_{s=1}^{\infty}\left[\frac{1}{(z_i - sL)^4} - \frac{1}{(z_i + sL)^4}\right]\right) \tag{4}$$

To simulate the hard spheres and hard walls, we introduce a short-range repulsive force between two particles

$$\mathbf{f}_{ij}^{rep} = \frac{3p^2\mathbf{e}_r}{\epsilon_f \sigma^4} \exp[-100(r_{ij}/\sigma - 1)] \tag{5}$$

and a short-range repulsion between a particle and the electrodes

$$\mathbf{f}_i^{wall} = \frac{3p^2\mathbf{e}_z}{\epsilon_f L^4}\{\exp[-100(z_i/\sigma - 0.5)] - \exp[-100((L - z_i)/\sigma - 0.5)]\}. \tag{6}$$

Now $\mathbf{F}_i$ in Eq.(1) is given by

$$\mathbf{F}_i = \sum_{j \neq i}[\mathbf{f}_{ij} + \mathbf{f}_{ij}^{rep}] + \mathbf{f}_i^{self} + +\mathbf{f}_i^{wall} \tag{7}$$

The random force $\mathbf{R}_i(t)$ has a white-noise distribution,

$$\langle R_{i,x} \rangle = 0, \quad \langle R_{i,\alpha}(0) R_{i,\beta}(t) \rangle = 6\pi k_B T \sigma \eta \delta_{\alpha\beta} \delta(t) \tag{8}$$

where k_B is Boltzmann's constant and T is the temperature. In our simulation, we replace $\mathbf{R}_i(t)$ by $\mathbf{R}_i(t, \Delta t)$ which is the average of $\mathbf{R}_i(t)$ over a short time step Δt, $R_{i,\alpha}(t, \Delta t) = \frac{1}{\Delta t}\int_t^{t+\Delta t} R_{i,\alpha}(t')dt'$. $\mathbf{R}_i(t, \Delta t)$ has a normal distribution,

$$< R_{i,\alpha}(t, \Delta t)^2 >= \Omega^2 \tag{9}$$

where $\Omega = \sqrt{6\pi k_B T \sigma \eta / \Delta t}$ is the Brownian force scale.

The intrinsic time scale in Eq.(1) is $t_0 = m/(3\pi\sigma\eta)$. We take $t = t_0 t^*$, $\mathbf{F}_i = F_0 \mathbf{F}_i^*$ where $F_0 = 3p^2/(\epsilon_f \sigma^4)$, $\mathbf{R}_i = \Omega \mathbf{R}_i^*$, and $\mathbf{r}_i = \sigma \mathbf{r}_i^*$ In Eq.(1). The scaling transformation produces a new equation

$$\ddot{\mathbf{r}}_i^* + \dot{\mathbf{r}}_i^* = A(\mathbf{F}_i^* + B\mathbf{R}_i^*) \tag{10}$$

where $A = F_0/[3\pi\sigma\eta(\sigma/t_0)]$ and $B = \Omega/F_0$. It is clear that $1/A$ is related to the Mason number, a ratio of the viscous force to the dipolar force. B is a ratio of the Brownian force to the dipolar force.

Equation (10) indicates that the final structure of our system is related to the two constants A and B. For a real ER system, such as alumina particles in petroleum oil, $\epsilon_f \sim 2$, $\epsilon_p \sim 8$, $\eta \sim 0.2$ poise, $\sigma \sim 10\mu$m, and the particle mass density $\rho \sim 3g/cm^3$. At $E_0 = 3$KV/mm and T=300 K, the ER fluid will be solidified in experiments. Under these conditions, we estimate $t_0 \sim 8.33 \times 10^{-7}$ s and $A \sim 10^{-4}$. As $\Delta t = 0.4t_0$, $B \sim 10^{-1}$. Our simulation just takes the above values for A and B.

An adaptive step-size control for Runge-Kutta method [18] is applied to integrate the motion equation (10). We specify a criterion δr_c. Let the largest position change among all particles during the time step δt be δr_x. If $\delta r_x > \delta r_c$, in the subsequent simulation, we reduce the time step to $\delta t/d$ where d is a controlled constant, greater than one. If $\delta r_x < \delta r_c$, in the next step, we increase the time step to $c\delta t$ where c is also a controlled constant, greater than one. Through constants c and d whose selection will be specified shortly after, we control the step size in our integration. It is also clear that δr_x is relatively big at the initial stage and gets smaller and smaller at the final stage. Accordingly, we will reduce δ_c in the course.

Our simulation has N=122 particles in a box with $L_x = L_y = 5\sigma$ and $L_z = 14\sigma$. Theoretical calculation has already shown that if $L_z \leq 6\sigma$, the single-chain structure has a lower energy than that of thick columns.[8] Therefore, we take a relatively big L_z. A periodic boundary condition is imposed in the x and y directions. This corresponds to a volume fraction $\phi = 0.183$. At every step, we apply the following three order parameters to characterize the structure[19],

$$\rho_j = \left| \frac{1}{N} \sum_{i=1}^{N} \exp(i\mathbf{b}_j \cdot \mathbf{r}_i) \right| \tag{11}$$

where three reciprocal lattice vectors of the bct lattice $\mathbf{b}_1 = (2\pi/\sigma)(2\mathbf{e}'_x/\sqrt{6} - \mathbf{e}_z)$, $\mathbf{b}_2 = (2\pi/\sigma)(2\mathbf{e}'_y/\sqrt{6} - \mathbf{e}_z)$, and $\mathbf{b}_3 = 4\pi\mathbf{e}_z/\sigma$. Among the three unit vectors, $\mathbf{e}_z$ is in the field direction, but $\mathbf{e}'_x$ and $\mathbf{e}'_y$ should be along the intrinsic axes of the bct lattice. In the structure formation, the ER system may rotate around the z axis, a phenomenon already observed in experiments.[4] Therefore, when measuring ρ_1 and ρ_2, we always rotate $\mathbf{e}'_x$ and $\mathbf{e}'_y$ about the z axis to find a position which maximizes $\rho_1\rho_2$.

The order parameter ρ_3 characterizes the formation of chains in the z direction. The other two parameters ρ_1 and ρ_2 characterize the structure in the x-y plane. All these three order parameters are unity if the ER system is the ideal bct lattice, When the dielectric particles are randomly distributed as in a liquid state, they are all vanishing.

3. Results and Discussions

The initial state is shown in Fig.2 where the dielectric particles are randomly distributed and the three order parameters are vanishing. At $t = 0$, a strong electric field is turned on to reach $A = 10^{-4}$ and $B = 10^{-1}$ and the particles begin to move. For the first 5000 steps, we take $\delta r_c = 10^{-3}\sigma$, $\delta t = 0.5t_0$, $c = 2$ and $d = 2$. It can be noticed from Fig.3 that after the first 5000 steps (about 10 milliseconds), single chains are in shape but the lateral ordering is very weak. The structure has $\rho_3 = 0.617$, while $\rho_1 = 0.32$ and $\rho_2 = 0.16$.

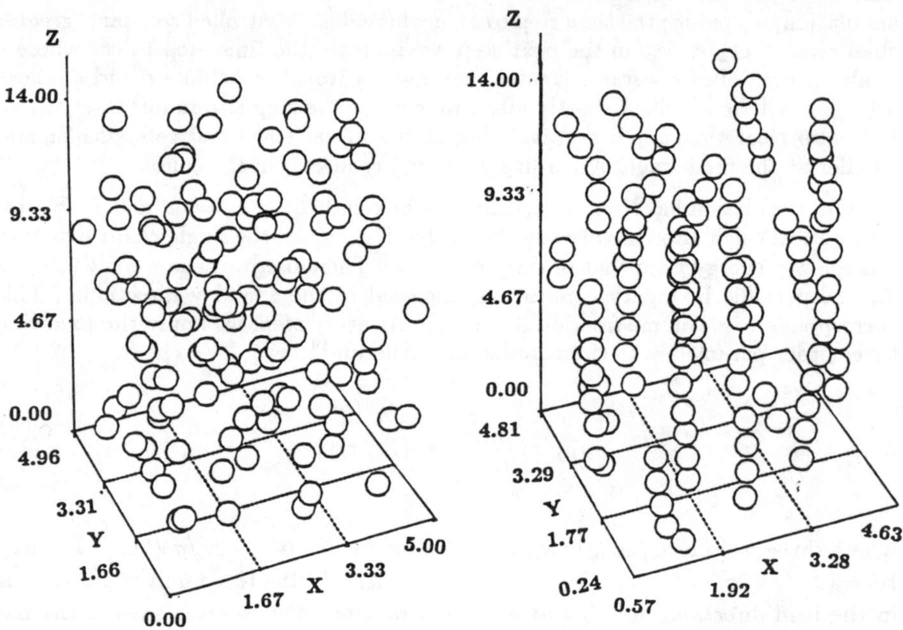

Fig. 2 In the initial state, dielectric particles are randomly distributed.

Fig. 3 The configuration after first 5000 time steps.

For the next 15000 steps, we change c to 1.1 and d to 3.0. At the end of this time interval, we have $\rho_3 = 0.92$, but ρ_1 and ρ_2 remain around $0.4 \sim 0.5$. The structure is in Fig.4. Clearly, the ordering in the field direction is almost perfect and the lateral ordering is building.

After 20000 steps, we reduce δr_c to 0.0005σ and continue the simulation. The structure obtained at 90000 steps in Fig.5 clearly shows excellent orderings both in the field direction and in the x-y plane. Single chains have been aggregated into a bct lattice structure. Three order parameters are $\rho_3 = 0.991$, $\rho_1 = 0.915$, and $\rho_2 = 0.850$. Further time evolution shows little improvement and the process becomes extremely slow.

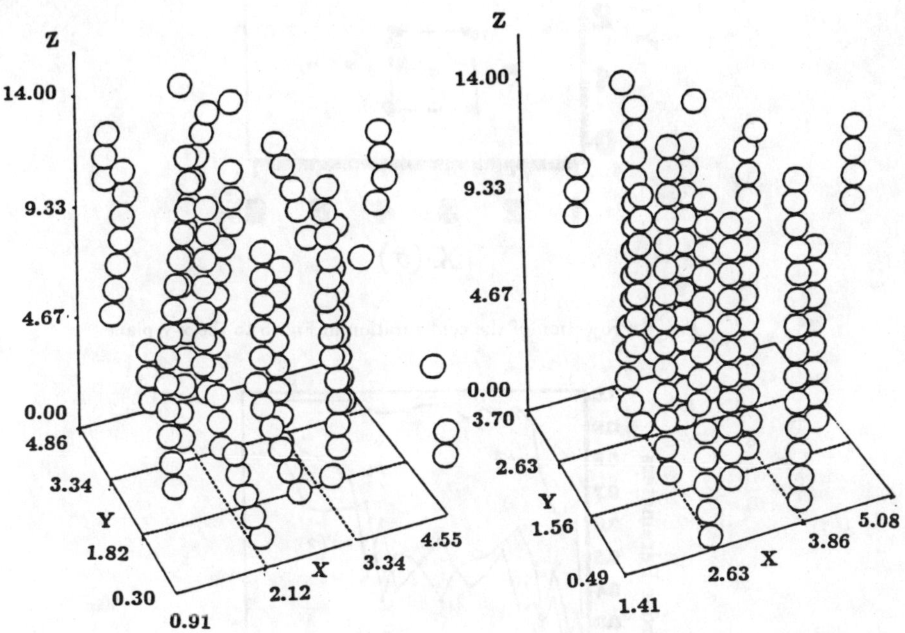

Fig. 4 The configuration after 20000 time steps.

Fig. 5 The configuration after 90000 time steps.

The projection of the 3-D structure in Fig.5 on to the x-y plane has a square lattice on the x-y plane (Fig.6). For example, the marked square has its side $\simeq \sqrt{1.5}\sigma = 1.225\sigma$, marking an ideal bct lattice. The four chains at the corners have 14 particles straight along the field direction. The chain at the center has 12 particles close-packed with the four neighbor chains. The structure is an ideal bct lattice if the chain at the center does not have one particle missing. The thick dots in Fig. 6 also indicate that the chains are straight in the field direction.

Fig. 7 shows the three order parameters develop with time. The formation of single chains is rapid. In about several milliseconds, ρ_3 reaches 0.6. After 0.4s, ρ_3 is above 0.9. The building of lateral ordering is relatively slower than the formation of chain. It takes about several seconds to have ρ_1 and ρ_2 reached 0.8.

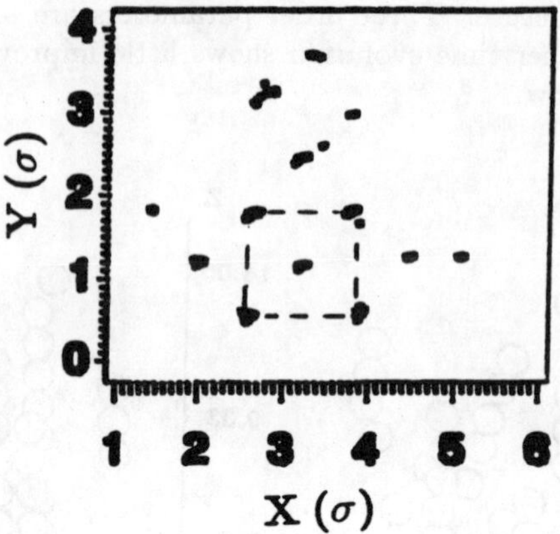

Fig. 6 Projection of the configuration in Fig. 5 to the x-y plane.

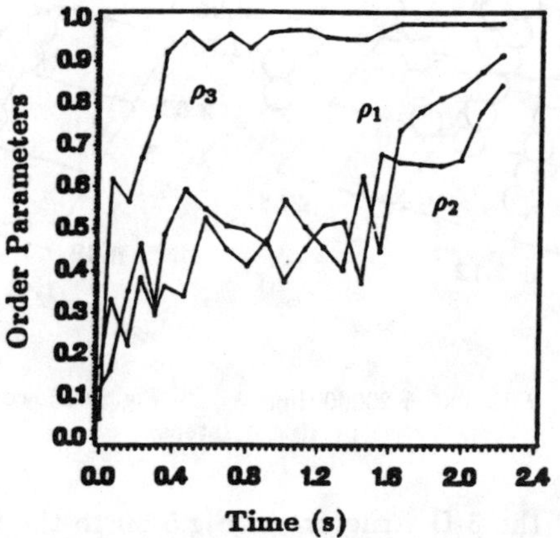

Fig. 7 The order parameters change with time after the electric field is applied. The time unit is second.

We have also done some tests to verify the importance of non-equilibrium Brownian force in development of ER solid structure. If we ignore the Brownian motion completely, or let B be extremely small, such as $B \sim 10^{-6}$, then the ER fluid has a rapid chain formation followed by a kinetic trapping into a complicated structure. After a long time, the structure has ρ_3 stayed about 0.7 but ρ_1 and ρ_2 remain 0.3 or below. In another words, the system is trapped in to a local energy-minimum state, but cannot get out to reach the global energy-minimum state.[17] The structure in such a case needs a further study because it may be related to directed polymers.[9] When B is very big ≥ 1, such as the case of very small dielectric particles, the ER system has too much vibrations which prevents from formation of a stable structure. However, for a quite a wide range of B, we find that the ER fluid has a rapid chain formation, followed by an aggregation of chains to form the bct lattice structure with three order parameters close or above 0.9.

Finally, we compare our simulation with experiments. The rapid chain formation in our simulation takes about 10 milliseconds or longer. This result is consistent with other simulations.[12] However, the experiments on ER fluids, such as alumina particles in petroleum oils, find this chain formation time is only about several milliseconds, shorter than our result.[20] The reason for this difference may lie in the fact that we ignore the contributions from higher multipoles, charges, and currents in ER fluids which begin to draw attention recently.[21]

Acknowledgements.

This research is supported by the Office of Naval Research grants N00014-90-J-4041 and N00014-93-1-0582 and a grant from Materials Technology Center of Southern Illinois University at Carbondale.

References

1. For example, see, *Electrorheological Fluids*, edited by R. Tao (World Scientific Publishing Comp., Singapore, 1992).
2. H. Block and J. P. Kelly, US Patent 4,687,589 (1987).
3. F. E. Filisko and W. E. Armstrong, US patent No. 4 744 914 (1988).
4. T. J. Chen, R. N. Zitter, and R. Tao, Phys. Rev. Lett. **68**, 2555 (1992).
5. J. E. Martin, J. Odinek, and T. C. Halsey, Phys. Rev. Lett. **69**, 1524 (1992).
6. J. M. Ginder and L. D. Elie in Ref.1, 23.
7. T. C. Halsey and W. Toor, Phys. Rev. Lett. **65**, 2820(1990).
8. R. Tao and J. M. Sun, Phys. Rev. Lett. **67**, 398(1991); Phys. Rev. A, **44**, R6181 (1991).
9. R. D. Kamien and D. R. Nelson, J. of Stat. Phys. **71**, 23 (1993).
10. R. D. Kamien and D. R. Nelson, Phys. Rev. A **45**, 8727 (1992).
11. R. P. Feynman, *Statistical Mechanics* (Benjamin/Cummins, Reading, MA, USA, 1972).
12. D. J. Klingenberg, F. van Swol, and C. F. Zukoski, J. Chem. Phys. **91**, 7888 (1989); **94**, 6170 (1991).
13. N. K. Jaggi, J. Stat. Phys. **64**, 1093 (1991); W. R. Toor, J. of Colloid and Interface Science, **156**, 335 (1993).

14. M. Whittle, J. Non Newtonian Fluid Mechanics, **37**, 233 (1990).

15. D. M. Heyes and J. R. Molrose, Mol. Simul. **5**, 293 (1990).

16. R. T. Bonnecaze and J. F. Brady, J. of Chem. Phys. **96**, 2183 (1992).

17. K. C. Hass, Phys. Rev. E, **47**, 3362 (1993).

18. For example, see, W. H. Press and et al, *Numerical Recipes* (Cambridge University Press, Cambridge, 1987), 554-560.

19. R. Tao, Phys. Rev. E **47**, 423 (1993).

20. K. D. Weiss and J. D. Carlson in Ref. 1 264.

21. L. C. Davis, J. Appl. Phys. **72**, 1334 (1992); N. J. Félici, J. N. Foulc, and P. Atten in the Proceedings of the Fourth International Conference on ER Fluids, edited by R. Tao (World Scientific Publishing Comp. Singapore,in press)

A CONDUCTION MODEL OF ELECTRORHEOLOGICAL EFFECT

N. FELICI, J. N. FOULC and P. ATTEN

Laboratoire d'Electrostatique et de Matériaux Diélectriques
CNRS and Université Joseph Fourier de Grenoble
B.P. 166 - 38042 Grenoble Cedex 9 (France)

ABSTRACT

A conduction model is proposed to explain the electrorheological effect under D.C. or low frequency A.C. fields. Firstly, an approximate analysis taking into account the bulk conduction of the solid and the non linear conduction properties of the liquid is confirmed by experiments on large scale spheres. Secondly, the predictions are extended to the case of surface conduction of the solid spheres.

1. Introduction

In the 1940's, Winslow[1] working with silica in kerosene discovered that applying a D.C. or A.C. field of the order of 1 kV/mm resulted in a drastic change in the rheological properties of the suspension : its apparent viscosity was increased by several orders of magnitude and a transition to a "solid" phase was obtained when applying the field to the suspension at rest; a static yield stress was obtained, i.e. a stress threshold at zero shear rate. These phenomena are manifestations of what is now called the electrorheological (ER) effect and the suspensions exhibiting this ER effect are named ER fluids[2]. The latter are mixtures of micrometer size particles of rather insulating materials and dielectric insulating liquids with the volume fraction of solid material usually ranging from 10 to 40%.

A great number of ER fluids[3], and particularly the early studied ones, are suspensions of more or less hydrophilic particles like corn starch, silica gel, cellulose, etc... : water or other additives adsorbed on or within the particles promote dramatically the ER effect. Unfortunately, these necessary additives make the rheological behavior of these fluids much dependent on temperature. Recently the use of anhydrous particles as semi-conducting polymers allows to obtain steady rheological behavior in a larger temperature range[4]. The main characteristics of ER fluids[5] are the yield stress (a few kPa), the ratio of dynamic viscosity with and without electric field (>100), the response time (~ ms) and full reversibility : the fluid recovers its original state and properties when the electric field is removed.

Numerous applications of ER fluids exist and a lot of devices have been proposed[6]. Among them are electromechanical coupling devices (active damper, brake, clutch), hydraulic actuators (ER valve and bridge), robotic devices (compliant wrist), servo-control systems (position, rate, force), acoustic apparatus, etc... To-day, the major drawback of commercial ER fluids is the operating temperature range limit.

Although ER effect is presently an active field of research, the physical mechanisms explaining the ER phenomena are not yet entirely understood[7]. Generally the ER effect is explained by the competition between the phenomenon of electrically induced aggregation of particles in fibers and the desaggregation due to hydrodynamic forces. Conventional wisdom ascribes the aggregation of particles to the mismatch of permittivities, ε_S (solid) and ε_L (liquid). As ε_S is always larger than ε_L, the particles attract one another and form chains, like iron filings in a magnetic field.

This cannot be the whole story. As mentioned by Filisko, titanium oxide ($\varepsilon/\varepsilon_o=80$) appears to be very active, but becomes inert in the dry state[8]. The role of water remained for a long time a mystery; it is plausible, however, that water merely gives the particles some sort of conductivity. Water can be dispensed with if the particle material exhibits an intrinsic conductivity in the dry state. On the other hand, it must be realized that most liquids, even purified hydrocarbons, are comparatively poor insulators. Their relaxation time $\tau_L=\varepsilon_L/\sigma_L$ (σ_L: conductivity) rarely exeeds 10^{-1} s, while many solids can keep their charge for minutes ($\tau_S>10^2$ s).

The classical condition for the field distribution not to depend on time (and, therefore, to remain identical to that given by permittivities) is $\tau_L=\tau_S$, a condition which is not satisfied by most ER mixtures, particularly in the absence of water. Thus, at low or zero frequency, the field distribution and the forces must be controlled by the conductivities of both phases.

On the basis of this fundamental fact we recall first the calculus of the attraction force between two spherical particles as a function of both the magnitude of electric field and the ratio of particles and carrying liquid conductivities, taking into account the non linear conduction properties of liquids. Our model is tested on a large scale experiment using slightly conducting polymeric spheres and controlling the liquid conductivity. The measurements fully confirm the occurrence of two regimes, quadratic and linear, of variation of the force as a function of the field. Finally we extend the interaction force estimation to the case of particles having only a surface conductivity.

2. Conductive model

The mechanisms proposed until recently[9,10] do not satisfactorily account either for the order of magnitude of the ER effect or for the drastic influence of added water. When retaining only the dipolar interaction due to polarisation of particles, most authors neglect the role of very small but non zero conductivity of dielectric materials which are never perfectly insulating. Now the conduction effects play the major role for D.C. and low frequency A.C. applied fields[11,12] and we proposed a new explanation based on the conductivity of both phases controlling the field distribution and therefore the interaction between particles[13]. We briefly recall below the main points.

2.1. A necessary condition of the ER effect

When particles with permittivity ε_S and conductivity σ_S immersed in a liquid with ε_L and σ_L are sujected to a D.C. uniform electric field $\mathbf{E}_o$, they acquire a dipole moment $\mathbf{p}$ determined by σ_S and σ_L. For a sphere :

$$\mathbf{p} = 4\pi\varepsilon_L R^3 \mathbf{E}_o (\sigma_S-\sigma_L) / (\sigma_S+2\sigma_L) \quad (1)$$

The dipole moment is parallel to the field if $\sigma_S > \sigma_L$ and antiparallel if $\sigma_S < \sigma_L$. In both cases, particles would seem to be able to attract one another and to build chains.

In the case of the prolate ellipsoid, the electric torque always tends to align the major axis parallel to the field. Thus, attraction and orientation would seem to be present in all cases, though weaker when $\sigma_S < \sigma_L$. This conclusion, however, is unwarranted, for the dynamics of the system does not allow stability in the latter case. In a nutshell, particles take a chaotic rotary motion[14] which prevents aggregation and alignment when the relaxation time τ_L of the liquid is larger than the time for the particles to rotate by a substantial angle under the pull of the electric torque. The condition for this may be roughly expressed by :

$$\tau_L > \eta_L/\varepsilon_L E_o^2 \quad (2)$$

where η_L is the dynamic viscosity of the liquid. This condition being usually satisfied in practical settings, no ER activity is to be expected whenever $\sigma_S < \sigma_L$.

2.2. Attraction force between two spheres

Let us consider two spheres of radius R of a solid material much less insulating than the liquid ($\sigma_S \gg \sigma_L$). Each sphere roughly behaves as a conductor immersed in a much more insulating liquid and takes the potential corresponding to the location of its center. Between two spheres at a distance d, the order of magnitude of the field is :

$$E \sim E_o(1+2R/d) \quad (3)$$

and as d tends to zero, E would increase dramatically. Indeed for high electric fields the conductivity of non polar liquids to a rough approximation increases exponentially with $\sqrt{E}$ [15]. This field enhanced conductivity results practically in a saturation of the field between the spheres. For two spheres in contact the field intensity is limited in the liquid close to the contact point, where the layer of liquid is very thin.

This basic remark allows to estimate the attraction force between the two spheres. In the case $\Gamma=\sigma_S/\sigma_L \gg 1$, the equipotential character of the sphere surface of course fails in the contact zone and we can distinguish two regions (Fig. 1-a) : for $x > \delta$ we can state that the sphere surface is equipotential and the current escaping from the sphere is negligible; for $x < \delta$, the field in the liquid more or less saturates due to the enhanced conductivity of the thin liquid layer and the major part of the current crosses the liquid in this region.

With this picture, the attraction force can be easily calculated. For $x>\delta$, the voltage is almost entirely applied to the liquid. The local field is $E \approx U/y$, where U is the potential difference between the spheres ($U \cong 2RE_o$) and y is the distance between them (Fig. 1-b). By replacing the sphere by paraboloids of same radius of curvature at $x=0$, y is approximated by x^2/R. The force per unit area is $\varepsilon_L E^2/2$ and in an annulus between x and x+dx, we get :

$$dF_1 \cong \pi\varepsilon_L U^2R^2dx/x^3 \tag{4}$$

By integration between δ and R one obtains ($\delta^2 \ll R^2$) :

$$F_1 \cong (\pi/2)\, \varepsilon_L U^2 \,(R/\delta)^2 \tag{5}$$

For $x<\delta$ we assume the field strength to be roughly constant and equal to the value at $x=\delta$: $E_\delta \cong RU/\delta^2$. The contribution of the inner zone then writes : $F_2 \cong \pi\delta^2\, (\varepsilon_L/2)\, E_\delta^2 \cong (\pi/2)\, \varepsilon_L U^2\, (R/\delta)^2$ and is equal to F_1. The total attraction force is :

$$F = F_1 + F_2 \cong \pi\varepsilon_L U^2\, (R/\delta)^2 = 4\pi R^2\, \varepsilon_L E_o^{\,2}\, (R/\delta)^2 \tag{6}$$

The next step to derive this force is to determine the transition radius δ.

2.3. Low applied fields

Coming back to the splitting of the sphere surface into two regions, it is clear that only a small part of the current escapes from the sphere in the first region ($x>\delta$), the

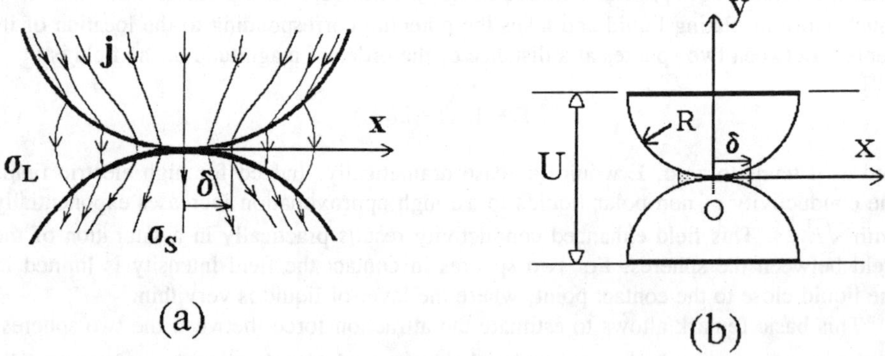

Fig. 1. Schematic view of the two half-spheres in contact. (a) sketch of current lines in both media ; (b) definition of notations.

resistance offered by the liquid being much greater than that of the solid phase. Very close to the contact point where the two spheres are touching each other, conversely, the resistance offered by the very thin layer of liquid (the conductivity of which can be enhanced) is negligible compared with the one of the solid material where the current density has increased drastically.

The distribution of current in such a system is given by Onsager's minimum dissipation theorem, which is the counterpart of Riemann's minimum energy theorem for static fields. One readily sees that the conductance C_s of the sphere increases with δ, while the opposite is true for the conductance C_L of the liquid. Thus, it is plausible that dissipation is minimum when both conductances are nearly equal, and this gives an equation for δ.

The conductance C_s of the solid spheres can be approximated by assuming that the current passes through the disks with radius δ (see Fig. 1). Using i) the capacitance $c=8\varepsilon\delta$ of an conductor disk (radius δ) isolated in an infinite medium of permittivity ε, ii) the relation $rc=\varepsilon/\sigma$ (r resistance, σ conductivity) and iii) the following approximation : two half spheres in contact are equivalent to two disks in serie (both of them isolated in an half infinite medium), we obtain for the conductance C_s :

$$C_S \cong 2\sigma_S\delta \tag{7}$$

As concerns the liquid phase, the conductance of the layer comprised between cylinders of radii x and x+dx is :

$$dC_L = 2\pi\sigma_L\, x dx/y \cong 2\pi\sigma_L R\, dx/x \tag{8}$$

and by integrating :

$$C_L = 2\pi R\sigma_L \ln(R/\delta) \tag{9}$$

This is the conductance of the part of the liquid which competes with the sphere for carrying the current (note that the conductance of the liquid within the δ-disk is very large but plays no role in our picture because this part of the liquid is in series with the sphere). Balancing both conductances C_s and C_L leads to the condition :

$$(R/\delta)\ln(R/\delta) = \Gamma/\pi \tag{10}$$

which determines δ (Γ is the conductivity ratio : $\Gamma = \sigma_S/\sigma_L \gg 1$). In the low electric field case, R/δ is a constant only depending on Γ. This gives the classical quadratic behaviour for the attraction force :

$$F \cong 4\pi R^2 \varepsilon_L K_\Gamma^{\;2}\, \Gamma^2\, E_o^{\;2} \tag{11}$$

where K_Γ stands for $[\pi \ln(R/\delta)]^{-1}$. Note that Eq. (11) also predicts a quadratic dependence on Γ.

2.4. High applied fields

In the high fields case, there are regions in the liquid ($x > \delta$) where the conductivity enhancement becomes significant. The field dependence of σ_L is described theoretically by Onsager's theory[16] which gives for the dissociation constant K_d of a solute :

$$K_d(E) = K_d(0)\ H(E/E_{Ons}) \tag{12}$$

where $E_{Ons}=(8\pi\varepsilon_L/e)(kT/e)^2$ (e: electronic charge; kT: thermal energy) and the function H expresses in term of the Bessel function J_1 [16]. The elementary conductance $dC_L=2\pi R\sigma_L(E)dx/x$ now involves the field dependent conductivity (with $E \approx UR/x^2$). In order to get qualitative insight rather than a precise quantitative description, $\sigma_L(E)$ is assumed to vary proportionally to the dissociation constant K_d. In order to integrate Eq. (8) and obtain analytical expressions, it is worthwhile to use the simplified expression :

$$\sigma_L(E) = \sigma_L(0)\ \{(1\text{-}A) + A\exp[(E/E_c)^{1/2}]\} \tag{13}$$

where A and E_c are constants depending on the considered liquid. For instance, in the case of a non polar liquid with $\varepsilon_L = 2.2\ \varepsilon_o$, $A=0.1$ and $E_c=0.335$ kV/mm give values differing only by a few percent from Onsager values for E in the range 15 to 50 kV/mm [17]. Substituting E by UR/x^2 and integrating after a change of variable and a further approximation finally leads to :

$$C_L \cong 2\pi\sigma_L\delta\ (E_c/2E_o)^{1/2}\exp[(R/\delta)\ (2E_o/E_c)^{1/2}] \tag{14}$$

Stating again that δ is determined by balancing C_L and C_S, one obtains the following expression for the attraction force

$$F = 2\pi R^2\varepsilon_L E_c E_o\ \{\ln[(10\Gamma/\pi)\ (2E_o/E_c)^{1/2}]\}^2 \tag{15}$$

The law is now linear in E_o to a first approximation and exhibits a very weak dependence on Γ ($\Gamma \gg 1$).

A very simple approximate formula for F can be obtained by simplifying further the model in this case of high applied field and $\Gamma \gg 1$. When $\Gamma \gg 1$ the spheres are practically equipotential and the radius δ may be determined by the condition that for $x \leq \delta$, $\sigma_L \cong \sigma_S$ thanks to the Onsager effect. This is achieved for a local field $E(x) \cong 30$ to 40 kV/mm which we denote by E_m. Inserting this condition in the expression $E(x) = UR/x^2$, we obtain $\delta^2 = UR/E_m$ and the force :

$$F = 2\pi R^2\ \varepsilon_L E_o E_m \tag{16}$$

The tensile strength of a chain of such spheres $F/\pi R^2 = 2\varepsilon_L E_o E_m$ gives values comparable to those of ER mixtures in the gel-like state.

3. Large scale experiment

The interaction processes between particules have been modelled using spheres of radius much larger than that of particles in an ER fluid. A first experiment aimed at measuring the interaction force between a slightly conducting polymeric sphere and a metallic plate[17]. We report here on results relative to sphere-sphere attraction.

3.1. Experimental set-up and procedure

A set-up was designed (Fig. 2) for measuring the force between two large scale half spheres (R=0.7 cm). The test cell comprises two duralumin electrodes 9 cm in diameter spaced by 1.4 cm. The planar section of the upper half sphere, solid with a scale of the balance, has been metallized and is grounded. The planar section of the lower half sphere, fastened to the lower electrode, is brought to potential U. The vertical position of the cell can be adjusted by a screw. The two spheres being in contact, the lower one is very slowly sunk until the force exerted by the balance causes the spheres to separate : the force is thereby measured. The sensitivity of the balance is a few 10^{-4} N.

The half spheres are made of polyamid having a conductivity $\sigma_S \cong 8\times10^{-10}$ S/m. The surrounding liquid is a mineral oil (Univolt 52 from ESSO) with permittivity $\varepsilon_L \cong 2.2\ \varepsilon_o$ and conductivity $\sigma_L \cong 6\times10^{-13}$ S/m at low fields. Adding the petroleum antistatic additive AOT to the oil allows σ_L to be increased in a controlled way.

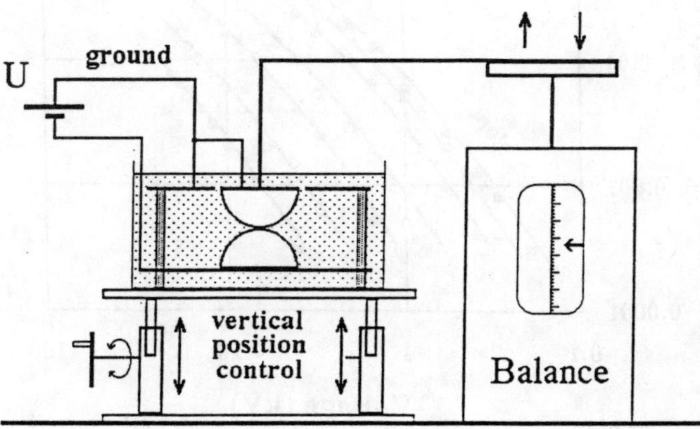

Fig. 2. Experimental set-up for the measurement of the attraction force between two half-spheres.

3.2. Force measurements

Fig. 3 shows the attraction force F between the two half spheres as a fonction of U for various values of the ratio Γ of solid and liquid conductivities. For high enough values of Γ, the force is proportional to U^2 for low fields and to U for strong fields. This is in full qualitative agreement with Eqs. (11) and (15).

Fig. 4 shows the same measured force as a function of the conductivity parameter Γ for different applied voltages. There appears a rather fast decrease of F when Γ decreases below 10. This is consistent with the first conclusion that the ER effect requires σ_S to be larger than σ_L. A quantitative test of Eq. (11) for U = 1 kV and 2 kV reveals a fair agreement for Γ ranging from 15 to 50 (see Fig. 4). A good agreement is also obtained between calculations from Eq. (15) and experimental results for U = 10 kV and 20 kV (Γ in the range 100 to 2000). Indeed these agreements appear surprisingly good if we remember the rough character of the estimation due to the various assumptions and approximations.

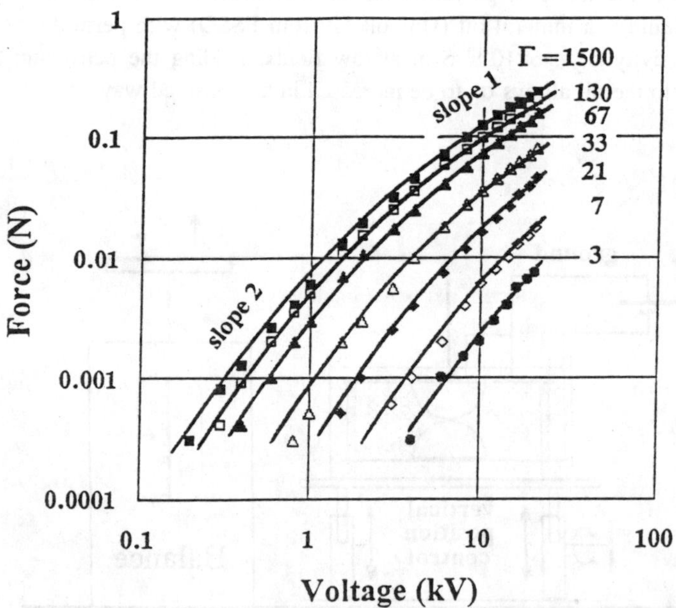

Fig. 3. Dependence of the attraction force F between spheres on the applied voltage U for various values of the conductivity ratio Γ.

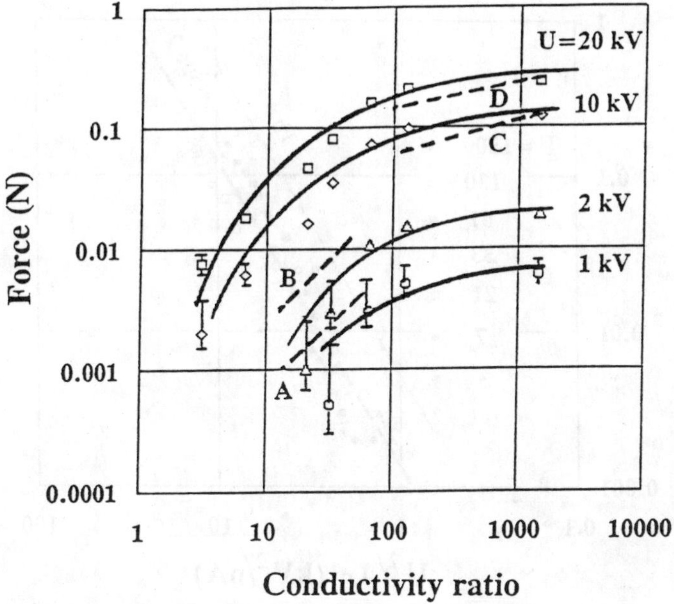

Fig. 4. Variation of the attraction force F between spheres as a function of the conductivity ratio Γ for various values of applied voltage U. Dotted curves A and B correspond to Eq. (11), and dotted curves C and D to Eq. (15).

3.3. Correlation between force and current

The good agreement obtained between predictions and results suggests that the basic approximations made in deriving Eqs. (6), (11) and (15) are sound. Now the current I_S flowing through the equatorial plane sections of the spheres (I_S is easily measured) provides a mean to test these approximations further. The key variable in the picture we propose is the radius δ of the contact zone. Eq. (6) expresses the force in terms of δ : $F \cong \pi\varepsilon_L U^2(R/\delta)^2$. Another basic point was to state that the conductance of the spheres in contact is equivalent to that of the disks of radius δ : $C_S \cong 2\sigma_S\delta$. By definition $I_S = C_S U$ and eliminating δ leads to :

$$F \cong 4\pi\varepsilon_L (\sigma_S R)^2 U^4/I_S^2 \tag{17}$$

Fig. 5 shows the variations of F as a function of (U^2/I_S) for the various values of Γ. Clearly there is a universal law, the dispersion appearing rather limited when we keep in mind the variations of Γ from 3 to 1500. Qualitatively the experimental results very satisfactorily agree with the prediction of F varying as U^4/I_S^2 (see Fig. 5). The factor of

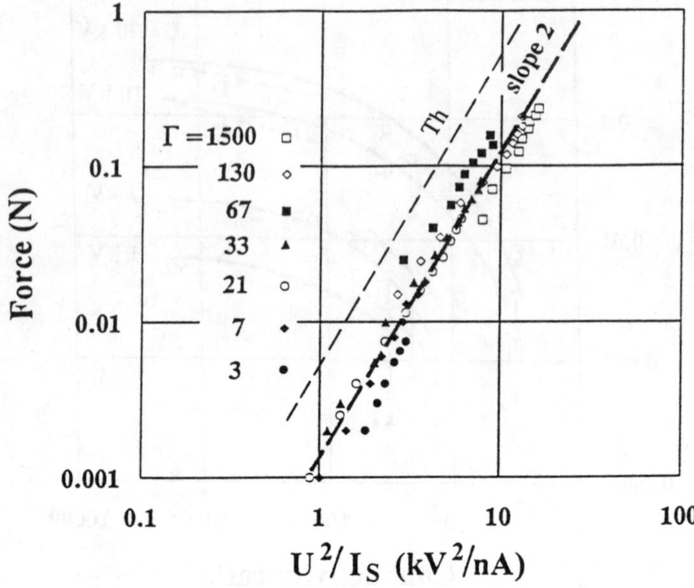

Fig. 5. Correlation of the attraction force F between spheres with the expression (U^2/I_S). Dotted line Th corresponds to Eq. (17).

about 5 between predictions of Eq. (17) and experimental results can be ascribed to the overestimate of the conductance of the two half spheres when approximating it by the conductance of the disks of radius δ. We can conclude that this test strongly validates the overall picture which is the basis for the force estimation.

An important consequence of this correlation between attraction force and current through the spheres, in D.C. fields, is that the leakage current is not a mere nuisance but a necessary by-effect of ER activity. Indeed ER forces are brought forth by the very strong fields prevailing near the contact points between particles, where the conductivity of the liquid is very much enhanced by the Onsager effect. This requires the particles to be even more conducting, and the chains they build are the channels of a leakage current. Thus, current drain and heating appear to be unescapable consequences of ER activity.

More precisely, it is possible to derive a semi-quantitative relation between tensile force and current. Assuming the simplified derivation of §2 valid for $\Gamma \gg 1$, we have as tensile force for a chain $F = 2\pi R^2 \varepsilon_L E_o E_m$ and as radius of the contact zone $\delta = (UR/E_m)^{1/2}$. The current flowing through the chain is $I_s = UC_s = 2\sigma_s U\delta$. Thus, we obtain for the ratio F/I_s :

$$F/I_s = (\pi/2\sqrt{2})\,(\varepsilon_L/\sigma_s)\,(E_m/E_o)^{1/2}\,E_m \tag{18}$$

Inserting $E_o = 3$ kV/mm, $E_m = 30$ kV/mm and $\sigma_S = 10^{-9}$ S/m gives : $F/I_s \equiv 2\times10^6$ N/A. For a tensile strength of 2 kPa we find a current drain of 10^{-3} A/m^2 or 0.1 μA/cm^2.

Finally, It is useful to notice that a related phenomenon has been already described as "Johnson-Rabeck effect".[18] When two nearly plane pieces of a moderately conducting material are allowed to touch each other, a strong attraction develops if a voltage of a few kV is applied to them. This is due to the strong fields appearing around the discrete contact points. Fully insulating plates give no effect, no more than metallic ones. The actual force was found not to depend very much on the fluid filling the gap, air or (insulating) liquid.

4. Role of surface conductivity

In numerous electrorheological fluids, the effect strongly depends on the water (or other additives) amount adsorbed by the solid particles (this is the case for instance of cellulosic or silica gel particles). In some cases the additive mainly concentrates on the surface and promotes a noticeable surface conductivity. Qualitatively the surface conductivity acts in a way quite similar to the bulk conductivity. We aim here at deriving quantitative estimates of the attraction force resulting from the surface conductivity of particles. We assume the spheres to be perfectly insulating in the bulk and to have a slightly conducting surface. The spheres are characterized by the surface conductivity γ_S which is the inverse of the surface resistivity R_S (measured in ohms per square).

The force can be estimated in a way very similar to the previous case of bulk conductivity of spheres if we can express the sphere conductance as a function of δ. With the notations of Fig. 6 the current flowing through a sphere surface is :

$$I = 2\pi\gamma_S R\, E'(\theta)\, \sin\theta \tag{19}$$

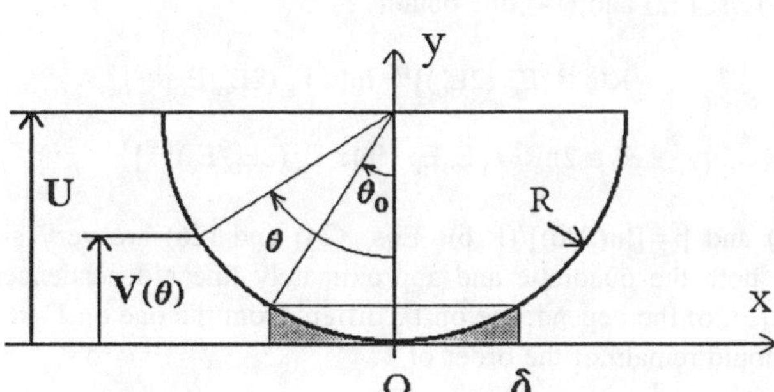

Fig. 6. Definition of notations used for calculating the attraction force between two spheres with surface conductivity.

where $E'(\theta)$ is the field component along the sphere surface. (Eq. (19) implies that there is no current escape through the liquid). The variation of this field is :

$$E'(\theta) \cong (U/R) / \sin\theta = E_o / \sin\theta \quad (20)$$

Integrating (19) between $\theta_o \cong \delta/R$ and $\pi/2$ and approximating $tg(\delta/2R)$ by $(\delta/2R)$ give:

$$U-V(\theta_o) \cong RE_o \ln[(2R/\delta)] \quad (21)$$

with $U=V(\pi/2)=RE_o$ (the contact point is at the potential $V(0)=0$). The conductance of the sphere is C_S', with : $C_S' = I/[U-V(\theta_o)]$. Using (19) and (21) we obtain for the conductance of the two half spheres :

$$C_S' \cong \pi\gamma_S/[\ln(2R/\delta)] \quad (22)$$

Balancing this conductance with the liquid conductance given by Eq. (9) leads, in the low field case to :

$$\ln(R/\delta) \cong (\Gamma_S/2\alpha)^{1/2} \quad (23)$$

which defines δ (here $\alpha = \ln[2R/\delta]/\ln[R/\delta]$ and $\Gamma_S = \gamma_S/(R\sigma_L)$). Applying Eq. (6) finally gives the attraction force :

$$F = 4\pi R^2 \varepsilon_L E_o^2 \exp[(2\Gamma_S/\alpha)^{1/2}] \quad (24)$$

Note that the exponential dependence on Γ_S is not really significant because Eq. (24) is only valid for low applied fields (and low force values).

For high applied fields, the field enhanced dissociation has to be taken into account. From Eqs. (22) and (14) one obtains :

$$R/\delta \cong [E_c/(2E_o)]^{1/2} \ln[c\,\Gamma_S\,(2E_o/E_c)^{1/2}] \quad (25)$$

$$F \cong 2\pi R^2 \varepsilon_L E_o E_c \{\ln[c\,\Gamma_S\,(2E_o/E_c)^{1/2}]\}^2 \quad (26)$$

with $c=5/(\alpha\beta)$ and $\beta=[\ln(R/\delta)]/(R/\delta)$. Eqs. (24) and (26) are very similar to Eqs. (11) and (15), both the quadratic and approximately linear dependences being again obtained. In Eq. (26) the dependence on Γ_S differs from the one on Γ in Eq. (15) by a factor which should remain of the order of 1.

5. Conclusion

Measurements of the attraction force between two weakly conducting spheres performed on a large scale model confirm the analysis based on the picture of electric field and interfacial charge distributions being controlled by the conduction properties of both media and not by their permittivities. Under D.C. (and low frequency A.C.) fields this has been clearly shown for non polar dielectric liquids. In this case, the quasi-linear increase of the force for high applied fields reflects the importance of non linear conduction processes in the liquid phase.

The analysis of the field distribution and resulting attraction force for spheres, characterized by only a surface conductivity, immersed in a non polar dielectric liquid, leads to very similar laws, quadratic and quasi linear in E_o. This shows that the sphere-sphere interaction is not very sensitive to the details of the conduction mechanism of the spheres.

This conduction model has now to be tested in the conditions of ER fluids. This implies to relate theoretically the global properties measured on the fluid, the yield stress for instance, to the interaction force between particles. It will presumably also be necessary to address the question of applying or adapting the present model to the conditions prevailing for particles of micrometer size. Nevertheless a very encouraging indication on the pertinence of the present conduction model was recently obtained for a particular ER fluid (cellulose/mineral oil). The static yield stress exhibits a quadratic dependence on E_o at low applied fields and a linear one for $E_o \geq 3$ kV/mm[19].

In conclusion, it appears that linear and non linear conduction properties of the suspensions must be taken into account in order to progress in the understanding of the basic mechanisms at work in ER fluids.

Acknowledgments

Part of this work was developed with the financial support of the french "Centre National de la Recherche Scientifique" and of the Company Elf-Aquitaine (under contract CNRS/SNEA No 50.8016).

References

1. W. M. Winslow, *J. Appl. Phys.*, **20**, 1137 (1949).
2. A. P. Gast and C. F. Zukoski, *Adv. Colloid Int. Sci.*, **30**, 153 (1989).
3. H. Block and J. P. Kelly, *J. Phys. D: Appl. Phys.*,.**21**, 1661 (1988).
4. Y. Z. Xu and R. F. Liang, J. Rheol. **35**, 1355 (1991).
5. H. Conrad and A. F. Sprecher, *J. Stat. Phys.*, **64**, 1073 (1991).
6. Z. P. Shulman, R. G. Gorodkin, E. V. Korobko and V.K. Gleb, *J. Non-Newton. Fluid Mech.*, **8**, 29 (1981).
7. T. C. Jordan and M. T. Shaw, *IEEE Trans. Elec. Insul.*, **24**, 849 (1989).
8. F. E. Filisko, in Proc. 3rd Int. Conf. on ER Fluids, R. Tao ed., (World Scientific, Singapore, 1992), p. 116.
9. D. L. Klass and T. W. Martinek, *J. Appl. Phys.*, **38**, 67 (1967).

10. D. J. Klingenberg, D. Dierking and C. F. Zukoski, *J.Chem. Soc. Faraday Trans.* **87**, 425 (1991).
11. R. A. Anderson, in Proc. 3rd Int. Conf. on ER Fluids, R. Tao ed., (World Scientific, Singapore, 1992), p. 81.
12. L. C. Davis, J. Appl. Phys., **72**, 1334 (1992).
13. J. N. Foulc, N. Felici and P. Atten, *C. R. Acad. Sci. Paris,* **314,** Série II, 1279 (1992).
14. A. O. Tsebers, Magnitnaia Gidrodinamika, No 3, 17 (1991), *in russian.*
15. A. Alj, A. Denat, J.P. Gosse and B. Gosse, *IEEE Trans. Elec. Insul.*, **20**, 221 (1985).
16. L. Onsager, *J. Chem. Phys.*, **2**, 599 (1934).
17. J. N. Foulc, P. Atten and N. Félici, submitted to J. of Electrostatics (1993).
18. R. W. Waring, in Proc. CNRS International Colloquium 102 (Ed. CNRS, Paris, 1961) p. 117.
19. J. N. Foulc and P. Atten, in this volume.

Theoretical Analysis of Field Induced Structures in ER and MR Fluids

G. Bossis, H. Clercx*, Y. Grasselli, E. Lemaire
Laboratoire de Physique de la Matière Condensée
Universite de Nice Sophpis Antipolis, Parc Valrose, 06108 Nice Cedex 2

ABSTRACT

The rheological properties of ER and MR suspensions are determined by the strength of the structure built under the application of an electric (respectively magnetic) field. We present a general multipolar method which allows to obtain the stress-strain curve for any sheared lattice. This is applied to infinite periodic chains of spheres arranged on three different lattices: simple cubic, simple hexagonal and body centered tetragonal. The predicted yield stresses and shear moduli are compared to the predictions of simpler models involving interactions between two or three particles. The field induced structures experimentally observed show that the chains of spheres aggregate into thick columns; we present a model which allows to calculate the diameter of these aggregates, at least for MR fluids, on the basis of a minimization of the free energy.

I - INTRODUCTION

The growing interest in electrorheological and magnetorheological fluids has motivated a new research which aims to predict the changes of structure and of rheological properties of a suspension which is submitted to an external field [1]-[4]. The chaining of the particles and the phase separation which are due to attractive forces between the dipoles induced on each solid particle by the external field, is at the heart of the E.R or M.R effect. It seems reasonable to think that, depending on the precise microscopic arrangement between the particles, we could have large differences in the macroscopic properties of interest such as the dielectric permittivity, the conductivity, the shear modulus or the yield stress of the suspension. These two problems : the prediction of the field induced structure and, for a given structure, the prediction of the macroscopic properties are tackled in this paper. We shall begin by the presentation of a numerical treatment of electrostatic interactions between spherical particles which allows to calculate the dielectric permittivity and the electrostatic forces for either a finite or an infinite-periodically replicated- set of spheres. Then, in section III, we shall present some results obtained with this method for two and three spheres and in section IV the shear modulus and the yield stress predicted for different lattices. In the last section we shall present a model which aims to predict the size of the columnar aggregates formed at equilibrium in the presence of a magnetic field. . This study is only devoted to structural models which do not depend on the presence of a flow . Flow induced structures are beyond the scope of this paper but are of course very important to know if we want to be able to understand phenomena like hysteresis or slow drift of rheological properties [5], [6].

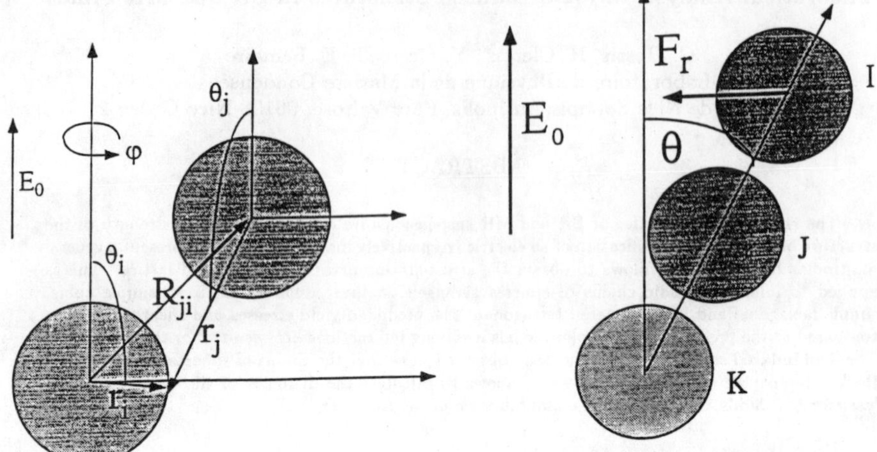

Fig.1. Representation of two spheres i and j with the notation used for their respective coordinates.

Fig.2. Three spheres initially touching and moving along the velocity lines in a shear flow. The maximum of the restoring force F_r on the particle I as function of the angle θ is calculated exactly and with a pair superposition approximation.

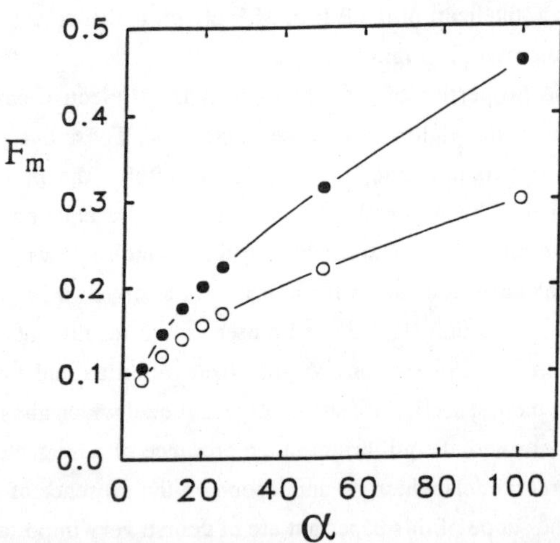

Fig. 3 : ●●● $F_m{}^3$: maximum restoring force acting on particle I calculated with three particle interactions versus the ratio $\alpha = \varepsilon_p/e_F$.
ooo $F_m{}^2$ idem as $F_m{}^3$ but calculated with a sum of two particle interactions.

II - MULTIPOLAR APPROACH OF ELECTROSTATIC INTERACTIONS

We consider a suspension of monodisperse solid spheres characterized by a dielectric permittivity ε_p suspended in a fluid of permittivity ε_f. The polarization is supposed to respond instantaneously to the applied field or in other words we do not consider a polarization mechanism whose characteristic response time could be larger than the inverse of the frequency of the applied field. Furthermore we do not consider the existence of currents or of interfacial polarization due to a non zero conductivity of the solid or of the liquid phase. These conditions apply principally to the category of E.R. fluids based on a large electronic polarizability of the constituent particles and acted on by electric fields whose frequency is high enough (> 10^2 - 10^3 Hz) to neglect ionic polarization and charge accumulation on the electrodes. On the other hand this model also applies to magnetic fluids in so far as the magnetic permeability does not vary too much with the intensity of the magnetic field. This is usually the case at low magnetic fields (< 10^2 O_e) but even if this condition is not fully respected, the predictions based on a constant magnetic permeability are still in good agreement with some macroscopic experiments on steel spheres [7].

The problem to solve is then to find the solution of the Laplace equation for the potential $\phi(\mathbf{r})$ with the appropriated boundary conditions on the surface S_i of a reference sphere i (cf fig. 1). Let us call $\phi_{in}{}^i$ the potential inside the particle i and ϕ_{out} the potential inside the liquid phase,then we must have:

$$\nabla^2 \phi(\mathbf{r}) = 0 \tag{1}$$

$$\phi^i_{in}(\mathbf{r}_i) = \phi_{out}(\mathbf{r}_i), \text{ with } \mathbf{r}_i \in S_i$$

$$(2)$$

$$\alpha \frac{\partial \phi^i_{in}(\mathbf{r}_i)}{\partial r_i} = \frac{\partial \phi_{out}(\mathbf{r}_i)}{\partial r_i} \qquad \text{with } \mathbf{r}_i \in S_i.$$

where $\alpha = \varepsilon_P / \varepsilon_F$

The general solution can be expanded in spherical harmonics :

⋆ **Present address: Department of Physics, Eindhoven University of Technology, P. O. Box 513, 5600 MB Eindhoven, The Netherlands**

$$\phi^{i}_{in}(\mathbf{r}_i) = \psi_i + \phi_o(\mathbf{r}_i) + \sum_{\substack{l \geq 0 \\ m}} \beta^{i}_{lm}\, r_i^{l}\, Y_{lm}(\theta_i, \phi_i), \qquad (3)$$

$$\phi_{out}(\mathbf{r}_i) = \psi_i + \phi_o(\mathbf{r}_i) + \sum_{\substack{l \geq 0 \\ m}} Q^{i}_{lm} \frac{4\pi}{2l+1} r_i^{-(l+1)}\, Y_{lm}(\theta_i, \phi_i),$$

$$+ \sum_{j \neq i} \sum_{\substack{l \geq 0 \\ m}} \frac{4\pi}{2l+1} Q^{j}_{lm}\, r_j^{-(l+1)}\, Y_{lm}(\theta_j, \phi_j) \qquad (4)$$

The function $\psi_i + \varphi_o(\mathbf{r}_i) = -\mathbf{E}_o.\mathbf{R}_i - \mathbf{E}_o.\mathbf{r}_i$ represents the potential of the external field. The vector $\mathbf{R}_i$ refers the position of the centre of the particle i. The quantities Q^{i}_{lm} are the multipole moments of the charge distribution $\rho(\mathbf{r})$ inside the particle i, defined in the usual way :

$$Q^{i}_{lm} = \int_{V_i} Y^{*}_{lm}(\theta, \phi)\, r^{l}\, \rho(\mathbf{r})\, dV \qquad (5)$$

with V_i the volume of the particle i.

The difficulty to apply the boundary conditions (2) with the equations (3) (4) comes from the part of the solution in (4) which is expressed relatively to the variables r_j, θ_j, φ_j defined in the frame of each particle j (see fig 1). Some previous solutions to this problem have been worked out in specific cases. Klingenberg [8] has obtained a numerical solution for two spheres but its method was not computationally usable for high values of α ($\alpha > 10$) when the particles are in or near contact. Chen, Sprecher and Conrad [9] have solved for the radial force on a sphere included in a chain of spheres parallel to the direction of the field. This case is much easier to solve due to the symmetry axis which allows to neglect the azimuthal angle φ.

Our aim is to extend the multipolar method to a general configuration of N spheres, either isolated or replicated on a network, with the possibility to take into account a multipolar degree (indice l in Q_{lm}) as high as 1000 without numerical problems. The method is detailed in Clercx and Bossis [10] and we shall only give here the main ideas. First the change of referential from particle j to i in (4) is obtained with the help of the Hobson relation (see for instance ref [11]).

$$r_j^{-(l+1)} Y_{lm} (\theta_j , \varphi_j) = \sum_{\substack{s \geq o \\ -s \leq t \leq +s}} \frac{n_{st}}{(s+t)!} M^{ji}_{lm;st} (\mathbf{R}_{ij}) \, r_i^s \, Y_{st} (\theta_i , \varphi_i) \qquad (6)$$

In this relation n_{st} is a numerical factor and

$$M^{ji}_{lm\,;\,st} = C \, \frac{Y_{l+s,\, m-t} (\xi_{ij} , \eta_{ij})}{R_{ij}^{l+s+1}} \qquad (7)$$

where C is an other numerical constant [10]. It appears that the use of Eq (6) allows to express the potential on the surface of the sphere i with only the variables r_i, θ_i, φ_i, relatively to the referential of the particle i. The price to pay is the introduction of a new sum of spherical harmonics (Eq 7) which only depends on the relative positions of the centres of the particles i and j (vector $\mathbf{R}_{ij}$ defined by R_{ij}, ξ_{ij}, η_{ij} in the frame of the particle i).

With the use of this relation in Eq (4) it is then possible to multiply each side of (2) by $Y_{pq} (\theta_i, \varphi_i)$ and to integrate on the surface of the particle i. In that way the boundary conditions lead to a set of linear equations whose general form is :

$$\begin{bmatrix} E_1 \\ \cdot \\ \cdot \\ \cdot \\ E_N \end{bmatrix} = \begin{bmatrix} Z_{11} \ldots\ldots Z_{1N} \\ \\ Z_{N1} \ldots\ldots Z_{NN} \end{bmatrix} \begin{bmatrix} Q_1 \\ \cdot \\ \cdot \\ \cdot \\ Q_N \end{bmatrix} \tag{8}$$

with every $E_i = (0, E_0, 0....)$ and

$Q_i = \left\{ Q^i_{lm} \; ; \; l \geq 0, \; |m| \leq l \right\}$

represents the set of multipoles belonging to the particle i. In the following we take $Q_{00} = 0$: no global charge on the particles; Thus the matrix **Z**, which depends only on the relative positions of the N particles, has the dimensions L (L + 2) N where L is the larger value of l used in the calculus. The linear system is numerically solved and we obtain the dipole $\mathbf{M}^\alpha$ on each particle α which is related to the first multipole as follows:

$$Q^\alpha_{10} = \sqrt{\frac{3}{4\pi}} \; M^\alpha_z \; ; \; Q^\alpha_{11} = \sqrt{\frac{3}{8\pi}} \; (-M^\alpha_x + i\, M^\alpha_y)$$

$$Q^\alpha_{1-1} = \sqrt{\frac{3}{8\pi}} \; (M^\alpha_x + i\, M^\alpha_y)$$

Actually the linear system is not solved under the form presented in (8) but rather we have used the variables $(A^\alpha{}_{lm})^+$ and $(A^\alpha{}_{lm})^-$ which are respectively proportional.
to $(Q^\alpha_{lm} + (-1)^m Q^\alpha_{l,-m})$ and $(Q^\alpha_{lm} - (-1)^m Q^\alpha_{l,-m})$

The physical meaning of these new variables is more obvious, (for instance we see that each component $A^+{}_{10}$, $A^+{}_{11}$ or $A^-{}_{1-1}$ is respectively proportional to M_z, M_x and M_y) and above all their use simplifies the numerical calculus.

At last, before to look at some results, it remains to deal with lattices. In that case each component of the configuration matrix (Eq(8)) contains infinite sums over all the replicaes of the particles contained in the unit cell. These sums over the lattice sites { L} appear as :

$$\sum_{\substack{j \in L \\ j \neq i}} M^{ji}_{lm;st}$$

and are sums over spherical harmonics divided by $(R_{ij})^{l+s+1}$ with $l \geq 1$ and $s \geq 1$ (cf Eq. 7). An efficient way to calculate these sums consist to separate them into two fast converging sums, one in the real space and the other in the reciprocal space. The application of this technic -known as Ewald sum- to the solid spherical harmonics appearing in Eq.(7) has been treated by Nijboer and de Wette, both in the case of converging sums [12] and conditionally converging sums ($l = 1, s = 1$) [13] . We refer the reader to these two papers for the lattice sums of long ranged interactions.

III. RESULTS FOR A FEW SPHERES

If we want to predict the shear stress as function of the strain in the presence of an electric field we need to specify the positions of the spheres at zero deformation and then we have to do an hypothesis on the way the spheres are moving when submitted to an external stress. The reference model consists of chains of spheres which are aligned on the direction of the field and then sheared affinely, the trajectory of each sphere following the velocity lines. Using this structural model, Klingenberg et al [14] were able to calculate the yield stress on the basis of a numerical solution of a two-spheres model. Indeed if we assume that only the interactions between two neighbouring spheres are important, then the electrostatic restoring force (perpendicular to the field) cancels on each sphere except on the last one which is supposed to be fixed on the upper plane. They write for the restoring force F_r (see fig. 2) as function of the angle θ :

$$F_r^{(2)} = 12\,\pi\;\varepsilon_0\,\varepsilon_F\,a^2\,\beta^2\,f_r\,E^2 \qquad (9)$$

with $f_r = \left(\frac{a}{r}\right)^4 \left[(f_{//} + f_\Gamma)\sin 2\theta \cos\theta - f_\perp \sin^3\theta\right]$

and $\beta = \dfrac{\alpha - 1}{\alpha + 2}$

The quantities $f_{//}$, f_Γ, $f_\perp$ are some functions of the normalized distance r/a between the spheres, and of α . The restoring force is proportional to the square of the radius a of the particles and to the square of the electric field, E. The shear stress for a given angle θ is the product of the number of chains per unit surface multiplied by $F_r^{(2)}$:

$$\tau = \frac{\phi}{\frac{2}{3}\pi a^2} F_r^{(2)} \qquad (10)$$

For conductive spheres ,the force in Eq (9) diverges when the particles come into contact and the multipolar approach will fail for r=2a. In this case the analytical expressions for $f_{//}$, f_Γ and $f_\perp$ have been obtained by Arp and Mason [15] following an analysis done by M.H. Davis[16] in bispherical coordinates. For instance they obtain for $\delta = r/a - 2 << 1$ the following expression for $f_{//}$:

$$f_{//} = \frac{1}{6}\left(\frac{r}{a}\right)^4 \left[\frac{\pi^4}{18} \frac{1}{\delta\,(\ln(\delta/12.869))^2} - 1.09\right] \qquad (11)$$

For $\delta = 2\ 10^{-4}$ this expression gives $f_{//} = 587.42$. Taking 600 multipoles in our development, we obtain $f_{//} = 587.23$ which is in fair agreement with the value obtained from (11). We have also been able to extend the results reported by Klingenberg to higher values of α and closer distances. For instance for $\alpha = 100$ we have a converged value at contact : $f_{//}(r = 2) = 182.2$.

In order to test the hypothesis of additivity of pair interactions which is used in Eq.(10) we have compared the cases of two and three spheres. The interesting quantity to compare with, is the maximum of the restoring force ,F_m ,as function of the deformation $\gamma = \mathrm{tg}\theta$, since this maximum is, in the chain model, related to the experimental yield stress through Eq (10). Let us call $F^{(3)}{}_m$ the maximum restoring force calculated with the multipolar approach applicated to 3 spheres in an external field and $F^{(2)}{}_m$ the force obtained by adding separately the interactions between i and j and between i and k alone (pair superposition approximation) (cf. Fig. (2)). These two forces, obtained for an affine deformation are plotted in Fig. (3) versus the ratio α of the permittivities. It appears than the "exact" force is always larger than the one obtained with a two body approximation. For $\alpha = 100$ the ratio $F^{(3)}/F^{(2)}$ is about 1.6. The limit of this ratio when $\alpha \to \infty$ can be studied because, even if the forces become very large, their ratio remains finite and seems to converge to 1.9. This value is to be taken cautiously but in any event it indicates that the two body superposition approximation strongly underestimates the yield stress. It is worth noting that this result is not completely intuitive since we could expect that the stronger the field inside the gap between two spheres the less it will be influenced by the presence of the third one. Actually the mutual attraction between opposite induced charges in one gap contributes to reinforce the opposite charge on the opposite pole of the central particle , which in turn will increase the attraction with the third particle. This important effect of three body interactions indicates that, in order to have a reliable reference model for the stress-strain behavior of ER of MR fluid, it is better to work with infinite chains of spheres disposed on some given lattices.

IV. SHEARED LATTICES

The study of the dielectric permittivity of dielectric spheres located on the nodes of a cubic lattice has been addressed by several authors [17] [18] [19] with different methods.

We have tested our multipolar method against the available results and found a good agreement [10] . The calculus of the effective permittivity of a regular arrangement of spheres is not only important for the determination of the electrical properties of the material but also for the determination of its mechanical properties. Actually due to the inversion symmetry, the force on each sphere is zero for an infinite lattice and we can only get the stress-strain relation through the derivatives of the electrostatic energy. In a real experiment of course the translational invariance is broken by the walls of the container and the force does not cancel on the last layers of particles.

The stress calculated from the forces on a finite system and the one obtained from the electrostatic energy of an infinite lattice will only differ by a surface term in the electrostatic energy whose importance decreases as a/L where L is the thickness of the sample.

We determine the effective permittivity tensor, ε_{eff} , of the suspension from its definition :

$$\left(\frac{\varepsilon_{eff}}{\varepsilon_F} - \mathbf{I}\right) . \mathbf{E} = 4\,\pi\,\mathbf{P} \qquad (12)$$

Where $\mathbf{E}$ is the Maxwell field and $\mathbf{P} = \mathbf{M}/V$ is the average polarization.The total dipole $\mathbf{M}$ of the unit cell is obtained from the solution of the set of linear equations (cf. Eq. (8)). After a partial inversion giving the higher multipoles as function of the dipoles we obtain :

$$\beta\, a^3\, \mathbf{E} = \mathbf{C} \,.\, \mathbf{M} \qquad (13)$$

The field which intervenes in this last equation is the Maxwell field **E** and not the external field $\mathbf{E_0}$ which appears in the left hand side of Eq. (8). The difference between the two fields comes from the conditionally convergent sums which have been transferred from the right hand side of Eq.(8) to the left hand side in order to write Eq.(13). The combination of Eqs. (12) and (13) gives :

$$\frac{\varepsilon_{eff}}{\varepsilon_f} = \mathbf{I} + 3\beta \ \phi \ \mathbf{C}^{-1} \qquad (14)$$

C is a configuration dependent matrix which does not contain the conditionally convergent contribution to the lattice sums and ϕ is the volume fraction of spheres.

The lattices we have mainly studied are composed of chains of touching spheres along the direction of the field (Z axis) and the projection of these chains on the perpendicular (OX,OY) plane is either a square lattice (Fig. 4 A) a triangular lattice (Fig. 4 B) or a more complicated pattern (Fig. 4 C) where the grey and black spheres overlap.The variation in volume fraction is obtained by changing the spacing between the chains in the OX,OY plane ; we have represented in Fig. 4 the close packed situation. In the pattern of Fig. 4C the grey chains are shifted by a radius in the Z direction : a side view of this figure either along X or along Y would give the triangular pattern of Fig. 4 B ; this lattice known as the BCT lattice has been demonstrated to be the more stable one, for a lattice of dipoles [1]. On the other hand for conductive spheres Davis[22] has shown that the three lattices had the same energy per particle. We have calculated the exact permittivity for a close packed BCT lattice with the multipolar approach for α between 1 and 100. These results for the permittivity are compared with the ones obtained in the dipolar approximation in fig. 5. We see that for high values of α the effect of higher multipoles considerably increases the value of the permittivity. The same effect occurs for other lattices and finally the conclusion that the BCT lattice is the more stable remains valid but the difference of energy per particle between the FCC and the BCT is much lower than in the dipolar approximation. For instance for $\alpha = 100$ we have $(W_{bct}/W_{fcc})_{dip} = 1.082$ but $(W_{bct}/W_{fcc})_{mult.} = 1.008$.

Let us now come back to the stress-strain curve for lattices of chains of spheres. We can shear with the displacement of each plane in the X direction increasing linearly with the distance along Z. (For the BCT lattice, due to the imbrication of the chains,the lattice can't be sheared affinely in the X direction; the easier direction of shear is at an angle of 45° relatively to the X or Y axis and even in this direction a continuous affine deformation is not possible for $\phi > \pi/9$ [21]). The change of electrostatic energy with the strain γ is given by :

$$W(\gamma) = -\frac{1}{2}\ \mathbf{E}\,.\,\varepsilon\,.\,\mathbf{E} = -\frac{1}{2}\ \varepsilon_{zz}(\gamma)\ E^2$$

The electrostatic energy is calulated for different strains of the lattices then a first numerical differentiation gives the shear stress : $\tau = \partial W / \partial \gamma$ and a second differentiation gives the shear modulus : $G = (\partial \tau / \partial \gamma)_{\gamma=0}$

Previous calculations of the shear modulus on a simple cubic lattice have been done by Davis [20] with the help of a finite element method. Instead of a sheared lattice they took only the "stretching" component of the shearing motion; it allows to keep the cubic symmetry and so to decrease the computational time . We have found that the stretching approximation becomes accurate for $\alpha > 10$; nevertheless the size of the mesh taken in [20] for the finite element method was too large for $\alpha > 10$, resulting in an underestimation of G which grows with α (already by a factor of 2 for $\alpha = 20$) [21]. Bonnecaze and Brady [17] have also addressed this problem for a configuration where the chains were touching in the y direction and for different spacing in the x direction. Their method is similar to the one used for hydrodynamic interactions in Stokesian dynamics [23] : an exact solution for the three first multipoles and a pair superposition for higher mulipoles. A comparison with their results and an extensive calculus of the variations of G and τ_S with the volume fraction for different lattices will be reported in a forthcoming paper [21]. We want here to focus on the comparison between the two spheres model of Klingenberg and the lattice chain model.

The two important quantities : the shear modulus G and the static yield stress τ_S (normalized by $2\, \varepsilon_0\, \varepsilon_F\, \beta^2 E^2$)are plotted versus the volume fraction respectively in Fig. 6 and in Fig. 7 for a ratio of the permittivities $\alpha = 10$. The strain is in the X direction for the SC and SH lattices (Fig 4A and 4B) and midway between the X and Y axes for the BCT lattice(Fig. 4C). The straight lines in both figures represent the prediction of the two spheres model which ,by construction (addivity of the effects of each chain), scales linearly with the volume fraction. For the two spheres model the value of G is obtained from the derivative of the restoring force relatively to the strain at small value of γ. Three things are worth noting on these plots : firstly the three lattices-except the BCT lattice above $\phi = 0.45$- give similar values for G as well as for τ_S which means that the mechanical properties are more sensitive to the number of chains per unit surface than to their precise relative positions in the XY plane. This is not true for BCT lattice at high volume fraction because of steric hindrance which occurs at low strain when the volume fraction increases.Secondly, the yield stress and the shear modulus increase almost linearly with the volume fraction and this linear behavior is still more evident at higher values of α - (always with the exception of BCT lattice for $\phi > 0.45$). Thirdly the two body approximation largely underpredicts the values of τ_S and G and this difference is found to increase when we increase α . For instance with $\alpha = 10$ we obtain $(\tau_S)^{lattice} / (\tau_S)^{2spheres} = 1.6$ and with $\alpha = 20$ $(\tau_S)^{lattice} / (\tau_S)^{2spheres} = 2.1$.

It is interesting to note that the linear behavior with ϕ and the absence of sensitivity to the precise structure indicate that we can reason only with unidimensional chains of spheres but that, in this chain, a large number of particles must be taken into account. For instance with $\alpha = 10$ $(\tau_S)^{lattice} / (\tau_S)^{2spheres} = 1.6$ whereas $(\tau_S)^{3spheres} / (\tau_S)^{2spheres} = 1.2$ only and for α =20 these values are 2.1 and 1.3 respectively .

The comparison with experiment is not always meaningful since this model only applies to electronic polarization and very often ionic polarization is present. Nevertheless, for the monodispersed silica particles we have used, the agreement between the theoretical and the experimental yield stresses are rather good [6]. Actually this model could more or less apply when the electric field is switched on rapidly since in that case we have the formation of a loose gel-like structure which can behave mechanically as the lattice model. But if the structure is formed at equilibrium by rising the field very slowly, we expect the formation of aggregates having the internal structure of a close packed BCT lattice. Such a lattice can't be sheared at constant volume and will developp high normal stresses and wall slip before to flow. A large increase of shear stress and of shear modulus associated with a densification of the gel like structure has been observed experimentally [6] ,[24] and, if the lattice chain model can still be useful as reference model, an approach based on the existence of isolated aggregates (whose internal structure would ideally be the BCT structure) could be an alternate way to model E.R. or M.R. fluids.

V. MODEL OF PHASE SEPARATION AT EQUILIBRIUM

Magnetorheological fluids as well as E.R. fluids exhibit a phase transition from the liquid to a solid-like state when the attractive forces between dipoles become larger than the entropic forces ($\lambda = F^{dip}\, a / kT > 1$). The experimentally observed structure does not consist of equally spaced chains of particles but of thick columns. For magnetic fluids the average radius, b, of these aggregates has been measured and has been found to increase with the thickness L of the sample with a rate which depend on the range of L [25],[26]. The mechanism of formation of these aggregates - do we have first isolated chains and then a gathering of these chains into columns or a simultaneous growth in two directions (parallel and perpendicular to the field) but at different rates imposed by the rate of increase of the field? - is an interesting question but will not be discussed here. We only want to deal here with the determination of the final radius of the columns, obtained after the completion of a phase separation which is supposed to have been performed slowly enough in order to allow for thermodynamic equilibrium. In that case, we can use a minimization of the free energy relatively to the size of the aggregates in order to find the equilibrium size. The model we use should only apply to MR fluids because we are considering the effect of depolarizing fields generated by the cutoff of the polarization on the walls of the cell. Strictly speaking this effect does not exist in ER fluids where the electrodes bring the charges needed to cancel the local field generated by the cutoff of the polarization. In other words the average field is constant in the sample for any distribution of matter since the potential is imposed on the conductive electrodes.

Let us see the main interactions which enter into play and how we can model them. First we assume that the aggregates have an ellipsoïdal shape, characterized by the ratio k =b/L (b is the transverse semi-axis and L the total length of the major axis); then the total moment $M_a{}^h$ of the aggregate in the direction of the external field H_o will be :

$$M_a^h = \alpha^{ell} . H_o \quad \text{with} \quad \alpha^{ell} = \frac{\mu^{ell} - 1}{4\pi} \left[\frac{1}{1 + n_z (\mu^{ell} - 1)}\right] V^{ell} \tag{15}$$

μ^{ell} is the internal permeability of the aggregate of volume V^{ell} and n_z is the depolarizing field factor of an ellipsoid[27] which for k << 1 is given by :

$$n_z \sim -4 (1 + \text{Log } k) k^2 \tag{16}$$

A thin aggregate correponds to small value of the depolarizing field ($n_z \to 0$) and so to large value of $M_a{}^h$ and of the magnetostatic energy given by :

$$W = - 1/2 \ \mathbf{M_a} . \mathbf{H_o} \tag{17}$$

(The moment M_a is equal to $M_a{}^h$ in an effective medium theory where the finite size of the particles do not intervene).This effect alone should contribute to the formation of individual chains of particles. But ,on the other hand, these chains are going the repel each other because the cut-off of the polarization P is equivalent to the effect of charges of opposite sign located at the extremities of each chain. Actually an uniform distribution of these equivalent surface charges whose surface density is σ (r) = **P** (r) . **n** (**n** unit vector normal to the surface) would give rise to the usual average field : H = Ho - 4 π P . But if we consider that each aggegate is located on the nodes of a lattice we shall have for the field acting on the reference aggregate i :

$$\mathbf{H}_i = (\mathbf{H}_0 - \sum_{j \neq i} T_{ij} . \mathbf{M}_j) \tag{18}$$

where $\mathbf{T}_{ij} . \mathbf{M}_j$ is the field coming from the total dipole $\mathbf{M}_j$ of the aggregate j and averaged on the length of the aggregate i .Owing to the cylindrical symmetry, we can only consider the z component of **T** which is :

$$T_{ij} = \frac{1}{L^2} \left[\frac{2}{d_{ij}} - \frac{2}{(d_{ij}^2 + L^2)^{1/2}}\right] \tag{19}$$

Using in (15) the local field H_i instead of H_o gives the result :

$$M_a^H = \alpha^{ell} Ho - n_r M_a^H \ ; \qquad n_r = \alpha^{ell} \sum_{j \neq i} T_{ij} \qquad (20)$$

The sum on the particles can be replaced by an integral between the radius d_0 of an equivalent cavity and infinity; we then obtain:

$$n_r = \left[\frac{\mu^{ell} - 1}{1+n_z(\mu^{ell} - 1)}\right] \frac{\phi}{\phi^{ell}} \left[\left(1+ \frac{d_0^2}{L^2}\right)^{\frac{1}{2}} - \frac{d_0}{L}\right] \quad \text{with} \quad d_0 = b\sqrt{\frac{2\,\Phi^{ell}}{3\,\Phi}}$$

In this expression ϕ is the average volume fraction of solid particles and ϕ^{ell} is the volume fraction inside the aggregates.The value of n_r decreases when the distance between the aggregates increases which means that, regarding to the repulsive interaction between agregates, the energy will be lower with a small number of thick aggregates than with a large number of thin aggregates. If we do not consider any surface tension, the energy is only a function of the ratio $k = b/L$ (through n_r and n_z) so the equilibrium size obtained by minimizing (20) relatively to b will scale linearly with L.

On the other hand we have to add to the energy a surface tension whose origin is the difference between the local field on a particle situated on the surface of an aggregate relatively to the local field on a particle situated inside the aggregate. This change in local field will lead to a change of the total moment from $M_a{}^h$ to M_a which can be incorporated in the factor n_σ :

$$M_a = M_a^h\,(1 - n_\sigma) \quad \text{where} \quad n_\sigma = C\,a\,\frac{S^{ell}}{V^{ell}} \qquad (21)$$

The constant of proportionnality C depends on the precise structure and on the value of α [28] ; it is equal to $\beta\,\phi^{ell}$ if we approximate the surroundings of a particle by an homogeneous medium. A good approximation for n_σ is :

$$n_\sigma = C\ 0.75\,\pi\left(\frac{a}{L}\right)\frac{1}{k} + \frac{a}{L}\,0(k) \qquad (22)$$

If the equality $n_z\,(\mu^{ell} - 1) << 1$ is verified, the polarizability tensor α^{ell} in Eq.(15) can be developped and using (16) and (22) we obtain :

$$\frac{M_a}{H^0} = \alpha^{ell}\ (1 - n_\sigma) \cong \left((\mu - 1)/4\pi + c'\,k^2\right)\ (1 - c''\ \frac{a}{L}\ \frac{1}{k}) \qquad (23)$$

In this equation c' and c" are some constants and we have neglected the logarithmic dependence of n_z so it is only valid for a restricted domain of k. Looking for the minimum of the energy is equivalent to take the derivative of Eq.(23) relatively to k and then to solve for k which will give : $b \propto a^{1/3} \; L^{2/3}$.

This result holds in the absence of repulsive energy ($n_r = 0$) . The general case ($n_r \neq 0$ and $n_\sigma \neq 0$) can be obtained in a self consistent way by writing instead of Eq.(20) :

$M^h{}_a = \alpha^{ell}\; Ho - n_r\; M^a$ with M^a given by Eq.(21) The resulting energy per particle is then :

$$\frac{U}{N} \propto \alpha^{ell}\,(1 - n_\sigma)\;/\;(1 + n_r\,(1 - n_\sigma))$$

It is not easy to obtain, in the general case, an analytical relation between b and L. We have plotted in fig. 8 the numerical solution for the radius of the aggregates versus the thickness of the cell, using a standard set of values ($a = 1\ \mu m$; $\mu = 10$; $\phi = 0.1$) and assuming an internal volume fraction $\phi_{ell} = 0.64$ (corresponding to a random packing)inside the aggregates . We can see on this log-log plot that there is no power law for b versus L if we take the whole range of variation of L. Such a result has been found by Liu et al [26] on droplets of ferrofluids. For a different set of parameters corresponding to polystyrene spheres charged with magnetite we have found a good agreement between theory and experiment with a slope close to 2/3 in the range $L = 130\ \mu - 700\ \mu$ [29].

VI CONCLUSION

We have seen with the help of a multipolar development that a reference model for the calculus of the yield stress or of the shear moduls cannot be based on the consideration of two or three spheres, especially for high values of α. The calculus of these quantities on different lattices have shown that their values increase linearly with the volume fraction for $\alpha > 10$ and that they are not very sensitive to the relative disposition of the infinite chains. Nonetheless for $\phi > 0.45$ hindrance effects result in a sharp decrease of G and τ_s for the BCT lattice. In opposition to the reference model where all the chains are equally spaced, we have demonstrated that, at least for MR fluids, the surface tension and the repulsive interactions between aggregates lead to the formation of aggregates whose radius increases with the thickness of the cell. The rate of increase is ultimately linear ($b \propto L$) but decreases continuously when the thickness of the cell decreases.

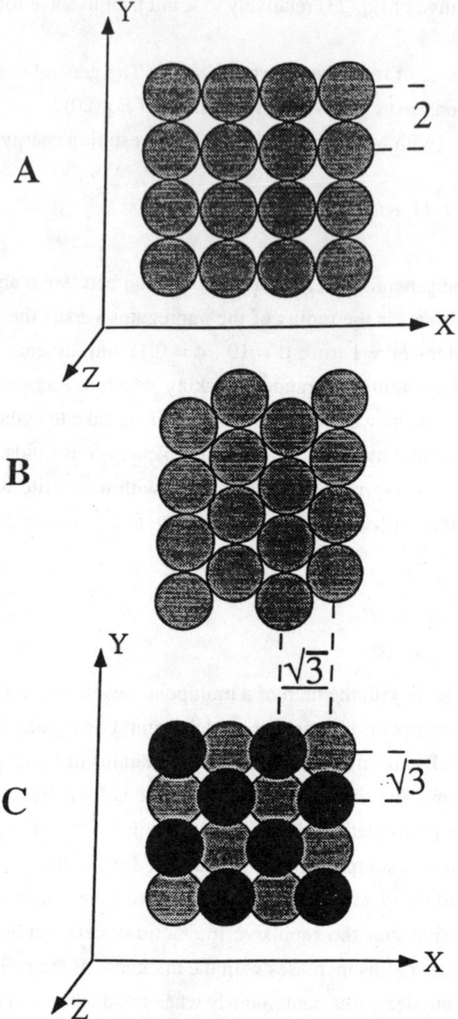

Fig. 4 : Different lattices studied ; the spheres are touching along Z (field direction). The projection on XY plane is represented for the close packed structure but the spacing between chains in the XY plane depends on the volume fraction.
4A simple cubic structure (SC)
4B simple hexagonal structure (SH)
4C body centered tetragonal structure (BCT).

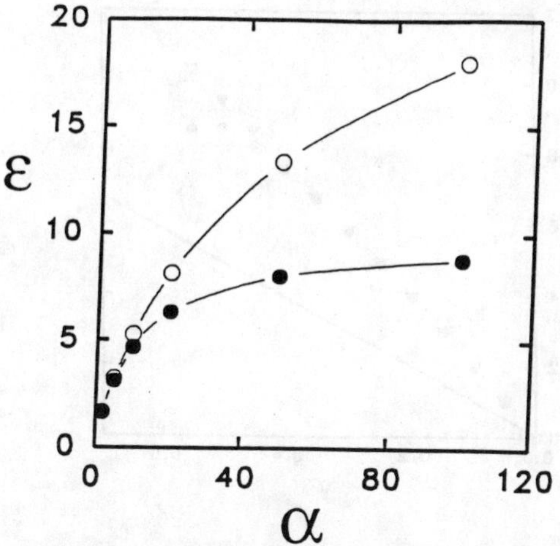

Fig.5 : Permittivity of the BCT lattice versus the ratio the permittivities : $\alpha = \varepsilon_p/\varepsilon_F$

• • • - dipolar approximation

o o o - multipolar development.

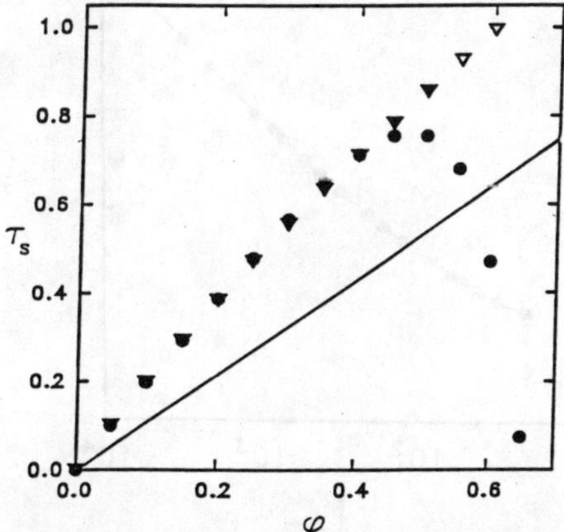

Fig. 6 : Theoretical yield stress (normalized by $2\,\varepsilon_0\,\varepsilon_F\,\beta^2\,E^2$) with α=10 versus the volume fraction for different lattices and for the two-spheres model.

▼▼▼ SC lattice • • • BCT lattice

▽▽▽ SH lattice ——— two spheres model

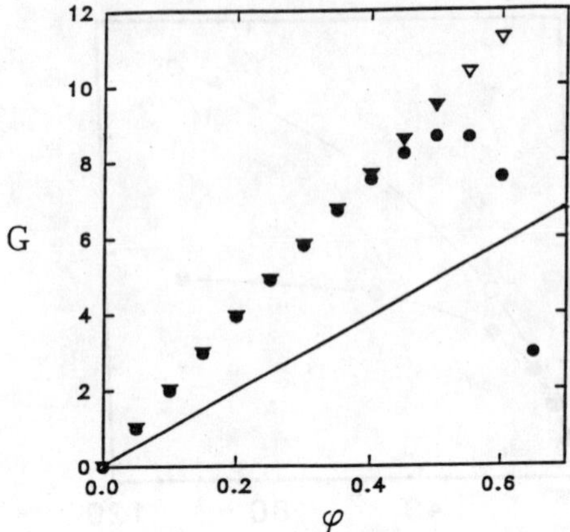

Fig. 7 : Theoretical shear modulus (normalized by $2\,\varepsilon_0\,\varepsilon_F\,\beta^2\,E^2$) for $\alpha=10$ versus the volume fraction - same legend as fig. 6.

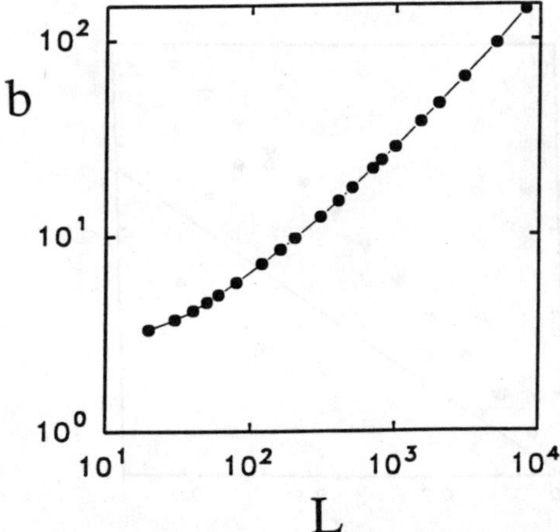

Fig. 8 : Theoretical prediction of the radius of the aggregates in a M.R. fluid versus the thickness of the cell;the lengths are in microns.

REFERENCES

[1] R. Tao and J.M. Sun, Phys. Rev. Lett. , 67, (1991) 398.

[2] T. Chen., R.N. Zitter and R. Tao, Phys. Rev. Lett, 68, (1992), 2555.

[3] E. Lemaire, Y. Grasselli and G. Bossis, J. Phys. II, France, 2, (1992), 359.

[4] T.C. Halsey and W. Toor, J. Stat. Phys., 61 (1990), 1257.

[5] E. Lemaire, G. Bossis, Y. Grasselli, Langmuir, 8, (1992), 2957.

[6] E. Lemaire, G. Bossis, Y. Grasselli, A. Meunier,Fourth International Conference on E.R. Fluids, 20-23 July 1993, Feldkirch.

[7] G. Bossis, E. Lemaire, J. of Rheology , 35, (1991), 1345.

[8] D.J. Klingenberg, Ph. D. Thesis University of Illinois, 1990.

[9] Y. Chen, A.E. Sprecher and H. Conrad, J. Appl. Phys., 70, (1991) 6796.

[10] H.J.H. Clercx, G. Bossis, Phys. Rev. E, (1993), in press.

[11] E.W. Hobson "The theory of spherical and ellipsoidal harmonics". (Cambridge University Press, Cambridge, 1931).

[12] B.R.A. Nijboer and F.W. de Wette, Physica 23, (1957) 309.

[13] B.R.A. Nijboer and F.W. de Wette , Physica 24, (1957) 422.

[14] D.J. Klingenberg and C.F. Zukoski, Langmur 6, (1990) 15.

[15] Arp P.A., Mason S.G., Coll. Polym. Science, 255, (1977), 56

[16] Davis M.H.,Quart.Journ.Mech.and Appl.Math. ,17,(1964),499

[17] R.T. Bonnecaze ,J.F. Brady, J. Rheol. 36, (1992), 73.

[18] R.C. Mc. Phedran and D.R. Mc Kenzie , Proc. Roy. Soc. Lond. A 359, (1978) 45.

[19] L.C. Davis, Appl. Phys. Lett. 60, (1992), 319.

[20] L.C. Davis, J. Appl. Phys. , 72, (1992), 1335.

[21] H.J. H. Clercx, G. Bossis, in preparation.

[22] L.C. Davis, Phys. Rev. A ,46 ,(1992), R719

[23] J.F. Brady and G.Bossis, Annu. Rev. Fluid Mech. 20,(1988),111

[24] G. Bossis, Y. Grasselli, E. Lemaire, J. Persello, L. Petit, Eur. Phys. Lett. , submitted.

[25] G. Bossis,H.J.H. Clercx, Y. Grasselli, E. Lemaire, Communication at the Fourth International Conference on E.R. fluids, 20-23 July 1993, Feldkirch.

[26] J. Liu.,T.Mou,E.Lawrence,M.Ivey ,Communication at the Fourth International Conference on E.R. fluids, 20-23 July 1993, Feldkirch.

[27] L.D.Landau,E.M.Lifshitz, Electrodynamics of Continuous Media ,2nd edition,Pergamon Press 1984 , Chap 1

[28] H.J.H. Clercx, G. Bossis, J. Chem. Phys.,98, (1993),8284.

[29] Y. Grasselli, G. Bossis, E. Lemaire, in preparation.

Field-Induced Structure of Confined Ferrofluid Emulsion

Jing Liu, Eric M. Lawrence, Mark L. Ivey, and George A. Flores
Department of Physics and Astronomy,
California State University, Long Beach, CA 90840

J. Bibette
Centre de Recherche Paul Pascal, Avenue A. Schweitzer, F-33600
Pessac, France.

J. Richard
Rhone-Poulenc Recherches, 52 Rue de la Hale Coq, 93308 Aubervilliers,
France.

ABSTRACT

Field-induced phase behavior of a confined monodisperse ferrofluid emulsion was studied using optical microscopy, light transmission, and static light scattering techniques. Upon application of magnetic field, randomly dispersed magnetic emulsion droplets form solid structures at $\lambda = 1.5$, where λ is defined as the ratio of the dipole-dipole interaction energy to the thermal energy at room temperature. The new solid phase consists of either single droplet chains, columns, or worm-like clusters, depending on the volume fraction, cell thickness and rate of field application. For the column phase, an equilibrium structure of equally sized and spaced columns was observed. Our measurements taken for cell thickness $5\mu m \leq L \leq 500\mu m$ and volume fraction 0.04 show the column spacing to be reasonably described by $d = 1.49\, L^{0.34}$.

1. Introduction

Phase behavior of colloidal dispersions with anisotropic interactions (mainly dipole interactions) strongly depends on the confinement of the colloidal system. It has already been shown that a single layer of colloidal particles interacting through long-range repulsive dipole forces forms a crystalline structure [1, 2]. In that case, each particle is well separated from its neighbors as a result of the strong in-plane repulsion arising from the confined parallel dipoles. In the limit of an extremely large cell thickness, it has been predicted [3, 4] and partially observed [5, 6] that the system undergoes a macroscopic phase separation; a liquid and a dense solid phase or, in some cases a liquid and a gas phase, may coexist [7, 8]. Indeed since dipole-dipole interactions may also be attractive, dense phases are expected to form once there is less confinement to restrict exploration of attractive configurations. Nevertheless, the resulting equilibrium structure should arise from a very subtle interplay between attractive and repulsive forces, leading to substantially more complex equilibrium patterns. Though experiments [9, 10] and theories [4, 9-11] have already dealt with this question, controversy between existing theories persists. Our experiment is designed to test the predictions of these theories.

In this paper we examine the behavior of a confined colloidal system in which particles exhibit magnetic dipole interactions upon application of a magnetic field. We demonstrate that at equilibrium column structures are formed which are strongly reminiscent of well-ordered structures in the two-dimensional case. On the other hand, compared to the unique macroscopic dense phase in an extremely large cell, the confined system shows many equally spaced and sized dense columns even for a cell thickness as large as about 1000 particle diameters. We show that the spacing d between these columns increases with the cell thickness L. Based on experimental data, the behavior is reasonably described by the power law $d \sim L^{0.34}$.

2. Emulsion Characterization

In these experiments we chose to use small monodisperse ferrofluid emulsion [12]. This is because the emulsion droplets act as hard spheres in the absence of an applied magnetic field and acquire magnetic dipole moments proportional to the magnetic field applied. The sample was made in Exxon by one of us (J. Bibbette) using kerosene-based ferrofluid (obtained from Dr. Ron Rosensweig) mixed with appropriate amount of water and surfactant sodium dodecyl sulfate (SDS). The result is an kerosene-in-water emulsion where each kerosene droplet contains about 6% by volume of iron oxide grains. Each magnetic grain has an average diameter of 90Å and a permanent dipole moment corresponding to a single magnetic domain. The droplets have an average diameter of 0.51 μm and a standard deviation from the mean of less than 10%, and are stabilized by the surfactant SDS. The droplets are dispersed in a density-matched aqueous continuous phase containing a suitable mixture of D_2O and water. The SDS surfactant concentration is set at 0.015 mole/ℓ which is about twice of the value of the critical micellar concentration (CMC = 0.008 mole/ℓ) for this surfactant. The droplets are monodisperse both in terms of their size and magnetic moment. The monodisperse emulsion is produced by applying a fractionated crystallization scheme to an initial polydisperse crude emulsion [13]. The emulsion droplets are stable over long periods, as long as the surface charge is maintained. This may be accomplished by keeping the continuous phase surfactant concentration at or above the CMC [14]. The repulsion between droplets due to the surfactant coating has a very short interaction range compared to the diameter of the droplet. The droplets are also almost non-deformable in the limit of small osmotic pressure at a low surfactant concentration [14-15]. Therefore, as long as the ionic surfactant concentration is also kept near the CMC, the droplets can be viewed as hard spheres.

One major feature of these droplets comes from their superparamagnetic behavior. In the absence of an external magnetic field, the droplets have no permanent dipole moments, since the small magnetic grains within each droplet are randomly oriented due to thermal motion. An external magnetic field induces a dipole moment in each droplet, whose magnitude increases with the strength of the applied field up until saturation of magnetization is reached. Consequently, these droplets interact through dipole forces which can be controlled by the magnitude of the external field [16]. The strength of this interaction can be conveniently described by the coupling constant λ [17], which is the ratio of the contact dipole energy of two adjacent dipoles to the thermal energy:

$$\lambda = E_d/kT = \pi\mu_0 a^3\chi^2H^2/18kT$$

where μ_0 is the magnetic permeability in a vacuum, a is the droplet radius, k is the Boltzmann constant, T is the temperature in Kelvin, and χ is the ferrofluid susceptibility. χ is defined as M/H and is equal to $\phi_m M_d L(\alpha)/H$ [18], where ϕ_m is the magnetic grain volume fraction, $L(\alpha)$ is the *Langevin function* with $\alpha = mH/KT$ where m is the magnetic grain dipole moment, M_d is the bulk saturation magnetization of magnetite solid. At low field limit $\chi = \pi\phi_m\mu_0 d_m^3 M_d^2/18kT$ where d_m is the magnetic grain diameter. For the ferrofluid used in our sample, ϕ_m = 5.8%, d_m = 8.9 nm, T = 300 K, M_d = 0.5605 T or 4.46×10^5 A/M for magnetite of Fe_3O_4 [19]. Therefore, χ = 0.43 to 0.37 in SI units for the applied field range from zero to 40 mT.

3. Experiment

The behavior of the ferrofluid emulsion is studied using optical microscopy, light transmission and static light scattering technique. The sample is loaded into a few flat wedge cells covering the thicknesses ranging from 5 μm to 500 μm. The homogeneous external field is set normal to the cell surface and is produced by a coil to avoid the formation of image dipoles at the magnetic boundary. Both static light scattering and the light transmission measurements are performed using an air-cooled Ar ion laser directed perpendicular to the surface of the cell. Scattered light is displayed on a screen behind the cell and is collected by a CCD camera. The signal is finally transmitted to a computer for processing. Transmitted light is detected by a photo diode.

3.1 Optical Microscopy

We have observed very rich phase behavior in these confined colloidal suspensions. The appearance of the different phases depends on several parameters including the magnitude of the applied magnetic field, the droplet volume fraction, the cell thickness, and the rate at which the field is applied.

As already noted, applied field induces a dipole moment in each droplet and generates dipole interactions between the droplets. When λ is less than 1, the droplets exhibit a random dispersed liquid phase. When λ approaches unity, the droplets follow a condensation process, merging together to form small chains. By further increasing the value of λ, the small chains combine to form long chains. Depending on chain length and volume fraction, chains can either stay separated or gradually coalesce to form columns. When λ approaches 2, there is still a visible exchange of droplets between the dense (chains or columns) phase and the dilute (disperse) phase. Before the completion of this condensation process, the ordering of the chains or columns becomes visible. This ordering sharpens and saturates when λ is further increased to about 10. For λ approaches 10, the chains or columns become rigid and are locked in position; the Brownian motion of each particle is reduced dramatically. Fig. 1 illustrates the phase transition for a sample of volume fraction $\phi = 0.03$ in a cell of thickness $L = 20$ µm.

When λ is larger than 1, the cell thickness controls the chain length and therefore the interaction between chains. Figure 2 shows the microscope image for droplet volume fraction $\phi = 0.007$ confined in a cell of thickness approximately 5 µm. The magnetic field H is set at 31.5 mT, oriented normal to the cell surface. The corresponding value of λ is 80. In this case, the chain length *L*, which expands to fill the cell, is 10 times the droplet diameter. It is clearly visible that the ferrofluid emulsion arranges itself into an organized array of chains of equal thickness *2r* and equal spacing *d*. The average spacing roughly measured from this picture is about 5 µm and the chain thickness is just one droplet diameter. For this thin cell, the interaction between droplets is mainly repulsive since the structure formed is very similar to that of a single layer of dipoles. However,

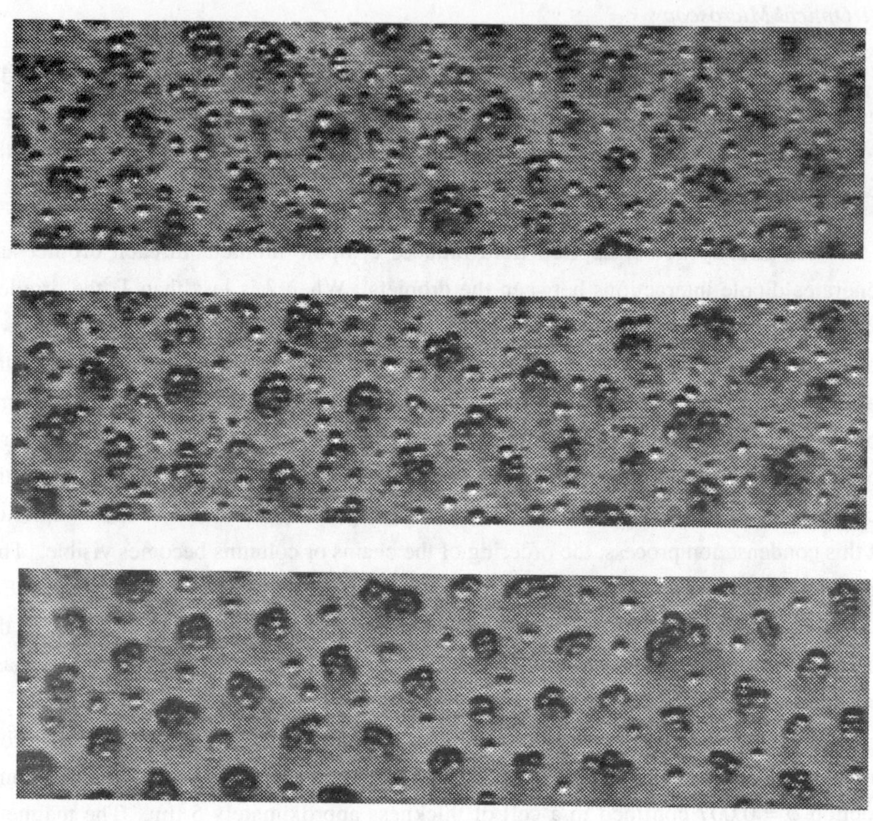

Fig. 1. Microscope image showing phase transition for ϕ = 0.03, L = 20 μm and magnetic field applied at a rate of 0.3 mT/min. The pictures, from top to bottom, correspond to 3.8 mT (λ = 1.4), 5 mT (λ=2.4) and 38 mT (λ=107).

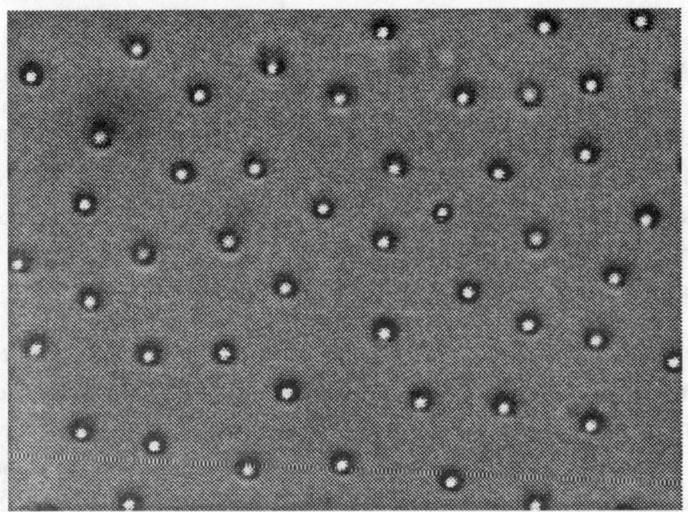

Fig. 2. Microscope image showing in cross section the field-induced structure when the droplet volume fraction is $\phi = 0.007$, the cell thickness is $L = 5$ μm, and $H = 31.5$ mT ($\lambda = 80$). The field direction is into the paper. Single chains are well separated with each chain width one droplet diameter, $2a = 0.51$ μm. The magnetic field is increased at a rate of about 800 mT/min from 0 to 31.5 mT. The horizontal width of the image corresponds to 40 μm.

the inter-chain repulsive interaction range and strength must be reduced compared to that of a single layer; otherwise, a crystal structure would be observed. Either decreasing the chain length or increasing the volume fraction results in a stronger repulsive interaction between chains and, therefore, a better ordering.

When the confinement is decreased, a different structure appears. Figure 3 shows a microscope picture where $\phi = 0.13$, $L = 65$ μm, $H = 38$ mT, and $\lambda = 107$. In this case, the chain length L is about 130 droplet diameters. The ferrofluid emulsion again arranges itself into a fairly well ordered structure; now however, columns are formed that are much larger in diameter, about 3.0 μm, than single droplets, 0.5 μm. The chains aggregate to form larger column structures. For $L >> a$, repulsive forces are dominant between chains that are parallel and in register with each other. However, if chains are brought close together and are shifted a distance of one droplet radius along the chain direction, a net attractive interaction between the chains results. Therefore, columns form from chains attracting each other and "zippering" together. The columns are observed to taper at the ends, forming sharp tips at cell boundaries.

A dramatically different structure results if the magnetic field is applied rapidly to this system. Figure 3 corresponds to the rate of 0.1 mT/min. In order to check whether other structures are possible, we vary the applied magnetic field at a rate of up to 10^4 mT/min. Figure 4 shows the structure formed by the ferrofluid emulsion when the applied magnetic field changes at a rate of 240 mT/min. In this case, $\phi = 0.12$ and $L = 50$ μm. A worm-like structure is observed with a clear characteristic length scale. If the field application rate is increased to 10^4 mT/min., a random connection of labyrinthine structures will result. We also note that below a critical volume fraction, in this case 6%, the worm structure is always absent. We then always find a column structure with a slight difference in column separation compared to the one shown in Fig. 3.

A very interesting phenomenon is found when we maintain the rapidly applied field to its final value, corresponding to about $\lambda \sim 3$, for few minutes. The worm-like pattern disappears gradually and a columnar structure appears. However, if the strength of the final applied field is relatively high, corresponding to $\lambda > 10$, the worm or labyrinthine structure freezes and does not revert back to a columnar pattern.

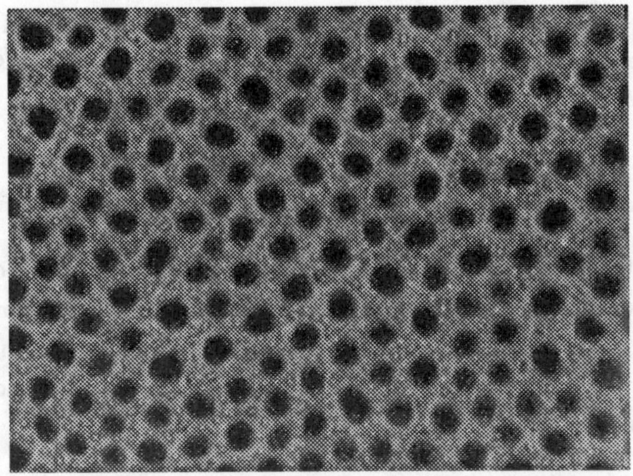

Fig. 3. Microscope image showing the cross section of a column structure taken at ϕ = 0.13, L = 65 µm, H = 38 mT (λ = 107). The magnetic field was increased at a rate of about 1 G/min from 0 to 38 mT. The horizontal width of the image corresponds to about 100 µm.

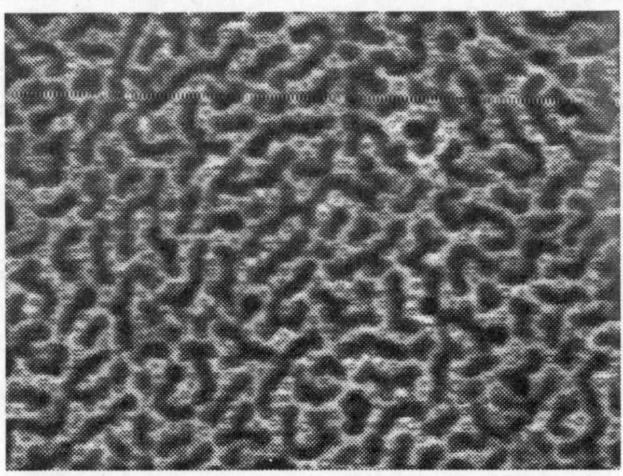

Fig. 4. Microscope image showing cross section of the "worm" structure for magnetic field increased at a rate of about 240 mT/min from 0 to 38 mT. Here ϕ = 0.12, L = 50 µm, H = 38 mT (λ = 107). The horizontal width of the image corresponds to about 100 µm.

This suggests that a columnar structure is the equilibrium phase and thermodynamic equilibrium can be reached if the applied field increases slowly; e.g., at 0.1 mT/min. This rate employed in Fig. 3, λ increases at a rate of about 0.1 min.$^{-1}$ in the range of λ = 1 to 6 in which structure formation occurs.

3.2. Light Transmission

The quantitative phase behavior under equilibrium conditions (low applied field rate) was studied using light-transmission measurements. When chains start to form, more light is allowed to pass through a sample which in the random phase absorbs and scatters light. Therefore, the critical field for phase transition can be determined by light transmission measurement.

For the light-transmission experiment, a laser beam parallel to the magnetic field is directed toward the sample surface and a photo diode is located behind the sample to detect the transmitted light. The result is shown in Fig. 5 for which $\phi = 0.10$, and $L = 100$ μm. The transmission starts to increase sharply when H approaches 4 mT (that is, λ exceeds 1.5). The transmission reaches its peak value when H approaches 8 mT corresponding to $\lambda = 6$. Transmission is increased by three orders of magnitude with a value of λ in the range 0 to 10. Reducing the field from 10 to 0 mT does not result in any hysteresis. When the droplets diameter is 2a = 0.41 μm, and the measurement is repeated under the same condition, we find a similar transmission curve; the transmission starts to increase at a field strength $H = 5.5$ mT. However, this field again corresponds to λ = 1.5. Our previous direct microscopic observations taken together with this measurement, lead to the conclusion that the condensation process into columns is very analogous to a first-order phase transition in a system where two different phases coexist with one condensing in the other. The critical field for the phase transition from randomly dispersed droplets to a columnar structure occurs in both cases at λ = 1.5.

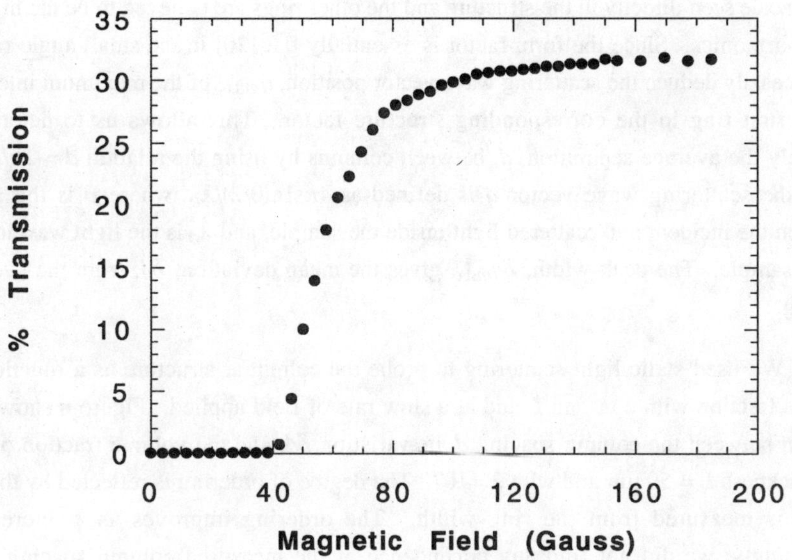

Fig. 5. Light transmission measured as a function of magnetic field applied at a rate of about 0.4 mT/min for $\phi = 0.10$ and $L = 100$ μm.

3.3 Light Scattering

In the static light scattering experiment, a laser light parallel to the magnetic field direction is again directed toward the cell surface. On the screen behind the sample, a typical scattering pattern of several concentric rings is observed. A number of rings as high as 7 has been observed in some cases. The rings indicate that a partially ordered structure is formed, which is consistent with what we found in the microscope images. The first inner ring, which is the most intense ring, corresponds to the characteristic length scale seen directly in the structure and the other rings are believed to be the higher-order harmonics. Since the form factor is essentially flat [20] in the small angle range, we can easily deduce the scattering wave vector position, q_{max}, of the maximum intensity of the first ring in the corresponding structure factor. This allows us to determine precisely the average separation, d, between columns by using the relation $d = 2\pi/q_{max}$. Here, the scattering wave vector q is defined as $4\pi\sin(\theta/2)/\lambda_o$, where θ is the angle between the incident and scattered light inside the sample, and λ_o is the light wavelength in the sample. The peak width, δq_{max}, gives the mean deviation, δd, from the average spacing.

We used static light scattering to probe the columnar structure as a function of volume fraction with constant L and at a slow rate of field applied. Figure 6 shows the relation between the column spacing d, its variation δd, and the volume fraction ϕ, at a cell thickness $L = 50\ \mu m$ and with $\lambda = 107$. The degree of ordering is reflected by the bar which is measured from the ring width. The ordering improves as ϕ increases. Surprisingly, we did not find any dependence of the measured column spacing d on volume fraction for $\phi \geq 0.04$. This result is confirmed for other thicknesses ranging from $10\ \mu m$ to $100\ \mu m$. This results are remarkable in that column spacing is independent of volume fraction. As ϕ increases, instead of forming more columns at reduced spacing, the column width increases as $\sqrt{\phi}$ so as to maintain a constant column spacing. The reason for this preferred structure is not clear yet.

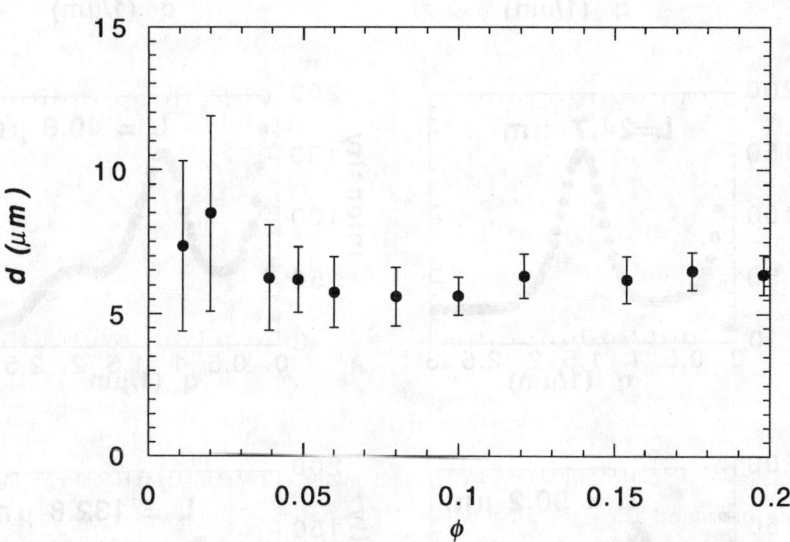

Fig. 6. From static light scattering, the column spacing d is measured as a function of ϕ at $L = 50$ μm and $H = 38$ mT ($\lambda = 107$). The magnetic field was increased at a rate of about 0.4 mT/min. The bar δd gives the variation of column spacing obtained from the scattering ring width.

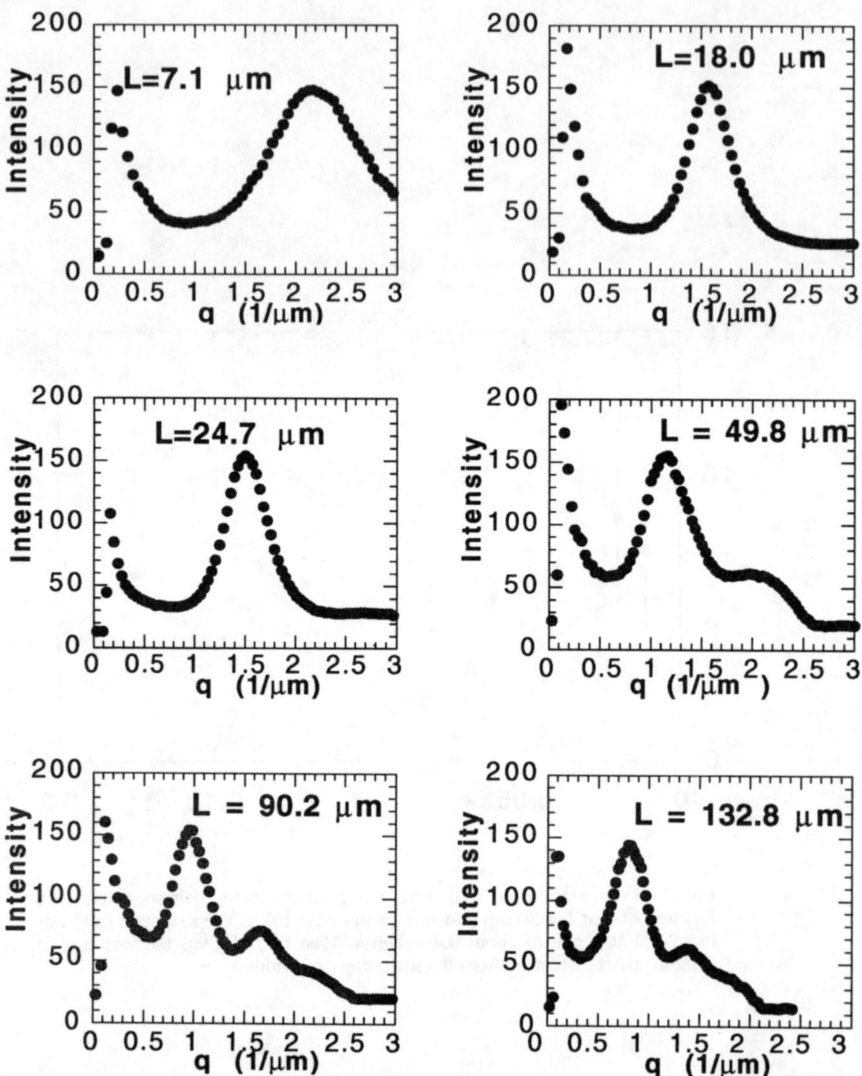

Fig. 7 Scattered light intensity (arbitrary unit) radially averaged at constant q and plotted as a function of scattering wave vector, q, at $\phi = 0.04$ and $H = 38$ mT. The scattering pattern was measured for wedge cells of thickness L ranging from 7.1 μm to 132.8 μm. The transmitted beam is blocked near the center and the magnetic field is increased at a rate of about 0.3 mT/min.

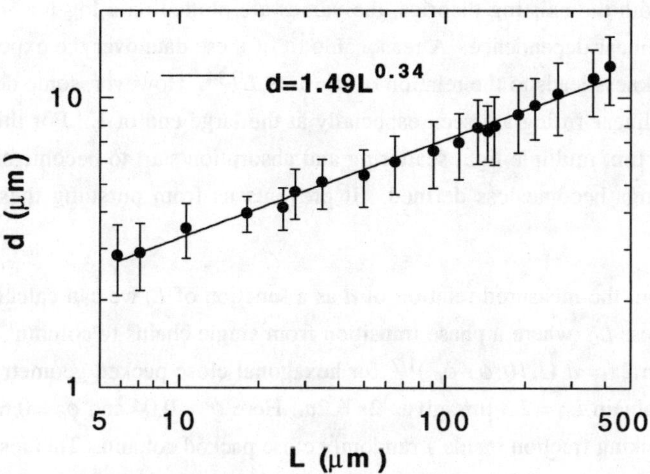

Fig. 8 Column spacing is obtained from the peak position, q_{max} in Fig. 7 and plotted on a log-log scale for 5 μm ≤ L ≤ 500 μm. The solid line is a best-fit power law given in the figure. The bar δd is again measured from the peak width.

A more interesting result is obtained when we vary the cell thickness L at constant ϕ and at a low rate of field application. Figure 7 shows the characteristic light scattering pattern obtained from the column structures for different values of L. As L increases from 7 μm to 130 μm, the peak in the intensity profile moves towards smaller angles and narrows; higher order harmonics are visible. This clearly shows that the column spacing increases with increasing cell thickness. From the first-ring peak position q_{max}, we can examine the dependence of the column spacing on cell thickness at constant droplet volume fraction $\phi = 0.04$ and with λ large. The result is presented in Fig. 8. In order to compare with the existing theories, the values are plotted on a log-log scale. The data exhibits a linear dependence. A reasonable fit of these data over the experimental range in cell thickness leads to the relation of $d = 1.49\ L^{0.34}$. However, some deviations in the data from linear fitting are seen especially at the large end of L. For thicknesses at or above 500 μm, multiple light scattering and absorption start to become significant such that the rings become less defined. It prevents us from pursuing these experiments further.

From the measured relation of d as a function of L, we can calculate the critical cell thickness L_0 where a phase transition from single chains to columns occurs. Using the relation $2r = d\ (1.10\ \phi / \phi_p)^{1/2}$ for hexagonal close packed geometry and $d = 1.49\ L^{0.34}$, we obtain $L_0 = 2.3$ μm, given $2r = 2a$. Here $\phi = 0.04$ and $\phi_p = 0.64$,which is the volume packing fraction inside a randomly close packed column. This result tells us that for $\phi = 0.04$, chains of 5 droplets or more will coalesce to form columns, whereas chains of 4 droplets or less will remain separate from other chains. In the other limit $L \to \infty$, from our measured relation $d = 1.49\ L^{0.34}$, $d \to \infty$. This means that a phase separation would occur for infinite chains, which is consistent with the theoretic predictions.

Our results differ from that of a previous experiment [10] for a similar system. The sample used in reference [10] was a polydisperse solid magnetic colloid and measurement were made for 100 μm $\leq L \leq$ 700 μm. The result of [10] showed a power-law relationship $d \propto L^{0.66}$, which differs significantly from the relationship $d \propto L^{0.34}$ we obtained. This difference is probably due to the dispersity of their sample and the different values of L used.

4. Conclusions

Several theories [4,9-11] have been developed to understand the field-induced structure in a confined system. All of them attempt to calculate the lowest free energy of a continuous system due to either two or all three of the magnetic interactions: (a) the dipole interaction within one column favors forming long single chains, (b) the surface tension term favors forming large columns to minimize the surface area, since it costs energy to form new surfaces, and (c) the interaction between neighboring columns, which is mainly repulsive and favors large column spacing. Competition between these interactions presumably settles columns at a finite width and spacing for a given column length. While some of theories predict a power-law dependence of column diameter on column length, the predicted power-law exponents are either 2/3 or 1. Only one theory [11] has predicted a phase change from columns to thin wall-like stripes as the volume fraction increases. Even though this theory suggests a phase transition from a column structure to a worm-like structure, its prediction does not agree quantitatively with our experimental results. Therefore, our experiment does not agree with any of the existing theories or previous experiments.

The dipoles induced by a magnetic field and their interactions in a ferrofluid emulsion are very similar to those incurred in an electrorheological (ER) fluid [17]. In an ER fluid, an external electric field is applied to a colloidal suspension in which the dielectric constant between particles and liquid is mismatched. Electric dipole moments are induced by the applied field in particles, which will interact and align the particles into chains [21]. However, the electrodes which apply electrical field to the system also induce image dipoles over the boundaries. Each formed chain connected with its image through interaction is equivalent to a infinitely long chain even if the cell thickness is finite. Therefore, the structure formed in the ER fluid corresponds to that of the large cell limit of our ferrofluid emulsion. The field-induced structure change in ER fluids turns an ER fluid into a Bingham solid, which leads to many potential applications [22]. The same is expected in this ferrofluid emulsion upon applying a magnetic field, and it is therefore called a magnetorheological (MR) fluid. Therefore, studying the phase behavior is very important for understanding the fundamental physics of an anisotropic interacting system and for understanding the rheological behavior for technological applications.

In summary, we have demonstrated that a confined ferrofluid emulsion exhibits a very rich phase behavior when subjected to a magnetic field. The anisotropic dipole interactions between droplets, induced by the external field, results in an anisotropic structure. The critical field for phase transition occurs at $\lambda = 1.5$ where random dispersed droplets begin chaining. When $\lambda >> 1$, in a highly confined system, a structure of partially ordered and well separated single chains form as a result of repulsive interaction between short chains. For a less confined system, equally spaced columnar structure forms with a degree of ordering depending on the volume fraction. A higher volume fraction gives a better ordering of the columns. Surprisingly, we have found that the rate at which the field is applied strongly affects the appearance of different phases. A low rate results in a columnar structure, and a high rate results in a worm-like or labyrinthine pattern if a critical volume fraction is reached. This is very similar to the labyrinthine structure observed in a ferrofluid confined between two plates and surrounded by another immiscible fluid, where the complexity of the pattern also depends on the rate at which the field is applied. In that case, the competition between dipolar interaction of magnetic grains and the intrinsic surface tension between two fluids controls the shape of the structure [18, 23]. However, in our ferrofluid emulsion, there is no intrinsic surface tension. It is surprising to see similar phenomena.

We have measured the equilibrium structure while applying the field at a very low rate. For a constant cell thickness in the range of $10\ \mu m \leq L \leq 100\ \mu m.$, we found that the column spacing is basically independent of volume fraction for $0.04 \leq \phi \leq 0.2$. At a constant volume fraction of 0.04, we obtained a power law between the column spacing and the cell thickness having the form: $d = 1.49\ L^{0.34}$, for $5\ \mu m \leq L \leq 500\ \mu m$. This result represents the first structure study in a confined ferrofluid emulsion system interacting through dipole force. A large gap between theory and experiment still exists. Therefore, more research, both experimental and theoretical, is strongly demanded in order to fully understand the complex phase behavior of this system.

5. Acknowledgments

We are grateful to the technical supports provided by Prof. E. Y. Wong and Prof. L. J. Eliason. We also acknowledge with gratitude the helpful discussions of Dr. Ron Rosensweig, Prof. E. Y. Wong, Prof. S. I. Salem, Prof. L.S. Learner, and Prof. E.L. Woollet. This research is partially supported by an award from Research Corporation. Acknowledgment is made to the Donors of The Petroleum Research Fund, administered by the American Chemical Society, for partial support of this research.

6. Reference

1. A. T. Skjeltorp, *J. Appl. Phys.* **57**, 3285 (1985).
2. B. Poulingny, R. Malzbender, P. Ryen and N.A. Clark, *Phys. Rev.* **42**, 988 (1990).
3. R. Tao and J.M. Sun, *Phys. Rev. Lett.* **67**, 398 (1991).
4. T.C. Halsey and W. Toor, *Phys. Rev. Lett.* **65**, 2820 (1991).
5. M. Fermigier and A. Gast, *J. of Coll. and Int. Sci.* **154**, 522 (1992).
6. J.E. Martin, J. Odinek and T.C. Halsey, *Phys. Rev. Lett.* **69**, 1524 (1992).
7. P. G. de Gennes and P. A. Pincus, *Phys. Kondens. Materie* **11**, 189 (1970).
8. P.C. Jordan, *Molecular Physics* **25**, 961 (1973).
9. E. Lemaire, Y. Grasselli and G. Bossis, *J. Phys. II France* **2**, 359 (1992).
10. E. Lemaire, G. Bossis and Y. Grasselli, to be published in *J. Phys. II France*, 1993.
11. T.C. Halsey, to be published in *Phys. Rev. E*, 1993.
12. J. Bibette, *J. Magnetism and Magnetic Materials,* **122**, 37 (1993).
13. J. Bibette, *J. Coll. and Int. Sci.* **147**, 474 (1991).
14. J. Bibette, "Ferrofluid emulsion as a tool for probing droplets forces" submitted for publication.
15. J. Bibette, D.C. Morse, T.A. Witten, and D.A. Weitz, *Phys. Rev. Lett.* **69**, 2439 (1992).
16. Landau and Lifshitz, *Electrodynamics of Continuous Media*, Pergamon Press, 1960.
17. A. P. Gast and C. F. Zukoski, *Adv. Colloid Interface Sci.* **30**, 153 (1989).
18. R. E. Rosensweig, *Ferrohydrodynamics,* Cambridge Univ. Press, New York, 1985.
19. R. E. Rosensweig, and J. Popplewell, *Electromagnetic Forces and Applications*, J. Tani and T. Takage eds., (Elsevier Science, B.V., 1992).
20 H.C. van de Hulst, *Light Scattering by Small Particles*, Dover, 1957.
21. T.C. Halsey, *Science*, **258**, 761, (1992).
22. *Proceedings of the Conference on Electrorheological Fluids , Carbondale, IL, 15-16 October , 1991*, R. Tao, Eds., (World Scientific, Singapore, 1992).
23. S.A. Langer, R.E. Goldstein and D.P. Jackson, *Phys. Rev. A,* **46**, 4894(1992).

The Evolution of Field-Induced Structure of Confined Ferrofluid Emulsions

Jing Liu, Tawei Mou, and George A. Flores
Department of Physics and Astronomy,
California State University, Long Beach, CA 90840

J. Bibette
Centre de Recherche Paul Pascal, Avenue A. Schweitzer, F-33600 Pessac, France.

J. Richard
Rhone-Poulenc Recherches, 52 Rue de la Hale Coq, 93308 Aubervilliers, France.

ABSTRACT

We report a real-time study of the evolution of structure of confined ferrofluid emulsions during the "liquid-solid" phase transition. A monodisperse oil-in-water ferrofluid emulsion is used. The structure evolution of the emulsion after rapid application of a magnetic field is probed by the static light scattering. The scattering pattern exhibits pronounced rings that reflect the formation of chains, columns, and possibly labyrinthine structures. The scattering ring is found to decrease in size and brighten in intensity with time. To monitor the structure evolution in time, both the ring peak position in scattering wave vector, q_{max}, and the peak intensity, I_{max}, are measured as a function of time. Both q_{max} and I_{max} saturate in less than 0.5 seconds after applying a magnetic field. At a constant cell thickness of 25 μm, the evolution of structure is essentially independent of volume fraction ranging from 0.015 to 0.13. In addition, a very good scaling is found in the scattered light intensity as a function of the scattering wavevector.

1. Introduction

The evolution of field-induced structure of a colloidal system through dipole interactions has been studied recently in cells with little or no confinement. For example, one study[1] was carried out in electrorheological (ER) fluids, where an applied electric field induces an electric dipole in each dielectric particle suspended in a liquid of different dielectric constant. The dipole interaction aligns the particles into chains, and different chains further coalesce to form columns. The columns' coarsening is believed to be due to thermal fluctuations in chains. The theory in ref. 1 has predicted that for an infinitely long chains the coarsening due to thermal fluctuation should follow a power law of the form $d \sim t^{5/9}$, where d is a characteristic length of the structure. The experimental result from the same study for a cell thickness of 0.7 mm shows that the chain coarsening varies in time in accordance with the power law $d \sim t^{0.4}$. A somewhat different result was obtained in a study[2] of a magnetic colloidal system where applied magnetic field induces a magnetic dipole moment in each particle suspended in a non-magnetic liquid. The dipole interaction results in a structure similar to that in ER fluids. The coarsening process shows a power law $d \sim t^{0.5}$ for a cell thickness of 1 mm. In both experiments, chain length is at least three orders of magnitude greater than the particle diameters. The data suggest a weak dependence, however, of the chain coarsening rate on the chain length (i.e. the height of the sample cell).

The chain coarsening process has not yet been studied either theoretically or experimentally in a cell with strong confinement. Presumably the thermal fluctuation amplitude of a short chain is much smaller than that of long chains, due to strong repulsion among chains. Therefore, the chain-chain coarsening is expected to be very slow, if it occurs at all. In this paper, we present our initial experimental study of the field-induced evolution of structure of monodisperse ferrofluid emulsions confined in a cell having a thickness of about 25 μm . The chain formation and coarsening are measured by monitoring the evolution of diffraction rings. The results show that the ring formation is complete or reaches saturation in less than 0.5 seconds after the application of the magnetic field. The formation of chains and their coarsening are indistinguishable for this strongly confined system. Furthermore, the saturation time of the structure formation was found to be basically independent of the volume fractions over the range studied, 0.015 to 0.13.

2. Experiment

In our experiments, small monodisperse ferrofluid emulsion[3] is used. The sample is an oil-in-water emulsion in which kerosene droplets contain about 6% volume of iron oxide grains. Each grain has an average diameter of 90Å and a permanent magnetic dipole moment corresponding to a single magnetic domain. The oil droplets have an average diameter of 0.51 μm with a standard deviation of less than 0.05 μm. Therefore, the droplets are essentially monodisperse in both size and magnetic moment. The advantage is that the emulsion droplets act as hard spheres in the absence of a magnetic field and acquire magnetic dipoles when a magnetic field is applied. Moreover, the interaction strength can be controlled by the magnitude of the applied field.

The oil droplets are stabilized by sodium dodecyl sulfate surfactant (SDS) and are dispersed in a density matched mixture of D_2O and water. The SDS surfactant concentration is set at 0.015 mole/l, which is about twice the critical micellar concentration (CMC = 0.008 mole/l) of the surfactant. The sample is prepared by applying the fractional crystallization scheme to an initial polydisperse crude emulsion[4]. The emulsion droplets are stable as long as the surface charge is maintained. This is accomplished by keeping the continuous phase surfactant concentration at CMC[5] or above. The repulsion between droplets due to the surfactant coating has an interaction range much shorter than the diameter of the droplet, as long as the ionic surfactant concentration is kept also around CMC. In addition, the droplets are almost non-deformable in the limit of small osmotic pressure[5,6]. Consequently, they can be viewed as hard spheres.

One major feature of these droplets arises from their super paramagnetic behavior. In the absence of external magnetic field, these droplets have no permanent dipole moments, as the small magnetic grains within each droplet are randomly oriented due to their thermal motion. However, once an external magnetic field is applied, the orientation of magnetic grains is slightly rotated towards the field direction, which results in a dipole moment in each droplet. Therefore, the magnitude of the magnetic dipole moment increases with the strength of the applied field until saturation is reached. Consequently these droplets interact through dipole forces which can be controlled by the magnitude of the external field[7]. The strength of this interaction can be conveniently described by the coupling constant λ [8], which is the ratio of the contact dipole energy of two adjacent dipoles to the thermal energy:

$$\lambda = E_d/kT = \mu_0 \pi a^3 \chi^2 H^2/18kT.$$

Here μ_0 is the magnetic permeability in vacuum, a is the radius of the droplet, k is Boltzmann's constant, and T is the temperature in kelvin, and χ is the ferrofluid susceptibility which is 0.4 in SI units for our sample and for an applied field ranging in strength from zero up to 400 G.

The behavior of the ferrofluid emulsion is studied using optical microscopy and static light scattering technique. The sample is loaded in a flat thin cell having a thickness of 25 ± 5μm. A homogeneous external field is set perpendicular to the cell surface and is produced by a coil to prevent the magnetic boundary from forming image dipoles. The applied magnetic field is varied from 0 to 380 G ($\lambda = 107$) with a rise time constant of 80 ms. A pair of car batteries is used to supply the current to the coil. The static light scattering measurement is performed using an air-cooled Ar ion laser directed perpendicular to the surface of the cell. Scattered light is displayed on a screen behind the cell, collected by a CCD camera, and recorded by a VCR. A LED light indicator, controlled by a switch which turns on the magnetic field, is put at the corner of the screen, which turns off at the instant when the coil becomes magnetized. The signal from the super VHS video tape is then sent one frame at a time to a Macintosh computer for processing.

3. Results

Figure 1 shows the evolution of the scattered light intensity measured as a function of the scattering wave vector, q, over a period of two seconds after the magnetic field is applied. In Fig. 1 the scattered light intensity is azimuthally averaged over constant q. The volume fraction is $\phi = 0.03$. The scattering wave vector q is defined to be $4\pi*\mathrm{Sin}(\theta/2)/\lambda_o$, where θ is the angle between the incident and scattered light inside the sample and λ_o is the light wavelength in the sample. As shown in the Fig. 1, at t = 0 the scattering pattern is essentially flat, corresponding to a liquid phase with droplets dispersed randomly. At t = 0.1 s, a ring appears in the scattering pattern. The peak intensity of the ring increases with time, whereas the peak position, q_{max}, of the ring decreases with time. This indicates the occurrence of the phase transition from a random to a partially ordered "solid" structure. The new structure has a characteristic length scale, $d = 2\pi/q_{max}$, which grows with time. After one second, a weak second ring at a higher q than q_{max} starts to develop in the scattering pattern. This ring is believed to be the second-order harmonic of the first, inner ring. After two seconds, no further change in the ring pattern is observed; the structure "freezes".

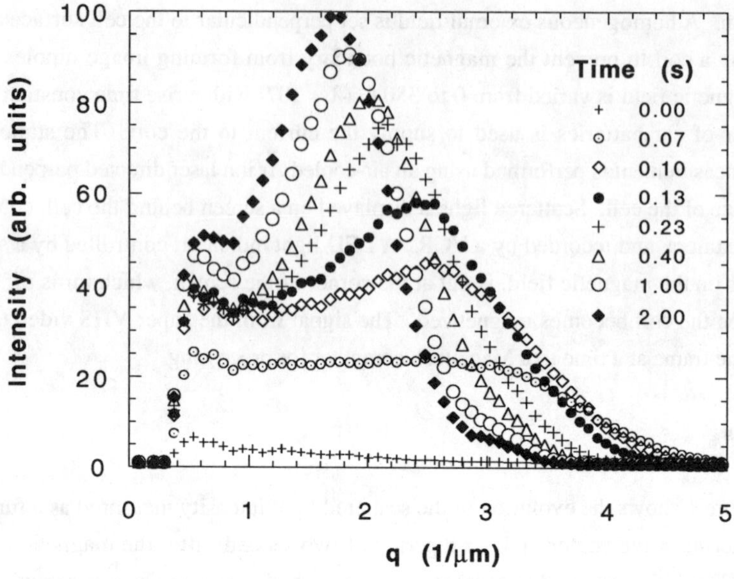

Fig. 1. Scattered light intensity (arbitrary units) is azimuthally averaged over constant q and plotted as a function of scattering wave vector, q, for various time after applying a magnetic field. The magnetic field H is applied in a step function from 0 to 380G (λ = 107) with a rise time less than 80 ms; the volume fraction ϕ used is 0.03 and the cell thickness L is 25 μm.

The growth of the solid structure is measured from the peak position, q_{max}, in scattering pattern of Fig. 1 as a function of time. Fig. 2 shows the results where the same measurement is performed for other volume fractions. These data are very different from results obtained by others in less confined or unconfined systems[1,2]. First, q_{max} decreases sharply and quickly reaches the asymptotic value q_s at $t \geq t_s$; here $t_s = 0.5$ s instead of the 10 min or more exhibited by the coarsening of very long chains[1,2]. This means that the phase transition from a randomly dispersed liquid phase to a partially ordered solid phase occurs within 0.5 s. Second, the formation of chains and the coarsening of chains to columns all occur within t_s. Therefore, from our data, the two processes cannot be distinguished. Finally, if we try to fit our results with a power law similar to those obtained in the systems of references 1 and 2, we find that the data in Fig. 2 can be fitted by a power law $q_{max} \sim t^{-0.15}$ only for $0 \leq t \leq 2$ s, that is, in the region before complete saturation. Since an asymptotic value of q_{max} is reached for $t > t_s$, a general function of the form $c_1/(c_2+t) + c_3$ fits the data much better over the entire time range measured. Where c_1, c_2, and c_3 are constants. c_2 is proportional to t_s and c_3 is the asymptotic value, q_s, as t goes to infinity.

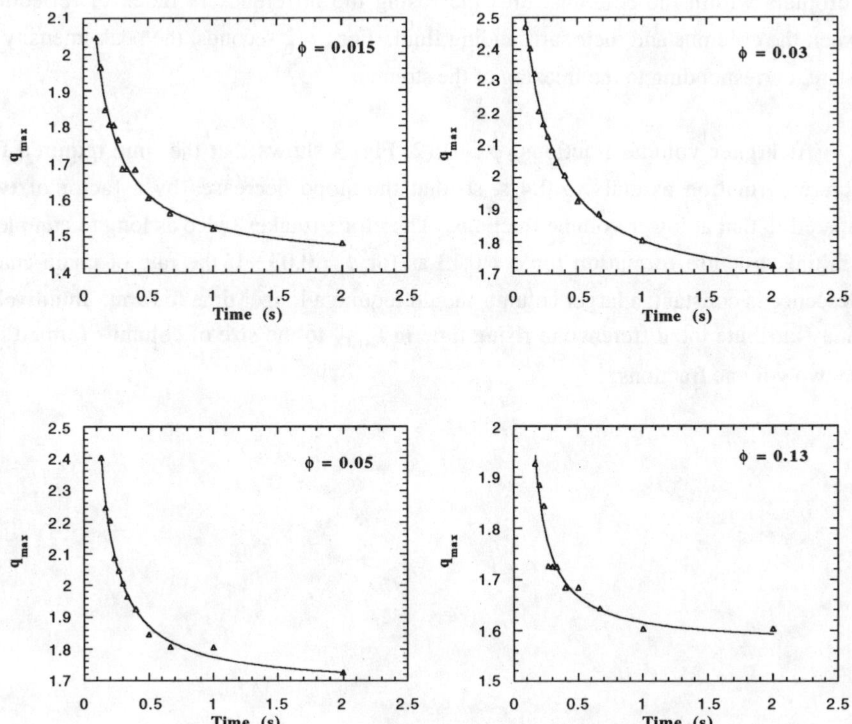

Fig. 2. The peak position, q_{max}, obtained from Fig. 1 is plotted as a function of time. It also shows results for other volume fractions obtained under the similar conditions. The solid lines are fitting of the function $q_{max} = c_1/(c_2+t) + c_3$.

Furthermore, by comparing the results obtained for different volume fractions in Fig. 2, we find that the saturation time t_s is independent of the volume fraction for $0.015 \leq \phi \leq 0.13$. The final value of q_s varies but slightly with volume fractions.

The growth of the columns can also be inferred from the peak intensity of the ring, I_{max}, as a function of time. The results are shown in Fig. 3 for the same range of volume fractions shown in Fig. 2. The peak intensity increases sharply with time and saturates even earlier than q_{max}. For $0.015 \leq \phi \leq 0.082$, the measured saturation time is about 0.2 s. Thus the formation of chains and columns must take place during that short period. Whereas the slight increase of I_{max} in the time period $t_s \leq t \leq 2$ s suggests the tightening of the droplets within the columns, thus increasing the difference in index of reflection between the columns and their surrounding fluid. For $t > 2$ seconds, the peak intensity is constant, corresponding to the freezing of the structure.

At higher volume fractions, $\phi > 0.12$, Fig. 3 shows that the time required for structure formation extends to 0.4 s, so that the slope decreases by a factor of two compared to that at lower volume fractions. Therefore, it takes twice as long to complete the initial structure formation for $\phi = 0.13$ as for $\phi = 0.03$. If the rate of chain-chain coalescence is constant, a larger column should require a longer time to form. Intuitively, we may attribute the difference in rising time in I_{max} to the size of columns formed for these two volume fractions.

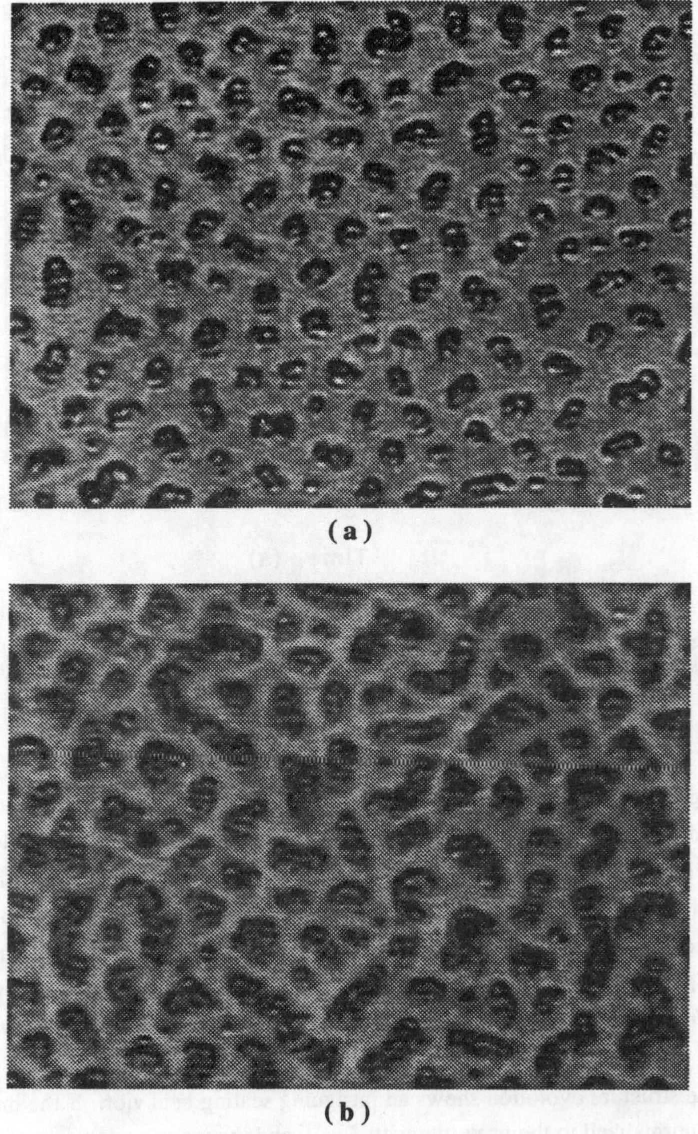

(a)

(b)

Fig. 4 a,b. Microscope image shows the cross section of the structure after the final magnetic field H = 380G ($\lambda = 107$) is reached and the structure freezes. Here $L = 25$ μm. and the horizontal width of the image corresponds to 80 μm. (a)$\phi = 0.03$, (b)$\phi = 0.12$.

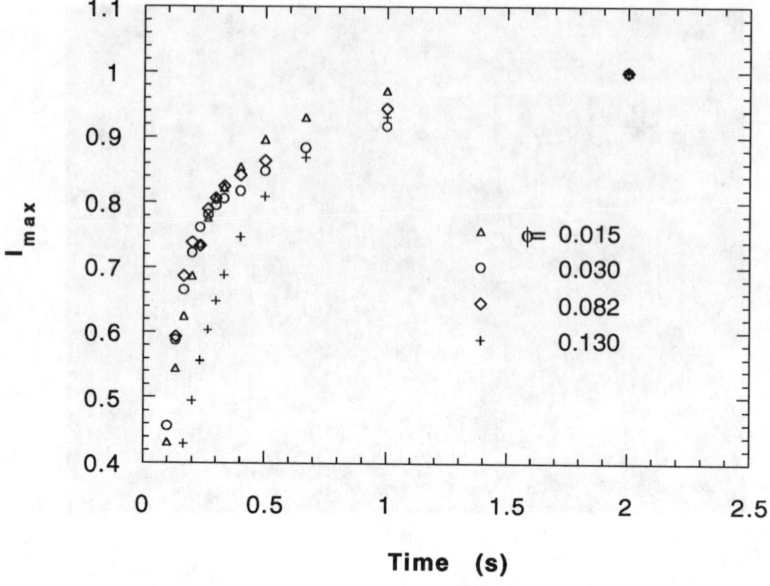

Fig. 3. The peak intensity, I_{max}, obtained from Fig. 1 is plotted as a function of time. Results for other volume fractions obtained under the similar conditions are also plotted for comparison.

Figure 4a and Fig. 4b shows the microscopic image of the final structure for (a) $\phi = 0.03$ and (b) $\phi = 0.12$ after rapidly applying a magnetic field to the sample. A cross section of columnar structure is observed at the lower volume fraction where the columns have equal size and are equally spaced. Similar columnar structure is observed for ϕ up to 0.082. In contrast, a worm-like pattern is observed for $\phi \geq 0.1$. The area of each "worm" in Fig. 4b is roughly twice the cross section of a column in Fig. 4a, which may account for the difference between the rise times of I_{max} for the two volume fractions shown in Fig. 3.

The structure evolution shows an intriguing scaling behavior. If the intensities of Fig. 1 are normalized to the peak intensity, I_{max}, and the wave vectors normalized to the peak position, q_{max}, the nice scaled curve of Fig. 5 results. An almost perfect scaling is obtained for the first ring over the period from 0.13 s to saturation. Good scalings are also found for all other volume fractions.

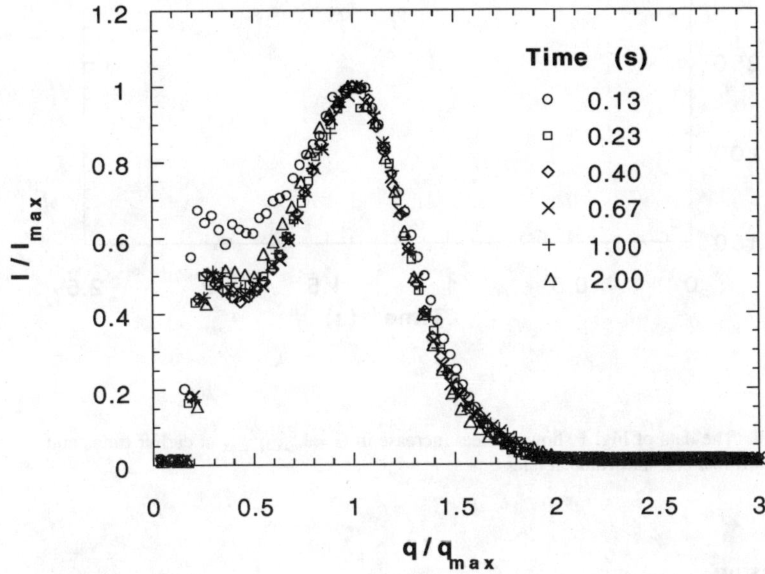

Fig. 5. The data of Fig. 1 replotted with I normalized to the peak intensity, I_{max}, and q to the peak position, q_{max}.

Another interesting scaling is found when the function G, defined as the peak intensity multiplied by the square of the peak position, is plotted as a function of time. Fig. 6 shows the results for $\phi = 0.03$. For $t < 0.2$ s, G increase linearly with time; thereafter, G is independent of time. The same results are also found in other volume fractions up to 0.082. At $\phi = 0.13$ the linear region extends to 0.5 s, after which a constant-G region is observed. Our result is very similar to the result found recently in colloidal aggregations[9].

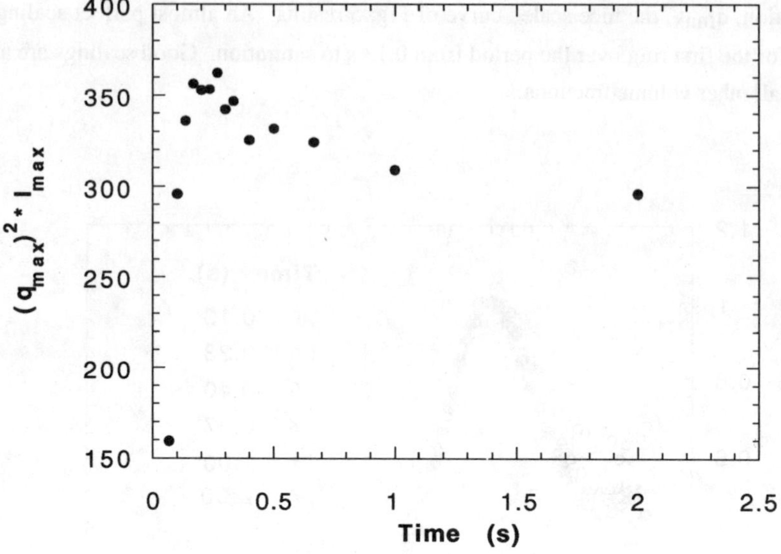

Fig. 6. The data of Fig. 1 show a linear increase in $G = I_{max}\, q^2_{max}$ at earlier times and level off to a constant value at later times.

4. Conclusion

Our initial experiments demonstrate that the confinement of ferrofluid emulsions in the applied field direction strongly affects the rate of structure evolution. In contrast to the long time (~ 10 min) required for chain-chain coarsening in others experiments [1,2] with less or no confinement, we do not observe the coarsening process to be the lone process in a thin cell of 25 μm. The strong repulsive interactions between short chains or thin columns prevents them from gathering. The time taken for completing the phase transition is found to be less than 0.5 s, which is much shorter than the 10 min observed in ER fluids[1] and in MR fluids in larger containers[2]. It is the relatively strong dipole repulsion between columns, which prevents the columns from coarsening further, that enables the sample in a thin cell to complete phase transition much faster than in a thick cell. This hypothesis is also confirmed by comparing the microscope images we obtain with different chain lengths. The column width increases with the chain length[10]. The rate of structure formation is essentially independent of volume fractions in the range of $0.015 \leq \phi \leq 0.13$ for fixed cell thickness. Very good scaling has been observed in the relation between scattering intensity and scattering wave vector during the structure formation.

5. Acknowledgments

We are grateful to the technical supports provided by Prof. E. Y. Wong and Prof. L. J. Eliason. We also acknowledge with gratitude the helpful discussions of Prof. E. Y. Wong, Prof. S. I. Salem, Prof. L.S. Lerner, Prof. A. Leung, and Dr. Ron Rosensweig. This research is partially supported by an award from The Research Corporation. Acknowledgment is made to the Donors of The Petroleum Research Fund, administered by the American Chemical Society, for partial support of this research.

6. References

1. J. E. Martin, J. Odinek, T.C. Halsey, *Phys. Rev. Lett.* **69**, 1524 (1992). T.C. Halsey, *Science*, **258**, 761, (1992).
2. M. Fermigier and A. Gast, *J. Coll. and Int. Sci.* **154**, 522 (1992).
3. J. Bibette, *J. Magnetism and Mag. Materials,* **122**, 37 (1993).
4. J. Bibette, *J. Coll. and Int. Sci.* **147**, 474 (1991).
5. J. Bibette, "Ferrofluid emulsion as a tool for probing droplets forces" Submitted for publication.
6. J. Bibette, D.C. Morse, T.A. Witten, and D.A. Weitz, *Phys. Rev. Lett.* **69**, 2439 (1992).
7. L. Landau and E. M. Lifshitz, *Electrodynamics of Continuous Media*, (Pergamon, Oxford, 1984), 2nd ed.
8. A. P. Gast and C. F. Zukoski, *Adv. Colloid Interface Sci.* **30**, 153 (1989).
9. Private communication with Prof. R. Klein, who reports the work performed on colloids of polystyrene spheres by M. Carpineti and M. Giglio.
10. E. M. Lawrence, M. L. Ivey, G. A. Flores, and J. Liu submitted to the *Int. J. Modern Phys. B.*

The Role of Suspension Structure in the Dynamic Response of Electrorheological Suspensions

M. Parthasarathy, K. H. Ahn, B. M. Belongia, and D. J. Klingenberg

Department of Chemical Engineering and Rheology Research Center
University of Wisconsin, 1415 Johnson Drive, Madison, WI 53706, USA

Abstract

The dynamic response of electrorheological (ER) suspensions has received little attention relative to the effort devoted to the study of the steady shear response. We report on simulation and experimental investigations of the dynamic oscillatory response of ER suspensions, in particular focusing on the relationship between suspension structure and the rheological response. We consider the response of monodisperse and polydisperse suspensions under linear deformation, as well as the response in the nonlinear regime. Dimensional analysis of the equations of motion predict that the linear rheological response obeys a time-field strength superposition principle, which is confirmed by experiment. The response is found to exhibit a sharp dispersion that is only broadened slightly by polydispersity. Nonlinear deformation is found to significantly broaden the observed dispersion.

1 Introduction

Since the early investigations of electrorheology (ER) by Winslow [1], considerable effort has been focused on understanding the steady-shear response of ER suspensions. Most proposed applications however, such as engine mounts, shock absorbers, brakes and clutches, actuators, and control valves, operate primarily in transient or periodic modes. Relatively little effort has been directed toward this aspect of ER, and as a result, our understanding of transient responses is lacking, inhibiting the development of ER technology to its full potential.

In this paper, we discuss our recent investigations of the dynamic response of ER suspensions. Experimental work with model alumina suspensions is complemented with particle-level simulations designed to reveal microscopic mechanisms responsible for observed behavior. The simulation method is outlined in Section 2, followed by a description of our experimental methods in Section 3. The results of these investigations are discussed in Section 4. The small amplitude response is found to exhibit a relatively sharp dispersion, due to a frequency-dependent structure produced by the competition between electrostatic and hydrodynamic forces acting on the particles. Polydispersity is found to have little influence on the breadth of the dispersion, whereas nonlinear deformation is found to significantly alter the frequency dependence of the moduli. Conclusions from this work are summarized in Section 5.

2 Model and Simulation Method

The microscopic model used here to describe ER suspensions has been discussed previously [2]. Here we outline the features required for the present discussion.

ER suspensions are treated as neutrally-buoyant, hard, class A dielectric spheres (dielectric constant ϵ_p) immersed in a class A dielectric, Newtonian continuous phase (dielectric constant ϵ_c, viscosity η_c). All the spheres possess the same dielectric constant, but can be of different size. Application of an electric field polarizes the spheres, inducing electrostatic interactions between them, and between each sphere and the electrodes. The continuous phase is treated as a continuum; its presence influences only the magnitude of the electrostatic interactions and the hydrodynamic resistance on each sphere. For large particles under large electric fields, colloidal and Brownian forces are negligible [3].

Neglecting inertia terms, the motion of sphere i not near an electrode is governed by

$$\mathbf{F}_i(\{\mathbf{R}_j\}) = 0, \tag{1}$$

where $\mathbf{F}_i(\{\mathbf{R}_j\})$ is the resultant force acting on particle i due to the contributions described above, which depends on the positions of all the spheres. Eq. (1) represents a coupled set of equations governing the motion of all the spheres not near electrodes (the motion of spheres near electrodes is treated below). Solutions for these equations requires explicit expressions for the forces.

The electrostatic force on sphere i (diameter σ_i) at the origin of a spherical coordinate system due to sphere j (diameter σ_j) at (R_{ij}, θ_{ij}) (see Fig. 1) is treated in the point-dipole limit [4]:

$$\mathbf{F}_{ij}^{el}(R_{ij}, \theta_{ij}) = F_s(\sigma_i, \sigma_j)\left(\frac{R_{min}}{R_{ij}}\right)^4 \left[(3\cos^2\theta_{ij} - 1)\mathbf{e}_r + \sin 2\theta_{ij}\mathbf{e}_\theta\right], \tag{2}$$

where $R_{min} = (\sigma_i + \sigma_j)/2$,

$$F_s(\sigma_i, \sigma_j) = \frac{3}{16}\pi\epsilon_0\epsilon_c\beta^2 E_0^2 \left(\sigma_j^2 \frac{16\lambda_{ij}^3}{(1+\lambda_{ij})^4}\right), \tag{3}$$

$\epsilon_0 = 8.8542 \times 10^{-12}$ F/m, $\beta = (\epsilon_p - \epsilon_c)/(\epsilon_p + 2\epsilon_c)$, and $\lambda_{ij} = \sigma_i/\sigma_j$. This force represents the interaction between class A dielectric bodies in a class A dielectric continuous phase, and is expected to be the leading order interaction for such systems (see ref. [5] for a discussion of accurate calculations of interactions between dielectric

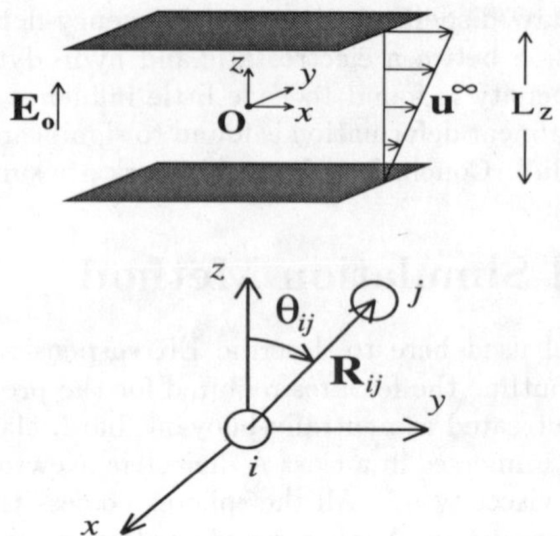

Figure 1: Schematic diagrams showing the geometries of the sheared suspension and sphere pairs. The bottom electrode is held fixed and the top electrode is displaced in the x direction. R_{ij} is the center-to-center separation and θ_{ij} is the angle between the line-of-centers and the applied electric field.

spheres). For systems that conduct, either due to the presence of mobile charge carriers or by electronic mechanisms, this representation is not likely to be correct, except in the limit of large frequencies where charge carriers can no longer respond to the varying electric field to an extent sufficient to alter the force described above. In this situation, the dielectric constants represent the values measured at the electric field strength and frequency employed.

The hydrodynamic force on sphere i is represented by the Stokes' drag,

$$\mathbf{F}_i^{hyd}(\mathbf{R}_i) = -3\pi\eta_c\sigma_i\left(\frac{d\mathbf{R}_i}{dt} - \mathbf{u}^\infty(\mathbf{R}_i)\right), \tag{4}$$

where $\mathbf{u}^\infty(\mathbf{R}_i)$ is the ambient fluid velocity.

The hard-sphere interaction is represented by a short-range repulsive force,

$$\mathbf{F}_{ij}^{rep}(R_{ij}) = -F_s(\sigma_i, \sigma_j)\exp\left(-\frac{R_{ij} - R_{min}}{0.01R_{min}}\right)\mathbf{e}_r. \tag{5}$$

The hard-sphere/hard-wall interaction is represented by a similar short-range function,

$$\mathbf{F}_{i,wall}^{rep}(\mathbf{R}_i) = \frac{3}{16}\pi\epsilon_0\epsilon_c\sigma_i^2\beta^2E_0^2\exp\left(-\frac{H_i - \sigma_i/2}{0.01\sigma_i}\right)\mathbf{n}, \tag{6}$$

where $H_i = L_z/2- \mid z_i \mid$, L_z is the electrode gap width, and $\mathbf{n}$ is the unit normal vector directed from the electrode into the suspension.

Incorporating these forces into Eq. (1), the equation of motion for sphere i is written

$$3\pi\eta_c\sigma_i\frac{d\mathbf{R}_i}{dt} = \sum_{j\neq i}\mathbf{F}_{ij}^{el}(R_{ij},\theta_{ij}) + \sum_{j}\mathbf{F}_{ij}^{el}(R'_{ij},\theta'_{ij}) + \sum_{j\neq i}{}^{''}\mathbf{F}_{ij}^{rep}(R_{ij},\theta_{ij}) + 3\pi\eta_c\sigma_i\mathbf{u}^{\infty}(\mathbf{R}_i). \tag{7}$$

The primes in the second summation indicate the electrostatic force between sphere i and the image of sphere j; these forces are summed over all images of all spheres, including those of sphere i. The double prime in the third summation indicates that repulsive forces are evaluated between sphere i and all spheres $j \neq i$, as well as between sphere i and the electrodes. Equation (7) governs the motion of spheres not near an electrode surface. Spheres within $\delta_w = 0.05\sigma_i$ of an electrode are considered stuck, and assume the lateral velocity of the electrode. This sticking condition is based on experimental observation, and is treated in the simplified model for monodisperse ER suspensions in ref. [6] . The equations governing the motion of stuck spheres in polydisperse suspensions is given below in dimensionless form.

The equation of motion is made dimensionless with the following scales

$$\ell_{scale} = \sigma_0, \qquad F_{scale} = \frac{3}{16}\pi\epsilon_0\epsilon_c\sigma_0^2\beta^2E_0^2, \qquad t_{scale} = \frac{16\eta_c}{\epsilon_0\epsilon_c\beta^2E_0^2},$$

where σ_0 is the diameter of a reference sphere, taken to be that of the largest sphere. The equation of motion in dimensionless form for sphere i not near an electrode is written

$$\frac{d\mathbf{r}_i^*}{dt^*} = \frac{1}{\sigma_i^*}\left(\sum_{j\neq i}\mathbf{f}_{ij}^{el*} + \sum_{j}\mathbf{f}_{ij}^{el*'} + \sum_{j\neq i}{}^{''}\mathbf{f}_{ij}^{rep*}\right) + \mathbf{u}^{\infty*}(\mathbf{r}_i^*), \tag{8}$$

where the asterisks denote dimensionless quantities. The equations governing the motion of stuck spheres are written

$$\frac{dx_i^*}{dt^*} = u_{x,elec}^* \tag{9}$$

$$\frac{dy_i^*}{dt^*} = u_{y,elec}^* \tag{10}$$

$$\frac{dz_i^*}{dt^*} = \frac{1}{\sigma_i^*}\left(\sum_{j\neq i} f_{z,ij}^{el*} + \sum_{j} f_{z,ij}^{el*'} + \sum_{j\neq i}{}^{''} f_{z,ij}^{rep*}\right) + u_{z,elec}^* \tag{11}$$

where u_k is the kth component of the electrode velocity. In this paper, we consider the following ambient flow profile:

$$\mathbf{u}^{\infty*}(r_i^*) = \omega^*\gamma_0\left(z_i^* + \frac{L_z^*}{2}\right)\cos\omega^*t^* \;\; \mathbf{e}_x; \tag{12}$$

where ω^* is the dimensionless frequency and γ_0 is the strain amplitude.

Equations (7)–(11) represent a microscopic model for ER suspensions. Dimensional analysis of these equations, described below, provides scaling information among various quantities, but further investigation of suspension behavior requires solutions of these equations to obtain the sphere positions as a function of time, and subsequently the rheological properties. Two methods of solution have been employed: dynamic simulation and regular perturbation. In this report, we use the dynamic simulation method. (See ref. [7], for discussions of the perturbation methods as well as other ambient flows.)

The dynamic simulation technique employs direct numerical integration of Eqs. (8)–(11). N spheres are placed between electrodes at $z^* = \pm L_z^*/2$, and within periodic boundaries at $x^* = \pm L_x^*/2$, $y^* = \pm L_y^*/2$. Simulations in two dimensions are performed by placing all spheres in a plane $y^* =$ constant. Spheres are placed randomly or in a prescribed, manufactured configuration. Equation (9)–(11) are integrated numerically for each sphere using Euler's method, with a time step $\Delta t^* \leq \sigma^*_{min} 10^{-3}$. Forces are evaluated within a cutoff radius $r_c^* = L_z^*/2$. For rheological simulations, a metastable structure is first obtained by integrating the equations of motion under no flow until motion ceases. After the metastable structure is obtained, simulation under the desired flow is begun. For oscillatory shear simulations, the runs are performed until all transients decay, and the motion of each sphere is periodic. The responses of structures simulated in this study were linear for strain amplitudes $\gamma_0 \leq 10^{-4}$ (and perhaps larger).

Once the sphere positions have been determined as a function of time, the rheological properties can be evaluated. The total stress is conveniently expressed as a sum of direct electrostatic and hydrodynamic contributions [4]

$$\boldsymbol{\sigma}^{total} = \boldsymbol{\sigma}^{E} + \boldsymbol{\sigma}^{H}. \tag{13}$$

This distinction between the contributions is arbitrary, but is chosen because the hydrodynamic contribution vanishes as the deformation rate vanishes or when the solvent is removed, whereas the electrostatic contribution may remain in both situations. These contributions are not independent, as both electrostatic and hydrodynamic forces influence the suspension structure and hence both contributions to the stress. Since elasticity arises from non-hydrodynamic forces [8], we consider only the direct electrostatic contribution to the stress, which is expressed

$$\sigma^*_{xz}(t^*) = -\frac{1}{V^*} \sum_i z_i^* f_{xi}^{total*}(\{\mathbf{r}_j^*(t^*)\}), \tag{14}$$

where $\sigma^*_{xz}(t^*)$ is the dimensionless time-dependent shear stress acting in the x-direction on a plane normal to the z-direction, $f_{xi}^{total*}(\{\mathbf{r}_j^*(t^*)\})$ is the total dimensionless electrostatic plus short-range repulsive force acting on sphere i in the x-direction, and V^* is the dimensionless suspension volume.

For linear oscillatory shear, the response can be represented by frequency-dependent storage $[G'^*(\omega^*)]$ and loss $[G''^*(\omega^*)]$ components of the time-dependent shear stress

$$\frac{\sigma^*_{xz}(t^*)}{\gamma_0} = G_e^* + G'^*(\omega^*) \sin \omega^* t^* + G''^*(\omega^*) \cos \omega^* t^*, \tag{15}$$

where G_e^* is the equilibrium modulus arising from a constant residual stress.

3 Experimental

Rheological experiments were performed using a 20 wt% suspension of alumina particles in poly(dimethylsiloxane) (PDMS, Huls, $\eta_c = 0.485$ Pa s, $\rho_c = 970.0$ kg/m^3, $\epsilon_c = 2.75$). The dielectric constant of alumina is found in the literature [9] to lie in the range 8–10; for this study, ϵ_p was assigned a value of 9. The precise value of ϵ_p ,does not influence the qualitative results presented in this report, since the same alumina particles were used in all of the suspensions. The alumina particles were ground with a mortar and pestle and then sieved to obtain a fraction of particles with characteristic dimensions in the range of 63–90×10^{-6} m. The alumina particles were spread on petri dishes and heated in a vacuum oven at 80°C for 6–8 hours to remove most of the water, and then added to the PDMS. Contact with air was minimized during suspension preparation. The suspensions were subsequently stored in a dessicator.

Rheological experiments were performed on a Bohlin VOR rheometer modified for the application of large electric fields across the parallel plates. This system evaluates the dynamic response by Fourier transforming the measured torque signal, and reporting the in phase and out of phase components at the fundamental frequency.

Potential differences were supplied by a Stanford Research Systems function generator (model DS345) and amplified with a TREK high voltage amplifier (model 10/10). An electric field frequency of 500 Hz was used for all experiments.

After loading the sample into the rheometer, the desired electric field was applied to the suspension for approximately 30 minutes to obtain a static structure. Shear moduli were then obtained by performing strain sweeps up to a strain amplitude of 0.02. From these curves, small strain plateaus could be identified, representing the linear response (see Fig. 13 for an example of one of these strain sweeps). Three replicates were performed for each experiment to ensure reproducibility. For consistency, experiments at the largest desired field strength and smallest desired oscillation frequency were performed first. Subsequent experiments were then performed in succession, without disturbing the sample.

4 Results

4.1 *Dimensional Analysis*

Dimensional analysis of Eqs. (7)–(13) provides useful information, without performing any simulations. Inspection of these equations reveals that for linear deformations (*i.e.*, particle trajectories that are linear in the strain amplitude γ_0), the dimensionless rheological response must be a function of only the dimensionless frequency:

$$\frac{G'}{\frac{3}{16}\pi\epsilon_0\epsilon_c\beta^2E_0^2} = f_1\left(\frac{16\eta_c\omega}{\epsilon_0\epsilon_c\beta^2E_0^2}\right), \tag{16}$$

$$\frac{G''}{\frac{3}{16}\pi\epsilon_0\epsilon_c\beta^2E_0^2} = f_2\left(\frac{16\eta_c\omega}{\epsilon_0\epsilon_c\beta^2E_0^2}\right), \tag{17}$$

i.e., $G'^* = f_1(\omega^*)$ and $G''^* = f_2(\omega^*)$. Dimensional analysis does not provide the functions f_1 and f_2, but only indicates which variables the dimensionless moduli must depend on.

To test these predictions, experiments were performed using a 20 wt% suspension of alumina particles in PDMS as described in the previous section. Shear moduli obtained in the linear limit are plotted as a function of frequency in Figs. 2 and 3, for electric field strengths ranging from 300 – 2000 V/mm. Uncertainties in moduli values are approximately ±10%. These data are plotted again in dimensionless form (as suggested by Eqs. (16) and (17)) in Figs. 4 and 5. These figures show that the dimensionless moduli at all field strengths collapse onto a single function of the dimensionless frequency, as predicted by Eqs. (16) and (17). As shown below, the shapes of these dimensionless plots are also in good agreement with predictions from simulations (See, for example, Fig. 9).

As mentioned above, dimensional analysis does not provide the functions in Eqs. (16) and (17), but only requires that these functions exist; these functions must be obtained by solution of the governing equations, *e.g.*, by numerical simulation. Simulation results for various model suspension structures are described below. However, the scaling relationships obtained from dimensional analysis provide another benefit for probing ER suspensions. Most rheological techniques permit investigation of only a limited range of frequencies, due to deformation and stress (torque) resolution, the natural frequency of the rheometer, and space limitations. Electric field strengths are also confined to specific ranges due to limitations of the power supply, dielectric breakdown, arcing, rheometer resolution, and electrode gap widths. However, dimensional analysis suggests that the field strength and oscillation frequency appear as independent variables only through the combination ω/E_0^2. Hence, dimensionless frequencies can be probed over a much larger range than the frequencies themselves through variation of the electric field strength. This is illustrated in Figs. 2–5 where frequencies are sampled in the range $0.6 < \omega < 30$ rad/s, while dimensionless frequencies span the range $0.3 < \omega^* < 6 \times 10^2$. This scaling is related to the Mason number scaling of the steady shear response of ER suspensions [3], and was referred to as "time-field strength superposition" in ref. [4].

4.2 *Simulation of Monodisperse Suspensions*

Consider first the response of a model structure represented by linear strands of spheres aligned with the electric field, connecting the electrodes ($N = 10$, $L_x^* = 10$, $L_z^* = 10$; see Fig. 6). For this structure, the storage modulus, $G'^*(\omega^*)$, is essentially independent of frequency. The loss modulus, $G''^*(\omega^*)$, is *negative* but negligible; the magnitude of the loss tangent, $|\tan\delta(\omega^*)| = |G''^*(\omega^*)/G'^*(\omega^*)| \leq 10^{-3}$ for all frequencies. This simple response is due to affine sphere motion—each sphere is displaced in concert with the ambient shear flow at all frequencies. Hence there is

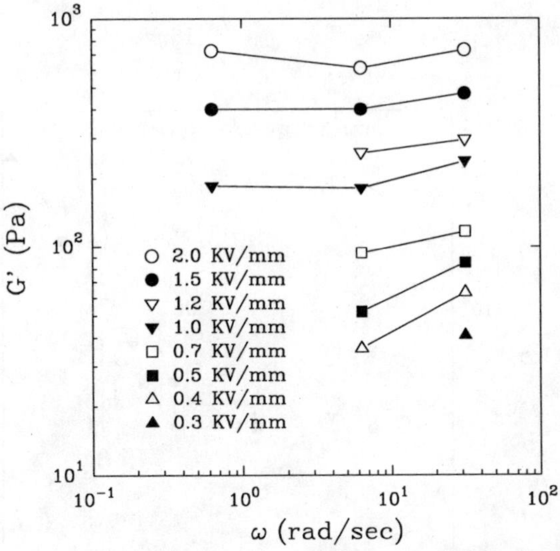

Figure 2: G' as a function of frequency, ω, and electric field strength, E_0.

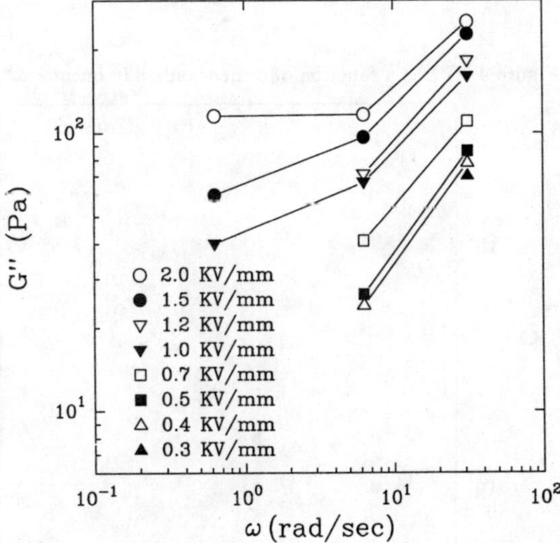

Figure 3: G'' as a function of frequency, ω, and electric field strength, E_0.

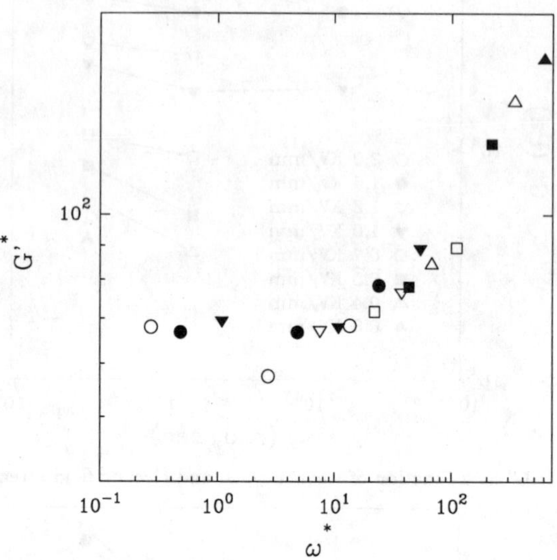

Figure 4: G'^* as a function of dimensionless frequency, ω^*.

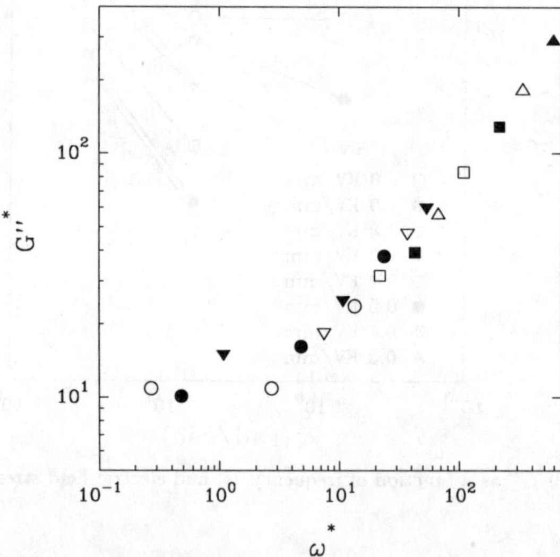

Figure 5: G''^* as a function of dimensionless frequency, ω^*.

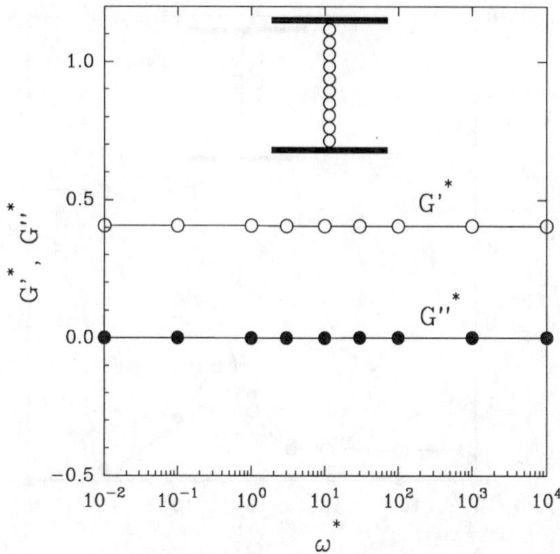

Figure 6: Dynamic oscillatory response for a single-sphere-width chain spanning the electrode gap.

no hydrodynamic resistance and no mechanism for relaxation. Since the suspension structure is independent of frequency, the electrostatic stress (Eq. (14)), and hence the moduli, are independent of frequency. [It is noted that there is a negligible relaxation associated with the slight nonaffine motion of the spheres attached to an electrode. These spheres are forced to translate with the electrode, while the ambient fluid displacement at the sphere centers differ slightly from that at the electrodes. This effect causes sphere motion to slightly lead the applied strain (at the top electrode), producing the negligible, negative loss modulus.]

The structure presented in Fig. 6 is an oversimplified representation of the structures commonly observed in moderate to highly concentrated ER suspensions; columns of particles, several particles wide, are observed to span the electrode gap, as opposed to single pearl-chain structures. Consider next the response for a suspension structure represented by thick clusters, each composed of the previous single-sphere-width strand with two additional strands placed along each side of the original strand ($N = 28$, $L_x^* = 10$, $L_z^* = 10$; see Fig. 7). The frequency dependent moduli for this structure are presented in Fig. 7. The loss modulus scales as $G''^*(\omega^*) \sim \omega^*$ at low frequencies and as $G''^*(\omega^*) \sim {\omega^*}^{-1}$ at large frequencies, passing through a maximum at a dimensionless frequency of $\omega^* \approx 25$. Near this frequency, the storage modulus increases from a small frequency plateau, $G'^*(0)$, to a large frequency plateau, $G'^*(\infty)$. Simulations performed at $\gamma_0 = 10^{-3}$ give identical results for this structure, indicating a linear response for $\gamma_0 \leq 10^{-3}$ (and perhaps larger values). The shapes of the small frequency portions of the curves presented in Fig. 7 ($\omega^* < 10$), are consistent with the shapes of the dimensionless experimental moduli curves presented in Figs. 4 and 5.

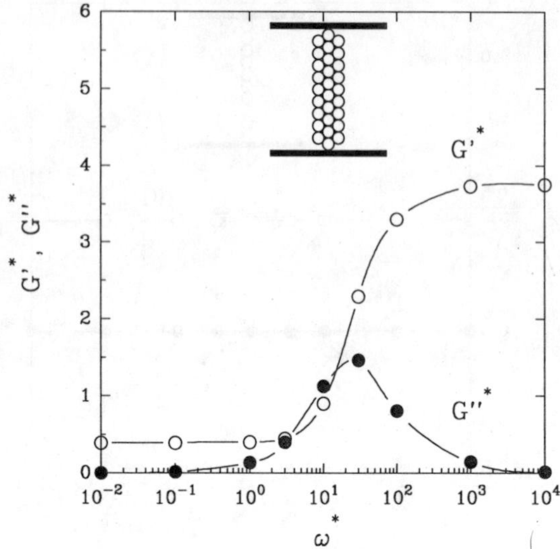

Figure 7: Dynamic oscillatory response for a thick 28-sphere cluster. The solid lines are cubic spline interpolations between the simulation data points.

This thick cluster exhibits a significant, sharp relaxation, whereas the structure represented by single-sphere-width strands exhibits none. Relaxation results from the addition of the side chains, as this is the only difference between the thin and thick clusters. The relaxation mechanism becomes apparent when the motion of individual spheres and the overall deformation of the cluster are considered in the small and large frequency limits.

At very small frequencies, deformation is completely determined by the spheres attached to the electrodes—hydrodynamic resistance to sphere motion is insignificant at these small deformation rates. As a result, the net electrostatic-plus-repulsive force on each sphere not attached to an electrode is zero at every instant (Eq. (14)). The extensional and rotational components of the quasi-static deformation require that spheres in stress bearing strands deform affinely (with a slight correction described above), while sphere displacement in the side chains possesses components in both the x and z directions. This motion is depicted in Fig. 8, where the displacement directions of a particular sphere in the right side chain are shown.

At large frequencies, the motion of spheres in the side chains is significantly different. Here, hydrodynamic resistance controls the displacement of each sphere. At any particular instant, there is still a net electrostatic-plus-repulsive force component in the z direction, but there is insufficient time during a half period of oscillation for the sphere to move in that direction. This time, proportional to $1/\omega^*$, vanishes as $\omega^* \to \infty$. As a result, each sphere displaces only in the x direction. This situation is also presented in Fig. 8, where the displacement of a particular sphere at large

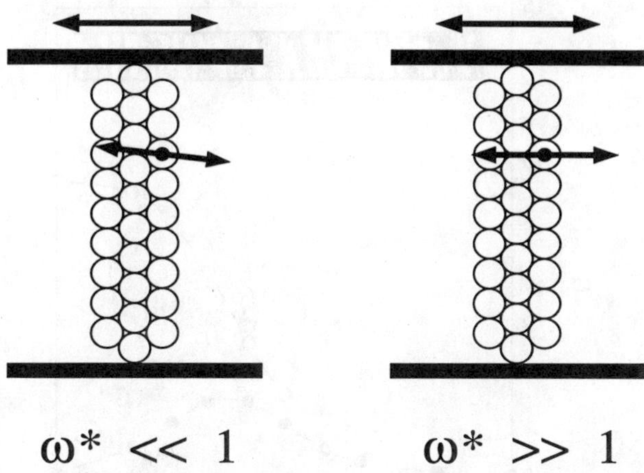

Figure 8: Illustration of the dynamic structures of the small and large frequency limits for the 28-sphere cluster. The arrows represent displacement directions of the electrodes and a particular sphere in the right side chain.

frequencies is shown. Motion is affine in this large frequency limit. However, note that this is only true for spheres interacting through continuous repulsive forces—for two hard spheres in contact with different z positions, at least one of the spheres must be moving relative to the ambient shear flow.

Thus the dispersion depicted in Fig. 7 is due to hydrodynamic relaxation of electrostatically driven sphere motion, providing a frequency-dependent dynamic structure. As the shear stress is simply a summation over position-dependent pair interactions, the stress, and hence the moduli, will be frequency dependent. The critical frequency marking the loss modulus maximum is directly related to the characteristic relaxation time, which in this case is just a fraction of the time scale t_{scale}. The peak in $G''^*(\omega^*)$ is pronounced because each sphere sees essentially the same environment, and therefore the response is dominated by a single relaxation time (factors that broaden the dispersion are considered below).

Although similar in some respects, this hydrodynamic relaxation mechanism is significantly different from that proposed by McLeish *et al.* [10], who attributed relaxation to the motion of "free strings" (single sphere-width chains attached to at most one electrode). In their theory, free strings deform affinely at large frequencies, but are able to relax at small frequencies due to the electrostatic forces, thus providing a relaxation mechanism. This mechanism is only likely to operate at small concentrations where thin, unattached chains are likely to exist. In contrast, the mechanism presented in this paper occurs within thicker clusters and hence will be

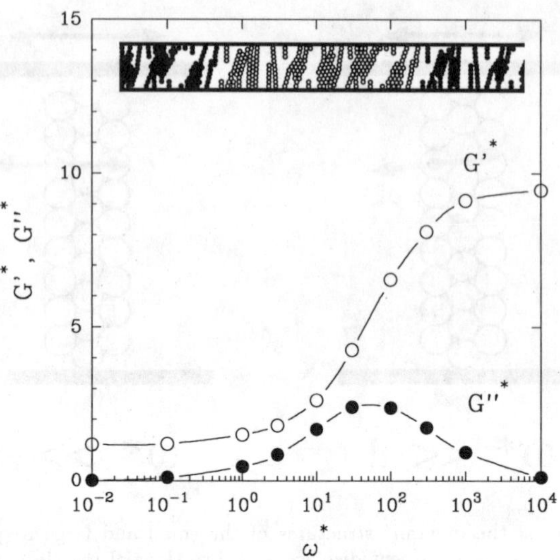

Figure 9: Dynamic oscillatory response for a 250-sphere structure. The filled spheres represent periodic images of the open spheres. The solid lines are cubic spline interpolations between the simulation data points.

active especially at moderate to large concentrations. It is significant, however, that both mechanisms arise from the electrostatic polarization model for ER suspensions, and *do not* rely on fiber rupture—dissipation is realized at infinitesimal as well as finite strain amplitudes. It is also important to note that relaxation of free strings is captured by the present model—however, only relaxation of thick clusters typically found in concentrated suspensions is considered in this paper.

The rheological response presented in Fig. 7 is characterized, to a first approximation, by a sharp dispersion with a single relaxation time. Experimental results reported in the literature typically do not exhibit this type of behavior—responses are sometimes reported to be insensitive to strain frequency [11,12] while in other situations moduli values increase or decrease with frequency [13,14,15,16]. We have simulated the response of various structures, in addition to those presented in Figs. 6 and 7, in order to determine if the variety of responses observed experimentally may be explained in terms of various cluster sizes and shapes in the linear response of monodisperse ER suspensions [4]. The results of these simulations are illustrated concisely in Fig. 9, where the simulated response for a suspension structure containing a distribution of cluster sizes and shapes in presented.

250 spheres were placed randomly in a two-dimensional simulation cell of dimensions $(L_x^*, L_z^*) = (50, 10)$. A fibrous structure was formed by integrating the equations of motion with $\mathbf{u}^{\infty *} = 0$ until motion ceased. This structure was then sheared slowly to a strain of 4.0, followed by relaxation to a metastable structure (see ref. [17] for a

description of simulations under steady shear relaxation). The frequency-dependent loss modulus does not become broader as a result of the various cluster sizes and shapes present (Fig. 9). Additional simulations in three dimensions also exhibit sharp dispersions, indicating that this observation is not an artifact of simulations in two dimensions. We conclude that monodisperse suspensions, conforming to the behavior respresented by our model, exhibit a sharp relaxation (*i.e.*, $G'^*(\omega^*)$ increases from a low frequency plateau to a large frequency plateau in 2-3 decades) under linear oscillatory deformation. Broad dispersions must arise from features not included in the present model. Below we consider two such features: polydispersity and nonlinear deformation. Other features that may influence the breadth of dispersion inlcude particle shape polydispersity, and particle interactions that do not follow the E^2 dependence (for instance, due to the presence of charge carriers and conduction).

4.3 *Influence of Polydispersity*

Simulating the dynamic oscillatory response of polydisperse suspensions is straightforward; spheres of various sizes are simply placed in the simulation cell, and size dependent interactions are evaluated as described in Section 2. The electrostatic interaction between spheres of different diameters is essentially the same as that for two spheres of the same size (in the point-dipole limit). The dipolar nature of the interaction is maintained for all possible size ratios $\lambda_{ij} = \sigma_i/\sigma_j$, but the magnitude of the interaction becomes a sensitive function of this diameter ratio. In fact, in the limit as $\lambda_{ij} \to 0$ or ∞, the magnitude of the interaction vanishes. Considering the role of electrostatic interactions in the nonaffine–affine transition that controls the relaxation process, we expect the presence of spheres of different size to produce a dispersion that is broader than that observed for a monodisperse suspension.

The response of several bidisperse suspension structures were investigated in ref. [7]. The response for one realization of a bidisperse suspension structure is presented in Fig. 10. Within the simulation cell, two large spheres ($\sigma^* = 1$) are placed near the electrodes. Thirteen small spheres ($\sigma^* = 0.2$) are placed between the two large spheres, forming a cluster similar in shape to the cluster presented in Fig. 7. The simulated oscillatory response indeed exhibits a broader dispersion; the loss modulus $G''^*(\omega^*)$ possesses two main peaks. By considering the various contributions to the electrostatic stress, we have determined that the peak at small frequencies arises from the electrostatic interactions between small and large spheres, while the peak at large frequencies arises from the interaction between small spheres [7]. As the difference in diameters gets larger, the peak arising from the interaction between small and large spheres moves to smaller frequencies. However, the magnitude of this peak also decreases as the diameter difference increases. As a result, the influence of diameter difference on dispersion broadening is restricted. By considering the response of several bidisperse structures, it was concluded in ref. [7] that the polydispersity does not significantly broaden the dispersion, relative to responses observed for monodisperse suspensions.

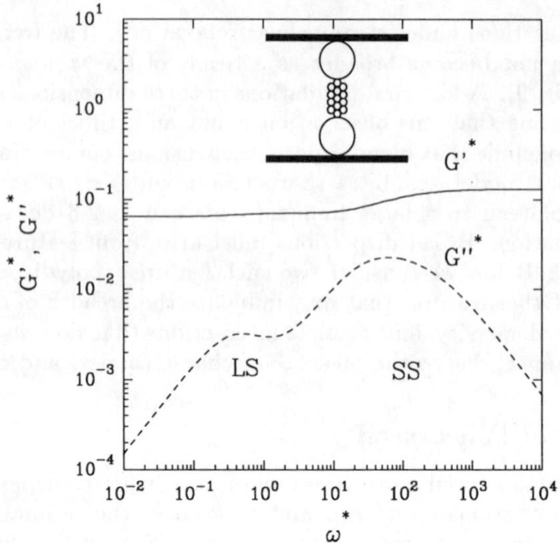

Figure 10: Dynamic oscillatory response for a bidisperse cluster.

4.4 *Nonlinear Deformation*

The simulated steady shear response for a two-dimensional monodisperse suspension is presented in Fig. 11. This curve was generated by performing simulations on ten randomly-generated configurations, each configuration containing $N = 50$ spheres in a simulation cell of dimensions $(L_x^*, L_z^*) = (10, 10)$. The equations of motion were integrated with $\mathbf{u}^{\infty *} = 0$ until all motion ceased. Simulations were then performed with a small shear rate of $\dot{\gamma}^* = 10^{-5}$. The shear stress-shear strain data presented in Fig. 11 represent the average over the ten configurations.

The response presented in Fig. 11 shows a linear region for shear strains less than a "critical" value of $\gamma \approx 4 \times 10^{-3}$. For larger shear strains, the response is nonlinear. The onset of nonlinearity is associated with slight rearrangements in the suspension structure; two such rearrangements are depicted in Fig. 12.

We note that the rearrangements we have seen in simulations can be either reversible or irreversible, depending on the ultimate strain amplitude experienced by the structure.

For these nonlinear deformations, the rheological response depends on the strain amplitude as well as the oscillation frequency. This is illustrated in Fig. 13 where experimental strain sweep curves $[G'(\omega, \gamma_0)$ as a function of γ_0 at fixed $\omega]$ are presented for two different situations. A strain sweep carried out at a large field strength of 2000 V/mm and a small frequency of 0.063 rad/s is represented by the circles. For this situation, we observe a linear region for strain amplitudes less than $\gamma_0 \approx 10^{-3}$, followed by a steady decrease in the apparent modulus with strain amplitude. At a

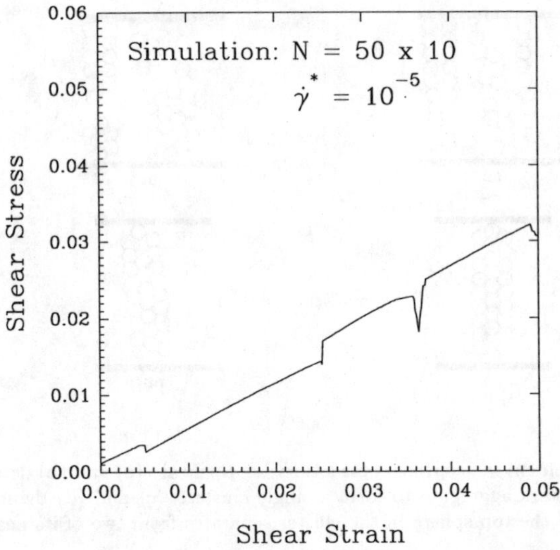

Figure 11: Stress-strain curve for a two-dimensional monodisperse suspension. The data averaged over the ten configurations.

strain amplitude of $\gamma_0 = 0.02$, the storage modulus is nearly 4 times smaller than the linear value. The inverted triangles represent a strain sweep performed at a smaller electric field strength of 300 V/mm and a larger oscillation frequency of 31.4 rad/s. For this situation, the response is essentially independent of strain amplitude, decreasing by only about 10% over the entire range of amplitudes. It is clear from this figure that for small frequencies, the dynamic response is sensitive to the strain amplitude outside of the linear regime.

The difference between the two sets of data in Fig. 13 can be explained in terms similar to our description of the nonaffine-affine deformation transition. At small frequencies and large field strengths, deformations in the nonlinear regime result in rearrangments in the suspension structure. These rearrangements are due to the electrostatic forces driving particles into positions where the net electrostatic-plus-repulsive force on each sphere is zero at every instant. At large frequencies and/or smaller field strengths, these driving forces still exist (weaker in the case of smaller field strengths), but there is insufficient time during a half period of oscillation for the rearrangement to occur. The precise set of conditions causing the transition to occur will depend on the relative positions of the particles as well as the strain amplitude, and the transition may not necessarily be sharp.

Dispersion can be interpreted as a measure of the difference in moduli values under different conditions. In Fig. 13, the difference in moduli at two different strain amplitudes is represented by the bold arrows. These arrows illustrate that the moduli difference, and hence the dispersion, is a function of strain amplitude. The influence

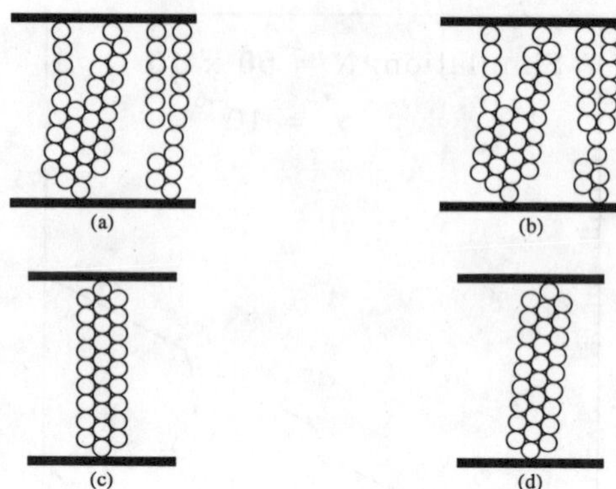

Figure 12: Rearrangement in suspension structure due to shear. (a) and (b) demonstrate rearrangement where two clusters aggregate to form a single cluster. (c) and (d) demonstrate a reversible rearrangement where the top sphere in the cluster separates from two of its nearest neighbors.

of strain amplitude on the breadth of dispersion is presented in Fig. 14, where the dimensionless storage modulus is plotted as a function of dimensionless frequency for experimental data. The open symbols represent the response for storage moduli measured in the limit of zero strain (*i.e.*, the linear response). The response exhibits a small frequency plateau extending over about 2 decades in dimensionless frequency, with the dimensionless modulus increasing with dimensionless frequencies for values above $\omega^* \approx 10$. The filled symbols represent the response where the moduli were measured at a value of $\gamma_0 = 0.02$. In this case, the dimensionless modulus appears to increase with dimensionless frequency over the entire range $0.3 < \omega^* < 6 \times 10^2$. Hence, the dispersion observed for experiments performed outside of the linear regime is significantly broader than that observed within the linear regime. In the previous section, polydispersity was found to have an insignificant influence on the breadth of dispersion for linear deformation. It is not clear at this point whether or not polydispersity will influence dispersion broadening under non-linear deformation.

5 Conclusions

This study has considered the dynamic response of ER suspensions, employing both simulation and experimental techniques. We find that relaxation is associated with a transition in the dynamic structure. At small frequencies, sphere motion is dominated

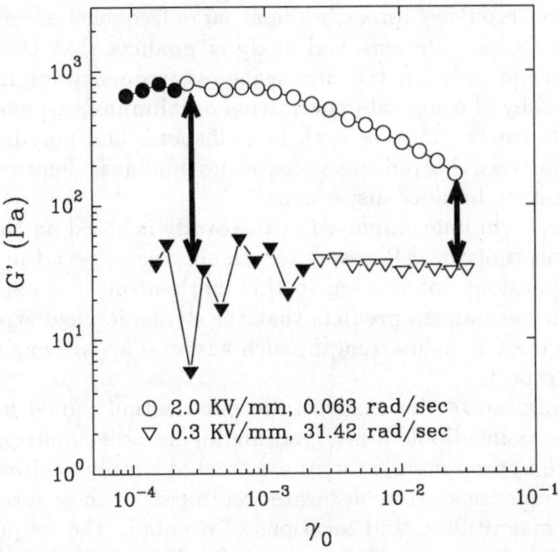

Figure 13: G' as a function of strain, γ_0, for specified electric field strength, E_0, and frequency, ω.

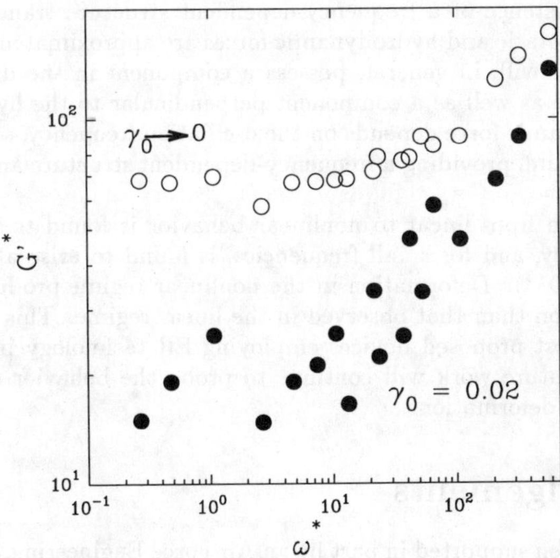

Figure 14: G'^* as a function of dimensionless frequency, ω^* , estimated at different strain amplitudes.

by electrostatic and repulsive forces, while at large frequencies, motion is dominated by hydrodynamic forces. Dimensional analysis predicts that the dimensionless response should depend only on the dimensionless frequency for linear deformation, which is confirmed by rheological experiments on alumina suspensions. Simulations predict that the linear response of both monodisperse and polydisperse suspensions should exhibit relatively sharp dispersions, while nonlinear deformation is capable of producing significantly broader dispersions.

The simulation technique employed in this study is based on the electrostatic polarization mechanism of the ER response. Hence, we expect the conclusions to be applicable to suspensions conforming to this mechanism. For example, the electrostatic polarization mechanism predicts that the dynamic yield stress should vary as the square of the electric field strength; such suspensions are expected to behave as described in this report.

The implementation of the electrostatic polarization model here treats electrostatic forces in the point-dipole limit, ignoring higher order multipole and multibody interactions, while hydrodynamic forces are treated simply as Stokes' drag, ignoring hydrodynamic interactions. More accurate treatments of these forces are expected to alter the moduli magnitudes, and locations of events in the frequency domain, but are not expected to alter the qualitative behavior discussed herein. For instance, this simplified model produces a relaxation mechanism embodied in a (dimensionless) frequency-dependent dynamic structure; the simplified forces produce a transition from nonaffine motion at small frequencies to affine deformation at large frequencies. However, the existence of a frequency-dependent structure transcends the level at which the electrostatic and hydrodynamic forces are approximated. The electrostatic force on a sphere will, in general, possess a component in the direction of the hydrodynamic force as well as a component perpendicular to the hydrodynamic force. As the hydrodynamic force depends on the oscillation frequency, so will the resultant force on each sphere, providing a frequency-dependent structure and a mechanism for relaxation.

The transition from linear to nonlinear behavior is found to depend on the oscillation frequency, and for small frequencies, is found to arise at very small strain amplitudes ($\leq 10^{-3}$). Deformation in the nonlinear regime produces a significantly different dispersion than that observed in the linear regime. This is particularly important since most proposed devices employing ER technology involve rather large deformations. Future work will continue to probe the behavior of ER suspensions under such large deformations.

Acknowledgements

This work has been supported in part by an Air Force Engineering Research Initiation Grant (RI-B-9207).

References

[1] Winslow, W. M., "Induced fibration of suspensions," J. Appl. Phys. **20**, 1137-1140 (1949).

[2] Klingenberg, D. J., F. van Swol, and C. F. Zukoski, "Dynamic simulation of electrorheological suspensions," J. Chem. Phys. **91**, 7888–7895 (1989).

[3] Marshall, L., J. W. Goodwin, and C. F. Zukoski, "Effect of electric fields on the rheology of nonaqueous concentrated suspensions," J. Chem. Soc. Faraday Trans. I **85**, 2785–2795 (1989).

[4] Klingenberg, D. J., "Simulation of the dynamic oscillatory response of electrorheological suspensions: demonstration of a relaxation mechanism," J. Rheol. **37**, 199–214 (1993).

[5] Klingenberg, D. J., F. van Swol, and C. F. Zukoski, "The small shear rate response of electrorheological suspensions: II. Extension beyond the point-dipole limit," J. Chem. Phys. **94**, 6170–6178 (1991).

[6] Klingenberg, D. J. and C. F. Zukoski, "Studies on the steady-state behavior of electrorheological suspensions," Langmuir **6**, 15–24 (1990).

[7] Ahn, K. H. and D. J. Klingenberg, "Relaxation of polydisperse electrorheological suspensions," in preparation (1993).

[8] Russel, W. B., D. A. Saville, and W. R. Schowalter, *Colloidal Dispersions* (Cambridge University Press, Cambridge, 1989).

[9] Von Hippel, A. R., (ed.) *Dielectric materials and applications; papers by twenty-two contributors*, (jointly published by the Technology Press of M.I.T. , Cambridge, and Wiley, New York, 1954).

[10] McLeish, T. C. B., T. Jordan, and M. T. Shaw, "Viscoelastic response of electrorheological fluids. I. Frequency dependence," J. Rheol. **35**, 427–448, (1991).

[11] Xu, Y.-Z and R.-F. Liang, "Electrorheological properties of semiconducting polymer-based suspensions," J. Rheol. **35**, 1335–1373, (1991).

[12] Yen, W. S. and P. J. Achorn, "A study of the dynamic behavior of an electrorheological fluid," J. Rheol. **35**, 1375–1384, (1991).

[13] Gamato, D. R. and F. E. Filisko, "High frequency dynamic mechanical study of an aluminosilicate electrorheological material," J. Rheol. **35**, 1411–1425, (1991).

[14] Korobko, E. V. and Z. P. Shulman, "Viscoelastic behavior of electrorheological fluids," in *Proceedings of the Second International Conference on Electrorheological Fluids*, edited by J. D. Carlson, A. F. Sprecher, and H. Conrad (Technomic, Lancaster, PA, 1990), pp. 3–13.

[15] Otsubo, Y., M. Sekine, and S. Katayama, "Electrorheological properties of silica suspensions," J. Rheol. **36**, 479–496, (1992).

[16] Shulman, Z. P., E. V. Korobko, and Y. G. Yanovskii, "The mechanism of the viscoelastic behaviour of electrorheological suspensions," J. Non-Newt. Fluid Mech. **33**, 181–196 (1989).

[17] Klingenberg, D. J., F. van Swol, and C. F. Zukoski, "The small shear rate response of electrorheological suspensions: II. Simulation in the point-dipole limit," J. Chem. Phys. **94**, 6160–6169 (1991).

Electric-Field-Induced Phase Separation in Electrorheological Fluids

Xin-Lu Tang, Ke-Qin Zhu, E Guan and Xiao-Ping Wu

Department of Modern Mechanics
University of Science and Technology of China
Hefei 230026, P. R. China

Abstract

The phase separation into a high-density and a low-density phase, which occurs after a strong electric field is applied to an ER fluid, is demonstrated in terms of the Coulomb interaction energy of the system. The phase separation is manifested by two-dimensional pattern observed in the field direction by means of a pair of transparent glass (TG) electrodes. The development of phase separation with an increasing field is recorded by an computer image processing system consisting of a CCD video camera. The patterns are irregular and their characteristic sizes are determined by two-dimensional auto-correlation approach. The dependency of those patterns on external electric field is analyzed based on statistical methods. Two critical fields are presented to characterize the electric-field-induced phase separation process in the ER fluid.

1. Introduction

The dramatic change in rheological properties of a suspension due to the application of an external electric field is known as the electrorheological (ER) response, whose potential value has recently attracted increasing attention [1]. An ER fluid, generally a suspension of small particles of high dielectric constant in a base liquid of low dielectric constant, can be switched from the fluid to the solid state when an applied electric field exceed a critical value. Both theoretical analysis and experimental observation have shown that under a high electric field, chainlike and columnar structures form parallel to the field direction. These structures arise from polarization forces produced by a mismatch in the dielectric constants of the disperse and continuous phases, and is responsible for the ER effect.

Halsey and Toor [2] argued that the thermodynamic ground state of ER fluids in a high electric field consists of a phase separation into a low-density and a high-density phase, and estimated that the columns have a width about $a(L/a)^{2/3}$, where a is the radius of dielectric particles and L the distance between two parallel electrodes. To understand the behavior of the phase separation in ER fluids, computer simulation methods seem to be an effective way, but fail to deal with a system consisting of a large number of particles [3]. Chen, Zitter and Tao investigated the phase separation and columnar structure by laser diffraction. Their observation was in the direction perpendicular to the electric field, and confirmed the predicted bct lattice structure within the columns [4].

Because branched columnar structures in ER fluids generally align with the electric field, the formation of the two-dimensional domains in the plane perpendicular to the field direction can supply important information to understand the phase separation. In the present paper, we study the phase separation based on the formation of the two-dimensional patterns observed in the field direction by means of a pair of transparent glass (TG) electrodes.

In the next section, we will demonstrate that phase separation into a low-density and a high-density phase should be an universal phenomenon in active ER fluids. And then in section 3 we will present our experimental method. In section 4, we will discuss the observation results and some statistical properties of the structure of ER columns. Furthermore, we also will demonstrate further application of TG electrodes to investigate the ER fluids.

2. Physical ground for the phase separation in ER fluids

In this section the physical ground for the electric-field-induced phase separation in ER fluids is analyzed in terms of the Coulomb energy of the system. Consider a model which has been widely used in the study of ER fluids. The model consists of spherical dielectric particles of radius a and dielectric constant ε_p suspended in a liquid of dielectric constant ε_f , with $\varepsilon_p > \varepsilon_f$. When a strong electric field **E** is applied, the spheres acquire dipole moments in the field direction, $\mathbf{p}_p = \alpha_p \varepsilon_f \mathbf{a}^3 \mathbf{E}_{loc}$, where $\alpha_p = (\varepsilon_p - \varepsilon_f)/(\varepsilon_p + 2\varepsilon_f)$, and $\mathbf{E}_{loc}$ is the local electric field. Similarly, the liquid is also polarized as $\mathbf{P}_f = 3\alpha_f \varepsilon_0 \mathbf{E}_{loc}$, where $\alpha_f = (\varepsilon_f - \varepsilon_0)/(\varepsilon_f + 2\varepsilon_0)$, according to the Clausius-Mosotti relation[5]. Denoting φ to be the local volume fraction of the particles in the ER fluid, the magnitude of the local polarization can be expressed as

$$P_{loc}^{erf} = \frac{\varphi}{\frac{4}{3}\pi a^3} p_p + (1-\varphi) P_f = \varepsilon_0 \chi_{loc}^{erf} E_{loc} \tag{1}$$

where the local electric susceptibility of the ER fluid is

$$\chi_{loc}^{erf} = \frac{3\alpha_p \varepsilon_f}{4\pi\varepsilon_0} \varphi + (1-\varphi) 3\alpha_f \tag{2}$$

Now consider local electric field $\mathbf{E}_{loc} = \mathbf{E} + \Delta\mathbf{E}$, where **E** is the external field and $\Delta\mathbf{E}$ is the field produced by polarization which is proportional to $\mathbf{P}_{loc}^{erf}$ according to the Lorentz's relation[5]. Introduce β such that

$$\Delta E = \beta \frac{P_{loc}^{erf}}{\varepsilon_0} \tag{3}$$

and then

$$E_{loc}=\frac{E}{1-\beta\chi_{loc}^{erf}} \tag{4}$$

$$P_{loc}^{erf}=\frac{\varepsilon_0\chi_{loc}^{erf}(\varphi)}{1-\beta\chi_{loc}^{erf}(\varphi)}E \tag{5}$$

The total Coulomb interaction energy of the system, including dipolar interaction and interaction between dipoles and external field, is given by

$$U_{co}=-\frac{1}{2}\int_V P_{loc}^{erf}\cdot E_{loc}dv=-\frac{1}{2}\varepsilon_0E^2\int_V\frac{\chi_{loc}^{erf}(\varphi)}{[1-\beta\chi_{loc}^{erf}(\varphi)]^2}dv \tag{6}$$

Next, we will demonstrate that in a fixed electric field, $\mathbf{U_{co}}$ will be always minimized when the ER system take the phase separation into a high-density and a low-density phase from a homogeneous density system. Rewrite eq.(6) in terms of probability function as

$$U_{co}=-\frac{1}{2}\varepsilon_0E^2V\cdot\int_0^1\rho(\varphi)F(\varphi)d\varphi \tag{7}$$

where $\rho(\varphi)$ is probability density function of the volume fraction and

$$F(\varphi)=\frac{\chi_{loc}^{erf}(\varphi)}{[1-\beta\chi_{loc}^{erf}(\varphi)]^2} \tag{8}$$

F(φ) is a convex function with respect to φ because

$$\frac{d^2F(\varphi)}{d\varphi^2}=\frac{\beta k^2(4+2\beta\chi_{loc}^{erf})}{(1-\beta\chi_{loc}^{erf})^4}>0 \tag{9}$$

where **k** is a constant determined by the dielectric constants of the particles and oil and generally $\beta > 0$ when a low-frequency electric field is applied on the ER system.

It is easy to prove that for a convex function F(φ) , the following inequality is always satisfied when $\rho(\varphi)\geq 0$ and $\int_0^1\rho(\varphi)d\varphi = 1$,

$$\int_0^1\rho(\varphi)F(\varphi)d\varphi\geq F[\int_0^1\rho(\varphi)\varphi d\varphi] \tag{10}$$

Defining $\Phi = \int_0^1 \rho(\varphi)\varphi d\varphi$ as the total volume fraction of dispersed particles in the ER fluid and noticing for a homogeneous density (HD) system

$$\rho(\varphi)=\delta(\varphi-\phi) \tag{11}$$

Therefore the total Coulomb energy of HD system is

$$U_{co}^{HD}=-\frac{1}{2}\varepsilon_0 E^2 V\int_0^1 \delta(\varphi-\phi)F(\varphi)d\varphi=-\frac{1}{2}\varepsilon_0 E^2 VF(\phi) \tag{12}$$

For a phase separation (PS) system, the total coulomb energy is

$$U_{co}^{PS}=-\frac{1}{2}\varepsilon_0 E^2 V\int_0^1 \rho(\varphi)F(\varphi)d\varphi<U_{co}^{HD} \tag{13}$$

It implies that when a fixed electric field is applied upon an ER fluid system, it will change from a HD system into PS system to minimize the total coulomb energy. In the next section, we will present an experimental system to observe the electric-field-induced phase separation phenomenon.

3. Experiment observation

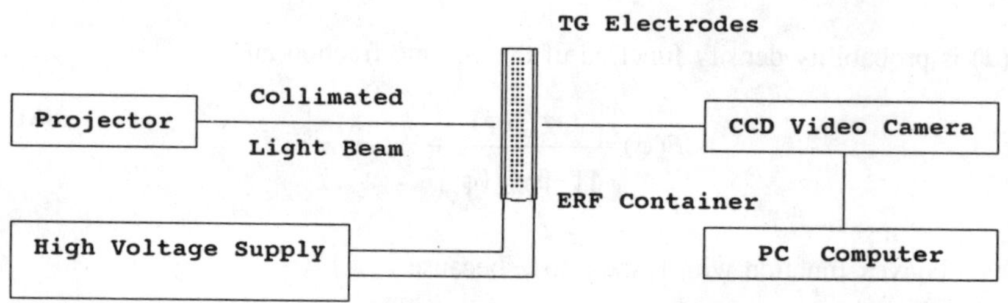

Figure 1 The Experiment Set-up

Our ER fluid consists of transformer oil containing fine diatomite particles (radius about 1 μm). The diatomite particles are mixed with oil to a volume fraction of about 0.2. The ER fluid is filled in a container, measuring 10 mm $\times$ 10 mm horizontally and 1.5 mm in vertical depth. Two sides of the container are a pair of parallel transparent-glass (TG) electrodes, on whose surfaces have the conductive film, which make it possible to observe the phase separation along the field direction. With no external electric field, the ER fluid in the cell is a homogeneous density suspension. When a high electric field is applied to the TG electrodes, the ER fluid separate into several high-density and low-density regions. Analyzing the phase separation patterns (observation area 5.5 mm × 5.5 mm) got from a CCD camera connecting a PC computer-image-processing system, one could obtain some statistical properties of the macrostructure changing with the electric field.

4. Results and discussions

(1). Figs. 2-7 are the pictures of the ER fluid taken under a series of increasing electric fields in the field direction at the room temperature, showing the development of the electric-field-induced phase separation in the ER fluid. Two critical fields $\mathbf{E}_{c1}$ and $\mathbf{E}_{c2}$ were found to describe this process. When **E** reached 0.8 kv/mm, the ER fluid had shown some ER effect such as the vanishing of the floatability. But it did not formed a macro-columnar structure, as shown in Fig. 3, until **E** rose by $\mathbf{E}_{c1}$ (1.41 kv/mm). Since then the columnar structure had formed and the whole picture separated into several high-density and low-density regions rapidly. With increasing field, those columns were becoming larger and larger, but still separated from each other, as shown

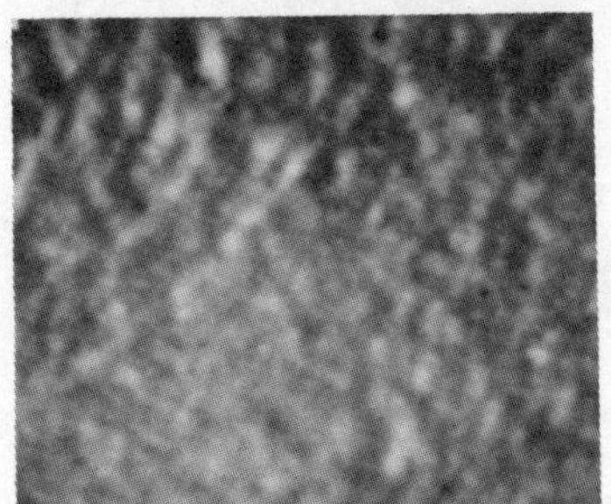

Figure 2 E=0.0 kv/mm

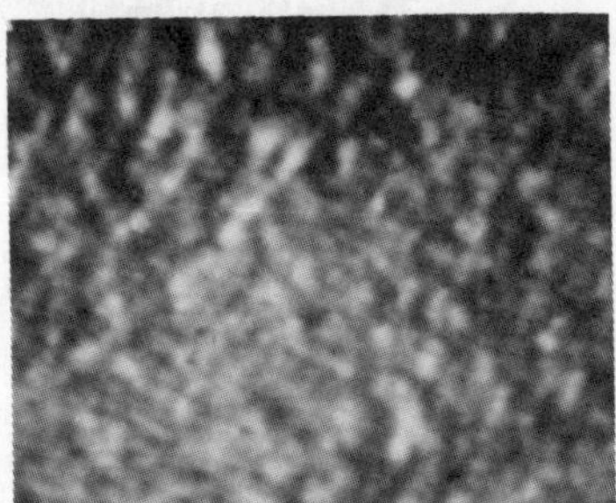

Figure 3 E=1.2 kv/mm

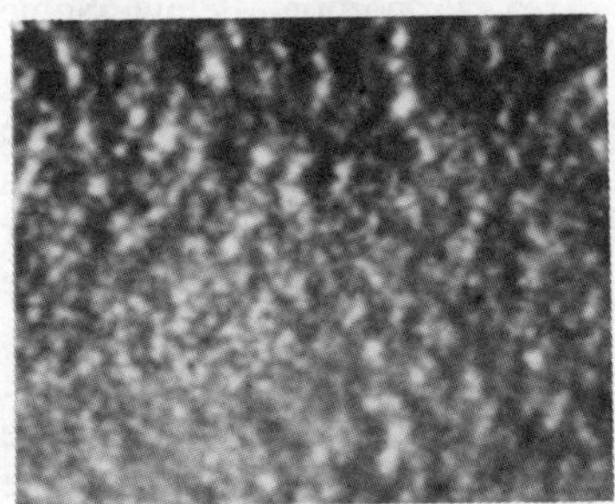

Figure 4 E=1.4 kv/mm

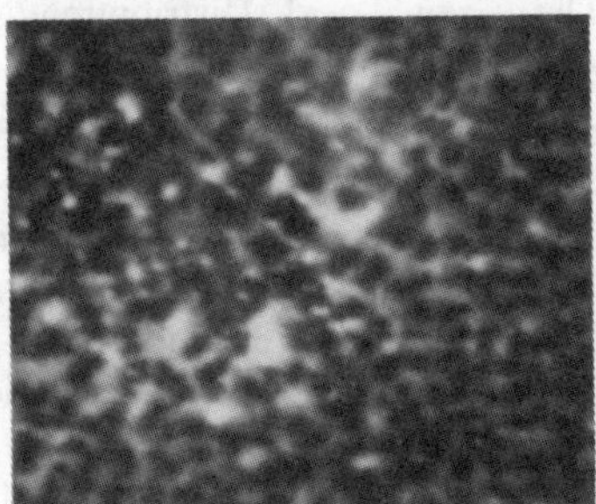

Figure 5 E=1.73 kv/mm

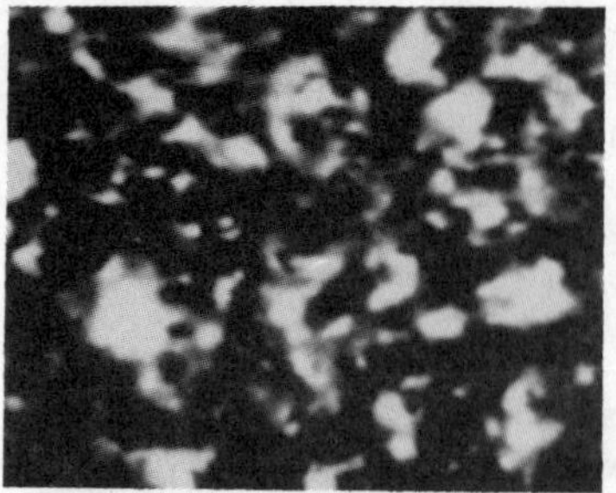

Figure 6 E=1.86 kv/mm

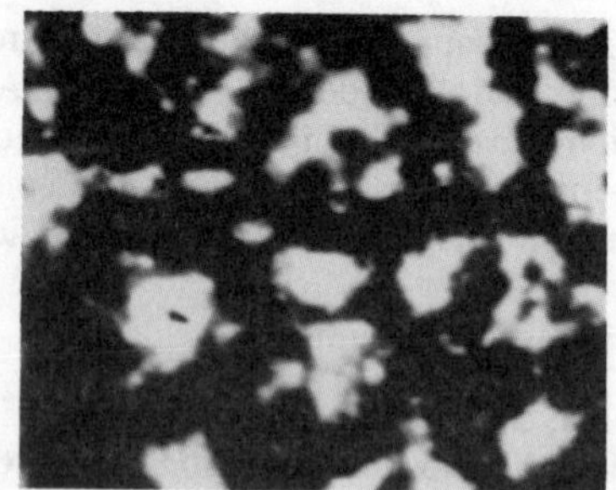

Figure 7 E=1.96 kv/mm

in Fig. 4 and 5. When **E** reached E_{c2} (1.86 kv/mm), those columns connected rapidly and formed a sieve-like structure, as shown in Fig.6 and 7.

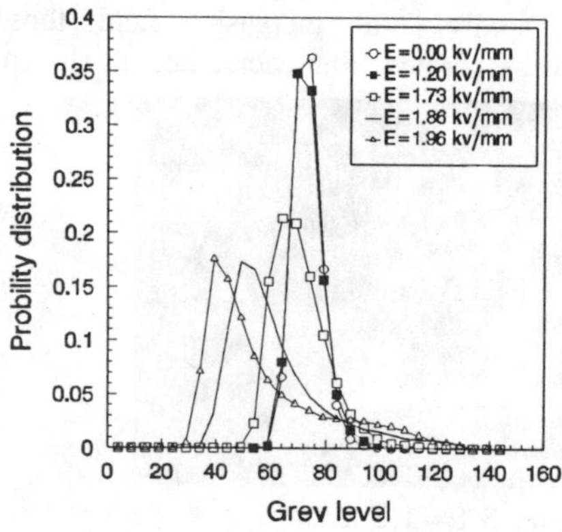

Figure 8 The Grey Level Distribution Changing with **E**

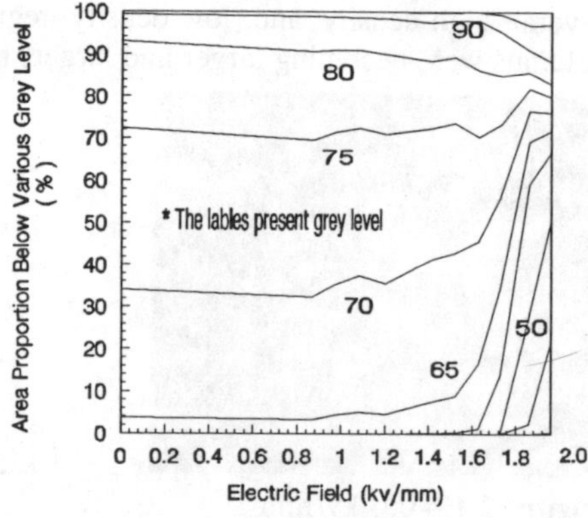

Figure 9 Area Proportion Below Some Specific Grey Levels Changing With **E**

(2) All above pictures were taken by a CCD camera. So the basic data we got were grey levels instead of the particles density in ER fluid. However, there should be a deterministic relation between those two physical values, which seemed to be a nonlinear monotone decreasing function in our experiment. The larger the density was, the less the grey level became at a same location.

Fig. 8 shows the grey level distribution of those pictures. The different curve indicates different electric field. As shown, when 0 kv/mm < **E** < 1.41 kv/mm, the distribution only changes slightly. But with **E** further increasing, the curves change greatly , whereas their peeks shift to the lower value and the distribution range become wider, which imply that the degree of the phase separation increases with **E**.

Fig. 9 demonstrates the area proportion changing with **E**, whose grey levels are below some specific values within the whole observation area. No obvious change has taken place in all these curves until **E** reaches the first critical field $\mathbf{E}_{c1}$. After **E** is over $\mathbf{E}_{c2}$, some of them level off again after a rapid change and the others are the new generated lower grey level curves that do not appear when **E** keeps a lower value, which implying that higher degree of the aggregation of the particles in ER fluid occur. If one divide the whole picture into three regions--region 1,2 and 3, whose grey levels are in the regimes [0, 62], (62, 86] and (86, 255] respectively--presenting the high-density, the intermediate-density and low-density phases, as shown in Fig. 10. he could find that the area of region 1 and 3 will increase but that of region 2 decreases while **E** rises.

(3) Consider the state of a fixed point in a picture to be determined by the region it occupies according to the mentioned region-dividing rule, i.e., if a point is within region 1, its state is called state 1. Let's gaze at a fixed point when **E** increases. Its state may transfer from one state to another or not. Making a census of all these points in a

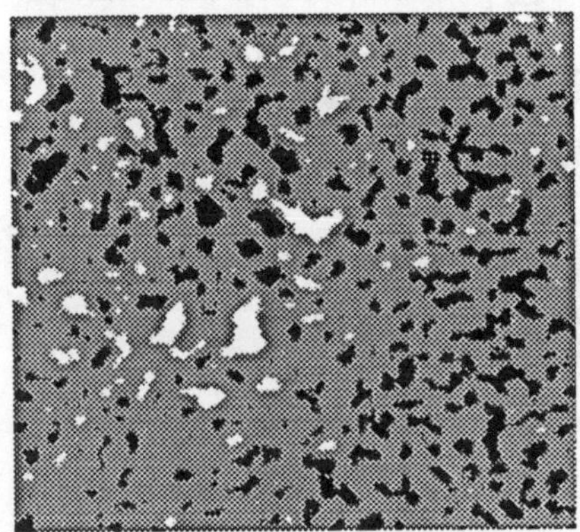

Figure 10 Region-dividing of Fig. 5 Black--Region 1, Grey--Region 2 and White--Region 3

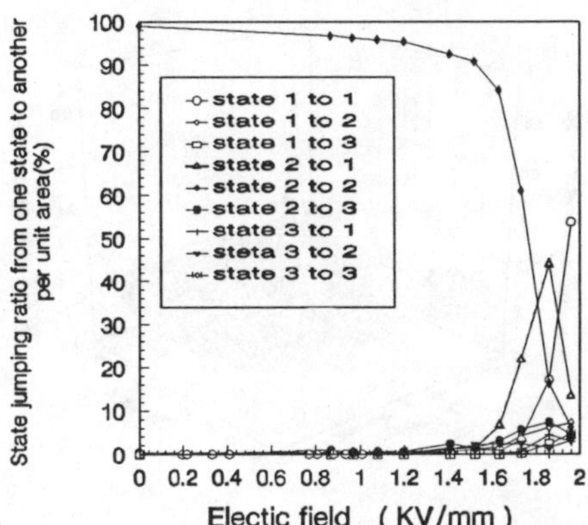

Figure 11 State jumping census with increasing **E**

observation area and denoting $\mathbf{T}_{ij}^{k}$(i, j=1,2,3) as the total number ratio of those points from state i to j when **E** increases from $\mathbf{E}_{k-1}$ to $\mathbf{E}_k$ per unit area, one could study the history of these three regions growth, i.e., where they are from and where they will come to. As shown in Fig. 11, two critical fields $\mathbf{E}_{c1}$ and $\mathbf{E}_{c2}$ exist and most area of region 1 and 3 come from state 2 . The fact that the jumping from state 1 to 3 or from state 3 to 1 occur occasionally at high **E** shows that reformation of the columnar structures has take place somewhere in the ER fluid.

(4) Calculate the auto-correlation function of those pictures as the following equation:

$$R_{cor}(\tau_x,\tau_y)=\frac{\int\int_{-\infty}^{\infty}[G(x-\tau_x,y-\tau_y)-<G>][G(x,y)-<G>]dxdy}{\sqrt{\int\int_{-\infty}^{\infty}[G(x,y)-<G>]^2dxdy\cdot\int\int_{-\infty}^{\infty}[G(x-\tau_y,y-\tau_y)-<G>]^2dxdy}} \quad \textbf{(14)}$$

Where G(x,y) is the grey level at location (x,y) and <G> is the average grey level of these pictures.

Define now the correlation length $\mathbf{L}^{cor}$ as the average radius of the root of the correlation peek, which can be considered as the characteristics scale of the ER macrostructure. $\mathbf{L}^{cor}$ is of meaningless When **E** is less than $\mathbf{E}_{c1}$ because no macrostructure has formed. After **E** exceed $\mathbf{E}_{c1}$, $\mathbf{L}^{cor}$ seems to grow with **E** in a exponential way in its initial growing stage in our experiment, as shown in Fig. 13.

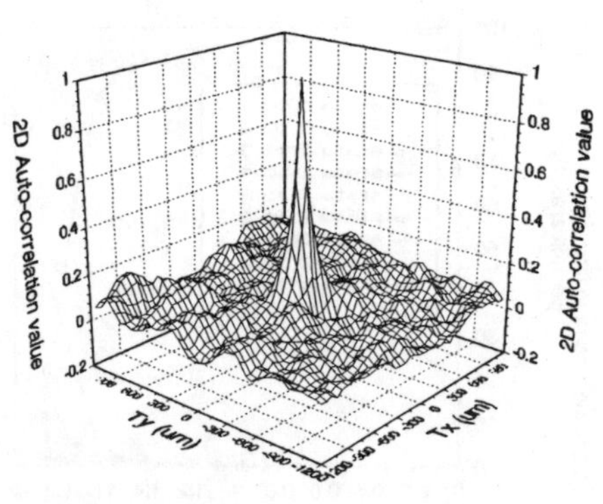

Figure 12 2D auto-correlation function of Fig. 4

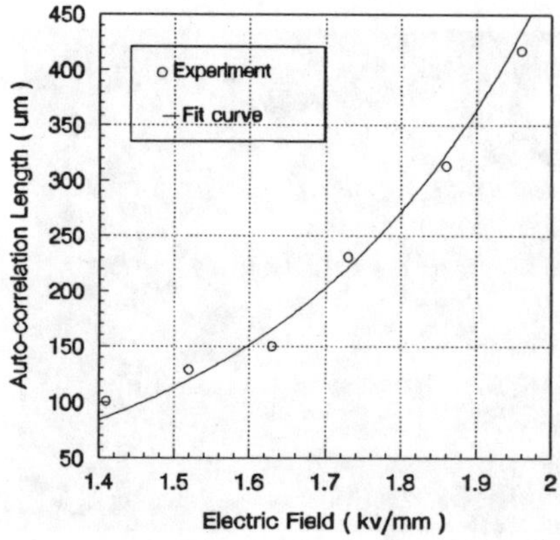

o Experiment
— Fit curve
Auto-correlation Length (um)
Electric Field (kv/mm)

Figure 13 The growth of the correlation length with **E** and the fitting curve.

The fitting curve in Fig. 13 is,

$$L^{cor}(E)=94.89e^{2.93(E-E_{c1})} \quad \textbf{(15)}$$

Where 1.41 kv/mm < **E** < 1.96 kv/mm.

(5) All above discussions are undertaken when **E** is less than 2.0 kv/mm. For some technical reason, further data were not obtained under higher **E**. However, one would also give some empirical predictions of the phase separation behavior of ER fluid when

higher **E** is applied:

a) Region 2 will disappear, and then region 1 will shrink again and region 3 will expand until the area ratio of region 1 to 3 reaches approximately the volume fraction of particles in ER fluid.

b) $\mathbf{L}^{cor}$ will decrease again in higher **E** after region 2 vanishes.

c) The macrostructure of ER fluid will change into some separated columns or even into one large cluster.

4. Conclusions and future works

In this paper, the electric-field-induced phase separation in ER fluid is investigated. A field-free ER fluid is assumed as a HD suspension consisting of a disperse and a continuous phases. It has shown that when an external electric field is applied upon an ER fluid, it will change from a HD state into a PS state to minimize the total Coulomb energy of the system. The phase separation into high-density and low-density regions can be discussed in terms of the two-dimensional patterns observed in the field direction because the disperse phase forms chainlike or columnar structures parallel to the field direction. To observe these patterns produced by electric-field-induced phase separation, two parallel arranged transparent electrodes are used. The pictures of these patterns are recorded by a CCD video camera connecting a PC computer-image-processing system, and have shown clearly the development of phase separation with increasing field. Two critical fields are presented in this experiment. The first one signs the lowest value of electric field to form the macro-columnar structure in ER fluid. Since then those structures are becoming larger and larger but still separated from each other. After the second critical field, the high-density domains connect rapidly and form a sieve-like structure. The patterns are irregular and their characteristic sizes are determined by two-dimensional auto-correlation approach and the correlation length grows with an increasing field. The relation between the correlation length and electric field are obtained, which seems to obey a exponential law.

The fundamental physics of ER fluid has just begun to be understood in recent decades. TG electrodes makes the behavior of ER fluid under direct observation. So, the application of TG electrode must have a promising prospect in further study of ER fluid.

(1) R. Tao [4] and his researchers predicted that for uniform spherical dielectric particles, the ER fluid formed a body-centered-tetragonal (bct) lattice structure, and later, this assumption was confirmed by means of the laser diffraction experiment. However, they failed to get a direct pattern along the field direction, which was as important as the other two directions. Apparently, the technique of TG electrodes may be a desired method for further experimental study of the microstructure of ER fluid as well as the macrostructure.

(2) With TG electrodes, flow visualization techniques may be employed, by adding some dilute fluorescent materials into ER fluid, to analyses the flow velocity field in

some ER devices, such as ER valves and ER clutches. The significant of this work lies in getting the firsthand data and verifying the theoretical model. Experiments can also be carried out to observe the phase separation phenomenon of ER fluid under flow conditions.

(3) The pictures got from our experiment seemed to have some similarity of the fractal analysis and percolation effect extensively studied in other critical phenomenon [6,7,8]. The finial shape of the macrostructure of ER fluid seemed to be dependent greatly on the electric field and its initial condition. However, for a certain ER fluid, it should exhibit same bulk properties under a fixed **E**. So , there must be some stable statistical values to characterize the global properties of the ER fluid behaviors, despite of the specific macro-shape formed under various initial conditions. Fractal analysis and percolation model may be a bridge to study ER fluid. It is clear that TG electrodes are necessities to get the statistical data in this case.

5. Acknowledge

The authors would like to thank Mr. Pang lin-yong, Mr. Zhang zheng and Mr. Lin wei-xing for developing some computer programs in this paper and their kindly help for preparing our experiments. Special thanks should be given to prof. Fu shao-jun and Dr. Wang bin for providing TG electrodes.

Reference

[1] Proceedings of the 3rd International Conference on ER Fluids, ed. by R. Tao (World Scientific, Singapore 1992)

[2] T. C. Halsey and W. Toor, Structure of ER fluids, Physical Review Letters, Vol. 65, No. 22 (1990) 2820-2823

[3] R. T. Bonnecaze and J. F. Brady, Dynamic simulation of an ER fluid, J.Chem. Phys., Vol. 96, No. 3 (1992) 2183-2202

[4] T. J. Chen, R. N. Zitter and R. Tao, Laser diffraction determination of the crystalline structure of an ER fluid, Physical Review Letters, Vol. 68, No. 16 (1992) 2555-2558

[5] M. A. Omar, Elementary Solid State Physics: Principles and Applications (Addison-Wesley, 1975)

[6] A. Sadig and K. Binder, Dynamics of the formation of two-dimensional ordered structures, J. of Stat. Phys., Vol. 35, Nos. 5/6 (1984) 517-585

[7] K. Binder, Percolation effects in the kinetics of phase separation, Solid State Communications, Vol. 34, (1980) 191-194

[8] D. Chowdhury, M. Grant and J. D. Gunton, Interface roughening and domain growth in the dilute Ising model, Phys. Rev. B, Vol. 35, No. 13 (1987) 6792-6795

SOME NEW EVIDENCE ON ELECTRO-RHEOLOGICAL MECHANISMS

Ruifeng Liang and Yuanze Xu
Polymer Physics Laboratory, Institute of Chemistry,
Academia Sinica, Beijing 100080, China

ABSTRACT

Some new evidence on electrorheological (ER) mechanisms for semiconducting polyacrylonitrile suspensions in silicone oil has been reported. Negative first normal stress difference was observed under an electric field. Effect of the particle size on ER properties was studied. The results have been related to the electrified fluid structure. Mason number was applied to dynamic responses by defining oscillatory Mason number M_n* (scaled as $\gamma\omega/E^2$) and it was found that the dynamic oscillation response and the steady shear response could be correlated with Cox Merz rule.

1. Introduction

As electrorheological (ER) effect was discovered in 1947, Winslow ascribed it to the field-induced fibrillation of small particles in the suspension[1]. Now it has been widely proved that the fibrous structure is a general characteristic for ER fluids[2-10], and interfacial polarisation mechanism is the major mechanism giving rise to ER responses[2-5,7,8,10-12]. Bingham plastic equation has been recognised as the most suitable rheological model for steady shear response of ER fluids[3,4,6,8,11]. Marshall, Zukoski & Goodwin[11] modified Bingham equation by introducing a dimensionless Mason number:

$$\eta/\eta_\infty = K/M_n + 1 \qquad (1)$$

where Mn is termed as Mason number, defined by:

$$M_n=\eta_\infty\dot{\gamma}/2\varepsilon_o\varepsilon_c(\beta E)^2 \qquad (2)$$

$$\rightarrow \text{scaled as } \dot{\gamma}/E^2$$

η is the suspension viscosity under field, η_∞ is the suspension viscosity at the high shear rate in zero field, K is the material constant, $\dot{\gamma}$ is the shear rate, E is the field strength, ε_o is the free space permittivity, ε_c is the dielectric constant of the medium, $\beta = (\varepsilon_p - \varepsilon_c)/(\varepsilon_p + 2\varepsilon_c)$ for the polarizability of the particles, ε_p is the dielectric constant of the particle. Mn is a measure of the relative importance of viscous shear forces to electric polarization forces acting on particles in the suspension. As seen from Eq.(1), the apparent viscosity could be correlated as a function of Mn. At small Mn, polarization forces dominate the ER response and the $\log(\eta/\eta_\infty) - \log M_n$ curve has a slope of -1 while at large Mn, viscous forces control the suspension structure and the term η/η_∞ approaches unity. In this paper, Mason number (M_n) will be applied to dynamic responses by defining oscillatory Mason number

$$M_n^* = \eta_\infty \gamma\omega / 2\varepsilon_0 \varepsilon_c (\beta E)^2 \quad \to \text{scaled as } \gamma\omega/E^2 \qquad (3)$$

where γ is the oscillation strain, ω is the dynamic frequency. The complex viscosity will be plotted against the oscillatory Mason number and compared with the apparent viscosity in steady shear.

On the other hand, the spanned strand model has often been used as a structural model for ER fluids[5]. The spanned strand model could be described as follows: Without electric field, an ER suspension behaves as usual suspensions do. In the presence of an electric field, the particles polarize and orient along the field direction to form fibrous bridge structures over the gap between electrodes. If a shear deformation is applied, the bridges will deform and incline to some angle. As the shear deformation proceeds, the bridges break down in the middle and flow occurs, but the broken chains still remain inclined. As the flow continues, the broken chains on the electrodes might build up again at a new position. In our previous paper[8], it was found that the rheological behaviors under different deformation modes could be approximately described by a rate insensitive stress, which was controlled only by the electric attractive force between the polarized particles. The characteristic ER responses were well explained using the spanned strand model.

In the present paper, some new evidence such as negative first normal stress difference, effect of the particle size and application of the dynamic Mason number will be discussed to further support the ER mechanism for semiconductive polymer-based suspensions.

2. Experimental

The dispersed phase used was the powder of a semiconducting polymer. The semiconducting polymer was made by controlled oxidation of polyacrylonitrile (PAN) fibre at high temperatures. Then, the powder was made by cutting these fibres in an agate ball mill. The different milling time was controlled to obtain particles with different sizes. The samples prepared from these powder were named in B-series as sample B1 - B5. The particle size in a suspension could be modified further by milling the prepared suspension in a colloid mill. This technology also provided the suspension with excellent stability against sedimentation. It was checked that the treatment in the colloid mill didn't bring about any contamination of the suspension. The samples treated with the colloid mill were called in C-series as sample C1 -C5. The particle size was determined with a Horiba CAPA particle analyser. The particle size data of B-series and C-series samples were given in Table 1. Typical particle size distribution was shown in Figure 1 for sample C2. The average size is about 5 μm. As seen from Table 1, different milling time corresponds to different particle size. The particle size decreases as the milling time increases. The effect of milling time on ER properties could be considered as an effect of the particle size.

Table 1. Particle size data of B-series and C-series samples

particles grade	mill type	milling time, hr	average size, μm	standard variation, μm	maximum size, μm
B - series:					
B1		0.58	7.52	4.62	20.0
B2	agate-	1.0	5.58	4.46	20.0
B3	ball mill	1.5	4.39	2.97	11.0
B4		2.0	3.46	2.14	9.0
B5		2.5	2.79	1.58	6.0
C - series					
C1		0.0	36.6	29.1	80.0
C2	colloid	1.0	4.62	3.20	10.0
C3	mill	1.5	4.26	2.54	10.0
C4		2.0	4.28	2.40	10.0
C5		2.5	4.11	2.25	9.0

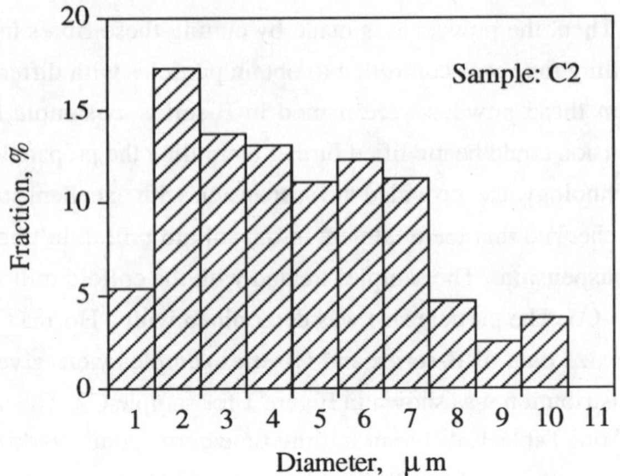

Fig.1. Typical particle size distribution for sample C2

The medium used was silicone oil with viscosity 46.9 mPa.s at 20^{0}C. The samples studied consisted of sample NC-2W with volume fraction 0.19, which was also studied in the paper[8], sample B1-B5 all with volume fraction 0.40 and sample C1-C5 all with volume fraction 0.30. In the preparation of sample B-series and sample C-series, the anhydride particles as well as the media were treated in an oven at 125^{0}C for hours to eliminate traces of water before mixing. Rheological properties such as yield stress, apparent viscosity, first normal stress difference and complex viscosity were measured by Rheometrics RMS-605 with parallel plates with gap =1mm and diameter 50mm at 25^{0}C. The main modification involved the insulation of the disc electrodes to the drive shaft and the torque transducer so that a DC electric field, up to 2.9 kV/mm in this study, could be applied to the sample. It should be mentioned that, because the shear rate is not homogeneous everywhere in the gap for the parallel plate geometry and no non-Newtonian corrections for stress were made, the measured material functions were apparent. In rheological measurements, different deformation modes might be applied to the sample. For the studied ER suspension, since the yield stress showed a rate insensitive dependence, electric field strength step change was often used at a given shear rate or at given strain and frequency in dynamic oscillation as well as shear rate sweep at a given electric field strength and dynamic frequency sweep at given strain and field strength. For more detail about the ER system and the experimental, see the previous paper[8].

3. Results and Discussion

3.1 Evidence 1: Negative Normal Stress

First normal stress difference (N1) data of ER suspensions was hardly mentioned in the literature[9]. In this study, special attention was paid to the measurement of first normal stress difference. It was found that the studied suspensions showed significant first normal stress difference with and without application of electric field. However, the electrified suspension was found to show negative normal stress while the suspension without voltage produced a positive normal stress. After having been checked carefully, the negative normal stress effect was understood to be a typical phenomenon for ER fluids although negative normal stress was also observed for liquid crystal systems[13] and some suspensions[14].

Figure 2 shows the first normal stress difference as a function of shear rate at different field strengths for sample NC-2W. A minus sign is added to make (-N1) positive, which will be used throughout the paper. As seen in Figure 2, the normal stress nearly remains constant over the range of the shear rate measured, but it increases with electric field strength. The curves for different field strengths run parallel to each other. The magnitude of the normal stress is found to increase approximately with the square of field strength to give an equation (-N1)=380 E^2, as shown in Figure 3.

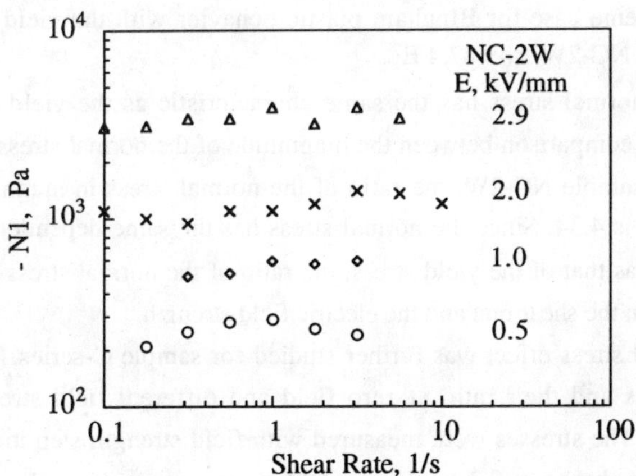

Fig.2. First normal stress difference as a function of shear rate at different field strengths for sample NC-2W

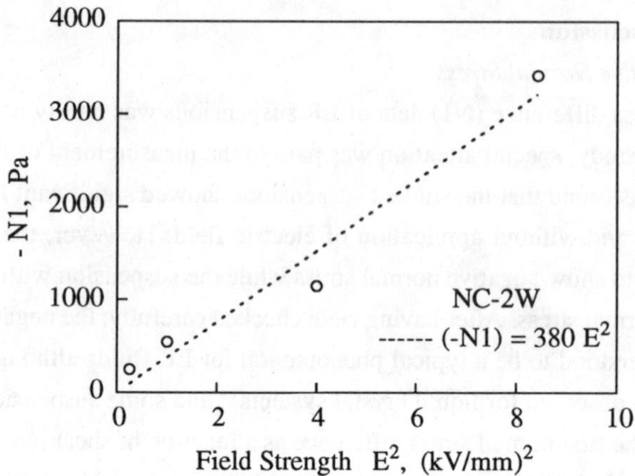

Fig.3. First normal stress difference as a function of field strength for sample NC-2W. The dashed line is the fit equation, $-N_1 = 380\ E^2$

In our previous paper[8], it was found that the shear stress showed an insensitive dependence on the shear rate but a strong dependence on the field strength. The yield stress was proportional to the square of electric field strength. The steady shear response could be considered as an extreme case for Bingham plastic behavior with the yield stress term dominant. For sample NC-2W, $\tau_y = 87.4\ E^2$.

Therefore, the normal stress has the same characteristic as the yield stress. It is meaningful to make a comparison between the magnitude of the normal stress and that of the shear stress. For sample NC-2W, the ratio of the normal stress in magnitude to the shear stress ($-N1/\tau_y$) is 4.34. Since the normal stress has the same dependency on field strength or shear rate as that of the yield stress, the ratio of the normal stress to the yield stress is independent on the shear rate and the electric field strength.

Negative normal stress effect was further studied for sample C-series. The normal stress, the yield stress and their ratio at zero field and different field strengths were presented in Table 2. The stresses were measured with field strength step increase from zero to 2.9 kV/mm at a shear rate 0.3 s^{-1}. Figure 4 - 6 were generated based on the data in Table 2.

Table 2. Normal stresses of ER suspension sample C-series

E, kV/mm		0.0	0.5	1.0	1.5	2.0	2.5	2.9
C1	$-N_1$, Pa	-8.30	21.4	248	710	1290	1970	2350
	τ_y, Pa	2.60	27.6	123	276	478	701	961
	$-N_1/\tau_y$	-3.20	0.77	2.03	2.58	2.69	2.81	2.45
	α, °	–	52.4	26.2	21.1	20.4	19.5	22.2
C2	$-N_1$, Pa	-20.4	40.0	210	623	1100	1740	2230
	τ_y, Pa	2.00	41.3	178	381	620	898	1140
	$-N_1/\tau_y$	-10.2	0.97	1.18	1.64	1.79	1.93	1.96
	α, °	–	45.9	40.3	31.4	29.2	27.4	27.0
C3	$-N_1$, Pa	-9.50	85.0	321	735	1240	1900	2270
	τ_y, Pa	3.80	44.3	169	354	585	832	1070
	$-N_1/\tau_y$	-2.50	1.92	1.90	2.07	2.11	2.28	2.11
	α, °	–	27.5	27.7	25.8	25.4	23.7	25.4
C4	$-N_1$, Pa	-31.9	-51.8	75.0	660	1220	1900	2380
	τ_y, Pa	2.10	48.2	176	355	570	800	991
	$-N_1/\tau_y$	-14.9	-1.07	0.43	1.86	2.14	2.37	2.40
	α, °	–	–	66.7	28.3	25.0	22.9	22.6
C5	$-N_1$, Pa	-76.4	8.50	320	697	1270	1840	2000
	τ_y, Pa	8.20	40.7	142	285	443	623	760
	$-N_1/\tau_y$	-9.32	0.21	2.25	2.44	2.87	2.95	2.63
	α, °	–	78.1	23.9	22.3	19.2	18.7	20.8

Figure 4 is the plot of the normal stress as a function of field strength E^2 for sample C-series. A common equation, $(-N1)=290E^2$, was got for all samples. Figure 5 is the plot of the yield stress as a function of field strength E^2 for sample C-series. In Figure 5,

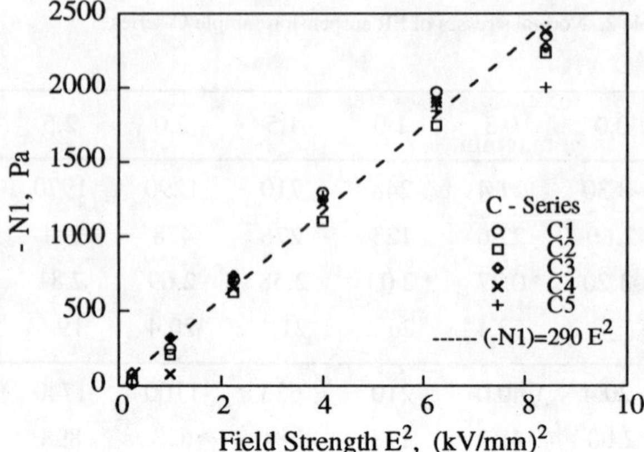

Fig.4. First normal stress difference as a function of field strength for sample C-series. The normal stress was measured at a given shear rate of 0.3 s^{-1}. The dashed line is the fit equation for all of the data, $- N_1 = 290\ E^2$

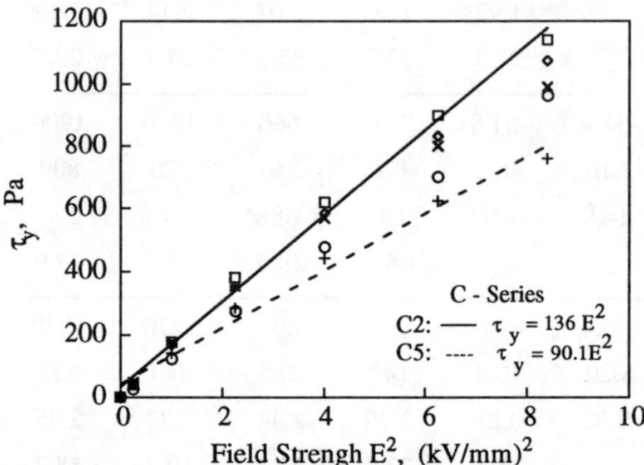

Fig.5. Yield stress as a function of field strength for sample C-series. Symbols as shown in Fig.4. The yield stress was measured at $\dot{\gamma}=0.3\ s^{-1}$. The dashed line is the fit equation for sample C5, $\tau_y = 90.1\ E^2$. The solid line is the fit equation for sample C2, $\tau_y = 136\ E^2$

however, there exists difference in field strength dependence among the samples, which represents effect of the particle size and will be discussed later. The fit equation for each sample could be obtained, for example, $\tau_y = 136\ E^2$ for C2 and $\tau_y = 90.1\ E^2$ for C5. The ratio of the normal stress in magnitude to the shear stress was also calculated for each sample. As seen from Table 2, most of the ratio values fall into the range of 2.0 - 3.0. The ratio value doesn't change very much indeed as the field strength increases. This fact is clearly shown in Figure 6 for sample C3.

The ER suspensions studied in this work were composed of fibrous powder dispersed in silicone oil. It was shown in Table 2 that in absence of electric field, the suspensions showed positive first normal stress difference as an usual fibrous rod-like suspension. Upon application of an electric field, the normal stress became negative. It would become positive again after the removal of the electric field. For the sake of completion, it should be mentioned that the positive first normal stress difference was also observed under electric field for high concentrated samples with volume fraction more than 0.40. In this case, the normal stress increased with the shear rate and was proportional to the field strength. The positive normal stress of high concentrated samples should be attributed to the dilatancy under the electric field, which was so strong that the electrified sample was found to squeeze out of the measuring cell.

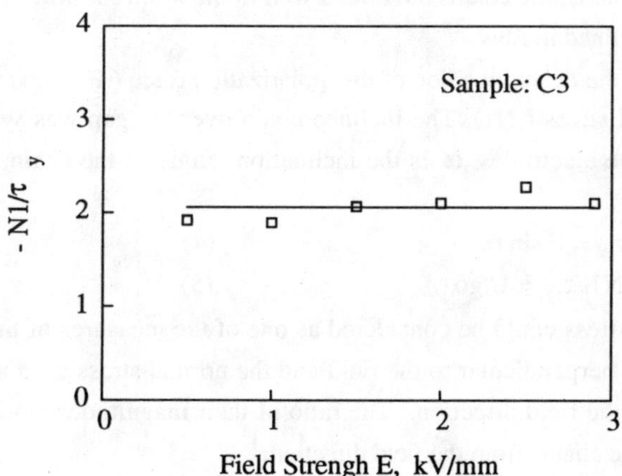

Fig.6. Ratio of first normal stress difference to yield stress as a function of field strength for sample C3. The solid line is the mean value, $-N_1/\tau_y = 2.07$

The ER suspension shows negative first normal stress difference under an electric field and the ratio of the normal stress in magnitude to the yield stress is independent on the shear rate and the electric field strength. For parallel plate geometry, negative normal stress implies existence of a compressing force which results in a tendency to pull the plates together. In the measurement of the electrified suspension, the large compressing force only could be ascribed to the polarization force. Based on the spanned strand model and the morphological observation result, a possible explanation might be proposed as follows: In the presence of the electric field, the particles polarize and orient along the field direction to form fibrous chain structures over the gap between the electrodes. Along these chains exists the interparticle electrostatic polarization force. Once a shear deformation is applied, the chains will deform to incline. At the same time, the deformed chains generate a resistance force against the shear deformation. That is the origin of the yield stress. On the other hand, the applied shear deformation makes the chains incline to a certain angle. Since the chains bind well with the electrodes, the produced deformation, that is, the inclination, will result in more or less elongation of the chains. However, the attractive polarization force transferred along the chains resists the chains to be elongated and pulls the chains to recover to their original positions. This gives rise to the negative normal stress. It should be noted that in the discussion two assumptions are used which are mainly based on the experimental observation results, that is, the chains are bound well to the electrode walls and the broken chains still remain inclined in flow.

Figure 7 shows the decomposition of the polarization force (F) into the yield stress (τ_y) and the normal stress (N1). The inclined chain over the gap was symbolised as a solid line between the electrodes. α is the inclination angle of the chain from the field direction. There exist:

$$\tau_y = F \sin \alpha \tag{4}$$

$$|N1|/\tau_y = 1/\mathrm{tg}\alpha \tag{5}$$

Therefore, the yield stress could be considered as one of the measures of the polarization force in the direction perpendicular to the field and the normal stress as a measure of the polarization force in the field direction. The ratio of their magnitudes would suggest the inclination angle of the chains from the field direction.

The inclination angle might be an important parameter related to the electrified fluid structure since the inclined chain structure is a general characteristic for flowing electrified suspensions. It could be understood as a critical inclination angle for chains to deform, at which the chains also break down in the middle as the applied shear deformation proceeds.

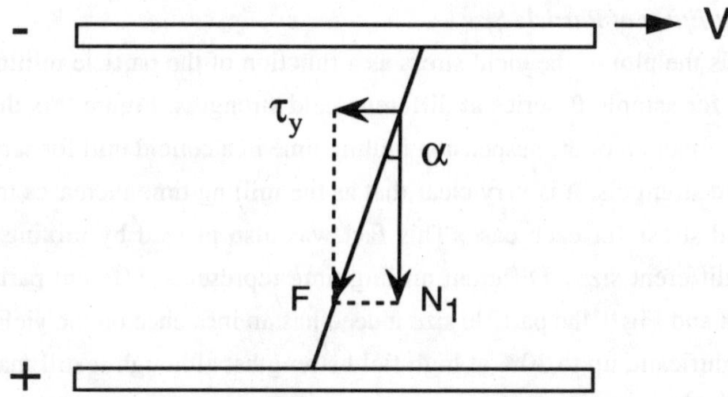

Fig.7. Schematic diagram for decomposition of polarisation force (F) into yield stress (τ_y) and normal stress (N_1). α is chain inclination angle. $-N_1 / \tau_y = 1/ \text{tg}\alpha$. If $-N_1 / \tau_y = 2 - 3$, $\alpha = 18 - 26^0$. A minus sign is added to make (-N1) positive since N1 is negative, in agreement with the text

According to the ratio $(-N_1)/ \tau_y$, this angle is independent on the shear rate and the electric field strength. It is mainly determined by the suspension composition parameters such as materials properties and particle concentration. For a particular suspension, the polarization force is certain. The yield stress increases with the inclination angle.

The $(-N_1) /\tau_y$ ratio is equal to 4.34 for sample NC-2W, implying that the deformed particle chains might incline 13^0 from the field direction. For sample C-series, the possible inclination angles of the chains at different field strengths were calculated and also listed in Table 2. Since the normal stress in value is 2-3 times the shear stress, thus, most of the particle chains over the gap should respectively incline 18.4^0 -26.5^0 from the field direction. The results agree well with the morphological observation. The inclination angle was indeed observed not to change much with the shear rate and the field strength. The angle ~ 20^0 seems to be a crucial position for flowing ER suspensions since it was also revealed by others[6,7,9,15,16] from computer simulation as well as experimental morphological observation. More work is desired, for example, to confirm the ratio of the normal stress to the yield stress from computer simulation of ER fluids[17] and to study effect of the concentration on the ratio.

3.2 *Evidence 2: Effect of Particle Size*

Figure 8 is the plot of the yield stress as a function of the particle milling time in an agate ball mill for sample B-series at different field strengths. Figure 9 is the plot of the yield stress as a function of the suspension milling time in a colloid mill for sample C-series at different field strengths. It is very clear that as the milling time increases there occurs a maximum yield stress for each case. This fact was also proved by mixing two type of particles with different sizes. Different milling time represents different particle size. As seen from Fig.8 and Fig.9, the particle size indeed has an influence on the yield stress. This influence is significant, up to 30% at high field strengths, although it still may be a small effect from an engineering viewpoint.

It should be mentioned that, in general, only when thermal forces dominate over polarization forces is there an influence of particle size on the ER effect [4,18]. The effect of particle size in the present study should be ascribed to the non-spherical fibrous geometry of the particles. This fact also means that the effect of the particle size includes both the influence of optimum size distribution of the particles and the influence of different length/diameter ratios (L/D < 10) of the particles. An influence of particle size on dynamic shear strength has also been reported at low shear rates for glass bead suspensions in silicone oil by Shih and Conrad[19].

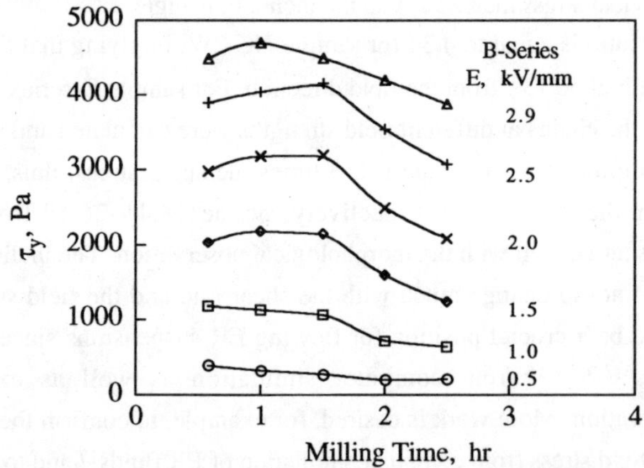

Fig.8. Yield stress as a function of particle milling time in an agate ball mill and applied field strength for sample B-series. The yield stress was measured at $\dot{\gamma}=0.3\ s^{-1}$

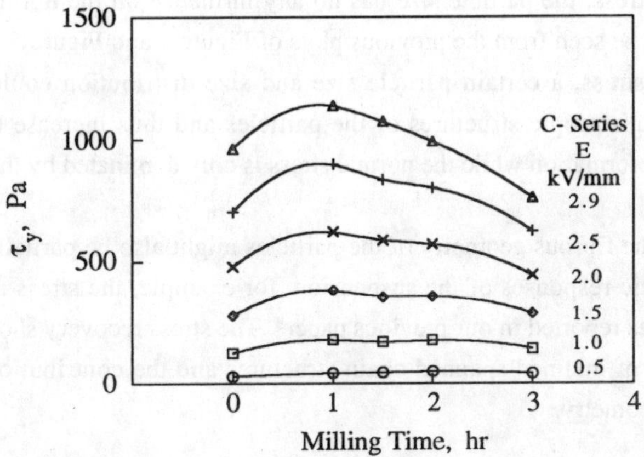

Fig.9. Yield stress as a function of suspension milling time in a colloid mill and applied field strength for sample C-series. The yield stress was measured at $\dot{\gamma}$=0.3 s^{-1}

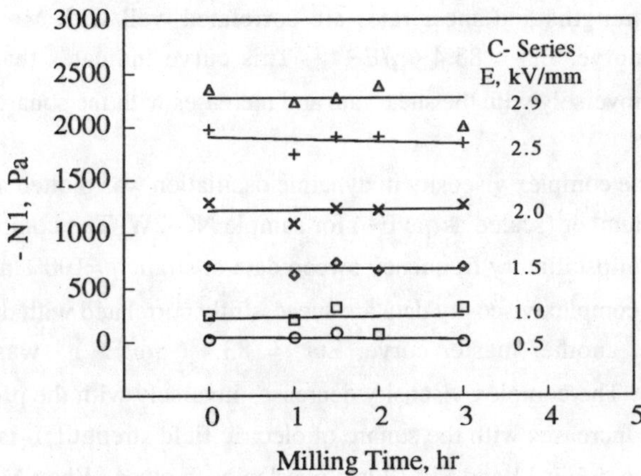

Fig.10. Normal stress as a function of suspension milling time in a colloid mill and applied field strength for sample C-series. The normal stress was measured at $\dot{\gamma}$=0.3 s^{-1}

Figure 10 shows the effect of the particle size on the normal stress. (-N1) was plotted against the milling time for sample C-series at different field strengths. Differently from the case of the yield stress, the particle size has no any influence on the normal stress. The difference can also be seen from the previous plots of Figure 4 and Figure 5. The reason is that for the yield stress, a certain particle size and size distribution could favours the formation of the anisotropic structures of the particles and thus increase the resistance against the shear deformation while the normal stress is only dominated by the interparticle polarization forces.

In addition, the fibrous geometry of the particles might also be partially responsible for the characteristic responses of the suspension, for example, the stress recovery after flow stop which was reported in our previous paper[8]. The stress recovery should be related to the reformation of inclined spanned chain structures and the contribution of the non-spherical particle geometry.

3.3 Evidence 3: Dynamic Mason Number

In Figure 11 the apparent viscosity in steady shear was plotted against Mason number (scaled as $\dot{\gamma}/E^2$) for sample NC-2W. The apparent viscosities were calculated from shear rate sweep data at different field strengths. It is found that all the steady shear viscosity data at different field strengths and shear rates are correlated well with Mason number to generate a master curve, Eta = 83.4 ($\dot{\gamma}/E^2$)$^{-1}$. This curve indicates that the apparent viscosity decreases inversely with the shear rate and increases with the square of the electric field strength.

In Figure 12 the complex viscosity in dynamic oscillation was plotted as a function of oscillatory Mason number (scaled as $\gamma\omega/E^2$) for sample NC-2W. The complex viscosities were calculated from oscillatory frequency sweep data at strain γ =100% and at different field strengths. The complex viscosity data are successfully correlated with dynamic Mason number. Therefore, another master curve, Eta* = 83.4 ($\gamma\omega/E^2$)$^{-1}$ was obtained for dynamic shear flow. The complex viscosity decreases inversely with the product of strain and frequency, and increases with the square of electric field strength. It is further found that these two curves in Fig.11 and Fig.12 can overlap each other. When Mn* equals Mn, the complex viscosity will equal the apparent viscosity, as described by Cox-Merz rule for polymer. Cox-Merz rule states that the complex viscosity approaches the apparent viscosity if the oscillatory frequency equals the shear rate[20].

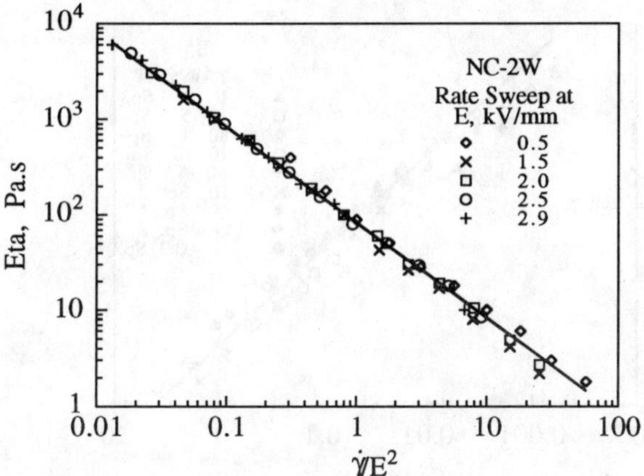

Fig.11. Apparent viscosity as a function of Mason number (scaled as $\dot{\gamma}/E^2$) for sample NC-2W. The apparent viscosities were calculated from shear rate sweep data at different field strengths. The solid line is the fit equation for all of the data, Eta = 83.4 ($\dot{\gamma}/E^2$)$^{-1}$

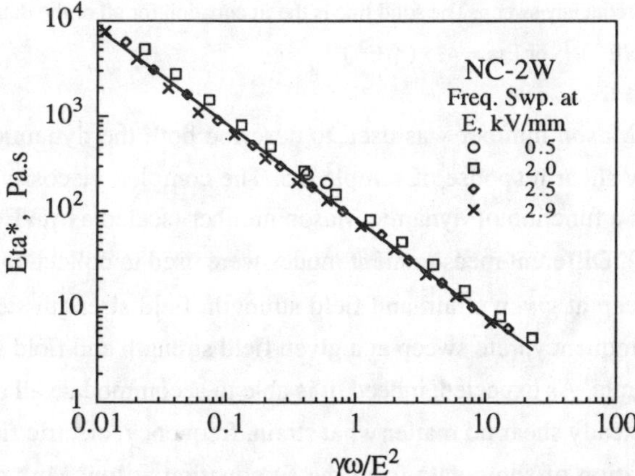

Fig.12. Complex viscosity as a function of dynamic Mason number (scaled as $\gamma\omega/E^2$) for sample NC-2W. The complex viscosities were calculated from oscillatory frequency sweep data at strain γ =100% and at different field strengths. The solid line is the fit equation, Eta* = 83.4 ($\gamma\omega/E^2$)$^{-1}$

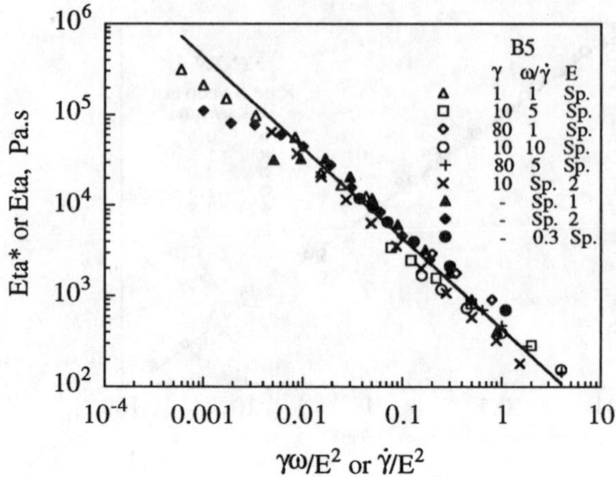

Fig.13. Complex viscosity or apparent viscosity as a function of dynamic Mason number (scaled as $\gamma\omega/E^2$) or Mason number (scaled as $\dot{\gamma}/E^2$) for sample B5. Open symbols for complex viscosity in oscillatory; Close symbols for apparent viscosity in steady shear. γ, % dynamic strain; ω, 1/s dynamic frequency; $\dot{\gamma}$, 1/s shear rate; E, kV/mm field strength; Sp., variable sweep, such as frequency sweep. The solid line is the fit equation for all of the data, Eta* = 435 ($\gamma\omega/E^2$)$^{-1}$ or Eta = 435 ($\dot{\gamma}/E^2$)$^{-1}$

In Figure 13 the Mason number was used to describe both the dynamic oscillation response and the steady shear response of sample B5. The complex viscosity or apparent viscosity was plotted as a function of dynamic Mason number (scaled as $\gamma\omega/E^2$) or Mason number (scaled as $\dot{\gamma}/E^2$). Different measurement modes were used to collect data, including dynamic frequency sweep at given strain and field strength, field strength step change at different strain and/or frequency, rate sweep at a given field strength and field strength step change at a given shear rate. As expected, indeed, it is able to accommodate all of the data in dynamic oscillation or steady shear no matter what strain, frequency, electric field strength, or shear rate. The deviation of some data from the fit equation at low Mn* or Mn range seems to be resulted from the yield behavior.

The significance of Mason number, in our understanding, is that it contains two most important field variables in ER responses: shear rate (or product of strain and frequency in dynamic oscillation) for a shear field and field strength for an electric field. Eta- Mn plot could summarize the rheological behaviors in steady shear while Eta* -Mn* plot could summarize the rheological responses in dynamic oscillatory. For the studied suspensions, the viscous flow resistance contributes little to the shear stress even at high Mn or Mn* range since the viscosity was not observed to approach the suspension viscosity at the high shear rate in zero field although the shear rate is up to 20 s^{-1}, comparable to that used in the paper[11].

It is surprising to find that the rheological responses of the electrified suspensions follow Cox-Merz rule. This result suggests not only that the polarization force scaled as E^2 dominates rheological quantities both in the dynamic oscillation and the steady shear, but also the microstructure within the suspension in dynamic flow might be the same as that in steady shear. One of the reasons for this equality is that the strains used in the oscillation measurements are all out of linear strain range. The linear strain amplitude range was found too small to be determined[8].

4. Conclusions

ER suspensions show large negative first normal stress difference under an electric field. For the semiconducting PAN suspensions, the characteristic shear rate independence and field strength dependence shows that the normal stress is also a rate insensitive stress which is controlled by interparticle electric polarisation forces. The yield stress could be considered as one of the measures of the polarization force in the direction perpendicular to the field and the normal stress as a measure of the polarization force in the field direction. The ratio of the normal stress to the shear stress, independent on shear rate and field strength, would suggest the inclination angle of the chains from the field direction. The particle size was found to have an important effect on the yield stress, but no influence on the normal stress due to the fibrous geometry of the particles which favours the formation of the anisotropic structure. Mason number(M_n) could be applied to dynamic responses by defining dynamic oscillatory Mason number M_n* (scaled as $\gamma\omega/E^2$). For the studied suspension, the characteristic ER behaviours in steady shear and dynamic oscillation are correlated with Cox Merz rule and, indeed, are dominated only by the electrically induced interparticle forces. The spanned strand model captures most features of ER phenomena, and can be used to explain the results very well. On the other hand, the results also provide evidence to the model itself.

5. Acknowledgments

We would like to acknowledge the Natural Science Foundation of China for the support for this work. We would also like to thank Mr Q.Zhang and Ms Y.Li for their help in rheological measurements and the referee for his suggestion to define dynamic Mason number. R.L. was most grateful to Dr M.Mackley at University of Cambridge, the British Society of Rheology and the Conference Committee for funding his attendance at the 4th Int.Conference on ER Fluids, Feldkirch, Austria, July 1993 where this paper was presented.

6. References

1. W.M.Winslow, *J.Appl.Phys.*, 20 (1949) 1137
2. Yu F.Deinega and G.V.Vinogradov, *Rheol.Acta*, 23 (1984) 636
3. H.Block and J.P.Kelly, *J.Phys.D, Appl.Phys.*, 21 (1988) 1661
4. T.C.Jordan and M.T.Shaw, *IEEE Trans.Elect.Insul.*, 24 (1989) 849
5. D.J.Klingenberg, F.van Swol and C.F.Zukoski, *J.Chem.Phys.*, 91(1989) 7888
6. D.J.Klingenberg and C.F.Zukoski, *Langmuir*, 6 (1990) 15
7. D.J.Klingenberg, D.Dierking and C.F.Zukoski, *J.Chem.Soc. Faraday Trans.*, 87 (1991) 425
8. Y.Xu and R.Liang, *J.Rheol.*, 35 (1991) 1355
9. T.C.Jordan, M.T.Shaw and T.C.B.Mcleish, *J.Rheol.*, 36 (1992) 441
10. K.M.Blackwood, H.Block, P.Rattray, G.Tsangaris and D.N.Vorobiov, Presented at the *4th Int.Conference on ER Fluids*, Feldkirch, Austria, July 1993
11. L.Marshall, C.F.Zukoski and J.W.Goodwin, *J.Chem.Soc.Faraday Trans.*, 85 (1989) 2785
12. D.J.Klingenberg, F.van Swol and C.F.Zukoski, *J.Chem.Phys.*, 94 (1991) 6160; 6170
13. C.E.Chaffey and R.S.Porter, *J.Rheol.*, 29 (1985) 281
14. K.W.Lew and C.D.Han, *J.Rheol.*, 27(1983) 263
15. M.Whittle, *J.Non-Newt.Fluid Mechanics,* 37 (1990) 233
16. G.L.Gulley and R.Tao, Presented at the *4th Int.Conference on ER Fluids*, Feldkirch, Austria, July 1993
17. K.R.Rajagopal and A.S.Wineman, *Acta Mechanica*, 91(1992) 57
18. P.M.Adriani and A.P.Gast, *Phys.Fluids*, 31(1988) 2757
19. Y.H.Shih and H.Conrad, Presented at the *4th Int.Conference on ER Fluids*, Feldkirch, Austria, July 1993
20. W.P.Cox and E.H.Merz, *J.Polym.Sci.*, 28(1958) 619

Effects of Electrode Morphology on the Electrorheological Response

Panos V. Katsikopoulos and C.F.Zukoski

Department of Chemical Engineering

Univerity of Illinois, Urbana, Illinois 61801.

INTRODUCTION

The electrorheological (ER) phenomenon has been known for more than 40 years (1) and is defined as the rapid and reversible change in the rheological behavior of a suspension under the action of large electric fields [O(1kV/mm)]. Specifically, orders of magnitude increases in the apparent viscocity and/or the development of some type of a yield stress in the otherwise Newtonian suspensions are observed for polarizable particles in a low permittivity continuous phase (2). The applied electric field induces dipolar interactions between the particles resulting in the development of particle structures which span the electrode gap. The formation and shear degradation of this structure is responsible for the dramatic change in the rheological properties of the ER fluids. Even though some understanding of the phenomenon has been achieved the details of the particle interaction potential are not fully understood. As a result, the range of stress transfer properties possible is not yet known. Due to the fact that the stresses that are achieved with currently existing ER fluids are lower than what is required and their power consumption is still high, commercial devices are not yet available. The objective of much current work is to increase the stress without increasing the current levels. However, a recent study (3) concerned with the feasibility of automotive devices based only on stress transfer considerations, concludes that damping applications become feasible provided increases in the yield stresses of a factor of two to three are realized.

A simple but successful model for the steady state rheological behavior of ER fluids is the Bingham plastic where the important parameter is the dynamic yield stress. Experimental and modelling efforts have focused on the measurement and prediction of the dynamic yield stress and the elucidation of its dependence on several parameters. The stress depends on the strength of the interparticle forces and on the structure of the suspension. The effect of external and material parameters like the electric field strength (E), the type of the field and its frequency (f), the permitivity (ε) and conductivity (σ) of the dispersed particles (p) and the continuous phase (c) and the volume fraction of the dispersed phase (φ) has been addressed. The majority of the work up to the present has dealt with increasing the interparticle forces by increasing the ratio $\alpha=\varepsilon_p/\varepsilon_c$ or E (2). Of course, there is an upper limit to which the electric field strength can be increased imposed by dielectric breakdown strength considerations as well as by power consumption ones. At the same time substances with large permittivities do not seem to produce large ER effects (2).

Another approach to increasing the dynamic yield stress lies in altering the suspension structure. When the elctric field is turned on, strands of particles form, aligned with the electric field, which extend across the electrode gap. In both simulations and experiments, shear is observed to produce dense clusters or strands. The resulting structures are observed to be insensitive to E or material properties. A transition from thin to thick clusters near a volume fraction of 0.3 is observed in simulations to accompany a saturation of the dynamic yield stress (i.e., the yield stress at a fixed field strength becomes insensitive to volume fraction). The saturation of the stress has been attributed to the formation of these thick strands. Bonnecaze and Brady (4) demonstrate that the saturation in the yield stress (static or dynamic) is observed (except for non touching conducting particles) even in an idealized, single particle width structure due to the increased importance of interchain interactions at high volume fractions. In addition, calculations by Kraynik show (5) that double particle width strands are less effective in transfering stress than single particle width stands. These calculations clearly indicate that the stress is sensitive to suspension microstructure and the maximum stress is transfered if all the particles are located in single particle width strands.

As a result, if an experimental manifestation of the idealized structure could be created, one would expect increases in stress transfer. To achieve steady state increases, these structures must be stabilized against the coarsening effect of the shearing motion; a process which requires localizing strands at particular points on the electrode surfaces. The experimental realization of this approach requires the existence of a no slip boundary condition (B.C.) for the particles at the electrode surfaces. This may be achieved by fixing particles to the surface, increasing the strength of particle-electrode interactions or increasing electrode roughness. In this paper, we explore the effects of creating positions on the electrode which have locally enhanced electric fields. This is accomplished by placing grooves in the otherwise smooth electrodes.

The effect of rough electrodes on the ER response has seen little systematic investigation. The only exception lies in the studies of Monkman (11) where he attached to one or both electrode surfaces one or more layers of a cotton fabric. The increases in the measured torque were small; at best they fell short of even a factor of two increase over the smooth electrode case. No visualization of the suspension during the shear was attempted. As a result, the actual effect of the cotton fabric is not clear. In summary, the importance of the boundary condition (which depends on the particle-wall interaction and the particle and electrode surface roughness) in the study of ER fluids has been largely overlooked experimentally and in simulations has either been completely ignored (8) or treated as a stick-slip situation (7). On the otherhand, to achieve the maximum yield stresses, a no slip B.C. has been assumed in all the theoretical calculations (4-6).

In magnetorheology the effect of the particle/shearing surface boundary condition has been addressed by Bossis and Lemaire (9,10) who have studied the apparent static yield stresses of colloidal polystyrene particles with magnetic inclusions dispersed in water. The magnitude of the yield stress was found to depend on the material used for the electrode (ferromagnetic or paramagnetic) and its surface roughness. Depending on the electrode material, two orders of magnitude differences in yield stress are reported for the same forces acting between the particles. Bossis and Lemaire concluded that the particle-wall interaction is very important. Unfortunately, in electrorheology we do not have the ability to change the magnitude of the particle-wall interaction without also altering the magnitude of the interparticle force. However, we can vary the average roughness of the surface of the electrodes, or engrave a particular pattern on them and therefore create sites with larger electric field strength. In this work, we report preliminary results towards this end.

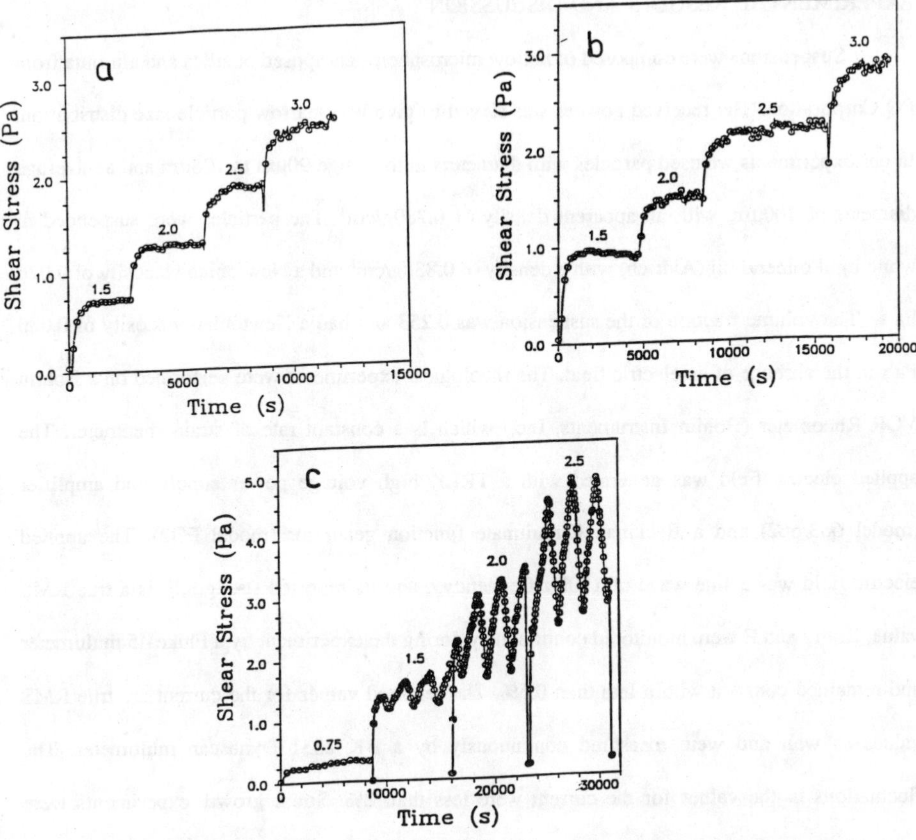

Figure 1a. The shear stress for the given ER fluid as a function of time at different electric field strengths using the typical smooth electrodes. The electric field strengths are given in kV/mm.

Figure 1b. The shear stress for the given ER fluid as a function of time at different electric field strengths using smooth electrodes with grooves parallel to the velocity direction. The electric field strengths are given in kV/mm.

Figure 1c. The shear stress for the given ER fluid as a function of time at different electric field strengths using smooth electrodes with grooves perpendicular to the velocity direction. The electric field strengths are given in kV/mm.

EXPERIMENTAL RESULTS AND DISCUSSION

Suspensions were composed of hollow microspheres composed of silica and alumina from PQ Corporation. The received powder was sieved to give very narrow particle size distribution. In our experiments we used particles with diameters in the range 90μm to 106μm and an average diameter of 100μm, with an apparent density of 0.780g/cm^3. The particles were suspended in white light mineral oil (Aldrich) with a density of 0.838g/cm^3 and a Newtonian viscosity of 0.026 Pa s. The volume fraction of the suspension was 0.253 and had a Newtonian viscosity of 0.080 Pa s in the absence of an electric field. The rheological experiments were performed on a Bohlin VOR Rheometer (Bohlin Instruments, Inc.) which is a constant rate of strain rheometer. The applied electric field was generated with a TREK high voltage power supply and amplifier (model 663/662) and a Beckman Circuitmate function generator (model FG2). The applied electric field was a sine wave of 100Hz frequency,f, and its reported strength,E, is a true RMS value. Both f and E were monitored continuously during the experiment by a Fluke 45 multimeter and remained constant within less than 0.5%. The reported values for the current are true RMS values as well and were measured continuously by a BK 2831 Dynascan multimeter. The fluctuations in the values for the current were less than 2%. Stress growth experiments were carried out at the lowest possible shear rate of 1.995x10^{-3} s^{-1} at different electric field strengths. The value for the shear stress at this low shear rate can be considered as a lower bound for the dynamic yield stress for the suspension. [The frequency was choosen to eliminate particle circulation between electrodes.] We used three different sets of stainless steel parallel plates : a) with smooth surfaces i.e., what is traditionally used, b) with smooth surfaces and regularly spaced grooves parallel to the velocity direction (i.e. concentric circles) and c) with smooth surfaces and regularly spaced grooves perpendicular to the velocity direction (i.e., along diameters of the circular plates). All the plates had a 30mm diameter and were held at a surface to surface separation of 1mm. The number of the grooves was 5 and 8 for case b and case c respectively. The width and the depth of the grooves were in both cases 300μm and 150μm pespectively.

Typical stress growth curves for the ER fluid are shown in figure 1. The approach to the steady state value is monotonic for smooth electrodes and for electrodes with grooves parallel to the velocity direction. The steady state values are approximately the same within experimental uncertainty. The same is observed for the stress growth curve for the electrodes with grooves perpendicular to the velocity direction for E=750 V/mm. On the contrary, for these electrodes at higher electric field strengths the shear stress demonstrates a very repeatable and periodic behavior and achieves an "oscillatory steady state".

The steady state values for the three different geometries are plotted in figure 2. For the first two geometries we report the steady state value and for the third the average stress over a number of periods. The typical dependence on the square power of the electric field is clearly demonstrated meaning that the suspension behaves like a classical ER fluid. Furthermore, the shear stress measured for the electrodes with the grooves perpendicular to the velocity direction is about two times larger, at all electric field strengths, than that for the two other sets of electrodes. In addition, the current passed through the suspension is the same for all the different electrodes (figure 3).

Understanding the reasons behind the different rheological behaviors is aided by visualization of the suspension during the shearing experiments. Observation with lens and a video camera shows that for smooth electrodes, clusters densify upon shearing and respond to deformation through a strand breaking and reformation mechanism. In addition, many clusters show a stick-slip motion where the strands remain intact but slide along one electrode surface. In system 3 where electrodes have grooves cut perpendicular to the velocity direction, there is accumulation of the particles next to the groove edges. These particles never move from their positions whereas particles "touching" the electrode surface far from the grooves are observed to slide along the surface of the electrode and occasionaly leave it. Single particle width strands are not observed for any of the three electrode configurations investigated. For system 3, walls of particles two or three particles thick next to the grooves and all along their edges are observed. In addition, columnar structures with one end lying next to a groove but the other being a particle attached to the smooth part of the opposite electrode have been observed. These columns strain under the action of the shear field but they do not reach the maximum strain. Instead the strand end lying on the smooth electrode slides (i.e., the strand has a kind of a stick-slip motion). In comparison to this behavior when only one end of a strand is pinned by a groove, when both ends of a strand are particles next to a groove, the strand never slips, instead it gets fully strained and always breaks approximately in the middle. Visualization also shows that the minimum in the shear stress coinsides with the grooves being in phase whereas the maximum corresponds to the grooves being out of phase. These observations suggest that under many circumstances, the weak link in the polarization force induced chains of particles lies at the electrode particle interface rather than between particles within the chain.

By intuition, we expect the disturbance to the electric field due to the grooves to be most prominent close to the edge of a groove. Preliminary finite element calculations for our particular groove geometry show that the electric field can increase up to a factor of two next to the edge compared to everywhere else along the electrode surface or across the gap. Also the relative position of the grooves at the two plates has no effect on the electric field (i.e., the disturbances appear to be decoupled). Based on this information we argue that the periodic stress trace for system 3 is not due to a change in the electric field experienced by the particles.

After eliminating variations of the electric field as being at the origin of the periodic stress curves, we are drawn to an explanation that this behavior of the stress is the result of the existence of two types of sites of attraction (strong and weak) on the surface of the electrode (which are allowed to interact). These two sites give rise to stands that deform 1) according to a no slip B.C. and 2) a stick-slip B.C respectively. The minimum stress corresponds mainly to the contribution of all the stands not associated with the edges of the grooves since at that time they are weakly strained. The difference between the maximum and the minimum stresses is a measure of the stress associated with the strands that are associated with the grooves and is an indication measure of the true dynamic yield stress of the suspension if all particles were incorporated into strands which experienced a no slip B.C..

The period of the stress pattern is inversely proportional to the imposed shear rate for a range of shear rates from 1.995×10^{-3} to 1.979×10^{-2} s^{-1}. Visual observations during oscillation experiments confirm the distinction of two types of strands indicated by the shearing experiments. Of course, the parallel plate geometry with the variable shear rate and distance between the grooves across the plate complicates the analysis. Experiments with polyaniline particles, of a much smaller and polydisperse size and irregular shape, in silicone oil have shown the same periodic behavior for the stress.

The lack of increases seen with electrodes with grooves cut in the direction of the velocity field are then understood in two ways. First, that the accumulation of particles at the edges of the grooves creates coarse structures at the very regions of the no slip B.C. and as a result what is gained because of its realization is lost because of the lower stress transfer capability of thicker structures. The reason that these electrodes do not show a periodic stress trace is that the two sites (strong, weak) do not interact with each other. Alternatively, it can be understood in terms of the lack of a blocking of particle slip. While grooves perpendicular to the velocity direction create local potential minima which resist particle displacements in the velocity direction, grooves parallel to the velocity direction offer no such blocking action. Thus while columns of particles are expected with grooves parallel to the velocity direction, these grooves offer no mechanism to produce a no slip boundary condition.

CONCLUSIONS AND FUTURE WORK

From the above described experiments on our generic ER fluid we can arrive at the following conclusions:

1. Single particle width strands will be very hard to realize experimentally even under conditions that provide for very localized structures.

2. Localization of the suspension structure can be achieved and can be visually observed when grooves are cut on the surface of the electrode perpendicular to the velocity direction.

3. The existence of two types of sites on the surface of the electrode with: a) strong attraction next to the edges of the grooves (stick) and b) weak attraction at the smooth surface (stick-slip) seems to be supported by visualization of the suspension under shearing and

oscillation experiments.

4. Columnar structures with both ends lying near the edges of grooves do not slip but strain and break in the middle of the gap. On the contrary, strands associated with the smooth part of the electrode follow a stick-slip pattern of deformation (motion). Therefore, increases in the value of the yield stress can be effected by reducing slip of the particles at the wall.

5. The above observations provide a reasonable explanation for both the twofold increase in the stress and its periodic nature with time.

6. Creating a pattern of grooves perpendicular to the velocity direction on the surface of an otherwise smooth electrode results in twofold increases in the dynamic yield stress while keeping the current the same.

Continuing investigations are underway in order to evaluate; 1) the influence of a uniform roughness on the surface of the electrodes of the order of the particle size (rough surfaces of roughness of about 100μm and 50μm), 2) the effect of varying the number of the grooves, and, 3) the effect of the particle diameter. We also plan to perform experiments with concentric cylinders where many complications that are present in the parallel plates are absent.

ACKNOWLEDGEMENTS

We would like to acknowledge the financial support of the National Science Foundation. We would also like to acknowledge Dr. Ralph Goodwin for the finite element calculations he performed for us.

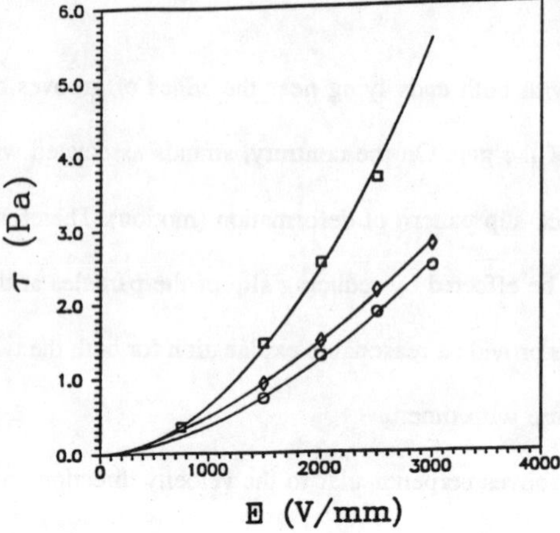

Figure 2. Average values for the shear stress as a function of the electric field strength. The squares are for the electrodes with grooves perpendicular to the velocity direction. The diamonds are for the electrodes with the grooves parallel to the velocity direction and the circles for the smooth electrodes.

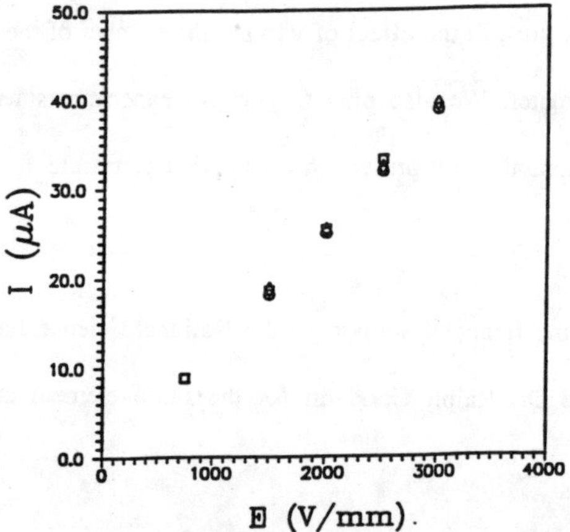

Figure 3. Average values for the current as a function of the electric field strength. The squares are for the electrodes with grooves perpendicular to the velocity direction. The diamonds are for the electrodes with the grooves parallel to the velocity direction and the circles for the smooth electrodes.

REFERENCES

1. W.M.Winslow "Induced Fibration of Suspensions", *J. Appl. Phys.*, **20**(12), 1137 (1949).
2. A.P.Gast and C.F.Zukoski "Electrorheological Fluids as Colloidal Suspensions", *Adv. Colloid Interface Sci.*, **30**, 153 (1989), H.Block and J.P.Kelly, "Electro-Rheology", *J.Phys. D: Appl. Phys.*, **21**, 1661 (1988).
3. D.L.Hartsock, R.F.Novak, and G.J.Chaundy "Electrorheological fluid requirements for automotive devices", *J.Rheol.*, **35**(7), 1305 (1991).
4. R.T.Bonnecaze and J.F.Brady "Yield Stresses in Electrorheological Fluids", *J. Rheol.*, **36**(1), 73 (1992).
5. A.M.Kraynik, R.T.Bonnecaze and J.F.Brady "Electrically Induced Stresses in ER Fluids: The Role of Particle Chain Structure", *Proceedings of the International Conference on Electrorheological Fluids*, editor Rongjia Tao, Southern Illinois University at Carbondale, October 15-16, 1991.
6. D.J.Klingenberg and C.F.Zukoski "Studies on the Steady-Shear Behavior of Electrorheological Suspensions", *Langmuir*, **6**, 15 (1990).
7. D.J.Klingenberg, F.van Swol and C.F.Zukoski, "The Small Shear Rate Response of Electrorheological Suspensions. I. Simulation on the Point-Dipole Limit and II. Extension Beyond the Point-Dipole Limit", *J. Chem. Phys.*, **94**(9), 6160 and 6170 (1991).
8. R.T.Bonnecaze and J.F.Brady "Dynamic Simulation of an Electrorheological Fluid", *J. Chem. Phys.*, **96**(3), 2183 (1992).

9. G.Bossis and E.Lemaire "Yield Stresses in Magnetic suspensions", *J.Rheol.*, 35(7), 1345 (1991).

10. E.Lemaire and G.Bossis "Yield Stress and Wall effects in Magnetic Colloidal Suspensions", *J.Phys. D: Appl. Phys.*, 24, 1473 (1991).

11. G.J.Monkman "Addition of Solid Structures to Electrorheological Fluids", *J. Rheol.*, 35(7),1385 (1991).

III. PROPERTIES

VISCOELASTICITY OF ELECTRORHEOLOGICAL FLUIDS: ROLE OF ELECTROSTATIC INTERACTIONS

J. M. GINDER and L. C. DAVIS
Research Laboratory, Ford Motor Company
P.O. Box 2053, M.D. 3028
20,000 Rotunda Drive
Dearborn, MI 48121-2053 (U.S.A.)

ABSTRACT

The dynamic stress response of a model electrorheological (ER) fluid system to oscillating shear strains has been measured. The model system, a suspension of barium titanate particles in a nonaqueous solvent, exhibits substantial ER activity only in alternating (ac) electric fields. A novel double-modulation technique, in which the component of the oscillating shear stress modulated at twice the frequency of the applied ac electric field is detected, was used to decompose the stress response into a modulated component that results directly from field-induced electrostatic interactions between the suspended particles and an unmodulated component that is the result of other interparticle interactions. Use of this technique reveals that the viscoelastic response of the barium titanate suspensions at small strains results from nonelectrostatic interparticle interactions, while for larger strains the directly field-induced interactions dominate. The measured field-dependent moduli and static yield stresses are compared with the results of finite-element calculations on ordered arrays of dielectric spheres embedded in a dielectric medium. The good agreement of the predictions of this model with these and other experiments demonstrates that the electrostatic model successfully describes the field-induced behavior of this class of ER fluids.

1. Introduction

Designing electrorheological (ER) fluids that will enable the development of adaptive or active electromechanical devices has motivated ER research from its inception[1-4]. Examples of these devices include automotive shock absorbers[5,6] and engine mounts[7,8]. However, some automotive and other applications of ER fluids are not feasible at present because of the relatively small field-induced shear stresses exhibited by current ER fluids[9]. A primary goal of ER materials research is thus to formulate fluid systems with higher field-induced shear stresses or, failing that, to understand the physical constraints that prevent their development.

Many rheological studies of ER fluids have focused on steady shear flows at relatively high shear rates, in part because these flow conditions are encountered in most devices. Nonetheless, the nature of the 'gels' or weak solids formed by the application of electric fields in ER fluids can be profitably studied by their dynamic viscoelastic response at relatively small shear strains and strain rates. For example, the electric field dependence of the viscoelastic moduli can indicate which interparticle interactions are responsible for the rigidity of ER gels. Unfortunately, little agreement on the field dependence has emerged: several studies have identified strongly electric-field-dependent shear moduli[10-12], while another found almost no dependence of the moduli on the magnitude of the field[13]. Remarkably large storage moduli have been observed both in parallel-disc[13] and bending-beam[14,15] experiments. Recent analytical[12,16] and simulation[17] results suggest that the frequency-dependent viscoelasticity of ER gels is a sensitive probe of their structure. The nonlinear viscoelastic response of ER gels for stresses beyond the static yield stress has been characterized by Fourier analysis[18] and by parametric modeling[19]; the rate of dissipation of mechanical energy has also been measured in this regime[20].

Important physical insight into electrorheological phenomena was provided by the dipole model[21-24], in which the primary role of the applied electric field $\mathbf{E_0}$ in an ER fluid is to induce a polarization $\mathbf{P} = \alpha\mathbf{E_0}$ in each suspended particle. The resulting dipole-dipole interparticle forces, which vary as $\mathbf{P} \cdot \mathbf{P} \sim E_0{}^2$, result in well-known ER phenomena like the formation of particle chains, columns[24], and more complicated gel structures[25,26], as well as the appearance of a nonzero shear modulus and yield stress[10,27,28]. While the observed E^2 dependence of the structure formation rate[29] and the yield stress in some ER fluids directly follow from this model, it severely underestimates the magnitude of the measured stresses[30].

One source of this discrepancy is the approximate nature of the dipole model, i.e., its neglect of higher-order multipole fields and local-field effects caused by the interactions of neighboring particles in these dense suspensions[28,30]. Approaches that have been used to obtain the induced electrostatic forces between particles in ER solids include series expansions[31,32] and finite element analyses[33-35]; in the latter, the decrease in the stored electrostatic energy produced by a shear strain applied to ordered arrays of particles is equated to the mechanical work required to deform the array quasistatically at constant applied voltage, allowing the electrostatic contribution to the shear stress and shear modulus to be calculated[31-36]. As these approaches presume quasistatic deformations of the gel structure, they are applicable to field-induced stresses at small strains and strain rates; to treat ER phenomena at high shear rates, they must be amended to include hydrodynamic forces[36,37]. They also do not include other interparticle interactions[38] such as screened Coulomb repulsive and van der Waals attractive forces[39].

These electrostatic models can be used to predict the electrorheological behavior -- the field-induced shear modulus and static yield stress -- of dense suspensions of dielectric particles in a dielectric fluid. Both the simple dipole and more sophisticated electrostatic models suggest that the strongest ER fluids are those with the largest ratio of particle to fluid dielectric constant. A *quantitative* test of this design rule for ER fluids is highly desirable, both to assess our present understanding of the physics of electrorheology and to suggest how one can make better ER fluids.

The dynamic viscoelasticity of a model anhydrous ER fluid at very small strains and strain rates is reported here. In order to extract that portion of the shear stress response that results directly from the induced particle polarization in this system, a double-modulation technique -- detection of the shear-stress modulation induced by sinuosoidal electrical excitation at a frequency higher than the mechanical oscillation frequency -- has been used. The field-controlled moduli and yield stresses obtained from this technique were compared with the predictions of a finite-element model for the stored electrostatic energy; the agreement obtained demonstrates that the electrically-induced contribution to the shear stress does indeed result from the storage of electrostatic energy in ER fluids consisting of dielectric particles in insulating liquids. However, the field-independent moduli observed at very small strains are apparently due to other interparticle interactions that are indirect consequences of field-induced structure formation.

In the following, the techniques used for the viscoelastic measurements will be detailed and results for a model fluid will be presented. The modeling techniques used to predict the field-induced moduli will be outlined and results will be compared with experiment.

2. Experimental Methods

The model fluids utilized in this study were composed of barium titanate particles (TAM Ceramics) having $K_p = 2000$[40], suspended at volume fractions $\phi = 0.20$ in dodecane (K_f=2.0) or silicone oil ($K_f = 2.5$). The notation that $\epsilon_i = K_i\epsilon_0$, where i = p or f for particle or fluid respectively, and where ϵ_i is the permittivity, K_i is the relative permittivity, and ϵ_o is the permittivity of free space, 8.85417 x 10^{-12} F/m, is used throughout this paper. The particles were dried *in vacuo* at temperatures greater than 120 °C for eight hours prior to dispersion in the suspending fluids. The equivalent spherical diameter of the particles, 1.3 ± 0.3 μm, was estimated by optical microscopy; the particles were irregular polyhedra in shape. As described by

Garino, et al.[40], a small amount of the surfactant OLOA 1200 (Chevron) was added to the dodecane suspensions to enable particle dispersal. The fluid samples were thoroughly mixed prior to the rheological measurements. All data were taken at room temperature, T = 22 ± 1 °C.

A controlled-strain rheometer system was constructed around a microstepping motor (Parker Compumotor DR 1008B) having 5.1×10^5 steps per revolution. The angular position of the motor was determined by an optical lever: the deflection of a He-Ne laser beam from a mirror attached to the motor was detected by a linear position sensor (Advanced Photonix SD 1166). The motor position was controlled by a closed-loop proportional feedback loop that compared the measured angular position with an analog reference signal derived from a function generator (Exact 627), allowing sinuosoidal or other oscillatory motion to be generated. The maximum frequency at which the motor oscillations could be controlled was about 20 Hz. Both Couette (cup and bob) and parallel disc fixtures constructed of stainless steel were utilized; the motor drove the grounded outer cup in the Couette geometry or the lower disc in the rotating disc geometry.

The torque transmitted through the sample to the nominally stationary fixture was measured using a strain-gauge-based sensor (Sensotec QWLC-8M) electrically isolated from the fixture by a length of machinable glass-ceramic. High voltage was applied to the upper fixture using a high voltage amplifier (Trek 10/10) driven by a second signal generator (Wavetek 75); high voltage sinuosoidal waveforms with a 20 kHz bandwidth were available. The angular compliance of the upper fixture/torque sensor assembly, ~ 1250 N-m/rad, was measured using a second optical lever, establishing the proportionality of the torque signal and the angular position of the fixture, and thus allowing the measurement and feedback control of the *net* angular deflection of the sample. Knowledge and control of the net shear strain was essential to characterize the ER gel viscoelasticity at small strain amplitudes[41]. The torsional resonance of the Couette fixture was identified at ~ 400 Hz. As Lou et al. have pointed out, this resonance places an upper limit on the frequency of the ac fields that can be applied without involving complications due to fixture dynamics[42]; consequently, the excitation frequencies used in this study were well below it.

Time series of the strain, stress, applied high voltage, and current were stored and averaged in a digitizing oscilloscope (Tektronix 11401). The in-phase (storage, σ_1') and quadrature (loss, σ_1'') components of the peak total stress $\sigma_1^{(tot)} \equiv [(\sigma_1')^2 + (\sigma_1'')^2]^{1/2}$ were measured via phase-sensitive detection using a lockin amplifier (Stanford Research Systems SR 850). The subscript '1' denotes the component of the stress at the fundamental frequency f_s of the oscillating applied strain; higher odd harmonics appear in the nonlinear viscoelastic regime[18]. Because *ac* electric fields are applied to the sample as it is sheared, a portion of the stress is modulated at twice the frequency f_e of the electric field: in effect, the polarization-induced interparticle forces responsible for the ER effect are turned on and off each time the electric field is reduced to zero. The double-modulation technique involves detecting this superimposed stress modulation with a second lockin amplifier; the components of that amplitude which varied with the applied strain were detected by a third lockin tuned to f_s. In order to minimize systematic errors due to the sensitivity of the sample rheology to excitation and strain history, the samples were 'conditioned'[13] by the application of high fields (typically 3 kV/mm peak) and nonzero strain amplitudes ($\sim 10^{-3}$) for several minutes before data taking at each new field.

3. Experimental Results

Electrification of ER fluids containing highly polarizable particles like barium titanate or other metal oxides by high *dc* electric fields leads to surprisingly small field-induced shear stresses[43,44]. However, *ac* excitation at frequencies as low as a few Hz gives rise to substantial ER activity[40,45]! These observations are significant, although not widely appreciated; they demonstrate that particle polarization is moderated by the particle-fluid conductivity mismatch in dc fields, and that only in ac fields does the dielectric mismatch control the polarization and the resulting ER activity[33,34,46]. They also reflect that electrophoretic deposition of the particles in dc fields can interfere with ER activity[47].

An example of the time-dependent shear stress induced by the simultaneous application of a f_s = 1 Hz sinuosoidal shear strain of peak amplitude $\gamma_{peak} = 10^{-2}$, Fig. 1a, and a f_e = 10 Hz sinuosoidal electric field of peak amplitude 3 kV/mm, Fig. 1b, is shown in Fig. 1c. Several interesting features can be discerned from these data. First, a portion of the shear stress is clearly modulated at *twice* the frequency of the applied electric field: this portion results directly from field-induced electrostatic interactions. To justify the observed frequency doubling, consider a time-dependent electric field $\mathbf{E}(t)$ that varies as $\mathbf{E_0}\cos(2\pi f_e t)$. Assuming that the interparticle polarization-induced force varies as $E^2(t)$, the resulting shear stress varies as $E_0^2\cos^2(2\pi f_e t) = E_0^2(1 + \cos(4\pi f_e t))/2$. Thus the frequency at which the directly field-induced stress varies is $2f_e$, as is found experimentally. Another possible source of stress modulation at $2f_e$ is the periodic distortion of the ER gel structure induced by the simultaneous application of shear and ac electric fields[48]. Second, inspection of Fig. 1c reveals that an underlying component of the shear stress is not modulated at $2f_e$. This unmodulated component may be plausibly ascribed to other interparticle

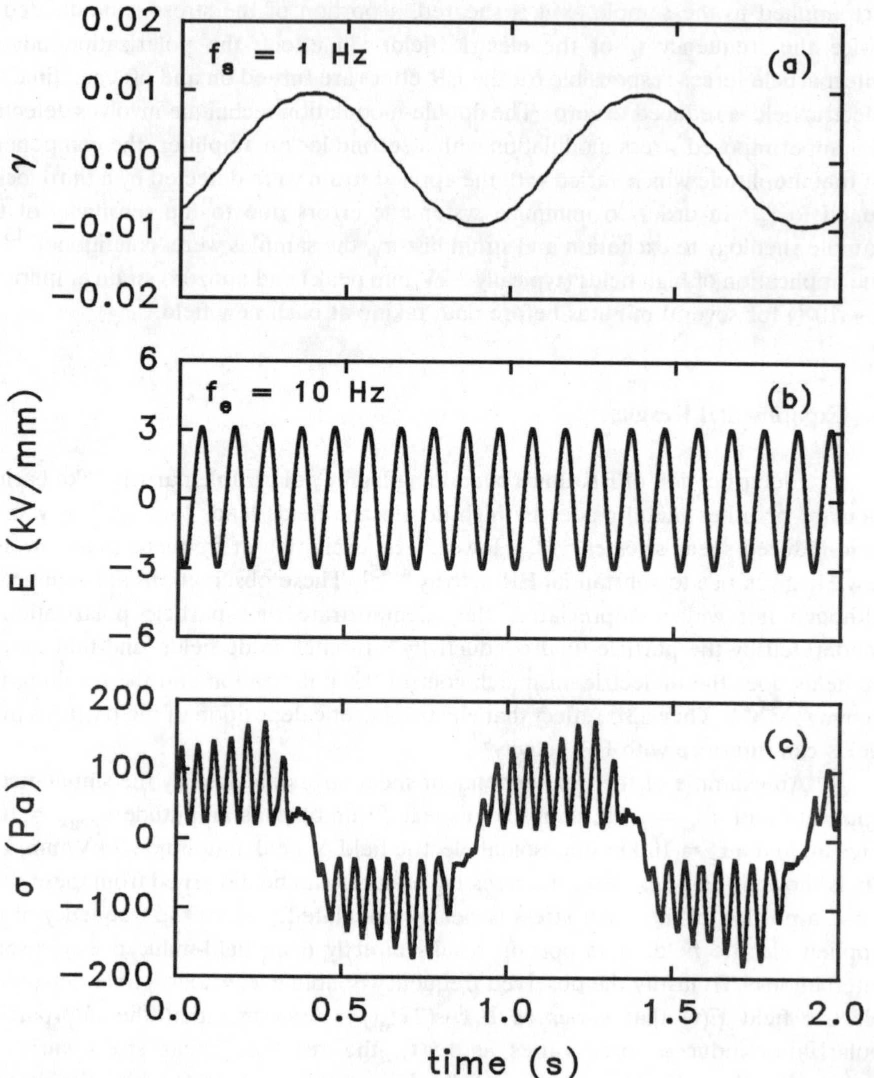

Figure 1: The time dependence of (a) the shear strain γ, (b) the electric field E, and (c) the resulting stress σ of a $\phi = 0.2$ $BaTiO_3$ suspension in a Couette geometry.

interactions, e.g., van der Waals attractive forces, that become relevant because of polarization-induced interparticle contact in the ER gel structure. Third, the overall stress envelope is not linearly related to the applied sinuosoidal strain: nonlinear or yield phenomena are thus encountered even at a strain amplitude of only 10^{-2} in this system.

Indeed, a linear viscoelastic regime is observed only for $\gamma_{peak} < 6 \times 10^{-4}$; in this regime, the storage modulus is virtually independent of field and is quite large, of the order of several hundred kPa[49]. In order to quantify the induced electrostatic contribution to the viscoelastic response of the model fluid, the magnitude of the total shear stress, $|\sigma_1^{(tot)}|$ (Fig. 2a), as well as the portion of the shear stress modulated at $2f_e$, $|\sigma_1^{(el)}|$ (Fig. 2b), were measured as a function of peak strain amplitude (f_s = 2 Hz) and root-mean-square (rms) electric field amplitude (sinuosoidal excitation with f_e = 20 Hz). Referring to Fig. 1c, it is noted that $|\sigma_1^{(tot)}|$ is qualitatively related to the smoothed amplitude of the total shear stress while $|\sigma_1^{(el)}|$ is comparable to the peak amplitude of the field-modulated portion of the stress. Fig. 2 reveals that the total shear stress is dominated by non-field-induced interactions for $\gamma_{peak} < 2 \times 10^{-3}$, while the direct contribution from particle polarization dominates for larger strains.

The linear variation of the polarization-induced stress with strain at small strains allows one to obtain a polarization-induced modulus whose magnitude $|G_1^{(el)}|$ is defined as $d|\sigma_1^{(el)}|/d\gamma_{peak}$ in the limit of small stresses. The field dependence of the measured $|G_1^{(el)}|$ is compared with a fit to the expected quadratic dependence on electric field in Fig. 3. A departure from the expected E^2 dependence may indicate, e.g., the presence of nonlinear conductivity effects[50].

The characteristic strain at which the electrostatic component of the stress saturates, ~ 10^{-2}, is qualitatively related to the dimensionless range of the relevant interparticle interactions; that this strain is two orders of magnitude smaller than predicted from the dipole model is consistent with the relevance of short-range multipole electrostatic interactions. The plateau in $|\sigma_1^{(el)}|$ observed for the largest strains can be associated with the polarization-induced contribution to the static yield stress σ_y. These polarization-induced stresses and moduli will now be compared with the predictions of an electrostatic model of the ER effect.

4. Theory

In this section, the finite element analysis (FEA) of the stresses in an ER fluid is sketched; the method has been described previously[33-35]. The previous analysis is modified by introducing an axisymmetric model, which is computationally more

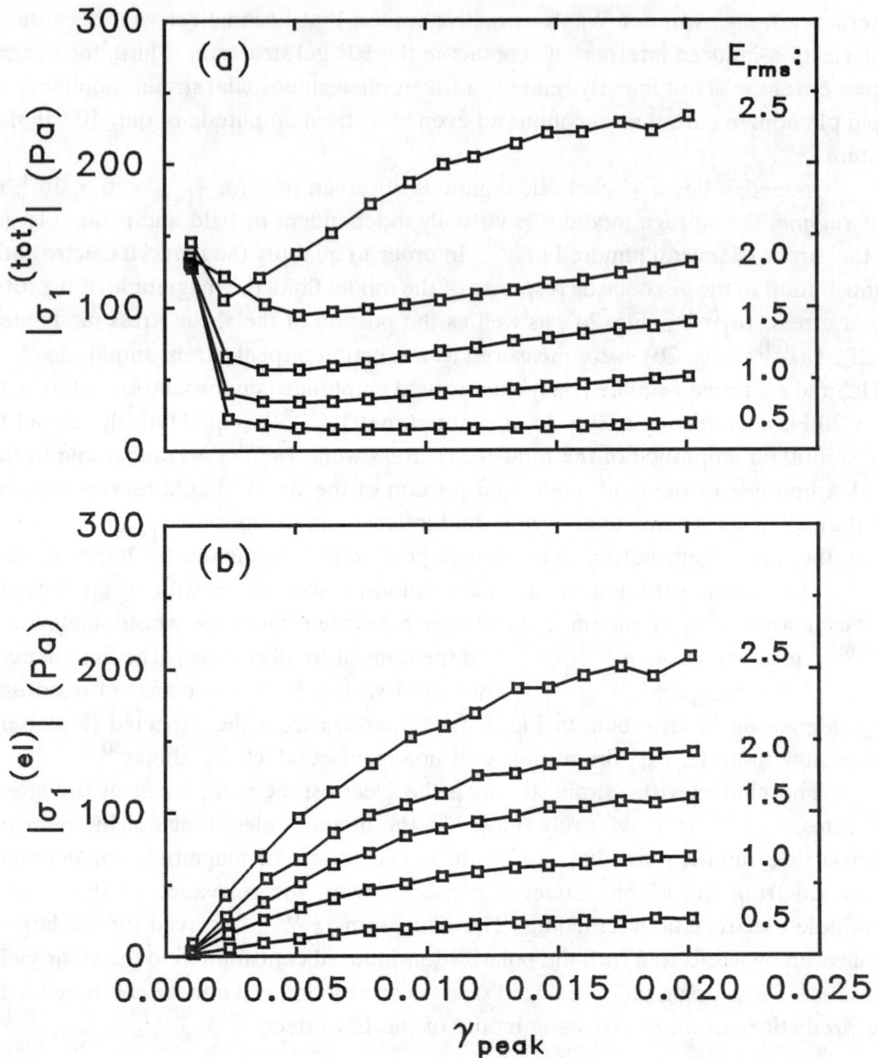

Figure 2: The dependence of (a) the total stress and (b) the electrostatic contribution to the stress on strain amplitude; $f_s = 2$ Hz and $f_e = 20$ Hz. The rms electric fields are in kV/mm.

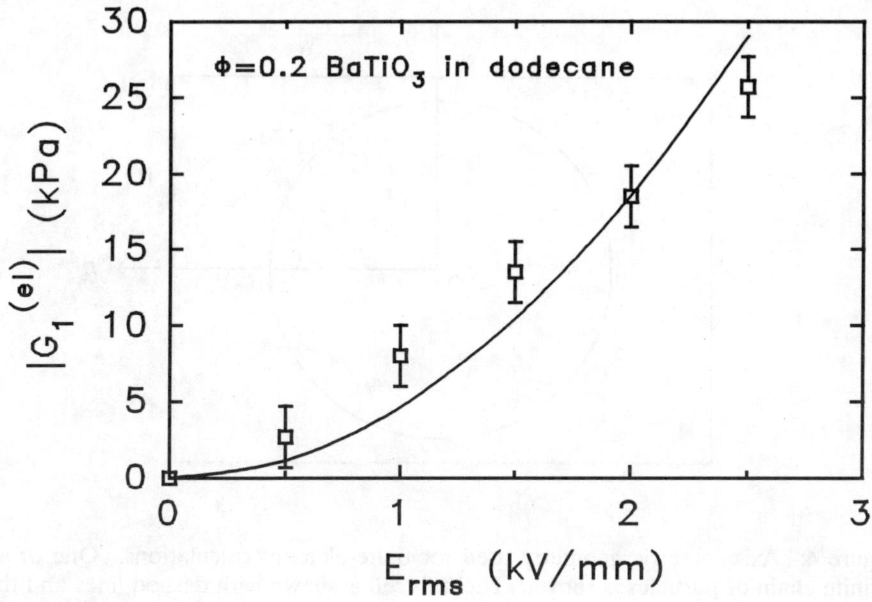

Figure 3: The variation of the electrostatic contribution to the shear modulus with rms electric field. The solid line is a fit to the expected E^2 dependence.

efficient than the full three dimensional (3D) model used in prior work. The analysis is also extended by explicitly calculating the static yield stress σ_y. As before, only single chains are considered; clustering of chains into columns is beyond the scope of this work.

A diagram of the model, Fig. 4, shows one spherical particle and its unit cell in the chain. The chain is aligned along the applied electric field E_0, which is taken to be the z direction. In the full 3D model, it was assumed that the chains were arranged on a square lattice in the xy plane. However, for dilute concentrations (ϕ = 0.2 was used in these calculations), the stresses should not be sensitive to the lateral arrangement of the chains. It is reasonable to assume that the chains are arranged somewhat randomly, suggesting that invoking cylindrical symmetry about the z axis is a good approximation. Thus, the boundary condition that the lateral component of the electric field (E_ρ) vanishes on the cylindrical surface defined by ρ = W is imposed. The size of the cylinder is determined by the volume fraction of particles: $W = (2R^3/3h\phi)^{1/2}$, where R is the particle radius. The center-to-center

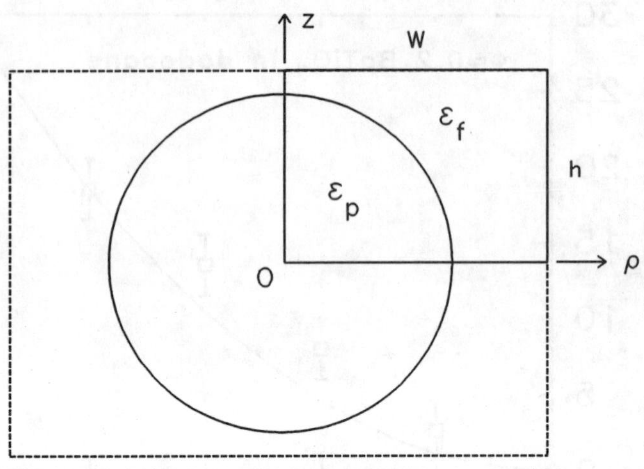

Figure 4: Axisymmetric geometry used for finite-element calculations. One of an infinite chain of particles is shown. The unit cell is shown with dashed lines and the computational region with solid lines. The applied electric field is in the z direction.

distance between adjacent particles in the chain is d = 2h. The remaining boundary conditions are: the potential $\Phi = -E_0h$ on the plane at z = h, $\Phi = 0$ on the plane at z = 0, Φ and the normal component of the displacement $\mathbf{D} = \epsilon\mathbf{E}$ are continuous at the surface of the particle. Laplace's equation $\nabla^2\Phi = 0$ with these boundary conditions is solved by the FEA code ABAQUS (from Hibbit, Karlsson, and Sorenson, Inc., Providence, RI) to find the potential and electric field everywhere in the unit cell.

An effective dielectric constant, $K_{eff}\epsilon_0$, is calculated from the integral of D_z over the upper surface of the unit cell (z = h):

$$K_{eff}\epsilon_0\pi W^2 = \int_0^W D_z 2\pi\rho d\rho. \tag{1}$$

Representative plots of K_{eff} vs d = 2h have been displayed in Ref. 34.

The analysis from this point on is based upon the approximation that the dominant effect in the shear stress σ is due to the stretching of the chains, as has been justified in Ref. 34. We take $h = h_0 \sec\theta_0$, where θ_0 is the shear angle and $h_0 \geq R$ (If particles touch the equality holds.). The shear stress is given by[33,34]

$$\sigma=-\frac{1}{2}\epsilon_0\frac{\partial K_{eff}}{\partial\theta_0}E_0^2, \tag{2}$$

where

$$\frac{\partial K_{eff}}{\partial\theta_0}=\frac{dK_{eff}}{dh}h_0\sec\theta_0\tan\theta_0. \tag{3}$$

The derivative with respect to h is taken with the volume fraction ϕ held constant (thus W varies with h), because a shear strain preserves the volume of the unit cell.

The shear stress σ goes as $G\theta_o$ for small shear strains. The modulus G can be found from Eqs. (2) and (3) to be

$$G=-\frac{1}{2}\epsilon_0\frac{dK_{eff}(h_0)}{dh}h_0E_o^2. \tag{4}$$

The dependence of the shear modulus on interparticle spacing for representative ratios of dielectric constants is shown in Fig. 5. The derivative is evaluated at three different values of the gap 2δ between particles at zero shear and is plotted against $\delta/R = (h_0-R)/R$, h_0 being the minimum allowed value of h. For typical micron-sized particles, the smallest gap considered is only a few nanometers. As is clear from the figure, the modulus is sensitive to the assumed gap. Previous calculations[33] in effect assumed $\delta/R = 5.55 \times 10^{-3}$, leading to smaller values of G. For example, for $K_p/K_f = 25$, the full 3D calculation at this larger gap gave G = 445 Pa with $\phi = 0.2$ and $E_0 = 2$ kV/mm. Taking $K_p/K_f = 1000$ and a plausible value of $\delta/R \approx 3 \times 10^{-3}$, the predicted shear moduli are in good agreement with the experimental polarization-induced moduli in the barium titanate system, Fig. 3.

Bonnecaze and Brady[36] have suggested that the static yield stress σ_y can be determined by finding the maximum value of σ as a function of θ_0. Evaluating Eqs. (2) and (3) numerically from the FEA, the dependence of the yield stress on K_p has been determined; the results scale as E_0^2 and are proportional to K_f for *fixed* K_p/K_f. For dilute concentrations, σ_y is linear in ϕ. Results are plotted for $K_f = 2$, $\phi = 0.2$, and $E_0 = 1$ kV/mm in Fig. 6.

Experimental values of σ_y are also compared to this FEA theory in Fig. 6. The circles are data of Garino, et al.[40], while the square was obtained from the

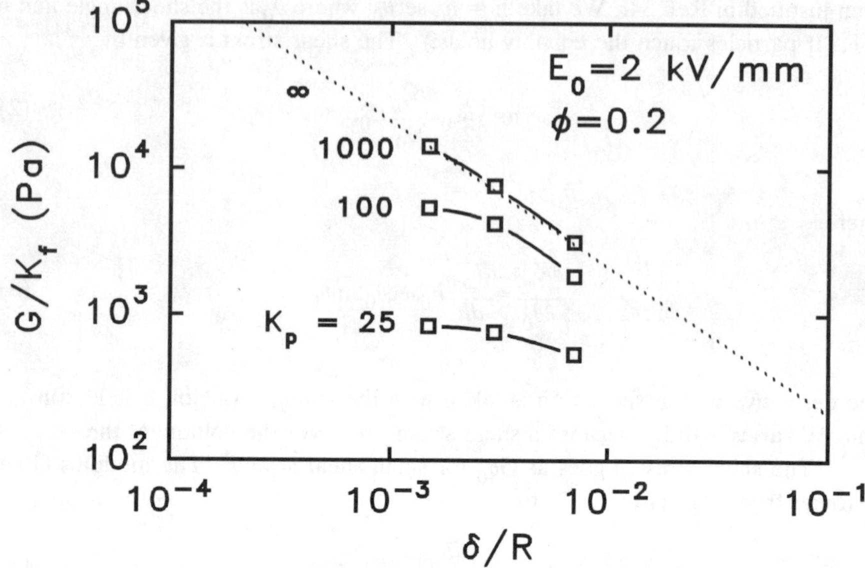

Figure 5: Calculated shear modulus vs. interparticle gap $2\delta = 2(h_0 - R)$, where h_0 is the minimum separation h. The dotted line is an analytic approximation for particles with infinite dielectric constant[35].

polarization-induced data in Fig. 2b. The fluids are comprised of TiO_2, $SrTiO_3$, and $BaTiO_3$ particles (K_p = 70, 270, and 2000, respectively) suspended at $\phi = 0.2$ in dodecane ($K_f = 2$). All data were collected with ac excitation to reduce or eliminate conductivity effects (See Refs. 34 and 46). *To the authors' knowledge, this is the first valid comparison of the dielectric mismatch theory and experiment.* Previous comparisons (e.g., Refs. 12, 31, and 51) have been made to dc data, for which a conductivity mismatch theory[34] is more applicable. Although the dielectric theory appears to overestimate the measured yield stresses, it accounts for both the observed order of magnitude and the trend with K_p. Since various factors such as incomplete chaining and slipping at the electrodes weaken the ideal single chain values, it is not surprising that the experimental values fall below theory.

The dielectric mismatch theory has essentially been verified for these fluids under ac excitation. Other factors that could influence the measured shear stresses include the influence of (rather substantial) particle roughness, particle shape effects, polydispersity, column formation, or other structural complications. The relevance of other colloidal interactions in ER rheology is suggested by the nearly field-independent moduli observed in the linear viscoelastic regime of the model systems

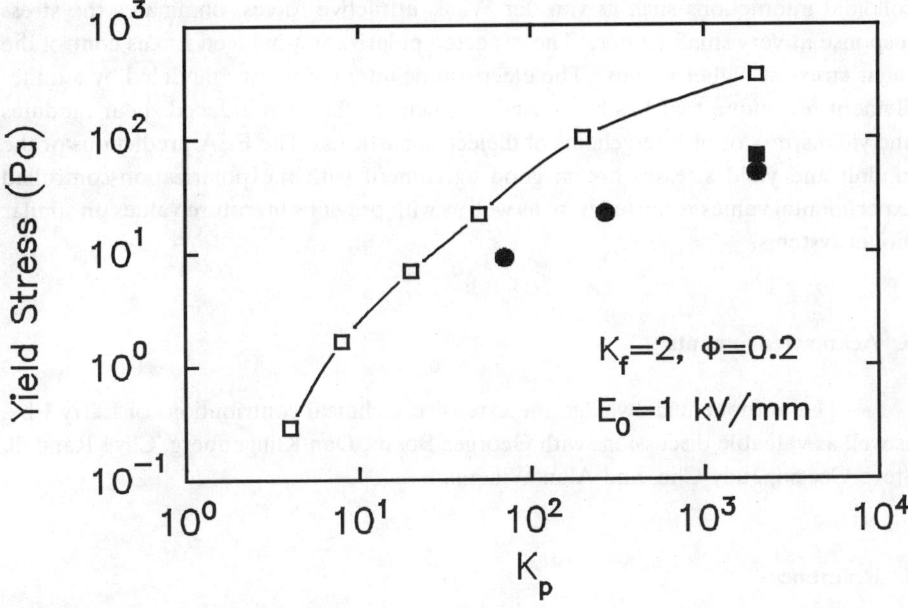

Figure 6: Yield stress vs. particle dielectric constant. Theoretical values (□) are from finite-element calculations; experimental data are from Ref. 40 (●) and the present work (■).

studied here and of quite different ER fluids[13]. Other ER fluids operated under dc electric fields, particularly those developed commercially, often possess larger shear stresses that could well involve different mechanisms than the dielectric polarization mechanism studied here.

5. Summary

The dynamic viscoelasticity of a model barium titanate ER fluid has been explored. A double-modulation technique, in which the samples are excited with sinousoidal electric fields at a frequency higher than the frequency of mechanical strain, has been used to decompose the shear stress response into one component that is directly controlled by field-induced particle polarization and another that is not modulated by the field. The unmodulated component, which may reflect other colloidal interactions such as van der Waals attractive forces, dominates the stress response at very small strains. The expected polarization-induced forces control the shear stress at higher strains. The electrostatic interactions are modeled by a finite-element technique that has been used to calculate the field-induced shear modulus and yield stress of ordered chains of dielectric particles. The FEA predictions of the moduli and yield stresses are in good agreement with the polarization-controlled experimental values reported here as well as with previous literature values on similar model systems.

6. Acknowledgements

The authors acknowledge the extensive technical contributions of Larry Elie, as well as valuable discussions with Georges Bossis, Dan Klingenberg, Clive Randall, Steve Ceccio, Judy Che, and Alan Wineman.

7. References

1. W. E. Winslow, *J. Appl. Phys.* **20** (1949) 1137.
2. *Proceedings of the First International Symposium on Electrorheological Fluids*, ed. by H. Conrad, A. F. Sprecher, and J. D. Carlson (North Carolina State University, Raleigh, NC, 1989).
3. *Proceedings of the Second International Conference on Electrorheological Fluids*, ed. by J. D. Carlson, A. F. Sprecher, and H. Conrad (Technomic, Lancaster, PA, 1990).
4. *Proceedings of the Third International Conference on Electrorheological Fluids*, ed. by R. Tao (World Scientific, Singapore, 1992).
5. N. Petek, *SAE Paper* No. 920275 (1992).
6. A. Pinkos, E. Shtarkman, and T. Fitzgerald, *SAE Paper* No. 930268 (1993).
7. S. Morishita and J. Mitsui, *SAE Paper* No. 922290 (1992).
8. T. Ushijima, K. Takano, and T. Noguchi, *SAE Paper* No. 881787 (1988).
9. D. L. Hartsock, R. F. Novak, and G. J. Chaundy, *J. Rheol.* **35** (1991) 1305.
10. D. Brooks, H. Goodwin, C. Hjelm, L. Marshall, and C. Zukoski, *Coll. Surf.* **18** (1986) 293.
11. D. R. Gamota and F. E. Filisko, *J. Rheol.* **35** (1991) 1411.
12. T. C. Jordan, M. T. Shaw, and T. C. B. McLeish, *J. Rheol.* **36** (1992) 441.
13. W. S. Yen and P. J. Achorn, *J. Rheol.* **35** (1991) 1375.

14. J. P. Coulter and T. G. Duclos, in Ref. 3, p. 300.
15. Y. Choi, A. F. Sprecher, and H. Conrad, *J. Intel. Mat. Sys. Struct.* **3** (1992) 17.
16. T. C. B. McLeish, T. C. Jordan, and M. T. Shaw, *J. Rheol.* **35** (1991) 427.
17. D. J. Klingenberg, *J. Rheol.* **37** (1993) 199.
18. D. R. Gamota, A. S. Wineman, and F. E. Filisko, *J. Rheol.* **37** (1993) 919.
19. P. J. Achorn and W. S. Yen, unpublished.
20. D. R. Gamota and F. E. Filisko, *J. Rheol.* **35** (1991) 399.
21. H. Block and J. P. Kelly, *J. Phys. D: Appl. Phys.* **21** (1988) 1661.
22. A. P. Gast and C. F. Zukoski, *Adv. Coll. Int. Sci.* **30** (1989) 153.
23. T. C. Jordan and M. T. Shaw, *IEEE Trans. Electr. Insul.* **24** (1989) 849.
24. T. C. Halsey, *Science* **258** (1992) 761.
25. J. R. Melrose, *Mol. Phys.* **76** (1992) 635.
26. K. C. Hass, *Phys. Rev. B* **47** (1993) 3362.
27. P. M. Adriani and A. P. Gast, *Phys. Fluids* **31** (1989) 2757.
28. A. M. Kraynik, R. T. Bonnecaze, and J. F. Brady, in *Proceedings of the Third International Conference on Electrorheological Fluids*, ed. by R. Tao (World Scientific, Singapore, 1992), p. 59.
29. J. M. Ginder, *Phys. Rev. B* **47** (1993) 3418.
30. D. J. Klingenberg, F. van Swol, and C. F. Zukoski, *J. Chem. Phys.* **94** (1991) 6160.
31. Y. Chen, A. F. Sprecher, and H. Conrad, *J. Appl. Phys.* **70** (1991) 6796.
32. H. J. H. Clercx and G. Bossis, *Phys. Rev. E* (1993), in press.
33. L. C. Davis, *Appl. Phys. Lett.* **60** (1992) 319.
34. L. C. Davis, *J. Appl. Phys.* **72** (1992) 1334.
35. L. C. Davis, *J. Appl. Phys.* **73** (1993) 680.
36. R. T. Bonnecaze and J. M. Brady, *J. Rheol.* **36** (1992) 73.
37. R. T. Bonnecaze and J. M. Brady, *J. Chem. Phys.* **96** (1992) 2183.
38. See, for example, W. B. Russel, D. A. Saville, and W. R. Schowalter, *Colloidal Dispersions* (Cambridge University Press, Cambridge, 1989).
39. J. T. Woestman and A. Widom, *Phys. Rev. E* (1993), in press.
40. T. J. Garino, D. Adolf, and B. Hance, in *Proceedings of the Third International Conference on Electrorheological Fluids*, ed. by R. Tao (World Scientific, Singapore, 1992), p. 167.
41. K. Koyama, at the *Fourth International Conference on Electrorheological Fluids*, July 20-23, 1993, Feldkirch, Austria.
42. Z. Luo, R. D. Ervin, and F. E. Filisko, *J. Rheol.* **37** (1993) 55.
43. Y. Otsubo, *Coll. Surf.* **58** (1991) 73.
44. F. E. Filisko, in *Proceedings of the Third International Conference on Electrorheological Fluids*, ed. by R. Tao (World Scientific, Singapore, 1993), p. 116.

45. C. A. Randall, T. R. Shrout, D. E. McCauley, and G. L. Messing, at the *Fourth International Conference on Electrorheological Fluids*, July 20-23, 1993, Feldkirch, Austria.
46. R. A. Anderson, in *Proceedings of the Third International Conference on Electrorheological Fluids*, ed. by R. Tao (World Scientific, Singapore, 1992), p. 81.
47. T. C. Halsey, J. E. Martin, and D. Adolf, *Phys. Rev. Lett.* **68** (1992) 1519.
48. H. See and M. Doi, *J. Rheol.* **36** (1992) 1143.
49. J. M. Ginder et al., unpublished.
50. N. J. Félici, J. N. Foulc, and P. Atten, at the *Fourth International Conference on Electrorheological Fluids*, July 20-23, 1993, Feldkirch, Austria.
51. H. Conrad, in *Proceedings of the Third International Conference on Electrorheological Fluids*, ed. by R. Tao (World Scientific, Singapore, 1992), p. 195.

Fluid Flow and Falling Ball Experiments in ER Fluids

R. N. Zitter, T. J. Chen, X. Zhang, and R. Tao
Department of Physics, Southern Illinois University
Carbondale, Illinois 62901, USA

ABSTRACT

The ability of the induced dipole-dipole model to predict various properties of an electrorheological fluid is tested in a series of experiments: the deformation of a single particle chain under fluid flow, the velocity of a falling ball at various electric fields, particle sizes, and concentrations, and flow valves operating at either constant pressure differential or constant flow rate. Analyses of these rather different experimental situations show that with proper application, the induced dipole model can give a fairly accurate description of observed characteristics.

1. Introduction

Electrorheological (ER) fluids have a variety of technological applications and it is important to have a good understanding of the forces and dynamics involved in these fluids.[1–3] The dominant force mechanism involved in the ER effect is generally believed to be the induced dipole-dipole interaction.[4–6] However, there have been relatively few quantitative experimental tests of the dipole model even in idealized laboratory configurations. The task of calculating electric fields and forces inside an ER fluid seems formidable because of the complex structures of particles of non-uniform size and shape.

We have carried out a unique sequence of experiments designed to test the ability of theory, based on the dipole model, to match experimental results. First we study the simplest of ER structures, a single chain of particles deformed by fluid flow. The shape of the chain is determined by the balance between external force (fluid drag) and internal force (dipole interaction), and unlike simple tension as in a rope or cable, the dipole-dipole force vector is generally not tangent to the chain. Therefore, the ability to predict the observed chain shape is a strong test of the dipole model.

In a second set of experiments, we measure the velocity of a glass ball falling through an ER fluid as a function of the electric field. This resembles the classic falling ball technique for determining the viscosity of a simple fluid, but now there are additional retarding forces on the ball as it stretches and breaks the ER chains and columns in its path. Thus, we are effectively measuring the strength of the ER structures as a function of electric field. The field is increased up to the point where the ball ceases to fall, corresponding to the stress yield point of the columns. Additional variables in these experiments are the particle size and concentration, so that the effects of these parameters on the strength of the ER structures can be ascertained.

The third set of measurements is on the rate of flow through an ER valve. The object here is to see if the results obtained in the previous experiments can be used to fashion a successful model of ER valve flow which would, in effect, complete the route between the dipole model and device applications.

In all of these experiments, the ER fluid consisted of dried silica gel particles in vacuum pump oil. Measurements gave the silica gel dielectric constant (immersed in oil) as 3.9 and that of the oil alone as 2.1. US Standard sieves were used to sort the particles according to size.

Detailed accounts of the first two experiments, chain deformation and falling ball, are given elsewhere[7,8] and only highlights will be presented here. The final section concerning flow in ER valves has not been previously published and is new. Analyses of these rather different experimental situations show that with proper application, the induced dipole model can give a fairly accurate description of observed characteristics.

2. Chain Deformation

A single chain of particles is formed in a square flow tube of cross section 6×6 mm^2. The electric field E is 1.6 kV/mm and the average flow velocity is 4.2 mm/s. Since the Reynolds number is 0.1, laminar flow is assumed. The chain is observed and photographed through a transparent side of the flow tube.

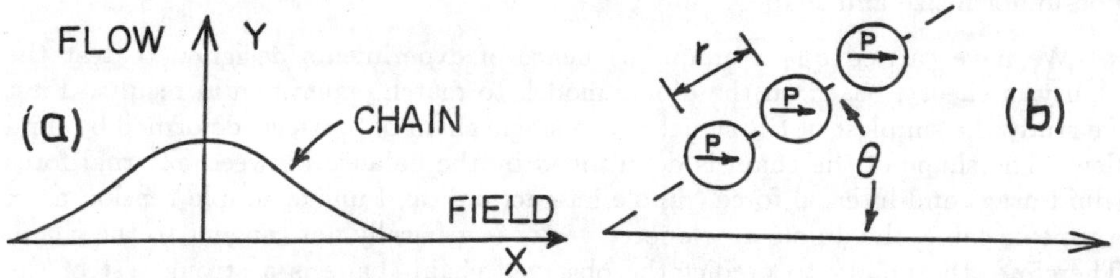

Fig.1. (a) Chain shape, with electric field along x and fluid flow along y. (b) Idealized chain segment of identical spherical particles.

Introducing dimensionless coordinates $X = x/L$, $Y = y/L$, and $Z = z/L$ where L is one-half of a tube side, we have the flow velocity in the y direction approximated as

$$v = v_0(1 - X^2) \tag{2.1}$$

where v_0 is the flow velocity at the tube center. In the steady state, the chain shape is determined by the balance between external force (fluid drag) and internal force (assumed to be dipole-dipole interaction),

$$d\mathbf{f}/ds - \mathbf{F} = 0, \tag{2.2}$$

where $d\mathbf{f}/ds$ and F are, respectively, the internal force and external force per unit chain length s. The drag force is in the y direction, given by

$$F_y = kv\partial x/\partial s, \tag{2.3}$$

where k is the proportionality constant or drag coefficient. We approximate the viscous flow past the chain by considering it as a thin cylinder and use a result already derived for this case.[9] The internal chain force is calculated with the following simplified assumptions: each particle is assigned an average induced point-dipole moment $\mathbf{p}$ aligned along the electric field, and only nearest-neighbor interactions are considered. A schematic illustrating the assumed system with coordinates is shown in Fig.1. The interaction potential between adjacent dipoles is then $p^2(1-3\cos^2\theta)/r^3$, from which the internal force components in the x and y directions can be calculated

$$\begin{aligned} f_x &= 3p^2\cos\theta(5\cos^2\theta - 3)/r^4 \\ f_y &= 3p^2\sin\theta(5\cos^2\theta - 3)/r^4. \end{aligned} \tag{2.4}$$

For the steady state, the equilibrium of all forces provides

$$df_x/ds = 0, \quad df_y/ds = -kv(\partial x/\partial s). \tag{2.5}$$

It follows that f_x is a constant. Thus, it is clear that the stretch r changes with θ. At the maximum angle θ, the chain has the minimal stretch. The experiments find the maximum angle $\theta_x = \pm 0.35$. If we assume that $r = D$ at θ_x, then we have

$$f_x = 4p^2/D^4 = A. \tag{2.6}$$

From Eqs.(2.3) and (2.4), we have

$$f_y = -kv_1L(X - X^3/3). \tag{2.7}$$

Since $\tan\theta = Y' = dY/dX$, we combine Eqs.(2.5), (2.6), and (2.7) to obtain an equation

$$Y'[4-(Y')^2]/[2-3(Y')^2] = -(kv_1L/A)(X - X^3/3). \tag{2.8}$$

There are three roots for Y' from Eq.(2.8), but only one of them has the proper symmetry. Subsequent integration gives the chain shape $y(x)$ which depends on only a single parameter $c = kv_0L/A$.

For $c = 2.0$ the calculated chain shape gives an excellent fit to that observed, as shown in Fig.2. The discrepancy at the chain end is due to the observed fact that there the chain becomes several particles in thickness (multiple chain). Since the fluid velocity approaches zero at the tube wall, there is a natural accumulation of particles at that portion of the chain. For the remainder of the chain, the match of observed and calculated shapes is actually a strong test of the dipole model because the shape depends critically on the internal force vector. Unlike simple tension as in a rope or cable, the dipole force vector is generally not tangent to the ER chain. If, in fact, we assume a tension-like internal force, the calculated shape differs significantly from the one observed.

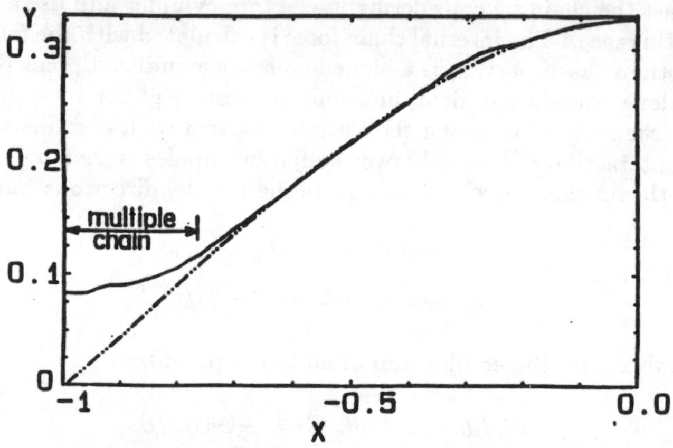

Fig.2. Comparison of observed (solid line) and calculated (dashed line) chain shapes. Half of the chain is shown, with an electrode at x=1.

The value of the parameter c that is calculated from experimental quantities of 8.6 as compared to the value 2.0 giving the best curve-fit. It is probably not unreasonable to ascribe the discrepancy to the various simplifications of the model (spherical particles of uniform size, only nearest-neighbor interactions, etc.). For example, a factor of two in the calculation of the magnitude of the average induced dipole moment would eliminate the quantitative discrepancy of the parameter c.

3. Falling Ball

A classic technique for determining the viscosity of a simple fluid relies on measuring a falling ball's terminal velocity, attained when the retarding force due to viscous drag equals the buoyant weight of the ball. In an ER fluid, there are additional retarding forces due to the bending and breaking of the ER structures (particle chains and columns) in the ball's path. Thus the experiment can provide insights into the strength and dynamics of those structures.

The falling balls here are dark-colored glass spheres 5 mm in diameter. Time-of-flight velocities are obtained from the interception of two laser beams sensed by photo-diodes. The silica gel particles for the ER fluid were passed through set of U.S. Standard sieves to give the following set of average particle diameters: 115, 138, 165 and 215 μm. For each, the particles were mixed with vacuum pump oil at the following volume concentrations: 1.5, 3.0 and 6.0%. For each size and concentration, the electric field was increased in a series of steps to the point where the ball ceased to fall.

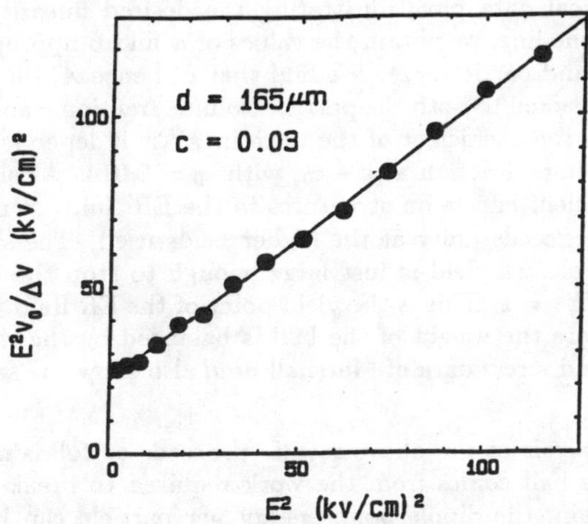

Fig.3. Typical falling ball data plotted in accord with Eq.(3.2). Symbols are as in the text.

To match the observed dependence of velocity v on the field E, it is necessary to invoke an equation of the form

$$m^*g = 3\pi D\eta_0 v + AE^2v + BE^2 \tag{3.1}$$

where m^*g is the ball's buoyant weight, D is its diameter, η_0 is the fluid viscosity at zero field, and A and B are some constants. The first term on the right side of this equation is the viscous drag force on the ball in the well-known Stokes form. The next term represents some force proportional to both the velocity and the square of the field, while the last term is a retarding force proportional to the field squared. After algebraic manipulation, Eq.(3.1) can be written as

$$(1 + \alpha E^2)v/v_0 + \beta E^2 = 1 \tag{3.2}$$

where $v_0 = m^*g/(3\pi D\eta_0)$ is the ball's steady falling velocity at zero field, $\alpha = A/(3\pi D\eta_0)$, and $\beta = B/(m^*g)$. Introducing $\Delta v = v_0 - v$, we write Eq.(3.2) as follows

$$E^2 v_0/\Delta v = (1+\alpha E^2)/(\alpha+\beta). \tag{3.3}$$

Experimentally, we measure the ball's velocity v at a series of values of electric field E. Therefore, according to Eq.(3.3), a data plot of $E^2 v_0/\Delta v$ vs. E^2 should give a straight line if the form of Eq.(3.1) is correct.

Fig.3 shows a typical data plot, illustrating the desired linearity. From the slope and intercept of the line, we obtain the values of α and β appropriate to that particle concentration and particle size. We find that α, hence A, the coefficient of the term vE^2, is proportional to both the particle volume fraction c and the particle size, while β, hence B, the coefficient of the term in E^2 is independent of particle size and varies with volume fraction as $c - c_0$, with $c_0 = 0.014$. As shown later, α is associated with transient effects on structures in the ER fluid. It may be noted that the value of αE^2 exceeds unity at the higher fields used. The significance of β is evident when the electric field is just large enough to stop the ball's motion ($v = 0$); in that case, $\beta E^2 = 1$. This is the yield point of the ER fluid (proportional to the yield stress) where the weight of the ball is balanced by the chain strength under the ball. The yield stress data of Marshall *et al* also show the same $c - c_0$ on volume fraction.[10]

Our theoretical analysis of the above results proceeds as follows. One of the retarding forces on the ball comes from the work required to break chains in its path. Within each chain, the dipole bond energy per particle can be written as $U = Ka^3E^2$, where K is a function of particle and fluid dielectric constants and a is the particle radius. When a ball encounters a chain, the chain is stretched until essentially all bonds break. For a ball of diameter D falling a distance Δs, the number of bonds broken is $nLD\Delta s$, where n is the number of particles per unit volume contained in chains and L is the chain length (inter-electrode spacing). The work ΔW required is then $nDLU\Delta s$, and since n is related to the particle volume fraction c by $c = (4/3)\pi n a^3$, the chain breaking force $F_c = \Delta W/\Delta s$ is

$$F_c = (3/4\pi)LDK(c-c_0)E^2 \tag{3.4}$$

where c_0/c is the particle fraction not contained in chains.

During stretching, the chains move through a viscous medium and experience drag force. The motion consists of two regimes: a brief initial acceleration and then motion at constant velocity until chain break. For the latter, the analysis already developed for the case of chain deformation under steady flow can be employed, i.e., the drag force is equal to the chain's internal dipolar force component along the falling ball's direction. Then the work required is approximately the integral of the dipole force component over deformations from zero to that of breaking. By considering the number of chains encountered by the ball in distance Δs and the number of bonds in a chain contributing to the dipole force, we obtain the retarding force F_d due to chain drag as

$$F_d = F_c \tag{3.5}$$

where F_c is given by Eq.(3.4).

The initial acceleration regime is treated by converting the steady state analysis of chain deformations to a time-dependent equation of motion. The situation is analogous to that of a flexible rod in a viscous fluid undergoing an impact at its center, where the initial deformation depends on the velocity of the impact as well as the stiffness of the rod. An exact solution of the motion equation for impact by a sphere is impractical, and so we make some crude approximations, e.g., replacing the sphere by a cube. Not only complete chains but also segments as short as two to three particles in length should be counted, since they also will be subject to the impact-acceleration process described, so if we assume that essentially all particles are either in complete chains or segments, then the result depends on the total volume fraction c, rather than $c - c_0$ as in the earlier cases. The expression obtained for the accelerative retarding force is then calculated as

$$F_a = (16/k)pD^2 KacvE^2 \tag{3.6}$$

where k is a chain's drag coefficient per unit length. Despite the crude approximations employed, the expression should be correct to an order of magnitude.

The total ER retarding force $F_c + F_d + F_a$ contains terms in E^2 and in vE^2. This is the form assumed in Eq.(3.1), which has been shown to fit the data. The analytic expression for the coefficients α and β are contained in Eqs. (3.4), (3.5) and (3.6). Their dependence on particle size and concentration is in agreement with the experimental results mentioned earlier. For quantitative comparisons of α and β with experiment, we need a value for K, the constant in the expression $U = Ka^3E^2$ for the bond energy per dipole interaction. Thus, K depends on dielectric constants and represents the local field correction. Magnified visual inspection of the chains in these experiments shows them to be a number of average particle diameters in thickness, and the silica gel particles are variegated in shape as well as having a spread of sizes. Since the particles will tend to be close-packed, a rough estimate of K can be obtained by viewing a thick chain as a continuous solid, with an energy density of $\epsilon_p E^2/8\pi$ where ϵ_p is the particle dielectric constant. Multiplication by the equivalent volume of a pair of particles $2(2a)^3$ gives the bond energy $U = 2\epsilon_p a^3 E^2/\pi$. Using this result, we find that for α, the ratio of experimental to calculated values is 4.5, and for β the corresponding ratio is 2.1. In consideration of the approximate nature of the calculations and also the fact that the effects of chains not directly in the ball's path have been ignored, the match of calculation and experiment seems satisfactory.

In summary, the falling ball experiments show that the electric field induces two types of retarding force, one varying as E^2 and the other as vE^2. The former is related to the shear stress and varies with volume fraction as $c - c_0$, independent of particle size; the latter is proportional to both particle size and total volume fraction. The forces in E^2 are associated with chain stretching and breaking at constant velocity, while the force varying as vE^2 comes from an impulse action on chains. This implies, for example, that in other situations such as the flow of fluid through an ER valve, there will be no force varying as vE^2 under steady flow.

4. Flow in ER Valves

The preceding sections have shown that the induced dipole model, properly applied, gives a fairly accurate account of particular laboratory-designed situations. As a final test, we now apply the model to a device with many important applications, the ER valve.

For simplicity the flow is assumed to be steady state, essentially laminar and unidirectional along an axis y, and unbounded in a transverse direction z; the electric field is along x. Then the Navier-Stokes equation takes the form

$$\eta_0(d^2v/dx^2) + F + G = 0 \tag{4.1}$$

where η_0 is the zero field viscosity, $v(x)$ is the flow velocity, F is any internal force component along y per unit volume, and G is the pressure gradient along y. In the present situation, F represents the viscous drag force on chains formed by the field. To calculate F, we employ the results obtained in our earlier study of chain deformation.

The drag force per unit length df_d/ds is directed along the flow direction y. In the steady state the drag force is balanced by the chain's internal dipole force. If the dipole force component in the flow direction is f_y, then per unit length the balance is $df_d/ds = df_y/ds$, or $df_d/dx = df_y/dx$. If there are N chains per unit area perpendicular to the flow, then the total force produced per unit volume in the flow direction is

$$F = -N(df_y/dx). \tag{4.2}$$

In terms of the symbols illustrated in Fig. 1, the dipole force components are given in Eq.(2.4) For the electric field direction x, we put the electrodes at $x = \pm L$. At the center ($x = 0$), the slope angle θ of the chains is zero and f_y vanishes. Then from Eqs.(4.1) and (4.2), integrations from $x = 0$ to x gives

$$\eta_0(dv/dx) - Nf_y + Gx = 0, \tag{4.3}$$

since $dv/dx = 0$ at $x = 0$ by symmetry. A second integration from $x = 0$ to $x = L$

gives, after re-arranging terms

$$\eta_0 v(0) = GL^2/2 - N \int_0^L f_y dx \tag{4.4}$$

because $v(L) = 0$ at the electrodes in Newtonian flow.

It is clear from the expression of f_y in Eq.(2.4) that a calculation of the integral in Eq.(4.4) would require a knowledge of the functions $r(x)$ and $\theta(x)$ along the chains. Instead, a reasonable estimate is obtained as follows. By the mean value theorem of calculus, the integral can be written as $\langle f_y \rangle L$, where $\langle f_y \rangle$ denotes an appropriate average value within the integration interval. Since $f(0) = 0$, we take $\langle f_y \rangle$ to be $f_y(L)/2$. In the section describing chain deformation, it was noted that at even moderate flow rates the chains break at their ends, sliding along the electrodes. According to Eq.(2.4), this means that the chain slope angle at $x = L$ is given by $\cos^2\theta = 3/5$. Furthermore, computer simulations of chain deformation show the particle spacing to be minimal at the chain ends, so that $r(L)$ is approximately equal to the average particle diameter D. Then by Eq.(2.4), $f_y(L) = -3.8p^2/D^4$ where the minus sign comes from the fact that $\theta(L)$ is negative. Since the dipole interaction energy is given by $U = p^2(1-3\cos^\theta)/r^3$, it follows that $f_y(L) = 4.8U/D$. As described in the preceding section on falling balls, we take $U = Ka^3E^2$, where a is the particle radius, $K = 2\epsilon_p/\pi$, and ϵ_p is the particle dielectric constant. Then $f_y(L) = 0.4\epsilon_p D^2 E^2$ and so the value of the integral in Eq.(4.4) is estimated to have the value $0.2\epsilon_p L D^2 E^2$.

A final point is that N, the number of chains per unit area perpendicular to the flow, can be written in terms of the volume fraction c of particles as $6(c - c_0)/\pi D^2$, where c_0 represents particles not contained in complete chains. With this and the preceding expression, Eq.(4.4) becomes

$$\eta_0 v(0) = GL^2/2 - 0.4L(c - c_0)\epsilon_p E^2 \tag{4.5}$$

which, it should be noted, is independent of particle size.

The next step depends on whether we are considering a valve operating at constant pressure differential or constant flow rate. The term $v(0)$ represents the flow velocity midway between electrodes and is proportional to the flow rate Q. For constant pressure differential G, Eq.(4.5) gives

$$Q/Q_0 = 1 - 0.8(c - c_0)\epsilon_p E^2/GL \tag{4.6}$$

where Q_0 is the flow rate at zero field. For constant flow rate we have the alternate expression

$$G/G_0 = 1 + 0.8(c - c_0)\epsilon_p E^2/G_0 L \tag{4.7}$$

where G is the pressure differential at zero field.

The relations (13) and (14) are in qualitative and quantitative agreement with experiments. For the case of constant pressure differential, we studied the flow of an 8% volume fraction of silica gel particles in vacuum pump oil through a simple horizontal ER valve of square cross section 6×6 mm^2 with a length of 5.5 cm. Pressure at the valve entrance was supplied by gravity feed over a constant height of 23.6 cm (the specific gravity of the ER fluid is close to unity). The valve exit was essentially at ambient air pressure, and flow rates were determined from the quantity of fluid collected over a ten second time interval. As illustrated in Fig.4, the data show the linearity predicted by Eq.(4.6) and the agreement between the experimental and calculated values of the slope is within 20%.

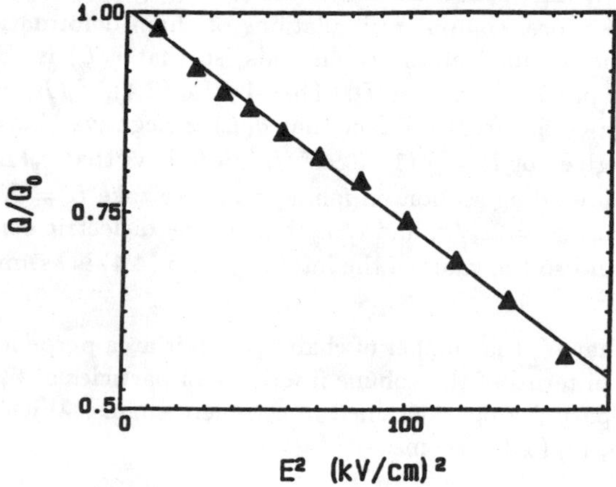

Fig.4. Flow rate as a function of the square of electric field in a simple ER valve.

For the case of constant flow rate, we used published data for a "commercial" grade ER fluid flowing axially in a valve of cylindrical geometry.[11] Since the gap between the cylindrical electrodes was very much smaller than the mean circumference, the metric in the gap is quite close to Cartesian and Eq.(4.7) can be used. Again the data show the required linearity in E^2 with a slope that is within 10% of the calculated value.

5. Summary

A series of studies, ranging from the deformation of a single chain under flow to the properties of an electrorheological device employing a "commercial" ER fluid, have shown that the induced dipole model can give a fairly accurate description of observed characteristics. It is important to point out that in all practical situations where thick ER chains are formed consisting of various particle sizes and shapes, each chain can probably by approximated as a continuous solid, and in that case the so-called problem of calculating the local field is trivially simple. Certainly the continuous solid supposition appears to give reasonably correct numerical values in the situations considered here. This and other analytic approaches described in the present report will be subject to continued test by further studies.

Acknowledgements

This work was supported by the U.S. Office of Naval Research under grant No. N00014-90-J-4041.

References

1. For example, see, *Electrorheological Fluids*, edited by R. Tao (World Scientific Publishing Comp., Singapore, 1992).
2. W. M. Winslow, J. Appl. Phys. **20**, 1137 (1949).
3. H. Block and J. P. Kelly, US Patent 4,687,589 (1987); F. E. Filisko and W. E. Armstrong, US patent No. 4 744 914 (1988).
4. T. C. Halsey and W. Toor, Phys. Rev. Lett. **65**, 2820(1990).
5. R. Tao and J. M. Sun, Phys. Rev. Lett. **67**, 398(1991); Phys. Rev. A, **44**, R6181 (1991).
6. T. J. Chen, R. N. Zitter, and R. Tao, Phys. Rev. Lett. **68**, 2555 (1992).
7. T.J. Chen, X. Zhang, R.N. Zitter and R. Tao, *J. Appl. Phys.* **74**, 942 (1993).
8. R. N. Zitter, X. Zhang, T.J. Chen and R. Tao, *J. Appl. Phys.* (Jan. 1994 issue).
9. L.D. Landau and E.M. Lifshitz, *Fluid Mechanics* (Pergamon, London, 1959), p. 68.
10. L. Marshall, C.F. Zukowski and J. Goodwin, *J. Chem Soc. Faraday Trans. 1*, **85**, 2785 (1989).
11. D. A. Brooks, in Ref.1, p. 367.

INFLUENCE OF PARTICLE SIZE ON THE DYNAMIC STRENGTH OF ELECTRORHEOLOGICAL FLUIDS

Yung-Hui Shih and Hans Conrad
Materials Science and Engineering Department
North Carolina State University
Raleigh, NC 27695–7517

ABSTRACT

The electrical properties, rheology and structure of model ER fluids consisting of glass beads in silicone oil were investigated as a function of electric field E (0–4 kV/mm), particle size D (6–100 μm) and shear rate $\dot{\gamma}$ (2–1000 s^{-1}). The conductivity of the suspensions was 3 orders of magnitude greater than that of the host oil at E $\gtrsim$ 1 kV/mm; their low-voltage d.c. permittivity was about 1.35 times larger. The flow stress of the the suspensions was given by

$$\tau = \tau_E + \tau_{vis} = A(\dot{\gamma}, D)\, E + (C_1 + C_2 / D)\,\dot{\gamma}$$

where τ_E is the polarization component and τ_{vis} the viscous component. The linear dependence of τ_E on E was attributed to dipole saturation. The observed opposing effects of D and $\dot{\gamma}$ on τ_E were concluded to result from their respective influence on the strength of the columnar structure normally produced by the electric field and its fragmentation during shear. The constant C_1 was in agreement with the Einstein equation for the effect of volume fraction of particles on the viscosity of suspensions. The parameter C_2/D was concluded to reflect either the effect of particle surface area on viscosity or a polydispersion effect. The present results did not correlate with the Mason number as normally formulated, but did when it was appropriately modified.

1. Introduction

It is now generally accepted that the shear resistance of an ER-active fluid reflects the combined action of polarization and viscous forces, giving for the flow stress

$$\tau = \tau_E + \eta_s \dot{\gamma} \tag{1}$$

where τ_E is the polarization contribution, η_s the viscosity of the suspension at zero electric field and $\dot{\gamma}$ the shear strain rate. Eq. 1 has the form for plastic flow of a Bingham material, which leads to a description of the relative viscosity η_a/η_s of an ER-active fluid in terms of the Mason number[1]

$$Mn = \frac{\text{Viscous force}}{\text{Polarization force}} = \frac{\eta_s \dot{\gamma}}{2\,\varepsilon_o\,K_f\,(\beta E)^2} \qquad (2)$$

$\eta_a = \tau/\dot{\gamma}$ is the apparent viscosity of the ER fluid with application of electric field, ε_o the permittivity of free space, K_f the relative permittivity of the host liquid, $\beta = (K_p - K_f) / (K_p + 2K_f)$ the dielectric mismatch parameter, K_p the relative permittivity of the particle and E the electric field. Formulation of the Mason number by the extreme right side of Eq. 2 is based on specific models regarding the polarization and viscous forces, which models have not been universally validated.

It follows from Eq. 1 that at low shear rates the polarization force dominates, whereas at high rates the viscous force becomes dominant. The low strain rate regime has been termed *quasi-static* and the high rate regime *dynamic*[2]. In general, the flow stress decreases with increase in $\dot{\gamma}$ in the quasi-static regime and increases in the dynamic regime; see for example Fig. 6 in Ref. 2. For hydrated zeolite particles in mineral oil the change from quasi-static to dynamic behavior occurred at $\dot{\gamma} \approx 10\ s^{-1}$ [2].

An identifying characteristic of ER suspensions under static conditions is that upon application of an electric field the particles align into a chain-like or fibril structure along the direction of the field[2–6]. This suggests that the magnitude of τ_E in the quasi-static regime represents the force to rupture the chains and that the shear process consists of the continued rupturing and reforming of the chains somewhat as shown in Fig. 1a. The observed decrease in τ_E with $\dot{\gamma}$ in this regime can then be explained by less time being available for reforming strong chains once they are broken. Particle size effects generally cancel out in theoretical treatments of τ_E in the quasi-static strain rate regime[2,6–8].

With increase in shear rate into the dynamic regime, the chain or fibril structure becomes disorganized and the suspension takes on a structure similar to that in Fig. 1b[4,6]. It now consists of densely-packed,

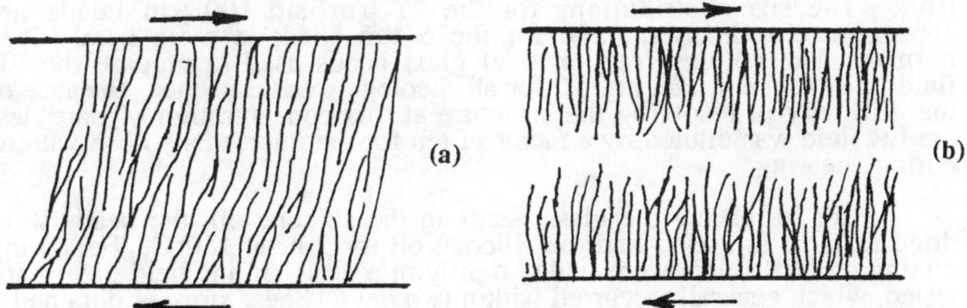

Fig. 1 Schematic of the typical structure of an ER fluid undergoing shear: (a) low shear rate and (b) high shear rate.

solid-like regions next to the electrodes separated by a more dilute, liquid-like region located either in the center of the electrode gap[4] or near one electrode[6]. The observed increase in τ with $\dot{\gamma}$ in the dynamic regime represents contributions from viscous flow of the liquid region and includes electrostatic and mechanical interactions between particles in this region and at the interface between the liquid-like and solid-like regions. As in the case of the quasi-static regime, theoretical treatment of ER fluid strength in the dynamic regime does not in general include a size effect[6].

Most theoretical considerations of the strength of ER fluids have dealt with the quasi-static regime, considerably less attention having been given to the more complex behavior in the dynamic regime. To better understand the mechanisms operative in the dynamic regime additional electrical, rheological and microstructural information pertaining to this regime is needed. The objective of the present research was therefore to provide such information, giving attention to a possible effect of particle size. Hence, electrical, rheological and microstructure studies were performed on model ER fluids consisting of several sizes of glass beads in silicone oil. Glass beads were chosen for the particles because: (a) they are spherical, (b) ER fluids containing them can be observed by optical microscopy[4,5] and (c) some pertinent basic properties and information on their ER response in silicone oil have been obtained in prior studies[5,9,10].

2. Experimental Procedure

2.1 ER Fluids

The host oil employed for our model ER fluids was Dow Corning 200 with specific gravity 0.97, relative permittivity 2.6 and viscosity 0.05 Pa-s at 25°C. The particles were soda-lime glass beads (specific gravity 2.5) provided by Potter Industries with the composition given in Refs. 5 and 9. Three sizes of glass beads were employed, having the linear intercept size distributions shown in Fig. 2. The mean sizes are 6 μm (with standard deviation $\sigma = 2.5$ μm), 27 μm ($\sigma = 7.7$ μm) and 100 μm ($\sigma = 24$ μm). The size distributions for the 27 μm and 100 μm beads are approximately Gaussian; that for the 6 μm beads is more nearly log-normal. The volume fraction ϕ of glass beads used to prepare the ER fluids was kept constant at 0.20 for all rheology measurements. Because of the difficulty in resolving the structure at this concentration of particles, the ER fluid was diluted by a factor of ten for the microscopy observations during shearing.

Prior to mixing the glass beads in the silicone oil, the beads were dried for 3 hr at 200°C and the silicone oil for 2 hr at 150°C. Following mixing, the ER fluids were sealed tightly in a glass jar and held there until tested, which generallyoccurred within two days. Shear stresses obtained 2 days after mixing were the same as those within the first hour.

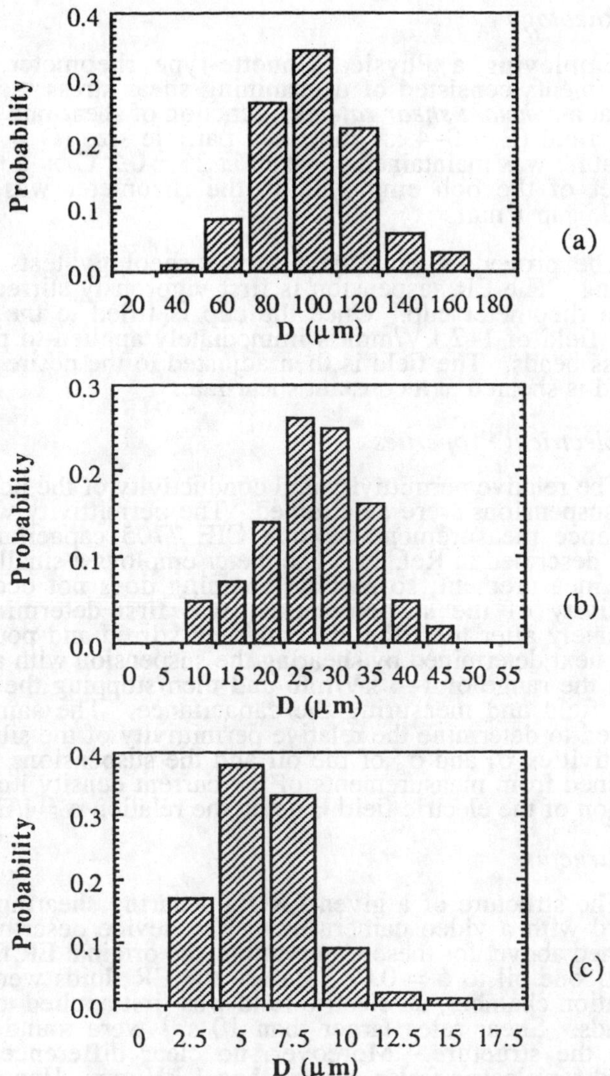

Fig. 2 Linear intercept size distributions of the glass beads with the following mean sizes: (a) 100 μm, (b) 27 μm and (c) 6 μm.

2.2 *Rheology*

Employing a Physica Couette-type rheometer, the rheological measurements consisted of determining shear stress τ vs shear strain γ curves at a *constant shear rate* as a function of shear rate ($\dot{\gamma}$ = 2–1000 s^{-1}), electric field (E = 0–4 kV/mm) and particle size (D = 6–100 μm). The temperature was maintained constant at 25 ± 0.2°C by a thermal bath. The diameter of the bob employed in the rheometer was 27 mm and the electrode gap 1 mm.

The procedure employed in the rheology tests consisted of the following. The ER suspension is first vigorously stirred prior to pouring into the rheometer cup. Once the cup is filled to the desired level, an electric field of 1–2 kV/mm is immediately applied to prevent settling of the glass beads. The field is then adjusted to the desired strength and the ER fluid is sheared at a constant shear rate.

2.3 *Electrical Properties*

The relative permittivity and conductivity of the host silicone oil and of the suspensions were determined. The permittivity was obtained from capacitance measurements with a CIE 7705 capacitance meter in the manner described in Ref. 11. The meter employs a small (1V) d.c. voltage for the measurement; so particle chaining does not occur. The relative permittivity of the suspension K_s was first determined at zero field immediately after the ER fluid had been stirred and poured into the cup. K_s was next determined by shearing the suspension with an applied electric field in the range of 1–4 kV/mm and then stopping the machine, turning off the field and measuring the capacitance. The same procedure was employed to determine the relative permittivity of the silicone oil K_f. The conductivities σ_f and σ_s of the oil and the suspensions respectively were determined from measurements of the current density j during shearing as a function of the electric field E using the relation $\sigma = j/E$.

2.4 *Structure*

The structure of a given ER fluid during shearing was viewed and recorded with a video camera using the device described in Ref 4. As mentioned above, for these observations the original ER fluids were diluted with silicone oil to $\phi = 0.02$. Before the ER fluids were poured into the observation chamber, an electric field was first applied to avoid settling of the beads. Shear rates larger than 10 s^{-1} were found to be too fast to resolve the structure. Moreover, no clear difference in the structure occurred for electric fields greater than 1 kV/mm. Hence, a field strength of 1 kV/mm and a shear rate of ~ 5 s^{-1} were chosen as the experimental conditions for observing the structure during shearing at 25°C.

3. Results

3.1 Electrical Properties

The relative permittivity of the silicone oil K_f was determined to be 2.6 independent of electric field, in accord with the value given by the supplier and measured earlier by Conrad et al[11]. Fig. 3 presents the relative permittivity of the suspension K_s vs prior electric field for the three particle sizes. It is seen that K_s initially increases slightly with electric field to ~ 1 kV/mm and then changes only little, if at all, in accord with previous observations[5] that the static fibril structure remains relatively constant for $E \geq \sim 1$ kV/mm. Although K_s is independent of particle size at zero field, it increases slightly with particle size with application of the field.

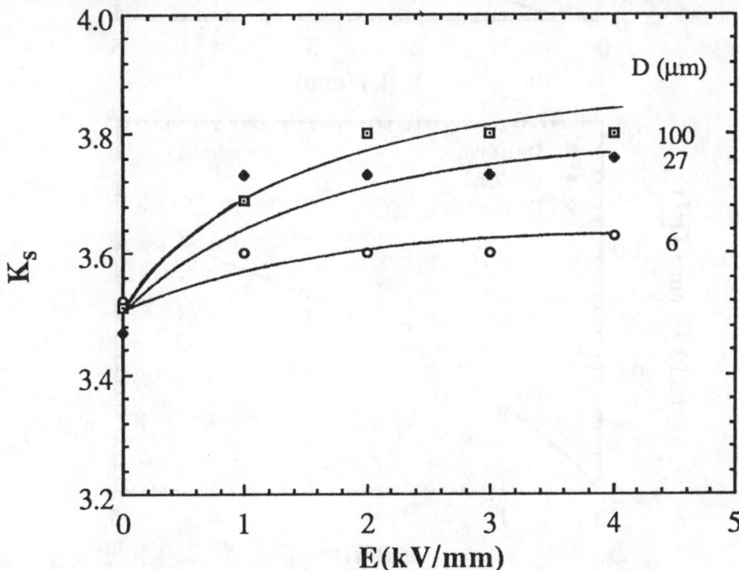

Fig. 3 Relative permittivity of the ER fluids $K_s^{d.c.}$ vs prior electric field E for the three particle sizes.

The conductivity of the silicone oil σ_f was ~ 10^{-13} S cm^{-1} for the range of electric fields employed here. Plots of the current density j and of the conductivity of the suspensions σ_s vs the electric field are given in Figs. 4a and 4b respectively. Evident is that j, and σ_s, increase with electric field, independent of particle size. Moreover, σ_s is about 3 orders of magnitude larger than σ_f, indicating that the conductivity of the particles σ_p is considerably greater than that of the host fluid at high electric fields, i.e. $\sigma_p \gg \sigma_f$ at $E \geq 1$ kV/mm.

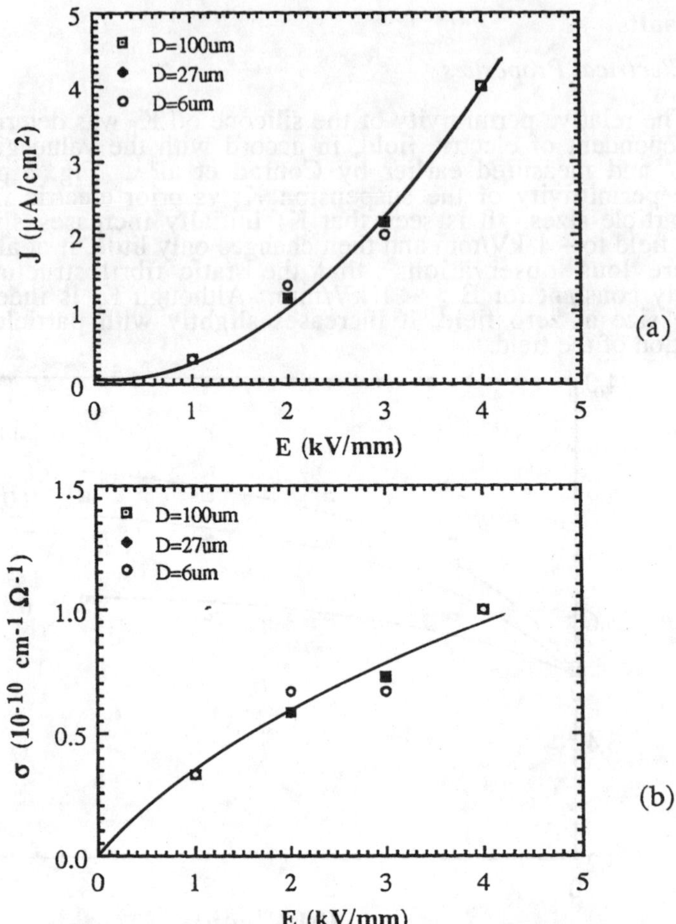

Fig. 4 Effect of electric field on: (a) current density and (b) conductivity of the suspensions containing three sizes of glass beads in silicone oil.

3.2 *Rheology*

Typical shear stress τ vs shear strain γ behavior as a function of shear rate $\dot{\gamma}$ is shown in the semilog plot of Fig. 5. At each strain rate the shear stress increased rapidly with strain to $\gamma = \sim 1.0$ and then remained relatively constant with further increase in strain. This gave a so-called steady-state or plateau stress τ, which varied with shear rate $\dot{\gamma}$, electric field E and particle size D. The effects of $\dot{\gamma}$, E and D on τ (taken as the flow

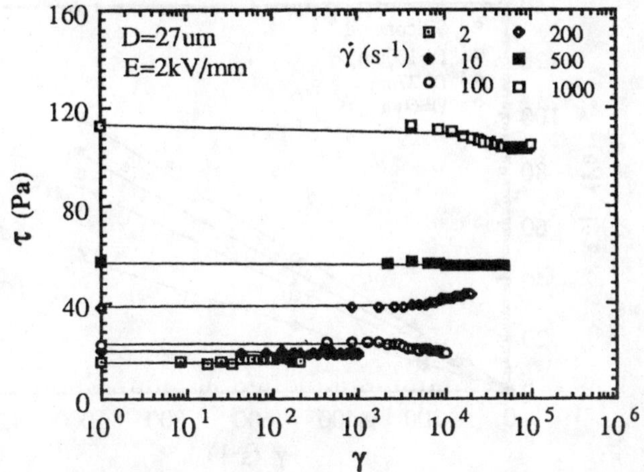

Fig. 5 Shear stress τ vs log shear strain γ at E = 2 kV/mm for various shear rates $\dot{\gamma}$ for the ER fluid with 27 μm particles.

stress at $\gamma = 5$) were considered in terms of their effect on each of the two components of the flow stress given by

$$\tau = \tau_E + \tau_{vis} \tag{3}$$

where $\tau_{vis} = \tau$ at E = 0 and τ_E includes all of the effects of the electric field on the flow stress. Eq. 3 is equivalent to Eq. 1 with $\tau_{vis} = \eta_s \dot{\gamma}$.

Fig. 6 presents a plot of τ_{vis} vs $\dot{\gamma}$ for the host silicone oil and for the ER fluids with differing particle size. In every case τ_{vis} is proportional to $\dot{\gamma}$, the proportionality constant (slope) giving the viscosity of the oil η_f or of the suspension η_s. Support for the validity of this method for determining the viscosity is the agreement in the value for silicone oil determined from the slope in Fig. 6 (0.05 Pa-s) with that provided by the supplier and measured by more conventional methods.

To be noted in Fig. 6 is that the slope for the suspensions decreases with increase in particle size. This is further revealed in Fig. 7, which gives a plot of η_s for the suspensions vs the particle size D. Also shown for comparison is η_f for the host silicone oil. The curve through the suspension data is given by

$$\eta_s = C_1 + C_2 / D \tag{4}$$

where $C_1 = 0.076$ Pa-s and $C_2 = 1.5 \times 10^{-7}$ N-s/m.

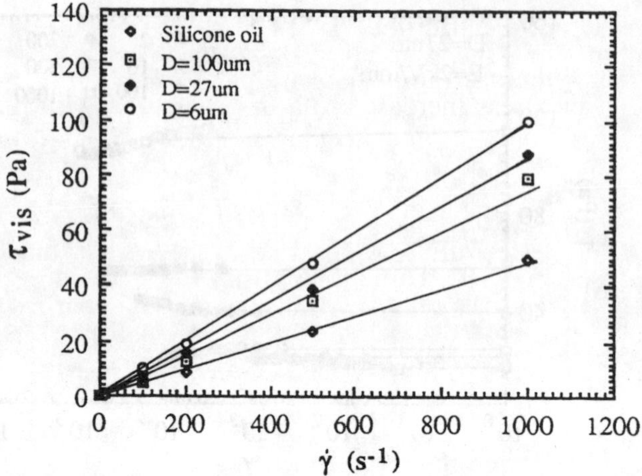

Fig. 6 Flow stress at E = 0 (= τ_{vis}) vs shear rate for silicone oil and for the three particle sizes.

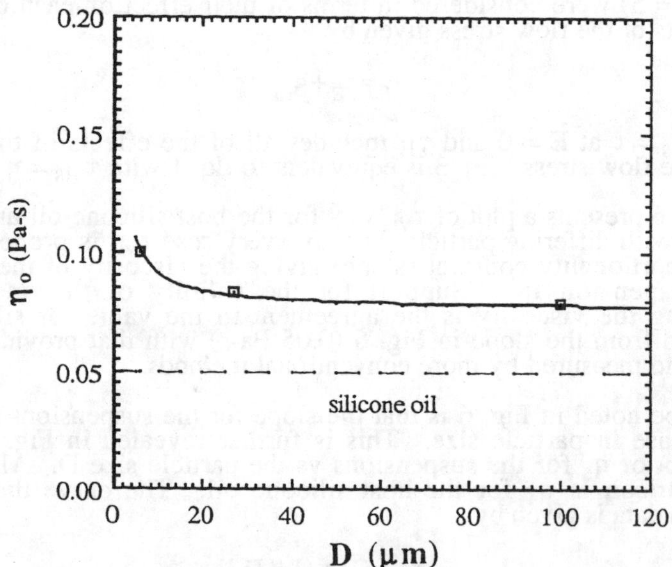

Fig. 7 Viscosity of the suspensions η_s vs particle size D for E = 0. Included for comparison is the viscosity of the silicone oil.

The effect of electric field on the flow stress as a function of particle size is illustrated in Figs. 8 and 9. Fig. 8 shows behavior typical of that at $\dot{\gamma} = 2$ to $10\ s^{-1}$, whereas Fig. 9 shows typical behavior at $\dot{\gamma} > 100\ s^{-1}$. In general, the flow stress increased in a linear fashion with the electric field, giving

$$\tau = A(\dot{\gamma}, D)\, E + \tau_{vis} \tag{5}$$

where $A(\dot{\gamma}, D) = d\tau_E/dE$ decreased with shear rate, but increased with particle size; see Fig. 10. To be noted from Fig. 10 is that the influence of particle size on $d\tau_E/dE$ is relatively independent of shear rate in the range $\dot{\gamma} = 2$ to $200\ s^{-1}$ and then decreases as the shear rate increases, with $d\tau_E/dE$ becoming independent of particle size at $\dot{\gamma} = 1000\ s^{-1}$. Moreover, $d\tau_E/dE$ approaches zero at $\dot{\gamma} > 1000\ s^{-1}$, i.e. the ER effect essentially vanishes at the highest shear rates.

Upon considering the shear rate regime ($\dot{\gamma} = 2$ to $200\ s^{-1}$) where the effect of the electric field is essentially independent of shear rate (i.e. $d\tau_E / dE \cong$ const), we find that (Fig. 11)

$$d\tau_E / dE = A'D^{1/2} \tag{6}$$

where A' is a constant independent of $\dot{\gamma}$ and D. Integration of Eq. 6 gives for the lower shear rate regime

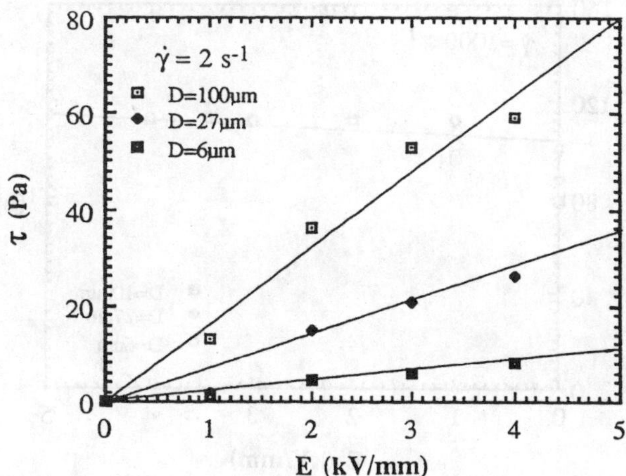

Fig. 8 Flow stress vs electric field at $\dot{\gamma} = 2\ s^{-1}$ for the three particle sizes.

$$\tau_E = A'D^{1/2}E \tag{7}$$

For $\dot{\gamma} > 200\ s^{-1}$, the effect of particle size on the polarization component τ_E is less than that given by Eq. 7, becoming essentially nil at $1000\ s^{-1}$ as τ_E approaches zero.

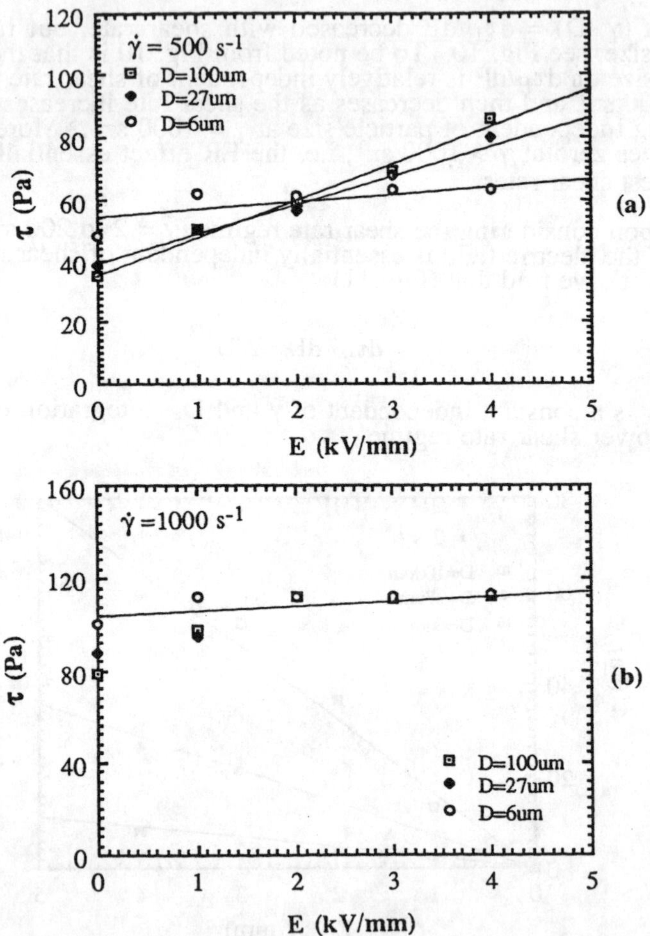

Fig. 9 Flow stress vs electric field for the three particle sizes at the shear rates: (a) $\dot{\gamma} = 500\ s^{-1}$ and (b) $\dot{\gamma} = 1000\ s^{-1}$.

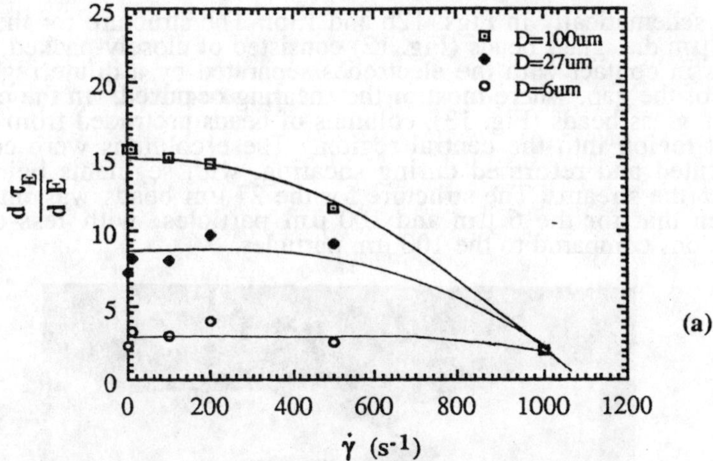

Fig. 10 $d\tau_E/dE$ (= $d\tau/dE$) vs shear rate $\dot{\gamma}$ for the three particles sizes.

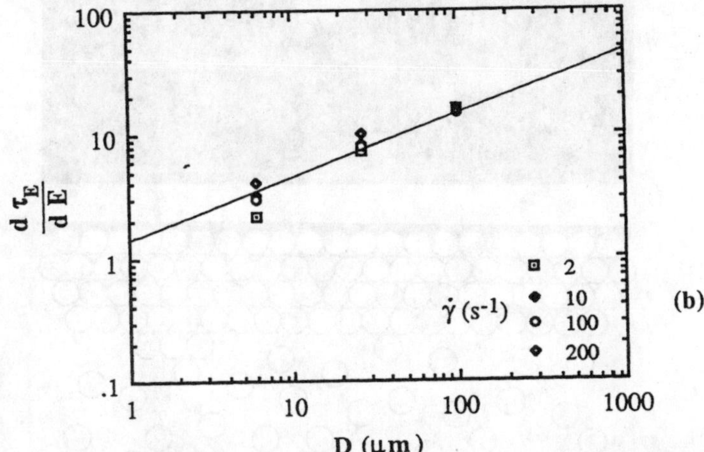

Fig. 11 Log-log plot of $d\tau_E/dE$ vs particle size D for the shear rate regime $\dot{\gamma}$ = 2 to 200 s^{-1}.

3.3 Structure

Photos taken from the video screen of the taped ER fluids undergoing shear are presented in Figs. 12a and 13a. These are not sufficiently clear to reveal the detailed structure, which by careful examination of many frames of the video tape was ascertained to be that shown schematically in Figs. 12b and 13b. The structure for the ER fluid with 6 μm dia. glass beads (Fig. 12) consisted of closely-packed, solid-like regions in contact with the electrodes separated by a dilute region in the center of the gap, where most of the shearing occurred. In the case of the 100 μm glass beads (Fig. 13), columns of beads protruded from the close-packed region into the central region. These columns were continually fragmented and reformed during shearing, with segments being carried away in the stream. The structure for the 27 μm beads was intermediate between that for the 6 μm and 100 μm particles, with less-developed protrusions compared to the 100 μm particles.

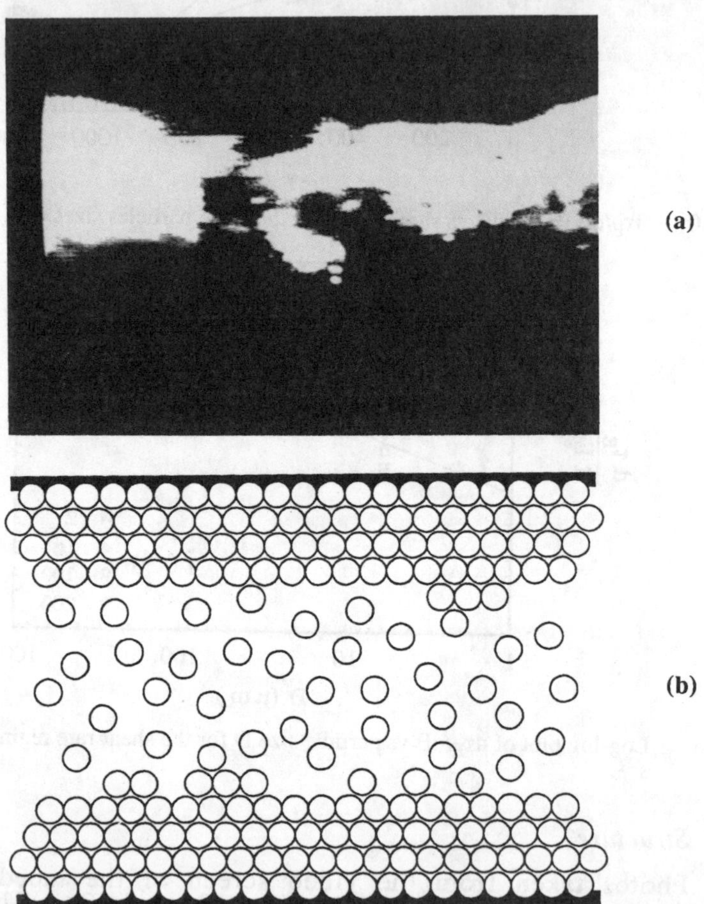

Fig. 12 Structure observed during shearing of the model ER fluid with 6 μm glass beads and E = 1 kV/mm: (a) photo from video tape and (b) schematic of the structure.

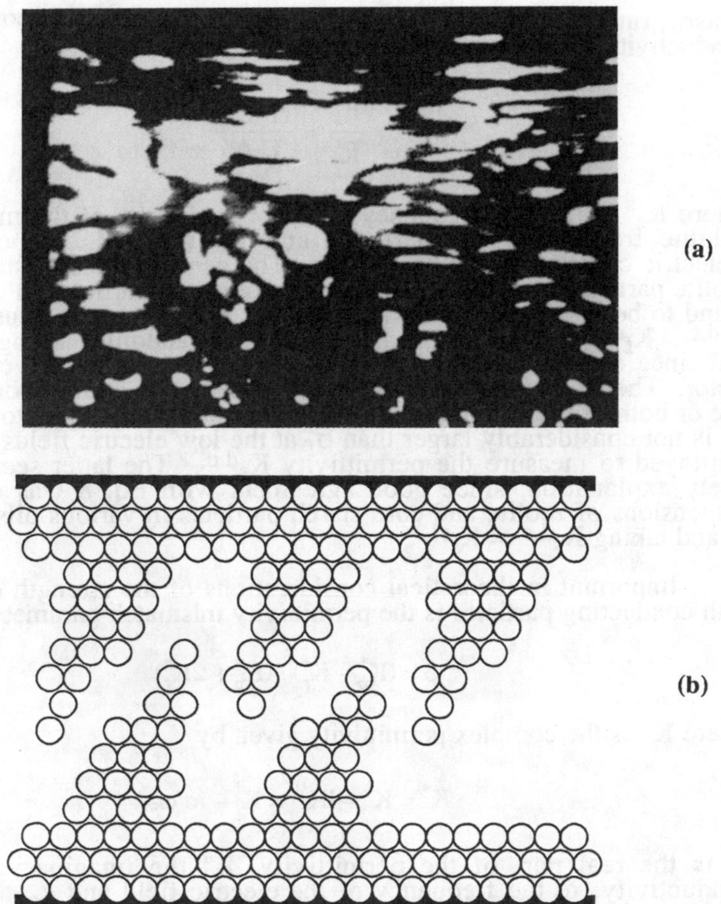

Fig. 13 Structure observed during shearing of the model ER fluid with 100 μm glass beads and E = 1 kV/mm: (a) photo from video tape and (b) schematic of the structure.

4. Discussion

4.1 Electrical Properties

For the case where the conductivity of the particle is considerably greater than that of the host oil (as is the case for the present ER fluids at E > 1 kV/mm), the Maxwell-Wagner equation for the permittivity of a two-phase, random mixture in terms of the dielectric constants and conductivities of the constituents reduces to[12]

$$\frac{K_s^s}{K_f'} = \frac{(1+2\phi)}{(1-\phi)} \tag{8}$$

where K_s^s is the low frequency (static) permittivity of the mixture, ϕ the volume fraction of the particles and K_f' the real component of the dielectric constant of the host phase. The permittivities of suspensions of zeolite particles in various oils and of corn starch particles in corn oil were found to be in good accord with Eq. 8[12]. The present measurements give $K_s^{d.c.} / K_f^{d.c.} = 1.35$ for the ER fluids with randomly arranged particles, and since $\phi = 0.2$ we obtain $(1+2\phi) / (1-\phi) = 1.75$ for the concentration factor. The lack of agreement between these two quantities could be due to one or both of the following: (a) $K_s^{d.c.}$ is not exactly equal to K_s^s, and (b) σ_p is not considerably larger than σ_f at the low electric fields (~ 1 V/mm) employed to measure the permittivity $K_s^{d.c.}$. The latter seems the more likely explanation, since good agreement with Eq. 8 was obtained for suspensions of zeolite and corn starch particles in various oils with $\sigma_p >> \sigma_f$ and taking $K_s^{d.c.} = K_s^s$[12].

Important in theoretical considerations of the strength of ER fluids with conducting particles is the permittivity mismatch parameter

$$\beta = (K_p^* - K_f^*) / (K_p^* + 2K_f^*) \tag{9}$$

where K^* is the complex permittivity given by

$$K^* = K' - iK'' = K' - i\sigma/\omega\varepsilon_o \tag{10}$$

K' is the real part of the permittivity, K" the imaginary part, σ the conductivity, ω the frequency of the electric field and ε_o the dielectric constant of free space. For $\sigma_p >> \sigma_f$, Eq. 9 becomes

$$\beta = (\sigma_p - \sigma_f) / (\sigma_p + 2\sigma_f) = 1 \tag{11}$$

For the present ER fluids we expect that $\sigma_p >> \sigma_f$ at $E \geq \sim 1$ kV/mm and therefore that $\beta = 1$.

4.2 Rheology and Structure

In keeping with general behavior of ER fluids, the flow stress τ of the present glass beads/silicone oil suspensions can be considered to consist of the sum of two components, namely a polarization component τ_E and a viscous component τ_{vis}. The effects of electric field E, shear rate $\dot{\gamma}$ and particle size D on these components gives the following expression for the flow stress

$$\tau = A(\dot{\gamma}, D)\,E + (C_1 + C_2 / D)\,\dot{\gamma} \tag{12}$$

where the first term on the RHS of Eq. 12 is τ_E and the second τ_{vis}. The viscous component exhibited Newtonian viscosity with the viscosity coefficient $\eta_s = (C_1 + C_2 / D)$ *increasing* with *decrease* in particle size. The polarization component τ_E is proportional to E, but in contrast to τ_{vis} *decreased* with *increase* in shear rate and with *decrease* in particle size. These opposing effects of $\dot{\gamma}$ and D on the two flow stress components can lead to the complex behavior shown in Fig. 9, where at low fields the flow stress increases with reduction in particle size, whereas at high fields it either decreases or is relatively independent of particle size. Moreover, because of the decrease in τ_E with increased $\dot{\gamma}$, the flow stress at high shear rates may be only little affected by E, as in Fig. 9b.

Let us now consider the effect of particle size on τ_{vis}. We note that the expression for the effect of D on η_s contains two parts, C_1 which is independent of D, and (C_2 / D) which various inversely with D. We will assume that C_1 represents the component of the viscosity $\eta_s{}^o$ which is given by the Einstein equation for the effect of concentration on the viscosity of suspensions

$$\frac{\eta_s^o}{\eta_f} = \frac{C_1}{\eta_f} = 1 + 2.5\,\phi \tag{13}$$

We then obtain for the present fluids $C_1/\eta_f = 1.52$ and $(1 + 2.5\,\phi) = 1.50$. This agreement suggests that C_1 represents the contribution of the concentration of the second phase to the viscosity.

The additional contribution to the viscosity of the suspensions given by (C_2/D) is in accord with an enhanced resistance to flow due to an increase in the total surface area A_s of the particles in the suspension, which is given by

$$A_s = 6\phi/D \tag{14}$$

If we assume that

$$C_2/D = \alpha\,A_s = \alpha\,6\phi/D \tag{15}$$

where α is a proportionality constant, we obtain $\alpha = 1.25 \times 10^{-7}$ N-s/m and $C_2 = 7.5 \times 10^{-7}\,\phi$ N-s/m for the present suspensions. The significance of these values needs further study. An alternate explanation for the effect of particle size may be that the size distribution of the particles is sufficiently wide to lead to a polydispersion effect[13].

We will now consider the effects E, D and $\dot{\gamma}$ on τ_E. Theoretical considerations[2,6–8] indicate that τ_E should be proportional to E^2 and independent of D. The E^2 behavior results from the fact that the dipole moment p induced on the particles is proportional to E and that the force between the particles is given by the product pE. A proportionality between the force and E rather than E^2 would be obtained if p were a constant, i.e. if saturation of the dipole moment on the particles occurred at small values of E. That saturation probably does occur for the present, dehydrated glass beads is indicated by the results shown in Fig. 14, which

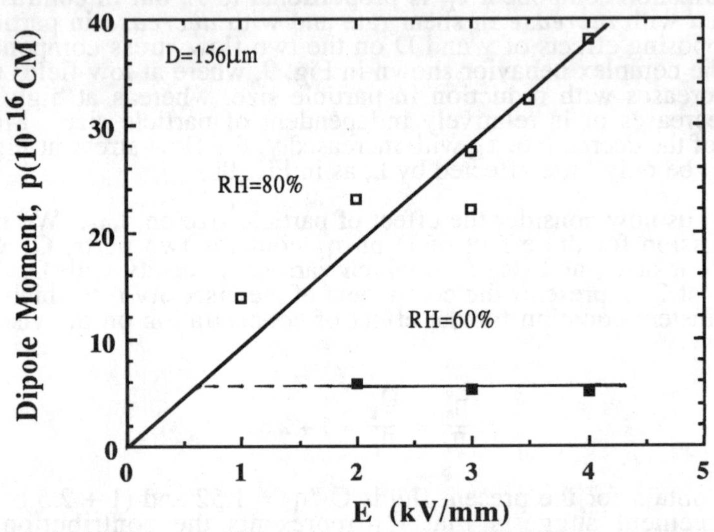

Fig. 14 Induced dipole moment vs electric field for 156 μm glass beads exposed to air environments with 60% and 80% relative humidity. From Conrad and Chen [12].

gives a plot of the dipole moment on 156 mm glass beads after exposure to environments of 60% and 80% humidity[12]. The dipole moment was determined from the interaction force between the glass beads in silicone oil using the apparatus described in Ref. 9. It is seen from Fig. 14 that at a relative humidity RH of 60% saturation of the dipole moment already occurs at low fields. Since the present beads were dehydrated by heating at 200°C prior to mixing with the silicone oil, it is expected that their water content will be less than that for beads exposed to 60% RH and in turn dipole saturation would occur at low fields. This would give τ_E proportional to E and not E^2.

Some insight into the effects of D and $\dot{\gamma}$ on τ_E can be obtained from the observations on the structure of the suspensions during shearing shown in Figs. 12 and 13. Theoretical considerations indicate that for a fixed distance of separation the force of interaction between particles is proportional to D^2. Hence, the resistance of a chain or column of particles to fragmentation during shearing should increase with particle size, as was observed. In turn this would lead to the observed higher resistance to shear, i.e. to an increase in τ_E. On the basis of this model, the proportionality between τ_E and $D^{1/2}$ at $\dot{\gamma} \leq 200\ s^{-1}$ may be attributed to the degree of resistance of the columns to fragmentation with increase in particle size due to the increased force of interaction between the beads. However, this increased resistance will diminish with increased shear rate, as even the stronger columns become fragmented at the higher shear rates.

Finally, as a result of the observed anomalous effects of E and D on the flow stress, the present data do not correlate with the Mason number as formulated in Eq. 2; see Fig. 15a. However, as seen in Fig. 15b good correlation is obtained by plotting η_a/η_s vs the experimental values for the ratio τ_{vis}/τ_E (= Mn'). This agreement of course simply reflects that the flow stress is given by Eq. 3. If we first divide Eq. 3 by $\dot{\gamma}$ and then by η_s we obtain

$$(\tau/\dot{\gamma}) / \eta_s = (\tau_E / \dot{\gamma}\eta_s) + 1 \qquad (16)$$

or

$$\eta_a / \eta_s = (\tau_E / \tau_{vis}) + 1 \qquad (16a)$$

Hence, a log-log plot of η_a/η_s vs τ_{vis}/τ_E should have a slope of –1 at the lower shear rates and approach a value of unity at the high rates.

5. Summary

1. The electrical properties, rheology and structure of model ER fluids consisting of glass beads in silicone oil were investigated at 25°C as a function of electric field E (0–4 kV/mm), particle size D (6–100 mm) and shear rate $\dot{\gamma}$ (2–1000 s^{-1}) in Couette flow.
2. At $E \geq 1$ kV/mm the conductivity of the suspensions σ_s was 3 orders of magnitude greater than that of the host silicone oil. The low-voltage, d.c. permittivity $K_s^{d.c.}$ of the suspensions was about 1.35 times that of the host oil. σ_s and $K_s^{d.c.}$ were relatively independent of particle size, with $K_s^{d.c.}$ for a chain structure being ~ 14% higher than for the random mixture.
3. The flow stress of the suspensions was given by

$$\tau = \tau_E + \tau_{vis} = A(\dot{\gamma}, D)\, E + (C_1 + C_2 / D)\, \dot{\gamma}$$

where τ_E is the polarization component and τ_{vis} the viscous component. The parameter A decreased with shear rate $\dot{\gamma}$, but increased with particle size D. For $\dot{\gamma} \leq 200\ s^{-1}$, A was proportional to $D^{1/2}$.

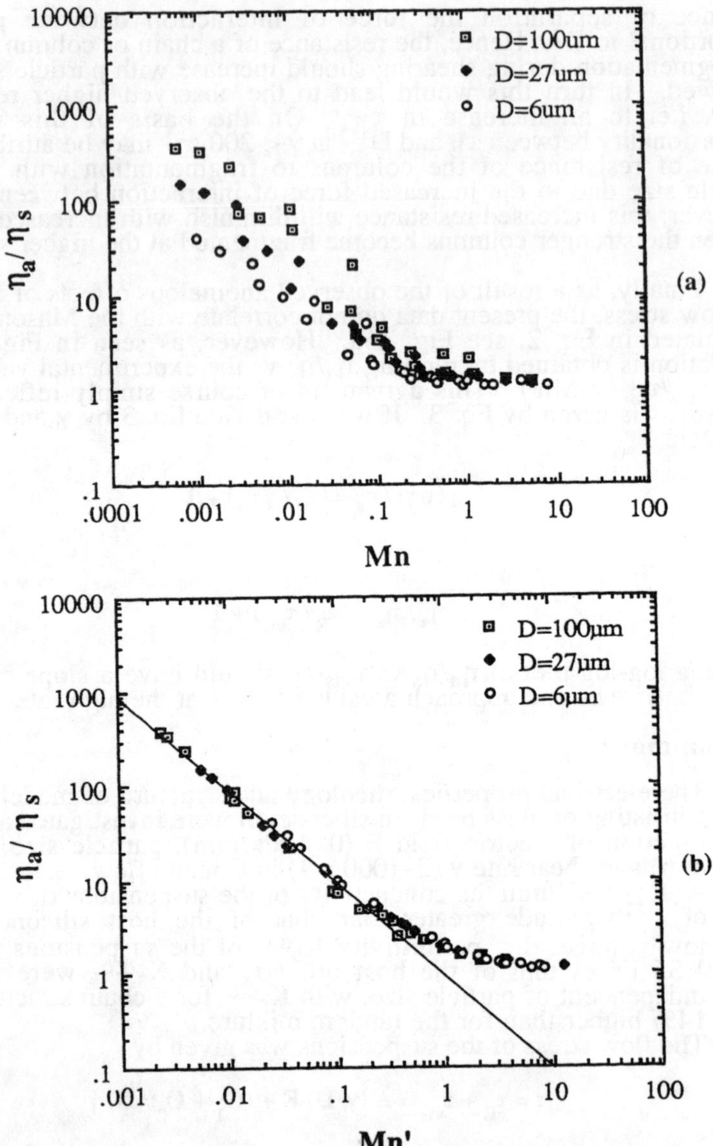

Fig. 15 Correlation of present results: (a) according to Mason number as formulated by Eq. 2 in text and (b) according to experimental values of the ratio τ_{vis}/τ_E.

4. The dependence of τ_E on E rather than E^2 is attributed to a saturation of the dipole moment on the glass beads. The effects of D and $\dot{\gamma}$ on τ_E are attributed to their influence on the strength and degree of fragmentation of the columnar arrangement of the particles normally produced by the electric field.
5. The constant C_1 in the equation for τ_{vis} is in accord with the Einstein equation for the effect of volume fraction of particles on the viscosity of suspensions. The parameter C_2/D is in accord with an effect of surface area of the particles on the viscosity or may reflect a polydispersion effect.
6. The present results did not correlate with the Mason number as normally formulated, but did when it was appropriately modified.

6. Acknowledgements

This research was jointly supported by NSF Award #CBT–8714515, the Ford Motor Company and the NCSU University/Industry Consortium on ER Fluids with Ford, Bridgestone, Physica, Texaco and Tokai Rubber, Ltd. participating. The authors also wish to thank Drs. A. F. Sprecher and Y. Chen for helpful discussions during the course of this work.

7. References

1. L. Marshall, J. W. Goodwin and C. F. Zukoski, Soc. Faraday Trans. **85** 2185 (1989).
2. H. Conrad, Y. Chen and A. F. Sprecher, in *Proc. 3rd Int. Conf. Electrorheological Fluids*, R. Tao, ed., World Scientific, Singapore (1992) p. 195; Int. J. Mod. Phys. B **6** 2575 (1992).
3. W. M. Winslow, J. Appl. Phys. **20** 1137 (1949).
4. A. F. Sprecher, J. D. Carlson and H. Conrad, Mat. Sci. Engr. **95** 187 (1987).
5. H. Conrad, M. Fisher and A. F. Sprecher, in *Proc. 2nd Int. Conf. Electrorheological Fluids*, J. D. Carlson, A. F. Sprecher and H. Conrad, eds., Technomic Publ. Co., Lancaster-Basel, (1990) p. 63.
6. D. J. Klingenberg and C. F. Zukoski, Langmuir **6** 15 (1990).
7. A. P. Gast and C. F. Zukoski, Adv. Colloid Interface Sci. **30** 153 (1989).
8. L. C. Davis, Appl. Phys. Lett. 60 319 (1992); J. Appl. Phys. **72** 1334 (1992); J. Appl. Phys. **73** 680 (1993).
9. A. F. Sprecher, Y. Chen and H. Conrad, in *Proc. 2nd Int. Conf. Electrorheological Fluids*, J. D. Carlson, A. F. Sprecher and H. Conrad, eds., Technomic Publ. Co., Lancaster-Basel (1990) p. 82.
10. J. Hill, N. Vaz, D. Klingenberg and C. F. Zukoski, in *Proc. 3rd Int. Conf. Electrorheological Fluids*, R. Tao, ed., World Scientific, Singapore (1992) p. 280.
11. H. Conrad, A. F. Sprecher, Y. Choi and Y. Chen, J. Rheology **35** 1393 (1991).
12. H. Conrad and Y. Chen, unpublished research, North Carolina State University (1993).
13. G. Bossis, oral discussion following presentation of the paper at the *4th Int. Conf. ER Fluids*, Feldkirch, Austria, July 21, 1993.

Static Rheological Properties of Electrorheological Fluids

G. L. Gulley and R. Tao
Department of Physics, Southern Illinois University, Carbondale, IL 62901

ABSTRACT

We have calculated the static shear stress of an induced electrorheological (ER) solid for a single-chain structure, double-chain structure, triple-chain structure, and body-centered tetragonal (bct) lattice. When the shear strain is small, all of these four structures prefer slanted configurations which will come back to the original configurations if the load is removed. As the shear strain exceeds a yield point, the structures break into parts which cannot return to the original configurations in a short time. The bct lattice is found to have the strongest shear modulus. The triple-chain structure is weaker than the bct lattice, but much stronger than the single-chain structure and double-chain structure. The single-chain structure has the Peierls-Landau instability if the chain is very long. A double chain is stronger than a single chain if the chains are quite long and the situation is reversed if the chains are short.

1 Introduction

An electrorheological fluid consists of a suspension of dielectric particles in a liquid of low dielectric constant [1–7]. When this suspension is exposed to an electric field, its viscosity increases dramatically. As the electric field exceeds a critical value, the suspension forms a solid whose shear modulus increases as the field is further strengthened. This phase transition is completed in about one millisecond and is reversible by reducing the electrical field below the critical field.

The understanding of the physical mechanisms of this phenomena is very important, since a wide variety of applications have been suggested for ER fluids. Some of these applications include suspension systems, valves, brakes, and clutches[1]. These alone will have a tremendous impact in the automobile and aerospace industries.

In this paper, we will discuss the static shear stress of the induced ER solid. The importance of shear stress is clear from the above applications. In experiments, it has been found that upon application of an electric field, the dielectric particles in ER fluids first form chains between two electrodes. Chains then aggregate to form thick columns[1]. Recent theoretical calculations and experiments have also shown that the ideal structure of the thick columns is a body-centered tetragonal (bct) lattice[3–5]. Since the static shear stress depends on the structure of the induced ER solid, we will calculate the shear stress for single chains, double chains, triple chains, and thick columns of the bct lattice.

Among the above four structures, which one is the strongest? This is an important issue to clarify. Some previous work claims that the double chains are weaker than the single chains and the structure consisting of single chains maximizes the shear stress[8]. This conclusion seems to contradict experimental results[9,10] and the well-known Peierls-Landau instability of a one-dimensional solid which implies a long single-chain is a weak structure[3,11].

To resolve the above controversy, we have carried out the calculations which eventually show that the thick-column structure is much stronger than the single-chain structure. The triple-chain structure is stronger than the single-chain and double-chain structures and the bct lattice is even stronger than the triple-chain structure. The comparison between a single chain and a double chain depends on the chain length L. If the chains are long, the double chain is stronger than the single chain. Otherwise, the situation is reversed. These theoretical results are consistent with recent experimental measurement[9,10].

Our calculation also finds that when the shear strain is small, all of these four structures prefer slanted configurations which will come back to the original configurations if the load is removed. As the shear strain exceeds a yield point, the structures break into parts which cannot return to the original configurations in a short time. We have found that the single-chain structure has its shear modulus tending to zero as the chain length L goes infinite, a conclusion consistent with the Peierls-Landau instability. The other thicker structures seem to be stable since their shear moduli do not extrapolate to zero as L goes infinite. The yield point and structure-breaking in the deformation of an ER solid have been observed in experiments[9,10]. The response force and yield point derived from our calculation seem to agree with these experiments reasonably well.

The present paper is organized as follows. In section 2, we will discuss the dipolar interaction. Section 3 is devoted to the calculation of shear stress and shear modulus. The results and discussions are in section 4 where we will also compare our theoretical results with experiments.

2 Dipolar Interaction

We consider a model of ER fluids consisting of spherical particles of dielectric constant ϵ_p suspended in a fluid of dielectric constant ϵ_f, $\epsilon_p > \epsilon_f$. This composite is subjected to an electric field inside a parallel plate capacitor whose two electrodes are planes at $z = 0$ and $z = L$ respectively. When the fluid is exposed to an electric field, the particles develop a dipole moment $\vec{p}$ in the direction of the electric field. Here $\vec{p} = \alpha\epsilon_f a^3 \vec{E}_{loc}$ where a is the radius of the particles, $\alpha = (\epsilon_p - \epsilon_f)/(\epsilon_p + 2\epsilon_f)$, and $\vec{E}_{loc}$ is the local electric field.

To calculate the energy of this system of particles, we consider the dipole-dipole interactions. Two dipoles at $\vec{r}_i$ and $\vec{r}_j$ have an energy of

$$u(\vec{r}_{ij}) = \nu(1 - 3\cos^2\theta_{ij})/r_{ij}^3 \tag{2.1}$$

where $r_{ij} = |\vec{r}_i - \vec{r}_j| = [\rho^2 + (z_i - z_j)^2]^{1/2}$, $\rho = \rho_{ij} = [(x_i - x_j)^2 + (y_i - y_j)^2]^{1/2}$,

θ_{ij} is the angle between their joint line and the direction of the applied field, and $\nu = \vec{p}^{\,2}/\epsilon_f$.

A dipole at (x,y,z) in a capacitor produces images at (x,y,-z) and $(x, y, 2Lj \pm z)$ for $j = \pm 1, \pm 2 \cdots$. The interaction between a dipole and its own image is $u(\vec{r}_{ij})/2$. The interaction between a dipole and a different dipole's images is $u(\vec{r}_{ij})$. All of these interactions must be summed to get the total energy of the interaction.

We introduce a function,

$$f(\rho, z) = \sum_n [(2nL - z)^2 + \rho^2]^{-3/2}. \tag{2.2}$$

Because of the periodicity, $f(\rho, z) = f(\rho, z + 2L)$, we can expand f into the form $f(\rho, z) = \sum_{s=-\infty}^{\infty} f_s(\rho) e^{-is\pi z/L}$ with

$$f_s(\rho) = \int_0^{2L} dz e^{is\pi z/L} f(\rho, z)/(2L) = \pi s K_1(s\pi\rho/L)/(L^2\rho) \tag{2.3}$$

where $K_1(\rho)$ is a modified Bessel function. Eq.(2.2) now reads as

$$f(\rho, z) = 1/(L\rho^2) + \sum_{s=1}^{\infty} 2\pi s K_1(s\pi\rho/L)\cos(s\pi z/L)/(L^2\rho). \tag{2.4}$$

A dipole at $\vec{r}_j$ and all its images interact with a dipole at $\vec{r}_i$ via

$$u_{ij} = -\nu(2 + \rho\partial/\partial\rho)[f(\rho, z_i - z_j) + f(\rho, z_i + z_j)]. \tag{2.5}$$

The formula $d(xK_1(x))/dx = -xK_0(x)$ enables us to write u_{ij} in Eq(2.5) as

$$u_{ij}(\rho_{ij}, z_i, z_j) = \nu \sum_{s=1}^{\infty} (4s^2\pi^2/L^3) K_0(s\pi\rho_{ij}/L)\cos(s\pi z_i/L) cos(s\pi z_j/L). \tag{2.6}$$

The interaction between a dipole inside the capacitor at (x, y, z) with its infinite number of images at $(x, y, 2Lj \pm z)$ $(j = \pm 1, \pm 2, \cdots)$, is given by

$$u_s(z) = -\nu\zeta(3)/(4L^3) - \nu f(0, 2z) \tag{2.7}$$

where the constant $\zeta(3) = \sum_{n=1}^{\infty} 1/n^3 = 1.2020569\cdots$.

Applying Eqs.(2.6) and (2.7), we have the total dipolar energies for a structure,

$$\sum_{j} u_s(z_j) + \frac{1}{2}\sum_{i \neq j} u_{ij}(\rho_{ij}, z_i, z_k). \tag{2.8}$$

3 Shear Stress and Shear Modulus

To apply a shear strain to the ER system, we move the electrode at $z = L$ along the x direction by a distance δ, while fixing the electrode at $z = 0$. The shear strain is then δ/L. We assume that there is no slipping between the electrodes and the induced ER structure. The slipping case has been studied in Reference 4.

For a single chain under a shear strain we consider three configurations as illustrated in Fig.1(a), Fig.1(b) and Fig.1(c). The slanted chain has each particle in the column moved along the x direction proportionally. The broken chain has the chain broken into two parts in the middle with each part in the field direction separated by an amount δ. The slanted-broken chain is broken in the middle but each part is still slanted. The separation of its two resulting parts, δ_0, is less than the deformation δ (Fig.1(c)). As we increase δ_0, the slanted-broken chain changes from the slanted chain to the broken chain.

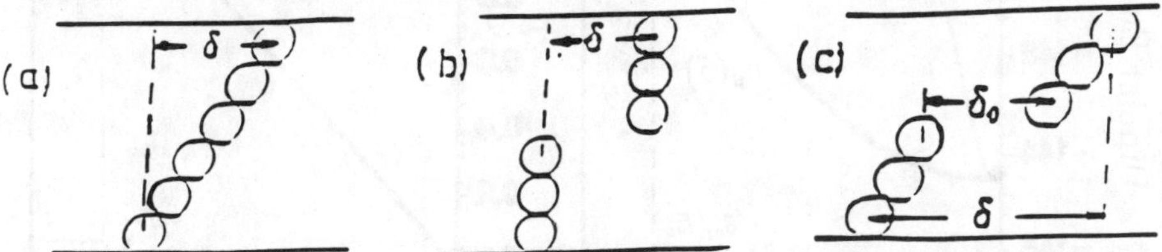

Fig.1. (a) Slanted single chain. (b) Broken single chain. (c) Slanted-broken single chain.

When the particle positions in Eq.(2.8) are for the slanted chain, slanted-broken chain, and broken chain, we obtain $U_s(\delta)$, $U_{sb}(\delta)$, and $U_b(\delta)$ respectively. As shown in Fig.2, $U_s(\delta) < U_b(\delta)$ until δ reaches a critical value δ_c where the curves for U_s and for U_b intersect. The slanted-broken chain has an energy interpolating between the slanted chain and the broken chain. This implies that the slanted chain is preferred when the shear strain is less than δ_c/L. However, when the shear strain exceeds δ_c/L, the broken chain has the lowest energy and the slanted-broken chain has its energy higher than that of the broken chain and lower than that of the slanted chain. Therefore, the competition is really between the slanted chain and the broken chain. When the shear strain is greater than δ_c/L, the chain suddenly breaks into two parts and the energy is given by $U_b$thereafter. There is no intermediate state. For this reason, U_{sb} is not plotted in Fig.2.

When $\delta < \delta_c$, the response force per chain is given by

$$\tau = -\partial U_s(\delta)/\partial\delta. \tag{3.1a}$$

When $\delta > \delta_c$, the response force per chain is given by

$$\tau = -\partial U_b(\delta)/\partial\delta. \tag{3.1b}$$

Fig.3 shows the behavior of the response force of a single chain versus the shear strain δ/L. When $\delta < \delta_c$, the slanted chain gives a response force τ_s almost linear with δ/L. When $\delta > \delta_c$, the preferred broken chain produces a much smaller and almost flat response force. The unit of τ in our calculation is $p^2/(\epsilon_f d^4)$ where $d = 2a$, the particle diameter.

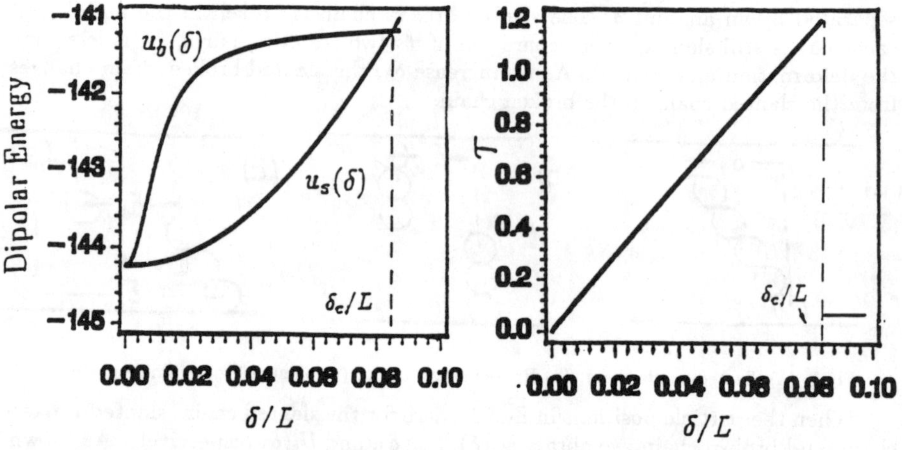

Fig.2. The dipolar energies of the slanted single chain and the broken single chain vs. the shear strain respectively. The single chain has 60 particles. The energy unit is $p^2/(\epsilon_f d^3)$.

Fig.3. The response force of a single chain of 60 particles. vs. the shear strain. The force unit is $p^2/(\epsilon_f d^4)$.

A single chain has $N = L/d$ particles. At a volume fraction ϕ, for the structure consisting of single chains, the number of chains per unit cross-area is $6\phi/(\pi d^2)$. The shear modulus of the single-chain structure at volume fraction ϕ is $6\phi S/(\pi d^2)$

where S is the response coefficient, given by

$$S = \tau L/\delta. \tag{3.2}$$

In Fig.4 we plot S versus shear strain δ/L up to δ_c/L. When $\delta > \delta_c$, it is clear from Fig.3 that the broken chain has a much smaller S.

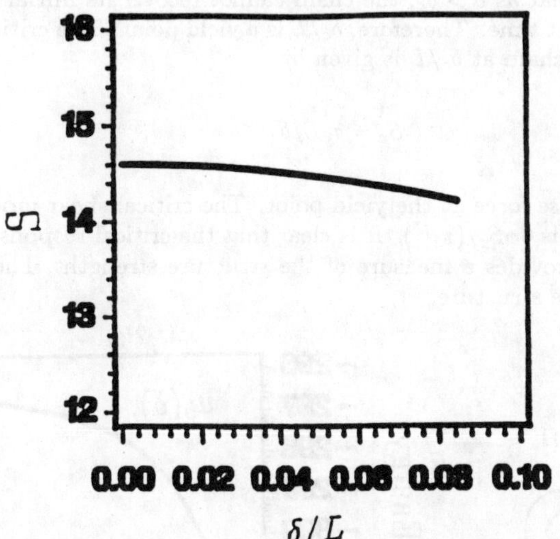

Fig.4 The response coefficient S of a single chain of 60 particles vs. the shear strain up to the yield point. The unit of S is $p^2/(\epsilon_f d^4)$.

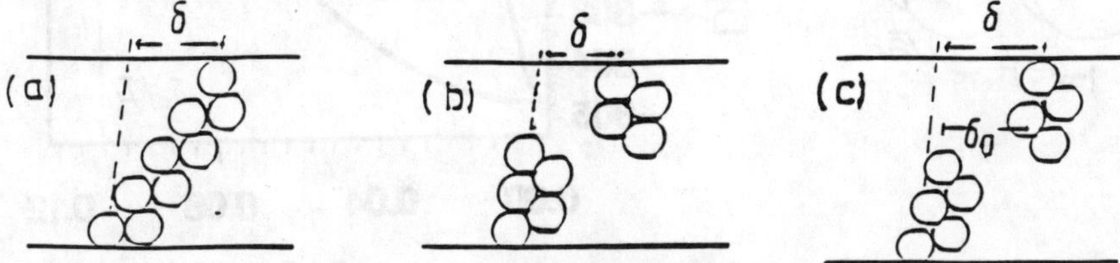

Fig.5. (a) Slanted double chain. (b) Broken double chain. (c) Slanted- broken double chain.

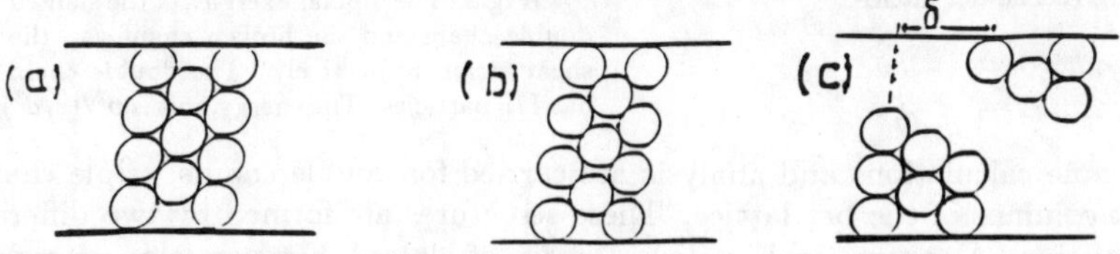

Fig.6. (a) Triple-chain structure. (b) Slanted Triple chain. (c) Broken double chain.

The quantity δ_c/L is of special interest. When the shear strain is small, the slanted chain represents a state which has a uniform deformation. The response force is approximately proportional to the shear strain (Fig.3) and the response coefficient is almost a constant (Fig.4). The single chain will recover its original shape as soon as the shear stress is removed. When δ becomes greater than δ_c, the chain suddenly breaks. The broken chain is the state in which there is structural damage. We expect that as $\delta > \delta_c$, the chain cannot recover its initial single chain configuration in a short time. Therefore, δ_c/L is a yield point. The critical response coefficient of a single chain at δ_c/L is given by

$$S_c = \tau_c L/\delta_c \tag{3.3}$$

where τ_c is the response force at the yield point. The critical shear modulus of the single-chain structure is $6\phi S_c/(\pi d^2)$. It is clear that the critical response coefficient per particle, S_c/N, provides a measure of the structure strength. The bigger the S_c/N, the stronger the structure.

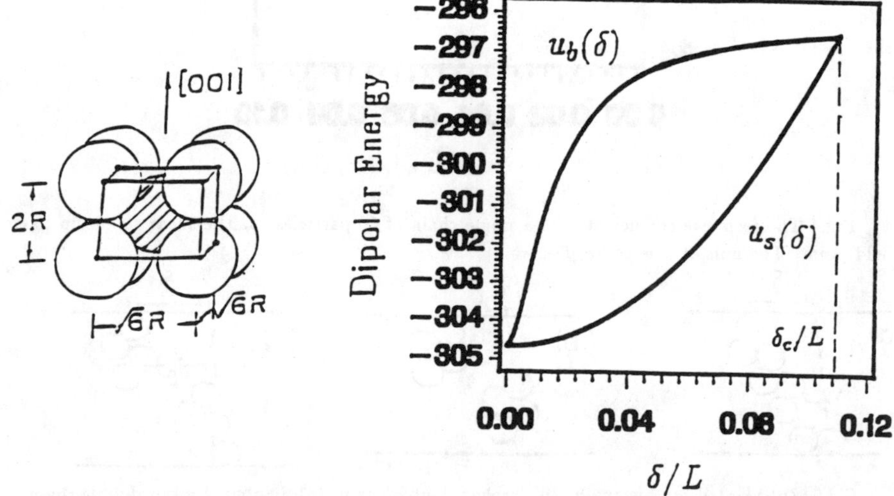

Fig.7. The bct lattice.

Fig.8. The dipolar energies of the slanted double chain and the broken chain vs. the shear strain respectively. The double chain has 119 particles. The energy unit is $p^2/(\epsilon_f d^3)$

The same calculations and analysis are carried for double chains, triple chains and thick columns of the bct lattice. These structures are formed by two different types of chains. As shown in Fig.5(a), a chain of class A has particles extending the full length of the capacitor. A chain of class B can be obtained by moving a chain of class A in the z direction by a sphere radius. In addition, because of the two electrodes, a chain of class A has L/d particles, while a chain of class B has $(L/d-1)$ particles. A double chain has one A chain and one B chain closely packed. Its three configurations under a shear strain, slanted double chain, broken double chain, and slanted-broken double chain are shown in Figs.5(a)-5(c).

As shown in Fig.6(a), the triple chains are formed from two A chains and one B chain closely packed. The slanted triple chain and broken triple chain are plotted in Fig.6(b) and Fig.6(c). A thick column of the bct lattice is a three-dimensional structure, shown in Fig.7. We denote its three conventional Bravais lattice vectors as $\sqrt{3}a(\hat{x}+\hat{y})$, $\sqrt{3}a(\hat{x}-\hat{y})$, and $2a\hat{z}$ (see Fig.7). This structure can also be considered as a compound of A chains and B chains[5]. There are four A chains around one B chain.

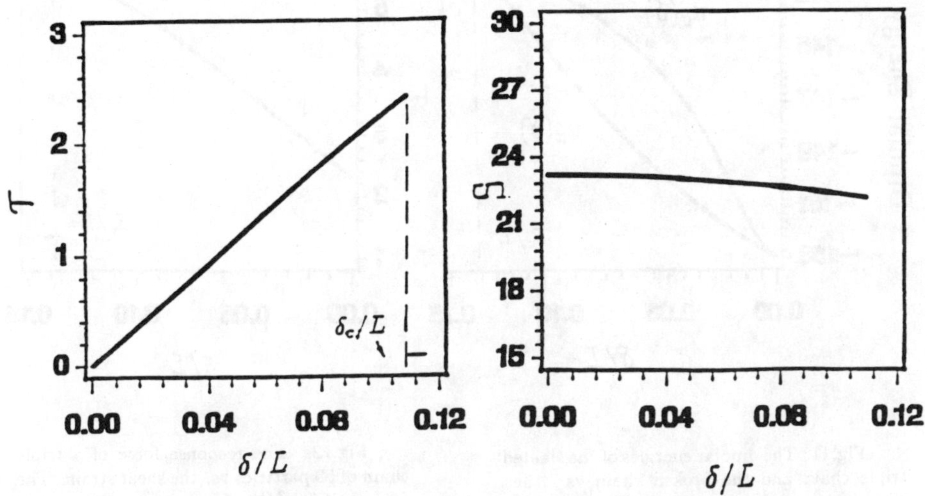

Fig.9. The response force of a double chain of 119 particles vs. the shear strain. The force unit is $p^2/(\epsilon_f d^4)$.

Fig.10. The response coefficient S of a double chain of 119 particles vs. the shear strain up to the yield point. The unit of S is $p^2/(\epsilon_f d^4)$.

Fig.8 gives the dipolar energies of the slanted double chains and broken double chains of 119 particles (60 particles in the A chain and 50 in the B). Fig.9 shows the response force τ of this double chain. Fig.10 is its response coefficient S. At volume fraction ϕ, the double-chain structure has $6\phi/[\pi d^2(2-d/L)]$ double chains per unit cross-section. Therefore, this double chain structure has shear modulus $6S\phi/[\pi d^2(2-d/L)]$.

Fig.11. gives the dipolar energies of the slanted and broken triple chain of 59 particles (20 particles in the A chain and 19 in the B). Fig.12 and 13 show the response force and response coefficient of this triple chain.

The bct lattice used in our calculation has nine A chains and four B chains. Since a bct lattice is a non-isotropic three-dimensional structure, we first compared deformations in different directions. It turns out that the shear strain in the x (or y) direction has the lowest response force for this bct lattice. In Fig. 14, we plot the dipolar energies of the slanted and broken bct lattice of 178 particles which has 9 A chains, 14 particles each, and 4 B chains, 13 particles each. Fig.15 depicts the response force τ for the bct lattice. Fig.16 shows the response coefficient S of this bct lattice. All deformations in Figs.14-16 are in the x direction.

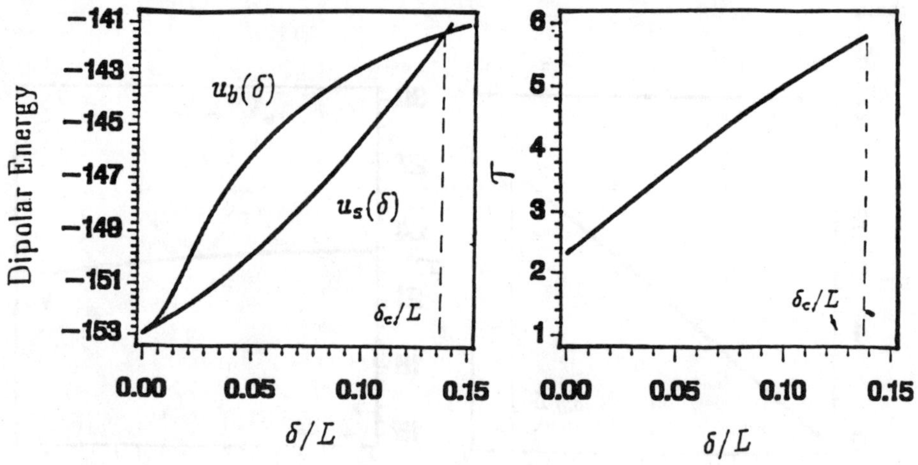

Fig.11. The dipolar energies of the slanted triple chain and the broken chain vs. the shear strain respectively. The triple chain has 59 particles. The energy unit is $p^2/(\epsilon_f d^3)$.

Fig.12. The response force of a triple chain of 59 particles vs. the shear strain. The force unit is $p^2/(\epsilon_f d^4)$.

Similar to the single-chain case, there is a yield point δ_c/L for all of these structures. When the shear strain is less δ_c/L, the slanted structures are always preferred. When the shear strain exceeds δ_c/L, the structures break into two parts. The slanted-broken structures always have an energy interpolating between the slanted structures and the broken structures. Therefore, the competition is really between the slanted structures and the broken structures. As seen in Figs.9, 12, and 15, similar to a single chain, when $\delta > \delta_c$, the preferred broken structures produce a much smaller and almost flat response force. In Figs.10, 13, and 16, we plot the response coefficient S of these thick structures up to their yield point. It is clear from Figs.9, 12, and 15 that all of these structures have a much smaller S beyond the yield point.

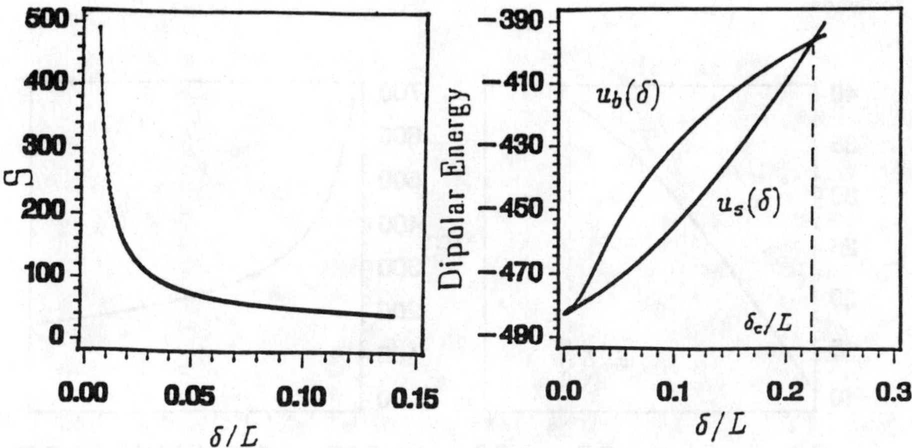

Fig.13. The response coefficient S of a triple chain of 59 particles vs. the shear strain. The unit of S is $p^2/(\epsilon_f d^4)$.

Fig.14. The dipolar energies of the slanted bct lattice and the broken bct lattice vs. the shear strain respectively. The bct lattice has 178 particles. The energy unit is $p^2/(\epsilon_f d^3)$.

As in Eq.(3.3), the critical response coefficient per particle, S_c/N, of the double chain, triple chain, and the bct lattice provides measurement of the strength of the induced structures.

4 Results and Discussions

To understand the stability of the solid structures and make a comparison, we plot the critical response coefficients per particle S_c/N versus d/L in Fig. 17. As stated earlier, the higher S_c/N, the stronger the structure.

Fig.17 first shows that a single chain is stronger than a double chain when L/d is small. This is consistent with the result reported in reference 8. However, as L/d increases (or as d/L decreases in Fig.17), the critical response coefficient of a double chain decreases slower than that of a single chain. When $L/d > 400$, a double chain becomes stronger than a single chain. Our numerical calculation also verifies this conclusion.

In addition, as $d/L \to 0$, the critical response coefficient per particle of a single chain tends to zero. This is the Peierls-Landau instability of an infinite one-dimensional solid. As seen from Fig. 17, the response coefficients per particle of a double chain, triple chain, and bct lattice do not extrapolate to zero as $d/L \to 0$. This implies that these thick structures are more stable than the single-chain structure. The single-chain structure may be stable only if the electrode spacing is not too wide.

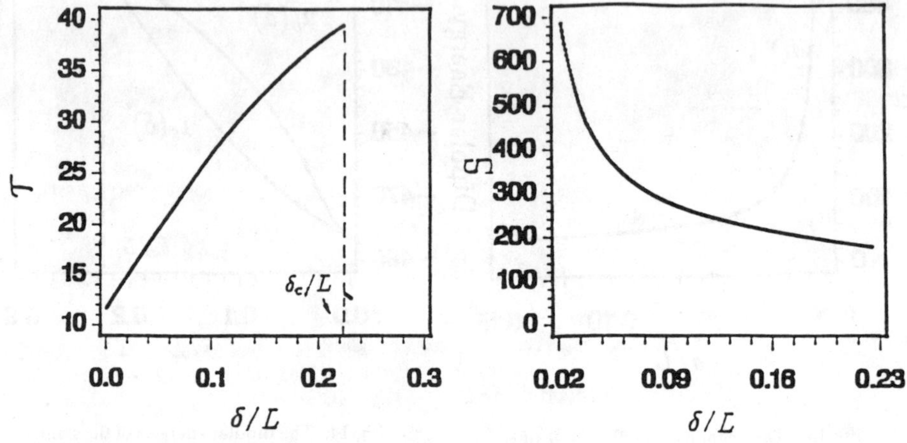

Fig.15. The response force of a bct lattice of 178 particles vs. the shear strain. The force unit is $p^2/(\epsilon_f d^4)$.

Fig.16. The response coefficient S of the bct lattice of 178 particles vs. the shear strain up to the yield point. The unit S is $p^2/(\epsilon_f d^4)$.

As seen from Fig. 17, the triple chain is much stronger than the double chain. When $L = 60d$, the critical response coefficient per particle of a triple chain is twice as big as that of a double chain. The bct lattice is even stronger than the triple-chain structure.

Fig.13 and 16 also show that the response force of a triple chain and a bct lattice is not linear with a small deformation δ. These structures are close-packed and have very limited room to move their particles in the initial part of the deformation (Fig.7). In responding to the initial deformation, they usually produce a quite large modulus to resist the deformation. For example, a triple chain in Fig.6(a) has two chains of class A and one chain of class B. During deformation, the positions of the second chain of class A is affected by the deformation of the middle chain of class B. If δ increases, the deformed structures are no longer close-packed, then there is more room to arrange particles and the response force becomes almost linear with the deformation. Therefore, the response coefficient of a triple chain and a bct lattice is very high for small δ, then it decreases with an increase of δ and tends to be stable as δ further increases until δ_c is reached.

Because the single-chain structure is not a close-packed structure (Fig.1), its response force is almost linear to the deformation from the beginning as seen in Fig.4. This is consistent with the results found by Conrad, Chen, and Sprecher[9]. Though a double chain has a single chain of class A and a single chain of class B close-packed, the geometric deformation of these two chains is not affected by each other as in a triple chain or bct lattice (Fig.5). Therefore, the response force is also almost linear to the deformation from the beginning.

In Fig.18, we plot the yield point δ_c/L versus d/L for all four structures. It is noted that as L/d increases, the yield point decreases. As $d/L \rightarrow 0$, the critical shear strain of a single chain does not tend to zero, This implies that to break a long singe chain, a minimum shear strain is still needed, though the shear modulus is vanishing.

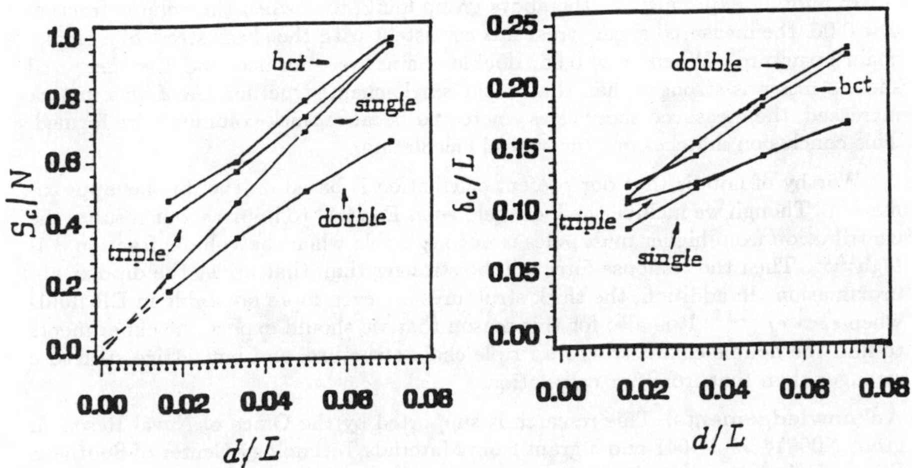

Fig.17. Critical response coefficient per particle versus d/L for the single chain, double chain, triple chain, and bct lattice.

Fig.18. Yield point δ_c/L versus d/L for the single chain, double chain, triple chain, and bct lattice.

To conclude our paper, we would like to make some comparison of our theoretical results with experiments[9]. Conrad, Sprecher, and Chen observed the single-chain structure when ϕ is low and L/d is not too big. Similar to our discussion, they applied a shear strain to the chain by sliding one electrode. The sliding electrode had a rough surface and there was sticking between the electrode and particles. The same as in our theoretical discussion, the chain first became slanted, then broke into two parts when the shear strain exceeded a yield point. For example, their experiment found the yield point δ_c/L was about 0.4 for $L/d = 3$. Our calculation finds $\delta_c/L = 0.31$ for $L = 3d$, close to the experimental results. Their experiment also found the response force was about 3.8×10^{-6}N for shear strain 0.21, $L/d = 3$, and $E = 2$KV/mm. Our calculation has the response force $3.0p^2/(\epsilon_f d^4)$ under the above condition. The dielectric particles in the experiment were moist glass spheres of $d = 150\mu$m. The liquid has $\epsilon_f = 2.5$. The effective dielectric constant of moist glass spheres was not measured in the experiment but was expected to be higher than 7.2, the dielectric constant of dry glass spheres. Applying the effective local field for a single chain[5],

$$E_{loc} = E/(1 - 0.601028\alpha), \tag{4.1}$$

we have found our response force close to 3.8×10^{-6}N when $\epsilon_p = 22.8$, a reasonable result for the effective dielectric constant of moist glass spheres.

In another experiment[10], the above group found that when the volume fraction $\phi < 0.06$, the measured shear stress was consistent with the shear stress of a single-chain structure. When $\phi > 0.06$, double chains were formed and the measured shear stress was stronger than that of the single-chain structure. As ϕ was further increased, the measured shear stress increased because thick columns were formed. This conclusion matches our theoretical calculation.

Worthy of note is that our present calculation is based on the dipolar approximation. Though we include the local field as in Eq.(4.1) to improve our results, the contribution from higher multipoles is not negligible when the volume fraction ϕ is high[12,13]. Then the response force will be stronger than that under the dipolar approximation. In addition, the thick structures are even more favorable in ER fluids when $\epsilon_p > \epsilon_f$ [12,13]. It is also for this reason that we should expect the experiments to find the double-chain structure, triple chain structure, and bct lattice structure stronger than that from our calculation.

Acknowledgements. This research is supported by the Office of Naval Research grant N00014-90-J-4041 and a grant from Materials Technology Center of Southern Illinois University at Carbondale.

REFERENCES

1. For example, see, *Electrorheological Fluids*, edited by R. Tao (World Scientific Publishing Comp., Singapore, 1992); P. M. Adriani and A. P. Gast, Faraday Discuss. Chem. Soc. **90**,1 (1990); W. M. Winslow, J. Appl. Phys. **20**, 1137 (1949); H. Block and J. P. Kelly, US Patent 4,687,589 (1987); J.D. Carlson, US Patent no. 4,772,407 (1988); F. E. Filisko and W. E. Armstrong, US patent No. 4 744 914 (1988); D. J. Klingenberg, F. van Swol, and C. F. Zukoski, *J. Chem. Phys.* **91**, 7888 (1989).
2. R. Tao, J. T. Woestman, and N. K. Jaggi, *Appl. Phys. Lett.***55**, 1844 (1989).
3. T. C. Halsey and W. Toor, *Phys. Rev. Lett.* **65**, 2820 (1990)
4. T. C. Halsey, J. E. Martin, and D. Adolf, *Phys. Rev. Lett.***68**, 1519 (1992).
5. R. Tao and J. M. Sun, *Phys. Rev. Lett.* **67**, 398 (1991).
6. R. Tao and J. M. Sun, *Phys. Rev. A*, **44**, R6181 (1991).
7. T. J. Chen, R. N. Zitter, and R. Tao, *Phys. Rev. Lett.* **68**, 2555 (1992).
8. A. M. Kraynik, R. T. Bonnecaze, and J. F. Brady, *Electrorheological Fluids* (edited by R. Tao, World Scientific Pub., Singapore, 1992) p59.
9. H. Conrad, A.F. Sprecher, and Y. Chen in *Electrorheological Fluids* edited by J.D. Carlson, A.F. Sprecher, and H. Conrad (Technomic Publishing Co., Lancaster, PA, 1990) p 82.
10. H. Conrad, A.F. Sprecher, and Y. Chen, in *Electrorheological Fluids* edited by R. Tao (World Scientific Pub., Singapore, 1992) p 195.
11. L. D. Landau, in *Collected Papers of L. D. Landau*, edited by D. ter Haar (Gordon and Breach, New York 1965) p209; R. E. Peierls, Helv. Phys. Acta Suppl. **7**, 81 (1934).
12. L. C. Davis, *Appl. Phys. Lett.* **60**, 319 (1992); J. Appl. Phys. **72**, 1334 (1992).
13. Y. Chen, A. F. Sprecher, and H. Conrad, *J. Appl. Phys.* **70**, 6796 (1991).

Dynamics of Structure Deformation and Rheology of Electrorheological Fluids

E. Lemaire, G. Bossis, Y. Grasselli, A. Meunier
Laboratoire de Physique de la Matière Condensée
CNRS, Universite de Nice, Parc Valrose
06108 Nice Cedex 2, France

ABSTRACT

We report some rheological feature of ER fluids observed on a suspension composed of silica particles dispersed in silicon oil. The experiments show a thixotropic behavior with two different characteristic times. We relate the shorter one to the separation time between two particles inside a chain formed by the electric field: the longer one originates from a mesoscopic organization of the suspension submitted to an electric field and to a flow. This structural change can occur as well in presence of a stationary flow as during an oscillatory solicitation. In this last case, the viscoelastic parameters are shown to be very sensitive to the kind of present structure. We explain this dependance by means of the calculus of interaction forces between two or three dielectric particles.

INTRODUCTION

Electrorheological fluids are known to present important changes in their rheological properties when they are submitted to an electric field of a few kilovolts per millimeter. This behavior comes from the polarization of the particles and from the resulting structural change. Generally, the ER effect is evaluated by means of the yield stress appearing under the application of an electric field. Indeed it is usual to describe the rheological behavior of ER fluids by a Bingham law so that the fluid is enterely characterized by the yield stress and the plastic viscosity which is generally taken equal to the viscosity in absence of the electric field.In order to predict the value of the yield stress, the suspension is often represented by an assembly of non interacting chains aligned in the direction of the applied electric field. Then, the yield stress is easily obtained from the knowledge of the electrostatic interaction force between two dielectric particles . In the second part of this paper (the first one reports experimental results), we compare this model to another based on the calculus of the restoring torque on an homogeneous ellipsoid tilted in a field. The next two parts are devoted to a tentative of an explanation of the experimental observations. Actually some experiments carried out on a silica based suspension show that the behavior of ER fluids could be much more complex than the one of a Bingham fluid; for instance, ER fluids can exhibit two distinct thixotropic features with two different characteristic times. In the third part we propose to relate the shorter characteristic time to the time evolution of the distance of separation between two particles. The second thyxotropic feature is shown to result from a new organization of the fibers of particles in stripes formed in the plane defined by the direction of the field and the flow.The same kind of structural changes occurs in the presence of an oscillating flow and influences considerably the viscoelastic behavior of the fluid. This point is discussed in the last section.

I - EXPERIMENTAL

In this first part, we present the suspensions and devices we have used to study the rheological behavior of ER fluids.

I.1. Composition of ER fluids

We have used two kinds of suspension which are both composed of micronic silica particles ($2a = 1\mu m$) dispersed in silicon oil. The solid volume fraction is equal to 12% and the viscosity of the silicon oil is either 20 cp or 500 cp. We are going to show the advantage to make use of two suspensions with rather different viscosities. Unfortunately the surface treatment used to stabilize the particles was different in these two samples, so the interactions between the particles are not the same in both systems and the experimental results obtained on each dispersion are not directly comparable.

The relative permittivities of the suspensions have been measured with a variable thickness cell. The permittivity of the less viscous dispersion has been found to be $\varepsilon_S = 21,6$. With the knowledge of this value and of the permittivity of silicon oil, $\varepsilon_p = 2,8$, we can deduce the internal permittivity, ε_p, of a particle from a mean field theory like Bruggeman's [1]. We get $\varepsilon_p = 62$. All these permittivity values are obtained for a frequency of 100 Hz. For the most viscous suspension the values are the following : $\varepsilon_S = 3.2$ and $\varepsilon_S = 7,7$ at 100 Hz.

I.2. Experimental devices

The steady flow measurements have been carried out on a controlled stress rheometer Carrimed CLS 100 either in cylindrical Couette geometry or in rotating parallel disks configuration. The external cylinder or the lower plate are connected to the high voltage and the internal cylinder or the upper plate are rotating and grounded through an electrode deeping in a mercury reservoir situated at the top of the rotation axis. Each of the two systems is convenient for a special experiment : with the Couette geometry we can assume that the shear rate within the gap is constant since the difference of radii of the inner and the outer cylinder is small ($\Delta R/R \approx 0,08$). So we can assume that everywhere in the cell the state of the suspension is the same. Concerning the second device the disks are made of glass and are coated with a transparent film of tin indium oxide. It allows to apply an electric field and to maintain enough transparency to observe the structure in a plane perpendicular to the field with a microscope.

The oscillating plate rheometer we have used is an original device [2]. The cell containing the ER fluid is composed of two parallel disks coated with tin indium oxide; the lower plane is motionless and the upper one is mounted on the arm of an electromagnetic vibrator; the applied sinusoIdal force is controlled by the electric current passing through the vibrator coil and the displacement of the upper disk is measured with an optic sensor whose precision is one tenth of micron [3]. A computer recordS simultaneously the force and the displacement, the motion of the upper disk is the one of an harmonic oscillator governed by the equation :

$$m\ddot{x} + \left(k + \frac{G'S}{d}\right) x + \left(\alpha + \frac{\eta' S}{d}\right) \dot{x} = F_o \cos \omega_o t \qquad (1)$$

m is the mass of vibrating part of the apparatus k and α are respectively the elastic and the damping factors of the vibrator, S and d, the surface of the electrodes and the gap between them.

At last, G' and η' are the viscoelastic characteristics of the suspension : storage modulus and viscosity . From eq.(1), we can deduce the phase shift, φ, between the displacement and the applied force :

$$\tan \varphi = \frac{1}{m} \frac{\left(\alpha + \frac{\eta S}{d}\right) \omega}{\omega_1^2 - \omega^2}$$

and (2)

$$\omega_1^2 = \frac{1}{m}\left\{k + \frac{G'S}{d}\right\}$$

Then, a measurement of the variation of φ versus the frequency allows to determine the storage modulus G'.

II.3. Preliminary experimental observations

Performing measurements of the static yield stress, we have observed that its value was dependant on the rate of increase of the shear stress. This time dependance is not so often reported since most of experiments are carried out with controlled shear rate rheometers. The second important observation is that, whatever the characteristic time, the loading and unloading curves are different. These observations were made under a field of 1710 V/mm on the more viscous fluid. They are reported in fig. 1 where are represented three loading curves obtained for different rates of increase of the shear stress and an unloading curve, which is independant of time in our experimental times range The time evolution of the loading behavior seems to be of the same order of magnitude as the time needed for the particles to separate of a diameter from each other. The experimental observations will be explained by means of a model based on two particles hydrodynamic and electrostatic interactions. The characteristic time evolution is proportionnal to the viscosity of the suspending fluid, so this thixotropic behavior is easier to observe on the more viscous fluid.

We also noted that the rheological behavior is history dependant. Indeed fig. 2 presents two rheograms obtained either when the suspension is stressed for the first time after the application of the electric field (lower curve) or when a prestress of 28 Pa ($\dot{\gamma} = 20\ s^{-1}$) has been applied during few minutes (upper curve). These last observations have been made with the parallel transparent disks on the less viscous fluid in order to be sure that the rheograms were equilibrium curves. We observed that during the prestress, isolated aggregates moved to form some concentric stripes in the direction of the flow and of the field. We shall explain the increase of the yield stress in the presence of stripes by their internal structure and the interactions of particles inside themselves.

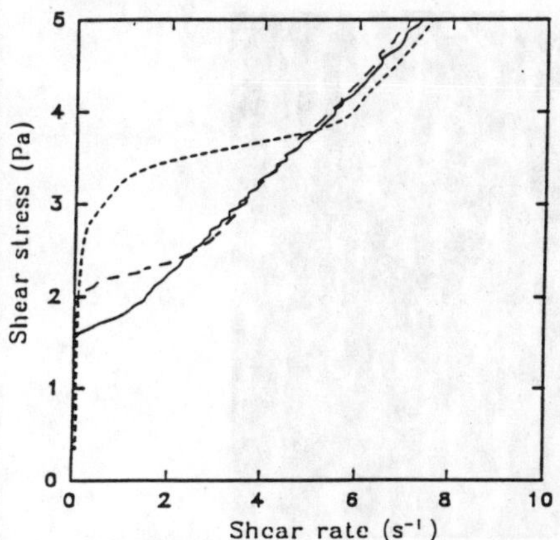

Fig.1. Rheograms obtained for different loading times: $c = 4 \times 10^{-3}$ Pa/s (—) , $c = 1.2 \times 10^{-2}$ Pa/s (- - -), $c = 0.33$ Pa/s (······). Whatever the characteristic unloading time the rheograms are the same.

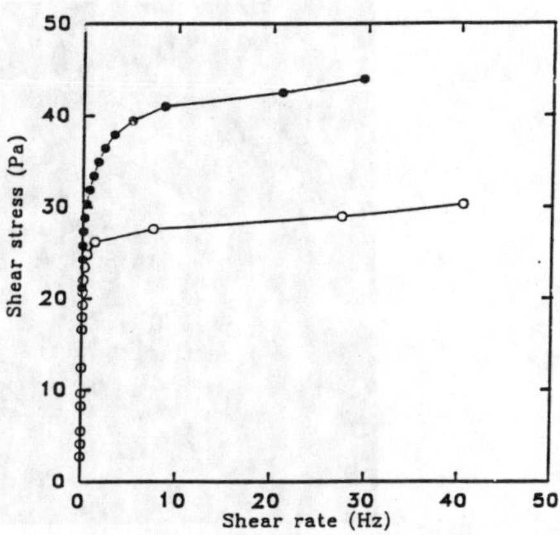

Fig.2. Equilibrium rheograms obtained without any prestress (• • •) and after a prestress of 28 Pa corresponding to a shear rate of 20 s^{-1} (o o o).

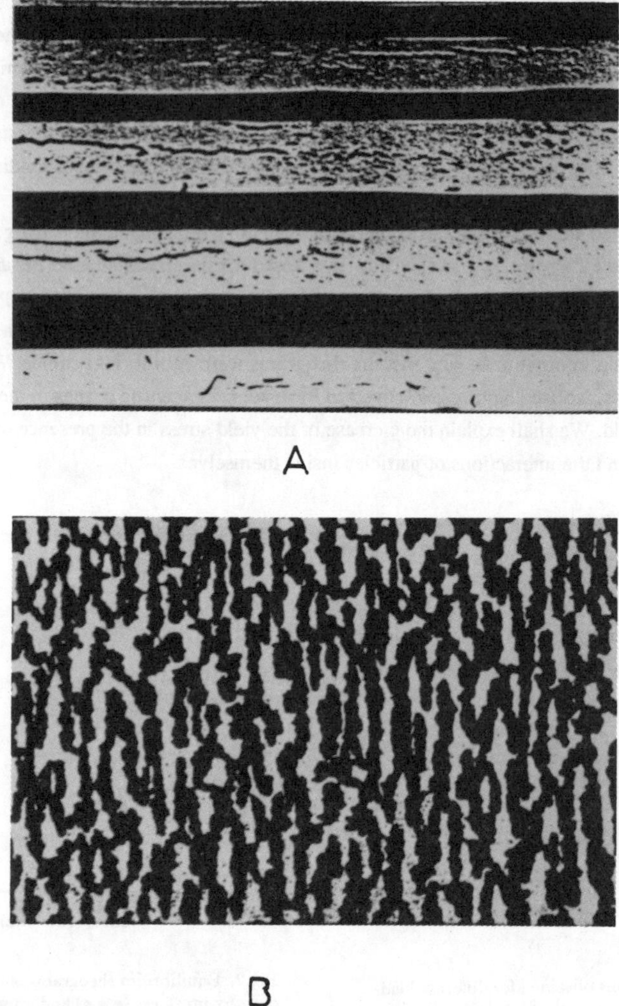

Fig.3 top views (in the plane perpendicular to the electric field)of the structure of the suspension submitted to a field and a flow; A :stripes parallel to the direction of the field and the flow observed either in a stationary flow or in oscillating shear at large amplitudes ($\gamma \approx 1$), B :sheets perpendicular to the flow obtained in oscillatory flow for intermediary amplitudes ($10^{-2}<\gamma<1$).

We can observe the same kind of phenomenon in sinusoidal oscillation experiments : when the amplitude of the strain is very small ($\gamma < 10^{-2}$), the aggregates remain isolated whereas, when the deformation is increased to about 1, they form a structure very similar to the one observed in steady flow (planes parallel to the direction of the flow and of the field, fig.3.A.). And, at last, under some intermediary strain amplitudes the aggregates group into sheets perpendicular to the direction of the flow (fig.3.B). The storage modulus, G', of the suspension under the field is very sensitive to the structure. Tab. 1 presents a summary of the variation of G' versus the structure. These results were obtained at a strain amplitude of 2.10^{-2} in the presence of a sinusoidal electric field of amplitude E_0 = 16.5 kV/cm and of frequency ν_E = 600 Hz. G' is practically independant of the oscillatory frequency in the range of frequencies we have used (50 < V < 80 Hz).

Structure	Isolated aggregates	Perpendicular sheets	Parallel stripes
G'(Pa)	170	74	1280

Tab. 1

As in the stationary case, the influence of the structure formation on the values of the storage moduli can be explained by the interactions between two or three particles inside the stripes.

II - STATIC YIELD STRESS

Generally,the rheological behavior of ER fluids is modeled by a Bingham law; indeed, in the presence of the field the particles form aggregates elongated in the field direction which can link the two electrodes. So, if we want the suspension to flow in direction orthogonal to the electric field, a finite stress is needed in order to break the fibers of particles. This is the static yield stress. Usually, the field induced structure is modeled by assembly of isolated chains of particles. This description is correct when the solid volume fraction is very low (less than 1%). At higher solid content, the chains group into thicker fibers. In this first part, we are going to show the influence of the choice of the structure model (chains of particles or prolate ellipsoidal aggregates) on the static yield stress predictions.

II.1. Chain model

This reference model is based on the hypothesis that the structure induced by the field is formed of isolated chains of spheres linking the two electrodes. In this case, the yield stress is simply related to the force needed to separate two particles initially aligned in the field direction and pulled tangentially to each other.

Klingenberg [4] has calculated this force for two dielectric spheres of permittivity ε_p surrounded by a dielectric fluid of permittivity is ε_f. The maximum of this restoring force F_{12}^m gives the yield stress :

$$\tau_S = \frac{\phi}{\frac{2}{3}\pi a^2} F_{12}^m \qquad (3)$$

where ϕ is the volume fraction of particles and a their radius.
F_{12}^m is the maximum of the restoring force which is given by :

$$F_{12}^m = 3\,\varepsilon_f\, a^2 \beta^2 E^2 f\left(\frac{\varepsilon_p}{\varepsilon_f}, \frac{r}{a}\right) \qquad (4)$$

with

$$\beta = \frac{\varepsilon_p - \varepsilon_f}{\varepsilon_p + \varepsilon_f}$$

and

$$f\left(\frac{\varepsilon_p}{\varepsilon_f}, \frac{a}{r}\right) = \left(\frac{a}{r}\right)^4 [\,(2f_{//} + 2f_\Gamma)\sin\theta\cos^2\theta - f_\perp \sin^3\theta\,]$$

Where $f_{//}$,f_Γ and $f_\perp$ depend only on the ratio of permittivities and are calculated for various ratios [4], [5].

II.2. Ellipsoids model

Since usually ER fluids are concentrated or semi diluted suspensions ($\phi > 10\%$) their structure under an electric field has to be modeled by elongated aggregates rather than by isolated chains. In previous works we have shown that the aggregates could be represented by prolate ellipsoids [6], [7]. We assume that, during the shear (before the yield stress is reached), the shape of the ellipsoids (ratio of the semi main axis, l, to the semi small axis, b) doesn't change. This hypothesis is not verified since the aggregates which connect the two plates of the rhometer are stretched; but we have verified that taking into account the stretch only modify the numerical results by a few percents. So we can calculate the yield stress from the torque needed to rotate the Na/S aggregates per unit area in the

presence of a field. The torque is given by the derivative of the energy with respect to θ, the angle of inclinaison :

$$\Gamma = -\partial U/\partial\theta \qquad (5)$$

with the electrostatic energy per unit volume given by

$$U = (-N_a/V)\, M_a . E_o \qquad (6)$$

In Eq. 6 N_a/V is the number of aggregates per unit volume, M_a their dipolar moment and E_0, the applied electric field.

So, from Eqs (5) and (6) we can express the torque per unit volume which is also the restoring force per unit area :

$$\Gamma = \frac{\varepsilon_o}{2}(\varepsilon_a - 1)\, E_o^2 \frac{\phi}{\phi_a} \frac{(\varepsilon_r - 1)(n_y - n_z)}{[1 + (\varepsilon_r - 1)\, n_a][1 + (\varepsilon_r - 1)\, n_y]} \qquad (7)$$

where ϕ_a is the solid volume fraction inside the aggregates which has been chosen equal to 0,64. ε_a is the relative permittivity of the aggregates, this value is calculated from ε_p with the help of Bruggeman's theory; $\varepsilon_r = \varepsilon_p / \varepsilon_F$ is the ratio of the permittivities .

At last, n_z and n_y are the depolarizing factors in the direction of the main axis of the ellipsoid and perpendiculary to it, respectively :

$$n_z = \frac{1 - e^2}{2e^3}\left\{ \ln \frac{1 + e}{1 - e} - 2e \right\}$$

and $n_y = (1 - n_z)/2$ **(8)**

where e is the exentricity :

$$e = \sqrt{1 - \frac{b^2}{l^2}}$$

II.3. Experimental results, comparison with theories

For this first experiment whose aim is to determine the static yield stress, we have used the less viscous suspension. Indeed, in the third part of this paper, we shall show that the rheological behavior of ER fluids is governed by a characteristic time proportional to the viscosity. So with the less viscous fluid, we can reach faster the equilibrium and measure the real equilibrium static yield stress.

The static yield stress, τ_S, is measured by increasing the applied shear stress from zero to a final value. τ_S is assumed to be reached when the velocity becomes non zero.

As expected theoretically, the variation of the yield stress versus the electric field is quadratic. This behavior is shown in fig. 4 where are also represented the theoretical predictions of the two models. Chain model underestimates the yield stresses while ellipsoid model gives larger values. In this last model, the possibility for the particles to separate from each other as in the model of isolated chains is not taken into account. This separation lowers the restoring torque and a more suitable model should consider both the real structure of the suspension (ellipsoidal aggregates) and the separation of particles inside the aggregates during the shear.

Yet, even if the model of isolated chains is not really suitable for concentrated ER fluids, it is interesting because it offers a microscopic description of the suspension. In particular we are going to see, in the next two parts, that this discrete modelization of the suspension allows a semi quantitative explanation of the thixotropy.

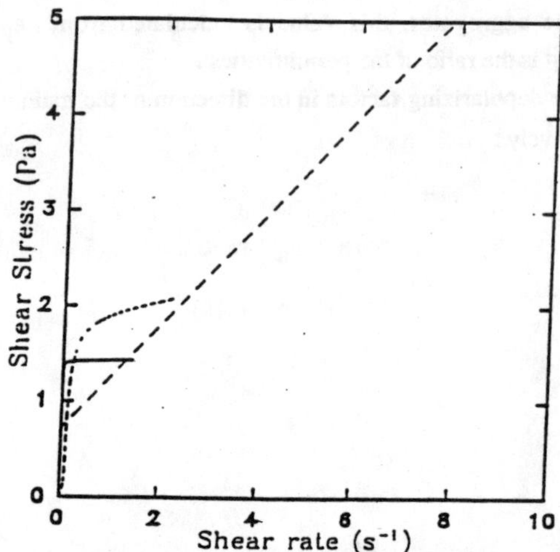

Fig.4 Yield stress versus electric field, experimental curve (o o o), chains model predictions (● ● ●) , ellipsoids model predictions (∇∇∇)

III - THIXOTROPY OF ER FLUIDS

In this part, we relate the thixotropic behavior of ER fluids to the microscopic scale. In particular, we are going to see that thixotropy has two distinct origins whose characteristic times are very different : an hydrodynamic origin with a short characteristic time corresponding to the separation of two spheres inside a chain and a structural origin where the caracteristic time is ruled by the displacement of entire fibers to form some planes in the direction of the field and the flow.

III.1. Hydrodynamic thixotropy : dynamical effect

Our purpose is to explain the dependance of the rheograms on the rate of increase of the shear stress τ. For a linear increase with time :

$$\tau = ct \qquad (9)$$

We shall model the suspension in presence of the electric field by a set of isolated chains linking the two plates of the rheometer. Each particle in a chain is submitted to a force F^a given by :

$$F^a = \frac{2\pi}{3}\frac{a^2}{\phi} ct \qquad (10)$$

and also to the electric restoring force, F_{12}, given by the formula (4). So we can obtain the relative velocity of a pair of particles in a chain by the product of the mobility matrix by the difference of forces acting on them :

$$\delta V = M\left[F^a - F_{12}\right] \qquad (11)$$

$$M_{ij} = \frac{1}{6\,\pi\eta a}\left\{G(\xi)\, e_i\, e_j + (d_{ij} - e_i\, e_j)\, H(\xi)\right\}$$

e is the unit vector along the direction of the tilted chain; its components are $\cos\theta$ in the direction of the field and $\sin\theta$ in the direction of the applied force. ξ is the normalized (by a) separation between particles. For small separations (ξ<0.1), the functions $G(\xi)$ and $H(\xi)$ can be approximated by :

$$G(\xi) = 2\xi + 1.8\xi^2\ \ln\xi - 4\,\xi^2$$
$$H(\xi) = 0.401 - 0.532 / \ln\xi$$

So, using eqs. (4), (10) and (11) we obtain the equation of relative motion of the two particles :

$$\frac{d\left[(\xi+2)\sin\theta\right]}{dt} = \frac{1}{6\pi\eta a^2}\left\{G(\xi)\sin^2\theta + H(\xi)\cos^2\theta\right\}.\left\{F^a(t) - F_{12}(\xi)\right\} \qquad (12)$$

In this model, we assume that the chain deformation is affine i.e. there is no motion along the field axis. Then θ and ξ are related by :

$$\cos\theta = (2+\xi_o)/(2+\xi) \tag{13}$$

where ξ_o is the initial normalized separation (when the chain is not strained) between particles. For this value, we have chosen 10^{-3}. Nevertheless, since the relative motion is tangential and the maximum restoring force is reached for a rather large angle ($\theta \approx 10°$), the choice of this parameter does not affect significantly the result of eq. (12). The left term in eq. (12) gives the relative normalized velocity δV which is related to the shear rate by :

$$\dot{\gamma} = \frac{\delta V}{(2+\xi_o)} \tag{14}$$

This model is valid as far as the chain does not break. Then after the rupture we have to consider the rebuilding of the chain but as can be seen experimentally , it is simpler to use a Bingham law :

$$\tau = \eta_o\dot{\gamma} + \tau_d \tag{15}$$

where η_o is the zero field viscosity and τ_d the dynamic yield stress. τ_d can be calculated by evaluating the extra viscous dissipation introduced by the presence of the electric restoring force [8]. Indeed when the relative motion of two particles is restricted by the electrostatic interactions the energy is stored. This energy is given back by means of an extra dissipation when, on the contrary, the chain breaks and reforms rapidly [9]. In this model τ_d can be evaluated by the following integral :

$$\tau_d = \frac{2a}{\langle d\rangle} \frac{\phi}{\frac{2}{3}\pi a^2} \int_0^{\langle d\rangle/2} F_{12}(y)\, dy$$

where <d> is the average distance between two chains.

Fig. 5, shows the theoretical rheograms obtained with this model for two different rates of increase the stress : $c = 4.10^{-3}$ Pa/s (lower curve) and $c = 0.33$ Pa/s (upper curve).

Comparing these results to the experimental one presented in fig. 1 we can first note that this simple two spheres model is able to explain the main features of the experimental results. The theory underestimates the values of the stress but this problem has been discussed in part II where we have shown that a chain model could not describe properly the rheology of a concentrated suspension in the presence of a field.

At last, it is worth noting that the experimental results are reproducible only if the increase of the stress is preceeded by an homogeneization of the suspension without electric field.Otherwise, the rheological behavior of the ER fluid is history dependant with a long characteristic time. In the next section, we relate this dependance to a mesoscopic structural change of the suspension.

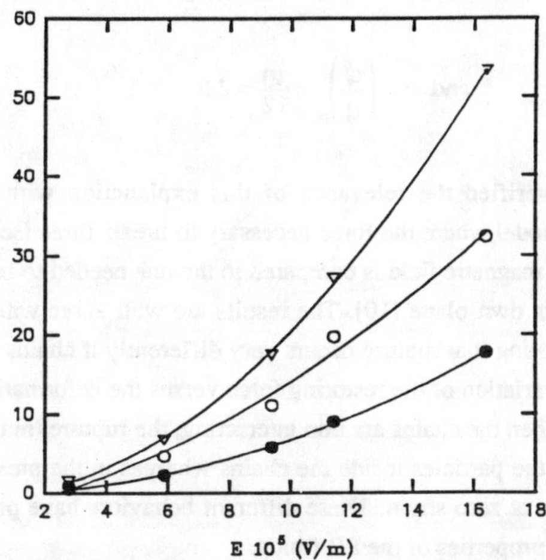

Fig.5 Theoretical rheograms obtained from a model based on the calculus of hydrodynamic and electrostatic forces between two particles, (—) laoding curve with c= 4 10-3 Pa/s ,(- - - -)laoding curve with c=0.33 Pa/s , (— — —) unlaoding curve.

III.2. Structural influence on rheology

Experiments show that the yield stress is larger when the suspension has been already sheared and is structured in stripes in the plane defined by the flow lines and the electric field (fig.2). We can model these stripes by a juxtaposition of chains of spheres shifted by a radius in the direction of the electric field (Fig.6). Then when this structure is being sheared, we have to consider not only the interaction between two spheres in a same chain (B1 and B2) but also the interaction between particles belonging to juxtaposed chains (B1 and A2). Assuming additivity of electrostatic forces - which is a not too bad approximation if the ratio of permittivities $\varepsilon_p/\varepsilon_f$ is small- the electrostatic restoring force per unit area is :

$$\tau_s^{//} = \frac{\phi}{\frac{2}{3}\pi a^2} F_m^{//} = \frac{\phi}{\frac{2}{3}\pi a^2} (\mathbf{F}_{B1B2} + \mathbf{F}_{B1A2}).\, \mathbf{e}_x \qquad (16)$$

This gives $\tau_s^{//} = 30$ Pa for the value of the yield stress to be compared to $\tau_s^i = 12$ Pa, for non interacting chains . We see that the experimental and theoretical ratios $\tau_s^{//}/\tau_s^i$ are very close :

$$\left(\frac{\tau_s^{//}}{\tau_s^i}\right)_{exp} = \frac{40}{25} = 2.6 \qquad \text{and} \qquad \left(\frac{\tau_s^{//}}{\tau_s^i}\right)_{th} = \frac{30}{12} = 2.5$$

Moreover,we have verified the relevance of this explanation with the help of a macroscopic experimental model where the force necessary to break three isolated chains of iron centimetric spheres in a magnetic field is compared to the one needed to break a stripe of three imbricated chains in its own plane [10]. The results are well agree with the proposed model. At last, it is worth noting that rupture occurs very differently if chains form planes or remain isolated. Indeed the variation of the restoring force versus the deformation is shown in figure 7 for the two cases. When the chains are non interacting, the rupture (maximum of F) is preceded by a separation of the particles inside the chains whereas in the presence of planar structures, the rupture occurs at zero strain. These different behaviors have probably a great influence on the viscoelastic properties of the ER fluids.

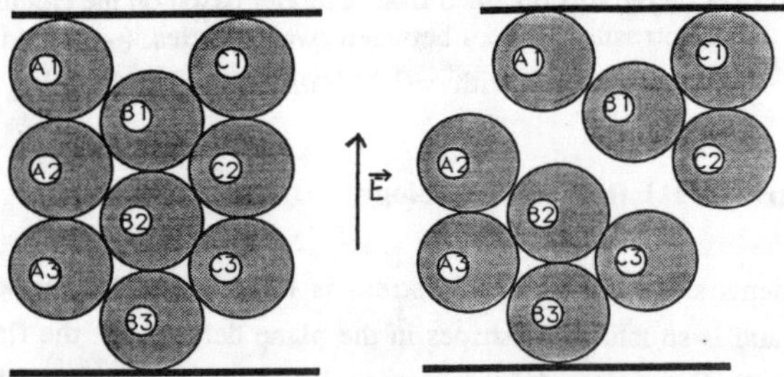

Fig.6 Shematic representation of the structure of a plane insine a stripe and of its rupture during a shear.

IV. VISCOELASTIC BEHAVIOR

Experiments show a strong dependance of the elastic moduli on the structure of the suspension (a difference of one order of magnitude depending on the stripes are parallel or perpendicular to the flow) (Tab.1). In the case of non interacting chains, the relation between the equilibrium stress, τ, and the deformaton, γ, can be obtained from equation (4) which can be rewritten for small strains :

$$\tau = \frac{\phi}{\frac{2}{3}\pi a^2} F_r \quad \text{with} \quad F_r = 24\pi\varepsilon_0\varepsilon_f a^2\beta^2 E^2\left(\frac{a}{r}\right)^4 (f_{//}+f_\Gamma)\,\gamma \qquad (17)$$

This expression leads to $G'(\omega=0)=\tau/\gamma=36$ Pa. This is to be compared to the experimental one obtained at a frequency of 60Hz : $G'(\omega=377))=170$Pa.

The theoretical value is about five times lower than the experimental one. This difference has likely several origins . We have already pointed out the first one in section II where we have shown that the non interacting chain model is not adapted to describe concentrated or even semi diluted supensions. On the other hand the non affine motion introduces a motion perpendicular to the velocity lines which can strongly increase the high frequency modulus [11]. This is also what is observed experimentally [12].

In order to explain the difference between moduli either in the presence of parallel stripes or isolated fibers, we can use on the same kind of argument than in the stationary case. Yet it is impossible to deduce the value of G' from the calculus of the interaction forces between three particles since the rupture of a plane occurs at zero deformation which would lead to an infinite modulus. To solve this difficulty , it would be interesting to build a model where the structures inside the planes are able to stretch and deform themselves

On the other hand, if we form parallel stripes at high shear amplitude and then measure the elastic modulus in the orthogonal direction we find about the same values of G' than the ones obtained with the perpendicular sheets. By this mean, we can evaluate the anisotropy of the elasticity modulus which itself gives an indication of the internal anisotropy of the structures.This observation would lead to propose a model for the structure inside the stripes : some monolayers of particles separated by a distance larger or of the order of magnitude of the particle diameter.This kind of structure has been partly observed by Brownian dynamics in a steady flow [13].

CONCLUSION

In this paper are summarized the main features of rheological behavior we have obseved on a silica suspension submitted to an electric field. Furthermore we have emphasized the thixotropic character of ER fluids with is two different characteristic times : a short one related to the separation between two particles and one much longer explained by the formation of a new mesoscopic order in the suspension. This effect due to a me soscopic change of the structure under a steady flow is also found in oscillating flow. In this last case the measured values of the elastic moduli differ of more than one order of magnitude depending on the presence of stripes either parallel or perpendicular to the velocity lines.

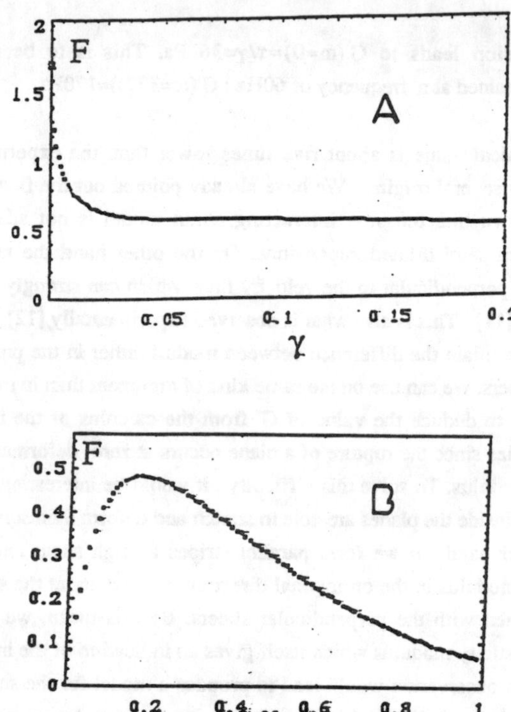

Fig.7 Variation of the restoring force versus the strain A : in the case of non Interacting chains (F^T=F_{A1A2}), B : in the case of planar structures sheared in their own direction (F^T= F_{B1B2}+F_{B1A2}).

REFERENCES

[1] BRUGGEMAN, D.A.G. Ann Phys. **1952**, 231, 779

[2] NOETINGER, B; PETIT,L.; GUAZZELLI,E.; CLEMENT,M.;Revue Phys. Appl. **1987**, 22, 1025

[3] BOSSIS,G.;GRASSELLI;Y.;LEMAIRE,E.; Submitted to Europhys.lett. 1993

[4] KLINGENBERG,D.J. Ph. D. Thesis, University of Illinois, 1990

[5] CLERCX H.J.H., BOSSIS G., Phys. Rev. E, in press

[6] LEMAIRE, E.; BOSSIS,G.;GRASSELLI,Y.;J. Phys D **1991**, 24, 1473

[7] GRASSELLI, Y.;BOSSIS,G.; LEMAIRE,E. et al.submitted to J.Phys. D1993

[8] BONNECAZE, R.T.;BRADY,J;F; J. RHEOL.**1992**, 36, 73

[9] LEMAIRE, E.;BOSSIS,G.; GRASSELLI,Y.; Langmuir **1992**, 8,2957

[10] LEMAIRE, E.;BOSSIS,G.; CLERCX, H.; GRASSELLI, Y.; MEUNIER, A.; Proceeedings of the international symposium "Advances in Structured and Heterogeneous Continua", Moscou August 1993.

[11] KLINGENBERG, D.J.; **1993**, 37, 199

[12] OTSUBO, Y.; SEKINE, M.; KATAYAMA, S; J. Rheol. **1992**, 35, 1411

[13] MELROSE, J.R. J. Chem. Phys. **1993**, 98, 5873

Determination of Rheological and Electrical Parameters of ER Fluids Using Rotational Viscometers

H. Janocha and B. Rech

Laboratory for Process Automation (LPA)
University of the Saarland
Postfach 1150
66041 Saarbrücken
Germany

ABSTRACT

The properties of an electrorheological fluid (ER fluid) strongly depend upon its particular composition. Hence, only limited information can be provided about an ER fluid's behaviour and its parameters without conducting experimental studies. The performance of an ER fluid actuator can be limited if the ER fluid used is not well known. That is why application-oriented measurement techniques are indispensable in the development of ER fluid actuators.

Rotational viscometers with an electrically insulated measurement arrangement can be implemented for determining rheological and electrical ER fluid behaviour. In this paper, the demands on the station used for measuring rheological and electrical ER fluid parameters are determined based on the essential ER fluid properties. Important problems concerning the measurements are also discussed. Moreover, the experience gathered by the Laboratory for Process Automation (LPA) during the construction of such a universal measuring station, which is based on a commercially available rotational viscometer, is described.

1. Introduction

The properties of an electrorheological fluid (ER fluid) strongly depend upon its particular composition. Therefore, only very general statements can be made about the behaviour of ER fluids. In addition, experiments are necessary to obtain exact ER fluid parameters while the relationship between the chemical composition and the physical properties of an ER fluid has not yet been established. These tests are unrenounceable for the technical application of electrorheological fluids since the functionability of an actuator can be limited through a lack of knowledge about the ER fluid used and/or the use of an inappropriate fluid. On this basis, the technique of measuring ER fluids is an important prerequisite for the development of ER fluid actuators.

2. Universal ER Fluid Measurement Station with Rotational Viscometer

The basic layout of the measurement stand installed at the Laboratory for Process Automation (LPA) for determining the rheological and electrical properties of ER fluids is shown in Figure 1. In addition to the ER fluid viscometer, the temperature control system and the high-voltage source, it consists of components for precision measurement of the current between and voltage across the electrodes. A computer equipped with D/A and A/D converters is implemented for recording the measurement values as well as for controlling the experiment. A software package which is capable of performing, for example, Fast-Fourier-Transforms (FFT) is used for signal analysis.

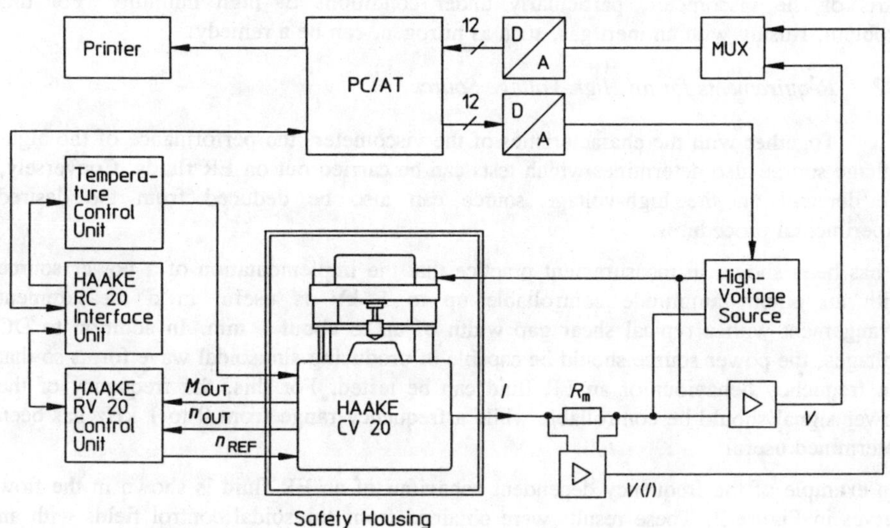

Figure 1: Layout of the ER fluid measurement station

2.1. Electrical Requirements for an ER Fluid Viscometer

A viscometer with electrically insulated rotational elements can be implemented in the determination of rheological and electrical ER fluid behaviour. By applying an electrical voltage potential, the test fluid can be exposed to an electric field. For cylindrical viscometers, the voltage potential exists between the outer and inner cylinders and for plate-plate viscometers, between the plates.

In addition to electrical insulation, it is important to prevent noise by shielding the sensors and electronic circuits from the ER fluid control signals. This is particularly important for viscometers with inductive torque and rotational speed sensors. In such viscometers, signal noise can be seen when ER fluids are activated with high amplitude alternating voltages if no special shielding is provided.

During measurement, electrical potentials with amplitudes of up to a few thousand volts can exist at the viscometer electrodes. To protect from personal injury, the viscometer should be further shielded. For the protective housing implemented on the LPA measurement station, the electrical control signal can be applied to the electrodes only when an additional high-voltage switch is activated by a closed housing door. Additionally, the operating conditions are indicated with signal lamps.

The control voltage amplitude may be limited by arcing between electrically activated parts of the viscometer, particularly under conditions of high humidity. For this problem, rinsing with an inert gas, such as nitrogen, can be a remedy.

2.2. *Requirements for an High-Voltage Source*

Together with the characteristics of the viscometer, the performance of the high-voltage source also determines which tests can be carried out on ER fluids. Conversely, the demand on the high-voltage source can also be deduced from the desired experimental procedures.

It has been shown in measurement practice that the implementation of a power source with an output amplitude controllable up to 5 kV is useful in a measurement arrangement with a typical shear gap width of up to about 1 mm. In addition to DC voltages, the power source should be capable of producing sinusoidal wave forms so that the frequency behaviour of an ER fluid can be tested. For this, the frequency of the power signal should be controllable while a frequency range from 0 to 1 kHz has been determined useful.

An example of the frequency dependent behaviour of an ER fluid is shown in the flow curves in Figure 2. These results were obtained from sinusoidal control fields with an effective (root-mean-square) field strength of E_{rms} = 2 kV/mm. For comparison, the curves obtained with no field applied as well as those with a DC field of strength E = 2 kV/mm are included. The measurements were performed with a modified HAAKE CV 20 viscometer with a cylindrical arrangement at a constant temperature of 25°C.

An undesirable behaviour is observed in the ER fluid controlled by a DC input signal. With an AC control voltage, on the other hand, the ER fluid exhibits good properties and demonstrates no reduction in the shear stress even at high shear rates. It can also be noticed from a comparison of the curves that the achievable shear stress increases up to a frequency of 250 Hz at which the maximum value is reached. The stress values become smaller again with higher frequencies.

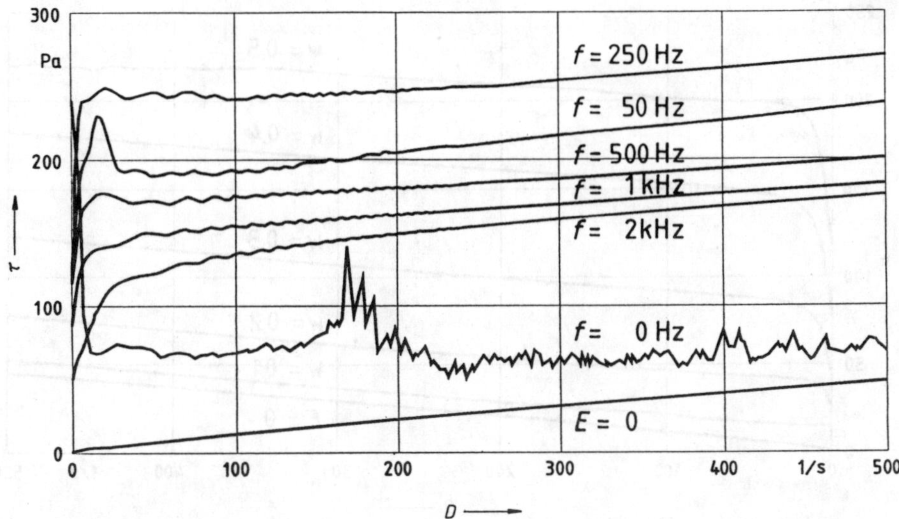

Figure 2: Flow curves for an ER fluid with sinusoidal control fields

The technical effort to produce square-wave alternating voltages is considerably smaller than for sinusoidal wave forms. The electronics for producing square-wave forms are also smaller and cheaper. For these reasons, the control of ER fluids with square-wave signals could play an important role in future applications. In automotive damping, for example, the control of motor mounts and shock absorbers can be performed by compact square-wave inverters coupled with high-voltage transformers at each damper. In this divided control concept, the inverter/transformer circuits can be supplied by a centrally located low-voltage DC source.

On this basis, it is useful to implement square-wave, in addition to DC and sinusoidal control signals, in determining the properties of ER fluids. This requires that the high-voltage source also be able to generate square-wave signals, of which the amplitude, the frequency and the pulse width of the signal should be controllable with a computer.

The flow curves demonstrating the influence of the change in the pulse width are shown in Figure 3. The ER fluid was controlled with a bipolar square-wave field signal with an amplitude of $E = 2$ kV/mm and a frequency of 300 Hz. Various pulse width ratios were tested. The pulse width ratio w is defined as the ratio between the pulse width t_1 and the period t_2 (see Figure 3). As in the previous example, the tests were performed with a cylindrical viscometer arrangement at a constant temperature of 25°C.

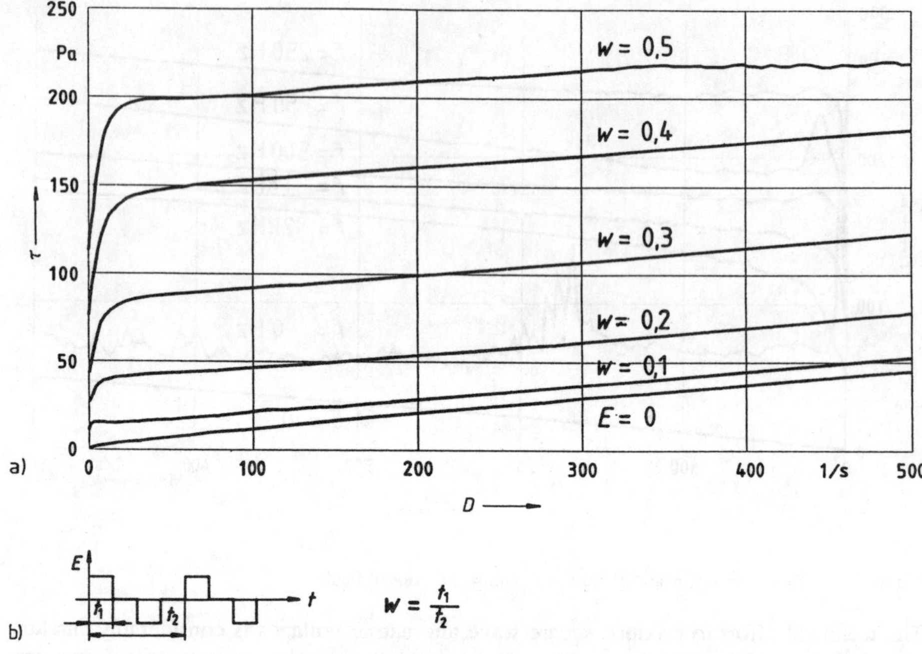

Figure 3: **a) Flow curves for an ER fluid with square-wave control of various pulse widths, b) Definition of the pulse width ratio**

A comparison between the curves shows that the highest shear stress value is obtained with the maximum possible pulse width ratio $w = 0.5$, for which the width of the pause between the pulses is zero. With narrower pulse widths, and therefore greater "pause widths," the curves are shifted further down. The flow curve corresponding to a pulse width ratio $w = 0.1$ lies only slightly higher than the flow curve without field ($E = 0$).

2.3. *Recommendations for the Control and Analysis Equipment*

The behaviour of electrorheological fluids is dependent on their particular composition and on a number of other factors including the temperature, shear rate and the amplitude, frequency and wave-form of the applied field. Control of a measurement process involving so many variables is aided through the use of a computer (e.g. PC/AT). In this case, not only should shear rate, shear stress and temperature be predetermined variables of time; the field strength should also be preprogrammed. In order to achieve this, the viscometer, high-voltage source and temperature control system should all be controllable.

A study of the electrical behaviour of ER fluids requires that the voltage across and the current between the electrodes be measured. It is recommended to incorporate the control and analysis of these measurements into the hardware and software of the computer rather than buying individual devices. The computer should be equipped with D/A and A/D converters for recording the measurement values as well as for controlling the experiment. A software package which is capable of performing, for example, Fast-Fourier-Transforms (FFT) is recommended for signal analysis.

2.4. *Testing of Electrical ER Fluid Properties*

The electrically equivalent circuit for an ER fluid actuator is needed for conceiving and dimensioning an optimum high-voltage source. In its simplest form, the model consists of an ohmic resistor in parallel with a capacitor, as shown in Figure 4. Both elements are determined by the electrode geometry and by the ER fluid properties, which are non-linearly dependent on the temperature, control field amplitude, field frequency and ER fluid shear rate.

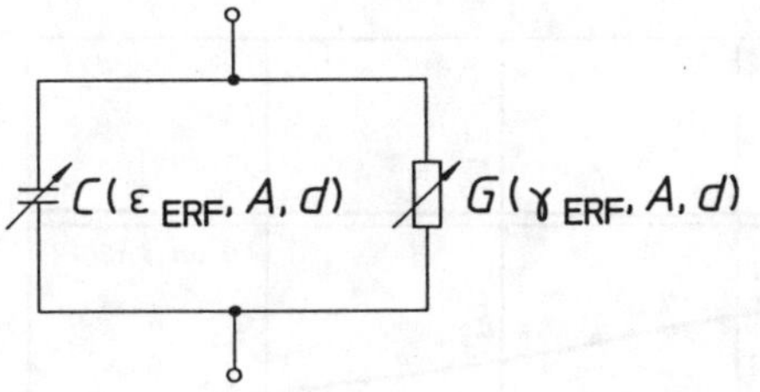

$C(\varepsilon_{ERF}, A, d)$ $G(\gamma_{ERF}, A, d)$

$$C = \frac{A}{d} \cdot \varepsilon_0 \cdot \varepsilon_{ERF}(T, V, D)$$

$$G = \frac{A}{d} \cdot \gamma_{ERF}(T, V, D)$$

Figure 4: Equivalent circuit for an ER fluid actuator

The ER fluid parameter which is simplest to measure is the conductivity. For measuring, a DC field is applied to the ER fluid and the current is determined with the help of a known resistance R_m. The specific conductivity of the ER fluid can be calculated from the current and the dimensions of the measurement apparatus.

The duration of the measurement is important in this test since the conductivity steadily increases from the beginning of the measurement. At the same time, the shear stress is seen to decrease. Figure 5 shows an example of the shear stress time dependence for field strengths of E_1 = 1 kV/mm, E_2 = 1.5 kV/mm and E_3 = 2 kV/mm. The measurements were performed during a time period of 60 minutes on a cylindrical measurement arrangement at a constant temperature of 25°C and a constant shear rate of $D = 100$ sec^{-1}.

A long-term test was also performed with the same ER fluid under the same measurement conditions with an alternating control field. Sinusoidal fields with a frequency of $f = 50$ Hz were applied. In order to permit a better comparison of the measurement results with those from the DC voltage tests, effective field strengths of $E_{4rms} = 1$ kV/mm, $E_{5rms} = 1.5$ kV/mm and $E_{6rms} = 2$ kV/mm were chosen. The test results are shown as well in Figure 5. A comparison of the shear stress-time curves shows that shear stresses of similar amplitude are achieved in the first few minutes for DC and alternating control voltages. In contrast to the DC control voltage tests, the measurement curves from the alternating control voltage tests shows no reduction in the shear stress.

This ER fluid behaviour can be explained by the effect of electrophoresis. Under DC field control, the suspended ER fluid particles migrate in the direction of one of the electrodes due to the force of the control field. Because of this, the originally uniform dispersion of the particles changes within the shear gap. At the one electrode there exists an accumulation of particles and at the opposite an impoverished zone. This results in an ever reducing achievable shear stress. These measurement results demonstrate, with respect to long-term stability in technical applications, that the application of alternating control fields without a DC component is favourable.

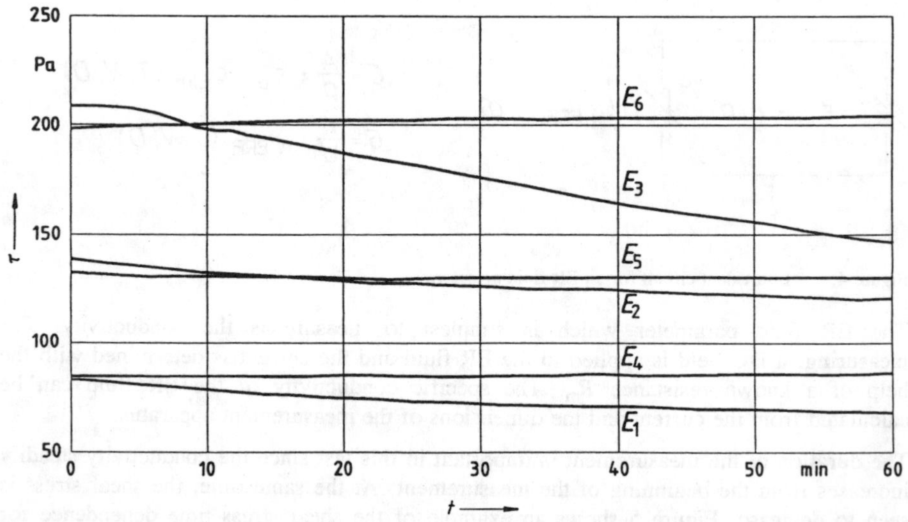

Figure 5: Shear stress-time behaviour of an ER fluid with DC and AC control fields

Distorted - instead of sinusoidal - current density curves were observed for sinusoidal control voltages due to the non-linear relationship between the ER fluid properties and the control field amplitude. This is illustrated in Figure 6 with control field strengths of

E_{1pk} = 0.5 kV/mm, E_{2pk} = 1 kV/mm and E_{3pk} = 2 kV/mm. The measurements were made with a cylindrical viscometer arrangement at a constant temperature of 25°C and a constant shear rate of $D = 100\ sec^{-1}$.

It is apparent that the distortion becomes greater with increasing field amplitudes. It is also recognizable in the power density spectrum of the measurement signal that additional frequency components occur in uneven multiples of the basic frequency under conditions of strong control fields. At the same time, the phase shift between the control and measured signals changes. With small control voltages, the phase shift φ is nearly -90° and tends towards zero with increasing field amplitudes.

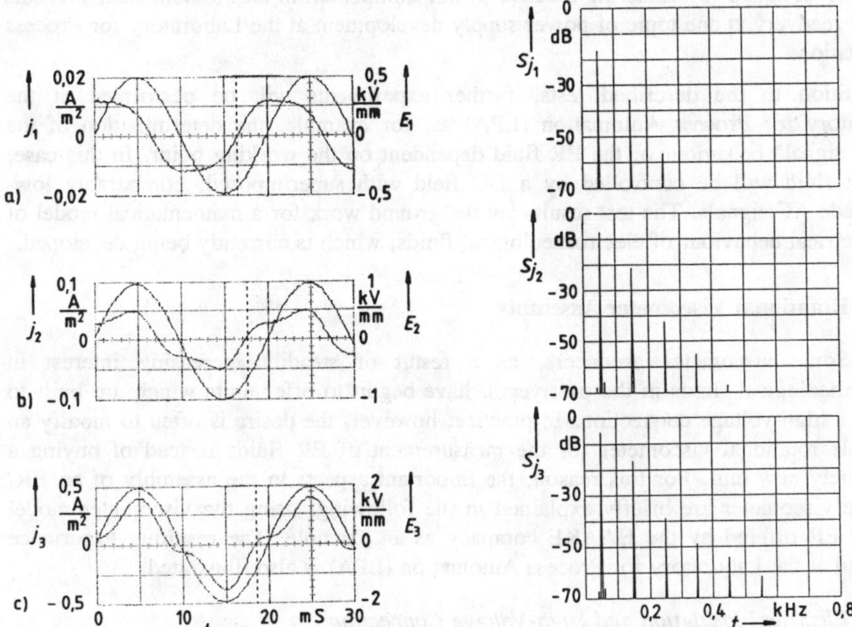

Figure 6: Time behaviour and power density spectrum of the current density with a sinusoidal ERF control signal with a) E_{1pk} = 0.5 kV/mm, b) E_{2pk} = 1 kV/mm, c) E_{3pk} = 2 kV/mm

The relationship between the active and the reactive components of the ER fluid control power changes with a change in the phase shift between the control field and the current density. In the example shown in figure 6, a phase shift of φ = -72° was achieved for a field strength of E_{1pk} = 0.5 kV/mm. This means that the high-voltage source must control a load with a very large capacitive and a small ohmic component. Therefore, the active power is very low in comparison to the reactive power. Control with a field strength of E_{2pk} = 1 kV/mm results in a phase shift of φ = -56°. In this case, the active power is almost as great as the reactive power. A phase shift of φ = -27° was measured when a field strength of E_{3pk} = 2 kV/mm was applied. That means the load has a larger ohmic component than a capacitive one. The voltage source should therefore mainly supply an active power.

The effectiveness of an ER fluid application as a whole depends on the rheological behaviour of the unit as well as on the design of the power supply and controller. The electrical efficiency of the system can be increased by selecting an ER fluid which has a low conductivity resulting in a low active power. It can be further increased if the power supply is designed to utilize the reactive power component in an efficient manner. This energy recovery is one topic of power supply development at the Laboratory for Process Automation.

In addition to the described tests, further experiments will be performed at the Laboratory for Process Automation (LPA) as, for example, the determination of the "small signal" behaviour of the ER fluid dependent on the working point. In this case, the ER fluid will be controlled by a DC field with superimposed, comparably low-amplitude AC signals. The test results lay the ground work for a mathematical model of the electrical behaviour of electrorheological fluids, which is currently being developed.

3. Rotational Viscometer Assembly

Some viscometer producers, as a result of steadily increasing interest in electrorheological fluids in the past years, have begun to offer units which are built to handle a high-voltage connection. In practice, however, the desire is often to modify an available rotational viscometer for the measurement of ER fluids instead of buying a completely new unit. For this reason, the important aspects in the assembly of an ER-suitable viscometer are briefly explained in the following, using the viscometer model CV 20 ER offered by the HAAKE company as an example. The resulting experience gathered at the Laboratory for Process Automation (LPA) is also illustrated.

3.1. Electrical Insulation and High-Voltage Connection

The HAAKE CV 20 is a shear-rate-controlled Couette-viscometer in which the lower cylinder/plate is driven by a DC-motor. A tacho-generator is implemented for the measurement of the rotational speed and is located, together with the drive motor, in the lower housing. The torque measurement system is installed in the vertically adjustable measurement system head. The torque is determined by sensing the twisting in the torsion element with the help of a differential transformer (LVDT).

The measurement system must be electrically insulated so that a voltage potential can be applied. The easiest way to achieve this is to insulate the torsion element and mounting clamp from the torque measurement system and the housing. Figure 7 shows a cross-sectional view of the measurement system head. The plastic electrical insulation is shown with crosshatching. The core of the differential transformer is held by an electrically insulating plastic arm. An insulating sheet lies between the differential transformer and the circuit board containing the signal amplifier.

The high-voltage is connected to a stationary portion of the torsion element via a short cable from the high-voltage socket (not shown in Figure 7) which is fixed directly to the measurement system head housing. The high-voltage connecting cable should be as short as possible to minimize signal noise.

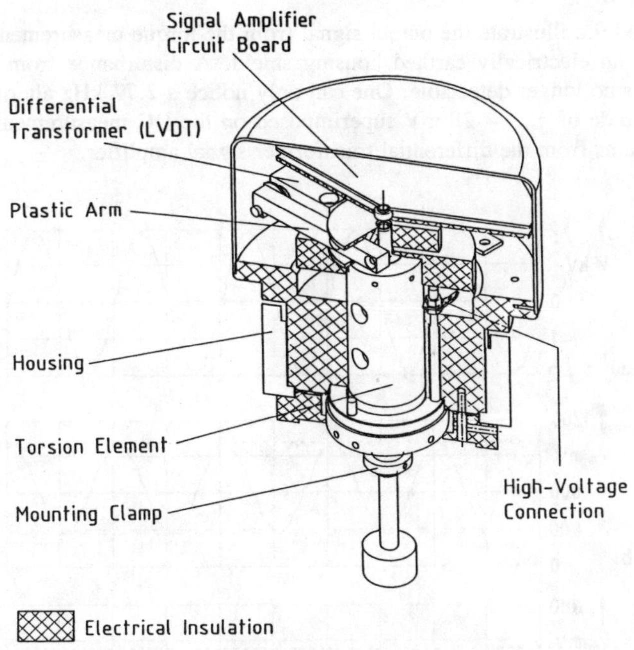

Figure 7: Cross-sectional view of the insulated HAAKE CV 20 measurement system head

3.2. *Shielding against Alternating Fields*

If special shielding is not implemented, signal noise can appear in viscometers with inductive torque and rotational speed sensors when a high-amplitude alternating ER fluid control voltage is applied. This was the case with the viscometer used at the LPA. Before shielding was implemented, signal noise was present on the output signal from the torque sensor unit.

Figures 8 and 9 illustrate examples of noisy torque signals. The results shown in Figure 8 were achieved by applying a 500 Hz sinusoidal voltage with an amplitude of V_{pp} = 4 kV (Figure 8.a). The torque signal V_m is made up of a DC component and a superimposed sinusoidal voltage with an amplitude of V_{pp} = 987.5 mV (Figure 8.b). The measurement results in Figure 9 were achieved using a 50 Hz square-wave signal to the ER fluid with an amplitude of V_{pp} = 4 kV (Figure 9.a). The resulting torque signal contains distinct voltage peaks. These peaks step up or down coincident with the edges of the square-wave signal and have an amplitude of 800 mV (Figure 9.b). In each measurement example, the noise component has an amplitude as high as the measurement signal itself.

Figures 8.c and 9.c illustrate the output signal from the torque measurement device after implementing an electrically earthed housing shield. A disturbance from the ER fluid control field is no longer detectable. One can only notice a 2.77 kHz alternating voltage with an amplitude of V_{pp} = 28 mV superimposed on the DC measurement signal. This disturbance stems from the differential transformer signal amplifier.

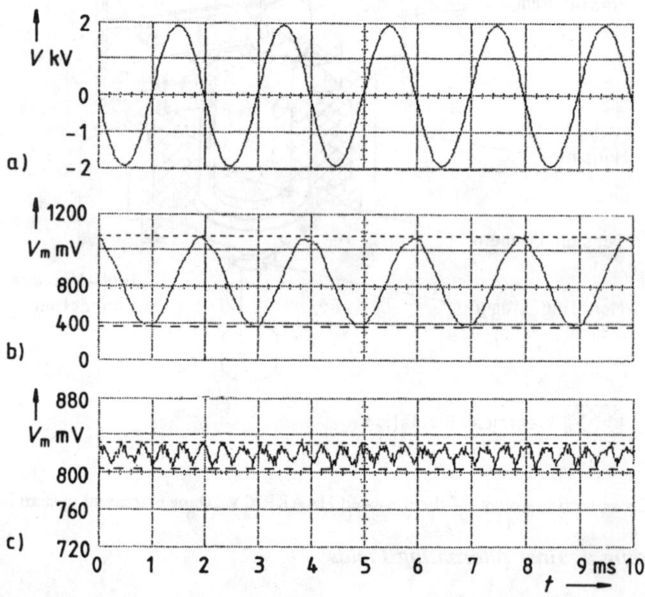

Figure 8: a) Sinusoidal ER input voltage, b) Unshielded torque signal, c) Shielded torque signal

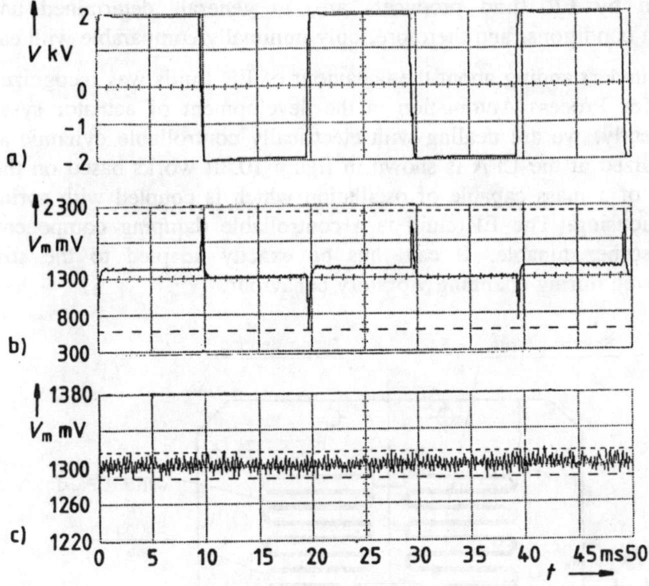

Figure 9: a) Square-wave ER input voltage, b) Unshielded torque signal, c) Shielded torque signal

The described signal disturbances result from an electromagnetic coupling between the ER fluid control signal and the torque measurement device. This coupling, and therefore the signal noise, was able to be considerably reduced through the implementation of shielding plates in the measurement system head housing. The desired result was achieved with an aluminium housing constructed at the LPA, in which an aluminium plate is installed next to the differential transformer and one beneath the amplifier circuit board. In this way, the high-voltage connection, the transformer and the amplifier are each located in their own chamber. The housing and the shielding plates are electrically connected and earthed.

4. Prospects

The rheological and electrical properties of ER fluids depend on a great number of factors with nonlinear relationships. This means that a comparison between different types of ER fluids is only possible when their characteristic values are taken under the same measurement conditions. At the present, no standardized specifications for the determination of ER fluid characteristic values exist. For this reason, the characteristic values given by ER fluid producers are, in general, determined under different measurement conditions, and therefore, only minimally comparable with each other.

The lack of understanding about the behaviour of ER fluids was recognized early at the Laboratory for Process Automation in the development of actuator systems with ER fluids. Currently, we are dealing with electrically controllable dynamic absorbers. An absorber realized at the LPA is shown in figure 10. It works based on the shear mode and consists of a mass capable of oscillation which is coupled with springs to the ER fluid-filled housing. The ER fluid is a controllable damping component making the dynamic absorber tunable. It can thus be exactly adapted to the structure to be dampened, even during changing vibratory behaviour.

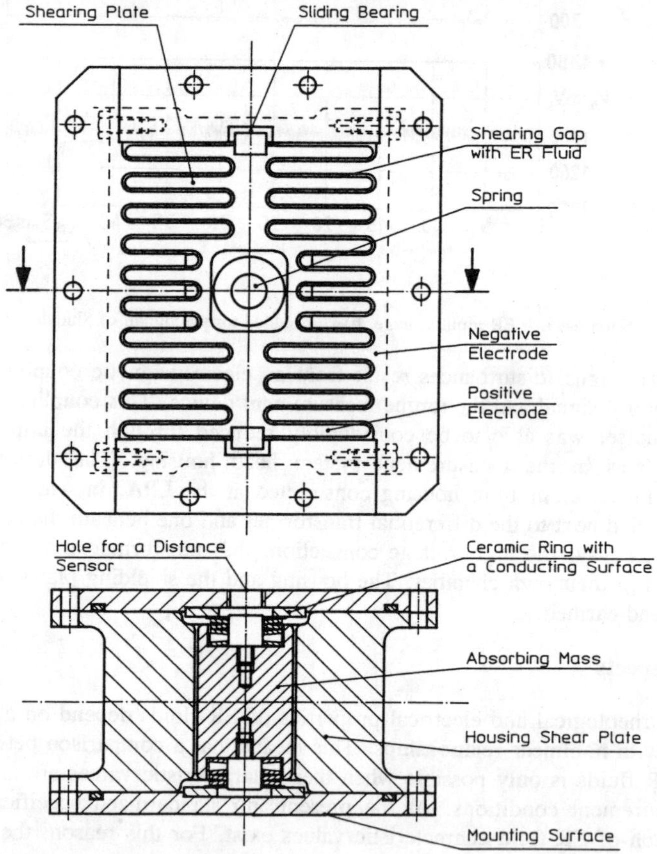

Figure 10: Construction of an ER fluid absorber

Here the actuator system comprises not only the damper but the whole control circuit consisting of the damper, the high voltage source and the controller. If the individual components are not co-ordinated, the results will be unsatisfactory. This might be the case if the output of the amplifier is not adjusted to the ER absorber or if the wave form and the frequency of the control signal are not adjusted to the ER fluid used.

Data about the frequency behaviour, square-wave response and electrical modeling of an ER fluid were not available in the data sheets provided by the producers of ER fluids. For this reason, producers and users of ER fluids should concentrate on the establishment of objective measurement methods and processes for characterising ER fluids. Those involved should agree to put together a catalogue of essential parameters and to define the conditions for their measurement.

5. References

1. H. Janocha (Hrsg.): Aktoren, Springer-Verlag, Berlin, 1992.
2. H. Janocha and B. Rech: Measurements on electrorheological liquids with rotational viscometers. Rheology 93 (March 1993), Vincentz Verlag, Hannover, p. 39-47.

EXPERIMENTAL STUDY OF YIELD STRESSES IN ELECTRORHEOLOGICAL FLUIDS

J. N. FOULC and P. ATTEN
Laboratoire d'Electrostatique et de Matériaux Diélectriques
CNRS and Université Joseph Fourier de Grenoble
B.P. 166 - 38042 Grenoble Cedex 9 (France)

ABSTRACT

The rheological behaviour of electrorheological (ER) fluids under steady shear flow can be characterized by a dynamic yield stress τ_{od} and a apparent viscosity η_{app}. Measurements at small shear strains exhibit another limiting stress defined as a static yield stress τ_{os}. We have studied ER fluids based on a mixture of cellulose and mineral oil. The variations of static and dynamic yield stresses have been investigated as a function of D.C. electric field E and volume fraction ϕ of cellulose. Static yield stress depends not only on electric field E and volume fraction ϕ but also on a time t_c of application of the electric field prior to the shear. The time t_c corresponds to an electrical conditioning of the suspension, necessary to obtain reproducible measurements; we present results which lead to estimates for a time t_c. Finally, we compare the yield stresses measured with the qualitative laws of interaction force between particules as given by a conduction model and note a qualitative relation between structuration of the medium and current flowing through it.

1. Introduction

One of the basic properties of ER fluids is the appearance of strength when they are subjected to an electric field. This strength is usually charaterized by static (no flow) and dynamic yield stresses[1]. Experimental studies have first focused on the measurement of the dynamic yield stress τ_{od} as a function of the shear rate $\dot{\gamma}$ and of the volume fraction ϕ [2,3]. More recently a detailed study pointed out the importance of the shear rate on the determination of the static yield stress τ_{os}[4]. The measurements generally mention that yield stress exhibits a quadratic dependence on the applied electric field [5,6]. Only recently, Sprecher et al. pointed out a more complex phenomenon; they found that the increase of shear stress as a function of E departs from the quadratic law at high enough applied fields[7].

In order to model the ER effect, the authors generally agree to consider that the electrically induced attraction force between particles arises from the electric polarization of the media, as reflected by the permittivities[8,9]. Attempts at estimating the values of τ_{os}

however reveal that this explanation of the particle-particle interaction does not account for the observed yield stresses[10]. One possible reason for this disagreement is that the postulated basic mechanism is not the true one at least at D.C. conditions. Indeed a recent conduction model confirmed by large scale experiments for D.C. fields provides a new way of explaining the particle-particle interaction and for estimating the attraction force between them[11]. This explanation is based on the fact that the electric field distribution is controlled by the conduction properties of both components at times clearly greater than the charge relaxation times of both media. Moreover the attraction force was found to vary proportionally to E for high enough fields[12].

The confirmation of this new approach on large scale spheres encourages us to investigate the behaviour of ER fluids with particles of much smaller diameter. One aim of the present work is to examine closely the field dependence of the yield stresses in D.C. conditions. Another point we address is to examine the current flowing through the fluid and to look at a possible correlation with its structuration. Prior to presenting results concerning these two points, we recall the definition of yield stresses in ER fluids and describe the experimental procedure.

2. Experimental

2.1 Definition of yield stresses

We retain here the definition of yield stresses in ER fluids as given by Conrad et al.[13]. From a curve representing the shear stress as a function of the shear strain, for very low shear rate, the static (or quasi-static) yield stress τ_{os} is the highest value reached by shear stress (Fig. 1-a). The dynamic yield stress τ_{od} corresponds to the threshold of a Bingham model. It is obtained by approximating the variations of shear stress as a function of shear rate $\dot{\gamma}$ by a linear relation : τ_{od} is the value extrapolated at $\dot{\gamma}=0$ (Fig. 1-b).

2.2 Experimental set-up

We used a modified Couette rheometer with a special arrangement using a pin-mercury contact. This apparatus (Mettler 115 A) is a controlled-rate rheometer. The shear rate could be varied from 0.1 to 1,000 s^{-1}. The test cell consisted of two concentric cylinders with gap 1 mm, mean radius 13 mm and height 37.5 mm. The inner cylinder is electrically insulated up to 5 kV and the outer cylinder is grounded.

The D.C. high voltage was imposed from a 5 kV HV power supply (Mesco) and the current was monitored using a resistor (50 kΩ) and a X(t) recorder.

The ER fluid used is a mixture of cellulose microcrystalline powder (typical size ≈ 30 µm, water content $W_w \approx 6\%$) and a mineral oil (TF 50 Elf) of characteristics : electrical conductivity $\sigma \approx 10^{-12}$ S/m, relative permittivity $\varepsilon_r \approx 2.2$, dynamic viscosity $\eta \cong 35$ cP (T=23°C). Cellulose powder is dispersed in oil by strongly mixing during approximately ten minutes before beginning experiments. The volume fraction of cellulose ranged from 10 % to 30 %.

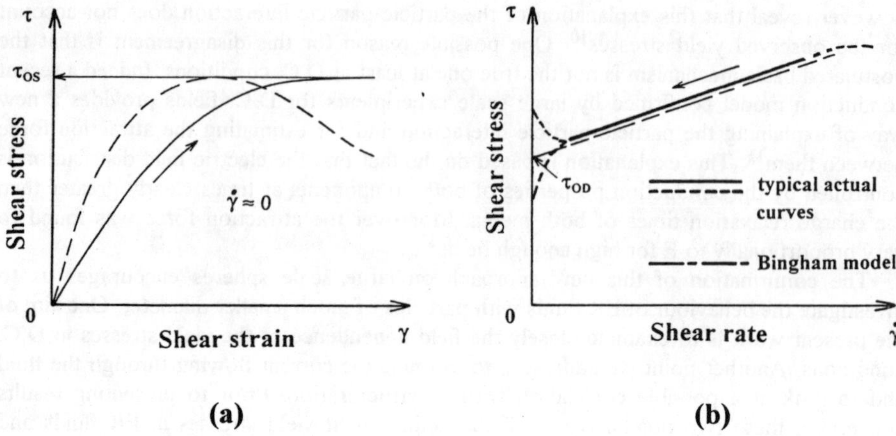

Fig. 1. Schematic curves defining yield stresses in ER fluids. (a) static yield stress τ_{os} ; dotted curve exhibits a "yield point" and solid curve indicates a smooth variation. (b) dynamic yield stress τ_{od} ; dotted curves show some typical flow situations.

2.3 Experimental procedure

An electrical conditioning of the cell and the ER fluid (for a freshly prepared suspension) is required in order to obtain reproducible measurements. Fig. 2 illustrates the typical behaviour of the fluid when starting immediately the measurements. It shows the values of static yield stress (for a detailed description see § 3) as a function of the used shear rate which was varied from 0.1 to 1 s^{-1}. Reproducibility of the results apparently occurs only after a full cycle. In fact the crucial parameter is the time t_c of application of the electric field (at least 5 to 10 minutes in the present case).

The raison for this "conditioning" effect presumably is an evolution of the electric conduction of the ER fluid. In dielectric liquids it has long been observed that the D.C. current drastically depends on minute traces of electrolytic impurities. It generally decreases with complex kinetics and usually several characteristic times can be identified[14]. The pure liquid used to elaborate our test fluid (mineral oil) exhibits such a behaviour (Fig. 3). Of course in the ER fluid, various electrochemical processes occur and the whole thermodynamic equilibrium is altered by application of D.C. field and passage of current. The practical consequence of these intricate phenomena is that a minimum delay of about five minutes must be observed between the first application of the field and the beginning of the rheological measurements.

3. Static yield stress

In order to investigate the influence of applied electric field E and of volume fraction ϕ of particles on the static yield stress it is necessary to first measure the shear stress against

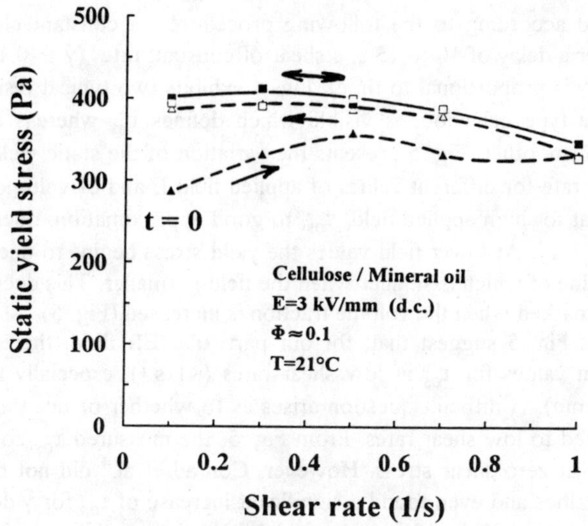

Fig. 2. Behaviour of an freshly prepared ER fluid (cellulose in mineral oil) when starting immediately the static yield stress measurements (each measurement lasts one minute). The conditioning time t_c required for reproducible results is about five minutes.

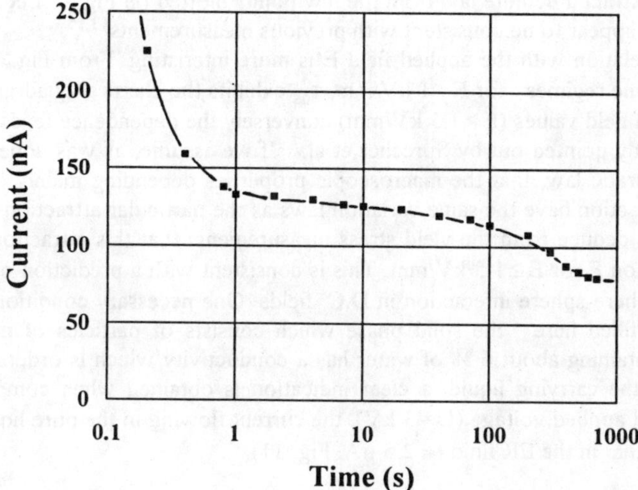

Fig. 3. Current response to a 3 kV step voltage in a mineral oil TF 50 Elf: the decrease of current revels several characteristic times (short-time ≈ 1 second, long-time ≈ 3 minutes...).

time. We worked according to the following procedure : a constant electric field E is applied and, after a delay of 10 to 15 s, a shear of constant rate ($\dot{\gamma} \geq 0.1\ s^{-1}$) is imposed (the shear strain γ is proportional to time). Fig. 4 exhibits two typical responses : on Fig. 4-a, a yield point-type behaviour is visible which defines τ_{os} whereas on Fig. 4-b the variation is much smoother. Fig. 5 presents the variation of the static yield stress τ_{os} as a function of shear rate for different values of applied field E and of volume fraction ϕ. We observe firstly that for high applied field, τ_{os}, to good approximation, does not depend on $\dot{\gamma}$ between 0.1 to 1 s^{-1}. At lower field values the yield stress begins to decrease above the shear rate, the value of which is smaller when the field is smaller. This decrease appears to be slightly more marked when the volume fraction is increased (Fig. 5).

The results of Fig. 5 suggest that, for our particular ER fluid, there are rather well defined saturation values for τ_{os} at low shear rates (<1 s^{-1}), especially for high applied fields ($E \geq 2$ kV/mm). A difficult question arises as to whether or not the constant value can be extrapolated to low shear rates. From Fig. 5, the measured τ_{os} could be expected to give the value at zero shear stress. However, Conrad et al.[4] did not obtain such well defined plateau values and even found a significant increase of τ_{os} for $\dot{\gamma}$ decreasing below $10^{-1}\ s^{-1}$. As the latter work is the only one wich investigated this question, the problain remains open (note that our rheometer does not allow to operate at shear rates $\dot{\gamma}<10^{-1}\ s^{-1}$).

Taking for granted that the static yield stress τ_{os} we have determined is a significant variable characterizing the behaviour of ER fluids, we can correlete it with other parameters. Fig. 6 shows that τ_{os} increases as the function of the volume fraction. It is difficult to extract a definite law from the few points plotted on Fig. 6. Let us remark that these results appear to be consistent with previous measurements[13,15].

The correlation with the applied field E is more interisting. From Fig. 7 there appear two well define regimes : for $E<1$ kV/mm, τ_{os} exhibits the classical quadratic dependence on E; at high field values ($E>1.5$ kV/mm) conversely the dependence tends to be linear, a feature already pointed out by Sprecher et al.[7]. If we assume, as was done usually in the case of quadratic law, that the macroscopic properties depending mainly of the particle-particle interaction have the same variation laws as the particular attraction force between particles, we deduce from the yield stress measurements that this attraction force should vary linearly on E for $E \geq 1.5$ kV/mm. This is consistent with a prediction of a conduction model for sphere-sphere interaction in D.C. fields. One necessary condition of the model is clearly fulfilled here : the solid phase which consists of particles of microcrystalline cellulose containing about 6 % of water has a conductivity which is orders of magnitude higher than the carrying liquid; a clear indication is obtained when comparing, for the same cell and applied voltage ($U=3$ kV), the current flowing in the pure liquid ($\approx 0.1\ \mu A$, Fig. 3) with that in the ER fluid ($\approx 2.5\ \mu A$, Fig. 11).

4. Dynamic yield stress

Most often the shear stress τ measured at decreasing shear rate $\dot{\gamma}$ is lower than that measured at increasing shear rate. This hysteresis explains why there are two yield stresses,

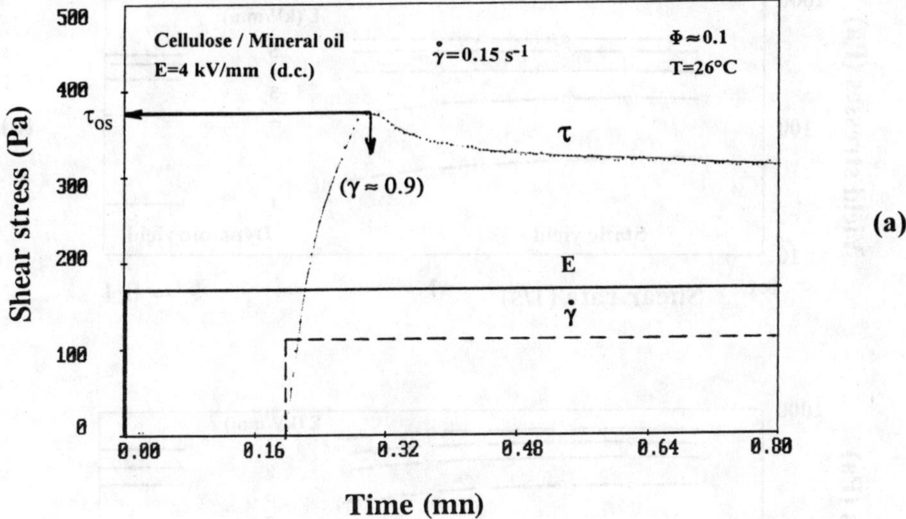

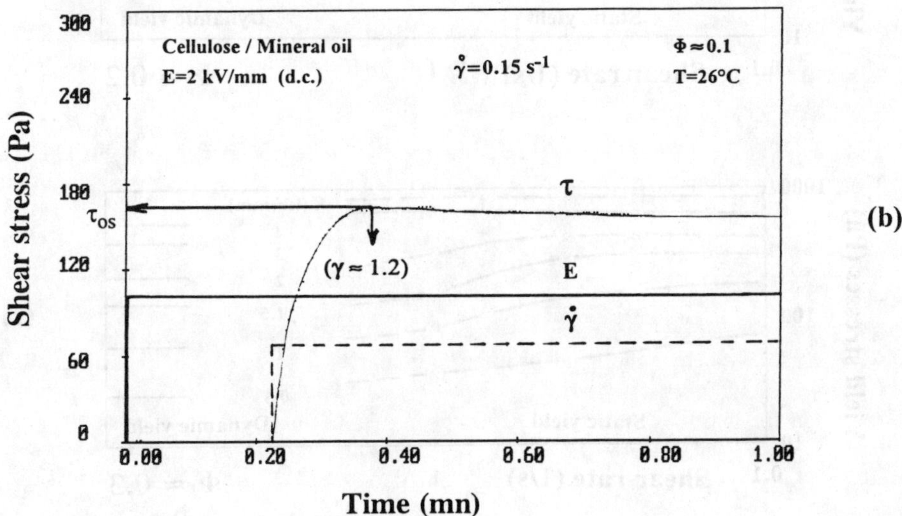

Fig. 4. Variation of the shear stress τ as a function of time at a shear rate of 0.15 s^{-1} ($\phi = 0.1$). Electric field E is applied prior to shearing $\dot{\gamma}$. The maximum value of shear stress defines the static yield stress τ_{os} (a) E = 4 kV/mm, (b) E = 2 kV/mm.

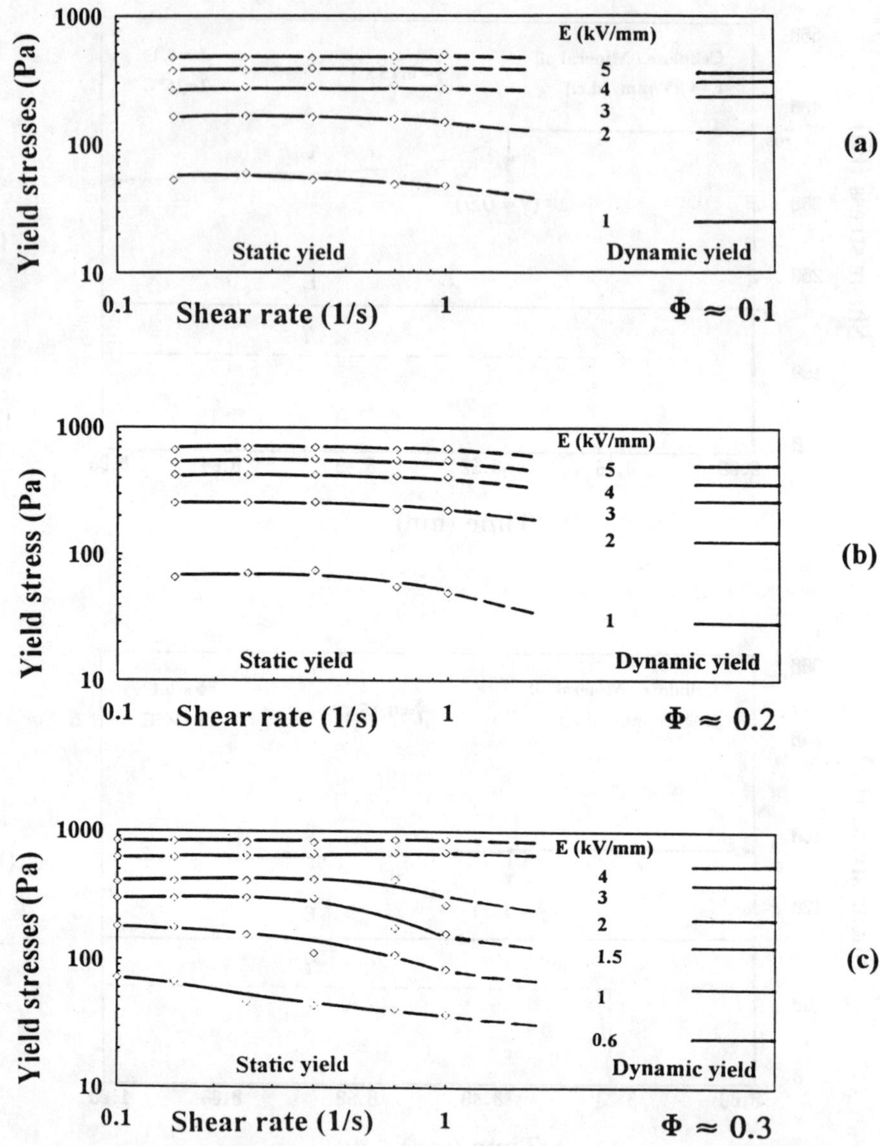

Fig. 5. Dependence of the static yield stress τ_{OS} on the shear rate in the ER fluid (cellulose in mineral oil) for various values of electric field E : (a) $\phi = 0.1$; (b) $\phi = 0.2$; (c) $\phi = 0.3$.

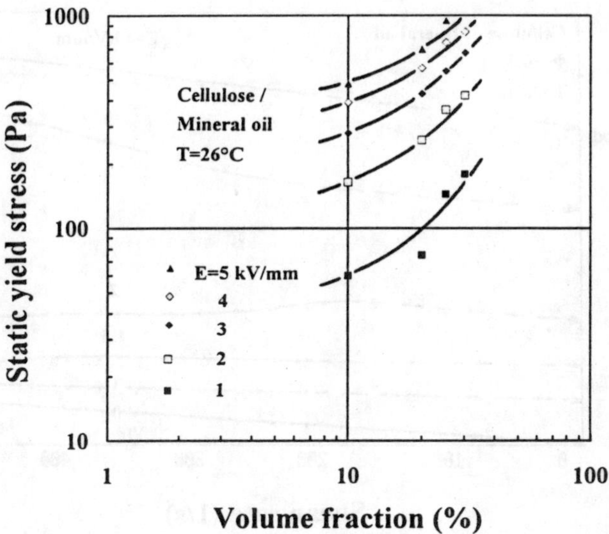

Fig. 6. Effect of the cellulose volume fraction ϕ on the static yield stress τ_{os} as a function of electric field E.

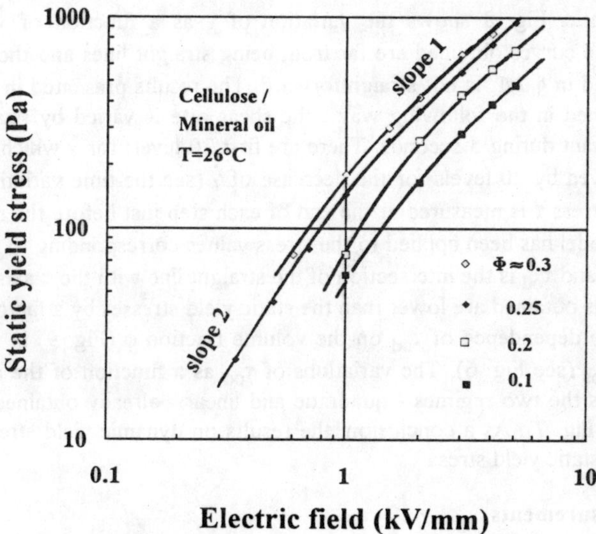

Fig. 7. Dependence of the static yield stress τ_{os} on electric field E for various values of volume fraction ϕ. The curves show a quadratic dependence on E at low fields and a linear dependence at high fields.

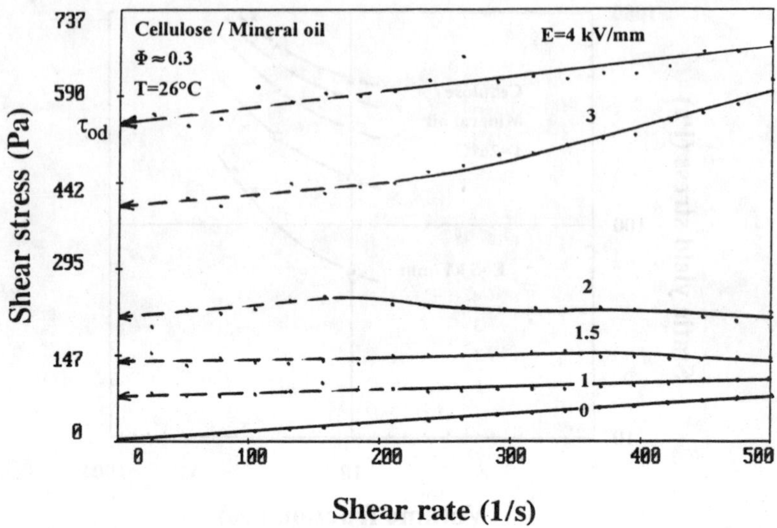

Fig. 8. Variation of the shear stress τ as a function of decreasing shear rate $\dot{\gamma}$ under shear rate-control mode ($d\dot{\gamma}/dt = 5\ s^{-2}$) for various values of electric field E.

dynamic and static. Fig. 8 shows the variation of τ as a function of $\dot{\gamma}$, measured at decreasing $\dot{\gamma}$. The curves obtained are far from being straight lines and the determination of τ_{od}, as defined in § 2.1, is not straightforward. The results presented in Figs. 9 and 10 have been obtained in the following way : the shear rate is varied by regular steps and maintained constant during 5 seconds. There are first 20 levels for $\dot{\gamma}$ which passes from 0 to 500 s^{-1} followed by 20 levels for the decrease of $\dot{\gamma}$ (see the time variation of $\dot{\gamma}$ in Fig. 11). The shear stress τ is measured at the end of each step just before the next jump of $\dot{\gamma}$. The Bingham model has been applied to the stress values corresponding to $\dot{\gamma}$ ranging from 20 s^{-1} to 200 s^{-1} and τ_{od} is the intersection of the straight line with the τ axis. The dynamic yield stresses thus obtained are lower than the static yield stresses by a factor of 1.5 to 2.5 (see Fig. 5). The dependence of τ_{od} on the volume fraction ϕ (Fig. 9) is very similar to that found for τ_{os} (see Fig. 6). The variations of τ_{od} as a function of the applied field E (Fig. 10) exhibits the two regimes - quadratic and linear - already obtained for the static yield stress (see Fig. 7). As a conclusion, the results on dynamic yield stress confirm the observations on static yield stress.

5. Current measurements

The conduction model proposed recently for explaining the ER effect[16] establishes that, in D.C. conditions, this effect can occur only when the conductivity of the solid particles is greater than that of the liquid. In the fluid we used, the solid phase consists in

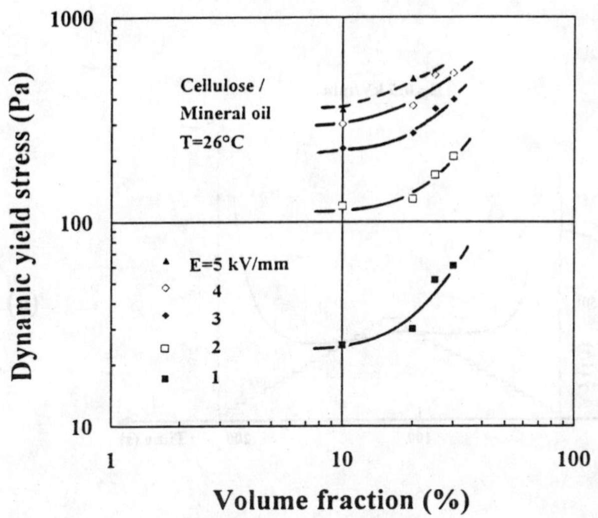

Fig. 9. Effect of the volume fraction ϕ of the cellulose on the dynamic yield stress τ_{od} as a function of electric field E.

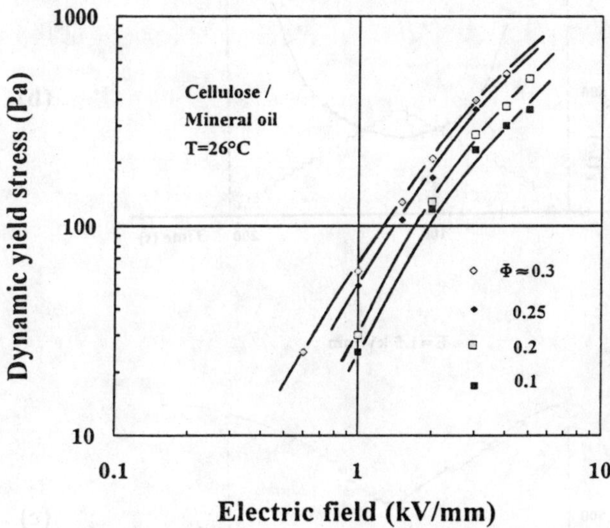

Fig. 10. Variation of the dynamic yield stress τ_{od} as a function of electric field E for different values of volume fraction ϕ. τ_{od} (E) exhibits the two regimes -quadratic and linear- obtained for a static yield stress (see Fig. 7).

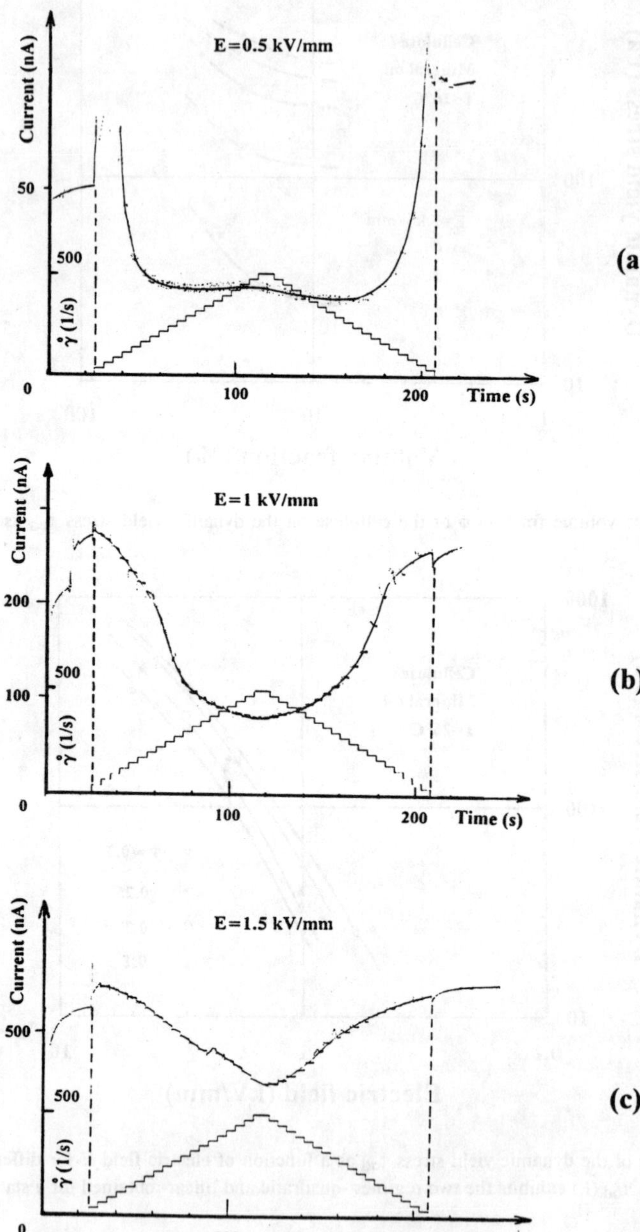
E=0.5 kV/mm
Current (nA)
50
500
γ̊ (1/s)
0
100
200
Time (s)
(a)
E=1 kV/mm
Current (nA)
200
100
500
γ̊ (1/s)
0
100
200
Time (s)
(b)
E=1.5 kV/mm
Current (nA)
500
500
γ̊ (1/s)
0
100
200
Time (s)
(c)

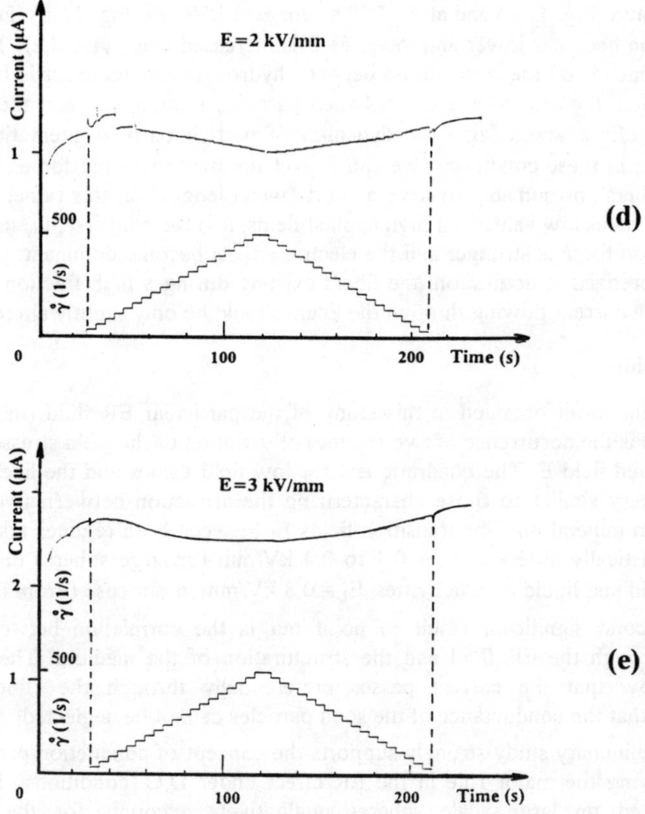

Fig. 11. Dependence of the current flowing through the ER fluid on the shear rate $\dot{\gamma}$, at volume fraction 0.25 and temperature 27°C, with increasing electric field : (a) E = 0.5 kV/mm, (b) E = 1 kV/mm, (c) E = 1.5 kV/mm, (d) E = 2 kV/mm, (e) E = 3 kV/mm.

cellulose containing about 6 % of water; its conductivity is several orders of magnitude higher than that of the mineral oil. As concerns the conductivity of the ER fluid itself, we can give the following naive picture : when the particles are randomly dispersed, the apparent conductivity is controlled by the liquid phase and should remain low; if we assume a structuration of the particles in fibers bridging the two electrodes, the current flows mainly through the fibers and the apparent conductivity is substantially enhanced.

This picture allows to explain qualitativily the variations of the current flowing through the ER fluid as a function ot time during a cycle : $\dot{\gamma}$ passes from 0 to 500s^{-1} and than to 0 (Fig. 11). For low applied D.C. field (E $\leq$ 1 kV/mm) there is a marked difference (ratio >2) for the current between low and high shear rates, the transition is at $\dot{\gamma} \sim 100$ s^{-1} for

E=0.5 kV/mm (Fig. 11-a) and at $\dot{\gamma} \sim 200\ s^{-1}$ for E=1 kV/mm (Fig. 11-b). For higher fields, the variation becomes lower and lower as E is increased (Fig. 11-c,d,e). The structure in the fluid depends on the competition between hydrodynamic and electric forces[17]. At low enough fields, the attraction force between particles is moderate and the electroviscous time $t_{ev}=\eta/\varepsilon E^2$ characterizing the dynamics of particle-particle interaction takes rather high values; in these conditions, we can expect the hydrodynamic forces to be dominant and the "fibers" presumably to have a short "mean length"; in this rather dispersed state the current takes low values. At high applied fields, it is the converse, t_{ev} takes low values, the attraction force is stronger and the electric effects become dominant; we can expect a well characterized structuration, the fibers existing during a high fraction of the time. In this case the current flowing through the fibers should be only slightly altered.

6. Conclusion

The main result obtained in this study of the particular ER fluid (moist cellulose in mineral oil) is the occurrence of two regimes of variation of the yield stresses as a function of the applied field E. The quadratic law for low field values and the linear one for high fields are very similar to those characterizing the attraction between two large spheres immersed in mineral oil. The transition fields E_t between both regimes take values which are not drastically different: E_t = 0.1 to 0.4 kV/mm for large spheres depending on the ratio of solid and liquid conductivities, $E_t \cong 0.8$ kV/mm in our case (From Fig. 7).

The second significant result to point out is the correlation between the current flowing through the ER fluid and the structuration of the medium. The measurements clearly show that the current passes preferentially through the "fibers". This fact underlines that the conductance of the solid particles cannot be neglected.

This preliminary study strongly supports the concept of conduction properties of both phases playing the major role in the ER effect under D.C. conditions. The conduction model tested on large scale spheres qualitatively accounts for the yield stresses dependence on E and work both experimental and theoretical has to be pursued in order to obtain more quantitative tests.

Acknowledgements

This work was performed with the joint financial support of the Université Joseph Fourier de Grenoble, the Centre National de la Recherche Scientifique and the compagny Elf-Aquitaine (under contract CNRS/SNEA No 50.8016). The authors are grateful to the compagny Elf-Atochem for providing samples of dielectric liquids and solid particles.

References

1. R. T. Bonnecaze and J. F. Brady, J. Rheol., **36**,73 (1992).
2. L. Marshall and C. F. Zukoski, J. Chem. Soc. Faraday Trans. I, **85**, 2785 (1989).
3. Y. Otsubo, M. Sekine and S. Katayama, J. Rheol., **36**, 479 (1992).

4. H. Conrad, Y. Chen and A. F. Sprecher, in Proc. 3rd Int. Conf. on ER Fluids, R. Tao ed. (World Scientific, Singapore, 1992) p. 195.
5. D. L. Klass and T. W. Martinek, J. Appl. Phys.,**38**, 67 (1967).
6. H. Uejima, Jpn J. Appl. Phys., **11**, 319 (1972).
7. A. F. Sprecher, J. D. Carlson and H. Conrad, Mat. Sci. Engr., **95**, 187 (1987).
8. D. J. Klingenberg, F. V. Swol and C. F. Zukoski, J. Chem. Phys., **91**, 7888 (1989).
9. A. P. Gast and C. F. Zukoski, Adv. Colloid Int. Sci., **30**,153 (1989).
10. T. C. Halsey, Science, **258**, 761 (1992).
11. N. Félici, J. N. Foulc and P. Atten, in this volume.
12. J. N. Foulc, P. Atten and N. Félici, submitted to J. Electrostatics (1993).
13. H. Conrad and A. F. Sprecher, J. Stat. Phys., **64**, 1073 (1991).
14. A. Denat, B. Gosse and J. P. Gosse, J. Electrostatics, **12**,197 (1982).
15. T. Garino, D. Adolf and B. Hance, in Proc. 3rd Int. Conf. on ER Fluids, R. Tao ed. (World Scientific, Singapore, 1992) p. 167.
16. J. N. Foulc, P. Atten and N. Félici, C. R. Acad. Sci. Paris, **317**, 5 (1993).
17. D. J. Klingenberg, D. Dierking and C. F. Zukoski, J. Chem. Soc. Faraday Trans., **87**, 425 (1991).

THE EFFECT OF SOLID FRACTION CONCENTRATION ON THE TIME DOMAIN PERFORMANCE OF AN ER FLUID IN THE SHEAR MODE

J Makin, W A Bullough, R Firoozian, A R Johnson and A Hosseini-Sianaki,

Department of Mechanical and Process Engineering
University of Sheffield
Mappin Street
PO Box 600
Sheffield S1 4DU
England

ABSTRACT

Some aspects of electro-stress and electron-hydraulic time delay in an electro-rheological fluid are examined for their dependence on the volume fraction of solid in the mixture. Early results show that the density matching of the solid and liquid phases of an ERF brings little improvement in the serviceability of a suspension. Subsequent testing is based on non density matched mixtures which are examined over a range of 0 $\leq$ shear rate $\leq$ 19000s^{-1}, 15 $\leq$ temperature °C $\leq$ 50 and 20 $\leq$ % volume fraction of solid $\leq$ 40%. The effect of prolonged operation on the durability of the fluid is also reported. The results are used to provide pointers towards fluid selection, a control strategy and to help in the determination of operating parameters in order to optimise clutch performance. Testing took place on a concentric clutch (Couette) type apparatus which had an electrode separation of 0.5 mm.

1. Introduction

To date the analysis of run up to speed time ($t^{**}_m + t^*_m$) for an ER clutch has been based on the assumptions that the electron-mechanical time delay (t^*_m) is relatively short, that the effect of any driving torque created within that duration has an insignificant effect on the prediction of driven rotor speed versus time, that parasitic stresses are small and that the electro-torque T_e is reasonably uniform during the subsequent (constant rate of) acceleration. The duration of t^{**}_m can then be determined from $T_e = I\,\alpha$. However, the requirement for more substantial fluid electro-stress τ_e and the quest for rotational stability of the ER mixture under centrifugal loading has fuelled an investigation into the effects of volume concentration of solid on these precepts and its density matching. There exists also a need to investigate the properties of high concentration (%C) fluid at high shear rates ($\dot{\gamma}$).

In the present paper, following brief tests to determine the effect of matching the density of solid and liquid phases, the effect of varying solid concentration is examined in shear, for its effect on the form of ERF response. Both the time response and short term 'steady state' are monitored for a wide range of shear rates, fluid temperatures (θ) and step input electric field strengths (E). At the same time, the opportunity is taken to monitor the effects of %C on the near static yield stress as an indication of the hysteretic performance of the fluid and also to throw light on which parameter to use for the control of torque in a device - voltage or current i; temperature, voltage step magnitude V and shear rate are important interrelated parameters.

For simplicity and ease of analysis the concentric clutch device was selected as the test instrument. Details of this and its supply and instrumentation systems are given elsewhere[1]. Essentially the apparatus is as shown in Figs. 1.

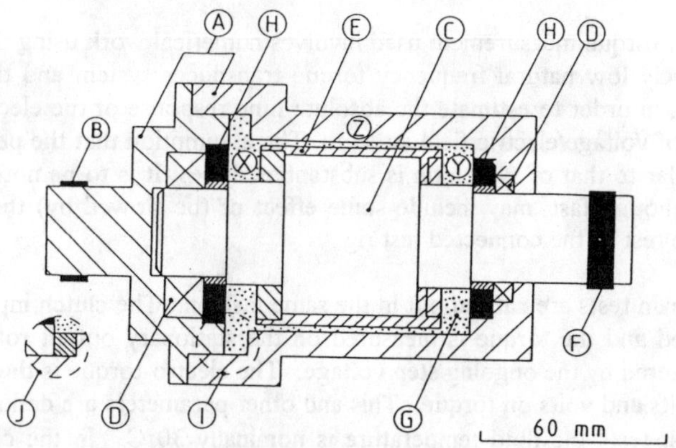

A - Input Member (Stainless Steel)
B - Earthing and Thermocouple Slip-Rings
C - Output Member (Glass Reinforced Nylon)
D - Rolling Element Bearings
E - High Voltage Electrode (Stainless Steel)
F - High Voltage Slip-Ring
G - Electro-Rheological Fluid
H - Viton Lip Seals
J - Nylon Insert
XYZ Thermocouple Positions

Figure 1a Cross section of test ER clutch.

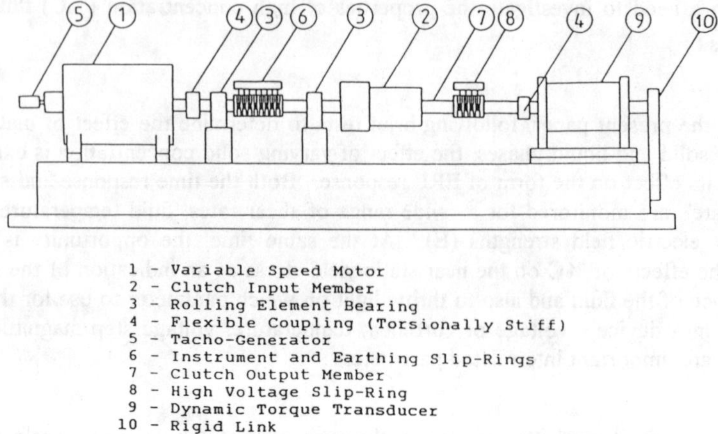

Figure 1b Schematic layout of the experimental apparatus.

The means of torque measurement used involves numerical work using the transfer function of a relatively low natural frequency torque transducer system and the reading from this transducer, in order to estimate the absolute time response of the electro-torque to a step switch on of voltage/electric field strength. The assumption that the performance on switch off is similar to that of switch on is substantiated in[2]. It is to be noted that the transformed event, though fast, may include some effect of (be slowed by) the response characteristics of the rest of the connected test rig.

All time domain tests are carried out in the same fashion. The clutch input rotor is run at a steady speed and the torque is measured on the stationary output rotor, results gathering being triggered by the ongoing step voltage. The electro-torque is the difference between the zero volts and volts on torque. This and other parameters are defined in Figs. 2a. and 2b. In most tests the fluid temperature is nominally 30°C. In the centrifuging tests, between monitoring events the clutch rotors are locked together for prolonged periods of high speed running .

In previous tests[3] the polarity of the excitation has been seen not to be an important factor in respect of any change of fluid behaviour under its endurance of electric field and rotation. The inner rotor is therefore always at -ve HT with the outer rotor earthed. Some verification of the test procedure is provided by the closeness of the low shear rate steady state test results to the fluid manufacturers calibrations. With the confirmation of a near continuum approach to the design of a clutch for one fluid[4], the suitability of different fluids for clutch application comes into focus.

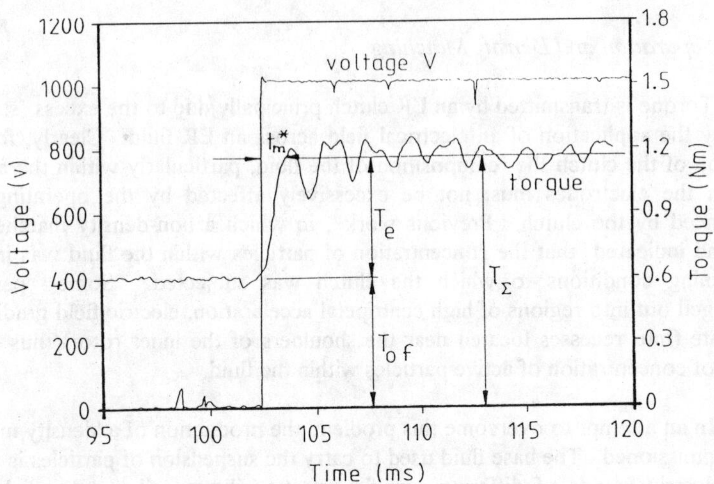

Figure 2a **General torque response to step input of voltage.**

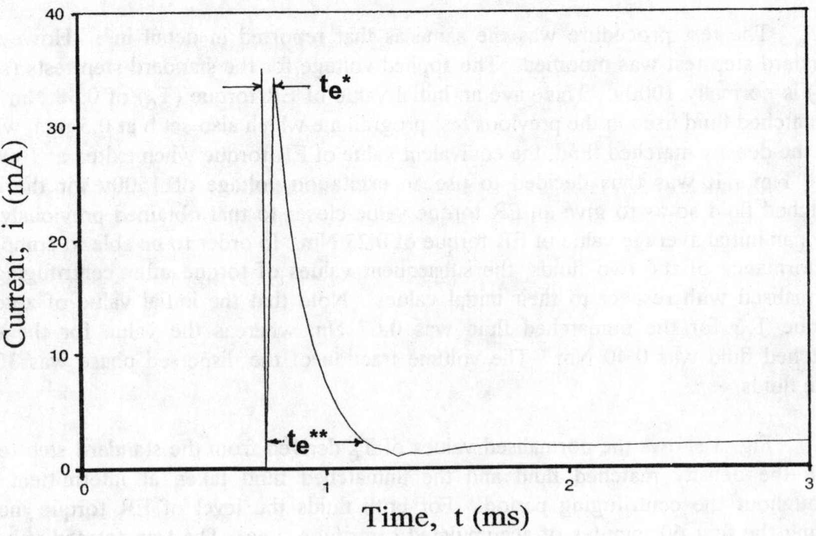

Figure 2b **General current response to step input of voltage.**

2. Test Results and Comments

2.1 *Separation and Density Matching*

Torque is transmitted by an ER clutch principally due to the excess stress brought about by the application of an electrical field across an ER fluid. Clearly, for consistent operation of the clutch the composition of the fluid, particularly within the active region between the electrodes must not be excessively affected by the operating conditions experienced by the clutch. Previous work[3], in which a non-density matched fluid was used, had indicated that the concentration of particles within the fluid was influenced by the running conditions to which the clutch was subjected. Some particles were centrifuged out into regions of high centripetal acceleration, electric field gradient and low shear rate (into recesses located near the shoulders of the inner rotor) thus leading to a change of concentration of active particles within the fluid.

In an attempt to overcome this problem, the production of a 'density matched fluid' was commissioned. The base fluid used to carry the suspension of particles is a mixture of two dielectric liquids of differing specific gravity (silicone oil and Fluorolube), of net density roughly equal to the density of the suspended active particles at 30°C. The dispersed phase was lithium polymethacrylate containing 12.5% free water and particle size 1 to 5 μm.

The test procedure was the same as that reported in detail in[3]. However, the standard step test was modified. The applied voltage for the standard step tests (see Fig. 2a.) is normally 1000v. This gave an initial value of ER torque (T_e) of 0.38 Nm for the unmatched fluid used in the previous test programme which also set h at 0.5 mm, whereas, for the density matched fluid, the equivalent value of ER torque when exited at 1000v was 0.10 Nm. It was thus decided to use an excitation voltage of 1500v for the density matched fluid so as to give an ER torque value closer to that obtained previously. This gave an initial average value of ER torque of 0.23 Nm. In order to be able to compare the performance of the two fluids, the subsequent values of torque after centrifuging were normalised with respect to their initial values. Note that the initial value of zero volts torque T_{of} for the unmatched fluid was 0.67 Nm whereas the value for the density matched fluid was 0.40 Nm. The volume fraction of the dispersed phase was 30% for both fluids.

Fig. 3 shows the normalised values of T_e derived from the standard step tests for both the density matched fluid and the unmatched fluid taken at intermittent stages throughout the centrifuging period. For both fluids the level of ER torque increased during the first 60 minutes of accumulated centrifuge time. The transformed value t^*_m, the electronic to shear stress time delay remained within 200μs of 1 ms throughout, for both fluids. Since $t^*_e << t^*_m$ (Figs 2a and 2b) this apparently limiting condition is unlikely to be imposed by the source of excitation. It is higher than the likely polarisation time.

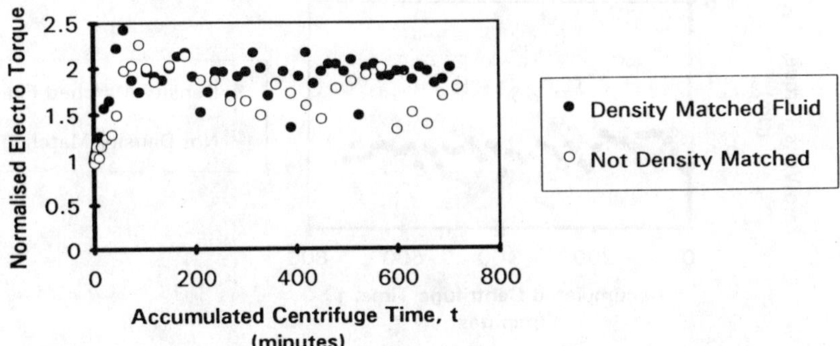

Figure 3 Normalised ER torque with running time. for 30% Concentration Fluid, density matched and unmatched

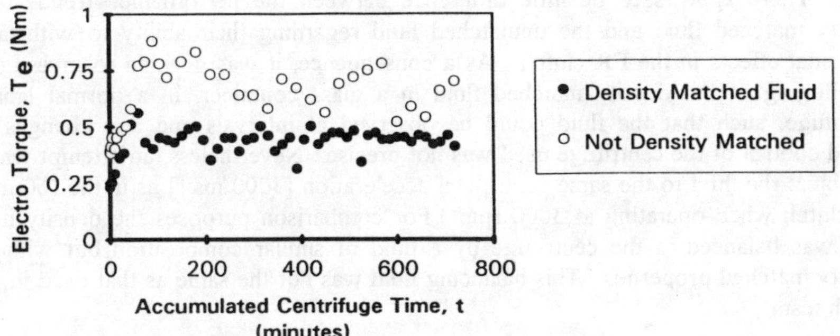

Figure 4 ER torque with running time. 30% Concentration Fluid, density matched and unmatched fluid

Figure 4 shows the values of T_e before normalisation for both fluids. The level of T_{of} also increased initially and then fluctuated about a steady mean, see Fig 5.

When the clutch was stripped down after 12 hours of accumulated centrifuge time it was found that some of the particles had separated out in the regions X and Y of Figure 1a, as with the unmatched fluid. In both cases the annular cavity had been blocked off before operation commenced. However, the shoulders of the inner rotor though smoothed off still presented relatively sharp (electrical) edges and high local field gradients would be present in the vicinity of the accumulations.

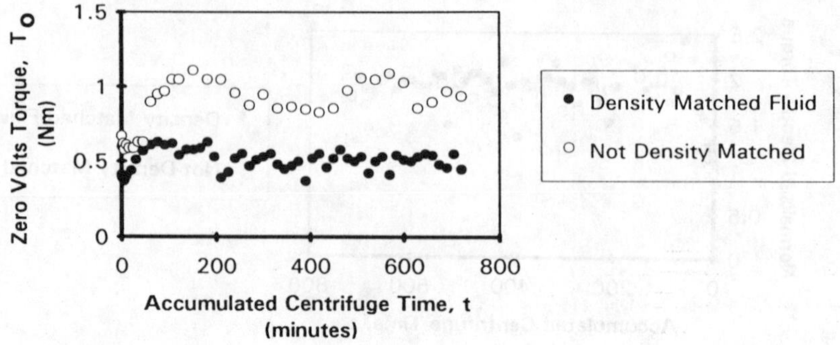

Figure 5 Zero volts torque with running time.
for 30% Concentration Fluid, density matched and unmatched

There appears to be little difference between the performance trends between density matched fluid and the unmatched fluid regarding their ability to withstand the rotational effects in the ER clutch. As a consequence, it was decided to investigate the centrifuging of the density matched fluid in a glass container, in a normal laboratory centrifuge, such that the fluid could be observed at intervals and any changes noted. Speed control of the centrifuge used was not precise. Nevertheless, an attempt was made to subject the fluid to the same centripetal acceleration [3000 ms^{-1}] as in the (60 mm dia) ER clutch when operating at 3000 rpm. For comparison purposes the density matched fluid was balanced in the centrifuge by a fluid of similar composition but without the density matched properties. This balancing fluid was not the same as that used in the ER clutch tests.

After one minute of centrifuging, the none density matched fluid had separated out substantially, i.e. the top 20% of the sample approximately was clear liquid, whereas the density matched fluid had shown no tendency to separate out. After ten minutes of centrifuging the density matched fluid did show signs of separation. Some clear liquid was seen at the top. After 20 minutes the top 10% was clear liquid and after 40 minutes the top 20% was clear. This centrifuging took place at room temperature. The fluid was balanced at 30°C.

Whatever the cause of the change in electro-torque T_e (and hence τ_e) with running duration the performance trends of both fluids are similar. The non-density matched fluid enjoys a higher electro-stress because its base oil properties are superior to those of the fluid containing the high density Fluorolube. The results tend to indicate that there is little advantage in density matching under the conditions experienced. Perhaps the improvement measured following the initial running of the fluid involves the generation of permanent dipoles? The phenomenon has been noted before[5]. The results of further separation effects testing of non density matched fluid are brought together (in so much as they can) in Fig 6.

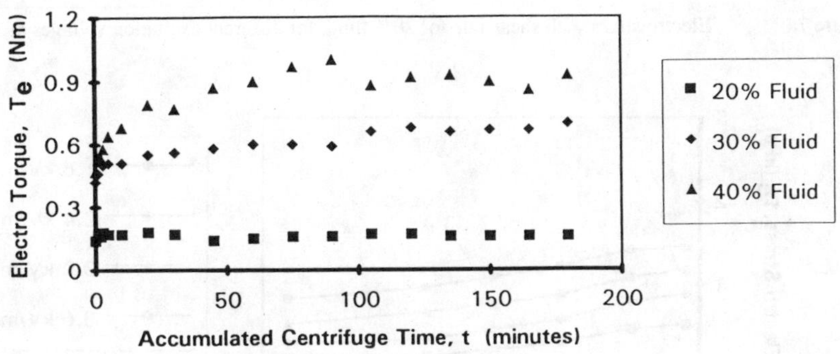

Figure 6 ER torque with running time.
for different volume fraction fluids, non density matched, 1500 volts excitation.

2.2 τ_e *Related High Shear Rate Characteristics*

In Figs 7 a, b, c, data is given for 20, 30 and 40% C volume fraction fluids respectively. At high shear rates the electro-stress is seen to continue to fall but, at a rate which is less severe than at lower shear rates. Fig 8 shows the same data normalised on the $\dot{\gamma}$ = 9500s^{-1} results for the 1800V excitation. The zero volts torques (seal friction prevents conversion to shear stress values) for the various volume fractions is shown in Fig 9. The definitions of T_{of} is given in Fig 2a. At high $\dot{\gamma}$ the rate of fall off of τ_e seems independent of %C. The low shear rate results indicate a potentially severe hysteretic tendency at 40%C. However, the result is not conclusive. Comparative manufacturers data adds confidence to the results.

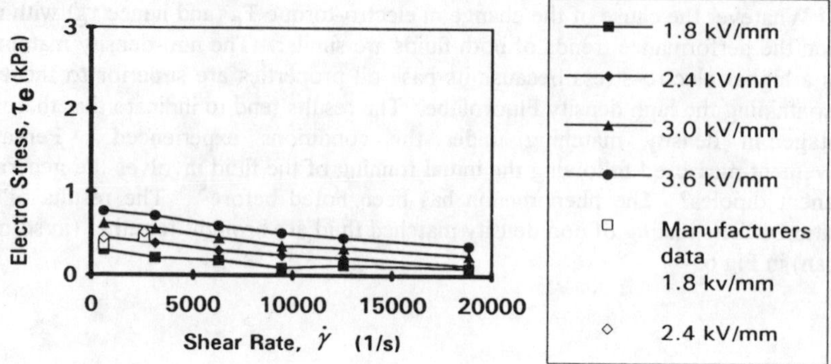

Figure 7a Electro stress with shear rate for 20% fluid, for different excitation voltages at 30°C.

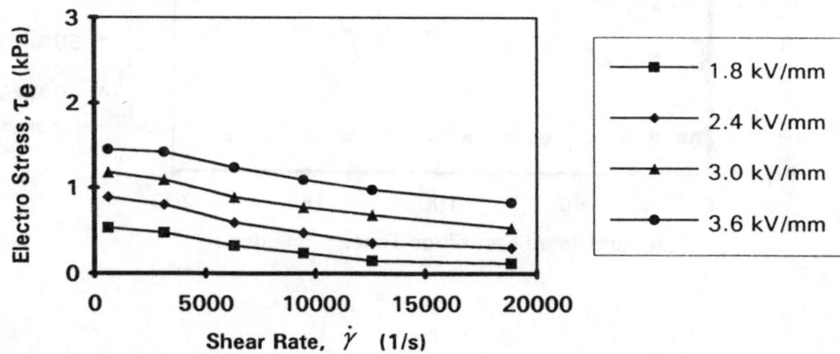

Figure 7b Electro stress with shear rate for 30% fluid, for different excitation voltages at 30°C.

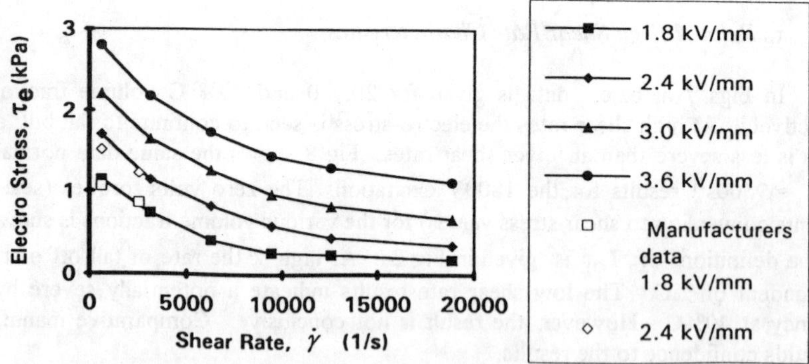

Figure 7c Electro stress with shear rate for 40% fluid, for different excitation voltages at 30°C.

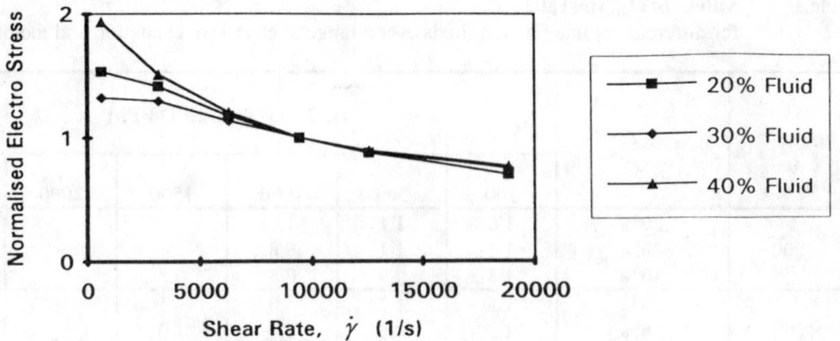

Figure 8 Electro stress normalised to values at 9500 s^{-1} shear rate, for different volume fractions at 30°C and 1.8 kV excitation.

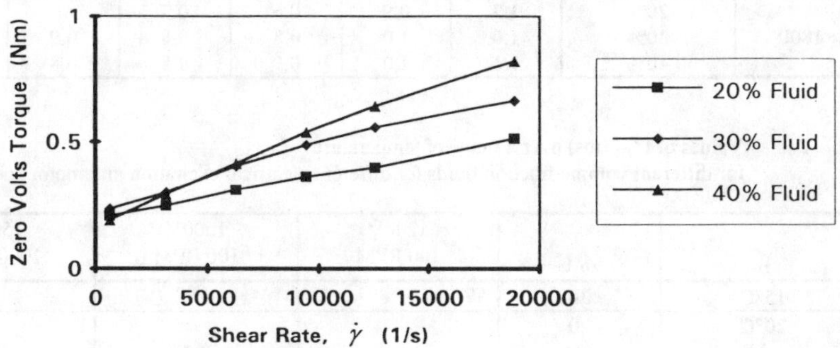

Figure 9 Average values of zero volts torque with shear rate, for different volume fraction fluids.

2.3 t^*_m Related High Shear Rate Characteristics

In previous work[6] it has been shown that for valve flow t^*_m is dependant on $\dot{\gamma}$ and to a lesser extent on the magnitude of the applied voltage step in valve tests conducted at 30°C and 30% volume fraction, and is similar when switching excitation on and off. In Table 1 these trends are confirmed in the clutch for $20 \leq \%C \leq 40$. Some slight lack of repeatability was encountered in the tests from which the tables are drawn and average values are given. The effect of %C is seen not to greatly effect t^*_m at high shear rate or, at least not within the accuracy of the tests.

The results for the effect of fluid temperature on t^*_m are again susceptible to error - in the precise control of temperature. However, despite the lack of comprehensive data some reduction in t^*_m is always evident as the temperature increases Table 2; shear rate is the dominant factor here.

Table 1 Values of t^*_m (ms) at 30°C
for different volume fraction fluids over a range of electrical excitation and motor speed.

V	%C	ω, MOTOR SPEED RPM					
		100	500	1000	1500	2000	3000
900	20%	1.6	1.1	1.0			
	30%	1.2	1.1	0.8			
	40%	1.3	0.9	0.8	0.8		
1200	20%	1.5	1.2	1.0			
	30%	1.2	0.8	1.0	1.0		
	40%	1.2	0.9	0.8	0.7	0.7	0.7
1500	20%	1.4	1.0	0.8	0.9		0.9
	30%	1.0	0.8	0.9	0.9	0.6	1.0
	40%	1.1	1.0	0.9	0.8	0.8	0.7
1800	20%	1.2	0.9	0.6	0.7		1.0
	30%	1.0	1.0	0.8	0.8	0.9	1.2
	40%	1.1	1.0	0.9	0.8	0.8	0.8

Table 2 Values of t^*_m (ms) over a range of temperatures
for different volume fraction fluids for different electrical excitation and motor speed.

θ	% C	1200V 100 RPM	1500V 100 RPM	1500V 1000 RPM
15°C	30		3.2	
20°C	20			
	30		2.1	0.9
	40	1.0		
25°C	20	1.1		
	30	1.2	1.3	0.5
	40	.8		
30°C	20	1.2		
	30	0.6	1.3	0.8
	40	1.0		
35°C	20	1.1		
	30	0.5	0.9	0.7
	40	1.0		
40°C	20	0.9		
	30	0.9	0.9	0.6
	40	0.9		
45°C	20	0.8		
	30	0.7	0.7	0.7
	40	0.9		
50°C	20			
	30		0.8	
	40	0.7		

Table 3 **Values of t^{**}_e (ms) at 30°C**
for different volume fraction fluids over a range of electrical excitation and motor speed.

V	%C	ω MOTOR SPEED RPM					
		100	500	1000	1500	2000	3000
900	20%	2.2	0.86	0.63	0.48	0.38	0.24
	30%	2.1	0.86	0.60	0.43	0.40	0.31
	40%	2.2	0.92	0.78	0.51	0.40	0.37
1200	20%	1.95	0.69	0.54	0.33	0.44	0.36
	30%	1.84	0.76	0.57	0.41	0.40	0.32
	40%	2.70	0.92	0.59	0.45	0.40	0.35
1500	20%	1.93	0.72	0.58	0.46	0.42	0.24
	30%	1.63	0.69	0.51	0.41	0.37	0.27
	40%	1.90	0.84	0.52	0.41	.034	0.30
1800	20%	1.90	0.60	0.46	0.38	0.34	0.28
	30%	1.76	0.67	0.53	0.40	0.34	0.27
	40%	1.60	0.70	0.50	0.43	0.34	0.30

Table 4 **Values of t^{**}_e (ms) over a range of temperatures**
for different volume fraction fluids for different electrical excitation and motor speed.

θ	% C	1200V 1000 RPM	1500V 100 RPM	1500V 1000 RPM	1500V 3000 RPM
15°C	20	0.98			
	30	0.75	2.60	1.70	1.10
	40	0.83			
20°C	20	1.10			
	30	0.76	2.44	0.95	0.36
	40	0.83			
25°C	20	0.84			
	30	0.80	2.21	0.60	038
	40	0.77			
30°C	20	0.56			
	30	0.50	1.82	0.48	033
	40	0.60			
35°C	20	0.40			
	30	0.43	1.44	0.40	
	40	0.45			
40°	20	0.32			
	30	0.33		0.35	
	40	0.44			
45°C	20	0.26			
	30			0.25	0.24
	40	0.37			
50°C	40	0.24			

The data in Table 3 shows how t^{**}_e (as measured from the test record) varies with %C, excitation voltage and $\dot{\gamma}$. The results are somewhat arbitrary because of the difficulty of judging the termination of the overshoot process. Nevertheless , it can be seen that t^{**}_e is reduced at high shear rates whilst it decreases with the magnitude of the voltage step. The results are in sympathy with valve electrical characterisations[7]. The effect of %C on t^{**}_e in the present tests is not significant or, inconclusive.

In the valve tests[7] the extent of t^{**}_e was found to be closely related to the electrical time constant (r x c) of the model shown in figure 10. This factor is also seen (in Table 4) to be reduced as the temperature increases in Table 4 and is consistent with increased conductivity. Separate values of r and c have not yet been determined. Capacitance is generally accepted to have but a secondary dependence on temperature. Data already published[4] shows that r x c does not vary much with h at constant temperature for given E and $\dot{\gamma}$. Throughout the range of the tests reported in this paper t^*_e varies within less than 100μs. This may yet be seen to be dependent on the power supply.

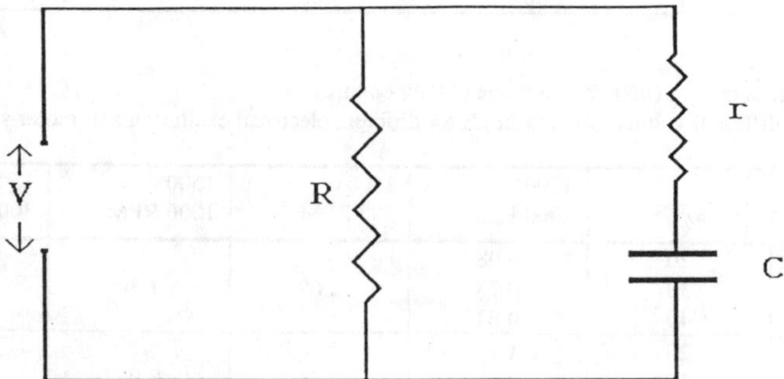

Figure 10 Approximate equivalent circuit for the Electro-rheological device.

2.4 *Controller Indications*

It is well known that τ_e is dependant on E (and h secondarily), $\dot{\gamma}$ and fluid temperature[8,9] Fig 11. This makes its precise control by voltage difficult[9]. In this section, the possibility of control by current is examined.

In Figs 12c, b and c the steady state currents drawn per unit of rotor surface area (as defined in Fig 2b) are shown for one temperature (30°C). It will be noted that the isothermal current and τ_e curves (Figs 7a, b and c) have the same trend in variation with shear rate and volume fraction, as well as voltage. The data is normalised to the values at 9500 s^{-1} shear rate in Fig 13, for a range of shear rates at one temperature, 30°C.

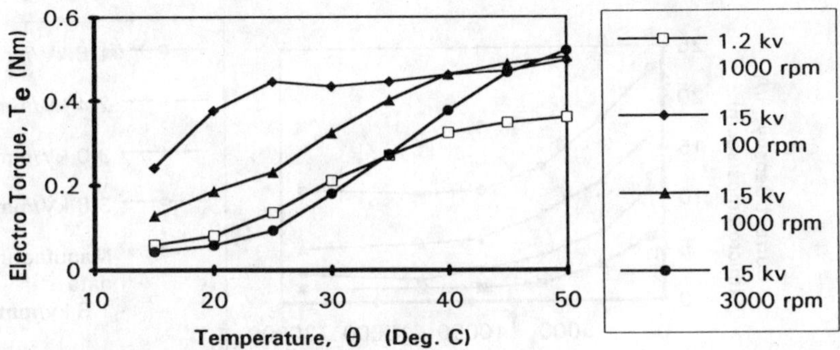

Figure 11 ER torque with temperature for different excitation voltages and motor speed.

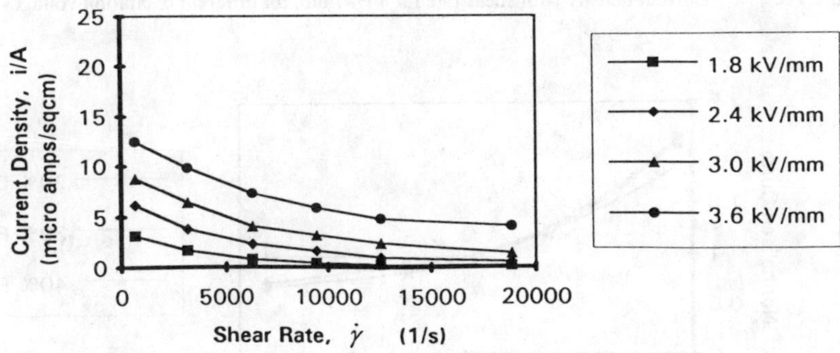

Figure 12a Current density with shear rate for 20% fluid for different excitation voltages at 30°C.

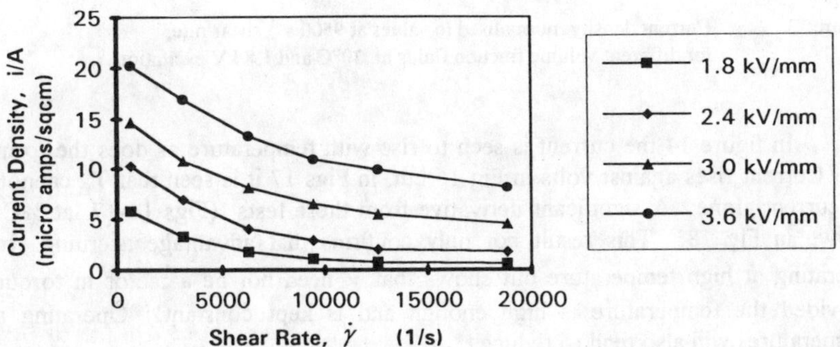

Figure 12b Current density with shear rate for 30% fluid for different excitation voltages at 30°C.

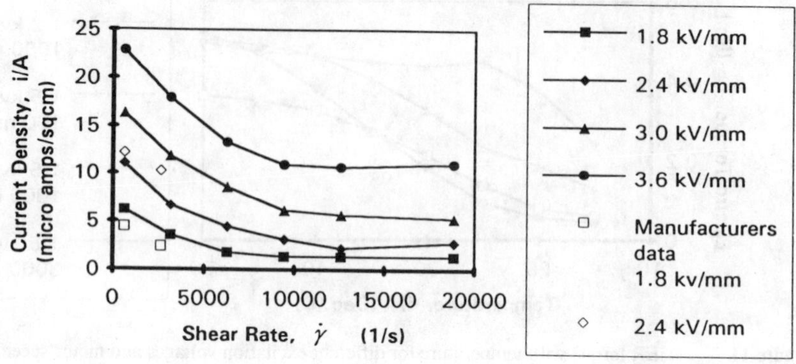

Figure 12c Current density with shear rate for 40% fluid, for different excitation voltages at 30°C.

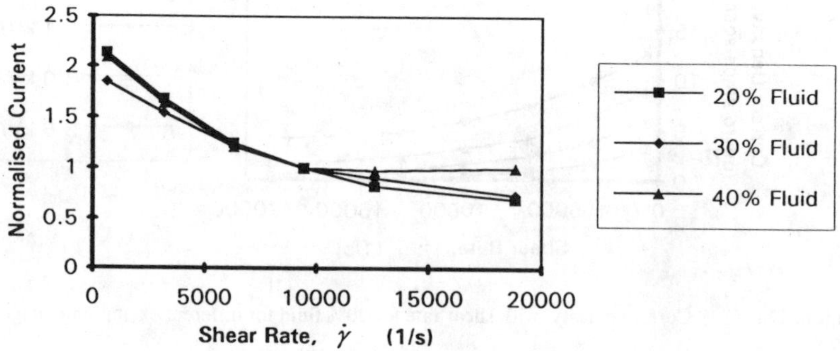

Figure 13 Current density normalised to values at 9500 s^{-1} shear rate, for different volume fraction fluids at 30°C and 1.8 kV excitation.

In figure 14 the current is seen to rise with temperature as does the torque in Fig 15. Current rises against volts in Fig 16 but, in Figs 17 it is seen that T_e cannot be fixed by current alone. A significant derivative from these tests (Figs 14-17, at %C = 30) is shown in Fig 18. This result not only confirms that advantage accruing to t^*_m in operating at high temperature but shows that $\dot{\gamma}$ need not be a factor in torque control provided the temperature is high enough and is kept constant. Operating at higher temperatures will also tend to reduce t^*_m.

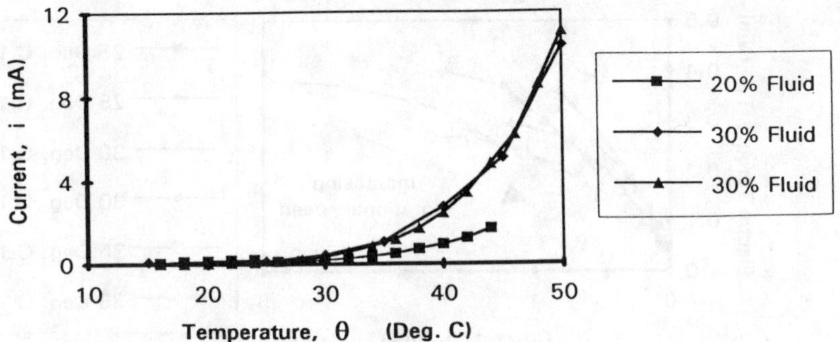

Figure 14 Current with temperature for different volume fraction fluids, 1.2 kV excitation and motor speed 1000 rpm.

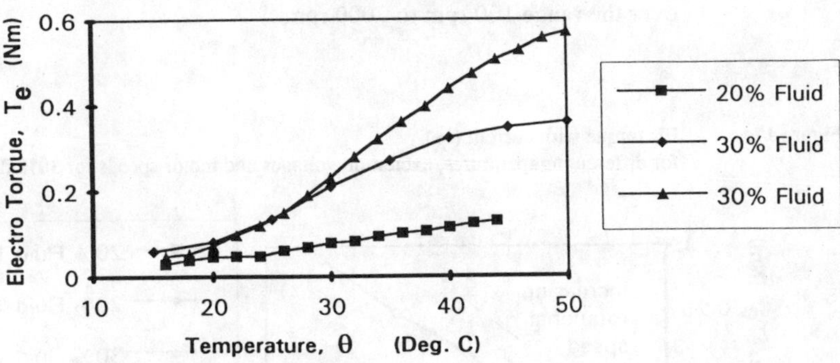

Figure 15 ER torque with temperature for different volume fraction fluids, 1.2 kV excitation and motor speed 1000 rpm.

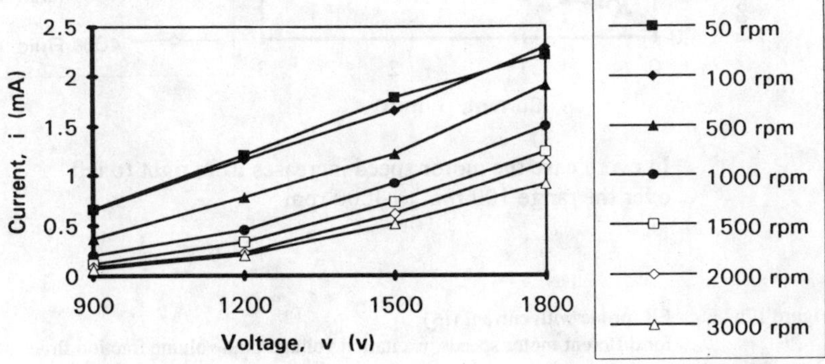

Figure 16 Current (i_R) with excitation voltage for different motor speeds, 30% fluid at 30°C.

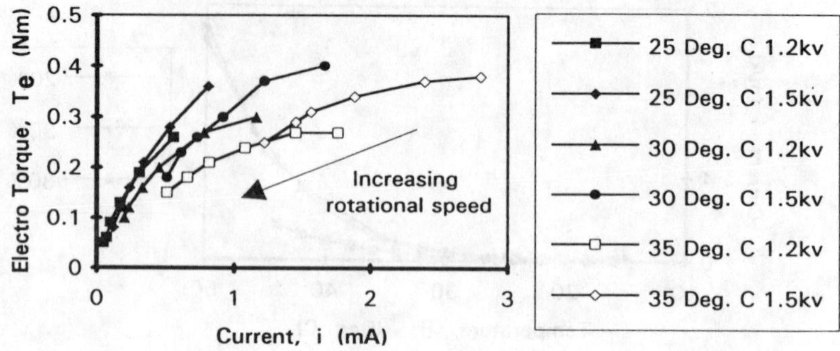

In every case the motor speed increases from right to left over the range 100 rpm to 3000 rpm

Figure 17a ER torque with current (i_R) for different temperatures, excitation voltages and motor speeds for 30% fluid.

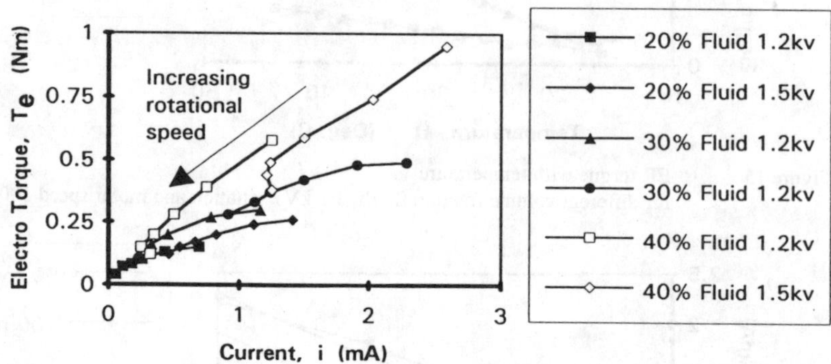

In every case the motor speed increases from right to left over the range 100 rpm to 3000 rpm

Figure 17b ER torque with current (i_R) for different motor speeds, excitation voltages and volume fraction fluids at 30°C.

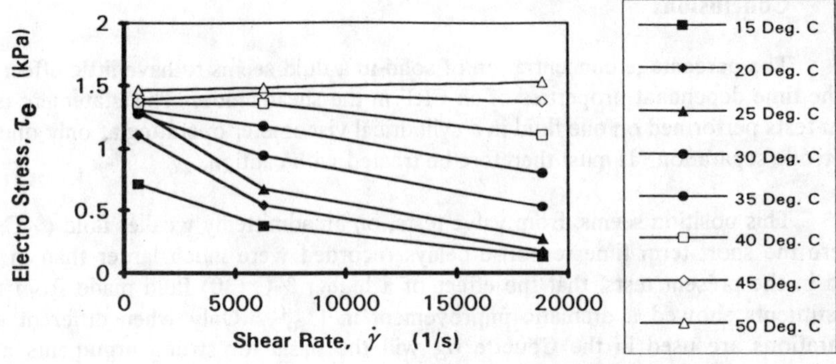

Figure 18 **Electro stress with shear rate for different temperatures, 30% volume fraction fluid and 1.5 kV excitation.**

3 Discussions

There are stated limitations on the accuracy of the measurements recorded in this paper. Nevertheless the main trends in performance are repeatable and established.

A running in period on a new fluid seems to improve its τ_e capacity. Beyond this there is little to choose with respect to density matched and non density matched fluid re their rotational stability and durability.

No catastrophic failure in τ_e (as $\dot{\gamma} \rightarrow 20{,}000\text{s}^{-1}$) is encountered, whatever the %C, and the gradual decline of τ_e with $\dot{\gamma}$ can be counteracted when operation is at industrial operating temperatures (circa 50°C in the present fluids). The corresponding level of τ_e is linked with the effective conductivity yet, current density alone does not control the level of shear stress.

The values of t^{**}_e and t^*_m encountered at significant shear rates are always small. Hence the simple procedure for calculating run up speed put forward previously[7] is validated. Some discretion is called for as τ_e is not precisely fixed by E, $\dot{\gamma}$ and θ for different electrode separations[4] and in the present test T_0 was not always insignificant.

Apart from the well known effects of %C (on τ_e and steady state current) it seems to have little effect elsewhere save possibly in the levels of charging current, hysteresis and elasticity which have yet to be fully investigated. An increase in current takes place whenever τ_e rises, whatever the value of θ°C. In most respects high $\dot{\gamma}$, E and θ conditions lead to optimal catch performance.

4. Conclusions

The percentage concentration of solid in a fluid seems to have little effect on most of the time dependant properties of an ERF in the shear mode. This statement is derived from tests performed on one fluid in a cylindrical viscometer operating at only one value of electrode separation. It must therefore be treated with caution.

This position seems, from valve tests, on an admittedly weaker fluid (%C = 20%), where the short-term time response delays recorded were much larger than in the fluid used in the present tests, that the effect of a higher %C (30) fluid made from the same constituents showed a dramatic improvement in t^*_m[6]. Only when different electrode separations are used in the Couette rig will the basis for strong arguments about the explanations for this be established (continuum properties of ERF vis a vis flow field effects). These and further results may then throw light on the different time response properties of valve (flow mode) and clutch (shear mode) situations that apparently accrue for the same fluid[2,6].

From what data is made available from the tests in a clutch, there is a case for moving to higher %C (circa 40%); this hardly affects t^*_m yet, τ_e enjoys a well-known increase as %C is raised. However, as %C rises so does zero volts viscosity and, if the inter electrode gap h has to be increased in order to control T_o, the ensuing fall in $\dot{\gamma}$, (for the same slip speed) may impair the operating t^*_m. Fortunately, at higher temperature levels (circa 50°C) there may be some alleviation of the problem since τ_e at that condition hardly falls with $\dot{\gamma}$.

The long term performance of the fluid seems to be little improved by density matching. Again this result must be qualified. Like the short term time domain results, though tested over a wide range of $\dot{\gamma}$, θ, %C and E, the geometrical situation was fixed.

The steady state control problem is not clear cut. Voltage or current alone will not suffice[10]. A more considered approach will be needed.

Acknowledgements

Financial support by the SERC/DTI, LINK High speed machines program is gratefully Acknowledged.

Nomenclature

A	-	effective surface area of a rotor $\pi D\ell$
%C	-	volume fraction of solid in ERF
D	-	nominal diameter of clutch
ERF	-	electro-rheological fluid
E	-	field strength V/h
h	-	inter electrode spacing
i	-	electric current
I	-	driven rotor inertia
c	-	capacitance
ℓ	-	nominal length of excited rotor surface
R, r	-	resistances
t	-	time
t^*_m	-	electron- hydraulic time delay
t^{**}_m	-	approximate time in acceleration
t^*_e	-	current rise time
t^{**}_e	-	approximate duration of surge current
T	-	torque measured on meter
V	-	magnitude of voltage step
α	-	constant rate of acceleration of driven rotor
$\dot{\gamma}$	-	shear rate (πD/2h)
τ	-	shear stress
θ	-	temperature of ER fluid
ω	-	rotational speed of clutch outer rotor

suffixes

f	-	due to parasitic friction
e	-	due to electric field application
o	-	due to zero volts fluid friction
R	-	resistive current
c	-	charging current

References

1. W A Bullough, R Firoozian, A R Johnson A H Sianaki, J Makin and X Shi, *Proc. I.Mech.E., Eurotech. 91.* (1991) Paper C414/070.

2. W A Bullough, R Firoozian, A R Johnson, A H Sianaki, J Makin and S Xiao, *Mechatronic Systems Engineering,* **Vol. No. 4,** (1992) pp. 315-327.

3. A R Johnson, W A Bullough, J Makin, A H Sianaki, S Xiao and R Firoozian. To be published in *Jnl. Intel. Matl. Systems and Structures,* (1993), 'Fluid Durability in a High Speed Electro-Rheological Catch'.

4. A Hosseini-Sianaki, W A Bullough, R Firoozian, J Makin, A R Johnson and S Xiao, *Proc Actuator 92, Bremen,* VDI/E, (German Association of Engineers), (1992) pp. 118-122.

5. W A Bullough, *Proc 2nd Int Conf. ER Fluids, Raleigh, North Carolina,* (1989) pp. 115-123, (Technomic Publishing Co. 1990).

6. D J Peel and W A Bullough, *Jnl. Intel. Matl, Systems and Structures,* **Vol 4. No. 1**, (1993) pp 54-64.

7. M Whittle, R Firoozian, D J Peel and W A Bullough, *Jnl. Mod. Phys. B,*.(1991) pp. 343-366.

8. F E Filisko, *Proc. 3rd. Int. Conf. ER Fluids, Carbondale, Illinois* (1991) pp. 116-128, (Technomic Publishing Co. 1992).

9. K D Weiss, J D Carlson and J P Coulter. *Jnl. Intel. Matl. Systems and Structures.* **Vol 4. No. 1,** (1993) pp. 13-34.

10. R Firoozian, W A Bullough, A R Johnson, A H Sianaki, J Makin and Xiao Shi. *Proc. Mechatronics Conf. Florence* (1991) pp. 479-487, (ISATA).

ELECTROSTATIC INTERACTIONS FOR PARTICLE ARRAYS IN ELECTRORHEOLOGICAL FLUIDS: I CALCULATIONS

Yung-Hui Shih, A. F. Sprecher and H. Conrad
Materials Science & Engineering Department
North Carolina State University
Raleigh, NC 27695–7517

ABSTRACT

A simple effective charge method and the principle of superposition were used to calculate the total system interaction energy of two-dimensional rectangular and hexgaonal arrays of particles in electrorheological suspensions. The calculations gave for *static conditions (no shear)* that a close-packed hexagonal array of particles was energetically favored over a rectangular array and that for very dilute suspensions a structure consisting of separated single chains was favored over one consisting of clusters. For *shearing* and assuming that no rotation of the polarization vector takes place, no difference in the maximum force per chain occurred for a two-chain system compared to a single-chain system in either the rectangular or close-packed hexagonal array. This was also the case with rotation of the polarization vector in the rectangular array, although the force was less with rotation compared to without rotation for this array. However, in the case of the close-packed hexagonal array with rotation of the polarization vector there occurred an order of magnitude enhancement of the maximum force for the two-chain system compared to the single-chain system. This enhancement is comparable to the ratio of the measured strength of ER fluids to that predicted for a structure consisting of separated single chains.

1. Introduction

A characteristic feature of electrorheological (ER) suspensions is the alignment of the particles into a chain-like or fibril structure along the applied electric field; see for example Fig. 1. At a low concentration of particles, the structure consists largely of single-row chains, whereas at high concentrations clusters or columns of particles span the electrode gap.

The polarization force of interaction between particles in a single-row chain has been calculated by Chen et al[2] and found to be in reasonable accord with that measured on a chain of humidified glass beads in silicone oil[3]. However, the measured shear strength of some common ER fluids is about an order of magnitude larger than that predicted based on the forces between particles in single chains[4–6]. The difference between the measured and predicted strength was attributed to the fact that in these ER

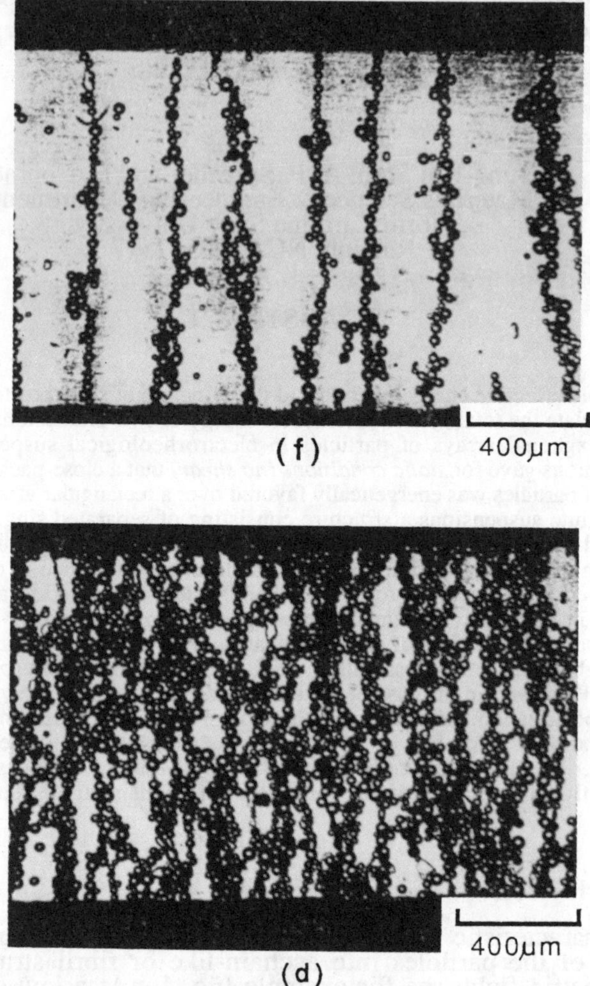

Fig. 1 Particle arrangements in a model ER suspension consisting of 27 μm dia. glass beads in silicone oil subjected to an electric field E = 1 kV/mm: (a) dilute suspension and (b) concentrated suspension. From Conrad et al [1].

suspensions the fibril structure consists of columns or clusters of particles spanning the electrode gap rather than single chains[4,5]. With such columns, contact between neighboring particles is maintained longer during shear than in the case of particles in a single chain. Since the force of interaction between particles decreases rapidly with their separation, the columnar structure would thus have a higher resistance to shear than would be the case for a structure consisting of separated, single chains. Support for this idea was provided by the concentration dependence of the flow stress of zeolite/mineral oil dispersions[4,5]. At low concentrations, where a single chain structure was prevalent, there occurred good agreement between the predicted and measured stress. The measured stress however became increasingly larger than that predicted for single chains as the volume fraction of particles was increased and the particles clustered to form a columnar structure.

The observed increase in strength with change in particle arrangement is in contrast to the calculations by Kraynik et al[7], which indicate that the strength of a double chain is less than that of two single chains. The present calculations were therefore undertaken to examine further the interaction energy and force between particles in multiple chains compared to single-row chains in ER fluids.

2. Approach

To simplify the calculations, an effective charge method is employed to calculate the interaction energy U of a two-dimensional system containing particles suspended in a dielectric liquid and subjected to an electric field. If there are infinite particles per chain and infinite chains in a two-dimensional periodic space lattice, every particle will have the same polarization, because all of the particles are equivalent in the periodic lattice. The charge distribution in every particle can then be represented by two effective charges +q and –q separated by a distance 2h as shown in Fig. 2. The relationships between the effective charge q and the electric

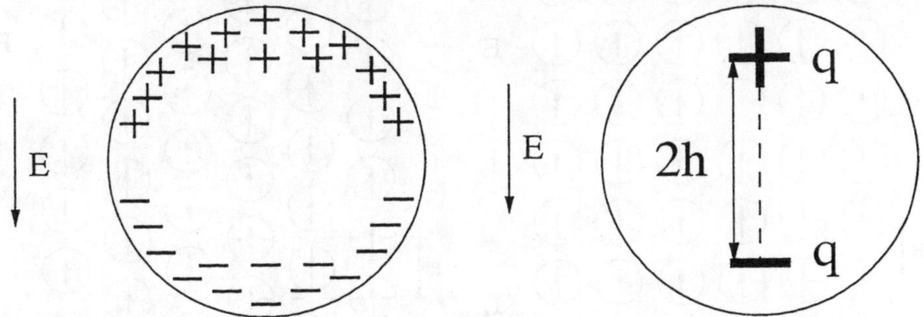

Fig. 2 Schematic of the charge distribution: (a) actual charge distribution and (b) effective charge representation.

field E and the relative permittivities of the particles K_p and of the fluid K_f are ignored in the present calculations. Thus, although the present method *can not* give the magnitude of the strength of an ER fluid in terms of E, K_p and K_f, it *can provide an estimate of the relative effects of different structures on the strength.*

The following assumptions are made regarding the calculations:

a. The particles are spherical and are all of the same size.
b. The electrostatic interaction is the dominant interaction, i.e. all other interactions are neglected.
c. Polarization is the same in every particle of an array.
d. Complete and perfectly straight chains of particles span the electrode gap.
e. When the array is sheared, this is so slow that a static or quasi-static state exists.
f. The separation of the particles during shearing is the same for all particles.

Two arrangements of the particles in two-dimensional space are considered, namely a rectangular array and a close-packed hexagonal array; see Fig. 3.

Since every sphere has two effective charges +q and −q, which are separated by the distance 2h, the total system interaction energy U of the two-dimensional space is calculated employing the principle of superposition, which gives

$$U = \frac{1}{2}\sum_{i}\sum_{j \neq i} U_{ij} = \frac{1}{4\pi\varepsilon}\left(\frac{1}{2}\sum_{i}\sum_{j \neq i}\frac{q_i q_j}{d_{ij}}\right) \tag{1}$$

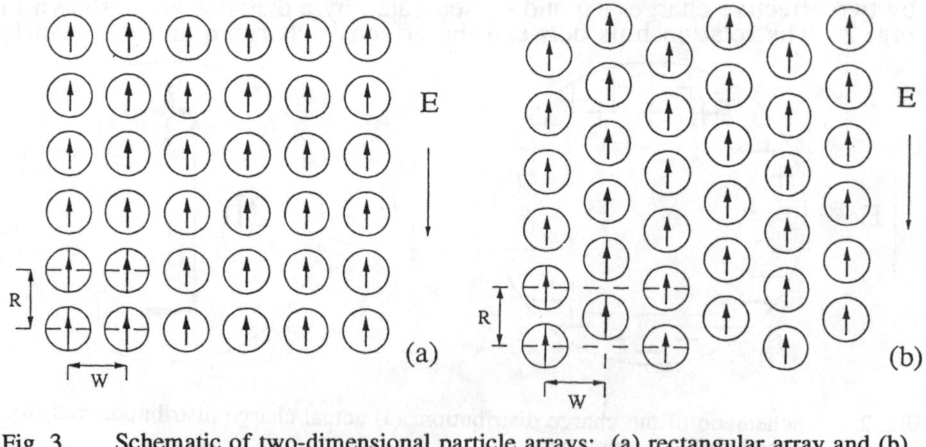

Fig. 3 Schematic of two-dimensional particle arrays: (a) rectangular array and (b) close-packed hexagonal array.

where U_{ij} and d_{ij} are respectively the system interaction energy and the distance between any two effective charges q_i and q_j. Here ε is the permittivity of the dielectric liquid. For example, if a **rectangular** array is considered, calculation is started from a single chain. Then the interactions between chains are considered. Assuming that N is the number of particles per chain and S the number of chains, the total system interaction energy for a single chain U_a that has 2N total effective charges can be written as

$$\begin{aligned} U_a &= U_{21} + (U_{31} + U_{32}) + (U_{41} + U_{42} + U_{43}) + \ldots\ldots \\ &\quad + (U_{2N.1} + U_{2N.2} + U_{2N,3} + \ldots\ldots + U_{2N,2N-1}) \\ &= \frac{1}{4\pi\varepsilon} \sum_{i=2}^{2N} \sum_{j=1}^{i-1} \frac{q_i q_j}{d_{ij}} \end{aligned} \tag{2}$$

By using the same procedure, the interaction energy between two chains U_b can be derived as a function of N, R, W, q, 2h. Here R represents the distance between the centers of two neighboring particles in the same chain and W is the separation between the center lines of two neighboring chains; see Fig. 3. The general expression of the system interaction energy for a two-dimensional particle array with N particles per chain and S chains can thus be divided into two components. The first is due to the interaction of the single chain itself U_a and the second is the interaction of any two chains in the array U_b. Therefore, the general expression for the total system interaction energy of a two-dimensional particle array is

$$\begin{aligned} U &= S\,U_a + (S-1)\,U_b\,(W) + (S-2)\,U_b\,(2W) + (S-3)\,U_b\,(3W) \\ &\quad + \ldots\ldots + 3\,U_b\,[(S-3)W] + 2\,U_b\,[(S-2)W] + U_b\,[(S-1)W] \\ &= S\,U_a + \sum_{i=1}^{S-1} (S\text{-}i)\,U_b\,(iW) \end{aligned} \tag{3}$$

For example, the normalized interaction energy of a single chain U_a is

$$U_a / \left(\frac{1}{4\pi\varepsilon}\frac{q^2}{2h}\right) = -N + \sum_{i=1}^{N-1} (N-i)\left(\frac{2}{ib} - \frac{1}{ib+1} - \frac{1}{ib-1}\right) \tag{4}$$

The normalized interaction energy of any two chains U_b in the **rectangular** array is

$$U_b(mW) / \left(\frac{1}{4\pi\varepsilon}\frac{q^2}{2h}\right) = \sum_{i=0}^{N-1} \sum_{k=i-(N-1)}^{i} \left(\frac{2}{\sqrt{m^2c^2+k^2b^2}} - \frac{1}{\sqrt{m^2c^2+(kb+1)^2}} - \frac{1}{\sqrt{m^2c^2+(kb-1)^2}} \right) \tag{5}$$

where the constants b and c are the ratios of R/2h and W/2h, respectively, and m is a dummy index number. Thus, the total normalized system interaction energy for a two-dimensional **rectangular** array is finally

$$U / \left(\frac{1}{4\pi\varepsilon}\frac{q^2}{2h}\right) = -SN + S \sum_{i=1}^{N-1} (N-i) \left(\frac{2}{ib} - \frac{1}{ib+1} - \frac{1}{ib-1} \right)$$

$$+ \sum_{m=1}^{S-1} \sum_{i=0}^{N-1} \sum_{k=j-(N-1)}^{i} (S-m) \left(\frac{1}{\sqrt{m^2c^2+k^2b^2}} - \frac{1}{\sqrt{m^2c^2+(kb+1)^2}} - \frac{1}{\sqrt{m^2c^2+(kb-1)^2}} \right) \tag{6}$$

The interaction energy for a two-dimensional hexagonal array is more complicated, but the calculation procedure is the same. The expression of the total normalized system interaction energy for a **hexagonal** array is

$$U / \left(\frac{1}{4\pi\varepsilon}\frac{q^2}{2h}\right) = -SN + S \sum_{i=1}^{N-1} (N-i) \left(\frac{2}{ib} - \frac{1}{ib+1} - \frac{1}{ib-1} \right)$$

$$+ \sum_{m=1,\,odd}^{[S]} \sum_{i=0}^{N-1} \sum_{k=i-(N-1)}^{i} (S-m) \left\{ \frac{2}{\sqrt{m^2c^2+(k-\frac{1}{2})^2b^2}} - \frac{1}{\sqrt{m^2c^2+[(k-\frac{1}{2})b+1]^2}} - \frac{1}{\sqrt{m^2c^2+[(k-\frac{1}{2})b-1]^2}} \right.$$

$$\left. + \sum_{m=2,\,even}^{[S]} \sum_{i=0}^{N-1} \sum_{k=i-(N-1)}^{i} (S-m) \left\{ \frac{2}{\sqrt{m^2c^2+k^2b^2}} - \frac{1}{\sqrt{m^2c^2+(kb+1)^2}} - \frac{1}{\sqrt{m^2c^2+(kb-1)^2}} \right\} \right. \tag{7}$$

Here [S] means the maximum integer which is small or equal to the chain number S and is dependent on the odd or even restriction of the dummy index.

Theoretically, it is necessary to input infinite numbers for N and S in order to satisfy the assumption of equal polarization for every particle in the array, but this is impossible for the calculation. If only a comparison of particle arrays in ER fluids is considered, finite numbers of N and S can be used to obtain the same properties as would be calculated from large numbers of N and S. Choosing large numbers of N and S will only change the scale; the trend would not be changed. Therefore, it is reasonable to assume finite number of N and S for the comparison of particle arrays of ER fluids.

In calculating the system interaction energy of particle arrays, N and S are fixed to satisfy particle conservation and only the particle arrays are changed. If a particle array is sheared, the shear force required to shear the particle array is obtained from the derivative of the total system interaction energy U with respect to shear strain γ.

$$F = \frac{1}{L}\frac{dU}{d\gamma} \tag{8}$$

F is thus a function of shear strain γ. The gap length of the two electrodes L is used to keep the units consistent. In the following, normalized quantities are used, including normalized interaction energy and normalized shear force, which are written as

$$U/(\frac{1}{4\pi\varepsilon}\frac{q^2}{2h}), \quad F/(\frac{1}{4\pi\varepsilon}\frac{q^2}{2Lh})$$

3. Results

3.1 Static Particle Arrays (No Shear)

The arrangements of particles shown in Fig. 1 can be classified into two general types: (a) single chain structure in dilute suspensions and (b) columns or clusters of chains in concentrated suspensions. Let us first consider static chain groupings in very dilute ER suspensions. Possible groupings are illustrated schematically in Fig. 4. For the calculations pertaining to these groupings we will assume that there are 20 particles per chain and a total of 32 chains in a very large two-dimensional space. The number of chains per grouping or bundle is chosen to be 1, 2, 4, 8, 16 and 32. In view of the large two-dimensional space, the interactions between bundles can be neglected.

The results of the calculations for very dilute suspensions are

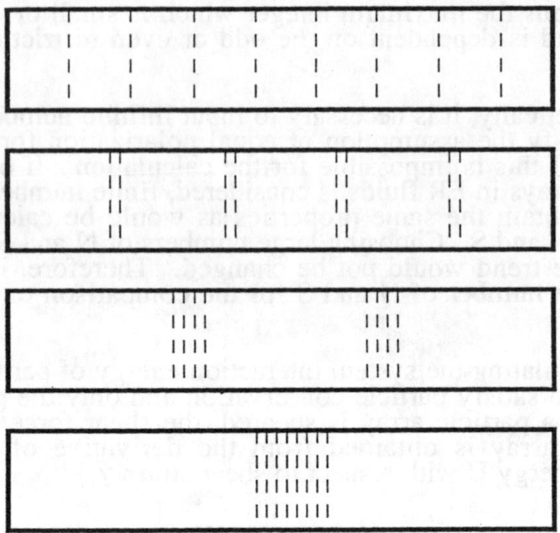

Fig. 4 Schematic of possible chain groupings in very dilute ER fluids.

presented in Fig. 5. It is here seen that the interaction energy is nearly the same for the rectangular and the hexagonal arrays and that the lowest interaction energy for both arrays occurs when the number of chains per bundle is one. Thus, in very dilute ER suspensions the most stable structure is predicted to consist of separated single chains, in accord with what is observed in actual ER suspensions.

For concentrated static suspensions we assume that the particles are arranged in either the rectangular or close-packed hexagonal array shown in Fig. 3 and calculate the resulting interaction energies. The results (Fig. 6) reveal that for the same separation parameters, the hexagonal array has a lower interaction energy and thus is more stable than the rectangular array, in accord with calculations by Tao and Sun[8] and Davis[9].

3.2 *Sheared Particle Arrays*

Four cases of motion with differing particle arrays and polarization orientations were considered in calculations pertaining to shearing of the particle arrays: (a) rectangular array without polarization rotation (Fig. 7), (b) rectangular array with polarization vector rotation along the row of particles (Fig. 8) , (c) close-packed hexagonal array without polarization rotation (Fig. 9) and (d) close-packed hexagonal array with polarization rotation (Fig. 10). Shearing of the rectangular array is taken to follow a rectangular planar motion, whereby the changes in particle position are in the direction of shear with no change in the direction of the applied electric

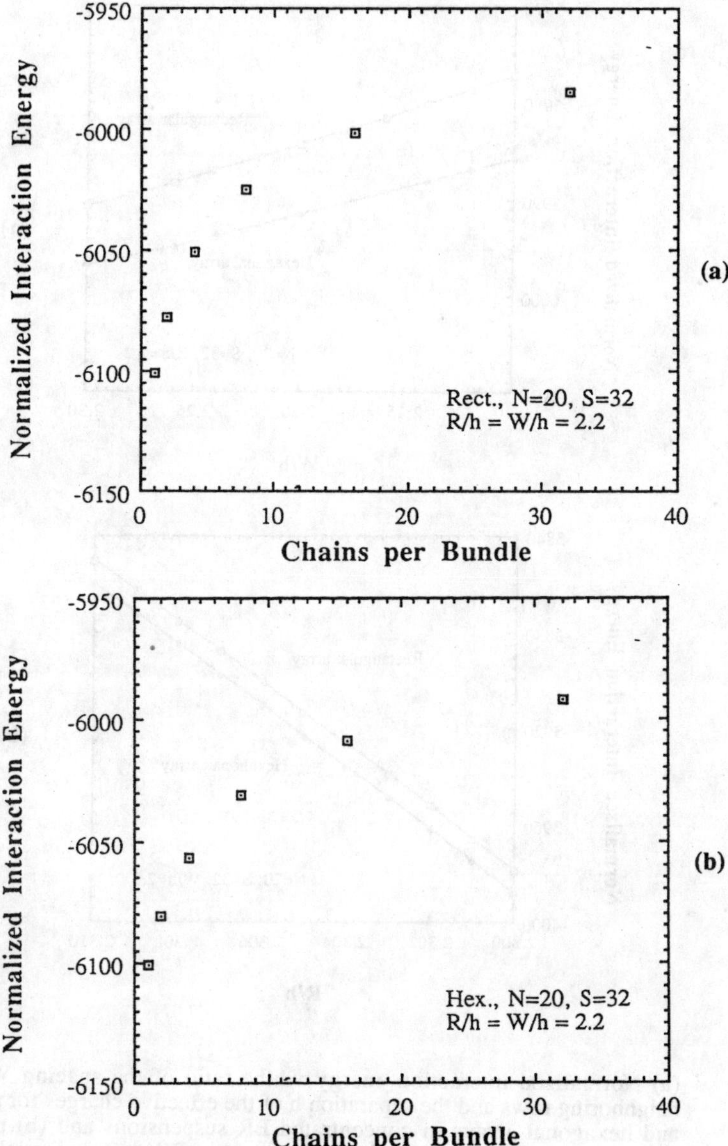

Fig. 5 Normalized interaction energy vs number of chains per cluster in a rectangular or hexagonal array in very dilute ER suspensions.

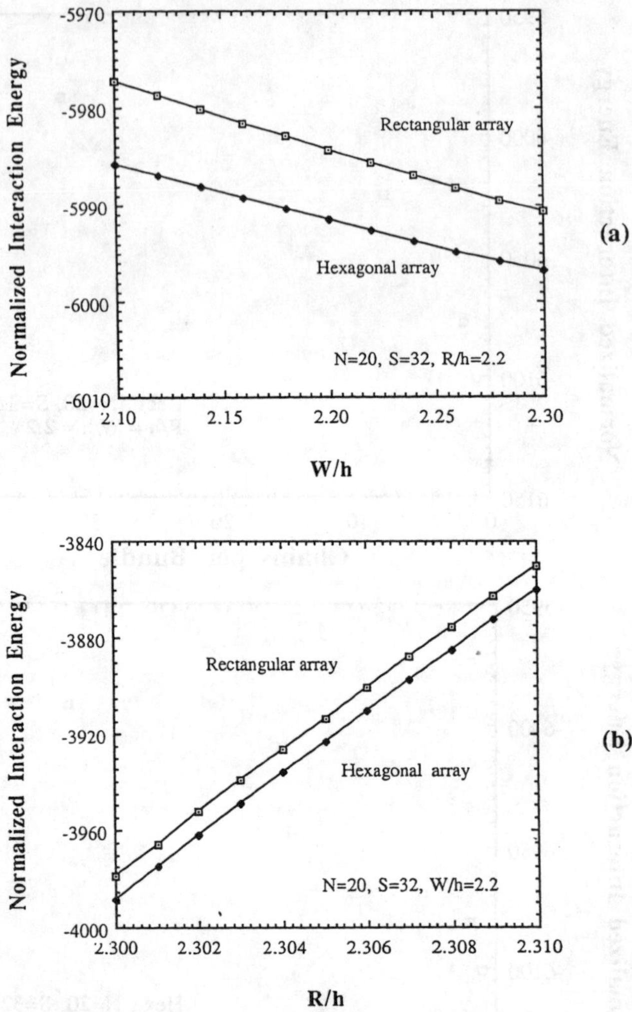

Fig. 6 (a) Normalized interaction energy vs the ratio of the spacing W between neighboring rows and the separation h of the effective charges for rectangular and hexagonal arrays in concentrated ER suspensions and (b) normalized interaction energy vs the ratio of the spacing R between particles and the spacing h of the effective charge for rectangular and hexagonal arrays in concentrated ER suspensions.

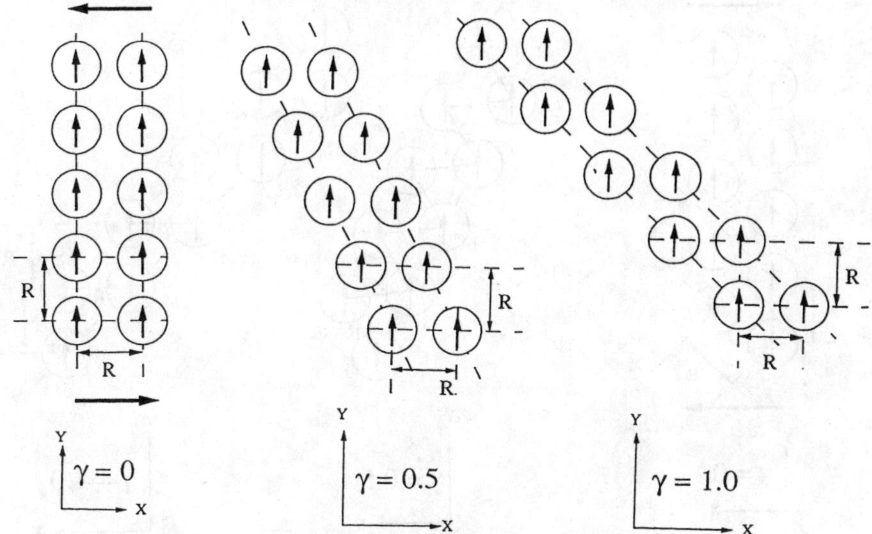

Fig. 7 Shear of two-chain rectangular array *without* polarization rotation.

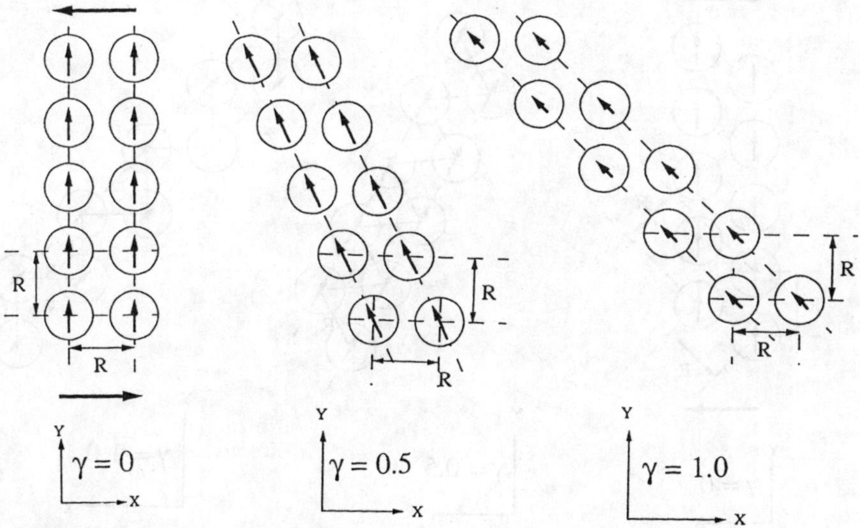

Fig. 8 Shear of two-chain rectangular array *with* polarization rotation.

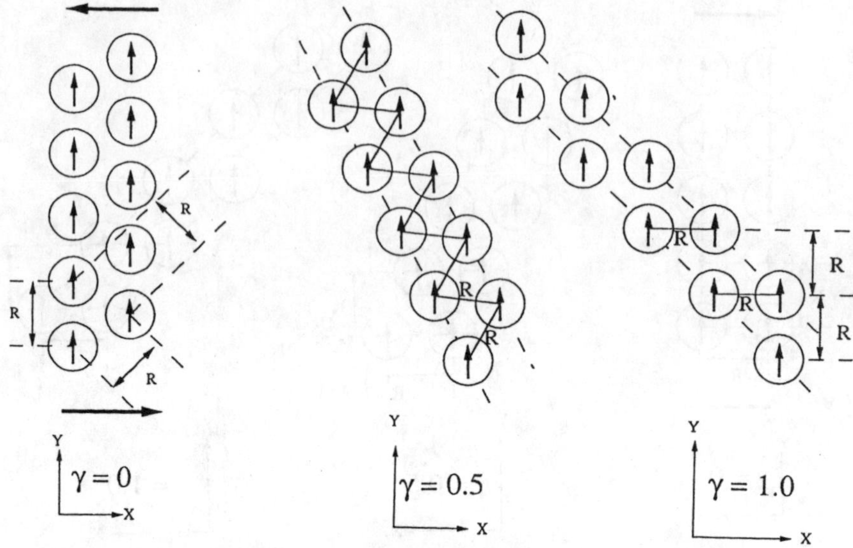

Fig. 9 Shear of two-chain hexagonal array *without* polarization rotation.

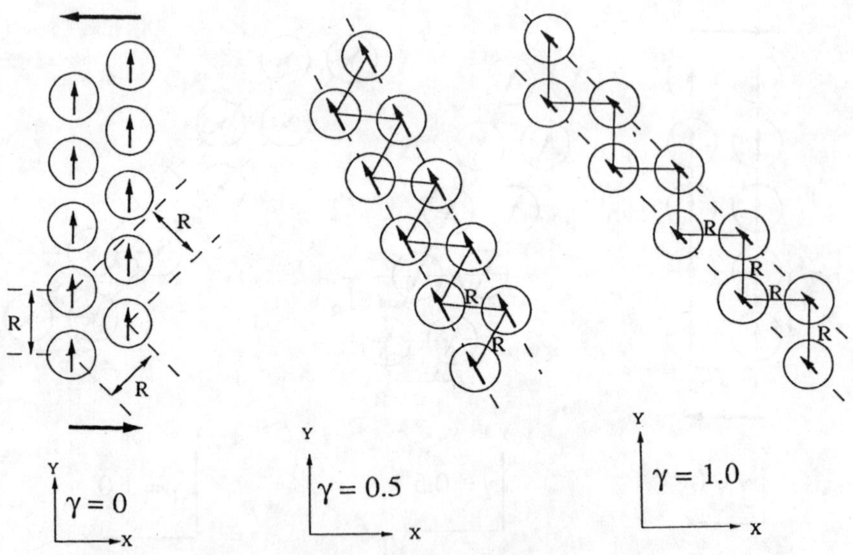

Fig. 10 Shear of two-chain hexagonal array *with* polarization rotation.

field. During shearing the hexagonal array it is assumed that the particles in the second chain move into the space produced during separation of the particles in the first chain. It is further assumed that the particles of the first chain change positions only in the direction of shear, while those of the second chain move in both the direction of shear and the applied electric field to maintain a structure of close-packed linkages of the particles in the two chains.

Taking N as the number of particles per chain, b the ratio R/2h and c the ratio W/2h, a normalized shear force can be expressed in terms of N, b, c and shear strain γ, using a procedure similar to the calculation of the interaction energy for the static condition without shear described above. The shear force of a two-chain system is considered to consist of two components: (a) the force due to a single chain itself, F_a and (b) that resulting from the interaction between the two chains, giving

$$F = 2F_a + F_b \tag{9}$$

Figs. 11 and 12 present the normalized shear force vs shear strain for the shearing of one-chain and two-chain **rectangular** systems with, and without, polarization rotation respectively. Note that for both polarization conditions, the two-chain system has a maximum force that is essentially twice that for a single chain, i.e. the strength of the two-chain system is the same as that of an equivalent number of single chains. However, in each case the force without polarization rotation is about three times that when rotation occurs.

The results for shearing the **hexagonal** array are shown in Figs. 13 and 14. In Fig. 13 it is seen that if no rotation of the polarization vector occurs, the maximum force for the two-chain system is equal to that for an equivalent number of single chains, similar to the situation for a rectangular array. Moreover, the two geometric arrays have the same maximum force. However, if we allow for rotation of the polarization vector in the hexagonal array, we see in Fig. 14 that the maximum force for the two-chain system is about 10 times that for an equivalent number of single chains. This is further shown in Fig. 15, which gives the ratio of the force per chain for the two-chain system compared to the single-chain system vs R/h. It is here seen that the force enhancement associated with the shearing of the close-packed hexagonal array with polarization rotation is about an order of magnitude for $R/h \leq 2.2$. This degree of enhancement is similar to the ratio of the measured strength of ER fluids to that predicted assuming that the structure consists of single chains[4–6].

5. Summary and Conclusions

A simple effective charge method and the principle of superpositron were used to calculate the total system interaction energy of two-dimensional arrays of particles in ER suspensions. The calculations gave the following results:

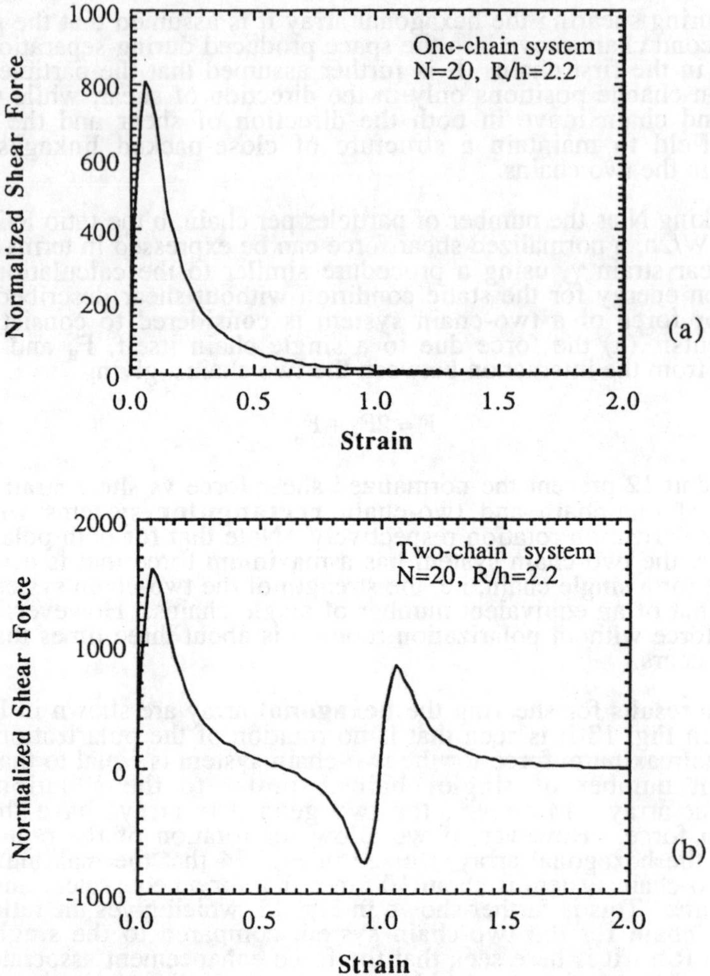

Fig. 11 Normalized shear force vs shear strain for a rectangular array *without* polarization rotation: (a) one-chain system and (b) two-chain system.

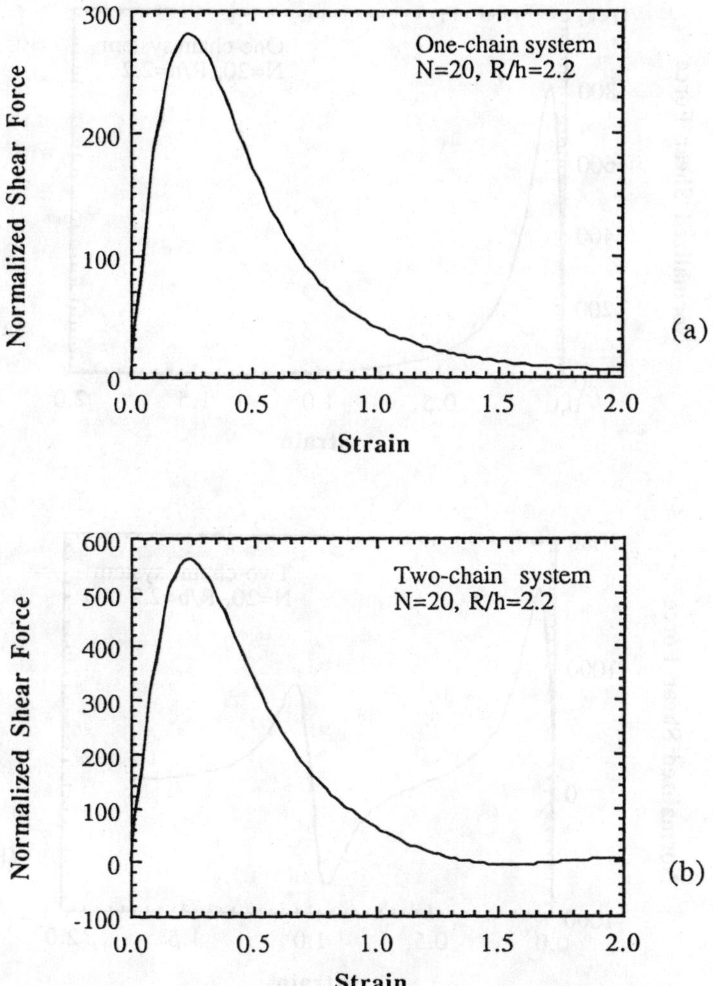

Fig. 12 Normalized shear force vs shear strain for a rectangular array *with* polarization rotation: (a) one-chain system and (b) two-chain system.

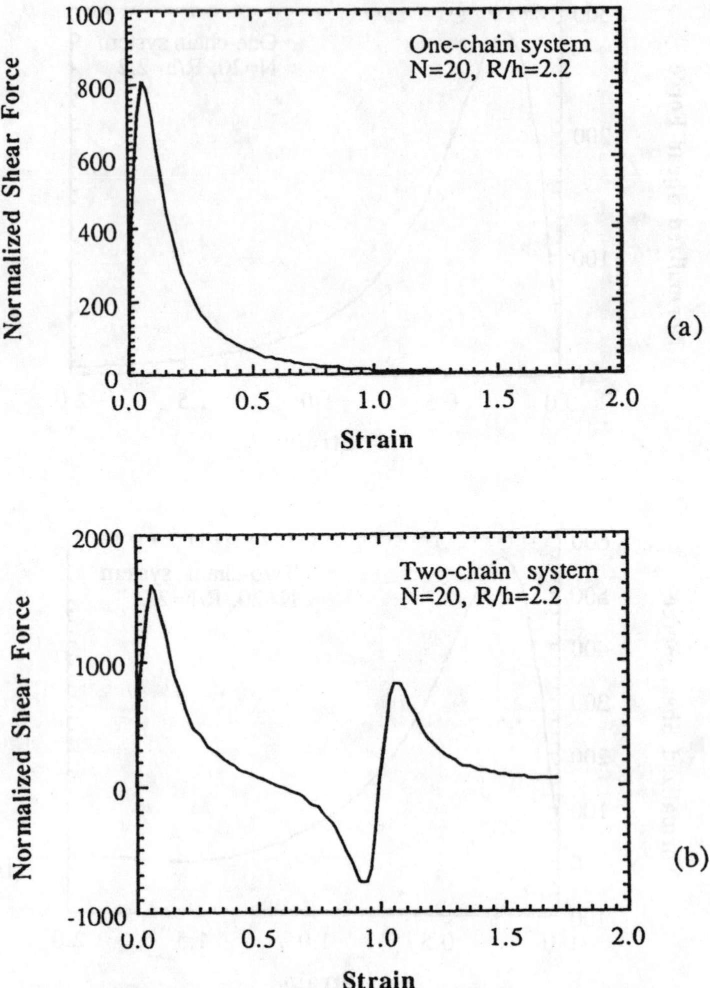

Fig. 13 Normalized shear force vs shear strain for hexagonal close-packed array *without* polarization rotation: (a) one-chain system and (b) two-chain system.

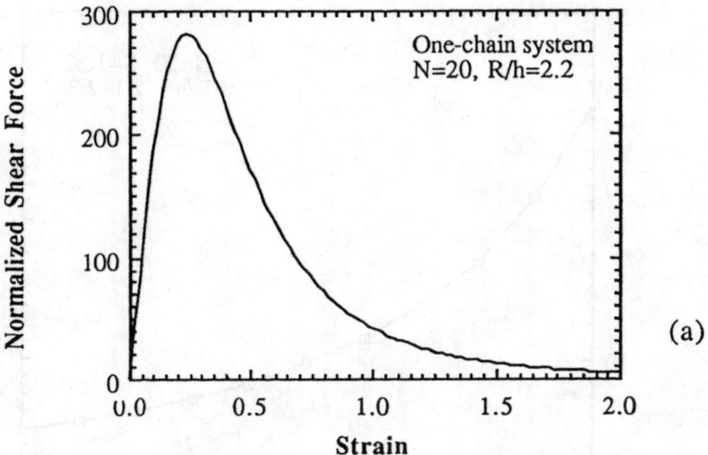

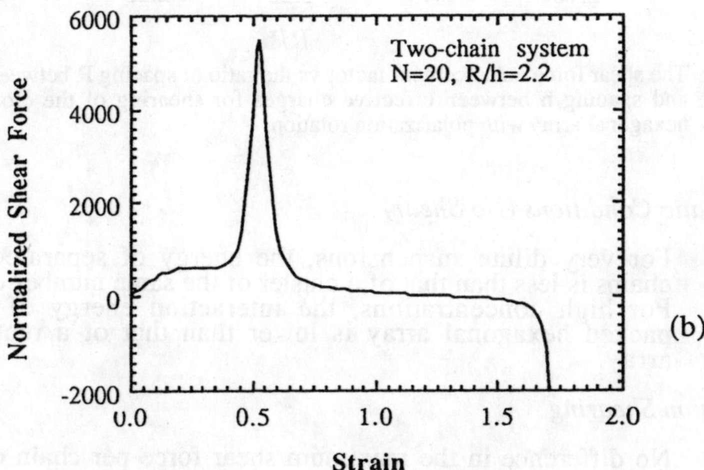

Fig. 14 **Normalized shear force vs shear strain for hexagonal close-packed array *with* particle rotation: (a) one-chain system and (b) two-chain system.**

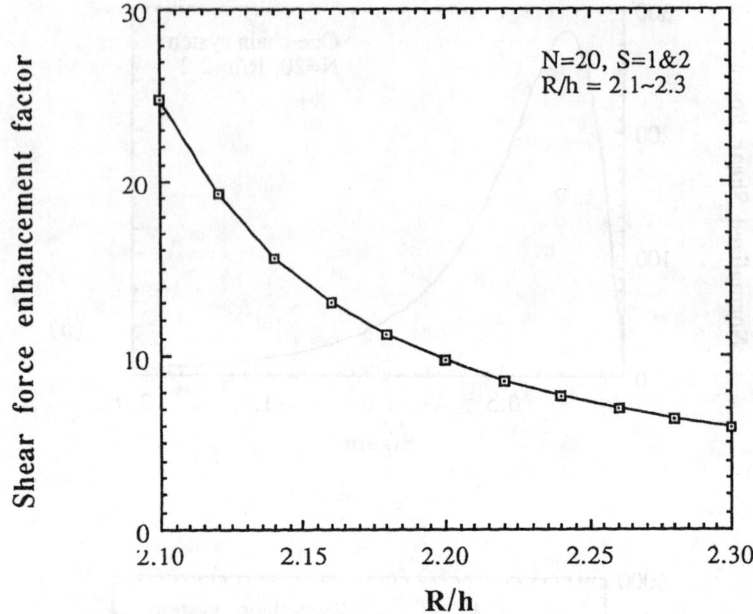

Fig. 15 The shear force enhancement factor vs the ratio of spacing R between particles and spacing h between effective charges for shearing of the close-packed hexagonal array *with* polarization rotation.

1) Static Conditions (No Shear)

a. For very dilute suspensions, the energy of separated single chains is less than that of a cluster of the same number of chains.
b. For high concentrations, the interaction energy of a close-packed hexagonal array is lower than that of a rectangular array.

2) Upon Shearing

a. No difference in the maximum shear force per chain occurred for a two-chain system compared to a single-chain system in either a rectangular or a close-packed hexagonal array of particles when the polarization vector does not rotate during the shear. This was also the case when rotation of the polarization vector occurred in the rectangular array; however, for this array the maximum force with polarization rotation was only about 1/3 that without rotation for both the single-chain and the two-chain systems.

b. With rotation of the polarization vector in the close-packed hexagonal array, there occurred an order of magnitude enhancement of the maximum force of a two-chain system compared to a single-chain system. This is comparable to the enhancement found for the measured strength of ER fluids compared to theoretical predictions based on a single-chain model.

6. Acknowledgements

This research was jointly supported by NSF Award #CBT–8714515, the Ford Motor Company and the NCSU University/Industry Consortium on ER Fluids with Ford, Bridgestone, Physica, Texaco and Tokai Rubber, Ltd. contributing.

7. References

1. H. Conrad, M. Fisher and A. F. Sprecher, in *Proc. 2nd Int. Conf. Electrorheological Fluids*, J. D. Carlson, A. F. Sprecher and H. Conrad, eds., Technomic Publ. Co., Lancaster, PA (1990) p. 63.
2. Y. Chen, A. F. Sprecher and H. Conrad, J. Appl. Phys. **70** 6796 (1991).
3. A. F. Sprecher, Y. Chen and H. Conrad, in *Proc. 2nd Int. Conf. Electrorheological Fluids*, J. D. Carlson, A. F. Sprecher and H. Conrad, eds., Technomic Publ. Co., Lancaster, PA (199) p. 82.
4. A. F. Sprecher, Y. Chen, Y. Choi and H. Conrad, in *Proc. 3rd Int. Conf. Electrorheological Fluids*, R. Tao, ed., World Scientific, Singapore (1992) p. 142.
5. H. Conrad, Y. Chen and A. F. Sprecher, ibid. p. 195.
6. Y. Chen and H. Conrad, "Effects of Water Content on the Electrorheology of Corn Starch/Corn Oil Dispersions", *ASME Symp. Developments in Non-Newtonian Fluids*, New Orleans, Nov. 28, 1993, in print.
7. A. Kraynik, R. Bonnecase and J. Brady, *Proc. 3rd. Int. Conf. Electrorheological Fluids*, R. Tao, ed., Word Scientific, Singapore (1992) p. 59.
8. R. Tao and J. M. Sun, Phys. Rev. Lett. **67** 398 (1991); Phys. Rev. A **44** R6181 (1991).
9. L. C. Davis, Phys. Rev. A **46** R719 (1992).

ELECTROSTATIC INTERACTIONS FOR PARTICLE ARRAYS IN ELECTRORHEOLOGICAL FLUIDS: II MEASUREMENTS

Ying Chen and H. Conrad
Materials Science & Engineering Department
North Carolina State University
Raleigh, NC 27695–7517

ABSTRACT

The force required to shear one-, two- and three-chain clusters of 230 μm dia. glass beads in silicone oil was measured. In each case the shear force was proportional to the shear strain, the proportionality constant increasing with electric field and number n of chains in the cluster. The derived shear modulus G also increased with n. An extrapolation of the present results suggests that a cluster of 4–5 chains would give the stress enhancement factor of 10–20 observed for real ER fluids.

1. Introduction

In the first paper of this series[1] it was pointed out that the measured flow stress of electrorheological (ER) fluids exceeds by an order of magnitude that predicted for a fibril structure consisting of separate, single-row chains of the particles. This has been attributed to the fact that the fibril structure in real ER fluids consists of columns or clusters of chains rather than separated, single-row chains[2,3]. However, calculations by Kraynik et al[4] indicate that the strength of a two-row cluster in close-packed arrangement is less than that of a single-row chain.

The calculations of the previous paper[1] give no difference in shear strength between a two-row chain cluster in rectangular or close-packed hexagonal array compared to a single-row chain, providing the polarization vector does not rotate during shear. However, with rotation of the vector an order of magnitude enhancement of the shear force occurs for the close-packed, two-row array compared to a single row. No such enhancement occurs for the rectangular array, even though a rotation of the polarization vector may take place.

The calculations in Refs. 1 and 4 reveal that the relative strength of clusters of chains compared to single chains depends on the specific geometric arrangement of the particles and whether or not rotation of the polarization vector occurs. Experimental measurements are needed to check the results of these calculations. The objective of the present work was to provide such measurements.

2. Procedure

The apparatus employed here to measure the strength of single chains and clusters of chains (Fig. 1) was the same as that described in Ref. 5. A single chain or cluster of glass beads was established between the two electrodes with application of an electric field. Shear was imposed by movement of the microscope stage. The shear force and displacement were determined from the motion of the pendulum bob. All tests were conducted at room temperature.

The host oil of the ER fluid was as-received Dow Corning 200 silicone oil, having 50 cs viscosity, 0.973 specific gravity and 2.6 dielectric constant. The soda-lime glass beads (230 μm dia.) were from Potter Industries and had a specific gravity of 2.5. Prior to mixing with the oil, the glass beads were held for more than 2 days in a 80% relative humidity chamber containing a $LiCl–H_20$ solution. The purpose of this treatment was to impart to the beads a specific amount of adsorbed water.

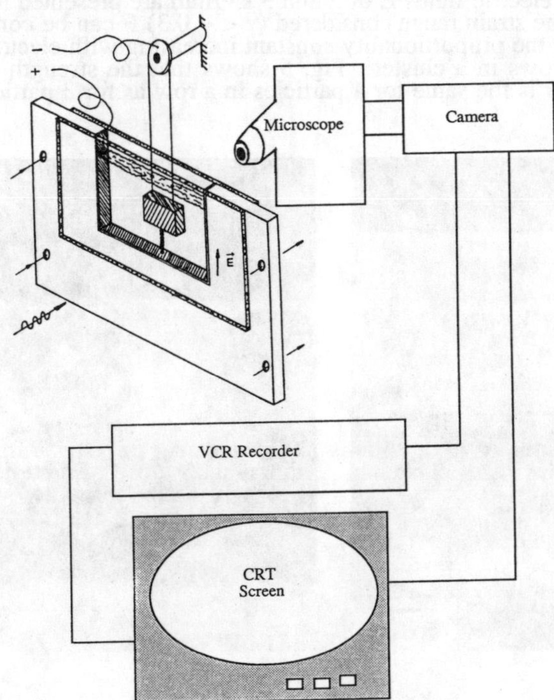

Fig. 1 Schematic of the force-displacement measuring system.

Since the force to shear a cluster depends on its configuration, it is important to characterize the arrangement of particles. In the present experiments we considered the strength of a single chain (Fig. 2a), a two-row rectangular array (Fig. 2b) and a three-row cluster (Fig. 3), which is part of a body-centered tetragonal array. In the case of the rectangular array, it was found that the results were reproducible if the structure of the array did not change during shearing. Occasionally during shearing, the particles in an adjacent chain slid into the gap which opened between the particles in the neighboring chain, leading eventually to a single chain. In this case, the force to continue shearing decreased to that of a single chain, and these results were not included. In all cases care was taken to avoid sliding of the chain or cluster along either of the electrode surfaces.

3. Results

The shear force F vs shear strain γ for the stable one-, two- and three-row clusters with d.c. electric fields E of 2 and 3 kV/mm are presented in Fig. 4. It is seen that within the strain range considered ($\gamma < \sim 0.3$) F can be considered to be proportional to γ, the proportionality constant increasing with electric field E and with number of rows in a cluster. Fig. 5 shows that the strength of a two-row rectangular cluster is the same for 4 particles in a row as for 3 particles in a row.

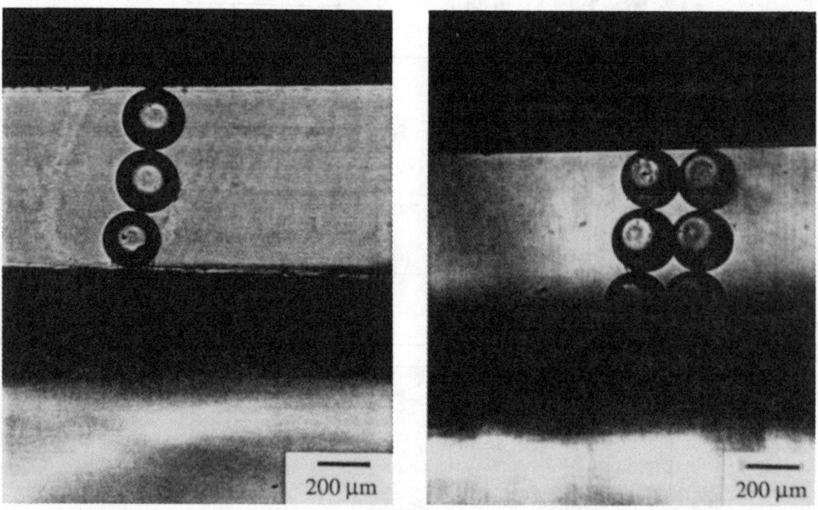

Fig. 2 Examples of: (a) single chain and (b) two-row cluster undergoing shear.

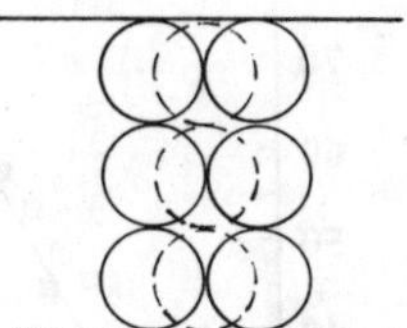

Fig. 3 Example of three-row cluster.

The shear modulus G of each cluster was derived from the slopes dF/dγ of the plots in Fig. 4 by taking

$$G = \left(\frac{dF}{d\gamma}\right) / n\pi a^2 \tag{1}$$

where n is the number of rows in the cluster and a is the particle radius. A plot of G vs number of rows (chains) in a cluster is given in Fig. 6. To be noted is that G is of the order of kPa and increases with electric field and with number of rows in a cluster.

4. Discussion

The present results show that for the configurations investigated the strength of a cluster of spherical particles increases with the number of rows n of chains comprising the cluster, the effect becoming especially pronounced for n > 2. Of further note is that the strength of the two-row rectangular array is only slightly larger than that for a single row chain, in reasonable accord with the calculations by Shih et al[1]. Also, the measured shear modulus for a single-row chain is in reasonable accord with values derived from the calculations for a single row by Chen et al[6], and from those of Bonnecaze and Brady[7] and Davis[8,9] for an ER fluid with a regular array of particles (volume fraction $\phi = 0.2$ and $K_p/K_f = 10$).

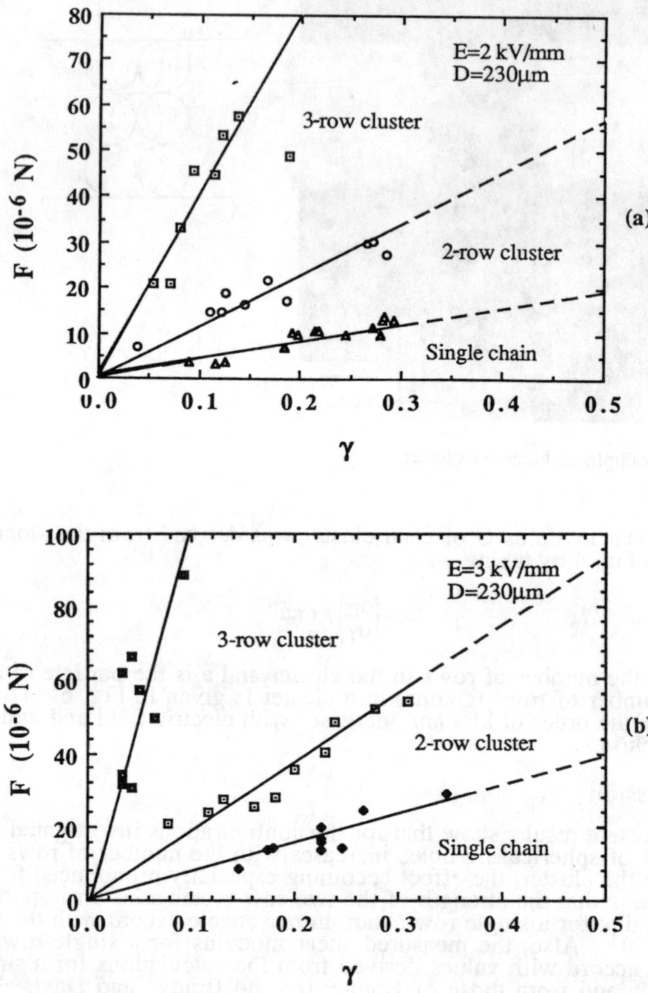

Fig. 4 Shear force F vs shear strain γ for the three particle arrays: (a) E = 2 kV/mm and (b) E = 3 kV/mm.

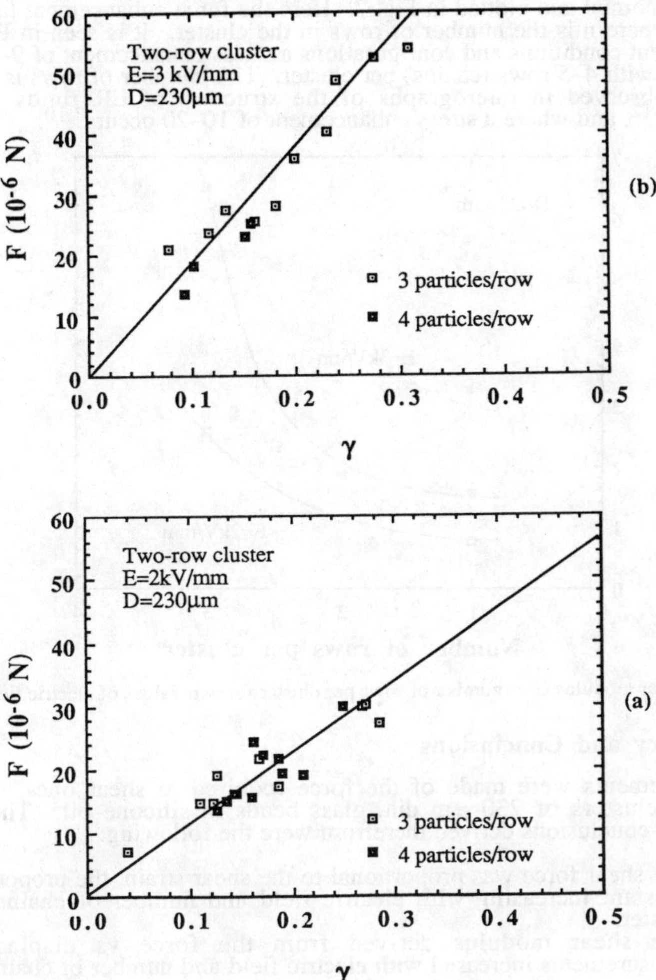

Fig. 5 Shear force F vs shear strain γ for the two-row cluster with 3 particles per row and with 4 particles per row: (a) E = 2 kV/mm and (b) E = 3 kV/mm.

Since the strength of a cluster increases with number of rows, it is of interest to estimate the number of rows which would give the stress enhancement factor of 10–20 found for ER fluids when compared with the strength predicted based on separated single chains[3,10]. To obtain some estimate the results in Fig. 6 were fit to a polynomial and plotted in Fig. 7. Here the force enhancement factor $A_s = G_n/G_{n=1}$, where n is the number of rows in the cluster. It is seen in Fig. 7 that for the present conditions and configurations a stress enhancement of 9–26 would be obtained with 4–5 rows (chains) per cluster. This number of rows is in accord with that observed in micrographs of the structure of ER fluids with ϕ = 0.2–0.4[3,10–12], and where a stress enhancement of 10–20 occurs[3,10].

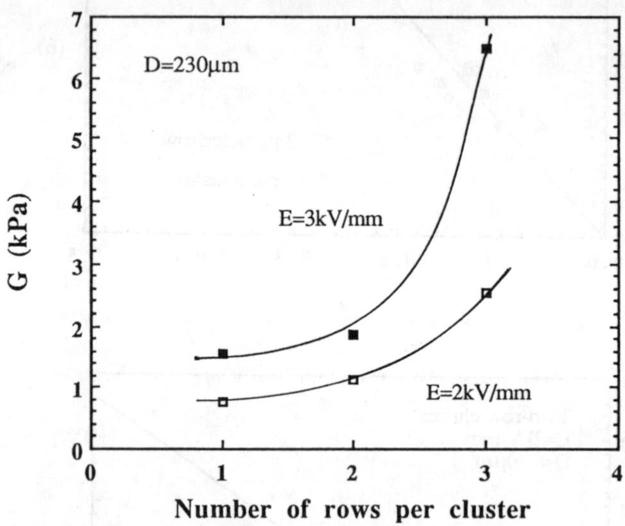

Fig. 6 Shear modulus G vs number of rows per cluster for two values of electric field.

5. Summary and Conclusions

Measurements were made of the force required to shear one-, two- and three-chain clusters of 230 μm dia. glass beads in silicone oil. The results obtained and conclusions derived therefrom were the following:

1. The shear force was proportional to the shear strain, the proportionality constant increasing with electric field and number of chains in the cluster.
2. The shear modulus derived from the force vs displacement measurements increased with electric field and number of chains in the cluster.
3. The measured modulus of the single chain and of the two-row

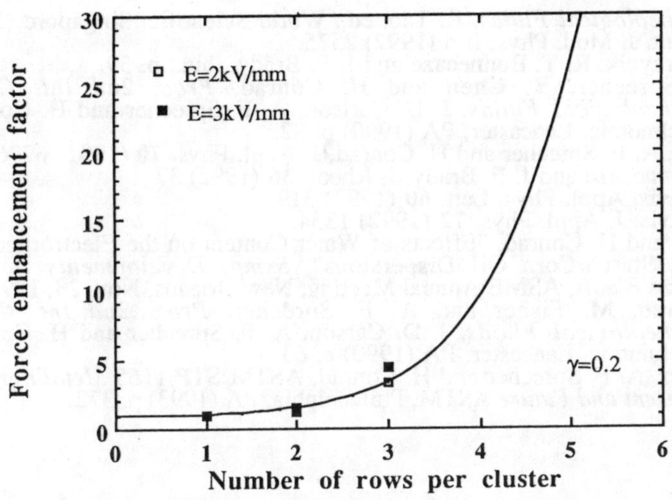

Fig. 7 Force enhancement factor A_s vs number of rows per cluster for E = 2 and 3 kV/mm. Solid curve is polynomial fit to the data.

rectangular cluster were in reasonable accord with theoretical predictions.

4. An extrapolation of the present results indicates that a cluster of 4–5 chains would give the stress enhancement factor of 10–20 observed for real ER fluids. This number of chains per cluster is in accord with microscopy observations on ER fluids.

6. Acknowledgements

The authors acknowledge support of this research by the NCSU University/Industry Consortium on ER fluids with Bridgestone Corp. and Texaco contributing.

7. References

1. Y. Shih, A. F. Sprecher and H. Conrad, "Electrostatic Interactions for Particle Arrays in Electrorheological Fluids: I Calculations", present volume.
2. A. F. Sprecher, Y. Chen, Y. Choi and H. Conrad, *Proc. 3rd Int. Conf. Electrorheological Fluids*, R. Tao, ed., World Scientific, Singapore (1992) p. 142;
3. H. Conrad, Y. Chen and A. F. Sprecher, *Proc. 3rd Int. Conf.*

Electrorheological Fluids, R. Tao, ed., World Scientific, Singapore (1992) p. 195; Int. J. Mod. Phys. B 6 (1992) 2575.
4. A. M. Kraynik, R. T. Bonnecaze and J. F. Brady, ibid., p. 59.
5. A. F. Sprecher, Y. Chen and H. Conrad, *Proc. 2nd Int. Conf. Electrorheological Fluids*, J. D. Carlson, A. F. Sprecher and H. Conrad, eds., Technomic, Lancaster, PA (1990) p. 82.
6. Y. Chen, A. F. Sprecher and H. Conrad, J. Appl. Phys. **70** (1991) 6796.
7. R. T. Bonnecaze and J. F. Brady, J. Rheol. **36** (1992) 37.
8. L. C. Davis, Appl. Phys. Lett. **60** (1992) 319.
9. L. C. Davis, J. Appl. Phys. **72** (1992) 1334.
10. Y. Chen and H. Conrad, "Effects of Water Content on the Electrorheology of Corn Starch/Corn Oil Dispersions", *Symp. Developments in Non-Newtonian Fluids*, ASME Annual Meeting, New Orleans, Nov. 28, 1993.
11. H. Conrad, M. Fisher and A. F. Sprecher, *Proc. 2nd Int. Conf. Electrorheological Fluids*, J. D. Carson, A. F. Sprecher and H. Conrad, eds., Technomic, Lancaster, PA (1990) p. 63.
12. M. Fisher, A. F. Sprecher and H. Conrad, ASTM STP 1165 *Metallography Past, Present and Future* ASTM, Philadelphia, PA (1993) p. 372.

Pressure Coupling in the Electrical Response of Electro-rheological Valves.

by

M. Whittle

Department of Mechanical and Process Engineering
The University of Sheffield, P.O.Box 600
Sheffield S1 4DU. United Kingdom.
Tel: (0742) 768555
Fax: (0742) 753671

R. Firoozian, W.A.Bullough and D.J.Peel .
Department of Mechanical and Process Engineering
The University of Sheffield.

ABSTRACT

An earlier model describing the electrical characteristics of an electrorheological (ER) valve is refined to include the effects of coupling to the pressure response. Expressions including additional linear terms in the pressure and time differential of the pressure are considered and it is found that a single term in the time derivative with a small time delay is a quite sufficient description of the data on the basis of goodness of fit.

1. Introduction

Electro-Rheological (ER) devices offer extremely rapid response times, these being often superior to those of the fastest electro-magnetic mechanisms. They can dissipate relatively little power making them potentially good candidates for interfaces between computers and hydraulic machines. Nevertheless, some current is drawn, and the control of these devices and the design of their power supplies requires a detailed understanding of the response characteristics. In earlier publications[1-3] we have separately described the electrical response and pressure response for a series of Electro-Rheological valve restrictors over a range of dimensions operating at a series of nominally steady flowrates. The radii of the valves are the same in each case and sufficiently large that the field between the electrodes is uniform and the system can be treated as a pair of parallel plates with width b, length l and gap h (Fig. 1). The valve dimensions are detailed in Table 1.

After allowing for the voltage rise time the current response to a voltage step consists of an effectively instantaneous rise sharp rise followed by a rapid decay with a time constant ~ 0.3 ms. This can be ascribed to the initial polarisation of particles and the

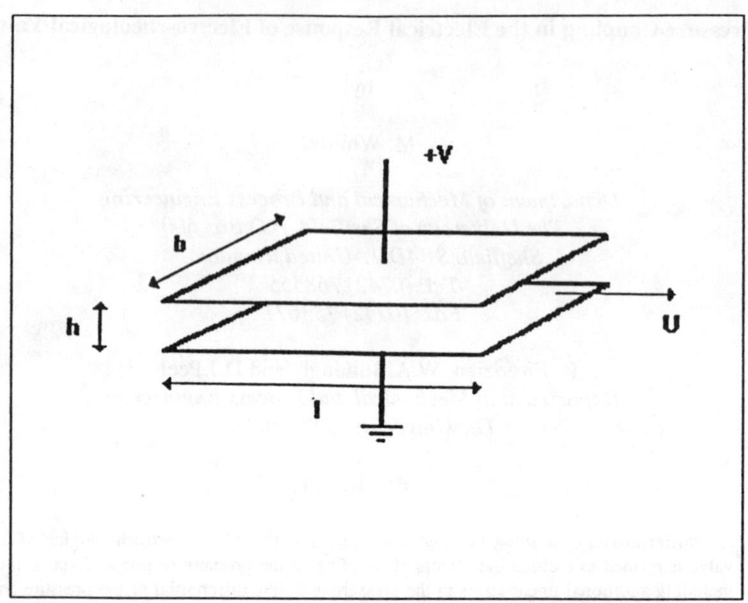

Fig. 1. Simplified valve geometry. ER fluid flows at mean velocity U Eq. (14) between plates length l, width b, gap h and held at voltage V.

Table 1. Valve dimensions: length l, gap h, surface area A.

Valve	l/mm	h/mm	A/m^2
A	100	0.5	.0362
B	50	0.5	.0181
C	100	1.0	.0362
D	140	0.5	.0507
E	100	0.75	.0362

observed relaxation time compares well with the expected Maxwell-Wagner relaxation time for this system. The pressure across the valve is related to the yield stress of the ER fluid[3] which is in turn related to the interaction force and thus (according to the widely accepted model) to the polarisation. The total particle charge is given by the time integral of the polarisation current and so the pressure initially rises on a comparable timescale to the charging process Figs. 2a & b. Underlying this current response is a constant

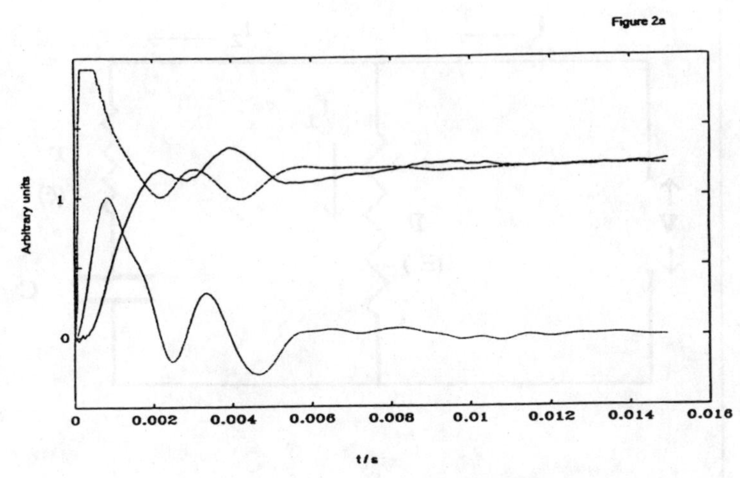

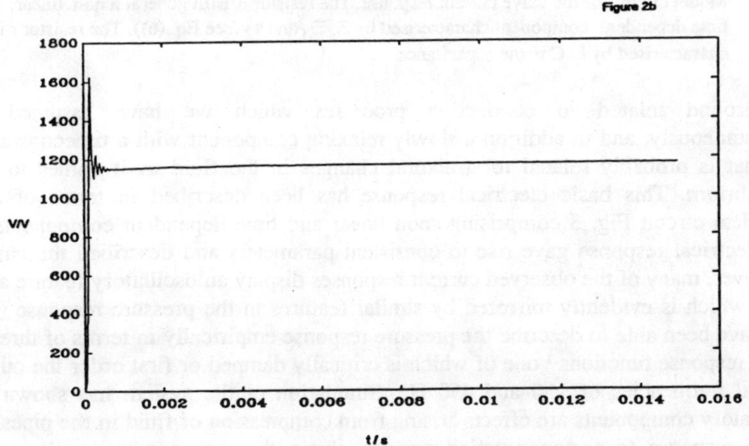

Fig. 2. a) Comparison of the current density, i ------ (1 arb. unit = 200 mA m^{-2}), pressure, P ——— (1 arb. unit = 10 bar), and dP/dt ·········· (1 arb. unit = 10 Kbar s^{-1}). This data is for valve D operated at ~1200 V and a flowrate of 9 lit min^{-1}.
b) Voltage input signal.

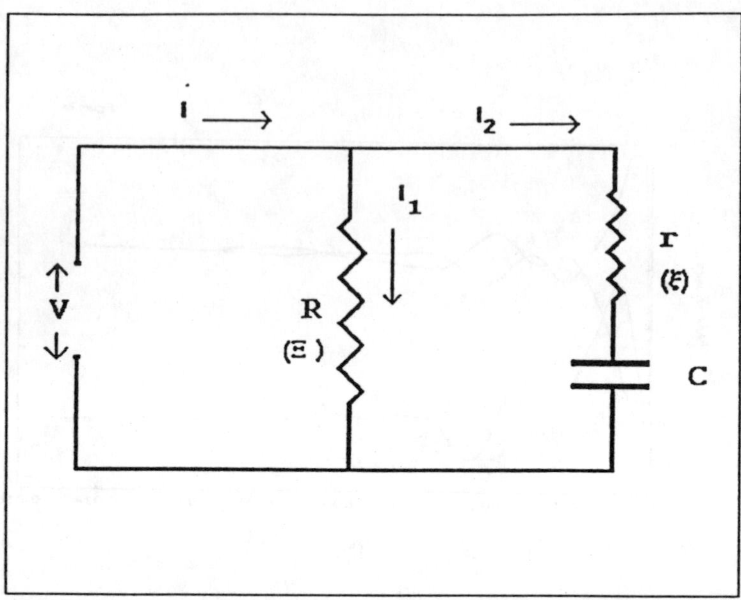

Fig. 3. Model circuit for the valve current response. The resistor R is in general a non linear, time dependent component characterised by Ξ, Ξ_0 and τ_2 (see Eq. (6)). The resistor r is similarly characterised by ξ. C is the capacitance.

background related to conduction processes which we have assumed appears instantaneously, and in addition a slowly relaxing component with a timeconstant 1 - 10 ms that is probably related to structural changes in the fluid as it comes to dynamic equilibrium. This basic electrical response has been described in terms of a simple electrical circuit Fig. 3 comprising non-linear and time dependent components. Fits to the electrical response gave rise to consistent parameters and described the basic form. However, many of the observed current responses display an oscillatory feature around 3-4 ms which is evidently mirrored by similar features in the pressure response (Fig. 2a). We have been able to describe the pressure response empirically in terms of three second order response functions[3] one of which is critically damped or first order the others with natural frequencies of 150 and 450 Hz. Simulation of the system has shown that the oscillatory components are effects arising from compression of fluid in the pipes[4]. If P(t) is the response to a step function we can obtain the response to a unit impulse by differentiation and then obtain the frequency response by one sided Fourier transform[5].

$$P(\omega) = \int_0^\infty \frac{dP(t)}{dt}\, e^{-j\omega t} \qquad (1)$$

Transformed in this way the data clearly show the two resonances at ~150 and 450 Hz and these become even more distinct when the data is plotted as a spectral density $S_p(\omega) = P(\omega)P^*(\omega)$ where the asterisk denotes the complex conjugate (Fig. 4). Treated in the same way the current also shows these peaks and in this work we have attempted to characterise this coupling.

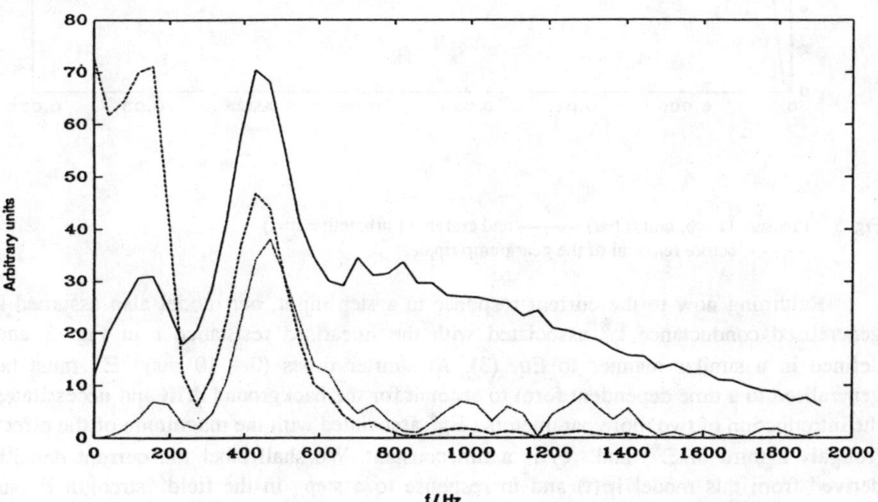

Fig. 4 Comparison of spectral densities obtained for valve A operated at 1200 V and a flowrate of 9 lit min^{-1}. Current I ——— (1 arb. unit = mA^2), Pressure P - - - - - (1 arb. unit = bar^2) and dP/dt ········ (1 arb. unit = 10^{-7} bar^2 s^{-2}).

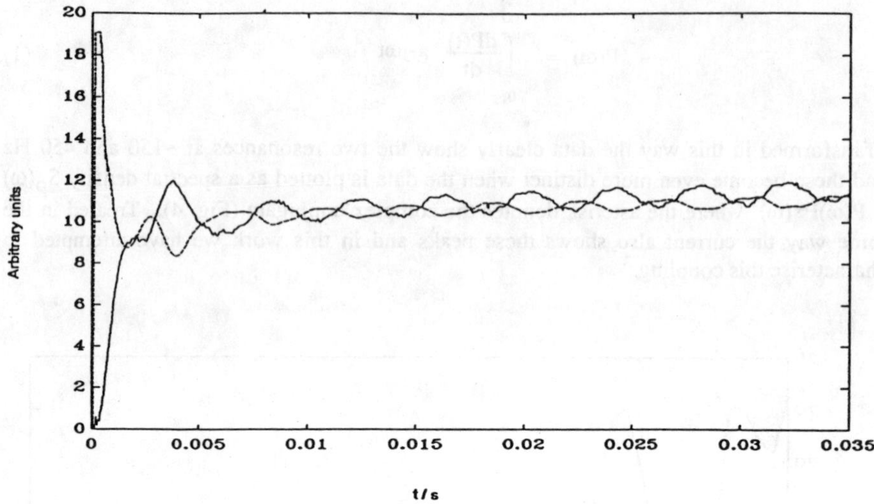

Fig. 5 Pressure (1 arb. unit = bar) ——— and current (1 arb. unit = mA) - - - - - before removal of the gear pump ripple.

Returning now to the current response to a step input, our model also assumed a generalised conductance ξ^* associated with the linearised resistance r in Fig. 3 and defined in a similar manner to Eq. (3). At shorter times (0 - 10 ms) Ξ^* must be generalised to a time dependent form to account for the background drift and necessitates the introduction of two more parameters : $\Xi_o{}^*$ associated with the magnitude of the effect (roughly a third of Ξ^*) and τ_2 is a timeconstant. We shall label the current density derived from this model $i_E(t)$ and in response to a step in the field strength E our original model[2] took the form

$$i_E(t) = (E - E_o)^2 \left[(\Xi^* - \Xi_o{}^* \exp(-t/\tau_2)) + \frac{\xi^* C^{*2}}{(\xi^*(E - E_o)t + C^*)^2} \right] H(0) \qquad (6)$$

where H(t') is the Heavyside step function and C* is the linear capacitance reduced by the valve dimensions

$$C^* = C \frac{h}{A} \qquad (7)$$

2. The Electrical Model

2.1 Original Form

The circuit in Fig. 3 has been found to account very well for the basic form of the current response in both the time and frequency domains[2]. For a voltage step input this circuit initially draws a large current to charge the capacitance C. The current then decays to a "steady state" determined by R at a rate controlled by C and the smaller resistance r. Although this circuit can qualitatively model the main features of the valve current response we have found that a more general description incorporating parameters invariant with valve size requires the use of non-linear resistances. In addition, the detailed form of the initial current decay has been found to be affected by this non-linearity. Furthermore, after the initial response the current increases slowly to a plateau and to account for this the main resistor R must be made time dependent. In spite of these complications, for small biased voltage steps or sinusoidal voltage inputs, linear resistances can be found that will adequately describe the response for most engineering purposes.

2.2 Pressure Coupling

We are primarily interested here in the coupling between pressure and current associated with the features around 150 and 450 Hz. However, we pause briefly to set the scene by examining the steady state and to establish the magnitude of relationships in this case. Steady state results show that the current and pressure both increase non-linearly and monotonically as the applied field increases[2,3]. Since the pressure difference across an ER valve is proportional to its length l an appropriate measure for this change is

$$K_o = l\frac{\Delta i}{\Delta P} \tag{2}$$

where the current density (current divided by valve area) changes by an amount Δi for a change in pressure ΔP. In our previous study[2], steady state measurements showed that above a threshold field E_o, determined from steady state results, the current drawn at long times (>0.75 s) depends upon the square of the applied field E. For the current density i this is characterised by a non-linear conductance coefficient Ξ^*.

$$i = \Xi^*(E - E_o)^2 \quad ; \quad E > E_o \tag{3}$$

Where the asterisk denotes quantities reduced with respect to the valve dimensions. The relationship between the steady state pressure and current density at constant flowrate can then be described by[3]

$$P = p_1^* \sqrt{\frac{i}{\Xi^*}} + p_2^* \left(\frac{i}{\Xi^*}\right) \qquad (4)$$

where p_1^* and p_2^* are coefficients. Neglecting the first term the value of K_o for high field strengths is then given approximately by

$$\frac{1}{}\frac{di}{dP} \sim \frac{\Xi^*}{p_2^*} \qquad (5)$$

With[3] $p_2^* \sim 10$ bar m MV^{-2} and[2] $\Xi^* \sim 70$ mA MV^{-2} this gives a steady state value of the constant as $K_o^{ss} \sim 7$ mA bar^{-1} m^{-1} at nominally constant flowrate (there is some dependence of flow on field due to pump slippage) and changing field. We also find that for a given field strength an increase in flow rate results in a decrease in the field induced pressure across the valve and a decrease in the current drawn. For valve A at E = 2.4 MV m^{-1}, ΔP is about 1 bar and Δi is roughly 25 mA m^{-2} giving a K_o value of 2.5 mA bar^{-1} m^{-1} for changing flow and constant field. In this case it is not possible to say what portion of the change in current is flowrate or pressure determined.

From Fig. 2 it can be seen that the interference from the pressure response is either negative or about 180° out of phase with the pressure. The flow to these valves was supplied by a gear pump and the raw data for these results contains a pressure ripple caused by engagement and disengagement of the gears. The current also contains this frequency but again the phase is shifted by a half cycle Fig. 5. This ripple was unimportant for the low flow rate and because our results are averaged over five individual steps the cycles sometimes cancel. Where important they were removed from the data (as in Fig. 2) by fitting a sinusoid at long times and then subtracting a curve with the same parameters from the whole response. In the frequency domain it could be seen that a peak at the pump frequency was removed while not disturbing the remaining shape. From the parameters obtained with pump frequencies at 156 and 255 Hz (corresponding to nominal fluid flow rates of 9 and 15 lit min^{-1} respectively) we find the phase difference $\phi = 2.96 \pm 0.33$ radians and the ratio of amplitudes rationalised by valve dimensions is $K_1 = 0.97 \pm 0.33$ mA bar^{-1} m^{-1}. There was no observable dependence on frequency. The dynamic coupling constant K_1 has the same units as steady state constant K_o but is significantly smaller than either of the values we have obtained for that constant. In addition the phase angle is near π radians and if set to zero the best fit magnitude of K_1 would be negative. This may indicate that a different mechanism is operating in the dynamic case.

For the non linear conductivities the comparable transformation is

$$\Xi^* = \Xi \frac{h^2}{A} \tag{8}$$

The second term in Eq. (6) is unfamiliar because of the non linear nature of the response. By analogy with the exponential response obtained from the linearised solution a time constant τ_1 can be defined by

$$\tau_1 = \frac{C^*}{2\xi^*(E-E_o)} \tag{9}$$

An initial attempt to include the pressure response involves a simple linear perturbation described by a coefficient K_1 which for our results (assuming zero phase shift) has been negative

$$i_{EP}(t) = i_E(t) + K_1 \frac{\Delta P(t)}{l} \tag{10}$$

where $i_E(t)$ is the original model current response. The modified current density $i_{EP}(t)$ is calculated according to this new model and compared with the observed response for each valve while the parameters are varied as part of a fitting procedure using a Nelder-Mead[6] routine until the best fit in a least squares sense is obtained.

This modified model gives some improvement in the fit over the original form but it can be seen from Fig. 2 that the oscillations in the current response are not always coincident with those in the pressure and the depth of the troughs are also incommensurate. Furthermore, K_1 obtained in this way is negative and since the pressure at long times is always positive this would imply a negative steady state contribution to the current. Our examination of the steady state above found that K_o was positive whether measured at constant field or constant flow and we found no evidence of a negative contribution. Alternative or additional contributions to the dynamic response were therefore sought. Since the pressure response may be described as a sum of three components[3] it is possible that the current is sensitive to only one or two of them or to some linear combination. Some improvement in the fit was obtained by using this method. However, more success was enjoyed by including a term in the time differential of the pressure. This has the effect of shifting the phase of any oscillatory components by 90° and it is also significant that this term makes no contribution at long times

$$i = i_E + \frac{K_1}{l}\Delta P + \frac{K_2}{l}\frac{d\Delta P}{dt} \tag{11}$$

The flow input to the system is controlled through the gear pump drive speed which is maintained constant to within $\pm$ 0.5% by means of a DC tach/thyristor. Nevertheless, when the valve is energised, fluid compresses in the upstream pipe (connecting valve to pump) and decompresses downstream (leading to the reservoir) resulting in transient changes in flowrate through the valve. Some compression also occurs in the valve itself altering the density and contributing to the flow change. The pipes and valve form a coupled system which responds in a complicated way[6]. However, these changes are all related to the pressure through expressions of the form

$$Q = \frac{V}{\beta}\frac{d\Delta P}{dt} \tag{12}$$

where β is the bulk modulus or the reciprocal compressibility and V is the volume of the component. There is thus some justification in introducing a differential term. Over the full range of valves studied, Eq. (11) is generally more successful than Eq. (6) in fitting the data. However a further improvement is obtained by including a small time delay δt_{sh} in the differential term giving

$$i(t) = i_E(t) + \frac{K_1}{l}\Delta P(t) + \frac{K_2}{l}\frac{d\Delta P(t+\delta t_{sh})}{dt} \tag{13}$$

The physical basis for this delay is presently obscure but the in many cases the fit is now exceptionally good. The inclusion of a time delay in the pressure component did not lead to any improvement. As we have seen, oscillations in the current are almost 180° out of phase with the pressure and differentiating the pressure gives a phase shift of 90°. The shortfall is taken up by δt_{sh}. Attempts to use the second differential of the pressure (which shifts the phase by 180°) were not so successful and required the inclusion of a small time advance in most cases. Having followed this route we found in practice that the term in the time derivative is quite sufficient and that inclusion of the pressure term is in fact unnecessary. When included it converges to a relatively large contribution but generally makes an insignificant improvement according to the least squares criterion indicating that the fit is overdetermined. For this reason and in view of the positive rather than negative pressure coupling coefficient seen at steady state as outlined above we quote results for $K_1 = 0$ in Table 2 and give for comparison only the mean values obtained for the full Eq. (13) in Table 3. Since $d\Delta P/dt = 0$ as $t \to \infty$ our original estimates of Ξ^* are undisturbed for $K_1 = 0$ while the inclusion of a pressure term leads to an interaction with Ξ^*. Examples of the fits obtained are shown in Figs. 6a and 6b. Further details of the procedure are given below.

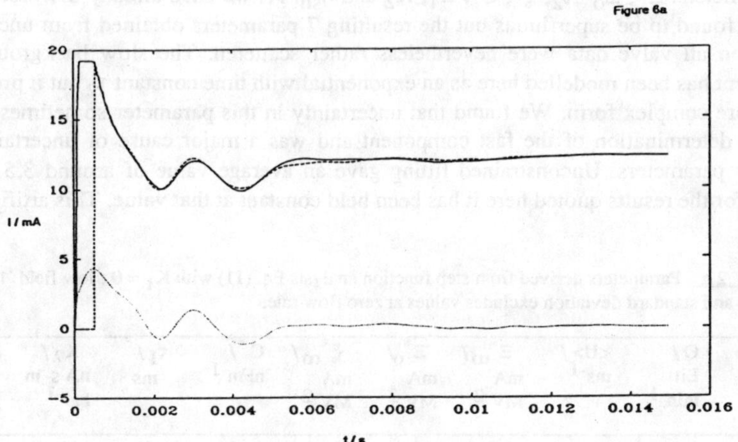

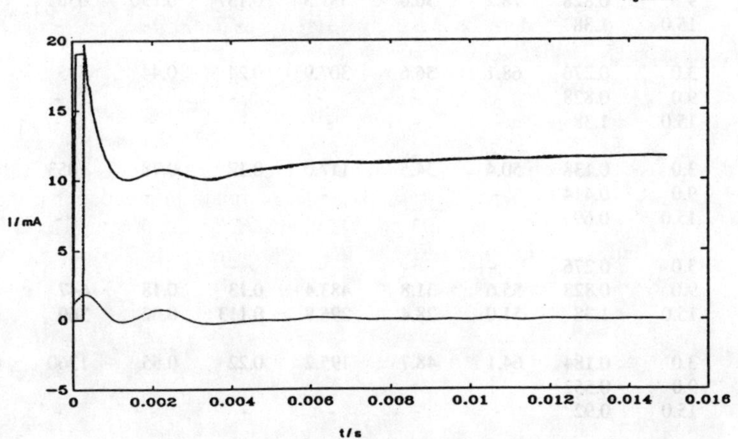

Fig 6. Fit to the current with $K_1 = 0$ (Eq. 13). Experimental data ———, fit - - - - - -, coupling term $(K_2 / l)\ (d\Delta P(t + \delta t_{sh})/dt)$ ········· .
a) Valve D operated at ~1200 V and a flowrate of 9 lit min^{-1}.
b) Valve A operated at ~1200 V and a flowrate of 15 lit min^{-1}.

3. Results

Fitting Eq. (13) at the same time as Eq. (6) involves the determination of 8 parameters: Ξ^*, $\Xi_o{}^*$, τ_2, ξ^*, C^*, K_1, K_2 and δ_{sh}. As we have already discussed, K_1 was later found to be superfluous but the resulting 7 parameters obtained from unconstrained fits on all valve data were nevertheless rather scattered. The slow background rise in current has been modelled here as an exponential with time constant τ_2 but it probably has a more complex form. We found that uncertainty in this parameter sometimes interfered with determination of the fast component and was a major cause of uncertainty in the other parameters. Unconstrained fitting gave an average value of around 3.5 ms for τ_2 and for the results quoted here it has been held constant at that value. This artifice

Table 2 a Parameters derived from step function analysis Eq. (11) with $K_1 = 0$: low field 1.2 MV m^{-1}. Mean and standard deviation excludes values at zero flow rate.

Valve	Q / Lit min^{-1}	<U> / ms^{-1}	Ξ^*_∞ / mA MV^{-2}	Ξ^*_o / mA MV^{-2}	ξ^*_∞ / mA MV^{-2}	C^* / nFm^{-1}	τ_1 / ms	K_2 / nA s m bar^{-1}	δ_{sh} / ms
A	0.0	0.0	43.9	7.04	339.1	0.104	0.128	0	-
	3.0	0.276	70.2	24.1	306.0	0.17	0.308	822	0.468
	9.0	0.828	78.2	36.0	331.5	0.137	0.190	656	0.329
	15.0	1.38	-	-	-	-	-	-	-
B	3.0	0.276	68.1	56.6	307.9	0.24	0.44	695	0.229
	9.0	0.828	-	-	-	-	-	-	-
	15.0	1.38	-	-	-	-	-	-	-
C	3.0	0.138	50.4	34.5	117.0	0.17	0.78	1053	0.225
	9.0	0.414	-	-	-	-	-	-	
	15.0	0.69	-	-	-	-	-	-	-
D	3.0	0.276	-	-	-	-	-	-	-
	9.0	0.828	55.6	31.8	483.4	0.13	0.18	647	0.225
	15.0	1.38	51.0	28.4	296.8	0.113	0.32	369	0.198
E	3.0	0.184	64.1	48.7	195.2	0.22	0.55	1360	0.523
	9.0	0.552	-	-	-	-	-	-	-
	15.0	0.92	-	-	-	-	-	-	-
Mean			62.5	36.7	291.2	0.169	0.40	825	0.314
S.D.			10.5	12.1	114.5	0.047	0.21	321	0.13

Table 2b Parameters derived from step function analysis Eq. (11) with $K_1 = 0$: high field $E = 2.4$ MV m^{-1} .Mean and standard deviation excludes values at zero flow rate.

Valve	Q / Lit min^{-1}	$\langle U \rangle$ / ms^{-1}	Ξ^*_∞ / mA MV^{-2}	Ξ^*_0 / mA MV^{-2}	ξ^*_∞ / mA MV^{-2}	C^* / nFm^{-1}	τ_1 / ms	K_2 / nA s bar^{-1}	δ_{sh} / ms
A	0.0	0.0	60.35	15.1	201.3	0.13	0.134	0	-
	3.0	0.276	72.4	18.4	147.8	0.134	0.321	1048	0.320
	9.0	0.828	76.5	22.9	202.3	0.124	0.147	607	0.322
	15.0	1.38	84.7	26.0	134.0	0.103	0.213	710	0.345
B	3.0	0.276	67.5	22.9	70.0	0.143	0.493	1698	0.375
	9.0	0.828	57.7	14.5	314.8	0.10	0.089	803	0.199
	15.0	1.38	49.3	20.2	334	0.144	0.14	608	0.197
C	3.0	0.138	88.6	1.37	183.9	0.09	0.119	917	0.325
	9.0	0.414	66.5	11.8	53.5	0.13	0.542	783	0.147
	15.0	0.69	60.2	5.49	130.3	0.093	0.159	830	0.219
D	3.0	0.276	61.0	14.6	345.1	0.117	0.078	1239	0.125
	9.0	0.828	65.3	22.1	113.9	0.105	0.239	1186	0.337
	15.0	1.38	66.3	22.1	450.4	0.092	0.057	996	0.231
E	3.0	0.184	69.3	21.4	98.0	0.119	0.283	1570	0.276
	9.0	0.552	62.8	22.7	230.0	0.152	0.162	1012	0.209
	15.0	0.92	58.9	21.0	233.4	0.126	0.143	736	0.222
Mean			67.0	17.8	202.8	0.118	0.212	983	0.256
S.D.			10.6	7.0	115.0	0.020	0.144	326	0.077

Table 3 Mean values obtained for fits of K_1 and K_2 (Eq. 13) .

	Field	Ξ^*_∞ / mA MV^{-2}	Ξ^*_0 / mA MV^{-2}	ξ^*_∞ / mA MV^{-2}	C^* / nFm^{-1}	τ_1 / ms	K_1 / mA m^{-1} bar^{-1}	K_2 / nA s m^{-1} bar^{-1}	δ_{sh} / ms
Mean	Low	94.7	33.4	322.4	0.131	0.213	-0.895	551	0.149
S.D.		18.4	10.8	134.2	0.050	0.073	0.18	347	0.099
Mean	High	81.1	17.3	254.0	0.089	0.136	-0.586	745	0.219
S.D.		8.09	6.18	205.8	0.027	0.104	0.17	220	0.078

considerably improved the grouping of other parameters. The pressure differential was obtained numerically which is an operation that can generate considerable noise. The result was therefore further processed using a low pass Butterworth filter with a cuttoff at 1000 Hz before use. Care was taken to ensure that this process did not introduce any spurious time delay.

The observed current response is of course dependent upon the form of the input voltage. This is in most cases a good approximation to a step function but includes a small rise time of around 0.15 ms (Fig. 2b). We have previously shown[(2)] how to deal with this strictly in the non-linear regime using a Runge Kutta solution to the differential equations but the procedure is laborious. Deconvolution of the observed response with the time derivative of the voltage, although not strictly applicable, is probably an acceptable approximation, but the voltage step is in most cases so near to ideal that we can account equally well for the imperfection by the incorporation of a small time delay of 0.15ms. This correction was not included in previous determinations. This and the addition of the new pressure coupling term has had the effect of reducing the value of ξ^* in most cases. The new values of ξ^* now agree more closely with those obtained previously from data in the frequency domain. Determination of the electrical constants is also hindered because the available data was truncated at a preset current value for operational reasons. This means that the maximum current drawn in the initial peak, which is directly related to ξ^*, is often unknown. Fitting to the remaining portion of the response does give an estimate of ξ^* but the additional pressure contribution sometimes drowned the available data in this region making a determination of this parameter difficult or impossible. Some results in the series have been omitted for this reason. There were also some results (in particular those for valve C - the widest gap) for which the pressure response is relatively smooth. The pressure coupled features in the current response for these traces were consequently inadequate for a reliable identification of the parameters and these results have also been omitted. Nevertheless, it is worth noting that good fits to these current responses could be obtained using average parameters derived from the other results. Values of all the parameters obtained are collected in Table 2. They are mostly comparable to those reported previously[2] but changes are discussed below. The estimates of ξ^*, where obtainable, are scattered but consistently higher for the lower field. These are plotted in Fig. 7 against the mean nominal (i.e. initial) flowrate estimated from

$$\langle U \rangle = \frac{Ql}{Ah} \tag{14}$$

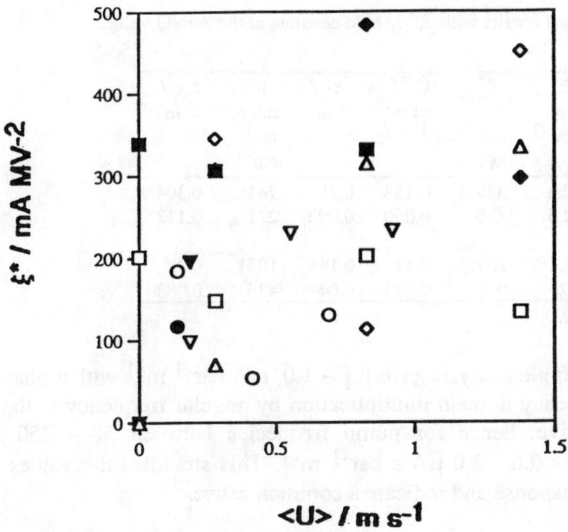

Fig 7. ξ* plotted against the mean flow rate <U>.
valve A , valve B, valve C, valve D, valve E.
Empty symbols E = 1.2 MV m^{-1}, filled: E =2.4 MV m^{-1}.

For valve A operating at zero flowrate the peak current was available and these estimates of ξ^* are probably the most reliable. It is not possible to detect any significant flow dependence of this parameter above the experimental noise and mean values agree reasonably well with the zero flow determination. In an attempt to overcome the problem of truncated data, fits were also carried out with ξ^* pegged at the zero flow value. However, this did not produce any significant changes or statistical improvement in the parameters. Mean values obtained in this exercise are presented in Table 4.

The new results of this study are values of the parameters K_2. Despite cosiderable scatter, these behave in a consistent way when plotted against the mean flow rate (Fig. 8) and, taken together, suggest that there is a small decrease in magnitude as the flow rate increases. Within the accuracy of this determination these parameters have no clear dependence on valve dimensions. The mean value of K_2 is about 1 μA s m^{-1} bar^{-1}.

Table 4. Mean values obtained for fits with ξ^*_∞ held constant at the zero Q value.

	Field	Ξ^*_∞ / mA MV^{-2}	Ξ^*_0 / mA MV^{-2}	ξ^*_∞ / mA MV^{-2}	C^* / nFm^{-1}	τ_1 / ms	K_2 / nA s m^{-1} bar^{-1}	δ_{sh} / ms
Mean	Low	64.5	32.6	339.0	0.158	0.29	741	0.304
S.D.		9.9	12.5	0.0	0.050	0.093	271	0.112
Mean	High	67.1	18.1	201.0	0.11	0.134	1031	0.22
S.D.		10.6	7.2	0.0	0.023	0.04	436	0.083

Results from the pump ripple analysis gave $K_1 \sim 1.0$ mA bar^{-1} m^{-1} with a phase shift of about 180°. In the frequency domain multiplication by angular frequency is the same as taking the time derivative, hence for pump frequecies between 50 - 250 Hz the equivalent value of K_2 is 0.6 - 3.0 μA s bar^{-1} m^{-1}. This straddles the values we have obtained from the time response and indicate a common cause.

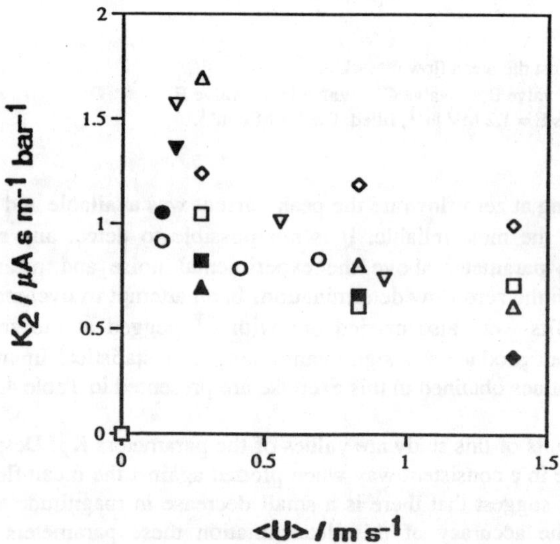

Fig. 8. Coupling parameter K_2 vs. <U>. Symbols as Fig 7.

Despite a large degree of scatter the time delay δt_{sh} appears to be roughly constant when plotted against mean flowrate (Fig. 9). The origin of this delay cannot be clearly ascertained from this study but it is probably related to the time taken for the new flow pattern to be established appearing through Q in Eq. (12). We have recently carried out simulations[4] which suggest a response time of this order for an incompressible fluid.

The pressure coupling causes some modification in values for the capacitance which are rather lower than those previously determined as are values of ξ^*. Decay times τ_1 are about the same and are generally faster at high field strength (Fig. 10).

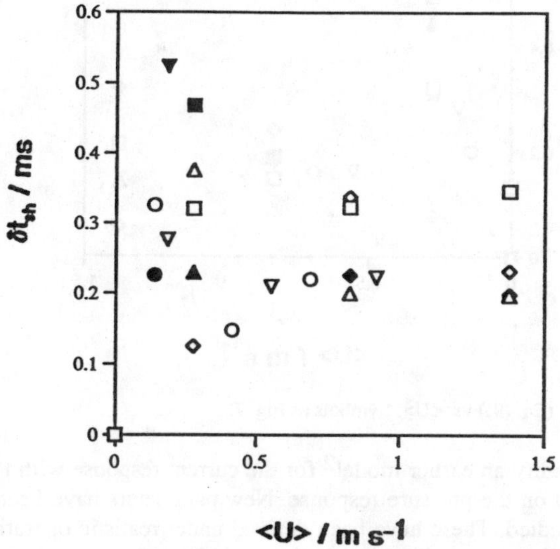

Fig. 9. Time delay δt_{sh} vs. <U> . Symbols as Fig. 7.

4. Conclusions.

We have demonstrated the existence of a dynamic coupling between pressure and electrical current in a flowing electrorheological system. We find that this can be characterised by a linear term in the time differential of the pressure (Eq. 13) which could be phenomenologically related to the changes in the flow rate. The inclusion of the term in the pressure leads to a negative value of K_1 which would be inconsistent with our findings for the steady state that indicate a positive coupling constant. We conclude that any coupling directly to the pressure is comparatively slow and that for short times K_1 = 0.

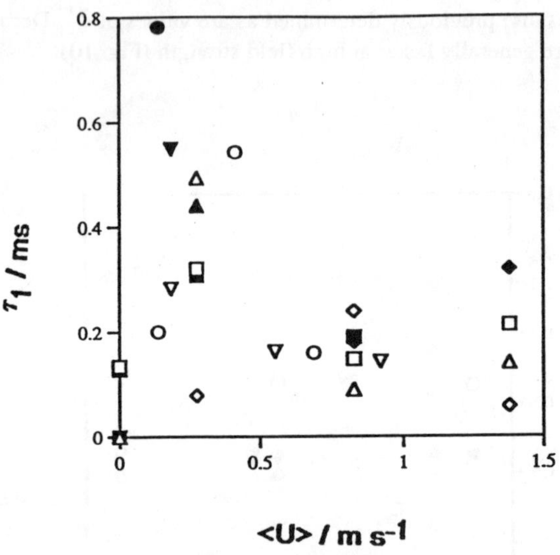

Fig. 10. Time constant τ_1 (Eq. (9)) vs. <U>. Symbols as Fig. 7.

Our findings modify an earlier model[2] for the current response with the addition of a new term dependent on the pressure response. New parameters have been obtained for the range of valves studied. These have been derived under realistic operating conditions and behave in a consistent way. Despite the large degree of scatter they should prove useful in the design and control of future electrorheological valves. In Table 2 we give the average values of these quantities for this purpose. A comparison of these results with the behaviour of valves using different ER fluids would now be a useful extension of this work.

The conductivity of ER fluids may depend upon a variety of phenomena[7]. In the water containing fluid considered here, charge transport is likely to be mediated by ionic charge carriers but the conductivity may also be affected by near-wall electrode processes, characteristics of the interfacial region and impurities in the dispersing fluid. In addition, the field induced fibrillated structure[8] of the fluid is certain to play an important role. Indeed, the square law current - voltage characteristics of most fluids are most easily, albeit superficially, explained in terms of a conductivity related to some aspect of the structure that depends linearly upon the field. Pressure and flowrate changes may affect the transport of electricity either through these structural effects or by inducing changes in any of the other contributing processes.

We have shown that the steady state current and pressure are linked and that the dynamic current response includes a component that reflects the pressure changes through the time differential dP/dt. Our evidence suggests that a direct link between the dynamic pressure P(t) and the current i(t) is less tenable. We have seen that pressure transients can be linked to transient changes in flowrate and it is difficult on the basis of this experiment alone to determine which effect is phenomenologically responsible for the coupling.

5. Acknowledgments

The authors thank the Science and Engineering Research Council for support of this work.

6. References.

1. R. Firoozian, W.A. Bullough, and D.J. Peel , *ASME Publication DE-Vol.* **18-4**, (1989) pp45-50, ASME.
2. M.Whittle, R. Firoozian, W.A. Bullough, and D.J. Peel. Int. Jnl. of Modern Physics B. **6** , 2683 - 2704, (1992).
3. M.Whittle, R. Firoozian, W.A. Bullough, and D.J. Peel. The Pressure Response in ER Valve Flow, *Proceedings of conference: Recent Advances in Adaptive and Sensory Materials and their Application.* Blacksburg, Virginia, April 27-29 (1992) pp495 - 505 (Technomic)
4. M.Whittle, R. Firoozian, D.J. Peel and W.A. Bullough. *Jnl. Non-Newt Fluid Mech.* Submitted for Publication.
5. J.A. Aseltine, *Transform Method in Linear System Analysis.* (McGraw-Hill, 1958)
6. W.H. Press, B.P. Flannery, S.A. Teukolsky and W.T. Vetterling, , *Numerical Recipes* (CUP 1986) p289.
7. R.A. Anderson , *Proceedings of the International Conference Electrorheological Fluids Carbondale Illinois.* (World Scientific Publishing Co. Singapore 1992) . pp 81-90.
8. M. Whittle, *J. Non-Newtonian Fluid Mech.*, 37, (1990) pp233-263.

Properties of Electrorheological (ER) Fluids Under Periodic Deformation

W.I.Kordonsky, A.D.Matsepuro, S.A.Demchuk, Z.A.Novikova

Luikov Heat and Mass Transfer Institute, Academy of Science, Minsk, Belarus

The progress in the overwhelming majority of the branches of science and technology is mainly determined by designing new host structural materials resistant to vibration, including composite ones that have the ability to change their mechanical characteristics in real time because of the altered operating conditions.

Of late, the considerable efforts of researchers have been aimed at designing composite materials, including the so-called "intelligent" fluids. Among these are the ER fluids capable of sharply changing their mechanical properties (effective viscosity, elasticity) at electric field for the time of an order of 10^{-4} s. The possibility to control the viscoplastic ER fluid properties over a wide range, on the one hand, and the simplicity in manufacturing an electric field inductor, on the other hand, allow simple and reliable vibration isolation designs containing ER fluid domains as a composite element to be developed. Among these is the so-called sandwich-element, whose schematic is shown in *Fig.1.* Such a device is composed of two flexible electrodes shaped as flat elastic-material plates, with an ER fluid-filled gap in between. Under the action of the external force there occurs elastic deformation of the electrodes, resulting in shear deformation of an ER fluid domain. For such a sandwich to operate profitably as a composite material element with adaptive vibration isolation parameters the fundamental mechanical properties of the ER fluid must be known under different deformation conditions. This has provided the subject matter for the present study.

In experiments, multicomponent ER fluids were studied when all their components were varied. Different natural and artificial adsorbents served as a dispersed phase. Optimal component proportions were chosen.

Experimental study of the mechanical properties of ER fluids was made under the conditions of periodic small-amplitude deformation using the classical scheme of a vibrating cantilever representing a sandwich composed of two elastic plates: electrodes and thin ER fluid domain in between. The plate size was (2x20x220)mm, the test ER fluid domain thickness was about 0.1 mm.

One end of such a sandwich was rigidly clamped by means of a special facility provided with electrode leads and with an electromagnetic actuator and inductive probes mounted near a free sandwich end. Thus, the "vibrating-reed method" was realized, i.e., use of a sample shaped as a flat long plate with one-side embedding and excitation of free end vibrations.

Facilities for exciting vibrations and recording a system response to them are the essentials of a design of sandwich-type systems.

The schematic of the experimental setup is shown in *Fig.2*.

Electromagnetic excitation of sandwich vibrations was performed by an electromagnetic actuator. A sandwich vibration amplitude was recorded by the inductive probes, whose amplified signals were sent to the oscillograph screen and the digit voltmeter. The assembled sandwich was placed into a heat-insulating chamber. This provided a possibility of making studies over a wide range of temperatures, both plus and minus.

In experiments, measurements were made of: shear storage modulus (G') and shear loss factor (η) of the ER fluid domain under periodic deformation. The behaviour and strength of an electric field, ER fluid composition, ER fluid domain thickness and temperature, vibration frequency and sandwich design were varied. The resonance frequency shift and resonance peak width were weasured, quantities G' and η were calculated by the method of A.

Nashif et al. (A.Nashif, D.Johnes. J.Henderson "Vibration Damping", John Willy Sons, 1985, New York).

In the first set of experiments, about 40 ER fluid samples were tested to determine an optimal ER fluid composition. Some results are cited in *Table 1.* Experiments were made at room temperature when the electric field strength and the force causing vibrations were varied. Having processed the obtained results, a more narrow set of compositions was determined for further studies. In addition to good efficiency (substantial resonance frequency shift and sandwich vibration amplitude decrease under the imposition of an electric potential) the selected compositions yielded good result reproducibility - return to the original (initial) state at field cutoff (return of a resonance frequency to its initial value before field cutin).

In the second set of experiments, the G' and η, dependences of the field behaviour and strength, ER fluid domain thickness, sandwich design, and vibration frequency over a wide temperature range were investigated (+50 ^{o}C to -30 ^{o}C).

Figures 3 through 6 present the graphical interpretation of the mechanical ER fluid-99 properties (see *Table 1*). In particular, Figure 3a,b plots G' and η as functions of electric field strength at room temperature and natural vibration frequency of 18 Hz (ER fluid domain thickness is 0.15 mm).

As seen, with growing the electric field strength in a gap shear storage modulus, G', increased 16.9 times simultaneously with decreasing shear loss factor, η, 3.6 times. The latter was caused by reinforcing the ER fluid structure with the field growth and shear flow energy dissipation reduction. As this took place, the system reversibly shifted in a resonance vibration frequency by 10-12 Hz at a vibration amplitude decrease and constituted 28-30 Hz, i.e., the system became more "stiff". The field cutoff brought the resonance frequency and vibration amplitude back to their original values. The effective action of d.c. and a.c. fields, all other things being equal, pointed to a great (10-12%) efficiency of the d.c. field.

Figure 4 illustrates the electric field action on the resonance frequency shift at different resonance vibration frequencies. As seen, the "more flexible" the system (larger sandwich length and smaller f_{res}) the more sensitive is the system to the field action and, vice versa, the "more stiff" the sandwich, the larger electric field strength is needed for f_{res} to shift by the same value.

Figure 5a,b plots G' and η vs. the ER fluids domain thickness (H_2) in the sandwich. With increasing the ER fluid domain thickness all things being the same, the value of the relative deformation of the ER fluid domain varies. As seen, relative deformation changes within 35% exert no essential influence on the shear storage modulus and shear loss factor (*Fig. 5a,b)*.

As for practical applications, of special interest is the environmental temperature influence on the mechanical characteristics of sandwich-elements. Measurements have been made in a cooling/heating chamber over a temperature range + 50 ^oC to -30 ^oC at a natural resonance frequency of 18 Hz and ER fluid domain thickness of 0.15 mm. The obtained values of G' and η at d.c. electric field are plotted in *Fig. 6a,b*.

As evident, the temperature influence on both the shear storage modulus and shear loss factor is not very essential although it takes place. In this case, the influence of high temperatures is more adverse: with decreasing G' and η the electrical conductivity of the ER fluid sharply grows. Apparently, this is associated with the thermal stability of some components of ER fluids and their dielectric properties, although the outward appearance of the ER fluids subjected to high temperatures did not change (the fluid and particles didn't separate, didn't change viscosity, colour, etc.).

What is more, the entire temperature range was examined for one sandwich assembly, i.e., the ER fluid domain was not replaced during the whole run of experiments. In this case, the electric field behaviour manifested itself better. If at room and minus temperatures the G' increase at d.c. field was by 10-15% greater as against at a.c. field, then at high temperatures when the electrical conductivity of the ER fluid domain sharply grew and this effect reduced, it retained at a.c. field with much smaller electrical conductivity of the ER fluid domain.

Also, it should be noted that over the entire considered temperature range the G' and η dependences of the field strength and temperature did not change qualitatively.

The performed study has evidenced that the fundamental possibility exists of significantly (up to 100 times) increasing the shear storage modulus of ER fluids at electric field. However, the problem on hysteresis of fluids exhibiting a maximal modulus increment still remains unsolved.

Over the investigated parameter range it is shown that at increasing electric filed the shear storage modulus tends to increase with a vibration frequency and practically does not depend on deformation and temperature values.

At increasing field the shear loss factor decreases, which is due to viscous energy dissipation reduction, as the ER fluid structure is reinforced.

In a general case, the studied design can be considered as an element with a natural electric field-tuned frequency of vibrations.

Thus, the idea of ER fluids for vibration control in distributed systems may be put in practice by embedding these fluids in host structural materials which are fabricated in either conventional or advanced composite materials. The imposition of an appropriate electric filed on the fluid domain in the structure permits the energy-dissipation characteristics and mechanical properties of the embedded fluid to be actively controlled, which enables the global properties of the structure containing the ER fluid domains to be controlled.

Table N1 Shear storage modulus G' (kPa)

Fluids	Alternating current field (kV/mm)							Direct current field (kV/mm)						
number	0	0.33	0.66	1.33	2.66	4	5.33	0	0.33	0.66	1.33	2.66	4	5.33
0	3	140	190	190	190	190	190	3	20	20	40	80	128	194
19	15	20	30	40	80	170	390	5	50	60	98	210	350	370
20	2	6	16	50	158	250	380	2	6	50	129	270	368	460
53	2	16	40	120	161	200	220	2	20	40	80	100	120	240
62	8	9	15	50	80	100	150	8	14	20	50	80	100	120
63	4	20	22	60	160	180	200	4	125	130	137	140	250	255
64	6	8	10	12	43	100	150	6	20	80	120	250	220	250
66	6	8	16	54	88	142	160	6	27	34	118	168	200	250
67	4	11	34	132	220	510	360	4	13	43	119	220	270	340
68	2	12	26	80	213	220	260	2	32	64	110	160	189	200
69	6	15	42	56	80	85	90	6	35	63	86	89	99	100
70	13	20	25	90	160	270	300	13	16	36	161	310	350	400
71	10	190	200	220	240	290	300	10	50	68	170	289	290	400
92	11	20	44	125	180	240	250	11	47	87	145	177	217	300
99	18	20	25	40	50	80	200	18	19	20	50	80	90	290
102	3	36	40	60	80	140	180	3	36	40	60	130	150	200

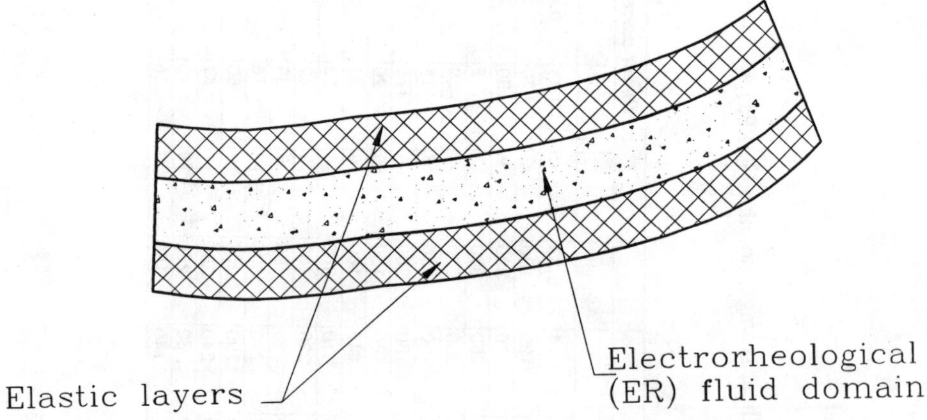

Figure N1 Typical view of a deformed sandwich (composite material)

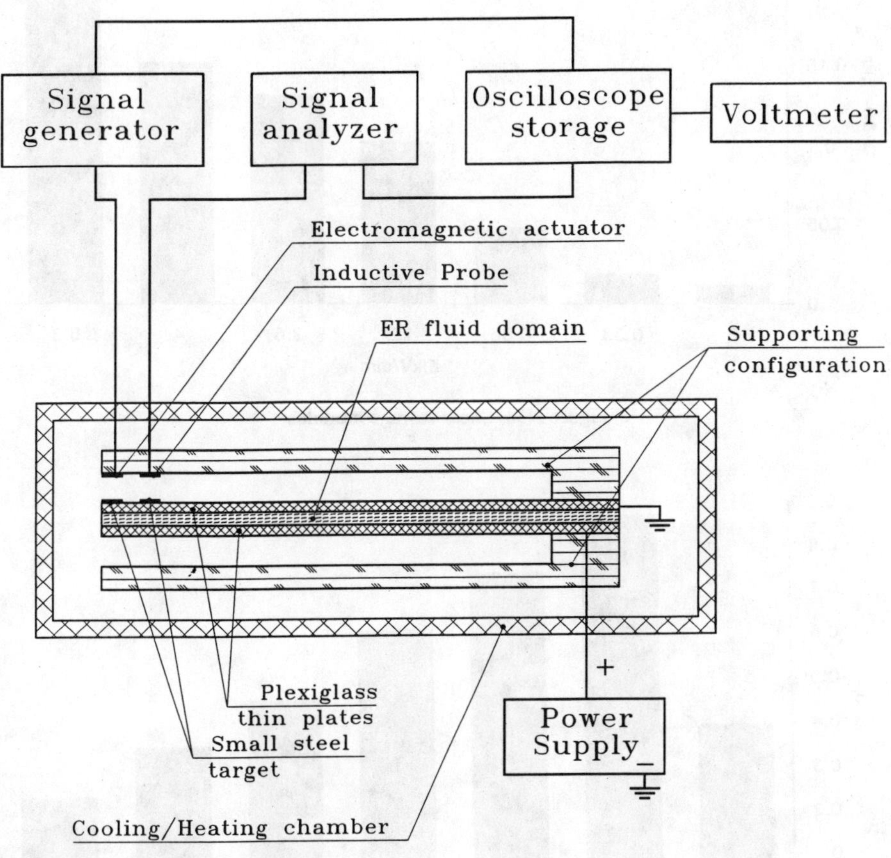

Figure N2 Schematic of the experimental setup based on vibration testing

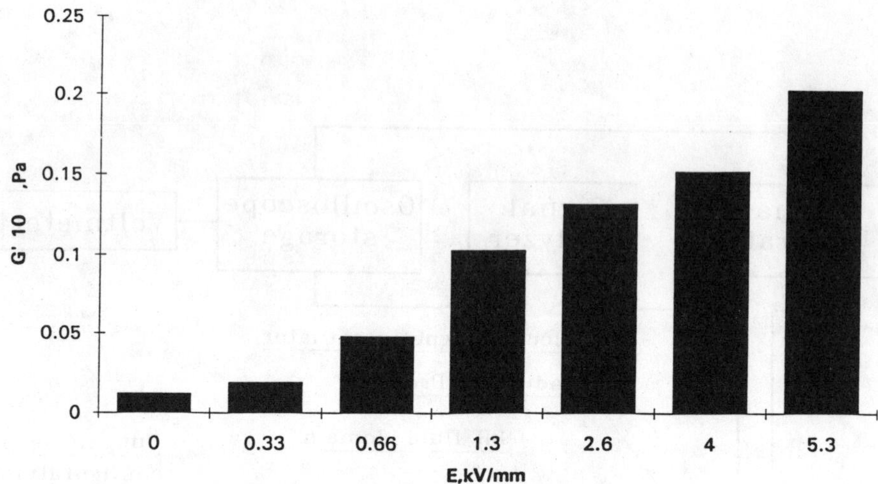

Figure N3a Shear storage modulus

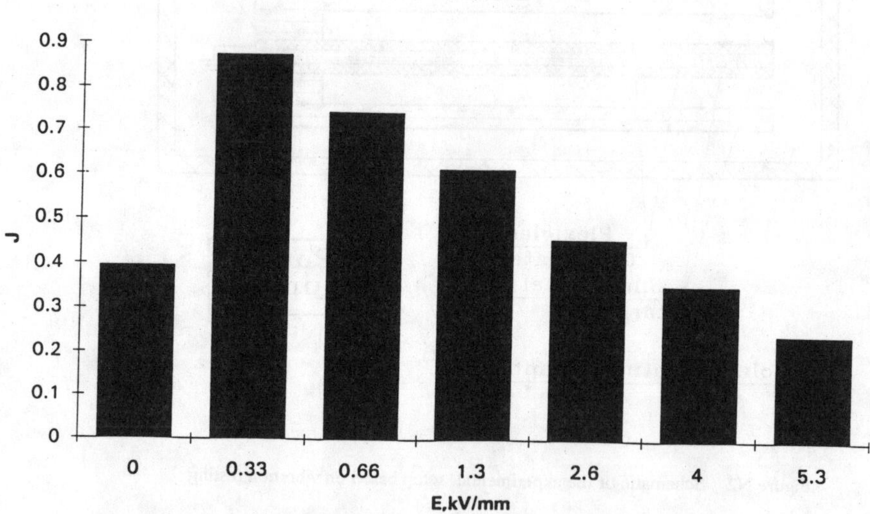

Figure N3b Shear loss factor

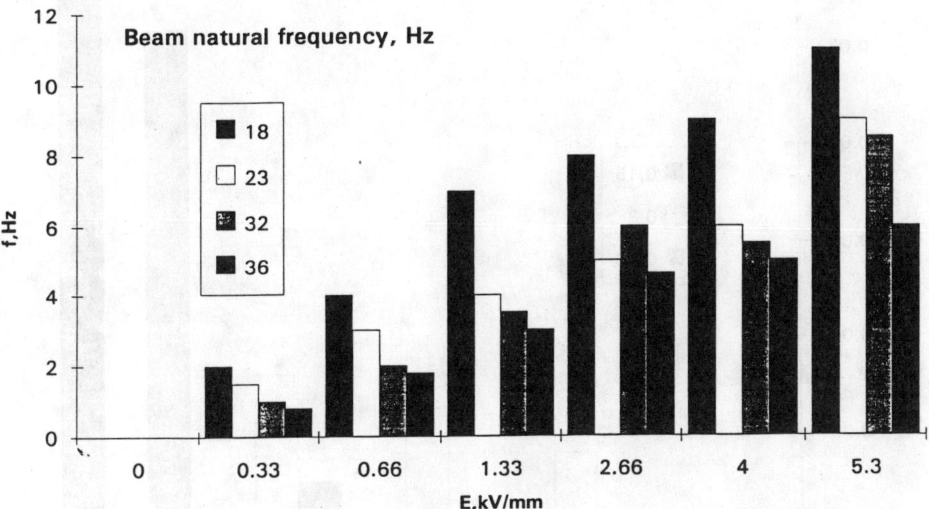

Figure N4   Resonance frequency shift

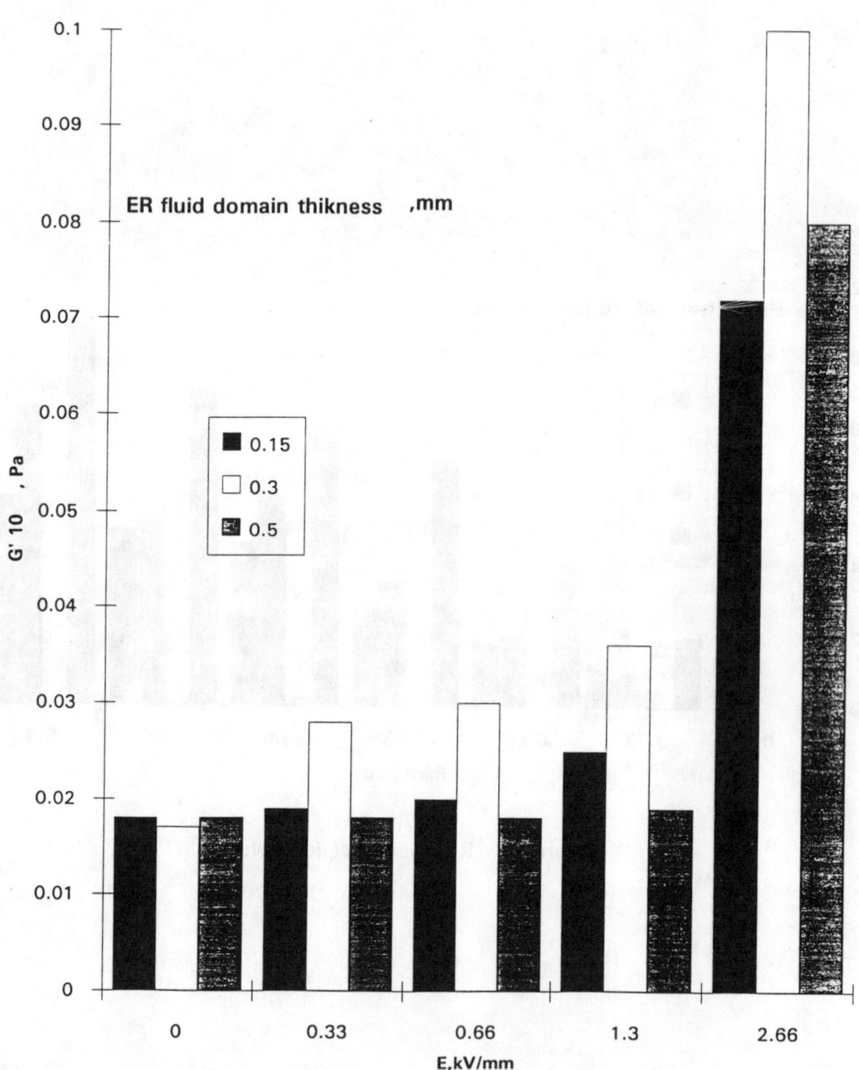

Figure N5a Shear storage modulus

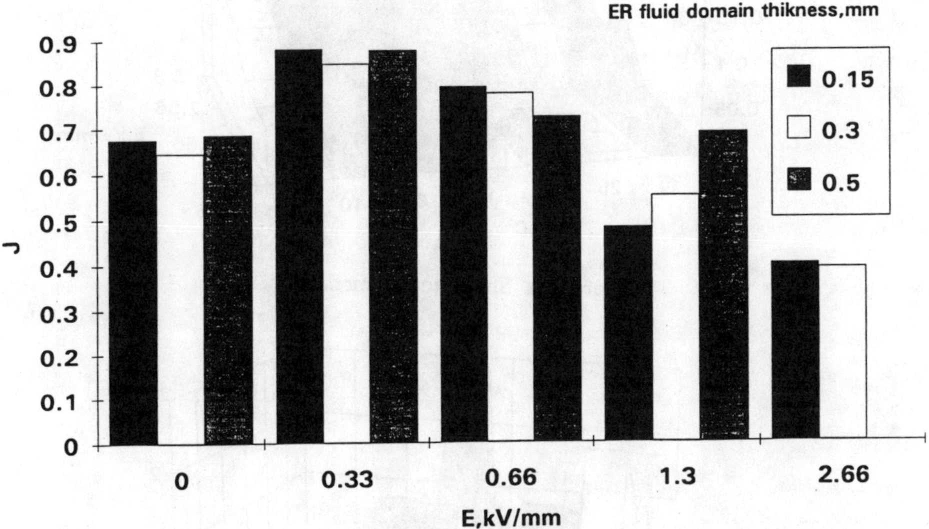

Figure N5b Shear loss factor

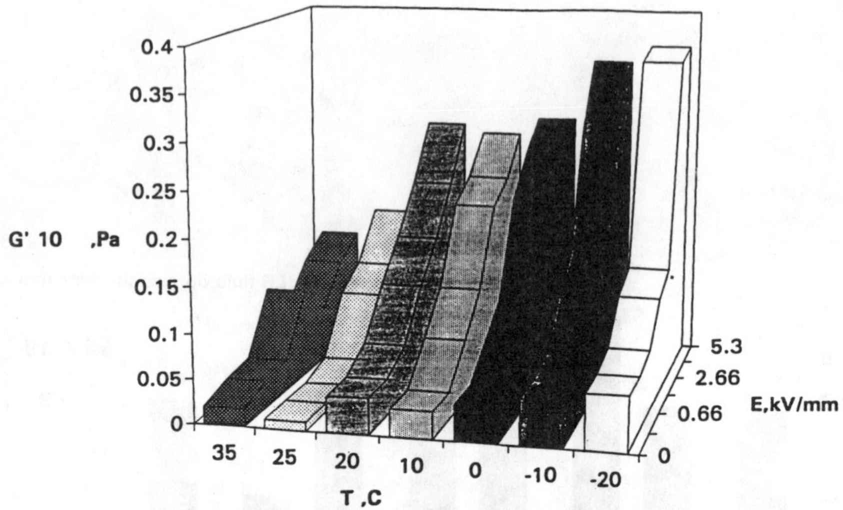

Figure N6a Shear storage modulus

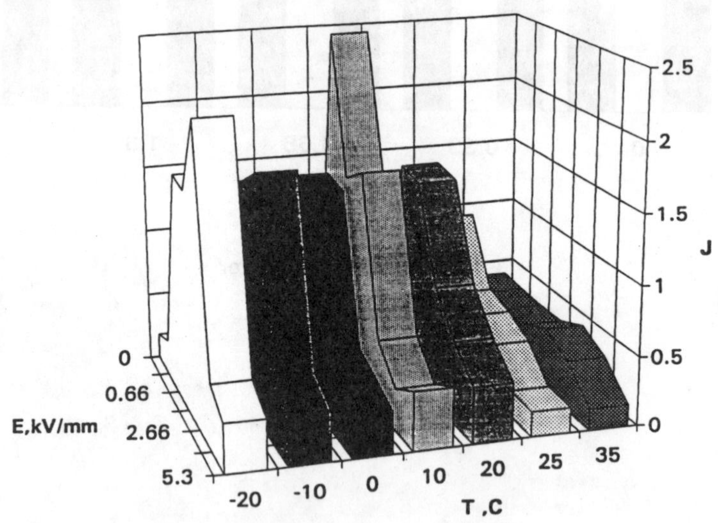

Figure N6b Shear loss factor

INFLUENCE OF THE ELECTRIC FIELD FREQUENCY ON THE ELECTRORHEOLOGICAL FLUIDS PROPERTIES

C. BOISSY, J.N. FOULC and P. ATTEN
Laboratoire d'Electrostatique et de Matériaux Diélectriques
CNRS et Université Joseph Fourier - Grenoble 1
B.P. 166 - 38042 Grenoble Cedex 9 (France)

ABSTRACT

The strength of hydrous ER fluids are usually decreasing when the frequency of the applied electric field is increased. The reason for such a response is probably due to the fact that the electric field distribution in the fluid is controlled at low frequencies by the conduction properties of the materials (solid and liquid) and, at high frequencies, by the dielectric properties (permittivity). We present preliminary results of current and force measurements on a large scale model using two half-spheres in contact. The dependence of phase-shift and magnitude of current through the half-spheres on frequency leads to a crude electrical model. The experimental results are discussed on the basis of this model. Finally, we compare the variation of the attraction force between the half-spheres with the shear stress of an hydrous ER fluid (cellulose in mineral oil) for different frequencies of the applied voltage.

1. Introduction

Even if the properties of ER fluids are more and more studied since the early 1980's, the physical mechanisms responsible for the electrically induced attraction between particles are not clearly understood. It is generally assumed that the particle-particle interaction is an attraction due to the induced dipole-dipole interaction. However this explanation is unsatisfactory, especially in the case of DC or low frequency applied voltages. In this case the electric field distribution is controlled by the conduction properties[1-3]. We recently proposed a model for DC electric field that seems to well describe the physical phenomena[4]. Nevertheless at high frequencies the electric field distribution is controlled by the permittivities, thus for the halfway values of frequency the effect of conductivity and permittivity are balanced. The electric field distribution then becomes more complex. It is also generally assumed that the materials are perfectly insulating but in fact, even if the conductivity is very low, it is not true. In particular, the liquid conductivity can strongly increase under high electric field[5] and thus modifies the field distribution. The phenomena that can influence the electrical properties of both liquid and particles are numerous. This is certainly why there is no general rule for the behaviour of ER fluids. Thus, it appears that the influence of field frequency on the strength of ER fluids varies according to the nature of the particles. For hydrous materials, the strength decreases when the frequency increases[6,7], but for anhydrous materials the behaviour seems to be opposite, the strength increases with frequency[8,9].

The purpose of this work is to see how the frequency of the applied field influences the interaction between particles, and to try to explain how the effect of conductance and capacitance are balanced.

2. Experimental set-up

2.1. Apparatus

The particles of an ER fluid are modelled using two half-spheres of radius much larger than that of actual ER fluid. The experimental apparatus (Fig.1) comprises two duralumin electrodes 9 cm in diameter spaced by 1.4 cm. The upper half-sphere can pass through a circular aperture in the upper electrode. The planar metallized section of the upper half-sphere and the upper electrode are grounded. The lower half-sphere, fastened to the lower electrode is brought to potential V. The cell is filled up with a transformer oil and put into a Faraday cage. The two half-spheres, of radius 0.7 cm, are made of polyamid having a conductivity $\sigma_s \cong 1.7\times10^{-8}$ S/m and a permittivity $\varepsilon_s \cong 24\varepsilon_o$ (at 50 Hz). The surrounding oil has a conductivity of $\sigma_L \cong 3\times10^{-13}$ S/m and a permittivity $\varepsilon_L \cong 2.2\varepsilon_o$.

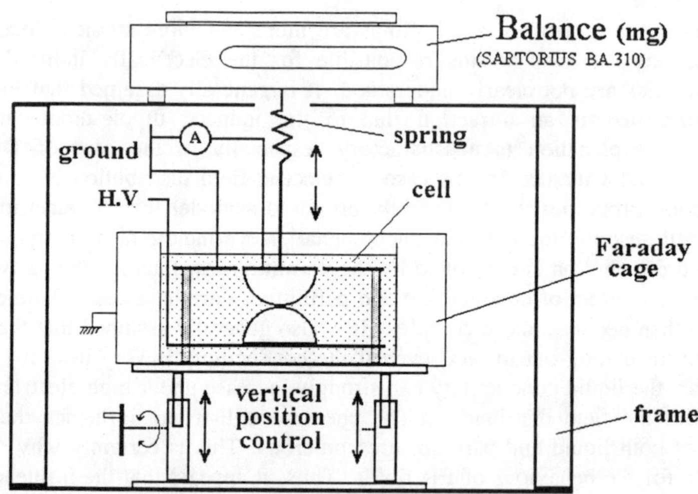

Fig.1 Experimental set-up for large scale model using two half-spheres. The vertical position of the lower half-sphere is controlled by a micrometric table.

2.2 *Measurements*

A sine-wave voltage V is applied on the two half-spheres in contact. The current passing through the upper half-sphere is monitored via an oscilloscope. The current magnitude and phase-shift (the phase reference is the voltage) are measured as a function of frequency. To measure, for fixed magnitude and frequency of the applied voltage, the attraction force between the two half-spheres in contact, we slowly lower the cell support. When the electric interaction force equal the restoring force of the spring the two half-spheres separate. The force measured with the balance at the moment of detachment is equal to the attraction force (its sensitivity is the milligram).

3. Results

3.1. *Electrical measurements*

We monitored the current in the upper half-sphere for different potentials V and for frequencies ranging from 0.01 to 2,000 Hz. We measured the phase-shift, between the current I and the voltage V, and the gain defined by $G=20Log(I/V)$. Results are presented in Fig.2 (phase-shift) and Fig.3 (gain) in a Bode plot representation. The phase-shift is positive (i.e. the current is in advance on the voltage) as can be expected with a capacitance system. We can note that the phase-shift and the gain are slightly dependent on the voltage magnitude. One of the possible electrical circuit that leads to the similar phase-shift and gain variations with the frequency is shown on Fig.4 (this equivalent circuit will be discuss in section 4). It exhibits three crossover frequencies ($f_c=1/2\pi\tau$). The values of the different RC components are difficult to estimate because we didn't explore a large enough frequency interval and also because the capacitances and resistances are not constant but vary with the frequency and magnitude of the applied electric field.

We notice that for frequencies above 1 Hz the gain doesn't depend on the applied voltage, but for low frequencies the gain depends on voltage; this means that the value of (R_1+R_2) is dependent of the applied voltage and therefore on the applied electric field E. It seems that (R_1+R_2) increases with increasing E. For the phase-shift, we notice the same behaviour but more accurate measurements are required before clearly interpreting the results.

3.2. *Force measurements*

The force between the two half-spheres has also been measured for frequencies above 10 Hz (see Fig.5). A few remarks can be pointed out. Firstly, with DC voltage the force increases with E (here, for relatively low fields, the force scales as E^2). Secondly, when comparing the shapes of the gain and force curves, we can notice that when the current increases, the force decreases. Thirdly, the decrease of the force with frequency is important, between one or two orders of magnitude, and it appears more marked for low electric fields.

It is interesting to compare the behaviour of the large scale sphere-sphere system with that of a actual ER fluid. The tested fluid is a suspension of cellulose particles (typical size 30 μm, volume fraction $\phi = 0.15$) in a transformer oil (ELF TF50). Fig.6 shows the dependence of the dynamic yield stress on frequency. The shape of this curve is similar to that of attraction force between spheres. This similarity suggests that our large scale system can, in a certain way, model the behaviour of an actual ER fluid.

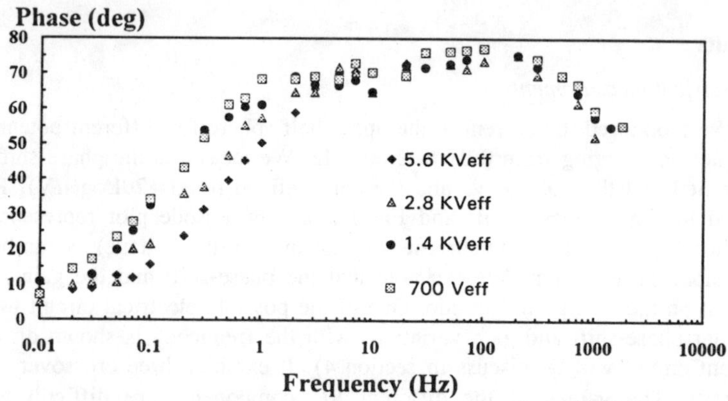

Fig.2 Phase-shift-frequency curves for different applied voltages.

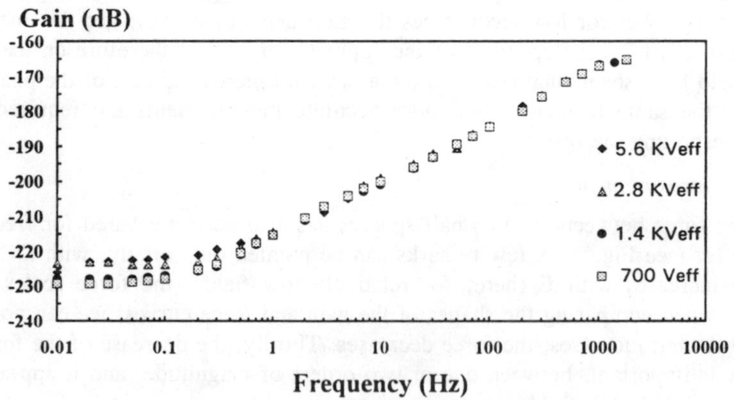

Fig.3 Gain-frequency curves for different applied voltages.

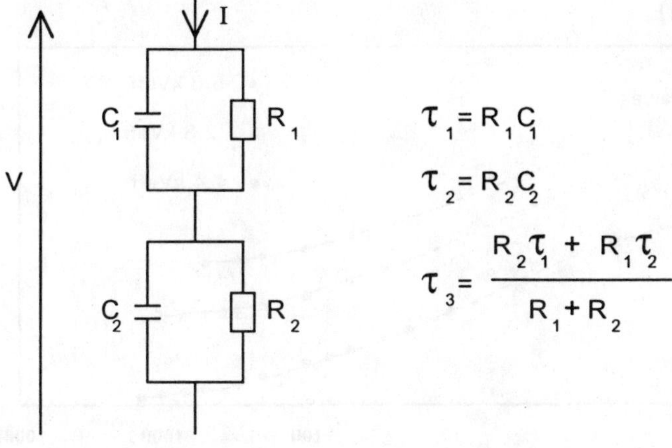

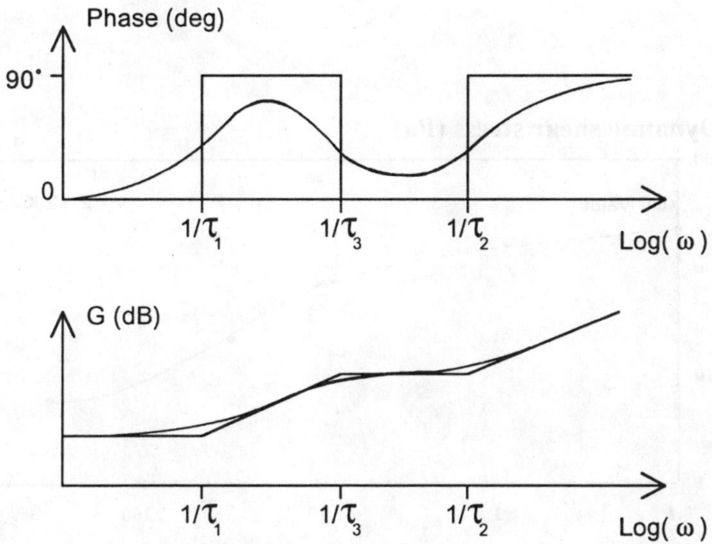

Fig.4 Equivalent RC circuit for the two half-spheres in contact and associated Bode curves. The time constants τ_1, τ_2 and τ_3 give three crossover frequencies ($fc_i = 1/2\pi\tau_i$).

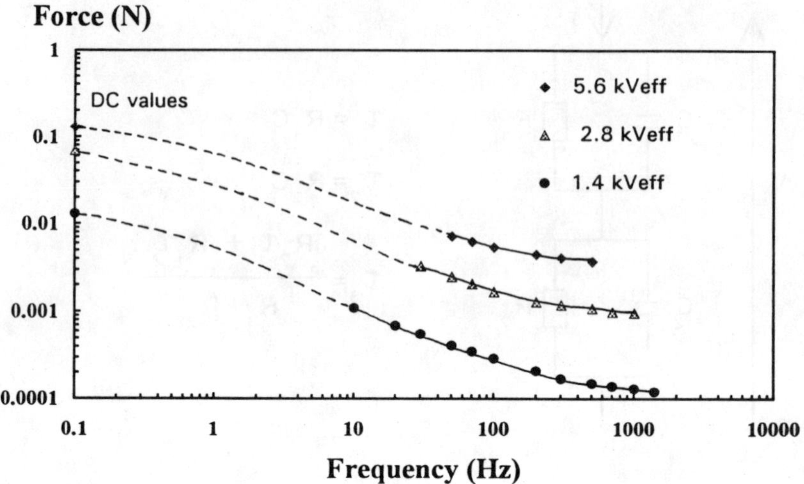

Fig.5 Variation of the attraction force between two half-spheres as a function of frequency for different values of potentials.

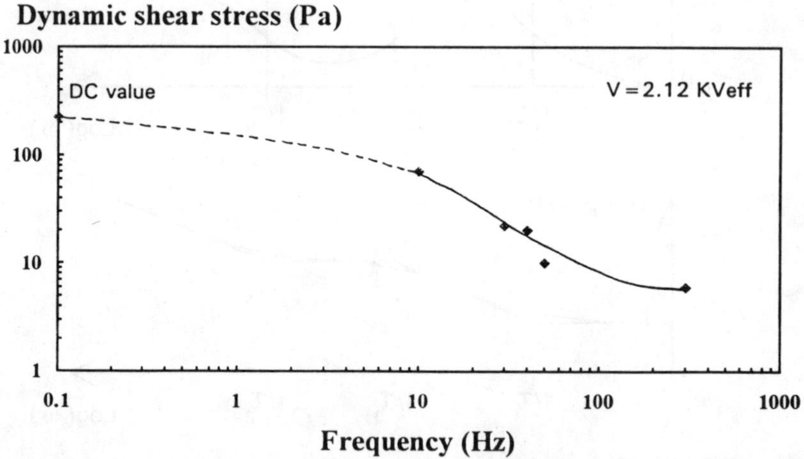

Fig.6 Effect of applied voltage frequency on the dynamic yield stress of the tested ER fluid (cellulose in transformer oil).

4. Discussion

The circuit of Fig.4 is a formal equivalent circuit. In order to progress in the understanding of this electrical model, it is worthwhile to see to what parts the various RC components correspond. Looking at one half-sphere (Fig.7), we can distinguish three parts : i) the upper part of the sphere with permittivity ε_s and conductivity σ_s, ii) the liquid with permittivity ε_L and conductivity σ_L, iii) the contact zone where ε and σ are very difficult to prescribe; let us note the capacitance of this zone C_c and its resistance R_c. The equivalent circuit is represented on Fig.8. On the basis of the conduction model[4], it appears that $R_L >> R_c$. We can also estimate that $C_L << C_c$. Therefore the electrical model of Fig.8 is equivalent to the circuit of Fig.4 and we have $R_1 = R_s$, $R_2 = R_c$, $C_1 = C_s$ and $C_2 = C_c$. The problem is to calculate the values of C_c, R_c, C_s, and R_s. As we can see on Fig.9, ε_s depends on frequency and, as we previously noticed, the conductivity σ_L of the liquid in the contact zone depends on the applied field. Therefore R_c, C_c and C_s vary with the frequency and the magnitude of the voltage. All these non-linearities make the problem very complicated and required further fine investigations.

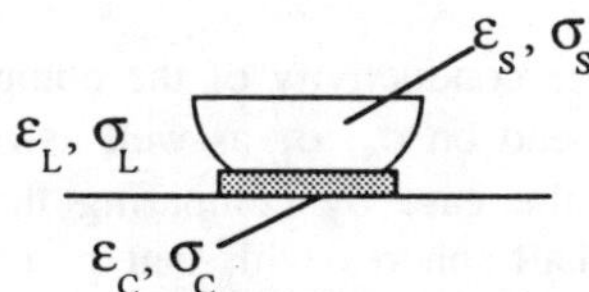

Fig.7 Different electrical parts of the half-sphere.

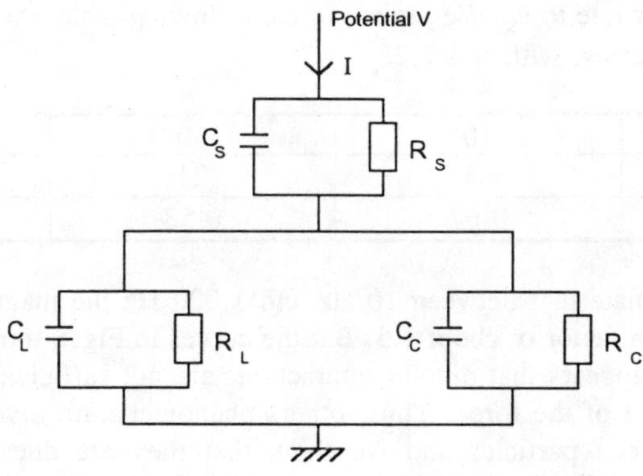

Fig.8 Electrical model equivalent to the half-sphere in contact.

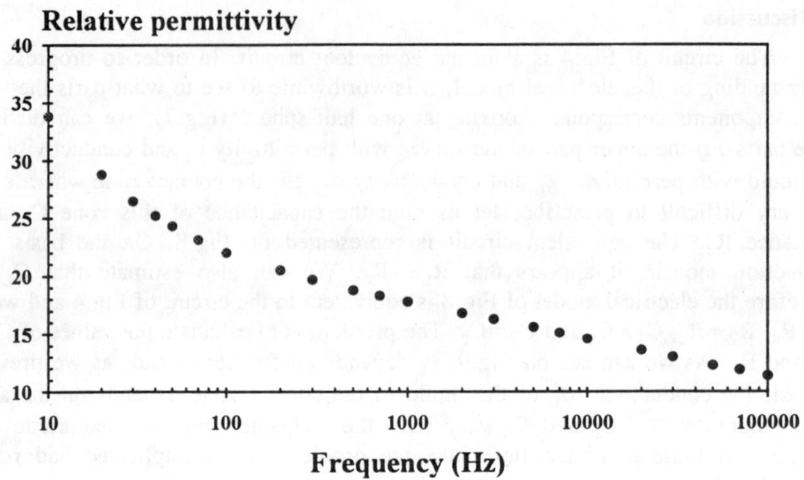

Fig.9 Variation of the relative permittivity ε_S of the solid material (polyamid) as a function of frequency.

In the above picture, the conductivity of the components plays a role and this implies that the force will depend on σ_S, σ_L as well as on ε_s and ε_L. We can get an argument that this is indeed the case by comparing the variation of the measured attraction force between the half-spheres with that given by the point-dipole model based on the polarisation[10]. This model indicates that the force F between particles is proportional to $\varepsilon_L.(\beta.E)^2$, with ε_L the permittivity of the liquid, $\beta=(\varepsilon_s-\varepsilon_L)/(\varepsilon_s+2\varepsilon_L)$ and E the applied field. At this range of frequency ε_L is constant and the variation of β on frequency is only due to ε_S. We report, in the following table, the value of ε_s and β^2 for different frequencies, with $\varepsilon_L \cong 2.2\varepsilon_o$.

Frequency (Hz)	10	50	1000
$\varepsilon_S/\varepsilon_O$	34	24	18
β^2	0.67	0.58	0.49

We can estimate that between 10 Hz and 1,000 Hz the magnitude of the force should decrease by a factor of about 1.5. But the curves in Fig.5 show that the factor is greater than 10. It appears that dipolar interactions are not sufficient to explain such a decrease in the value of the force. Thus, others phenomena are involve in the process of interaction between particles and we think that they are due to the conduction properties of the liquid and of solid particles. Establish a mathematical relation that describe the electrical behaviour is difficult because of non-linearities.

5. Conclusion

We present a preliminary study of the influence of the electric field frequency on ER fluids properties. Electrical measurements on a large scale model using two half-spheres, made in a weakly conducting polyamid and immersed in an insulating liquid, lead to an simplified equivalent circuit. This electrical circuit based on R-C components is discussed and connected to the conduction and dielectric properties of the different parts of the large scale model. The dependence on frequency of the attraction force between half-spheres is compared to that predicted by the polarisation model (point dipole approximation). As the comparison shows a great difference, we conclude that a possible raison is that, even in AC electrical field, the conduction properties of the liquid and the solid materials play a role in the ER effect.

These first results must be developed in order to understand the ER effect in AC conditions. In particular we have to investigate the range of frequencies above 2 KHz because it is not impossible that the force varies again. It is clear that liquid conductivity may play an important role and make the problem more complex. But if we manage to find how the distribution of the electric field varies, we will be able to calculate a more precise expression of the force. The study of the frequency dependence on ER fluid is important, as well on the theoretical point of view than on practical one. For future works we need to study the behaviour of others solid materials and/or others liquids.

One question remains : in what degree our large scale model agrees with real ER fluids ? The phenomena that can occur around particles of radius about centimeter can be very different than those around particles of micrometer size.

6. Acknowledgements

This work was performed with the financial and technical support of the company ELF-Atochem. The authors wish to express their thanks to Prof. N. Félici for useful discussions and advice.

7. References

1. J.-N. Foulc, N. Félici and P. Atten, *C.R. Acad. Sci. Paris* **t.314, Série II** (1992) 1279.
2. R.A. Anderson, in *Proc. 3rd Int. Conf. on ER Fluids*, R. Tao, ed. (World Scientific, Singapore, 1992) 81.
3. L.C. Davis, *J. Appl. Phys.* **72** (1992) 1334.
4. N. Félici, J.-N. Foulc and P. Atten, *in this volume.*
5. L. Onsager, *J. Chem. Phys.* **2** (1934) 599.
6. D.L. Klass and T.W. Martinek, *J. Appl. Phys.* **38** (1967) 75.
7. H. Uejima, *Japanese J. Appl. Phys.* **11** (1972) 319.
8. T. Garino, D. Adolf and B. Hance, in *Proc. 3rd Int. Conf. on ER Fluids*, R. Tao, ed. (World Scientific, Singapore, 1992) 167.

9. A. Inoue, in *Proc. 2nd Int. Conf. on ER Fluids*, J.D. Carlson, A.F. Sprecher and H. Conrad, eds. (Technomic Publ. Co., Lancaster-Basel, 1990) 176.
10. D.J. Klingenberg, F.V. Swol and C.F. Zukoski, *J. Chem. Phys.* **91** (1989) 7888.

PRELIMINARY OPTICAL STUDY ON ER FLUIDS

L.W. ZHOU, J.F. YE[+], R.B. TAO,
T.D. Lee Physics Laboratory, Department of Physics
Fudan University, Shanghai 200433, CHINA

Y. TANG, J.F. PENG, Z. GAO
Department of Chemistry, Fudan University, Shanghai 200433, CHINA

L.Y. LIU, S.H. MA, and W.C. WANG
T.D. Lee Physics Laboratory, Department of Physics
Fudan University, Shanghai 200433, CHINA

ABSTRACT

Linear and non-linear optical study on electrorheological (ER) fluids is reported. The ER fluids under the investigation were glass beads, zeolite and ferroelectrics. The linear optical responce of some ER fluids showed sharp changes near critical electric fields. An enhancement of electric field induced second harmonic generations (EFISH) was observed as the function of E^2, where E is the external electric field. The said enhancement is considered to be corresponding to a modulation of the material's refractive index associated with the electric field induced polarization of the delocalized electrons. The enhanced non-linear optical response on the transition between liquid and solid states can be related to the phase transition in ER fluids.

1. Introduction

After electrorheological fluids consisting of "dry" particles were invented several years ago, booming intensive theoretical and experimental studies on mechanism, materials and applications have been yielding a deep understanding in this area. Intensive studies on the mechanism of ER phenomena have benefit the material research and application exploration. Researchers in this area dream that, someday, they will be able to design ER fluids, either solid particles or base liquids, even single-phase materials, on a molecular level according to the requirements of applications. This is strongly depending on the better understanding the mechanism. In the sense of material design, the mechanism of ER fluids is far from complete awareness.

This work is the first part of a study to explore what is the nature of the transitions between solid and liquid states. It is expected to know how intra-

[+]Permanent address is Shanghai University of Engineering Technology, Textile College, Shanghai 200061, CHINA

molecular or inter-particle charges transfer near critical electric field, E_c. It is observed[1] that particles form a body-centered tetragonal lattice structure when $E>E_c$, and an order parameter was defined.

It is shown experimentally[2] and theoretically[3] that, near superconducting transition temperature, T_c, the power signals of second harmonic generation, $P^{2\omega}(T)$, and third harmonic generation, $P^{3\omega}(T)$, have a form of spikes of generation. $P^{2\omega}(T)$ is also magnetic field, H, dependent: $P^{2\omega}(T)$ shows peak structure only when H is not zero. Nevertheless, $P^{3\omega}$ shows peak structure even in the absence of magnetic field. It is evident that the strong enhancement of $P^{2\omega}$ and $P^{3\omega}$ is closely related to the phase transition.

To pursue the optical study we made several ER fluid samples, such as glass beads in silicone oil, zeolite, $BaTiO_3$-system and so forth with correspondent base liquid. The samples were first characterized by measuring their static yield stress. The linear optical measurements and the electric field induced second harmonic generation (EFISH) measurements were then conducted. A series of experiments were done to confirm that the ER fluids did show their intrinsic EFISH signal. We expect also to establish the qualitative relation between the microscopic structure and main characteristics of ER fluids.

2. Theory

Microscopic polarization $P(\boldsymbol{k},\omega)$ can be expressed as[4]

$$P(\boldsymbol{k},\omega)=P^{(1)}(\boldsymbol{k},\omega)+P^{(2)}(\boldsymbol{k},\omega)+\ldots, \tag{1}$$

where

$$P^{(1)}(\boldsymbol{k},\omega)=\chi^{(1)}(\boldsymbol{k},\omega)\cdot E(\boldsymbol{k},\omega) \tag{2}$$

and

$$P^{(2)}(\boldsymbol{k},\omega)=\chi^{(2)}(\boldsymbol{k}=\boldsymbol{k}_i+\boldsymbol{k}_j,\omega=\omega_i+\omega_j):E(\boldsymbol{k}_i,\omega_i)E(\boldsymbol{k}_j,\omega_j) \tag{3}$$

with $E(\boldsymbol{k},\omega)$ being the incident laser electric field of wavevector $\boldsymbol{k}$ and angular frequency ω, $\chi^{(1)}$, the first-order susceptibility tensor describing linear optical properties, such as refraction, absorption and so forth, and $\chi^{(2)}$, second-order susceptibility tensor. $\chi^{(2)}$ is zero for centra-symmetrical material; non-zero for the materials with lower symmetry yielding frequency doubling effect -- so called second harmonic generation (SHG). SHG is forbidden under the electric-dipole approximation in a medium with inversion symmetry. The inversion symmetry is broken at a surface interface.[5] The inversion symmetry is also broken when certain dc electric field is applied on a medium with such symmetry. This is why the technique of electric field induced second harmonic generation (EFISH) is commonly used.

In this case the macroscopic polarization $P^{2\omega}$ induced in an ER fluid by an incident laser field E^{ω} can be described as[6]

$$P_I^{2\omega} = \varepsilon_0 d_{IJK}(\mathrm{E}^0) E_J^{\omega} E_K^{\omega}, \tag{4}$$

where the component d_{IJK} of the non-linear optical susceptibility tensor is dependent on the strength of the applied dc electric field E^0, and ε_0 is the vacuum permittivity.

The microscopic dipole moment $p^{2\omega}$ induced in a particle by dc electric field E^0 and a laser field E^{ω} is as

$$p_i^{2\omega} = \varepsilon_0 (\beta_{ijk} F_j^{\omega} F_k^{\omega} + \gamma_{ijkl} F_j^{\omega} F_k^{\omega} \mathrm{F}_l^0), \tag{5}$$

where $F^{\omega} = f^{\omega} E^{\omega}$ i and $\mathrm{F}^0 = \mathrm{f}^0 \mathrm{E}^0$ are the local fields, i.e., the fields experienced by the molecules of particles and liquid, β_{ijk} and γ_{ijkl} are the second- and third-order polarizabilities, respectively, and f^{ω}, the local field factors.

The intensity of the output SHG is as

$$I(2\omega) \propto \left|\chi^{(2)}\right|^2 I^2(\omega). \tag{6}$$

The ratios of order of magnitudes of first three orders of susceptibilities can be expressed as following

$$\frac{\chi^{(2)}}{\chi^{(1)}} \sim \frac{\chi^{(3)}}{\chi^{(2)}} \sim \frac{1}{E_a}, \tag{7}$$

where E_a is Coulomb field inside an atom with an intensity of 10^8V/cm.

Each particle of an ER fluid can be considered as a source of nonlinear polarization and so the macroscopic response is the sum of the microscopic response averaged over all orientation. The small particles in an ER fluid with either zero or non-zero ground state dipole moments give no contribution to the second harmonic generation in the absence of external dc electric field E^0 because of random orientation of the dipole moments. Upon applying E^0, charges of polar molecules are oriented, non-polar molecules may be polarized, then the central symmetry is broken, and the SHG can be observed. In other words, if SHG is not observed before is turned on, but can be observed after E^0 is turned on, one can conclude that the charges in the particles are oriented with E^0 turned on.

3. Experiments and Analysis

3.1. Sample Preparation and Characterizations

The ER fluids used in the optical study were glass beads with silicone oil, ferroelectrics (whose static yield stress measured with electric field is given in Fig.1(a)) and zeolite with corresponding liquids.

A systematic study on zeolite ER fluids was conducted and the results of the study were summarized in Table 1. All of the samples were dehydrated at 250°C overnight. Firstly, the zeolites with the same framework type (say, type Y) but different cations were compared. Both the static yield stress and leaking currentof the zeolite samples increase in the order of $Ba^{++}>Ca^{++}>Mg^{++}>K^{+}\approx Na^{+}$, which is consistent with the decreasing order of their activation energy for conductance, ΔH in Y zeolite[7]. Y zeolite with divalent cations has smaller yield stress and leaking current than Y zeolite with univalent cations. The reason is that each divalent cation is bound by two negative potentials in the framework, but the univalent cation is bound by only one potential, so it is harder for the divalent cation to move than the univalent cation. Secondly, a comparison was made among the zeolites with different frameworks A, Y and M but the same cation, say Na. It is evident that Y zeolite has much higher static yield stress and leaking current density than A and M zeolites. This is because, as shown in Table 1, type Y has

Table 1. The static yield stress and the leaking current density of the zeolite samples.

Frame-work	$\frac{Si}{Al}$ Ratio	Pore Opening Diameter(s) (nm)	Type of Cations							Proposed Dominant Polarization Mechanism
			Na			K		Ca, Mg, Ba		
			wt.%	$\frac{Y.S.}{E}$	$\frac{J}{E}$	$\frac{Y.S.}{E}$	$\frac{J}{E}$	$\frac{Y.S.}{E}$	$\frac{J}{E}$	
A	1.0	0.41	16.2	$\frac{0.6}{2}$	$\frac{2.5}{2}$	$\frac{0.5}{2}$	$\frac{1.2}{2}$	-	-	Cation Movement
Y	2.5	0.74	9.6	$\frac{3.5}{2}$	$\frac{100}{2}$	$\frac{2.5}{2}$	$\frac{125}{2}$	$\frac{\sim 0.3}{2}$	$\frac{\sim 0.25}{2}$	
M	6.9	0.67 ×0.70	4.6	$\frac{0.6}{2}$	$\frac{\sim 5}{2}$	$\frac{0.8}{2}$	$\frac{\sim 25}{2}$	-	-	
HSA	62	0.54 ×0.56	0.6	$\frac{1.4}{2}$ $\frac{9}{5}$	$\frac{1.2}{2}$ $\frac{12}{5}$	-	-	-	-	Framework Polarization

Y.S./E -- static yield stress in kPa at the indicated electric field in kV/mm. J/E -- leaking current density in $\mu A/cm^2$ at the field in kV/mm.

the largest pore opening diameter and large enough number of cations among these three types of zeolites. Although Type A contains a largest number of cations, it has the least pore opening diameter which limits the mobility of the cations, and thus its yield stress. Type M is not able to contribute high static yield stress because of its least number of cations among these three zeolites. From the discussion above, the polarization of A, Y and M zeolites is conducted via the cation movement mechanism to which high cation content and large pore opening are favorable. Finally, a type of zeolite with high Si/Al ratio (HSA) was tested. Although both the diameter of pore opening and the number of cations of HSA are less than those of Y zeolite, it is found that HSA has strong static yield stress at high field, as shown in Fig. 1(b), and a very small leaking current density even at high electric field (5000V/mm). This indicates that cation movement is not responsible to the high static yield stress in HSA. On the other hand, shearing deformation of the crystal structure of HSA may occur in some circumstances, which strongly suggests that the framework polarization due to structure deformation may be a dominant reason for the high yield stress in HSA. This points out that the zeolite with high Si/Al ratio may be a promising material in some ER applications.

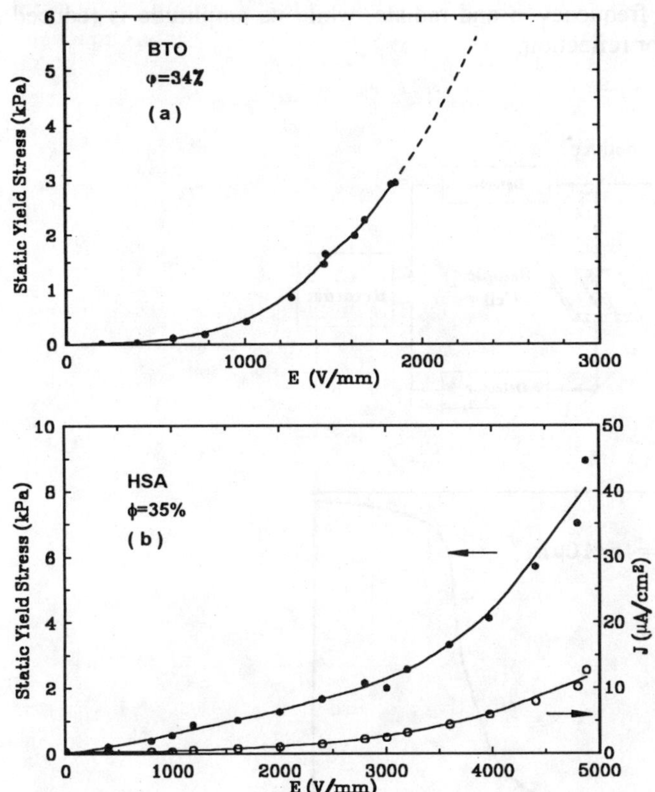

Fig. 1 Static yield stress vs. E for (a)sample BTO, a ferroelectrics ER fluid and (b)sample HSA, a zeolite ER fluid.

A homemade parallel-plate shear stress measuring system similar to that designed by Block[8] is used to determine static yield stress. To reduce friction, rolling needles are used. The gap between two electrodes is 0.5mm, a block-electrode has 2X2 cm^2 contact area with fluid to be tested, and copper-dish electrode has an area of 4X6 cm^2. A strain transducer was calibrated after the system is made.

3.2. Linear Optical Properties

Linear optical properties are refraction, absorption and so forth. For the first approximation, when a laser light beam of angular frequency ω is shining onto a sample, the macroscopic polarization of the media can be expressed as $P(\omega) = \chi^{(1)}E(\omega)$. In a classical point of view, one can image that the charges are the simple harmonic particles, which are forced by laser beam of angular frequency ω to oscillate also in angular frequency ω and radiate, while its amplitude is reduced due to either absorption or reflection.

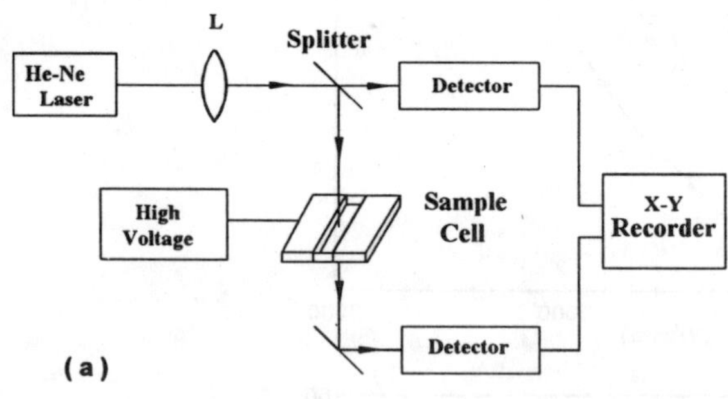

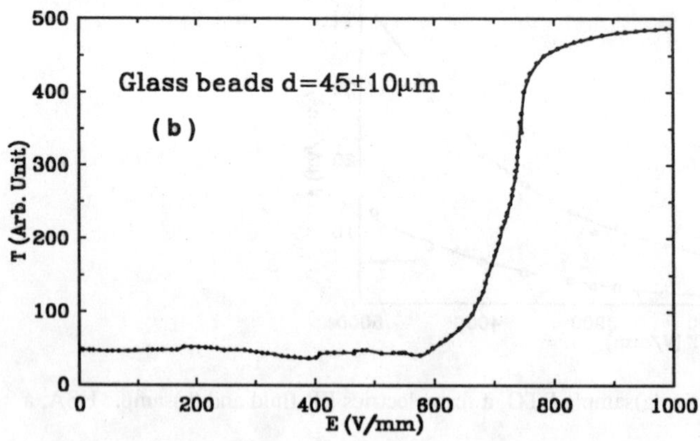

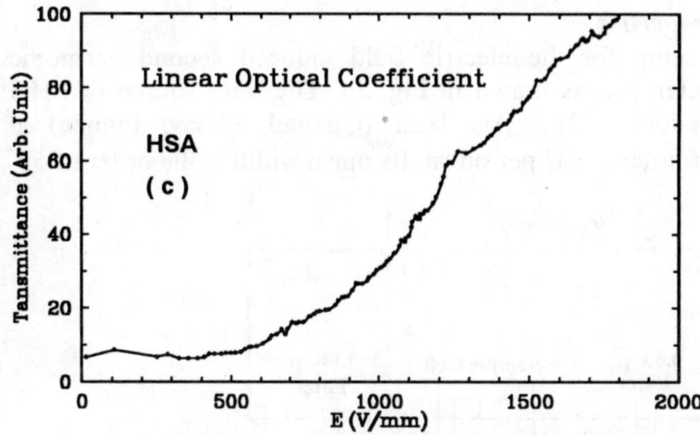

Fig. 2 (a) A schematic diagram of an experimental set-up for measuring linear optical coefficients of ER fluids. Experimental results of transmittance of ER Fluids of (b) glass beads with silicone oil and (c) HSA (a zeolite ER fluid).

Fig. 2(a) displays an experimental set-up for measuring the transmittance of ER fluids. The light source used is a He-Ne laser of power of 0.2W. A lens set L expends the diameter of the light spot to about 1 cm. One light beam after a splitter goes directly to a silicone photo cell detector, and another shines onto a sample cell, and then onto another silicone photo cell detector, which has been calibrated with the first one. The sample cell is constructed on a piece of optical glass, the distance between two parallel electrodes is 2.5 mm, and the fluid in the cell is about 0.5 mm thick and 10 mm long. The output signals from both photo cells are recorded through a two-pen X-Y recorder.

The experimental results of transmittance verses applied electric field is plotted in Fig. 2(b). The diameters of glass beads are 45 ± 10 μm, and a thin film of water is absorbed on their surfaces shortly before they are mixed with silicone oil for measurement. The volume fraction for glass-bead ER fluid is about 40%, and the lower volume fraction gave much weaker effect of transmittance change. The transmittance of the fluid shows an abrupt change when E reaches 700 V/mm, which is proposed to be critical electric field. This abrupt change arouses a possible way of making a light switch. When E is above E_c, the signal is very noisy, and that part of the curve is only an average value guided by eyes from the original X-Y recorder sheet. Similar behavior is also seen on other sample, when abrupt increase is observed. The noisy curve at high fields is probably from the severe hopping of particles between columns, a sign of a non-equilibrium state. After the field is turned off, the sample cell is observed under a microscope and a columned structure is observed. Apparently, more light can be transmitted through silicone oil when the columns are formed. The abrupt changes are also observed in zeolite samples as shown in Fig. 2(c).

3.3. Non-linear Optical Properties

The experimental setup for the electric field induced second harmonic generation (EFISH) measurements is drawn in Fig. 3a. The light source of SHG measurements is a mode-locked Nd:YAG laser (Quantel, Orsay, France) at λ=1.064μm with energy of order of mJ per pulse. Its pulse width is the order of 50

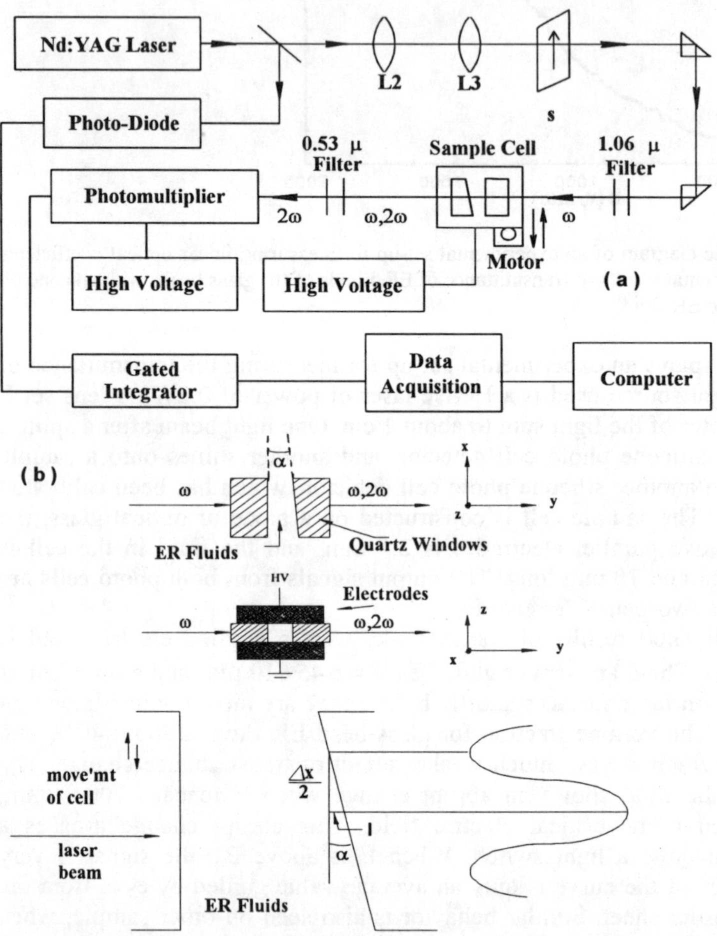

Fig. 3(a) A block diagram for the experimental set-up for measuring EFISH of ER fluids.(b) A sketch diagram of a sample cell used in EFISH measurements together with a definition of coherent length, l.

ps with 10 Hz of repetition rate, and the power is the order of 10 MW. Light is focused into a spot of diameter of about 1 mm. An S polarizeris put after lenses so that the electric field vector of the laser light is polarized along the same direction as dc electric field (in z axis). Detector is a photomultiplier sensitive to 2ω with wave length of 0.532μm. A boxcar (gate integrator) with gate width of 0.5μs is used to process the signal from photomultiplier. An X-Y recorder is used to record the results of the measurements.

The EFISH experiments are performed using wedge technique. The sample cell, shown in Fig. 3(b), consists of two quartz (JDS2) windows aligned along y axis, the direction of laser beam, two Teflon blocks (not shown) glued aside of quartz windows bilaterally, and two brass electrodes separated by the quartz windows. After electric field is applied, by translating the wedged liquid cell along x-axis, perpendicularly to the fundamental laser beam, Maker-fringe amplitude oscillations of the generated second-harmonic signal are obtained.

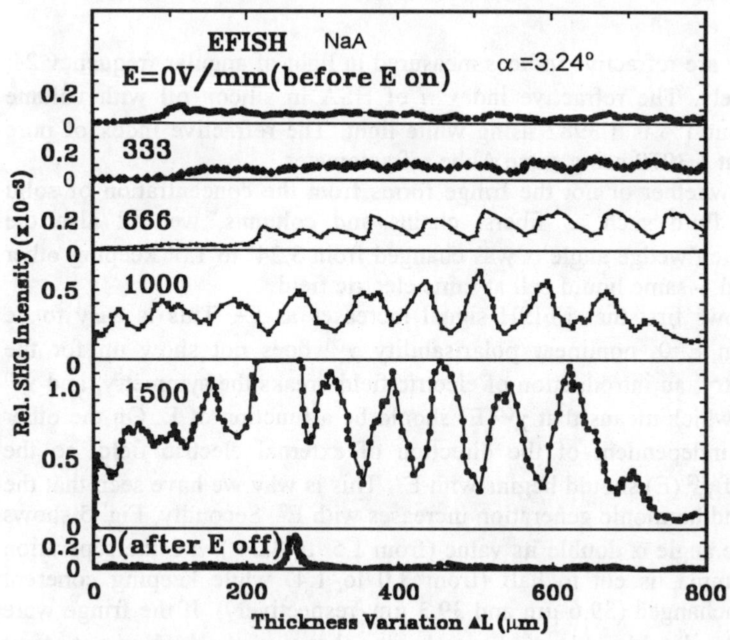

Fig. 4 Relative SHG vs. thickness variation for NaA, a zeolite ER fluid.

Fig. 4 shows how the SHG intensity relative to quartz (as unity) oscillates with thickness variation ΔL. ΔL is defined as the deference of the optical path in liquid cell before and after translation. ΔL=0 before the sample cell starts to translate. SHG intensities change when electric field increases from 0 V/mm to 1500 V/mm. When E=0, no fringe form can be detected; while E increases,

Maker's fringe appears and its amplitude increases dramatically. When E is cut off, the fringe disappears again. Coherent length l is defined, in this paper, as the difference of optical path corresponding to the scanning separation of Δx/2 (see Fig. 3(b)),

$$l = (\frac{\Delta x}{2}) \times \tan\alpha. \qquad (8)$$

It is named so because only two light beams, whose optical path difference is shorter than l, can interference each other. Coherent length is a material parameter. For a normally incident light with a given wave length λ, a coherent length of a material is determined as

$$l = \frac{\lambda}{4(n^{2\omega} - n^{\omega})}, \qquad (9)$$

where $n^{2\omega}$ and n^{ω} are refractive indices measured in light of angular frequency 2ω and ω, respectively. The refractive index n of HSA in silicon oil with volume fraction ϕ of about 1% is 1.3982 using white light. The refractive index of pure silicon oil is about 1.3970 using same Abbe refractometer.

To check whether or not the fringe forms from the concentration of solid particles in ER fluid such as fibers, chains and columns, we did also the measurements when wedge angle α was changed from 3.24° to 1.5° keeping other factors unchanged -- same liquid cell at same electric field.

Fig. 5 shows first that EFISH signal increases as E^2. This is easy to be understood: when E=0, nonlinear polarisability $\chi^{(2)}$ does not show up for the reason of symmetry; an introduction of electric field breaks the symmetry, and $\chi^{(2)}$ starts to appear, which means that $\chi^{(2)}(E)$ should be a function of E. On the other hand, $\chi^{(2)}(E)$ is independent of the direction of external electric field, so the principal terms of $\chi^{(2)}(E)$ should begins with E^2. This is why we have seen that the intensity of second harmonic generation increases with E^2. Secondly, Fig. 5 shows that, when wedge angle α double its value (from 1.5° to 3.24°), Δx, the separation between two minima, is cut to half (from 3.0 to 1.4) while keeping coherent lengths (Eq.8) unchanged (39.6 μm and 39.3 μm, respectively). If the fringe were from "column", Δx should not have changed since changing α only does not effect the formation of column.

One might suspect that the fringe form is from pure silicone oil itself. To rule out such suspicion, we measured the EFISH signal of pure silicone oil using same liquid cell at same electric field. The result shows that the pure oil has different EFISH behavior as one can see from that for ER fluid (Fig. 4).

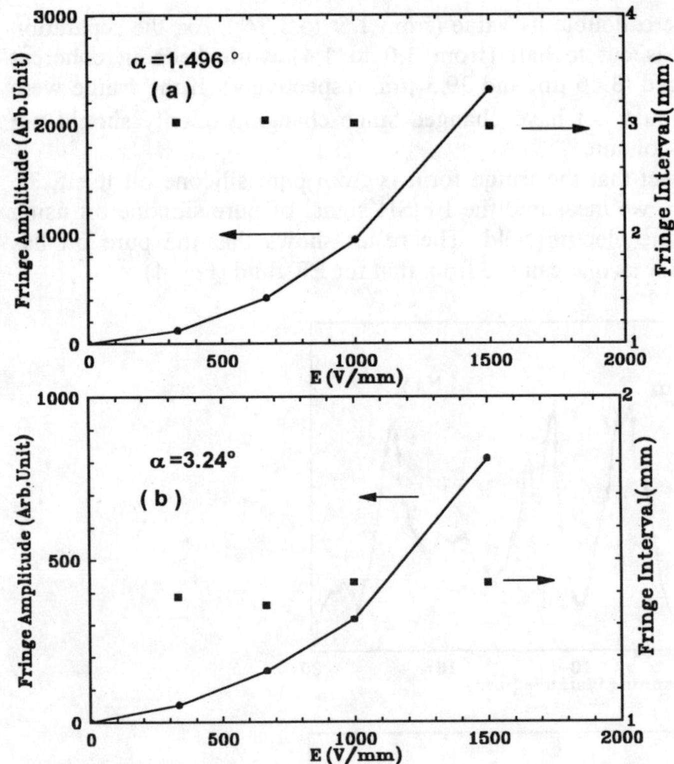

Fig. 5. Fringe amplitudes measured in EFISH experiments for NaA at wedge angle of (a) about 1.5° and (b)3.24°.

To give more information on the relation between the EFISH signal and ER effect, we did also the EFISH measurement on a zeolite with high silicone-to-aluminum ratio (HSA for short), which shows much lower ER effects, as shown in Fig. 6. Both samples were measured at same electric field but using different sample cells with very close wedge angle α (1.5° and 1.44°, respectively). Stronger ER effective sample does not necessarily show stronger EFISH signal although their coherent lengths are quite same (39.6 μm and 39.0 μm, respectively).

Fig. 5 shows first that EFISH signal increases as E^2. This is easy to be understood: when E=0, nonlinear polarisability $\chi^{(2)}$ does not show up for the reason of symmetry; an introduction of electric field breaks the symmetry, and $\chi^{(2)}$ starts to appear, which means that $\chi^{(2)}(E)$ should be a function of E. On the other hand, $\chi^{(2)}(E)$ is independent of the direction of external electric field, so the principal terms of $\chi^{(2)}(E)$ should begins with E^2. This is why we have seen that the intensity of second harmonic generation increases with E^2. Secondly, Fig. 5 shows that, when wedge angle α double its value (from 1.5° to 3.24°), Δx, the separation between two minima, is cut to half (from 3.0 to 1.4) while keeping coherent lengths (Eq.8) unchanged (39.6 μm and 39.3 μm, respectively). If the fringe were from "column", Δx should not have changed since changing α only should not effect the formation of column.

One might suspect that the fringe form is from pure silicone oil itself. To rule out such suspicion, we measured the EFISH signal of pure silicone oil using same liquid cell at same electric field. The result shows that the pure oil has different EFISH behavior as one can see from that for ER fluid (Fig. 4).

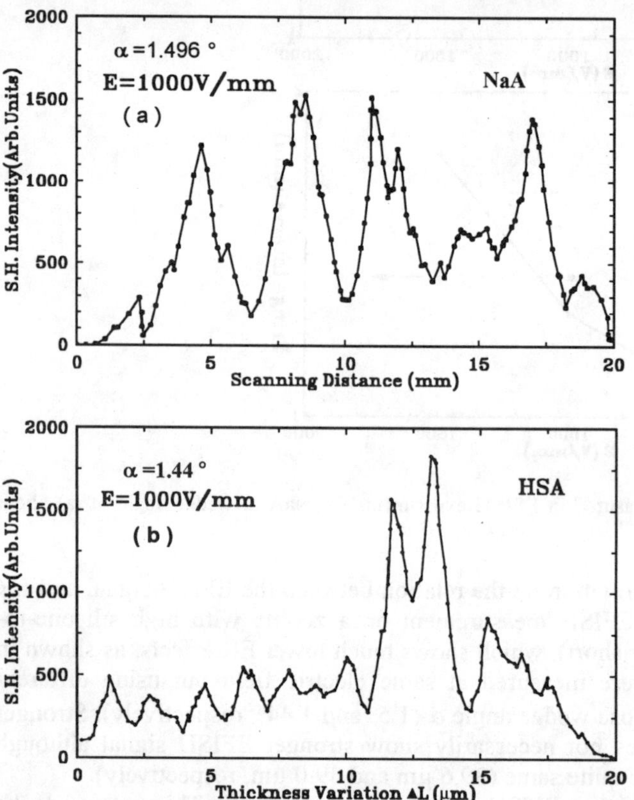

Fig. 6(a), (b) Quite different behaviors in intensities of second harmonic generation of two different zeolite ER fluids at about same wedge angle.

We did not observe a spike form of EFISH signals verses E, as that seen for high T_c superconductors in power signals of second harmonic generation $P^{2\omega}$ verses temperature near T_c when a phase transition occurs. The reasons may be (1) the sample we studied did not show strong enough phase transition, and/or (2) the frequency range we used may not be right for this kind of measurements.

4. Conclusion

In conclusion we would like to mention that the linear optical response of ER fluids consisting of glass beads, zeolite and ferroelectrics, such as $BaTiO_3$ system, show changes, sometimes sharp changes, upon applying electric field. Sharp changes in transmittance reveal a possibility of a suitable ER fluid being a photo switch.

The electric field induced second harmonic generation (EFISH) measurements showed that Δx, scanning distance between two adjacent minima of Maker's fringe, increases when wedge angle α decreases keeping coherent length $l = (\Delta x/2) \times \tan\alpha$ unchanged. The fringe amplitude increases as E^2, where E is the external dc electric field, pure silicone oil shows a different EFISH signal, and the EFISH signal is sample dependent. In other words, ER fluid does show its intrinsic second harmonic generation under electric field.

Future work will include similar measurements at different frequency range of incident electromagnetic field, such as visible light or millimeter microwave.

5. References

1. T. J. Chen, R. N. Zitter, and R. Tao, *Phys. Rev. Lett.* **68** (1992) 2555.
2. O. V. Abromov *et al.*, *JETP Lett.* **46** (1987) 546.
3. M. R.Trunin and G. I. Leviev, *J. Phys III France* **2** (1992) 355.
4. See, for example, Y. R. Seen, *The Principle of Nonlinear Optics* (John Wiley & Sons, Inc., 1984).
5. Y. R. Shen, *Annu. Rev. Phys. Chem.* **40** (1989) 327.
6. C. Bosshard *et al.*, *J. Appl. Phys*. **71** (1992) 1594 .
7. D. W. Breck, *Zeolite Molecular Sieves*, (John Wiley & Sons, New York, 1974), p. 400.
8. H. Block and J. P. Kelly, *US Patent*, No. 4687589 (1987).

PRESSURE RESPONSES OF ER FLUID IN A PISTON CYLINDER – ER VALVE SYSTEM

Masami NAKANO and Takuya YONEKAWA

Department of Mechanical Systems Engineering, YamagataUniversity
4-3-16, Jonan,Yonezawa, Yamagata, 992, JAPAN

ABSTRACT

The pressure responses of an electrorheological(ER) fluid containing ion exchange resin particles in silicone oil in a piston cylinder-ER valve system which is modeled after an ER damper, have been measured to investigate the steady and the transient responses of the ER fluid against the step inputs of electrical field to the ER valve. The ER valve has one channel of 0.6 mm in hight between two fixed parallel electrodes, where the flow establishes the Poiseuille-flow geometry. During testing, the piston is moved at a given constant speed to keep the mean flow rate of the ER valve constant. The measured step response of the pressure can be approximated by a first order response. From the saturated values and the rising time constants of the responses, we can estimate the characteristics of both steady and dynamic forces acting on the piston controlled by the ER valve, respectively. It is found that at a constant electrical field strength the saturated pressure caused by ER effect decreases as the flow velocity increases, and the rising time constants are fairly longer than the usual response time of ER fluid. These observations are discussed from the viewpoint of the damping performances and the system characteristics.

1. Introduction

Electrorheological(ER) fluids show intriguing properties of yielding solids in the presence of electric fields and of rapid response to the change of electric fields[1,2,3,4], and thus, are anticipated some applications of industrial devices such as a dumper, a valve, a clutch, an engine mount, and so on. In most of these ER devices except the clutch[5], the ER fluid flow forms a Poiseuille-flow configuration. So far, however, there are many researches[2,3,4] on the ER effects in Couette-flow configurations using a rotary type rheometer, but a few researches[6] on ER effects in Poiseuille-flow configuration, and it seems that the detailes of the Poiseuille-flow ER effects are not yet clarified. And also, in the closed ERF hydraulic systems such as an ER dumper[7], an ERF suspension system[8], and an ERF servo-actuator[9], the steady and transient pressure responses of ER fluid controlled by ER valve[10] are of significant importance in the estimation and the controllability of the induced force, respectively. The response characteristics might mainly depend upon the ER effects of ER fluid flowing through an ER valve and the compressibility of the pressurized ER fluid in the cylinder.

Therefore, in this research the authors have adressed two questions, that is, how the ER effects of ER fluid in Poiseuille-flow geometry are, and how the transient response of

ER fluid in the cylinder is. In order to answer these questions, a piston cylinder-ER valve experimental system was designed. Using this experimental apparatus, the steady and transient pressure responses of ER fluid in the cylinder have been measured against the step inputs of electric field to the ER valve. From these responses, we can estimate the characteristics of the steady and transient forces acting on the piston controlled by the ER valve.

2. Experiment

2.1 ER Fluid Properties

The tested ER fluid is a suspension of the stong acid ion exchange resin particles of about 5 μm in diameter dispersed into a 20 cSt silicone oil. The nonspherical rsin particles made by milling the particles of Amberlite IR-124 were dried and moisturized for a proper time before the dispersion. The mass fraction of particles is 30 wt%.

2.2 Experimental Apparatus and Procedure

Experiments were performed by using the experimental apparatus shown in Fig. 1, consisting of a piston of 230 mm stroke, a cylinder of 60 mm in inner diameter, and an ER valve which has a channel of 0.6 mm in hight between two parallel plate electrodes of 10 mm in length and 10 mm in width. The ER fluid is filled in the cylinder. The apparatus is designed for testing the ER effects in Poseuille-flow geometory, whose environment is similar to that expected for a variety of ER devices such as an ER damper for controlling mechanical vibrations and an ER valve for controlling a hydraulic servo actuator.

During testing, the piston is moved at a given constant speed by the speed controlled

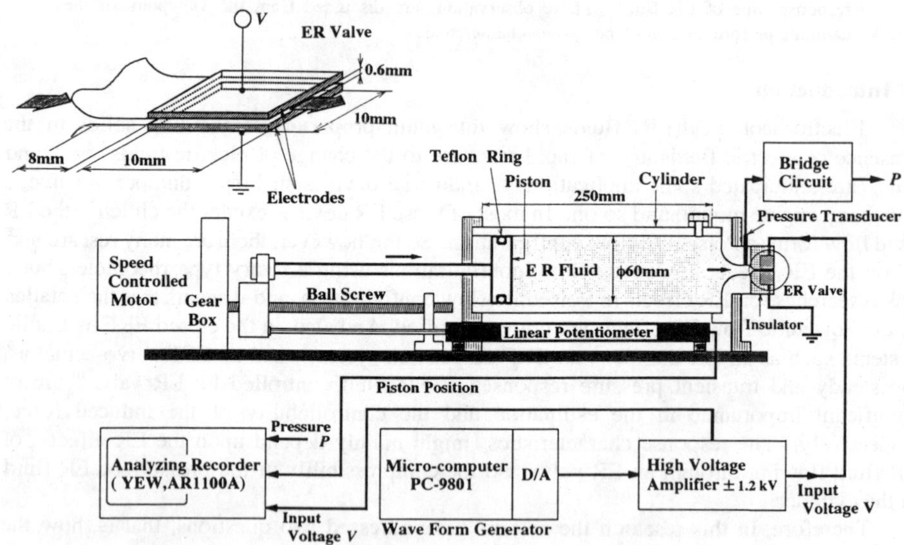

Fig. 1 Schematics of a piston cylinder-ER valve experimental apparatus for testing ER effects and the measuring system.

motor, so that the mean flow velocity between the electrodes of the ER valve should be constant. Under this condition, the step inputs of the electric field are applied to the electrodes through a high voltage amplifier of the frequency range from DC to 10 kHz , and the resultant pressure responses of the ER fluid in the cylinder are measured by a pressure transducer and are recorded on a degital recorder connected to a computer. From the saturated values and the rising time constants of the pressure responses obtained on the various conditions of mean flow velocity in the ER valve channel, the characteristics of both steady and transient pressures controlled by the ER valve are obtained as a function of the mean flow velocity.

3. Steady Characteristics of ER Effects

3.1 Theoretical Study Based on Bingham Plastic Model

Steady state ER fluid flows passing through a channel having a rectangular cross section between two parallel electrodes are investigated theoretically, assuming that the ER fluid behaves following a two-dimensional Bingham plastic model represented by eq. (1), which

$$\left.\begin{array}{lll} \tau = \tau_{ER} - \mu\,(du/dy) & (y_0 < y \leqq h/2) & : \text{post-yield} \\ \tau \leqq \tau_{ER} & (0 < y \leqq y_0) & : \text{pre-yield} \\ \tau = 0 & (y=0) & \multirow{2}{*}{\Big\}: \text{boundary condition}} \\ u = 0 & (y=h/2) & \end{array}\right\} \quad (1)$$

are defined in an orthogonal coordinate system, O-xy, with the x-axis of a channel center line and the y-axis crossing a channel, where τ is a shear stress, τ_{ER} is a yield stress associated with ER effects, u is a flow velocity in the x-direction, and h is the channel hight. Velocity profiles along the y-axis are determined by eq. (2) and (3), and depend on a

$$y_0 \leq y \leq \frac{h}{2}: \; u(y) = \left\{ \left(\frac{h}{2}\right)^2 + y^2 \right\} \frac{\Delta P_e}{2\mu L_e} - \left(\frac{h}{2} - y\right) \frac{\tau_{ER}}{\mu} \quad \cdots\cdots\cdots\cdots (2)$$

$$0 \leq y \leq y_0: \; u(y) = u(y_0) = \text{const.} \quad \cdots\cdots\cdots\cdots (3)$$

$$\text{where } y_0 = (L_e / \Delta P_e)\, \tau_{ER}$$

yield stress τ_{ER} and a pressure drop across the channel of the electrodes $\Delta P_e = \Delta P_{e0} + \Delta P_{ER}$, where ΔP_{e0} and ΔP_{ER} are given as eq. (6) and (7) respectively. And then, the integration of the velocity over the whole cross-sectional area of the cannel gives the flow rate Q of eq. (4), where B is the channel width of 10 mm and L_e is the channel length of the electrodes of 10 mm.

$$Q = B\left(\frac{h^3 \Delta P_e}{12\mu L_e} - \frac{h^2 \tau_{ER}}{4\mu} + \frac{L_e{}^2\, \tau_{ER}{}^3}{3\,\mu \Delta P_e{}^2} \right) \quad \cdots\cdots\cdots\cdots (4)$$

In the case of relatively high speed flow as the flow conditions of this experiments, a total pressure drop across the whole channel $\Delta P_T = P - P_a$, where P is a pressure in the cylinder and P_a is an atmospheric pressure, is approximated by eq. (5)[11]. That is, ΔP_T is devided into two components; one is a viscous component $\Delta P_0 = \Delta P_n + \Delta P_{e0}$ and the other is an electrorheological component ΔP_{ER}. L_n is the whole channel length of 18 mm.

$$\Delta P_T = \Delta P_n + \Delta P_{e0} + \Delta P_{ER} = \Delta P_0 + \Delta P_{ER} \quad \cdots\cdots (5)$$

$$\Delta P_n = \{12\mu(L_n - L_e) / Bh^3\}Q, \quad \Delta P_{e0} = (12\mu L_e / Bh^3)Q \quad \cdots\cdots (6)$$

$$\Delta P_{ER} = (3L_e/h)\,\tau_{ER} \quad \cdots\cdots (7)$$

Therefore, if the pressure increase ΔP_{ER} caused by the application of the electric field is measured under the condition of a constant flow rate, the yield stress τ_{ER} due to the ER effects can be estimated from the ΔP_{ER} using eq. (7).

Figure 2 shows the dependence of theoretical velocity profiles at a constant mean flow velocity U on an electric field strength E, which are obtained from eq. (2) and (3). The velocity profiles dramatically change with the electric field strength. That is, as E is increased, a plug flow region appears around a channel center and expands to both electrode walls, and the pressure drop ΔP_e increases to keep the mean flow velocity same. The increase of the mean flow velocity reduces the plug flow region as obvious on comparing Fig. 2(b) with Fig. 2(a).

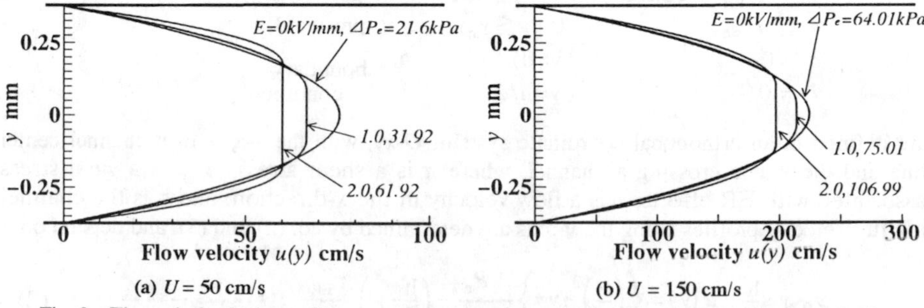

(a) $U = 50$ cm/s (b) $U = 150$ cm/s

Fig. 2 Theoretical velocity profiles at constant mean flow velocity U based on the Bingham plastic model, which strongly depend on electric field strength E and pressure drop ΔPe.

3.2 Pressure Drop ΔP_{ER} Due To ER Effects

Figure 3 shows the pressure responses of the ER fluid in the cylinder to a series of several square voltage V inputs having different levels, which were measured under the two conditions of the mean flow velocity U. The pressure P reversibly rises and falls against the on-off switching of the electric field, and the pressure increase ΔP_{ER}, corresponding to the pressure drop across the electrode channel due to the ER effects, increases with increasing input voltages. But, if the mean flow velocity becomes higher, ΔP_{ER} components are reduced as seen in Fig. 3(b). Thus, the levels of ΔP_{ER} components depend on the mean flow velocity between the electrodes.

The variations of the ΔP_{ER} components against the mean flow velocity are shown in Fig. 4 in terms of the electric field strength $E = V/h$. For each electric field strength E, as the mean flow velocity U is increased, ΔP_{ER} component tends to decrease almost exponentially from a maximum value $\Delta P_{ER,0}$ at $U = 0$ and to converge to a value $\Delta P_{ER,S}$ at a sufficiently high flow velocity. While, the increase of the electric field strength raises $\Delta P_{ER,0}$ and $\Delta P_{ER,S}$, weakening the decreasing rate to the flow velocity. The weakening of ER effects

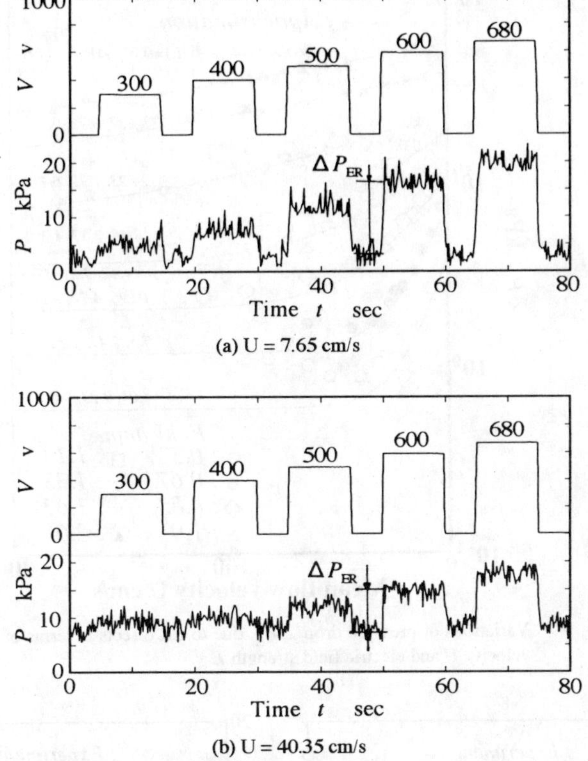

(a) U = 7.65 cm/s

(b) U = 40.35 cm/s

Fig. 3 Time traces of pressure response to a series of several square input voltage, which strongly depend on the mean flow velocity U.

with increasing flow velocity suggests that the cluster formation of dispersed particles may depend upon the flow velocity.

3.3 Approximate Function For Estimating ΔP_{ER} Component

By assuming the above mentioned decreasing characteristics of ΔP_{ER} component against the flow velocity, the following approximate function for estimating ΔP_{ER} component can be proposed in terms of the electric field strength and the flow velocity.

$$\Delta P_{ER} = \Delta P_{ER,0} - (\Delta P_{ER,0} - \Delta P_{ER,S})(1 - e^{-\beta(E)U}) \quad \cdots\cdots (8)$$

The application of a curve fitting method to the experimental data gives the values of $\Delta P_{ER,0}$, $\Delta P_{ER,S}$, $\beta(E)$, and fitted curves for each electric field strength E, which are indicated in Fig. 4. The fitted curves show a fairly good approximation of the experimental data, so it can be concluded that the approximate function of eq. (8) is suitable for estimating the levels of ΔP_{ER}.

Next, three parameters of $\Delta P_{ER,0}$, $\Delta P_{ER,S}$, and $\beta(E)$ obtained in Fig. 4 should be

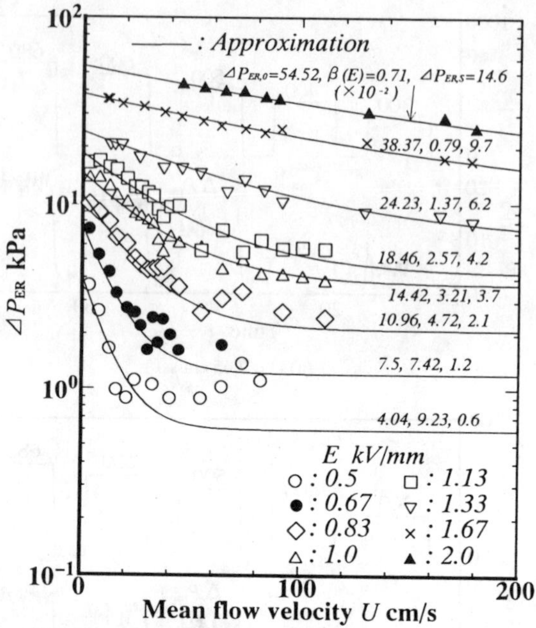

Fig. 4 Variations of pressure drop ΔP_{ER} due to ER effects in terms of mean flow velocity U and electric field strength E

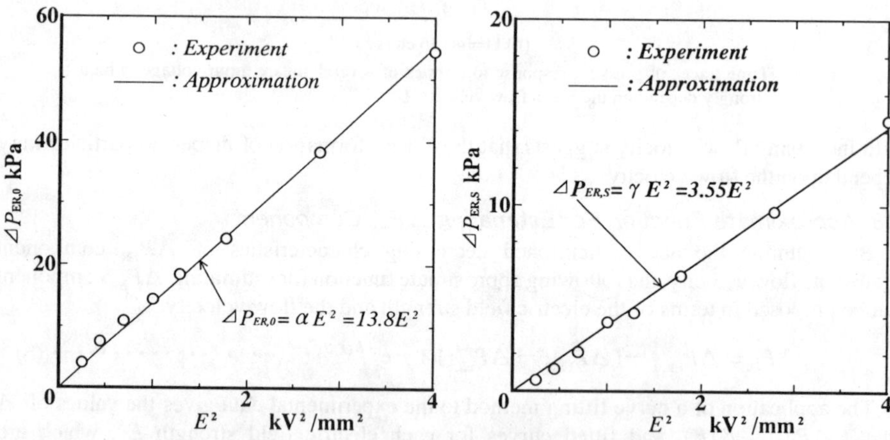

Fig. 5 Relation between maximum pressure drop $\Delta P_{ER,0}$ at U=0 and the square of electric field strength E^2.

Fig. 6 Convergent pressure drop $\Delta P_{ER,S}$ at sufficiently high flow velocity as a function ofthe square of electric field strength E^2.

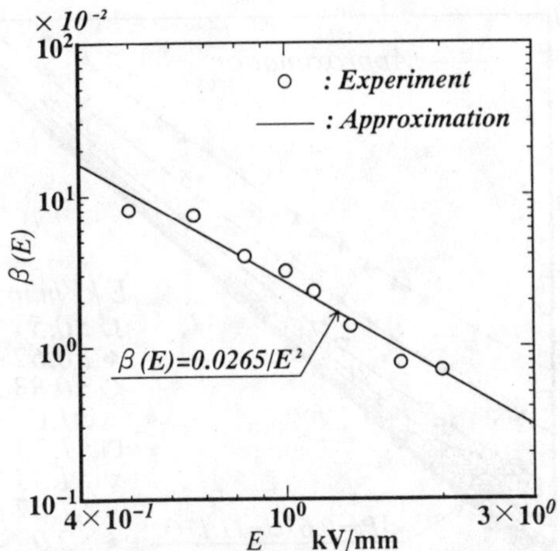

Fig. 7 Exponent β (*E*) as a function of electric field strength *E*.

arranged as a function of the electrical field strength. As shown in Fig. 5, $\Delta P_{ER,0}$ increases in proportion to the square of electric field strength, according to the relationship of $\Delta P_{ER,0} = 13.8\ E^2$. And also, the increase of the convergent pressure $\Delta P_{ER,S}$ has a proportional relationship to the square of the electric field strength; $\Delta P_{ER,S}=3.55\ E^2$, as seen in Fig. 6. Figure 7 shows the relation between the exponent $\beta(E)$ and the electric field strength. If both of them are represented on a log scale, a linear relationship between the logarithms of them are obtained. That is, the exponent $\beta(E)$ is found to vary inversely as the square of the electric field strength, according to the relationship; $\beta(E)=0.0265/E^2$. As a result, the approximate function of ΔP_{ER} component is given as follows;

$$\Delta P_{ER}(E, U)= \alpha E^2-(\alpha E^2-\gamma E^2)(1-e^{-\beta(E)U}) \qquad (9)$$

where, $\alpha=13.8$, $\gamma=3.55$, $\beta(E)=0.0265/E^2$.

3.4 Total Pressure Drop ΔP_T Across ER Valve and Yield Stress τ_{ER}

The total pressure drops ΔP_T across the ER valve , which correspond to the pressure P in the cylinder, are plotted in Fig. 8 in terms of the mean flow velocity U and the electric field strength E . The approximate curves of ΔP_T, which are calculated by adding the electrorheological component ΔP_{ER} of eq. (9) to the viscous component ΔP_0 represented by $0.67U$ experimentally, are drawn in the same figure, and have a fairly good agreement with the experimental results. Thus, it is possible to estimate the pressure drop ΔP_T across the ER valve with a fairly good accuracy.

Since, in this experimental system, the flow velocity U and the pressure drop ΔP_T are related to the piston speed and the force acting on the piston, respectively, Fig. 8 represents the relation between a piston speed and a damping force controlled by ER valve in an ER damper. Therefore, it is expected that a high level of the damping force is caused by ER effects in a range of relatively low piston speed, but as the speed becomes higher the damping force is weaken due to the decrease of ER effects.

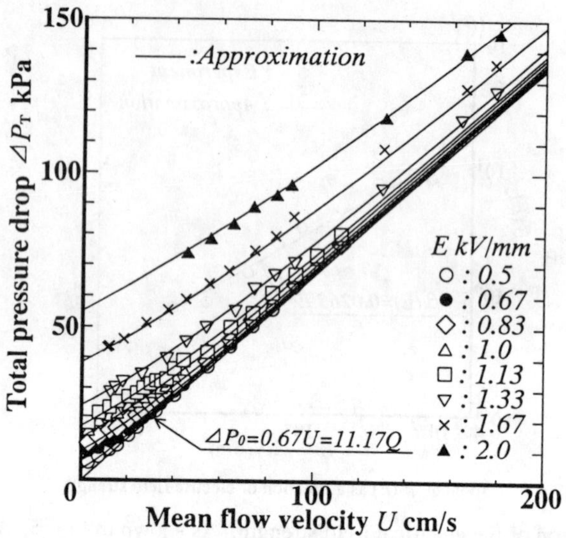

Fig. 8 Variations of total pressure drop ΔP_T with mean flow velocity U and electric field strength E, and the approximate curves represented by eq. (9).

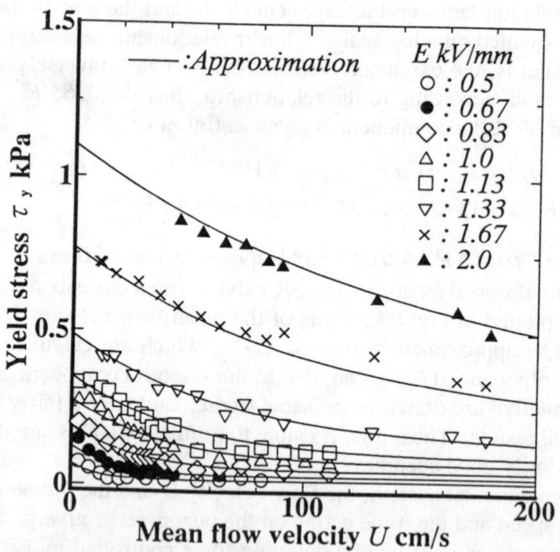

Fig. 9 Yield stress τ_{ER} estimated from the measured ΔP_{ER} against mean flow velocity U and electric field strength E and the approximate curves represented by eq. (10).

The yield stress τ_{ER} can be estimated from the measured electrorheological component ΔP_{ER}, according to the eq. (7). Figure 9 shows the estimated yield stress against the flow velocity in terms of the electric field strength and the approximate curves of τ_{ER} represented as follows;

$$\tau_{ER}(E,U) = \alpha_\tau E^2 - (\alpha_\tau E^2 - \gamma_\tau E^2)(1 - e^{-\beta(E)U}) \quad \cdots\cdots\cdots\cdots (10)$$

where, $\alpha_\tau = 0.28$, $\gamma_\tau = 0.07$, $\beta(E) = 0.0265/E^2$.

When the electric field strength E of 2 kV/mm is applied, the yield stress τ_{ER} caused by ER effects ranges from 0.5 to 0.9 kPa depending on the flow velocity.

In Fig. 10, the changes of the measured total pressure drop ΔP_T with U and E are compared with that of the theoretical one, which is calculated from eq. (4) and (5) using the relation of $\tau_{ER}(E, 0) = \alpha_\tau E^2$. The theoretical ΔP_T increases with increasing the electric field strength E as well as the measured ΔP_T, because of the increase of ΔP_{ER} components. Measured ΔP_{ER} components slightly decreases with increasing flow velocity , but even if the flow velocity is increased, theoretical ΔP_{ER} components are almost constant except a range of very slow flow velocity.

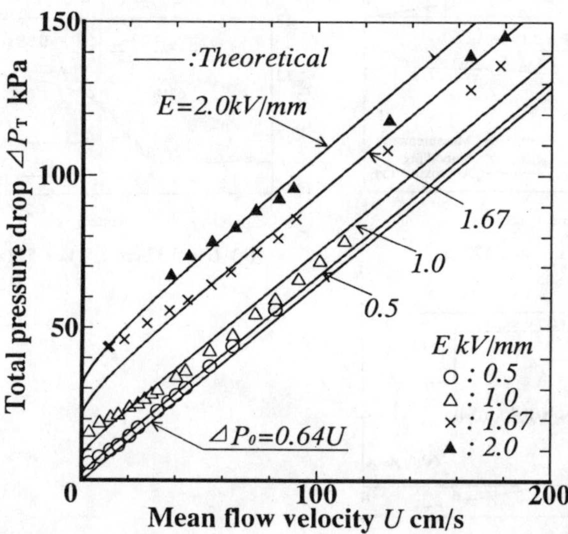

Fig. 10 Comparison of changes of theoretical total pressure drop ΔP_T against mean flow velocity U with that of measured one, in terms of electric field strength E.

4. Transient Pressure Response to Step Input of Electric Field

4.1 Transient Pressure Response

Figure 11 shows transient responses of the pressure in the cylinder against the step input of voltage to the ER valve, which can be approximated by a first order response. In the bottom side of the figure, the approximate first order responses are drawn in the enlarged manner, after smoothing the experimental responses to remove noisy components. The estimated time constant in the case of Fig.11(a) is 0.19 second, which is fairly longer than the usual response time (several millisecond) of ER fluid. When decreasing the flow velocity U as shown in Fig.11(b), the time constant is 0.38, which is longer than the previous one. The decrease of the volume of ER fluid in the cylinder under the same condition of the flow velocity as the case of Fig.11(b), shortens the time constant from 0.38 to 0.21 second as seen in Fig.11(c). Thus, the time constant T depends on both of the flow velocity U at the ER valve and the volume V_c of the ER fluid in the cylinder. Therefore, it is supposed that the transient responses of the pressure in this experimental system result from the compressibility of the ER fluid.

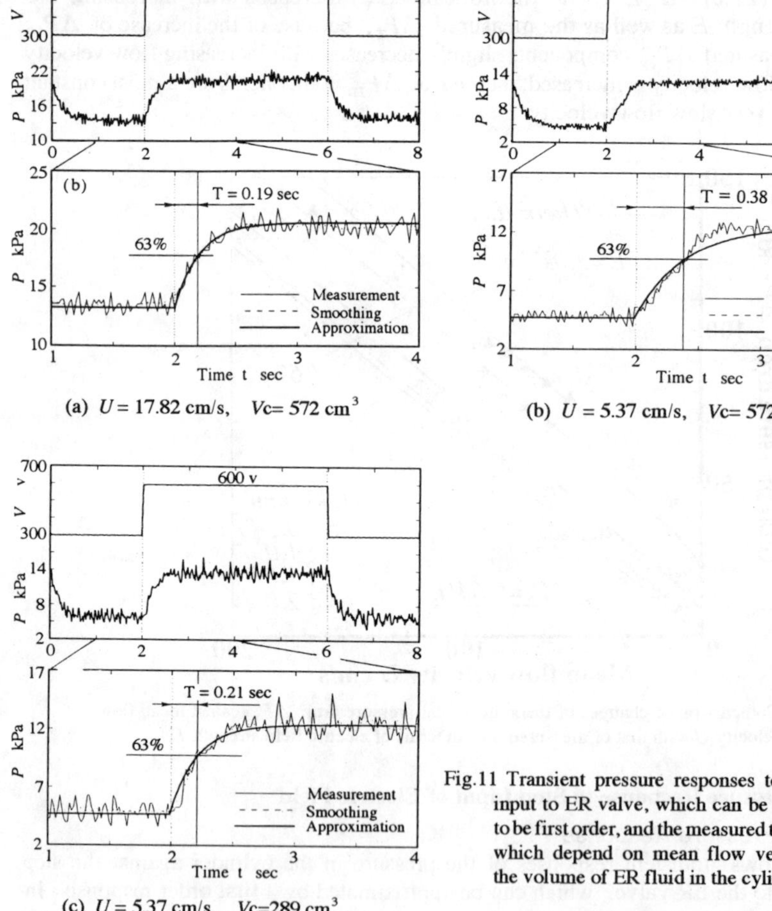

(a) U = 17.82 cm/s, Vc= 572 cm^3

(b) U = 5.37 cm/s, Vc= 572 cm^3

(c) U = 5.37 cm/s, Vc=289 cm^3

Fig.11 Transient pressure responses to step votage input to ER valve, which can be approximated to be first order, and the measured time constants which depend on mean flow velocity U and the volume of ER fluid in the cylinder Vc.

4.2 Theoretical Analysis on Transient Pressure Response

The transient pressure responses are investigated theoretically for the analytical model as shown in Fig.12, assuming the compressibility of ER fluid. The conservation equation is given as eq.(11), and the relation between the flow rate Q and the pressure drop $\Delta P_T = P - P_a$ across the ER valve is given as eq.(12), considering a fluid inertia and a flow resistance within the ER valve channel.

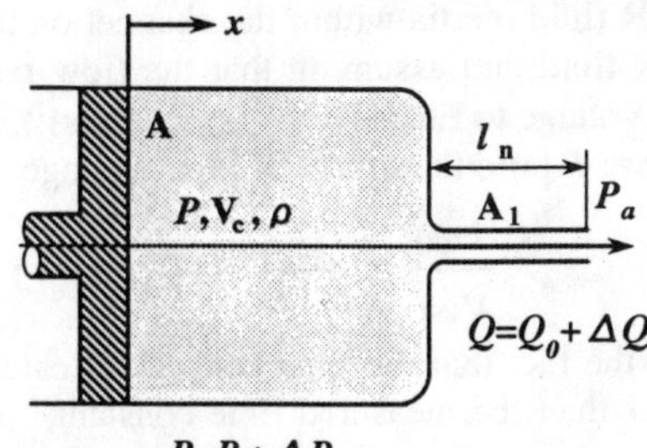

Fig. 12 Analytical model for transient pressure response.

$$\frac{V_c}{K}\frac{dP}{dt} = A\frac{dx}{dt} - Q \qquad (11)$$

$$\frac{\rho l_n}{A_1}\frac{dQ}{dt} + R_n Q = P - P_a \qquad (12)$$

where, K and ρ are the bulk modulus and the density of ER fluid, respectively, A is the area of the piston, and A_1 and R_n are the cross sectional area and the flow resistance of the ER valve channel respectively. Assuming very small fluctuations of the pressure P and the flow rate Q around the equilibrium state denoted by zero subscript , and considering the relation of $A(dx/dt)=Q_0$ in a steady state, transform the eq.(11) into the following equation.

$$\frac{V_c}{K}\frac{d}{dt}\Delta P = -\Delta Q \qquad (13)$$

And also, by considering the small change ΔR_n of the flow resistance caused by ER effects, the equation (12) is transformed as follows;

$$\frac{\rho l_n}{A_1}\frac{d}{dt}\Delta Q + R_{n0}\Delta Q = \Delta P - Q_0 \Delta R_n \qquad (14)$$

where $R_{n0} = (P_0 - P_a)/Q_0$. Eliminating ΔQ from the eq.(13) and (14) derives the following differential equation which represents the response of ΔP to ΔR_n .

$$\frac{\rho l_n}{A_1}\frac{V_c}{K}\frac{d^2}{dt^2}(\Delta P) + \frac{V_c R_{n0}}{K}\frac{d}{dt}(\Delta P) + \Delta P = Q_0 \Delta R_n \qquad (15)$$

And, by applying the Laplace transformation to eq.(15), the transfer function of the pressure P to the resistance R_n of ER valve is written as follows;

$$\frac{P(s)}{R_n(s)} = \frac{Q_0/C}{Ls^2 + R_{n0}s + 1/C} \qquad (16)$$

where, s is the Laplace operator, $L=\rho l_n/A_1$, and $C=V_c/K$. This transfer function is second order, but the measured response of the pressure is approximated to be first order as mentioned in section 4.1. Therefore, it can be considered that there is almost no effect of the ER fluid inertia within the channel on the pressure response. So, neglecting the inertia of ER fluid and assuming that the flow resistance varies in proportional to the square of input voltage to ER valve ($R_n(s) = a \cdot V(s)^2$), the following first order transfer function of the pressure $P(s)$to the square of input voltage $V(s)^2$ is given, whose time constant T is $R_{n0}V_c/K$.

$$\frac{P(s)}{V(s)^2} = \frac{a\,Q_0}{1+T\,s} \qquad \cdots\cdots (17)$$

From the fact that the time constant is calculated to be several millisecond which is fairly shorter than the measured time constant, it can be concluded that the measured pressure responses are independent of the inherent compressibility of ER fluid.

4.3 Transient Pressure Response Due to Compressibility of ER Fluid Containing Air

In this section, the measured transient pressure responses are understood, assuming that the tested ER fluid contains some fine air bubbles. The assumption of isentropic state change of air gives the bulk modulus of air; $K_{air} = \kappa P_0$, where κ is the specific heat ratio of air, and the time constant of the pressure response represented by the following equation.

$$T = R_{no} v_0 / K_{air} \qquad \cdots\cdots (18)$$

where, v_0 is the volume of air at the equilibrium pressure P_0.

Figure 13 shows the comparison of the measured time constants with the theoretical time constants which are estimated from eq.(18) by assuming that the volume fraction of air to the ER fluid is 0.5 %. Changes of the estimated time constants against the cylinder volume V_c and the flow velocity U are almost similar to that of the measured time constants.

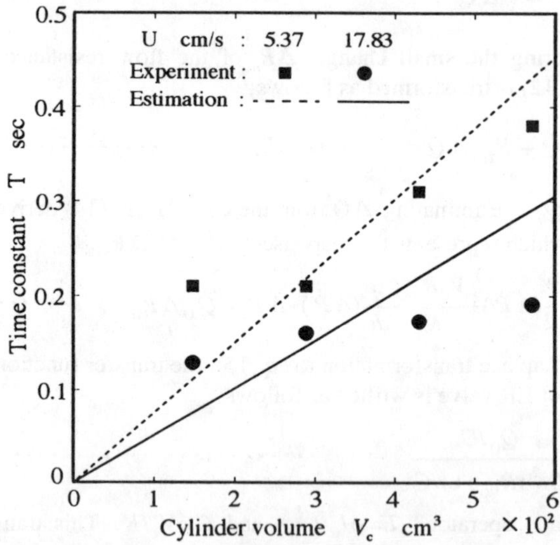

Fig. 13 Comparison of measured time constant T with theoretical time constant estimated from eq. (18), in terms of mean flow velocity U and volume Vc of ER fluid in the cylinder.

Therefore, it can be concluded that the transient response of the pressure in this experimental system mainly depends on the compressibility of ER fluid containing air. Thus, the above observations suggest that the deaeration of ER fluid and the stiffness of dispersed particles must be considered to improve the transient pressure responses of hydraulic ER devices.

5. Conclusions

The piston cylinder-ER valve system, in which the ER valve consists of two parallel electrodes, has been designed to investigated the steady and transient responses of the pressure in the cylinder to an input voltage. From the saturated values and the rising time constants of the step responses of the pressure, which were obtained on the various conditions of the flow velocity in the ER valve channel, we could estimate the characteristics of both steady and dynamic forces acting on the piston controlled by the ER valve, respectively.

The steady pressure drop ΔP_{ER} across the ER valve caused by ER effects increases in proportion to the square of electric field strength E^2, while, decreases exponentially with increasing flow velocity U and then converges to a certain value of the pressure drop at sufficiently high flow velocity which depends on the electric field strength. As a result, ΔP_{ER} was approximated by the following equation:

$$\Delta P_{ER}(E, U) = \alpha E^2 - (\alpha E^2 - \gamma E^2)(1 - e^{-\beta(E)U})$$

The above characteristic of ΔP_{ER} in terms of the flow velocity means that, in the case of an ER damper, the damping performance worsens as the flow velocity is increased.

The measured transient step responses of the pressure could be approximated to be first order. The measured rising time constants of the responses were fairly longer than the usual response time of ER fluid, and changed with the flow velocity and the ER fluid volume in the cylinder. The worsening of the transient characteristics is due to the compressibility of air containing in the ER fluid, and the pressure response in this system can be estimated from the dynamic analytical model which is led by considering the compressibility of ER fluid containing air.

The above observations obtained in this research are important for the design and the engineering application of ER devices.

6. References

1. D. L. Klass and T. W. Martinek, *J. Appl. Phys.*, **38** (1967) 67.
2. H. Uejima, *Jpn. J. Appl. Phys.*, **11** (1972)319.
3. N. Sugimoto, *Bull. of JSME*, **20** (1977)1476.
4. H. Block and J. P. Kelly, *J. Phys. D : Appl. Phys.*, **21** (1988)1661.
5. A. R. Johnson, *Proc. 3rd Conf. on ER Fluids, ed. R. Tao* (World Scientific Publishing, Singapore, 1992) 424.
6. J. E. Bures and J. D. Carlson, *Proc. 2nd Conf. on ER Fluids* (Technomic Publishing, Lancaster, 1990) 93.
7. R. Stanway, J. L. Sproston, and N. G. Stevens, *J. of Electrostatics*, **20** (1987) 167.
8. N. K. Petek, *SAE Paper* No. **920275**(1992)
9. Z. Lou, R. D. Ervin, and F. E. Filisko, *Proc. 3rd Conf. on ER Fluids, ed. R. Tao* (World Scientific Publishing, Singapore, 1992) 453.
10. M. Whittle, R. Firoozian, D. J. Peel, and W. A. Bullough, *ibid.* , 343.
11. R. W. Philips, *Ph. D Thesis, Univ. of California, Berkley*, (1969)

IV. APPLICATIONS

ELECTRO-RHEOLOGICAL CATCH/CLUTCH : INERTIAL SIMULATIONS

A. R. Johnson, J. Makin, W.A. Bullough

Department of Mechanical and Process Engineering
University of Sheffield
Mappin Street
Sheffield S1 3JD
England

ABSTRACT

A high-speed reciprocating mechanism is described in order to provide the basis of a general investigation into the required properties of electro-rheological fluids and associated materials for use in flexible, inertial mechanisms. The dynamic model of this, when run for realistic existing machine requirements clearly illustrates the need for a fully integrated approach to high speed machinery design. The work sets quantified targets and draws attention to the need for the continuing development of improved electro-rheological fluids which will have high yield stresses with acceptable viscosities and the conditions they must operate under: high shear rates, centrifugal loadings and accelerations.

1. Introduction

Industrial organisations are constantly seeking to improve the quality and reduce the cost of their output. This results in two, often conflicting, requirements for new processes and machines. Firstly, the speed of operation should be uprated, which increases the throughput. Secondly, the flexibility of the machine should be improved, which reduces the time spent in changing the machine to produce a modified or different output. Improved operational speeds can sometimes be achieved through the use of purpose-made machines, often using mechanical mechanisms such as linkages, gears and cams. However, these tend to be inflexible and very long delays can occur whilst the machine is re-configured in order to modify or change the output. On the other hand, electro-magnetic machines using devices such as controllable motors, solenoids and other electro-magnetic devices tend to be more flexible, thus reducing change-over times. However, they do tend to have significant inertias and also the fairly long time constants associated with electro-magnetic devices; both of these will limit the speed of operation of the machine. Thus whilst flexibility is improved, the output speed is often limited and is usually less than a purpose-made mechanical machine. This dilemma between output speed and flexibility is a common feature of high speed machine design. Further, it highlights the necessity of adopting a fully mechatronic approach to the design of any high speed machine. That is a full mechatronic model of the design is required to investigate the complex interactions between the mechanical and electrical parameters of the machine, in order to give resolution at high speed.

One technological area which is showing significant promise in the development of high speed machines involves devices which use Electro-Rheological fluids. One Electro-Rheological device, a cylindrical clutch, has demonstrated potential benefits in its speed of operation and controllability[1]. This paper will describe the use of the clutch to produce a high speed reciprocating mechanism. One aspect of a mechatronic model developed to describe its dynamic performance is described and results showing the performance are presented.

2. Reciprocating Mechanism

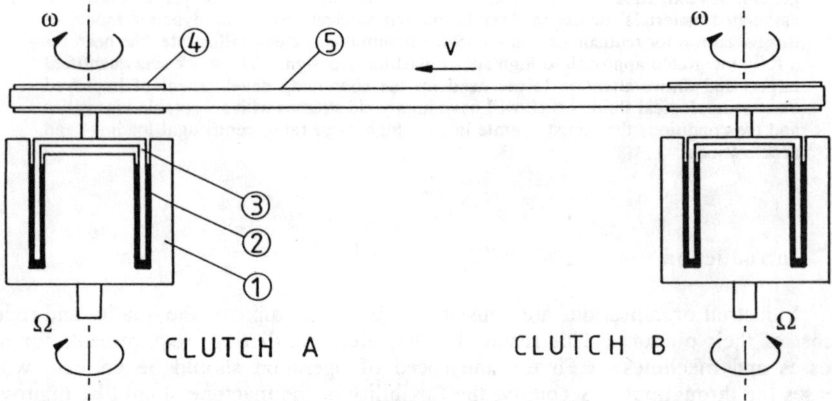

Figure 1. Schematic view of high speed electro-rheological reciprocating mechanism.

A schematic view of the high speed reciprocating mechanism is shown in figure 1. It consists of two vertically mounted cylindrical electro-rheological clutches A and B. The input rotors (1) of each clutch are driven in opposite directions by their own speed controllable electric motors. An important conceptual feature of the electro-rheological clutch used in this arrangement is that the input rotor (1) should run at a constant speed throughout the reciprocating cycle. This is effectively achieved by having the inertia of the input rotors and the electric motors large compared with that of the output rotors (3), pulleys (4) and belt (5). The belt could, for example, carry a positioning arm.

The vertical configuration is chosen in order to eliminate the need for bearings or seals in contact with the electro-rheological fluid (2) so as to minimise the frictional losses on the output rotor (3). The output rotor (3) is constructed using a plastic material with a conducting material, coated onto the high voltage electrode regions; for safety the input rotor (1) is earthed. The output rotor (3) is connected to a pulley (4) which drives the reciprocating belt (5). The materials and constructional details for the output rotor, pulley and belt need to be chosen to keep the inertia of these parts as low as possible. This enables the reciprocating mechanism to respond quickly to changes in the electro-rheological torque applied to the clutches.

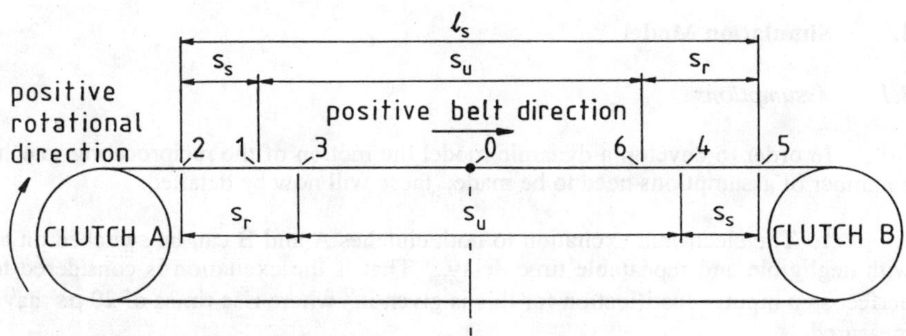

Portion of belt motion	Belt speed	Clutch A		Clutch B	
		Energised	Slipping	Energised	Slipping
0 to 1 - constant speed	$-u$	No	Yes	Yes	No
1 - A on, B off	$-u$	Yes	Yes	No	No
1 to 2 - slowing	$-u \le v \le 0$	Yes	Yes	No	Yes
2 - stopped	0	Yes	Yes	No	Yes
2 to 3 - accelerating	$0 \le v \le u$	Yes	Yes	No	Yes
3 to 4 - constant speed	u	Yes	No	No	Yes
4 - A off, B on	u	No	No	Yes	Yes
4 to 5 - slowing	$u \ge v \ge 0$	No	Yes	Yes	Yes
5 - stopped	0	No	Yes	Yes	Yes
5 to 6 - accelerating	$0 \ge v \ge -u$	No	Yes	Yes	Yes
6 - 0 - constant speed	$-u$	No	Yes	Yes	No

Figure 2. Reciprocating cycle.

The reciprocating cycle is defined in figure 2 and is produced by alternating which clutch, A or B, is energised. Important parameters defining the reciprocating cycle are the constant belt velocity, u; the constant velocity travel distance, s_u; the stopping distance, s_s; the run-up distance, s_r; the travel, l_s; the time at constant velocity, t_u; the stopping time, t_s; the run-up time, t_r; the turn-round time, t_{tr}; and the cycle time, t_c. Relationships between these quantities are given by

$$l_s = s_r + s_u + s_s \qquad t_{tr} = t_s + t_r \qquad t_c = 2(t_r + t_u + t_s) \tag{1}$$

3. Simulation Model

3.1 *Assumptions*

In order to develop a dynamic model the motion of the reciprocating mechanism a number of assumptions need to be made; these will now be detailed.

1. The electronic excitation to both clutches A and B can be switched on and off with negligible and repeatable time delay. That is the excitation is considered to be a perfect step input. Justification for this is given in[2] where rise times of 20 μs have been measured.

2. The electro-rheological torque follows this step electronic excitation change after a short electron-hydraulic time delay, t_m*. Typically this delay between switching on, or off, the electronic excitation and the development of the electro-rheological torque is of the order of 100μs and is repeatable for given conditions of operation. The effect can thus probably be allowed for by initiating the electronic excitation step before the electro-rheological torque is required. Additionally the full electro-rheological torque is assumed to reach its short time value[3] instantaneously, that is with no significant rise-time. This is a difficult value to measure due to the lack of suitably fast response torque transducers. The electro-rheological torque is therefore treated as a pure step function in both the on and off modes.

3. During the electron-hydraulic time delay it is assumed that no change in velocity of the output rotors takes place due to viscous drag.

4. The velocity profile across the electro-rheological fluid is assumed to be linear during the acceleration period, that is during clutch slippage. There is no known basis for this assumption since the acceleration of a plastic constitutive form has not been investigated in the time domain. It does, however, enable the viscous torque on the output rotor to be specified.

5. The effects of fluid inertia are ignored, that is no allowance is taken of the torque needed to accelerate the fluid surrounding rotors during slippage. Again no real

basis exists for this assumption, but some initial assessments suggest the inertias will be negligibly small, for inter electrode fluid gaps of between 0.5 and 1.0 mm.

6. Frictional losses on the output rotors are assumed to be negligible since there are no seals and the rotors are mounted on ball bearings.

7. Isothermal conditions only are considered, but see[4].

8. The electro-rheological shear stress τ_e is assumed to be invarient with the shear rate $\dot{\gamma}$ and the zero volts viscosity μ is assumed constant.

All dimensions, masses and inertias have been calculated from drawings used to produce an experimental rig which is currently undergoing proving trials.

3.2 *Model Development*

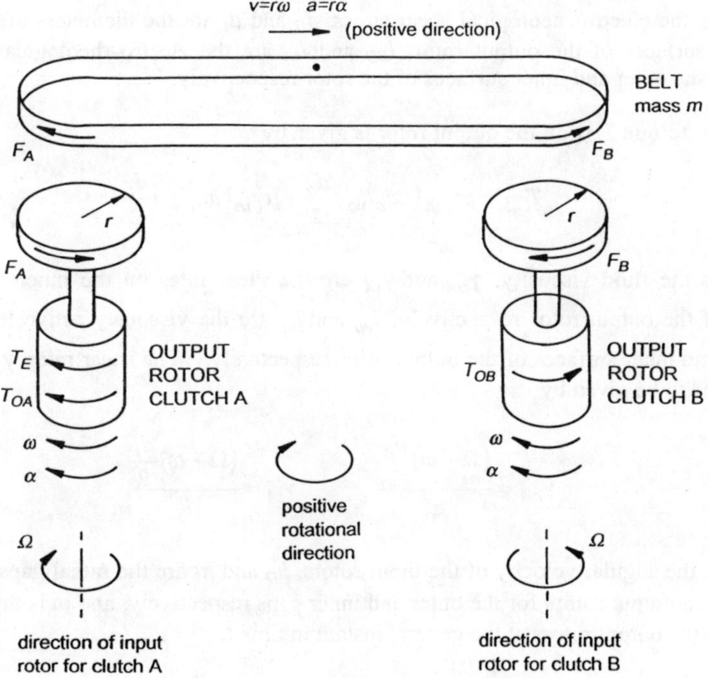

Figure 3. Free body diagrams for reciprocating mechanism (clutch A energised, clutch B not energised).

The free body diagrams for the two clutch output rotors with their associated pulleys, and also for the belt, are shown in figure 3, for a typical position when clutch A is energised and clutch B is not energised. The linear motion of the belt (displacement s, velocity v and acceleration a) are related to the angular motion of the pulleys/output rotors (rotation θ, angular velocity ω, angular acceleration α) by

$$s = r\theta \qquad v = r\omega \qquad a = r\alpha \tag{2}$$

where r is the effective radius of the pulley. Both s and θ are zero at the centre line position 0 in figure 2 when the time t is also zero.

Considering firstly the energised clutch A. The electro-rheological torque T_e on the output rotor is given by

$$T_e = \tau_e(\pi d_o l_{eo})\frac{d_o}{2} + \tau_e(\pi d_i l_{ei})\frac{d_i}{2} \tag{3}$$

where τ_e is the electro-rheological shear stress, d_o and d_i are the diameters of the outer and inner surfaces of the output rotor, l_{eo} and l_{ei} are the electro-rheologically active lengths on the outer and inner surfaces of the rotor respectively.

The viscous torque T_{OA} on the output rotor is given by

$$T_{OA} = \mu\dot{\gamma}_{oA}(\pi d_o l_{vo})\frac{d_o}{2} + \mu\dot{\gamma}_{iA}(\pi d_i l_{vi})\frac{d_i}{2} \tag{4}$$

where μ is the fluid viscosity, $\dot{\gamma}_{oA}$ and $\dot{\gamma}_{iA}$ are the shear rates on the inner and outer surfaces of the output rotor respectively, l_{vo} and l_{vi} are the viscously active lengths on the outer and inner surfaces of the output rotor respectively. The shear rates $\dot{\gamma}_{oA}$ and $\dot{\gamma}_{iA}$ are assumed to be given by

$$\dot{\gamma}_{oA} = \frac{(\Omega - \omega)\frac{d_o}{2}}{h_o} \qquad \dot{\gamma}_{iA} = \frac{(\Omega - \omega)\frac{d_i}{2}}{h_i} \tag{5}$$

where Ω is the angular velocity of the input rotors, h_o and h_i are the radial gaps between the input and output rotors for the outer and inner gaps respectively, and ω is the angular velocity of the output rotors at the general instant in time t.

Hence, putting (5) into (4)

$$T_{oA} = \pi\mu\left(\frac{l_{vo}d_o{}^3}{4h_o} + \frac{l_{vi}d_i{}^3}{4h_i}\right)(\Omega - \omega) \tag{6}$$

Applying Newton's second law to the output rotor A together with its pulley gives

$$T_e + T_{oA} - F_A r = I\alpha \tag{7}$$

where F_A is the force between the pulley and the belt and I is the moment of inertia of the output rotor and pulley assembly.

Now considering the un-energised clutch B. The viscous torque T_{OB} on the output rotor is given by

$$T_{OB} = \mu\dot{\gamma}_{oB}(\pi d_o l_{vo})\frac{d_o}{2} + \mu\dot{\gamma}_{iB}(\pi d_i l_{vi})\frac{d_i}{2} \tag{8}$$

the shear rates $\dot{\gamma}_{oB}$ and $\dot{\gamma}_{iB}$ on the outer and inner surfaces of the output rotor respectively, are given by

$$\dot{\gamma}_{oB} = \frac{(\Omega + \omega)\frac{d_o}{2}}{h_o} \qquad \dot{\gamma}_{iB} = \frac{(\Omega + \omega)\frac{d_i}{2}}{h_i} \tag{9}$$

Hence putting (9) into (8)

$$T_{oB} = \pi\mu\left(\frac{l_{vo}d_o{}^3}{4h_o} + \frac{l_{vi}d_i{}^3}{4h_i}\right)(\Omega + \omega) \tag{10}$$

Applying Newton's second law to the output rotor B, together with its pulley, gives

$$F_B r - T_{oB} = I\alpha \tag{11}$$

where F_B is the force between the pulley and the belt.

Finally, considering the belt and applying Newton's second law, gives

$$F_A - F_B = ma \tag{12}$$

where m is the mass of the belt, and using (2)

$$F_A - F_B = mr\alpha \tag{13}$$

Combining equations (7), (11) and (13) gives

$$T_e + T_{oA} - T_{oB} = \left(2I + mr^2\right)\alpha \tag{14}$$

Combining equations (3), (6), (10) and (14)

$$\pi\tau_e\left(\frac{d_o^2 l_{eo}}{2} + \frac{d_i^2 l_{ei}}{2}\right) - \pi\mu\omega\left(\frac{l_{vo} d_o^3}{2h_o} + \frac{l_{vi} d_i^3}{2h_i}\right) = \left(2I + mr^2\right)\alpha \tag{15}$$

but $\alpha = \dfrac{d\omega}{dt}$ and therefore

$$dt = \frac{\left(2I + mr^2\right)d\omega}{\left[\dfrac{\pi\tau_e}{2}\left(d_o^2 l_{eo} + d_i^2 l_{ei}\right) - \dfrac{\pi\mu\omega}{2}\left(\dfrac{l_{vo} d_o^3}{ho} + \dfrac{l_{vi} d_i^3}{hi}\right)\right]} \tag{16}$$

Equation (16) is the basic differential equation defining the motion of the clutch from the moment that clutch A is energised (point 1 on figure 2) until the instant when no slipping is occurring at clutch A (point 3 on figure 2). An identical equation can be derived to define the motion when clutch B is energised (point 4 to point 6 on figure 2). Between points 6 to 1, and points 3 to 4, the motion is simply constant velocity.

In order to simplify the analysis further, let

$$A = 2I + mr^2 \qquad B = \frac{\pi\tau_e}{2}\left(d_o^2 l_{eo} + d_i^2 l_{ei}\right) \qquad C = \frac{\pi\mu}{2}\left(\frac{l_{vo} d_o^3}{ho} + \frac{l_{vi} d_e^3}{hi}\right) \tag{17}$$

equation (16) then becomes

$$dt = \frac{A}{B - C\omega} d\omega \tag{18}$$

This equation is of standard form which can be interpreted to give the following equations which define the motion of the belt.

(a) **Belt position s, belt velocity v, pulley angular position θ and pulley angular velocity ω at the instant in time t_a after the clutch A is energised.**

$$v = r\omega = r\left[\frac{B}{C} - \frac{(B + C\Omega)}{C} e^{-\frac{C}{A}t_a}\right] \tag{19}$$

$$s = r\theta = \frac{r}{C}\left[Bt - \frac{A}{C}(B + C\Omega)\left(l - e^{-\frac{C}{A}t_a}\right)\right] \tag{20}$$

(b) **Stopping time t_S, stopping distance, s_S**

$$t_s = \frac{A}{C}\ln\left[\frac{B + C\Omega}{B}\right] \tag{21}$$

$$s_s = \frac{r}{C}\left[Bt_s - \frac{A}{C}(B + C\Omega)\left(l - e^{-\frac{C}{A}t_s}\right)\right] \tag{22}$$

(c) **Run-up time t_r, run-up distance, s_r**

$$t_r = \frac{A}{C}\ln\left[\frac{B}{B - C\Omega}\right] \tag{23}$$

$$s_r = \frac{r}{C}\left[Bt_r - \frac{AB}{C}\left(l - e^{-\frac{C}{A}t_r}\right)\right] \tag{24}$$

The complete motion is described using the appropriate combination of equations (1), (2), (16), (19-24). Typical results and their implications are discussed in the following section.

4. Reciprocating Motion

In this section the general features of the motion predicted by the analysis will be detailed. Consideration of the effects of changing geometrical parameters and fluid parameters will then be discussed.

4.1 General features of motion diagrams

Typical values of geometric and fluid properties are given in table 1; these are based on an experimental rig currently undergoing trials. The reciprocating cycle was defined in figure 2 and using the equations developed in section 3, the displacement, velocity and acceleration diagrams shown in figure 4 are derived. The main features are the steady velocity portions (points 6 to 0 to 1 and points 3 to 4) and the turn-round periods (points 1 to 3 and points 4 to 6). The analysis outlined in section 3 was for when clutch A was energised, that is during the portion of the cycle 1 to 4, the remainder of the cycle 4 to 0 to 1, is when clutch B is energised. Similar equations can be developed for this portion of the cycle, or the inherent similarity of the turn round periods can be used to generate the full cycle shown in figure 4.

TABLE 1 Typical values of geometric and fluid properties

<u>inner surface of output rotor</u>
- diameter $d_i = 47.0$ mm
- inter electrode gap $h_i = 0.5$ mm
- electro-rheologically active length $l_{ei} = 30.0$ mm
- viscously active length $l_{vi} = 30.0$ mm

<u>outer surface of output rotor</u>
- diameter $d_o = 51.0$ mm
- inter electrode gap $h_o = 0.5$ mm
- electro-rheologically active length $l_{eo} = 30.0$ mm
- viscously active length $l_{vo} = 40.0$ mm

effective radius of pulley $r = 39.55$ mm
constant speed belt velocity $v = 5$ m/s
input rotor angular velocity $\Omega = 1207$ rpm
moment of inertia of one output rotor and pulley assembly $I = 3.83 \times 10^{-5}$ kg m^2
mass of belt $m = 16$ grams
electro-rheological shear stress of fluid $\tau_e = 10$ kPa
fluid viscosity $\mu = 100$ m Pa.s

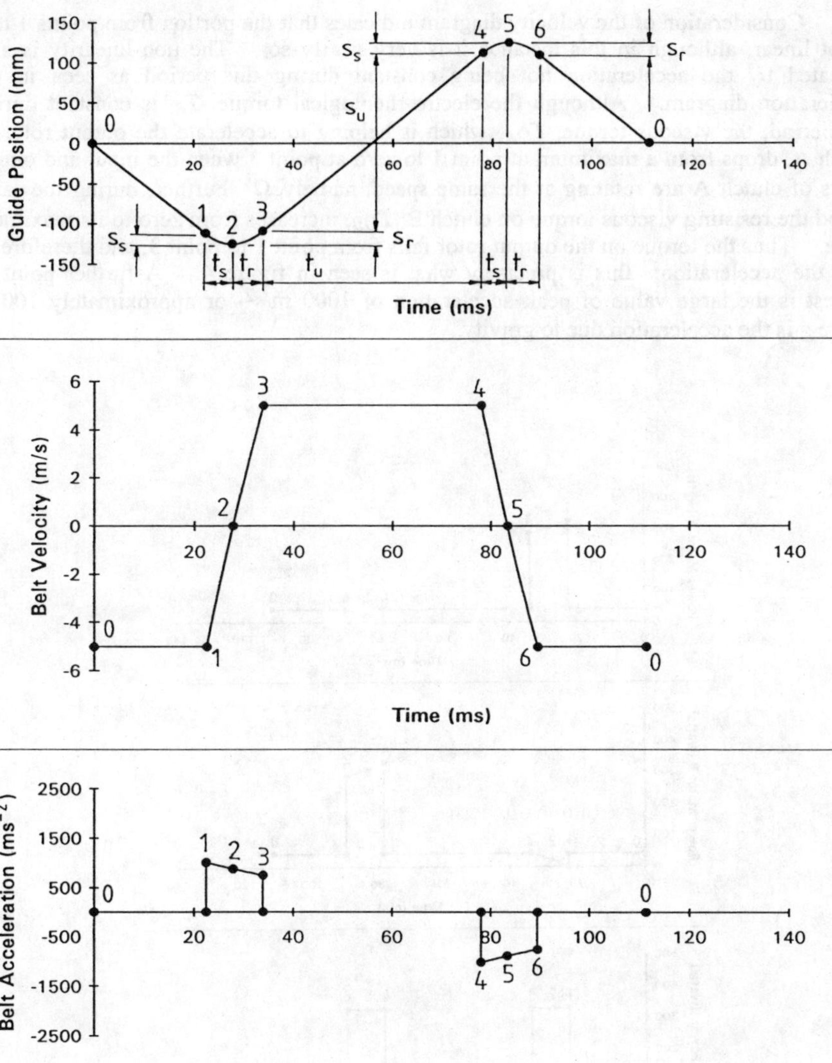

Figure 4. Typical motion diagrams for one complete reciprocating cycle (parameters and properties defined in table 1).

Consideration of the velocity diagram indicates that the portion from points 1 to 3 is not linear, although in this instance it is very nearly so. The non-linearity is also indicated by the acceleration not being constant during this period as seen in the acceleration diagram. Although the electro-rheological torque, T_e, is constant during this period, the viscous torque, T_{OA}, which is helping to accelerate the output rotor of clutch A, drops from a maximum at point 1 to zero at point 3 when the input and output rotors of clutch A are rotating at the same speed, namely Ω. Further, during the same period the resisting viscous torque on clutch B, T_{OB}, increases from zero to its maximum value. Thus the torque on the output rotor falls from point 1 to point 3, and therefore so does the acceleration; this is precisely what is seen in figure 4. A further point of interest is the large value of peak acceleration of 1000 m/s^2, or approximately 100 g, where g is the acceleration due to gravity.

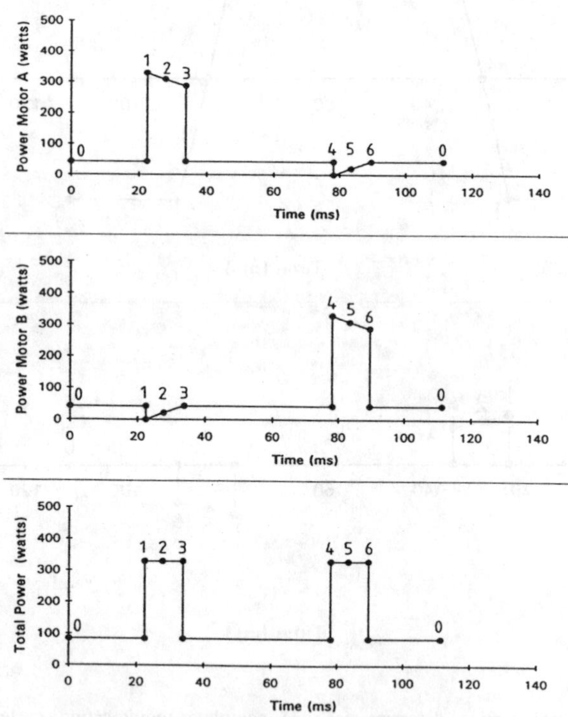

Figure 5. Power inputs for a typical reciprocating cycle (parameters and properties defined in table 1).

The power input to the reciprocating mechanism is important for two main reasons. Firstly sizing the mechanism, particularly the motors, and secondly in predicting the energy dissipated in the fluid which will result in temperature rises in the fluid. The power input to a clutch at any instant is found by multiplying the torque on the clutch rotor by the angular speed of the motor. Results are shown in figure 5 for the motors on clutch A, clutch B and the total power input to the mechanism. The main feature to emerge is the relatively large power needed during the "turn-round periods" compared with the "steady velocity" periods[5]. The former is needed to provide the large accelerations during the turn-round period, whereas the latter is required to overcome viscous losses whilst maintaining the velocity constant. Both these eventually result in a rise of internal energy of the fluid, and hence increased temperature.

4.2 *Effects of changing geometrical parameters*

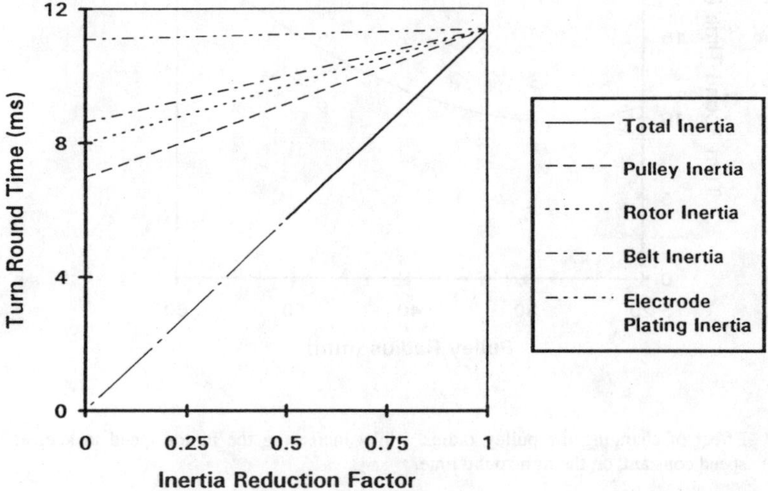

Figure 6. Effect of reducing inertias on the turn-round time.

Any changes to the geometric parameters will affect the torques, masses, moments of inertia and hence the turn-round speeds of the mechanism. One way in which to increase the speed of response of a mechanism is by reducing its inertia, either by using a lighter material or through detailed dimensional changes. The effects on the turn-round time (t_{tr}) of reducing the inertias of various parts of the mechanism is shown in figure 6. An inertia reduction factor of 1 is no change from the values given in table 1, a value of 0.5 is half the value, and a value of 0 is the utopian value of zero inertia.

Two important features emerge from figure 6, firstly, as would be expected, reducing the total inertia significantly reduces the turn-round time. Secondly, and perhaps more importantly, figure 6 gives an indication of what features are worthy of effort in reducing the inertia. For example, the electrode plating inertia even when reduced to zero, from the present level produces a negligible decrease in turn-round time.

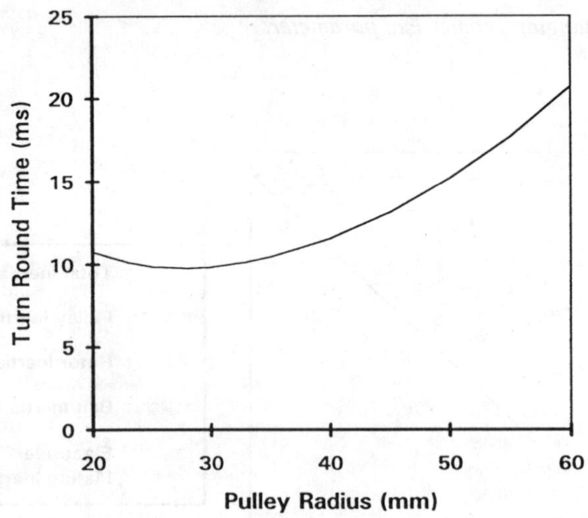

Figure 7. Effect of changing the pulley radius, whilst increasing the motor speed to keep the belt speed constant, on the turn-round time.

A major requirement of the traverse mechanism is that the belt should have a constant velocity, in this case 5 m/s. This requirement determines the relative values of the pulley radius (r) and the motor speed (Ω), the product $r\Omega$ giving the belt speed. Figure 7 shows the effect of varying the pulley radius whilst changing the motor speed to keep the belt speed constant, all other parameters remain unchanged. There is a minimum value of 28 mm. Below this gains in reducing the pulley's inertia are offset by the higher rotational speed of the motor which increases the viscous drag and the speed change of the rotor during the turn-round period. Above 28 mm the reverse situation applies.

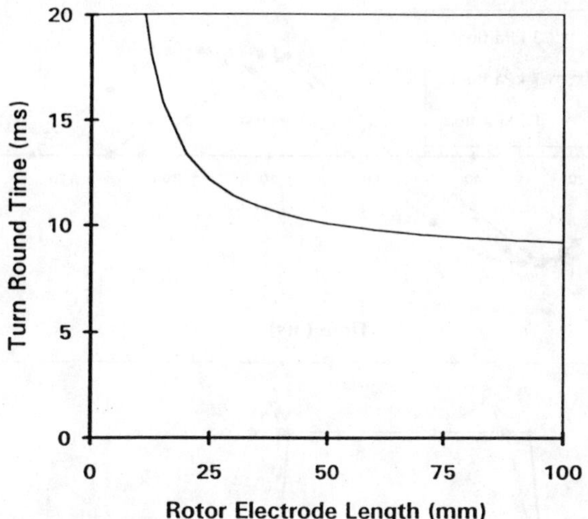

Figure 8. Effect of changing the length of the rotor, whilst keeping the electro-rheological torque constant by altering the diameters, on the turn-round time.

For a constant electro-rheological shear stress fluid, the electro-rheological torque depends on the area and diameter of the rotor. Figure 8 shows the effect of reducing the output rotor diameter whilst increasing the length to keep the electro-rheological torque constant, all other parameters are kept identical. The improvement approaches an asymptotic limit of a very long rotor with a very small diameter, that is when the inertia approaches zero. Clearly such a rotor would be impractical, nonetheless figure 8 does indicate that, for this example, whilst significant gains are made at low rotor lengths little is to be gained for rotors in excess of approximately 40 mm in length.

4.3 *Effects of changing fluid parameters*

The preceding section indicated that reductions in inertias from those of the prototype rig, given in table 1, would be advantageous in reducing the turn-round time. In view of this realistic reductions in the inertia will be included for the study of the effects of changing the fluid parameters detailed in this section. The two changes made are given in table 2, all the other parameters are left as given in table 1.

TABLE 2 Reduced values of inertia and mass for the fluid parameter study.

Pulley inertia reduced by 50% - new inertia of the rotor and pulley assembly is $I = 2.83 \times 10^{-5}$ kg m^2
Mass of the belt reduced to $m = 7$ grams.

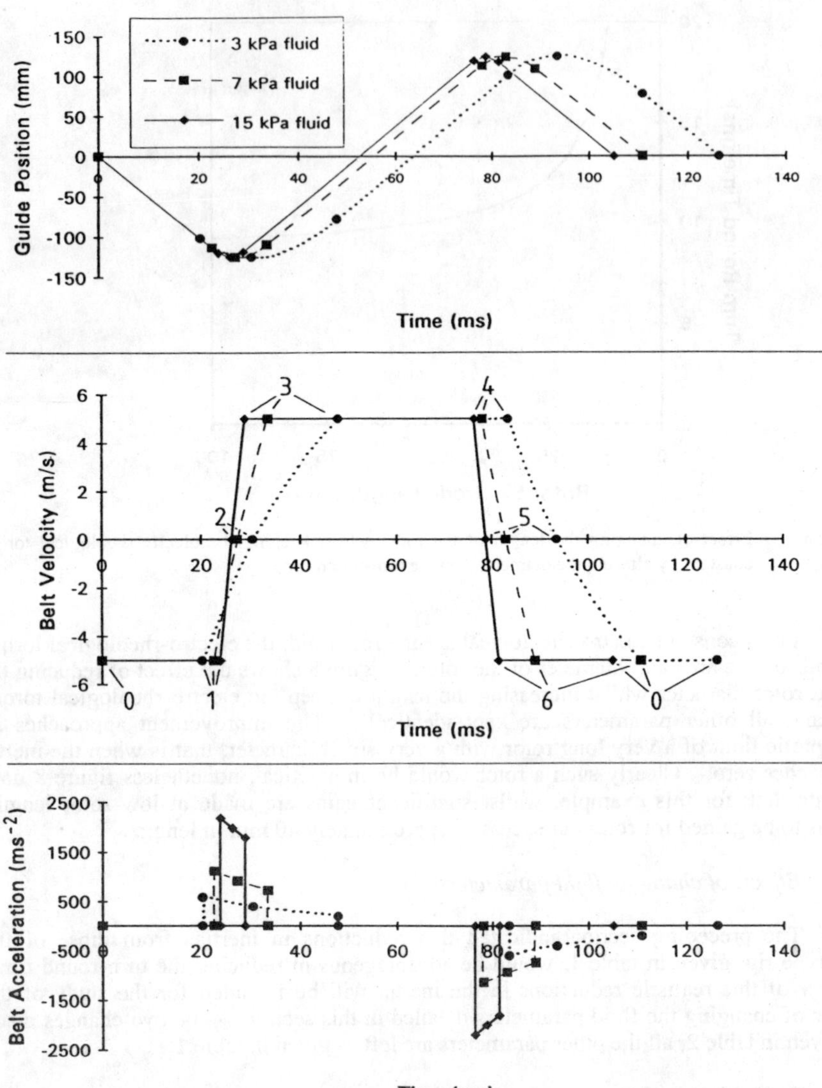

Figure 9. Motion diagrams for different electro-rheological shear stress fluids (parameters defined by table 1 and modified by table 2).

The motion diagrams for three different electro-rheological shear stresses, $\tau_e = 3$, 7 and 15 kPa, are shown in figure 9; each case has the travel length l_s kept constant at 250 mm. The important parameters determining these diagrams are given in table 3. A number of interesting improvements emerge from figure 9 and table 3. Firstly as the shear stress increases from $\tau_e = 3$ to 15 kPa, the cycle time reduces from $t_c = 126.3$ ms to 105 ms, whilst at the same time the length at constant velocity increases from $s_u = 178.1$ mm to 237.4 mm. Thus the useful constant velocity stroke length is significantly increased whilst the overall cycle time is reduced. Secondly the velocity changes virtually linearly during the turn-round period 1 - 3 at the higher shear stresses. On the converse side, the acceleration rises to a peak of 2184.6 m/s or 222.7 g at $\tau_e = 15$ kPa compared to 593.4 m/s^2 or 60.5 g at $\tau_e = 3$ kPa. Further, the corresponding peak power rises to 472.1 W from 128.3 W with its implications for motor size. The lower power required over the constant velocity portions of the cycles remains the same at 84.6 W. However, since the higher power during turn-round acts over a shorter period, the overall energy input to the fluid per cycle is similar, and so the heating of the fluid is not affected to a great degree.

TABLE 3 Parameters defining the traversing cycles shown in figure 9.

	Electro-Rheological Yield Stress τ_e kPa		
	3	7	15
stopping time t_s ms	10.2	4.89	2.40
run-up time t_r ms	17.3	6.05	2.64
turn round time t_{tr} ms	27.5	10.9	5.04
constant velocity time t_v ms	35.6	44.5	47.5
cycle time t_c ms	126.3	110.9	105.0
stopping distance s_s mm	23.9	11.8	5.90
run-up distance s_r mm	48.1	15.7	6.73
turn-round distance s_{tr} mm	72.0	27.6	12.6
constant velocity distance s_u mm	178.1	222.5	237.4
peak belt acceleration a_b m/s^2	593.4	1123.8	2184.6
peak total power P_T W	128.3	242.9	472.1
constant velocity power P_u W	84.6	84.6	84.6

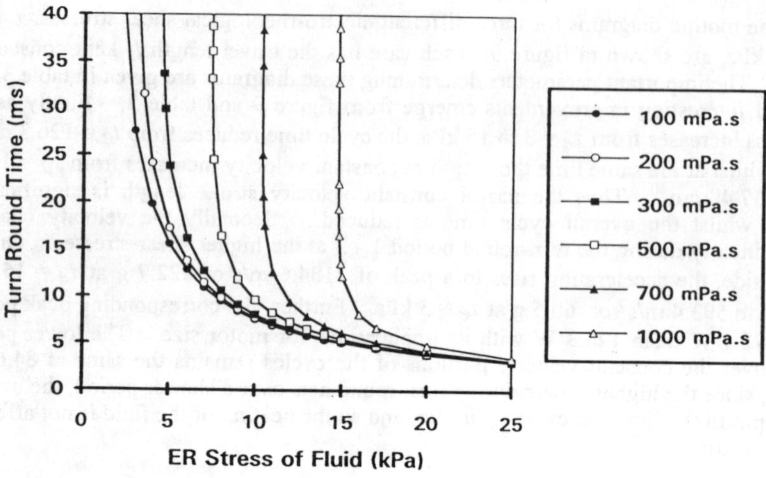

Figure 10. Effect of electro-rheological shear stress and fluid viscosity on the turn-round time.

The turn-round time, t_{tr}, gives a good indication of the usefulness of the cycle, and its variation with electro-rheological shear stress τ_e for a variety of fluid viscosities (μ) is given in figure 10. For high electro-rheological yield stress fluids the effects of viscosity on run-up time are small, whereas for lower yield stress fluids the effect is significant. Indeed, for lower electro-rheological yield stress fluids the higher viscosity fluids will never turn-round as indicated by the near vertical curves. This feature can be understood if the motion after clutch A is energised is considered in conjunction with the $\tau_e = 3$ kPa velocity and acceleration curves shown in figure 9. During the acceleration period (point 2 to 3) the electro-rheological driving torque has to overcome the viscous drag on clutch B. As clutch A approaches the motor speed (i.e. the slippage is very small) then the relative slippage at B approaches the maximum. If the viscosity is too high, then the viscous torque will reach the value of the electro-rheological torque being applied to clutch A and therefore the rotor will cease to accelerate. On figure 9 this reduced acceleration is clearly seen and results in the distinct curve in the velocity motion diagram as the belt speed approaches the steady state belt speed of 5 m/s. However, for the case shown in figure 9, the mechanism still turns round. Mathematically this feature can be seen in the run up time equation (23). The integral is only valid when $B>C\Omega$, that is when the electro-rheological torque T_e is greater than the peak viscous torque. For the case studied this can be further illustrated by plotting the electro-rheological shear stress against the fluid viscosity as shown in figure 11, which shows regions where the mechanism will or will not run-up.

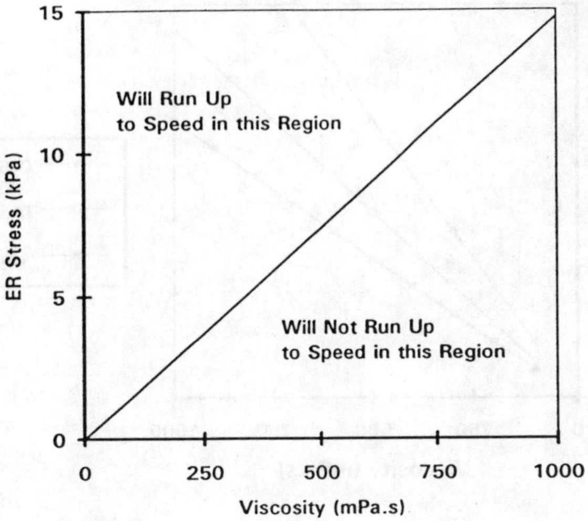

Figure 11. Regions for which the mechanism will, or will not, run-up for a belt speed of 5 m/s.

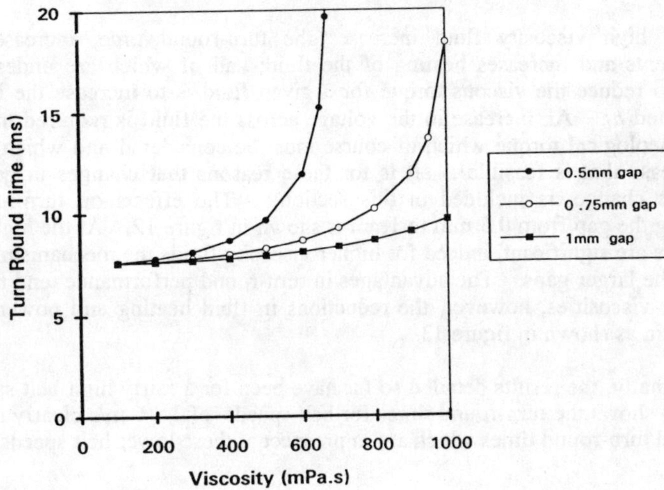

Figure 12. Effect of changing the inter electrode gaps on the turn round times ($h_o = h_i$).

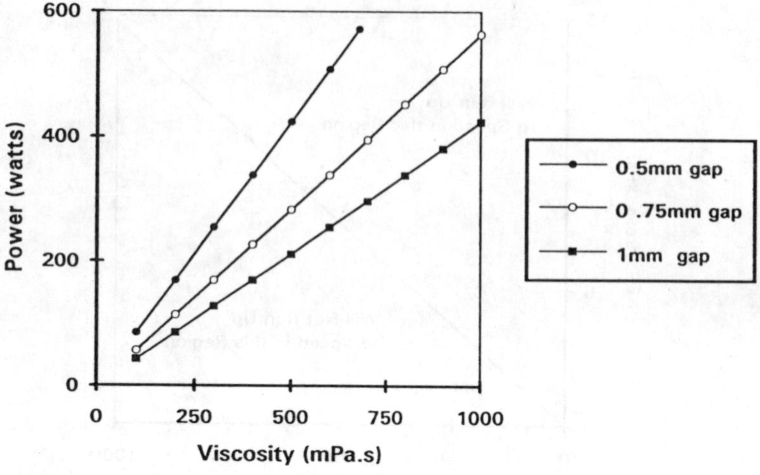

Figure 13. Effects of changing the inter electrode gaps on the power consumed during the constant velocity portions of the reciprocating cycle ($h_o = h_i$).

A high viscosity fluid increases the turn-round time, increases the power requirements and increases heating of the fluid; all of which are undesirable. One method to reduce the viscous torque for a given fluid is to increase the inter-electrode gaps h_o and h_i. An increase in the voltage across the fluid is required to maintain the electro-rheological torque which of course must be considered and which will limit the gap increase that is feasible[1]. It is for these reasons that changes in gaps, a purely geometric change, is included in this section. The effects on turn-round time, of increasing the gap from 0.5 mm to 1mm is shown in figure 12. At the higher viscosities the effects are significant, indeed for higher viscosity fluids the mechanism may only run up with the larger gaps. The advantages in turn-round performance tend to disappear at the lower viscosities, however, the reductions in fluid heating and power consumption will remain as shown in figure 13.

Finally, the results detailed so far have been for a fairly high belt speed of 5 m/s. Figure 14 shows the turn-round times for belt speeds of 1 - 6 m/s clearly illustrating the very rapid turn-round times which are in prospect at these lower belt speeds.

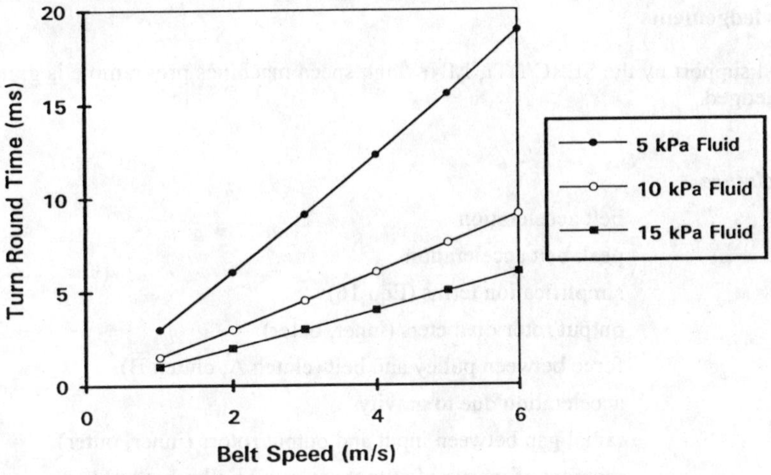

Figure 14. Effect of different constant belt velocities on the turn-round time.

Conclusions

A high speed electro-rheological reciprocating mechanism has been described and an ideal dynamic model describing the motion developed. Considering the effect that changing the variables has on the system performance, the need for a fully integrated mechatronic approach to high speed machinery design bringing together dynamic, heating and electrical design aspects is substantiated. Additionally, electrical supply, control of cooling, constructional materials and more accurate fluid properties need to be considered. The limiting upper viscosity may well be governed by the heating of the fluid[4] rather than by dynamic considerations. The effect of temperature on viscosity is likely to be a very important parameter, a rise in temperature will reduce viscosity but increase conductance. The requirements stated in this paper are only on the basis of ideal dynamical considerations.

The need for continuing fluid development has also been highlighted. The provision of high speed mechanisms such as the one described will require high yield stress fluids with acceptable viscosities. Further they must operate and be durable at the high shear rates, centrifugal loadings and accelerations illustrated by the mechanism described.

Acknowledgements

Financial support by the SERC/DTI, LINK/high speed machines programme is gratefully acknowledged.

Nomenclature

a	belt acceleration
a_b	peak belt acceleration
A,B,C	simplification terms (Equ 16)
d_i, d_o	output rotor diameters (inner, outer)
F_A, F_B	force between pulley and belt (clutch A, clutch B)
g	acceleration due to gravity
h_i , h_o	radial gap between input and output rotors (inner, outer)
I	moment of inertia of output rotor and pulley assembly
l_{ei}, l_{eo}	electro-rheologically active length on output rotor (inner, outer)
l_{vi}, l_{vo}	viscously active length on output rotor (inner, outer)
l_s	travel length
m	mass of belt
PT	peak total power
Pu	constant velocity power
r	effective pulley radius
s	distance of belt travel (zero at centre line position 0)
s_r	run-up distance
s_s	stopping distance
s_u	distance at constant velocity
t	time (zero at centre line position 0)
t_a	time after clutch A is energised at position 1
t_c	cycle time
t_r	run-up time
t_s	stopping time
t_{tr}	turn-round time
t_m*	electron-hydraulic time delay
t_u	time at constant velocity
T_e	electro-rheological torque

T_{OA}, T_{OB}	viscous torque (clutch A, clutch B)
u	constant belt velocity
v	belt velocity
α	angular acceleration of output rotor
$\dot{\gamma}_{oA}$, $\dot{\gamma}_{iA}$	fluid shear rate, rotor A (inner, outer)
$\dot{\gamma}_{oB}$, $\dot{\gamma}_{iB}$	fluid shear rate, rotor B (inner, outer)
θ	angular position of output rotor (zero at centre line position 0)
μ	fluid viscosity at zero volts
τ_e	electro-rheological shear stress
ω	angular velocity of output rotor
Ω	angular velocity of input rotor (constant)

References

1. W A Bullough, A R Johnson, A Hosseini-Sianaki, J Makin and R Firoozian, 'The Electro-rheological Catch: Design, Performance Characteristics and Operation. Proc. Inst. Mech. Engrs., Part 1: Journal of Systems and Control Engineering, Vol 207, pp 87-95, 1993.

2. R Tozer, C T Orrell and W A Bullough, 'On-Off Excitation Switch for ER Devices'. Proceedings of 4th Int. Conf. on Electro-Rheological Fluids, Feldkirch, Austria, July 20-23, 1993.

3. J Makin, W A Bullough, A R Johnson and A Hosseini-Sianaki, 'The Effect of Solid Fraction Concentration on the Performance of an ERF in the Shear Mode. Proceedings of 4th Int. Conf. on Electro-Rheological Fluids, Feldkirch, Austria, July 20-23, 1993.

4. R Smyth, K H Tan and W A Bullough, 'Heat Transfer Modelling of an ER Catch. Proceedings of 4th Int. Conf. on Electro-Rheological Fluids, Feldkirch, Austria, July 20-23, 1993.

5. C T Rogers, 'Intelligent Material Systems - The dawn of a New Materials Age', Jnl. of Intelligent Materials, Systems and Structures. Editorial p. 4-12, Vol 4, No. 1, 1993.

6. A Hosseini-Scanahi, R Tozer, M Whittle and W A Bullough, 'Experimental Investigation into the Electrical Modelling of Electro-Rheological Fluids in Shear Mode', submitted to Inst. Elect. Engrs. for publication.

DIELECTROPHORETIC ASSEMBLY: A NOVEL CONCEPT IN ADVANCED COMPOSITE FABRICATION

C.A. RANDALL, C.P. BOWEN, T.R. SHROUT,
G.L. MESSING AND R.E. NEWNHAM
Particulate Material Center
and Center for Dielectric Studies
The Pennsylvania State University
University Park, PA 16802, U.S.A.

ABSTRACT

Particle fillers can be assembled in an uncured polymer using the mutual dielectrophoretic effect. This chained microstructure can be "frozen in" by curing the polymer matrix to form advanced composites. Assembly alignment conditions are dependent on both electric field strength and frequency. The optimum electric field frequency for dielectrophoretic assembly may be determined via in-situ optical microscopy and electrorheological measurements. The potential of the dielectrophoretic assembly process is to intelligently fabricate a number of ceramic-polymer composites for various electronic applications such as micron-scale devices for packaging.

Introduction

Composites are multiphase materials whose components are mechanically separable, and which exhibit properties unattainable in any of the constituent phases. The properties of a composite material are controlled through the material selection, volume fraction of these materials, connectivity, percolation behavior and anisotropy. For the typical case of a polymer matrix and ceramic filler composite, the dielectrophoretic assembly technique can induce unidirectional alignment or chaining of the filler phase within the uncured polymer.[1-3]

The spatial redistribution of filler and the associated changes in physical properties are due to a connectivity transformation. Connectivity is essentially a means to describe the interconnection of the component phases in a composite material. Newnham developed a self-consisted nomenclature to describe the connectivity of composite materials.[4] For a given number of components (n) in a composite, there is a finite set of connectivity patterns given by the expression:

$$\frac{(n+3)!}{3!\,n!} \tag{1}$$

In the case of a diphasic composite material (n=2), there exists ten possible connectivity patterns. The first term in the notation refers to the dimensional interconnection of the filler phase, while the second term refers to the matrix phase dimensional interconnection. In the 0-3 connectivity case, Figure 1a, a filler phase is dispersed or self-connected and has, therefore, zero dimensional interconnection. The matrix phase is continuous in all three directions and, therefore, has 3-dimensional interconnection. This differs from 1-3 connectivity, Figure 1b, where the active filler phase is continuous in one direction, as is the case of parallel rods in a polymer matrix. With the dielectrophoretic assembly process,

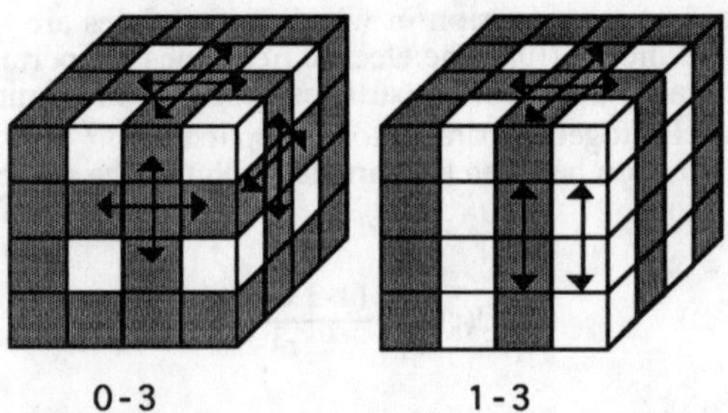

Figure 1. Schematic connectivity designs. In this diagram, the filler is represented by the colorless regions and the matrix is represented by the shaded regions. As can be seen, the 0-3 design consists of isolated particles in a continuous matrix. The 1-3 design shows the filler connected in one dimension in the continuous matrix.

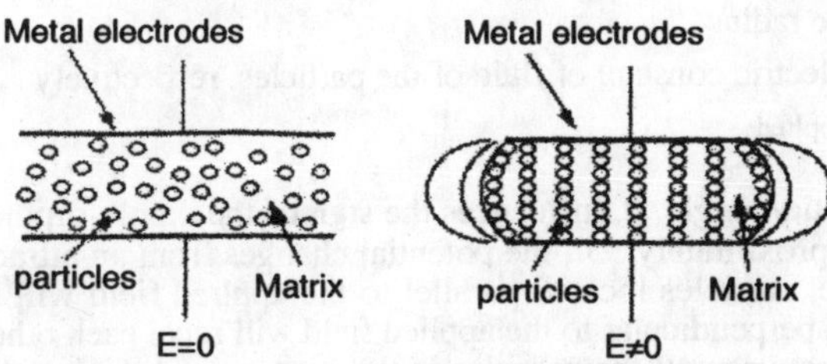

Figure 2. Schematic showing the unidirectional agglomeration phenomena. At zero applied field, the particles stay in a dispersed state. Upon the application of an external electric field, dipole-dipole interaction causes particle chains parallel to the applied field direction.

we have the opportunity of developing a 1-3 connectivity from a disperse 0-3 connectivity case (Figure 2).

The redistribution of particles under dielectrophoretic assembly can strongly influence the relative physical properties.[2,5,6] This occurs because of the relative change in the series and parallel property mixing in the composite. The result is an anisotropy of physical properties such as dielectric constant, resistivity, elastic stiffness, etc., within the composite.

Dielectrophoresis

In general, when an electric field is applied to a suspension of particles, two phenomena can occur; namely, electrophoresis and dielectrophoresis. Electrophoresis is defined as the translational motion of charged particles within a stationary fluid under an electric field. Dielectrophoresis is defined as the translational motion of neutral matter in a non-uniform electric field. This motion is caused by an induced polarization which interacts with a non-uniform field such that the particle migrates to the region of higher field intensity. In the case of a suspension in which the particles are of a higher dielectric constant than the surrounding fluid, the electric field lines are perturbed such that a non-uniform field is generated between the particles, resulting in a mutual dielectrophoretic effect that draws particles together parallel to the applied field.[7] This process is equivalent to a dipole-dipole interaction between the particles such that the attractive potential is given by the expression:

$$U(r) = \upsilon \frac{(1-3\cos^2\theta)}{r^3}$$

$$\text{where: } \upsilon = \frac{(\beta a^3 \varepsilon_f E_{loc})^2}{\varepsilon_f} \text{ and } \beta = \frac{\varepsilon_p - \varepsilon_f}{\varepsilon_p + 2\varepsilon_f} \qquad (2)$$

θ - orientation angle of particles relative to the applied electric field
r - interparticle distance
β - effective polarizability
a - particle radius
$\varepsilon_f, \varepsilon_p$ - dielectric constant of fluid of the particles, respectively
$E_{loc} \simeq E_{applied}$.

The orientation angle, θ, influences the sign of the dipole-dipole interaction. At a critical angle of approximately 55°, the potential changes from an attractive to a repulsive interaction. Hence, particles located parallel to the applied field will be drawn together while those located perpendicular to the applied field will repel each other. It is the dipole-dipole interaction between all the particles in the suspension which redistributes the filler particles into chains along the applied field direction, as shown in Figure 3.

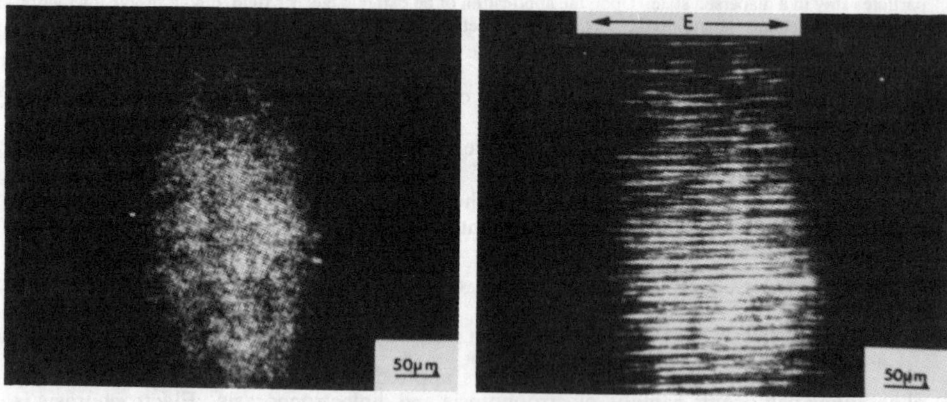

(a) (b)

Figure 3. Optical micrographs showing SrTiO3 dispersed in a silicone elastomer polymer. Figure (a) shows the zero applied field case while (b) shows the chained microstructure. (After Bowen, Bhalla, Newnham and Randall).

Matrix Materials

There exists several general requirements necessary for matrix materials in the dielectrophoretic assembly of composites. The first requirement is that the matrix must be insulating with both a lower dielectric constant than the filler and a low dielectric loss. This permits the induced polarization of the filler phase which, in turn, allows the dielectrophoretic chaining process to occur.

Secondly, the polymerization from the liquid state to the solid state must be a fast process in order to limit sedimentation effects. Sedimentation can also be limited by using a high viscosity fluid; however, the viscosity must not be high enough to limit the dielectrophoretic assembly.

A third requirement for dielectrophoretic assembly is that the uncured polymer matrix must have a high electrical breakdown strength. This will retard the crosslinking reaction through the evolution of carbon dioxide gas bubbles following the degradation of the carbon-based polymer materials.

Table I lists the polymers that have been successfully shown to support dielectrophoretic assembly. The Norland optical adhesive (a U.V. curing polymer) has the advantage of very rapid curing, but is only suited to low volume fraction filler applications. Limitations of the Norland polymer exist with higher filler concentrations owing to increased scattering and absorption of the U.V. radiation, therefore preventing full polymerization of the composite.

Table I. Thermoset Polymers and Suppliers

Thermoset Polymer	Trade Name and Supplier
Polyurethane	Hysol-Dexter U50048
Silicone elastomer	Sylgard-184, Dow Corning Eccogel, Emerson-Cummings
Eccogel epoxy	Eccogel, Emerson-Cummings 1365
Epon epoxy	EPON 865 shell
Norland optical adhesive	Norland-81

Filler Materials

Theoretically, any filler material with a dielectric constant higher than the chosen matrix can be assembled through the dielectrophoretic effect. Table II lists insulators, semiconductors and metals which have been successfully aligned in a thermoset polymer matrix. Insulating and large band gap semiconducting particles can be easily aligned. For small band gap semiconductors and conducting fillers, limitations associated with dielectrophoretic assembly exist. For example, the volume fraction of the filler material must be less than the percolation limit. If the conducting powder is percolated, the interconnected conducting pathways will prevent the polarization of the particles and no chaining phenomena can occur.

Conductive filler particles can, as they assemble into chains, act as a catalyst for the generation of large fields and field gradients in the insulating polymer gap between approaching particles. These field levels can exceed the breakdown strength of the uncured polymer which can result in a retardation of the curing process.

Table II. Dielectrophoretic Alignment of Various Filler Materials in Uncured Silicone Elastomer[1]

Insulator	Semiconductor	Metals
$BaTiO_3$ $PbTiO_3$ $Pb(Zr,Ti)O_3$ $SrTiO_3$ $Ba_2TiSi_2O_8$ ZrO_2 TiO_2 SiO_2 balls[(a)]	$YBa_2Cu_3O_{6.5+\delta}$ Graphite SiC fibers	Aluminum Powder Ag covered resin balls[(b)] Ag covered acrylic fibers[(b)]

Manufacturers
(a) Spheriglass — Potters Indus. Inc.
(b) Mitsubishi Metal Corp.

Determination of Ideal Dielectrophoretic Assembly Conditions for Composite Fabrications

An electric field will induce both electrophoresis and dielectrophoresis suspension in a polymer matrix-ceramic particulate suspension. If electrophoresis is induced, there is a net migration of the filler to one of the electrodes. This net migration prevents homogeneous chaining within the composite, hence limiting the advantages of the 1-3 connectivity. To override the electrophoretic effect and induce only the dielectrophoretic effect, alternating electric fields can be used. However, the frequency of the alternating electric field then becomes another variable in the assembly process.

The determination of the optimum frequency for dielectrophoretic assembly can be performed via two techniques. The first technique is the use of direct observation through optical microscopy. This allows the visual inspection of the chaining phenomena under various processing conditions. However, this technique is limited to low volume fractions of filler material (<0.05) owing to light scattering effects and to particle sizes $\geq 2.0\ \mu m$ in diameter. The second technique indirectly determines the ideal alignment conditions through the electrorheological effect. As thermoset polymers are continually undergoing polymerization during the experiment, a full determination of the Bingham behavior is not possible. It was therefore necessary to determine the field strength and frequency dependence of the shear stress at a fixed shear rate in the dynamic regime of Bingham plastic flow using a rotational viscometer.

From these measurements the relationship between shear-stress and field strength can be determined for a fixed volume fraction and alternating field frequency. As can be

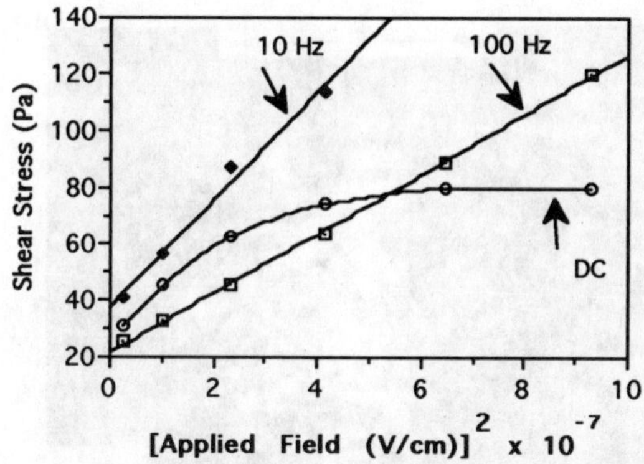

Figure 4. Plot of measured shear stress versus applied field squared. Plot shows a square law behavior for the ac fields while dc behavior shows evidence for the occurrence of electrophoretic phenomena.

observed in the typical case of silicone elastomer polymer and $SrTiO_3$ particles, Figure 4, a square law behavior exists for alternating fields. The direct field electrorheological behavior departs from the square law and is believed to be associated with electrophoretic migration of particles to one electrode. The strongest electrorheological effect is noted for a 10 Hz alternating field, which corresponds to the coarsest fibrils found by direct optical microscopy.

The epoxy thermosets investigated here showed ideal electrorheological effects at higher frequencies (~ 750 Hz). This frequency also corresponded to the thickest chains as observed with the optical microscope. Under direct or low frequency alternating fields, no dielectrophoretic alignment was observed in the epoxy thermosets. Particles flowed in a random and chaotic manner, leading to vortices at higher field strengths as shown in Figure 5. Figure 6 summarizes the viscometry data obtained on four thermoset polymers. The ideal frequency, which corresponds to the ideal alignment conditions and electrorheological effects, is plotted against the volume fraction of $SrTiO_3$ filler for two epoxies, a polyurethane and silicone elastomer. The ideal frequency conditions are volume fraction independent, but the polymers fall into two groups. Polyurethane and silicone elastomer optimally aligned at ≃ 10 Hz, whilst the epoxies aligned at ~ 750 Hz.

The primary difference in optimum alignment frequency is believed to be related to both the dielectric constant and the loss in the frequency spectra from 10^{-2} to 10^{6} Hz in each individual polymer. The epoxies at low frequencies 10^{-2} to 10^{2} Hz are dominated by ionic space charge polarization.[8] This gives rise to an extremely high dielectric constant and loss up to about 700 Hz. The silicone elastomer and polyurethane, on the other hand, do not have such a significant ionic space charge contribution to the dielectric properties; the dielectric loss and dielectric constant relaxes out at low frequencies (~ 0.1–1.0 Hz). Due to the extremely high matrix dielectric constant at low frequencies, it is improbable that the filler particles would polarize and undergo uniaxial agglomeration in this regime. The differences between the ideal alignment frequencies in the tested polymers will be discussed in more detail in future papers.

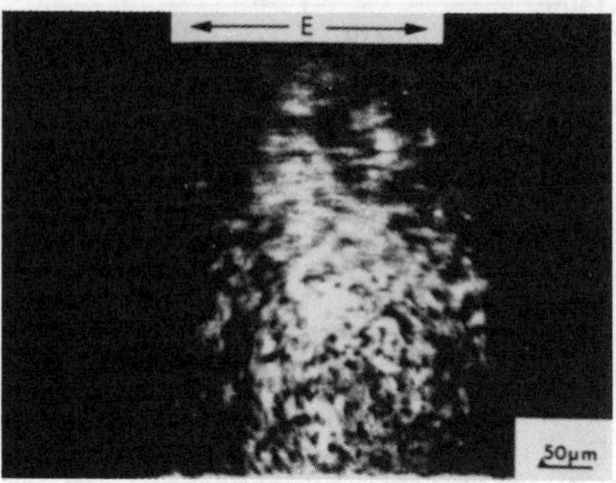

Figure 5. Optical micrographs showing $SrTiO_3$ dispersed in an epoxy polymer. Note that when assembly conditions are far from ideal, swirling vortices can form instead of a chained microstructure. (After Bowen, Bhalla, Newnham and Randall).

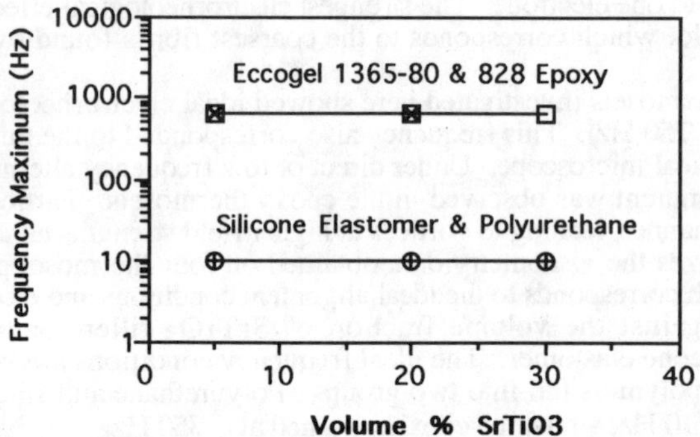

Figure 6. Plot of the optimum frequency versus volume % filler. It was found that the optimum alignment frequency was independent of the volume fraction filler used. Also, the behavior of the polymers fell into two distance regimes. Reasons for the differences in optimum alignment frequency are due to different dielectric responses of each polymer.

The electric field strength also has an influence on the coarseness of the chains. This can be observed directly in the scanning electron and optical micrographs in Figure 7. The field strength optimization is much more difficult to establish than frequency, as it is very much dependent on the application in question. Therefore, the applied field strength is best optimized for each individual composite.

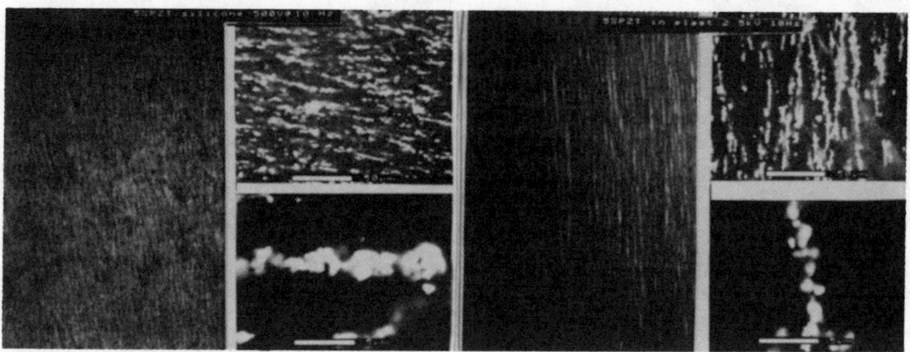

Figure 7. Optical/SEM micrographs of 5% PZT in silicone elastomer showing the field dependence of chaining. Note that in the 500 volt case (left), the chains are less defined than in the 2.5 kV case (right).

Applications for Dielectrophoretically Assembled Composite Materials

Composite materials assembled utilizing the dielectrophoretic effect can be exploited in a number of potential applications, as summarized in Figure 8. These materials can be utilized as thermal, mechanical and electronic devices at scales much smaller than can be achieved with conventional composite fabrication techniques. Here we outline two possible electronic composite applications for dielectrophoretic assembly.

Piezoelectric Hydrophone

Using the direct piezoelectric effect, piezoelectric ceramics based on $Pb(Zr,Ti)O_3$ can be utilized as pressure sensors. However, the performance of these materials is limited in hydrostatic applications such as underwater acoustic sensing devices (hydrophones) due to transverse piezoelectric contributions. The limitations of the monolithic ceramic sensitivity can be overcome by combining the ceramic with a polymer to form a composite material. In the composite design, higher sensitivity levels are obtainable due to enhancements of the effective d_{33} coefficient via stress transfer from the polymer to the ceramic.[9] The most sensitive hydrophones have been designed with 1-3 connectivities which have the advantages of large d_{33}, easy poling conditions and good impedance matching to the working environment.

Hydrophones have also been utilized as biomedical transducers which allow ultrasonic imaging of internal organs. These biomedical devices require high frequency operation ($\geq$ 1 MHz) in order to provide accurate and distinct images.[10] To obtain these high frequencies, small scale hydrophone composites are required. The dielectrophoretic assembly technique provides a means to fabricate such devices on a much smaller scale than is currently possible.

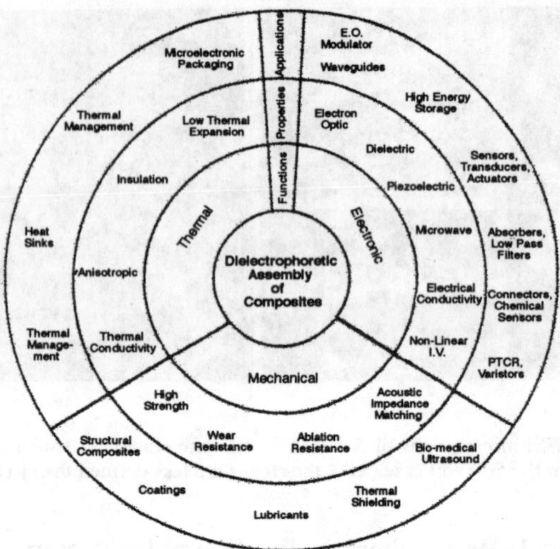

Figure 8. Potential applications for dielectrophoretically assembled composite materials. The circle chart is read from the center radially outwards with titles located at the top of each ring.

Decoupling Capacitor

Semiconductor chip packaging is a continually developing technology. The optimization of speeds, space and performance requires continued readjustment of processing techniques. There exists in the near future the need for high speed digital multichip modules (MCMs). One of the major problems with these packages is the suppression of electromagnetic noise, which continues to become a problem with higher speeds. Inductive noises are traditionally suppressed using surface-mounted decoupling capacitors which act as virtual power supplies close to the chips. Future designs propose locating the decoupling capacitors within the package between the power and ground planes.[11] Since the local decoupling capacitor actually supplies the current required to operate the integrated circuits during each switching cycle, the amount of local decoupling capacitance per unit area must be substantial. High capacitance per unit area may be accomplished with very thin dielectric layers and/or with high dielectric constant materials. In polymer-based MCMs, this requirement becomes a problem, owing to the intrinsically low dielectric constants available. The use of dielectrophoretically assembled high dielectric constant particles gives a possible solution to this materials problem using the induced parallel mixing.[2]

Summary and Conclusions

Dielectrophoretic assembly is proposed as a new novel processing technique for ceramic-polymer composites. It has been shown to be applicable to a wide range of matrix and filler materials with restrictions on volume fraction in some cases. The ideal assembly conditions differ from polymer to polymer due to intrinsic differences in the dielectric properties in the uncured state. Problems such as space-charge polarization within the polymer and electrophoretic migration can be overcome using alternating electric fields. Techniques for determining the optimum frequency are demonstrated using thermoset polymers. Dielectrophoretic assembly is believed to have wide implications in processing ceramic polymer-based composites. Some of these possible applications are outlined.

Acknowledgements

The authors wish to thank J. Mantz for typing this manuscript and Professor L.E. Cross for many useful discussions.

References

1. C. A. Randall, D. V. Miller, J. H. Adair, and A. S. Bhalla, *J. Mat. Res.*, **8** [4], 899 - 904 (1993).

2. C. A. Randall, S. Miyazaki, K. L. More, A. S. Bhalla, and R. E. Newnham, *Matl. Letters*, **15**, 26 - 30 (1992).

3. C. P. Bowen, A. S. Bhalla, R. E. Newnham, C. A. Randall, (submitted to *J. Mat. Res.*).

4. R.E. Newnham, D.P. Skinner, L.E. Cross, *Matl. Res. Bull.*, **13**, 525 (1978).

5. R. E. Newnham, Ann. Rev. Mat. Sci. **16**, 47 - 68 (1986).

6. R.E. Newnham, *Rep. Prog. Phys.*, **52**, 123 - 156 (1989).

7. H. A. Pohl, Dielectrophoresis, Cambridge University Press, Cambridge, London, New York, and Melbourne, 34 - 47 (1978).

8. D. Kranbuehl, S. Delos, E. Yi, J. Mayer, T. Jarvie, T. Hou and B. Wintree, *SAMPE Symp. Ser.*, **30**, 638 (1985).

9. W. Cao, Q.M. Zhang and L.E. Cross, *J. Appl. Phys.*, **72** [12], 15 December (1992).

10. W.A. Smith and A.A. Shaulov, *Ferroelectrics*, **87**, 309 - 320 (1988).

11. B.K. Gilbert and W.L. Walters, *The International Journal of Microcircuits and Electronic Packaging*, **15** [4], 171 - 181 (1992).

TWO DIMENSIONAL BINGHAM PLASTIC FLOW IN A CYLINDRICAL PRESSURISED CLUTCH

R J Atkin

Applied and Computational Mathematics Section
School of Mathematics and Statistics
The University of Sheffield, Sheffield S3 7RH, England

T J Corden, T G Kum and W A Bullough

Department of Mechanical and Process Engineering
PO Box 600, The University of Sheffield
Mappin St, Sheffield S1 4DU, England

ABSTRACT

An examination of the capability of an energised electro-rheological fluid to act as a rotating shaft seal is reported. A two dimensional laminar flow model based on a Bingham plastic flowing between parallel plate electrodes indicated that leakage would always occur whenever the shaft was in rotation, whatever the magnitude of the pressure gradient along the sealing annulus. This result was confirmed in experiments, the results of which show some discrepancy between theory and experiment in the magnitude of the leakage rate. However this error is accountable and is acceptable in the area of immediate interest-shaft speed of circa 1000/1500 rpm and at maximum yield or electro-stress.

1. Introduction

The High Speed Machines Unit at Sheffield University is developing a lightweight, fast response catch which uses an electro-rheological fluid as the controlled connecting medium. The performance of this horizontal axis cylindrical clutch, shown schematically in Fig. 1, is hindered by existing sealing problems. The rubber lip seal in current use causes undesirable heat generation and rotational friction. At high speeds (3000 rpm) the seal creates more drag than that due to the excited ER fluid. This sealing friction is restricting the speed range. It was proposed that this problem could be reduced with the use of an electro-rheological shaft seal (inspired by the success of ferro fluid seals [1]). Permanently excited ER fluid would hopefully restrain leakage and produce less rotational drag than the present seal.

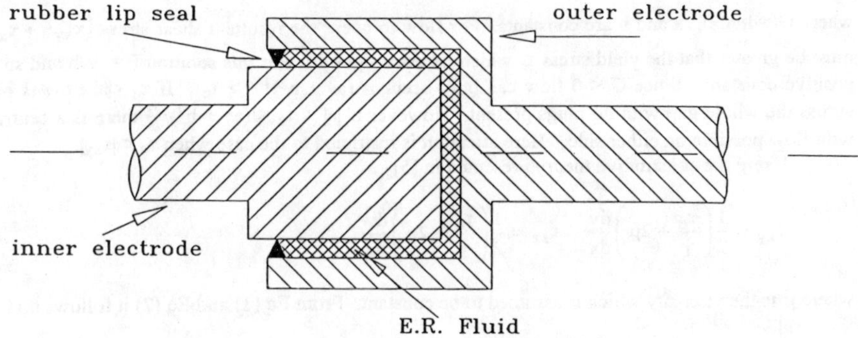

Fig 1 Schematic diagram of high speed clutch

Experience has shown [2,3,4] that ER fluids can be successfully modelled, albeit approximately, as a Bingham plastic. This is possible if electrode separation and the size of the plug flow region is large compared to the particle diameter. Also significant values of shear rate, $\dot{\gamma}$ and electric field strength, E must be involved. Throughout this paper the ER fluid is assumed to be a homogeneous incompressible continuum with Bingham plastic properties (when excited) the yield stress τ_e depending upon E. The flow is assumed to be fully developed and isothermal. The mathematical analysis outlined in section 2 shows that the shear (Couette) and pressure driven (Poiseuille) type flows are coupled, both flows in general depending upon the rotational speed and the axial pressure gradient. When there is flow due to rotation the theory predicts that leakage occurs. An expression for the leakage flow rate involving fluid parameters and seal geometry is derived. Theoretical and experimental values of leakage rate obtained from the test apparatus described in section 3 are compared and discussed in sections 4 to 6 and the feasibility of an electro-rheological fluid seal is evaluated.

2. Theoretical Analysis

In the test rig flow occurs in both the circumferential and axial directions, the former caused by rotation of the inner electrode and the latter due to the difference in pressure inside and outside the seal. Since the gap width is small compared to the radius of the inner electrode flow is considered between two flat plates, x = 0 which is earthed and moves with speed U in the y-direction and a fixed plate x = h which is raised to a potential V. The pressure gradient is in the z-direction. If v(x) and w(x) denote the velocity components in the y- and z- directions respectively the two rate of strain components 1/2 dv/dx and 1/2 dw/dx give an expression for the resultant two dimensional shear rate $\dot{\gamma} = \{(dv/dx)^2 + (dw/dx)^2\}^{1/2}/2$. From the equations of motion it follows that the non-zero shear stress τ_{xy} and τ_{xz} are given by

$$\tau_{xy} = a, \qquad \tau_{xz} = Gx + b, \tag{1}$$

where G(=dp/dz), a and b are constants. For flow to occur the resultant shear stress $\left(\tau_{xy}{}^2+\tau_{xz}{}^2\right)^{1/2}$ must be greater that the yield stress τ_e which depends upon E. For this solution E = V/h and so τ_e is a positive constant. Since G < 0 flow can only occur if $(a^2 + b^2)^{1/2} > \tau_e$. If $\tau_e < |a|$ flow is possible across the whole gap with no plugs present. However if $|a| < \tau_e < (a^2 + b^2)^{1/2}$ there is a central plug with flow possible on either side. Here attention is restricted to the case when $\tau_e < |\tau_{xy}|$.

Using the constitutive theory presented in [5]

$$\tau_{xy}=\frac{1}{2}\left(\frac{\tau_e}{\dot{\gamma}}+2\mu\right)\frac{dv}{dx},\quad \tau_{xz}=\frac{1}{2}\left(\frac{\tau_e}{\dot{\gamma}}+2\mu\right)\frac{dw}{dx},\tag{2}$$

where μ is the viscosity which is assumed to be constant. From Eq (1) and Eq (2) it follows that

$$\mu\frac{dv}{dx}=a\left\{1-\frac{\tau_e}{\left[a^2+(Gx+b)^2\right]^{\frac{1}{2}}}\right\},\tag{3}$$

$$\mu\frac{dw}{dx}=(Gx+b)\left\{1-\frac{\tau_e}{\left[a^2+(Gx+b)^2\right]^{\frac{1}{2}}}\right\},\tag{4}$$

from which the velocity components v and w can be found.
In particular

$$\mu w=\frac{G}{2}x(x-h)\ -\ \frac{\tau_e}{G}\left[\left\{a^2+(Gx+b)^2\right\}^{\frac{1}{2}}-\left(a^2+b^2\right)^{\frac{1}{2}}\right]\tag{5}$$

To satisfy the boundary conditions w(0) = w(h) = 0 it follows that b = -Gh/2. The boundary conditions v(0) = U, v(h) = 0 are satisfied for small h if $a=-(\tau_e+\mu U/h)$. Both components therefore depend upon G and U and so are affected by varying either the axial pressure gradient or the speed of the plate. If the plate moves, for any finite pressure gradient flow occurs in the z-direction. The leakage flow rate is given by

$$Q_z=\frac{-Gh^3\pi r}{6\mu}+\frac{2\pi r}{\mu}\left\{\frac{\tau_e h}{4G}(G^2h^2+4a^2)^{1/2}+\frac{\tau_e . a^2}{2G^2}\ell n\left[\frac{-Gh+(4a^2+G^2h^2)^{1/2}}{Gh+(4a^2+G^2h^2)^{1/2}}\right]\right\}\tag{6}$$

where r is a typical radius, and so Q_z has Newtonian and Bingham contributions, the latter always trying to reduce the former. When G = 0, Q_z = 0 and at high values of G the Newtonian term dominates.

3. Experimental Details

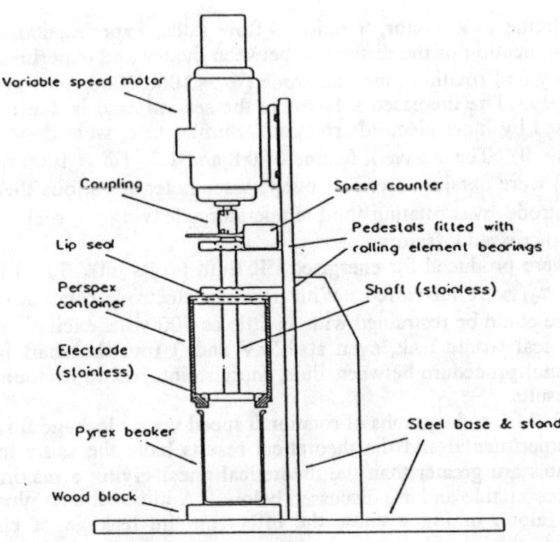

Fig 2 Schematic view of electro-rheological seal test rig

In the test apparatus shown in Fig 2 a substantial steel shaft is mounted vertically in two ball races. This shaft has the inner electrode/seal attached to its lower end which fits concentrically into a stationary outer electrode/seal giving 1/2 mm (±0.02mm) radial clearance. This arrangement gives a sealing length of 10mm on a 60mm diameter which is the same diameter as the clutch. Having such a large diameter seal ensures that measurable leakages will be obtained and also allows an analysis in which plane geometry is assumed and so any radial variation of the electric field is neglected.

The outer electrode/seal is supported in a clear plastic tube which acts as a reservoir within which up to 100mm head of fluid can be accommodated. The diameter of the inner shaft was arranged so that a minimum of fluid could be used. Both inner and outer electrodes were made from stainless steel. The rotating inner electrode was earthed while the stationary outer electrode was connected to a high potential stabilised DC source. The set but variable shaft speed was measured by an encoder/digital meter which was calibrated with a precision tachometer.

The fluid leaking from the seal was collected over a given time and then weighed so that the flow rate could be calculated. Two different fluids were used. Both were silicon oil/polymethacrylate mixtures and were characterised by Advanced Fluid Systems Ltd of Wandsworth, London. The density of the first fluid, type code S2030 made up of 20% fine suspension, was 1092 kgm^{-3}. The second fluid was a mixture made up of equal parts of two fluids, one having 30% and the other 40% fine suspension. Its density was 1485 kgm^{-3}.

4. Results

For a Newtonian fluid laminar flow caused by an axial pressure gradient through a stationary annular gap has a well established solution. Initial tests were done with oil (Nexus 68) known to be Newtonian to establish the accuracy of the apparatus and hence produce some control results to which the ER tests could be compared.

It is useful to define a 'k' factor, a ratio of flow rates, experimental divided by theoretical results. This gives an indication of the difference between theory and experiment. The oil tests were done with both stationary and rotating inner electrode ($\overline{\omega}$ = 1000/2000 rpm) giving k factors of 1.09 and 1.11/1.19 respectively. The increased k factor in the second case is due to lower fluid viscosity because of heating caused by inner electrode rotation. Similar tests were done with ER fluid and no excitation (volts off, $\tau_e = 0$). These gave k factors of 0.8 and 1.21 ($\overline{\omega}$ = 1000 rpm). Viscosity values (zero volts for ER fluid) were obtained from a Couette viscometer at various shear rates. It was found that when the inner electrode was rotating fluid temperatures leaving the seal were 2-3°C higher than that entering, hence the increased k factors.

Finally results were produced for energised ER fluid (volts on). Fluid head, rotational speed and voltage (and hence τ_e) were all varied. With no inner electrode rotation ($\overline{\omega} = 0$) with an axial pressure gradient leakage could be restrained with as little as 500 volts excitation but with the slightest amount of rotation the seal would leak, even at 2.5kV and 3 rpm the shaft being turned by hand. Changing the experimental procedure between fluid energisation before rotation and vice versa made little difference to the results.

For the fluid S2030 from the graphs of rotational speed verses leakage flow shown in Fig 3, it is evident that both the experimental and the theoretical results have the same form. In all cases the experimental leakage rates are greater than the theoretical ones, giving a maximum k factor of 1.87. These differences are accountable and are discussed below. Additional data gleaned from the plots of the theoretical analysis alone in Fig 4 show the effect on the leakage of changing the electrode separation h and the viscosity μ. Figs. 5 and 6 show the effects of yield stress level on leakage flow rate at a fluid head of 50mm and the effect of pressure for E = 2kV/mm. Further work involved checking both experimental and theoretical results for the second fluid. These results gave similar differences between theory and experiment (Figs 7, 8 method 1) indicating that the results are reproducible.

In order to produce all the theoretical results, values of τ_e and μ have to be substituted into Eq. (6). As this was thought to be one of the main sources of error various schemes for evaluating τ_e and μ were used. In Fig. 3 method 1 shown schematically in Fig. 9(a) is used. In all the methods the actual values of τ_e and μ were obtained by plotting the data supplied by Advanced Fluids Systems Ltd. Method 2 derives τ_e from $\tau_{eo} = \tau_e + \mu\dot{\gamma}$, where the local shear rate and voltage fix τ_e but μ is again the zero-volts viscosity see Fig. 9(b). In method 3 τ_e is calculated as in method 1 but the values of μ used are calculated from the slope of each constant voltage line as shown in Fig. 9(c). Method 4 is similar to method 2 except that the viscosity and yield stress values are both calculated for the local voltage or shear rate as shown in Fig. 9(d). It must be noted, see Figs. 7, 8 that the manufacturer's data for the second fluid is not available for theoretical methods 2 and 4 above 300rpm. Hence a direct comparison is not possible but it is immediately apparent that the choice of the Bingham parameters τ_e and μ for an imperfect fluid is an important factor. Also it should be noted that the values for the second fluid were interpolated from the manufacturer's data, which was given for the two fluids (30 and 40%) separately, according to the method in [6].

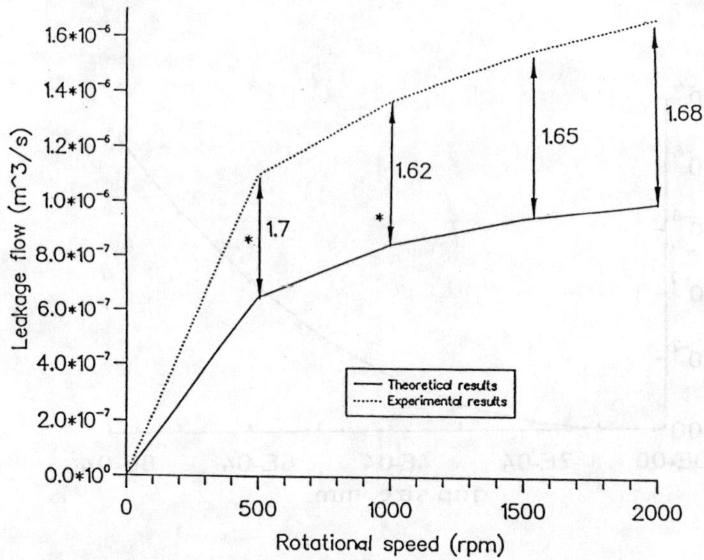

Fig. 3 Leakage flow rate v rotational speed of inner cylinder for 0.5mm gap. Fluid S2030, viscosity 0.096Pas, 5cm head.
(a) applied voltage 500 volts, yield stress 265Pa.

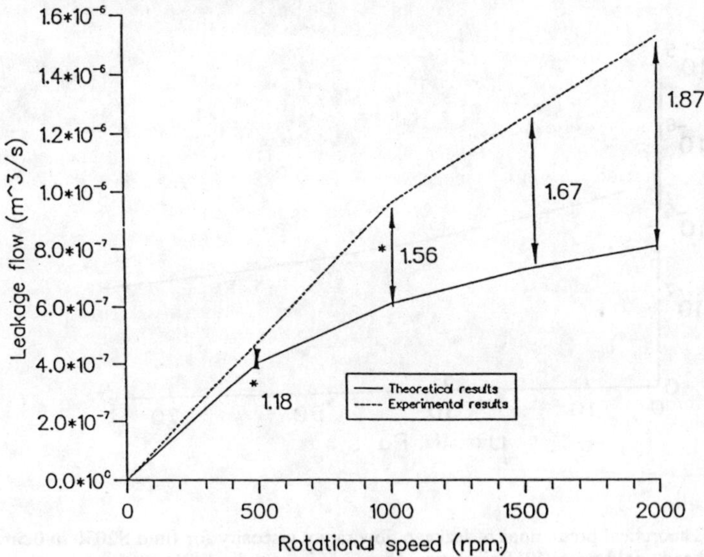

(b) applied voltage 1000 volts, yield stress 595Pa.

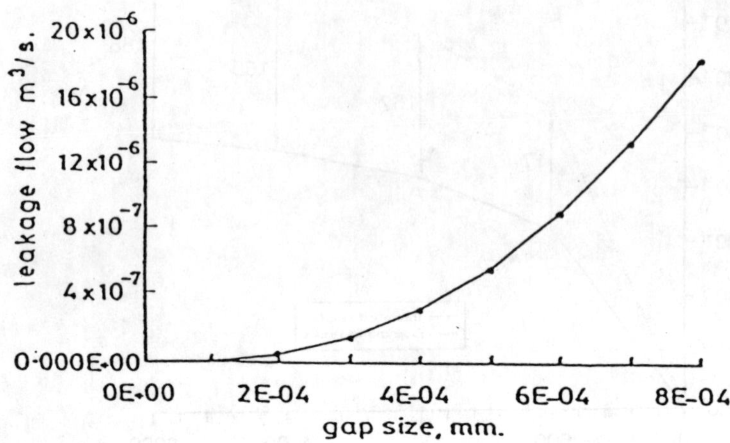

Fig. 4 (a) Theoretical predictions of leakage flow rate v gap size for fluid S2030 in 0.5mm gap at 1000 volts, 5cm head, rotational speed of inner cylinder 1000 rpm.

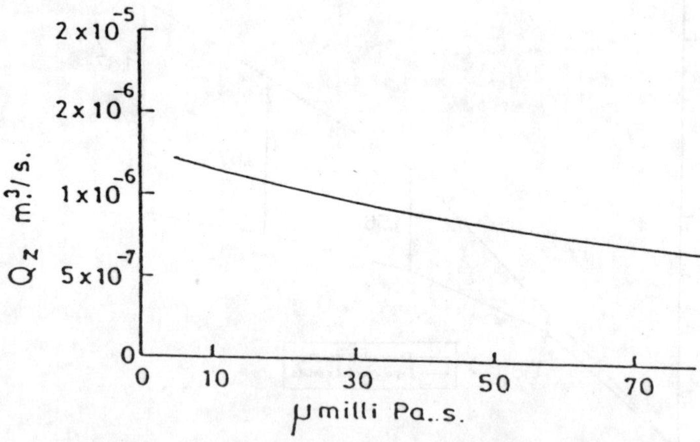

(b) Theoretical predictions of leakage flow rate v viscosity for fluid S2030 in 0.5mm gap, 5cm head, yield stress 595Pa, rotational speed of inner cylinder 1000rpm.

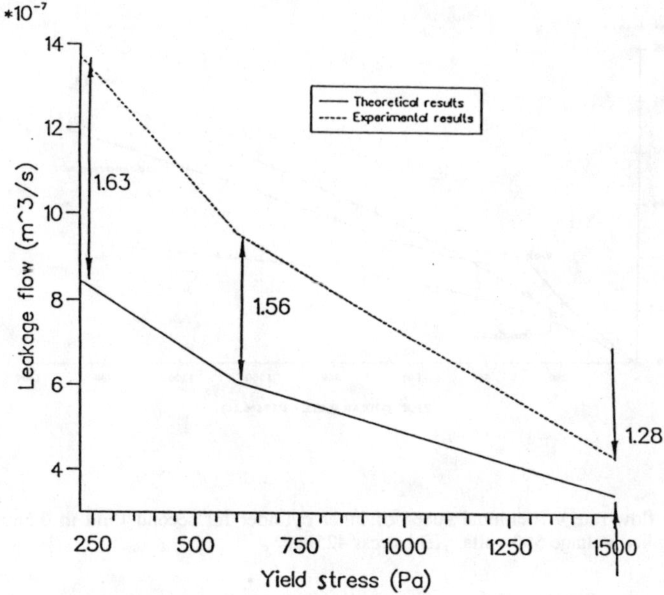

Fig. 5 Leakage flow rate v yield stress for fluid S2030 with 5cm head, rotational speed of inner cylinder 1000rpm.

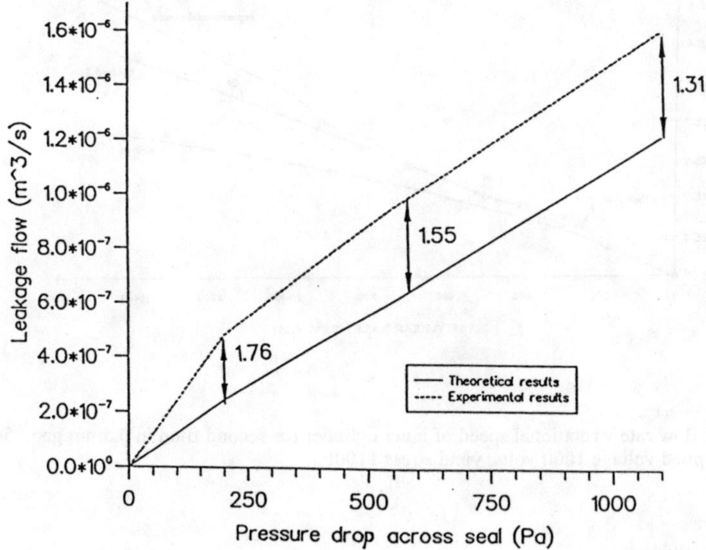

Fig. 6 Leakage flow rate v pressure drop for fluid S2030 at 1000 volts and rotational speed of inner cylinder 1000rpm .

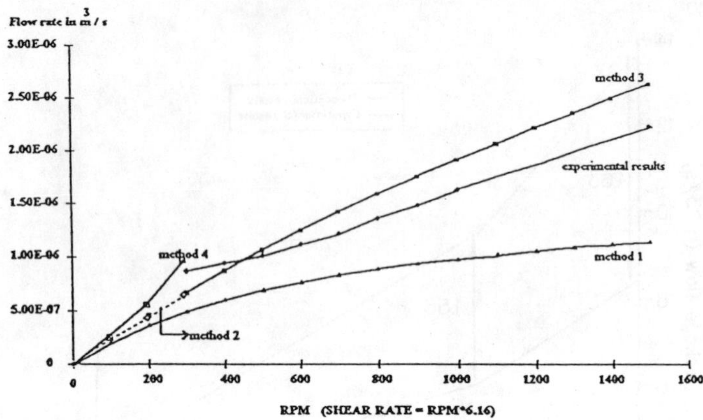

Fig. 7 Leakage flow rate v rotational speed of inner cylinder for second fluid in 0.5mm gap. 5cm head, applied voltage 500 volts, yield stress 425Pa.

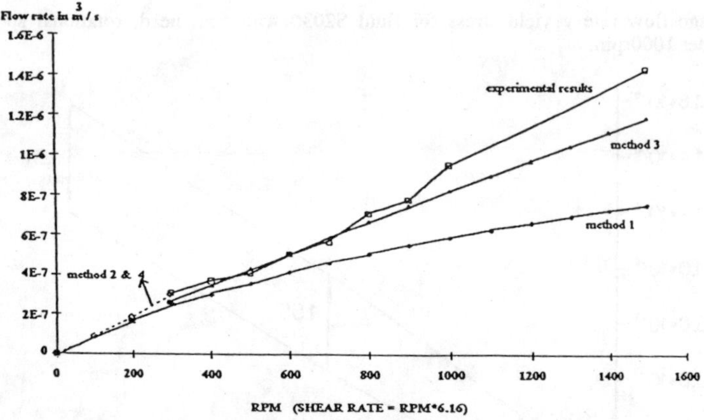

Fig. 8 Leakage flow rate v rotational speed of inner cylinder for second fluid in 0.5mm gap. 5cm head, applied voltage 1000 volts, yield stress 1100Pa.

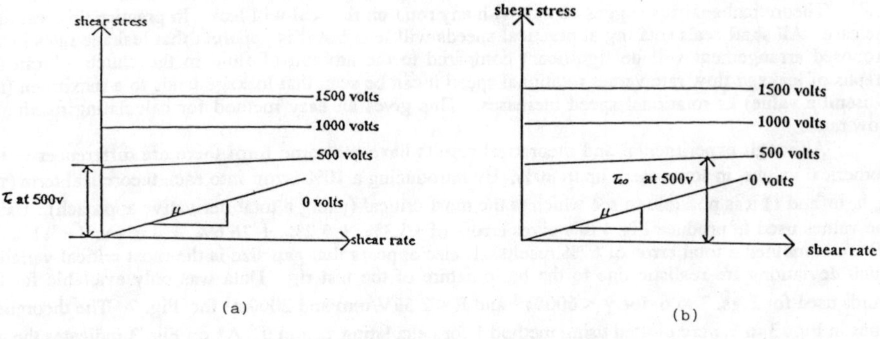

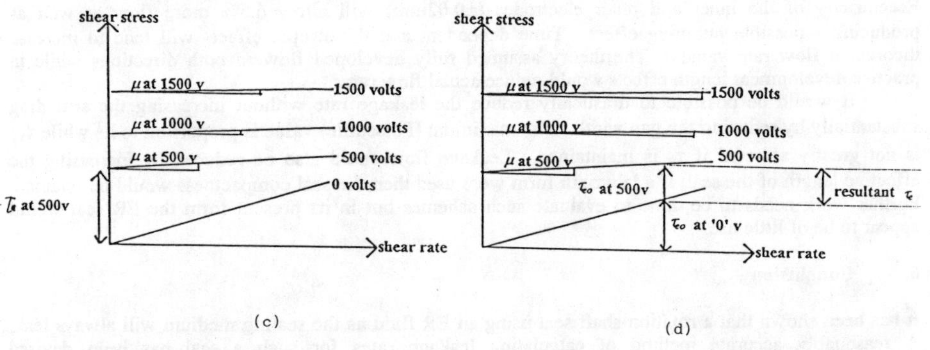

Fig. 9 Schematic diagrams indicating the methods used to calculate τ_e and μ
(a) method 1 (b) method 2 (c) method 3

(d) method 4

5. Discussion

Any success in applying the ER clutch will probably be due to its fast speed of response/operation and ease of electro/mechanical interfacing [7]. It is particularly suited to lightweight applications where a high torque/inertia ratio is desirable. At high speeds of operation (3000rpm) the combined viscous/drag stress due to unexcited fluid/rubber lip seal can be greater than the yield stress. This was the incentive for an electro-rheological seal. If axial leakage could be restrained whilst reducing the parasitic drag then further progress would have been made towards improving the clutch.

Theoretical analysis suggested that with any rotation the seal will leak. In practice this was also the case. All shaft seals rotating at practical speeds will leak but it is apparent that leakage rates in the proposed arrangement will be significant compared to the amount of fluid in the clutch. From the graphs of leakage flow rate verses rotational speed it can be seen that leakage tends to a maximum (the Poiseuille value) as rotational speed increases. This gives an easy method for calculating maximum flow rates.

Although experimental and theoretical results have the same form there are differences in the numerical values, in some cases up to 80%. By introducing a 10% error into each theoretical term (τ_e, μ, h, $\overline{\omega}$ and ℓ) it is possible to see which is the most critical (using a total derivative approach). Using the values used to produce Fig 4 this gives errors of +5.3%, + 5.2%, + 26.6%, + 4.6% and + 11%. If all are combined a total error of 67% results. It also appears that gap size is the most critical variable. Such deviations are realistic due to the basic nature of the test rig. Data was only available for the fluids used for Figs. 3 to 6 for $\dot{\gamma} < 6000s^{-1}$ and $E < 2.5kV/mm$ and $2000s^{-1}$ for Fig. 7. The theoretical lines in Figs 3 to 6 were plotted using method 1 for calculating τ_e and μ. A* on Fig. 3 indicates the use of method 2 when data was available. Figs. 7, 8 compare theoretical results using all four methods. The method of calculating τ_e and μ used in method 3 results in less difference between theory and the experimental results..

There are many other reasons for the significant k factors. Fluid temperature increased when passing through the seal by up to 3°C hence reducing viscosity and increasing leakage rates. Eccentricity of the inner and outer electrodes (±0.02mm) will allow 6.5% more flow as well as producing a possible pumping effect. Time dependent and thixotropic effects will tend to increase theoretical flow rate values. The theory assumed fully developed flow in both directions while in practice development length effects would reduce actual flow rates.

It would be possible to drastically reduce the leakage rate without increasing the seal drag substantially by reducing the gap width. The maximum (Poiseuille) value is proportion to h^3 while τ_{xy} is not greatly affected if τ_e is maintained. Leakage flow could also be reduced by increasing the effective length of the seal; if a labyrinth form were used then the seal compactness would be retained. Further work needs to be done to evaluate such schemes but in its present form the ER seal would appear to be of little use.

6. Conclusion

It has been shown that a rotation shaft seal using an ER fluid as the sealing medium will always leak. A reasonably accurate method of calculating leakage rates for such a seal has been devised demonstrating the validity of the Bingham plastic/continuum approach for modelling ER fluids. For the simple clutch the ER seal seems to be of little use. Further investigation will be necessary to evaluate the use of an ER seal in other situations.

7. References

[1] H Butcher, "Industrial Sealing Technology". (Wiley Interscience).

[2] R T Bonnecaze and J F Brady, "Dynamic Simulation of an Electro-Rheological Fluid", Jnl Chem Phys, 69,(3), 1 Feb 1992, 2183-2202.

[3] D J Klingenberg and C F Zukoski, "Studies on Steady Shear Behaviour of Electro-Rheological Suspensions" - Langmuir 6, 1990, 15-24.

[4] W A Bullough, "Electro-Rheological Fluids: an introduction for biomedical applications". Jnl Biomed Eng, 13, 1991, 234-238.

[5] R J Atkin, Xiao Shi and W A Bullough, "Solutions of the Constitutive Equations for the Flow of an Electro-Rheological Fluid in Radial Configurations", J Rheol 35 (7) (1991) 1441-1461.

[6] D A Brooks, "Applicability of Simplified Expressions for Design with Electro-Rheological Fluids". Recent Advances in Adaptive and Sensory Materials and their Applications (Technomic Press Lancaster Pa) April 1992.

[7] W A Bullough, "Innovative Electron Hydraulic Developments for The Fluid Power Industry". Fluid Power-Proceedings of the 9th International Symposium. BHRA Cambridge. 1990.

8. Acknowledgements

The authors are grateful to Messrs D Butcher and J Makin for their help in the preparation and running of the test equipment.

BINGHAM PLASTIC ANALYSIS OF ER VALVE FLOW

D J Peel and W A Bullough
Department of Mechanical and Process Engineering
The University, PO Box 600, Sheffield, S1 4DU, United Kingdom

ABSTRACT

Yield shear stresses and shear rates at a valve wall are derived from experimental results on a series of concentric cylinder valves. The fluid constitutive equation used for this purpose is that of a Bingham Plastic. Valve plates (being such that the radial gap is small compared to its mean pitch) are taken to be parallel so far as the derivation of the flow v's pressure v's geometry model is concerned. A range of electrode separations from 0.5 to 1.0 mm are used with flow velocities being limited to the region where the viscous pressure drop component is below that caused by the electro stress.

Results show that (away from the region of low shear rates and high voltages) the wall stresses for equivalent conditions are comparable for different valves, for a range of applied field strengths and mean flow velocities. Thus, provided the hysteretic region is avoided the fluid can be treated as a Bingham continuum with some stated reservations. However, this only applied with precision for the truly corresponding situations defined in the paper.

1. Introduction

Methods of modelling the flow of non-Newtonian fluids abound. For electro rheological fluids (ERF) multi-particle analyses are becoming refined to the point that it should soon be possible to include the effects of material properties (including conductance) in them. At the other extreme use has been made of apparent viscosity and indeed two viscosity models [1]. None of these analyses have so far been shown to give the applications engineer what he wants - an efficient and reasonably accurate technique for the sizing of a range of ER controllers from one set of fluid characteristic data, especially when time domain operation is called for [2].

A fundamental fact which is regularly overlooked is that the main use of ERF often accrues because of their fast time response and the ability to impose a pre yield zone electronically. These are usually the prime requirements of machines which can be flexibly operated on a variable configuration of displacement, velocity or force versus time basis. For valve control the positive slope of the ΔP v Q curve is necessary for stability of machine operation. The hold position should be as stiff as possible. In short, the <u>ideal</u> machine characteristic would posses a Bingham plastic form [3] with a response so fast that high resolution of the desired process output shape could be achieved. It follows from this and the Bingham form of torque versus speed difference arising from an idealised ER clutch that engineers will initially favour a Bingham plastic analysis for the valve/fluid/machine inter-action calculations provided that it does not cause too gross an error.

In valve flow the electron-hydraulic time delay value in ERF is often so short as to be of negligible effect when compared to transient flow/inertial considerations in the connecting circuit [4]. The choice of which τ_e or μ values to use in an analysis when as is often the case τ_e, $\tau_{eo} = f(E, \dot{\gamma})$ and $\mu \neq Kf(E)\}$ can cause problems, especially in two dimensional flows [5]. Nevertheless it is useful to have an insight into the value of the model over a range of valve sizes and flow conditions. The present paper purports to throw light on this scenario by a work back from the measured flow/pressure characterisation for a range of near parallel plate flow valves of electrode separation $0.5 \leq h \leq 1.0$ mm and length $50 \leq \ell \leq 100$ mm operating at mean velocities of $0 < \bar{u} < 2$ m/s in the laminar region and valve wall shear rates up to 40,000s^{-1}.

In some fluids, at a given operating temperature, $\tau_e \neq f(\dot{\gamma})$ and $\mu = K$[6]. So far as modelling is concerned such a situation would seem to be in need of little further improvement, provided the electron-hydraulic time delay was short and the hysteresis region was not encountered. However, in operation many fluids will have regions of $\tau_e = f(\dot{\gamma}, E)$, $\mu \neq K = f(E)$ since it is not always possible to predict the parameters a machine will work at (fluid temperature for instance depends on the specific duty of a flexible machine). It is this latter and more complex scenario that is considered here.

2. Apparatus

Parts of the test rig in question have been described in detail in previous papers viz: Electrical supply [7], pressure measurements [8], pump performance and flow monitoring [9], heat transfer and temperature control [10]. For the purpose of the present work the test rig is described in Figs 1 (general arrangement), Fig 2 (valve) and Fig 3 (drive layout etc).

A thyristor controlled electric motor using feedback from the pump shaft speed is used to drive the load within +0 to -1% of full speed (+0 to -15 rpm zero to full load) of 1500 rpm. This drives through a four speed reduction gear box, ratios 1:1, 2:1, 4:1 and 10:1, in order to control the regulation of the motor at low pump speeds. The tachometer used was a DC type. Using an order of magnitude technique slight falls of speed with valve loading are corrected for [9]. A typical recording (Fig 4) shows this fall of speed and the accompanying pressure response of the valve.

3. The Test Programme

With a particular ER valve geometric configuration selected, a straightforward test procedure was followed to determine firstly the relationship between pressure drop ΔP_{eo} across the valve, volume flow rate Q through the valve, and voltage V across the valve electrodes, under steady conditions as far they are attainable.

A test temperature $\theta = 30°C$ was maintained within limits of approximately ±0.25°C, and for a selected nominally steady volume flow rate, conditions were allowed to stabilise for zero electrode voltage. A selected electrode voltage was then applied as a step function, and valve pressure drop values ΔP_0 before the voltage step and ΔP_{eo} after the voltage step were observed. While the zero volts pressure drop ΔP_o was always steady, the pressure drop ΔP_{eo} developed rapidly (over a period of order 0.1 sec) to what is in general strictly a quasi-steady state - variations over a period of a few seconds normally amount to only a very few percent, can be neglected, and where appropriate the mean pressure drop is observed. The required data recorded, the electrode voltage was returned to zero, a new electrode voltage was selected if required and applied when zero volts conditions were stable once more. With all test values of electrode voltage completed for the selected volume flow rate, the volume flow rate was changed to a new value and tests for the different electrode voltage values repeated at the new volume flow rate.

Each valve test generally comprised pressure drop measurements for eight levels of electrode voltage at ten volume flow rate values within the range 0-20 lit/min. The electrode voltage range was 0 - 1000 V for the 0.5 mm valve electrode gap h, and 0 - 2000 V for h = 1.0 mm, so that the electric field strength E = V/h was in the range 0 - 2000 V/mm.

After each valve test was complete, a sample of ER fluid was taken from the test rig and tested in a small 0.5 mm fixed electrode gap clutch mode device [11] over the same range of electric field strengths as the valve test.

Outside the ER fluid/valve configuration tests reported here, the valve test rig has been examined in detail. Using the positive displacement flowmeter, the gear pump itself has been calibrated as a flowmeter, so that volume flow rate can be determined from pump speed and valve pressure drop; this is of value in applying corrections to measured flow rates arising from changes in valve pressure drop which affect slip around the gear teeth. Also, with the rig filled with a hydraulic oil of known Newtonian viscosity properties, detailed tests were carried out with each ER valve geometric configuration to determine the valve properties, end losses etc under predictable and known conditions.

The sequence of valve configurations tests was as follows:

(i) A : $h = 0.5$ mm x $\ell = 100$ mm, symbol o
(ii) B : $h = 0.5$ mm x $\ell = 50$ mm, symbol ▲
(iii) C : $h = 1.0$ mm x $\ell = 100$ mm, symbol □
(iv) Repeat of A above, symbol x

Each of the configurations of valves A, B, C is of two concentric channels of total $b = 361.5$ mm

3.1 *Presentation of the Test Data*

At each test point, a direct measurement is made of the valve pressure drop ΔP_{eo} corresponding to the electric field strength E, measured as voltage V across the known electrode gap h. Prior to the step applications of the electrode voltage, the no-field valve pressure drop ΔP_o and volume flow rate Q for the test point are measured directly. The rise of valve pressure drop from ΔP_o to ΔP_{eo} gives rise to an extra load on the pump motor, a slowing of speed and an increase of slip in the pump, and hence a decrease of volume flow rate overall and a decrease in the actual no-field pressure drop component of ΔP_{eo} below initial ΔP_o. These changes are calculated from the known, calibrated gear pump characteristics and are applied as corrections to the no-field measurements. Corrected plots of the entire valve test data, for four specimen constant voltage values for each valve configuration, are shown in Figs. 5, 6, 7 - the number of voltages shown has been reduced to keep the graphs clear; intermediate voltage data is straight forwardly interpolated .

Figs. 5, 6, 7, show the valve pressure drop ΔP_{eo} in bars, and opposite is the pressure scale translated into electrode wall shearing stress, through the simple force balance $\tau_{weo} = h\Delta P_{eo} / 2\ell$

3.2 *General Observations Related to the Test Data*

The following basic observations relate to the valve tests and the test data.

(i) On applying the electrode voltage as a step change disturbing an initial steady no-field flow, there is generally a rapid rise of valve pressure drop to a new, strictly quasi-steady state. The measured rise time is of order 0.1 secs, and deviation of the new quasi-steady valve pressure drop is generally within a very few percent over a period of, say, ten seconds. This time includes both ERF and test rig effects.

(ii) Doubling the valve length ℓ means doubling the electro-pressure component $\Delta P_e = \Delta P_{eo} - \Delta P_o$ to a first approximation.

(iii) Doubling the valve electrode gap h means halving the electro-pressure component ΔP_e to a first approximation, for constant field strength E = V/h.

(iv) To a first approximation, over the test range of volume flow rate, the electro-pressure component ΔP_e in any given valve is constant for constant electrode voltage V. More specifically, increasing flow rate tends to mean increasing electro-pressure for the wider electrode gap, and decreasing electro-pressure for the narrower electrode gaps tested

(v) For the larger valve electrode gap, at lower volume flow rates and higher valve electrode voltages, there is a tendency to unsteady electro-pressure; after the initial rapid development of electro-pressure common to all the test points, somewhat higher levels of electro-pressure are developed in a slower, unsteady fashion.

(vi) For longer valve lengths, after the initial quasi-steady electro-pressure development, there is a prolonged slow fall of the electro-pressure component with time which becomes cumulatively quite significant.

(vii) There are some slight non-Newtonian characteristics of the no-field flow through the valve, which are seen more clearly at very low flow rates.

(viii) The final repeat test of valve configuration A shows generally a very good repeat of the initial test data, which tends to confirm the distinctions between the three configurations A, B and C. However, it is the nature of the fluid tested that non-uniformity can be expected in any handling/mixing/sampling/time dependent procedure.

(ix) Practical operational difficulties which can occur with the fluid tested include compacting under high pressure e.g. in the pump journal bearings and electrical short-circuiting in the valve under high voltage/high pressure conditions, made worse by small electrode gaps.

(x) The ranges of test parameters chosen here stayed clear of these areas of difficulty - the fluid proved reliable and easy to work with.

4. First Steps in Analysis of Test Data

For ER fluid devices with a parallel flow channel similar to the ER valves tested here, there is a simple common-sense relationship borne out to a first approximation. The electro-pressure component is clearly approximately constant for a given electrode voltage, and dependent on electric field strength and valve length. It can be anticipated that the ratio $\Delta P_e h / \ell$ will be fixed for a given electric field strength V/h for any valve electrode gap h or length ℓ.

In Figs 8, 9, $\Delta P_e h / \ell$ is plotted against volume flow rate Q for four different constant values of the electric field strength V/h. In each case, the overall mean value of the data is shown, together with an indication of ± 20% bounds (dotted line). However, although the result is broadly as anticipated, and though the initial and repeat tests with valve configuration A give closely similar results, the fact is that looking at the field strength of E = 2000 V/mm at volume flow rate Q = 17 lit/min, the value of $\Delta P_e h / \ell$ for h = 1 mm at 55 mbars is about 35% above that for h = 0.5 mm at 41 mbars, valve length being ℓ = 100 mm in both cases. Furthermore, the data for the two valve configurations is clearly diverging as volume flow rate increases further. In the case of a fixed electrode gap of h = 0.5 mm, $\Delta P_e h / \ell$ is some 20% higher at Q = 10 lit/min for the valve length ℓ = 50 mm than it is for the valve length ℓ = 100 mm, although the trend of the data as volume flow rate increases is the same for the two configurations. These two particular examples are seen in Fig. 10. Much the same is true for all of the electric field strengths shown. Note that Q is in effect an indication of Reynolds Number.

4.1 Pointers to Further Analysis of Test Data

The tests reported here were part of an ongoing general engineering feasibility study with ER fluids. One aspect of feasibility of general interest is characterisation. Also, it is required to know e.g. is it possible to carry out a simple small scale laboratory test which will produce generalised design/application data translatable to other kinds of device? This is really the key question. Or, will generalised data apply to all valve flow situations only, rather than the shear mode also? For this reason, to serve as a crude comparator, simple shear mode tests were done on the ex valve fluids.

5. Bingham plastic analysis

Bingham plastic flow is well described in the literature [12], and is written here for the case of flow between parallel flat plates, which approximates the ER valve configuration of these tests. Following this, for any test point of Figs. 5, 6, 7, a Bingham yield stress τ_b for the fluid can be found from the equation

$$4\left(\frac{\ell}{h\Delta P_{\infty}}\right)^3 \tau_b^3 - 3\left(\frac{\ell}{h\Delta P_{\infty}}\right)\tau_b + \left(1 - \frac{12\mu\ell Q}{bh^3\Delta P_{\infty}}\right) = 0$$

where valve dimensions b, h, ℓ are known, ΔP_{eo} and Q are known for each test point by measurement, and Newtonian viscosity coefficient μ is found from the slope of the no-field pressure drop ΔP_o vs Q by measurement. Using this equation, the $\Delta P_{eo} h / \ell$ vs Q data of Figs. 8, 9 is translated into lines of τ_b vs Q at constant voltage/field strength in Figs. 10, 11. Inspection shows that the underlying trends of the data are not altered, whilst the calculated yield stresses for the two cases h = 0.5 mm and h = 1.0 mm at E = 2000 V/mm and Q = 17 lit/mm are even further apart than was the $\Delta P_e h / \ell$ data . On the other hand, the yield stress data for a given electric field strength coincide over much of the flow rate range for the two cases of valve B , h x ℓ = 0.5 x 50 mm, and valve C, h x ℓ = 1.0 x 100 mm, to some degree.

Also included on Figs 10, 11 is an indication of the yield stresses, τ_y, determined for the samples of fluid taken from the valve rig after the corresponding tests, in the rudimentary clutch device, at the corresponding electric field strength.

Although there is often quite good correlation between the measurement of yield stress with this clutch device and the levels of Bingham yield stress calculated for the flow mode in the valve, it is by no means clear that the plot of Bingham yield stress vs volume flow rate for different valve configurations at the same constant electric field strength represents a satisfactory or accurate generalisation of the valve data. The nature of the differences between the various valve configurations is made clearer by the following Figs. 12, 13, 14.

The similarity between general (radial effects neglected) clutch type and valve type fluid motion is in a portion of the valve flow field between parallel plates extending from one plate, or wall, out across the flow to an extent which is variable and can reach up to but not beyond the centre line of the flow. For the Bingham plastic flow in a valve (between parallel plates) there is always a plug at the centre, because without a yield stress, the shear stress and shearing rate are zero at the centre. In the clutch device there will normally be, say, a high shearing rate right across the flow channel with little variation of shear stress and shearing rate, with the exception that in a radial clutch if there is a sufficiently high yield stress, then the shearing rate will fall to zero there, a plug or solid body of fluid will be present at the outer wall and the region of shearing flow will be narrowed. Unless this occurs, the clutch situation is likened to the region immediately next to the walls in the valve situation, extending insufficiently to reach the central plug. Thus it is not immediately obvious what is the best generalised description of flow in the clutch and valve modes.

In Figs. 12, 13 the wall shear rate $(\partial u / \partial y)_{weo}$ is taken, this being found from:

$$\left(\frac{\partial u}{\partial y}\right)_{weo} = \frac{1}{\mu}(\tau_{weo} - \tau_b) \qquad \text{where} \quad \tau_{weo} = \frac{h\Delta P_{eo}}{2\ell}$$

A number of things stand out in the plot of Bingham yield stress vs wall shearing rate in Fig. 12. For the higher electric field strength of E = 2000 V/mm the wall shearing rate calculated is of a high value at all volume flow rates; there is considerable scatter of the data for the different valve configurations; and for the wide valve electrode gap h = 1.0 mm there is a clear minimum level of the calculated wall shear rate, with an increasing level accompanying the rising calculated Bingham yield stress as volume flow rate decreased to the lowest values. It seems quite clear that there are differing configurations of the flow field for the different valve configurations.

There are various potential sources of error and difference between the valves, but these are likely to lead only to differences in magnitude. For example, error in the electrode gap of 0.025 mm represents 5%, likely to lead to an 8% error in calculated yield stresses, and 15% in the no-field pressure drop through the valve. On the other hand, no-field pressure measurements for the two valve configurations A and B with gap h = 0.5 mm imply a viscosity coefficient differing by only 3% , hence only a very small gap difference of perhaps 1% in the two cases. Flow development effects for the fluid in the presence of the electric field may be present in the data; on the other hand, in the case of zero electric field with approximately Newtonian behaviour, the very low Reynolds Number of flow (of order 10 based on $\overline{u}$ and h, at most) implies flow development lengths of the order of the gap size, i.e., 1 or 2% of valve length ℓ.

The scatter of the wall shearing rate data is shown clearly in Fig. 13.

One of the most interesting pictures given directly by the Bingham plastic analysis is that of Fig. 14, showing a plot of the ratio plug width/gap width, δ/h, vs volume flow rate Q, for constant values of electric field strength. It can be clearly seen that for the two valve configurations A and B for electrode gap h = 0.5 mm, the data for a given field strength lie very close together, but for the gap h = 1.0 mm of valve configuration C at the same field strength, the data is very different. In fact the data at field strength 1000 V/mm. for gap h = 1.0 mm lies very close to the data for field strength 2000 V/mm. For gap h = 0.5 mm. This means that the flow fields must correspond in a dimensionless map. It also indicates that the plug of flow which will occur always in the flow mode will be more significant for the wider electrode gap than for the narrower gap at a given electric field strength - in other words, similarity of flow is not determined by electric field strength alone.

The electric field strengths selected provide four clear bands of (similar) flow on the plug/gap ratio vs volume flow rate plot of Fig 14.

In examining Fig. 14, it should be borne in mind that the particle size of the very wet Lipol 30% volume fraction fluid tested ranged approximately from 5 to 25 μm, i.e. from 1 to 5% of electrode gap h = 0.5 mm and from 0.5 to 2.5% of electrode gap h = 1.0 mm. For the electrode gap h = 0.5 mm, a plug/gap ratio of 0.9 will give a shearing flow channel width at each wall of the same order as the maximum particle size. For the electrode gap h = 1.0 mm, the equivalent plug/gap ratio is 0.95. These are the values of plug/gap ratio found for the high electric field strengths and low volume flow rates of the tests.

6. Dimensional Analysis

Given this scatter and the various trends of results so far encountered dimensional analysis is indicated as the next clear step in examining the data. In [12] Wilkinson describes Hedström's analysis for Bingham plastic flow in pipes, and this is readily applied to the case of flow in a rectangular channel, assuming unit width with no edge effects. Whereas in the case of Newtonian flow there are two significant parameters, the friction coefficient and the Reynolds Number (assuming smooth walls), for the case of Bingham plastic flow there is one more dimension - the yield stress τ_b - and one more dimensionless parameter, referred to as the Hedström Number $\tau_b \rho h^2 / \mu^2$ or Plastic Number $\tau_b . h / \mu \bar{u}$. Just as the incompressible, Newtonian fluid motion depends on the ratio of inertial to viscous forces, so for a Bingham plastic another factor enters in as the viscous force field is modified by the superposition of the yield stress on the viscous shearing stress.

In the absence of yield stress, the Hedström Number is zero and the friction coefficient vs Reynolds Number plot is that for a Newtonian fluid in laminar flow where $Q \alpha \Delta P$. A non-zero yield stress will increase the Hedström Number and lead to a higher friction coefficient at a given Reynolds Number: the 'plug flow' increases the shearing rate at the walls for a given mean fluid velocity $\bar{u}$, i.e. for a given volume flow rate Q.

If the electrode gap h is increased at constant yield stress τ_b and volume flow rate Q, then mean velocity $\bar{u}$ falls in proportion as h increases. The nominal shearing rate $\bar{u}/h$ will then fall as $\bar{u}^2$ - hence for the case $\tau_b = 0$, the friction coefficient C_f is fixed as Reynolds number R_e is fixed. If τ_b has some fixed value >0, the result of a fall in the value of $\bar{u}/h$ is that shearing forces fall in proportion as $\bar{u}^{-2}$ or h^2, and the yield stress becomes rapidly more significant than the shearing stress, and will contribute the same effect as an increase of yield stress.

If the Bingham plastic model is a good approximation for the ER fluid, then the data must fall reasonably clearly into the three dimensionless groups, and this is in fact the case. Data is shown in Fig. 15 for five reasonably constant values of the Hedström Number, including the zero case of no-field flow in all the test valve configurations. The data sets are exactly those shown in Fig. 14 where electrode voltage is held constant for a given valve configuration, the Hedström Number denoted π_3 is reckoned constant. For each data set of constant π_3, the relationship between friction coefficient $\tau_{weo} / \rho\bar{u}^2$, denoted π_1, and Reynolds Number, $\rho\bar{u}h / \mu$ denoted π_2, is represented by a linear regression equation of the form, where a, m are constants,

$$\ln\pi_1 = a + m \ln\pi_2$$

The degree of constancy of the Hedström Number π_3 is shown in Fig. 16, where it can be seen that π_3 is fixed to a reasonable degree of accuracy by the electric field strength and the electrode gap. Fig. 16 also shows the prediction of Hedström Number for the valve flow from static clutch mode test data determined for small samples of the fluid taken from the valve test rig at the end of each test.

The linear regression data of Fig. 15 is summarised in Fig. 17 where the constants a and m are plotted against π_3.

We can now attempt to work backwards. If Hedström Number π_3 is specified, then a plot of π_1 vs π_2, friction coefficient vs Reynolds Number, is specified to a first approximation. A simple clutch mode test of a fluid sample should predict valve mode performance to a first approximation, for any valve configuration within reasonable bounds considering the scope of the tests. This methodology will be the subject of a further paper.

7. Conclusion

The finding that single properties of fluid can probably be used to predict what happens in shear and flow over a range of conditions and sizes, and with some accuracy provided corresponding conditions are observed, is tentative. A further paper will however show the closeness of simple clutch and valve data over the ranges encountered in the present paper.

ACKNOWLEDGEMENTS

The Ministry of Defence are thanked for their kind permission to disseminate these results which are mainly drawn from Departmental Research Reports No 106A, The Control Valve in the Steady State, Nov 1970, and 104A, Characteristics of an ER Valve, Aug 1980, of the Dept. Mech. Eng. University of Sheffield.

Comparative yield test results taken on a representative sample of fluid were supplied by courtesy of the use of a test rig loaned by Dr J E Stangroom.

8. References

1. J L Sproston et al, 'Electrorheological Fluids in the Squeeze Mode: A Numerical Study'. To be published in Jnl. Phys(D).

2. Zheng Lou, R D Ervin and F E Filisko 'A Preliminary Parametric Study of Electro-rheological Dampers'. To be published in Jnl. Fluids ASME.

3. W A Bullough, Advanced Material for Optics and Electronics, Vol. 1, p. 159-171 'A Rheological Semiconductor'.

4. M Whittle, W A Bullough, D J Peel and R Firoozian, 'Electro-rheological Dynamics Derived from Pressure Experiments in the Flow Mode'. (Submitted to Jnl Newtonian Fluids).

5. T Corden, W A Bullough, R Atkin and T G Kum, 'Two Dimensional Bingham Plastic Flow in a Cylindrical Pressurised Clutch' - see these proceedings.

6. W A Bullough, J Makin, A R Johnson and A Hosseini-Sianaki, 'The Effect of Solid Fraction Concentration of the Performance of an ER Fluid in the Shear Mode', see these proceedings.

7. R Firoozian, W A Bullough and D J Peel, Proc. ASME Design Conf. Montreal, 'Time Domain Modelling of the Response of an Electro-Rheological Fluid in the Flow Mode', Vibration Analysis - Techniques and Applications ASME publication. DE-Vol. 18-4, p. 45-50.

8. D J Peel and W A Bullough, Jnl. Intel. Matl. Syst. and Struct. Vol 4, No 1, Jan 93, p54-64. 'The Effect of Flow Rate, Excitation level and solid content on the Time Response of an ER Valve'.

9. D J Peel, D Brooks and W A Bullough, Proc. 9th Conference Fluid Machines, Budapest, p.362-369, 'Experiences in the Pumping of Electro-Rheological Fluids', Academy of Sciences, Hungary.

10. M Whittle, R Firoozian, D J Peel and W A Bullough, Jnl. Mod. Phys. B. vol. 6 No 15. 'A Model for the Electrical Characteristics of an ER Valve', p93-108.

11. J E Stangroom - personal correspondence.

12. W L Wilkinson, Non Newtonian Fluids, Pergamon 1960.

9. Nomenclature

a - linear regression, constant

b - valve effective width = 2π x mean radius, (x 2) in double channel valves.

E - magnitude of electric field = V/h.

f( ) - a function of ( ).

K - general constant.

h - inter electrode spacing.

ℓ - length of valve.

ΔP - valve pressure drop over length ℓ. - direction of increasing flow rate.

Q - volumetric flow rate through valve, in direction of ℓ.

m - linear regression, slope.

u - velocity along the valve.

$\bar{u}$ - mean velocity.

V - voltage applied to inter electrode space.

y - direction across the valve.

δ - width of plug flow region.

μ - fluid viscosity at zero volts and plastic viscosity when excited.

υ - distance to point in shearing zone new from centre of channel.

π - dimensionless group.

$\dot{\gamma}$ - general shear rate.

θ - temperature of fluid.

τ - shear stress.

Cf - friction coefficient, $\tau_{weo} / \rho\bar{u}^2 = \pi_1$

Re - Reynolds Number, $\rho\bar{u}h / \mu = \pi_2$

Hd - Hedström Number, $\tau_b . \rho h^2 / \mu^2 = \pi_3$

Suffixes

b - derived from use of Bingham model.

e- - due to field effect alone.

o- - due to flow effect alone.

eo- - due to combined flow and field effects.

n - identification number.

w- - wall condition, on electrode.

y - a yield effect, in shear/clutch mode.

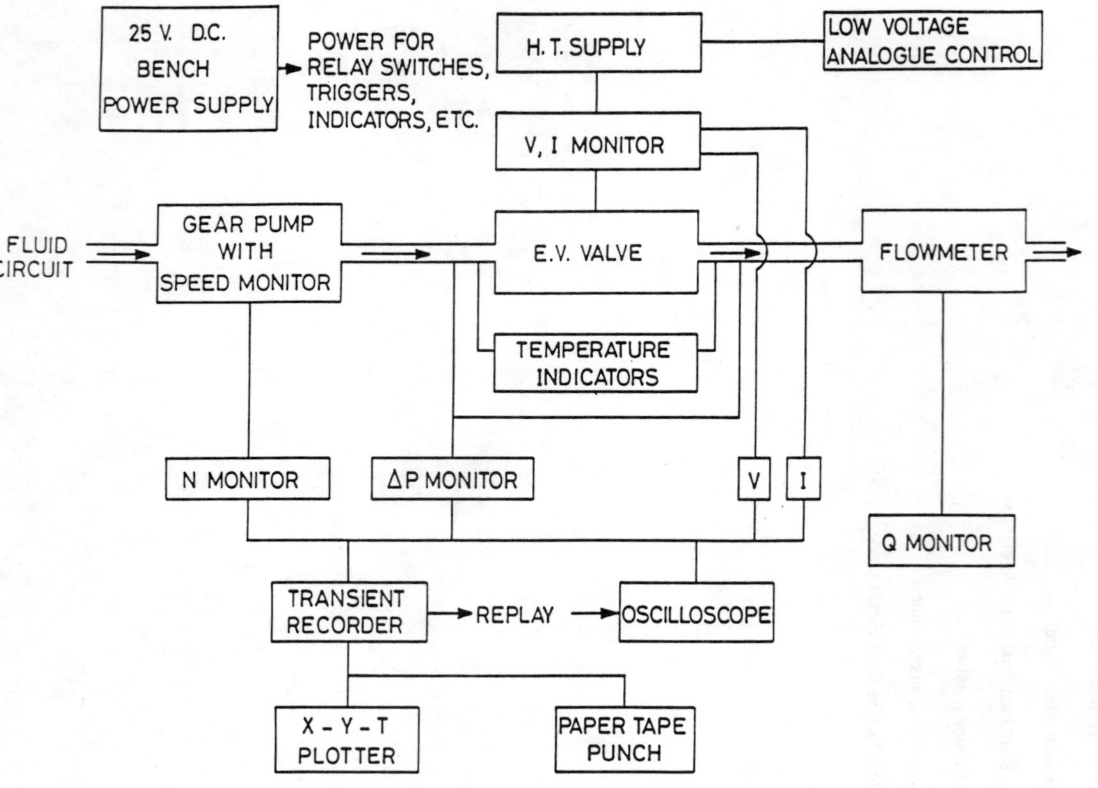

Fig 1. Test Rig

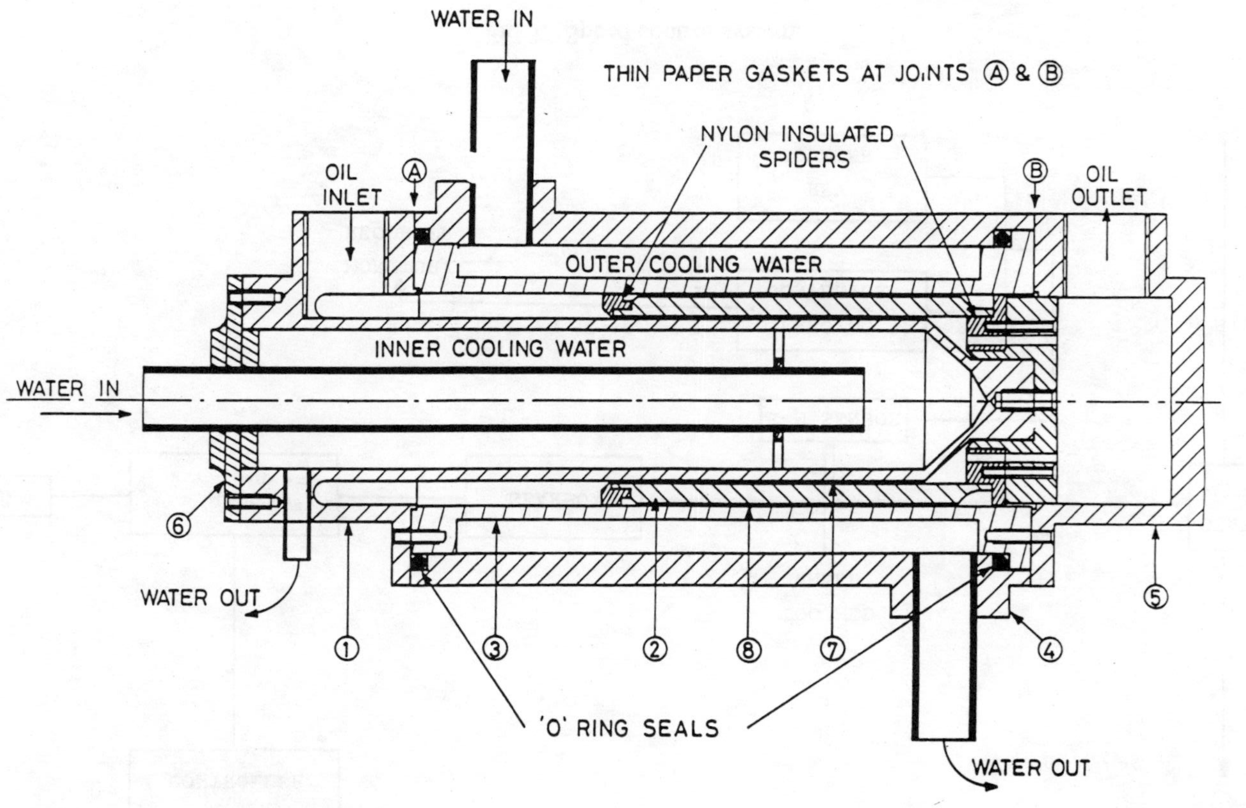

Fig 2. Test Valve Valve A : h = 0.5 mm ℓ = 100 mm Valve B: 0.5 x 50 mm
Valve C: 1 x 100 mm

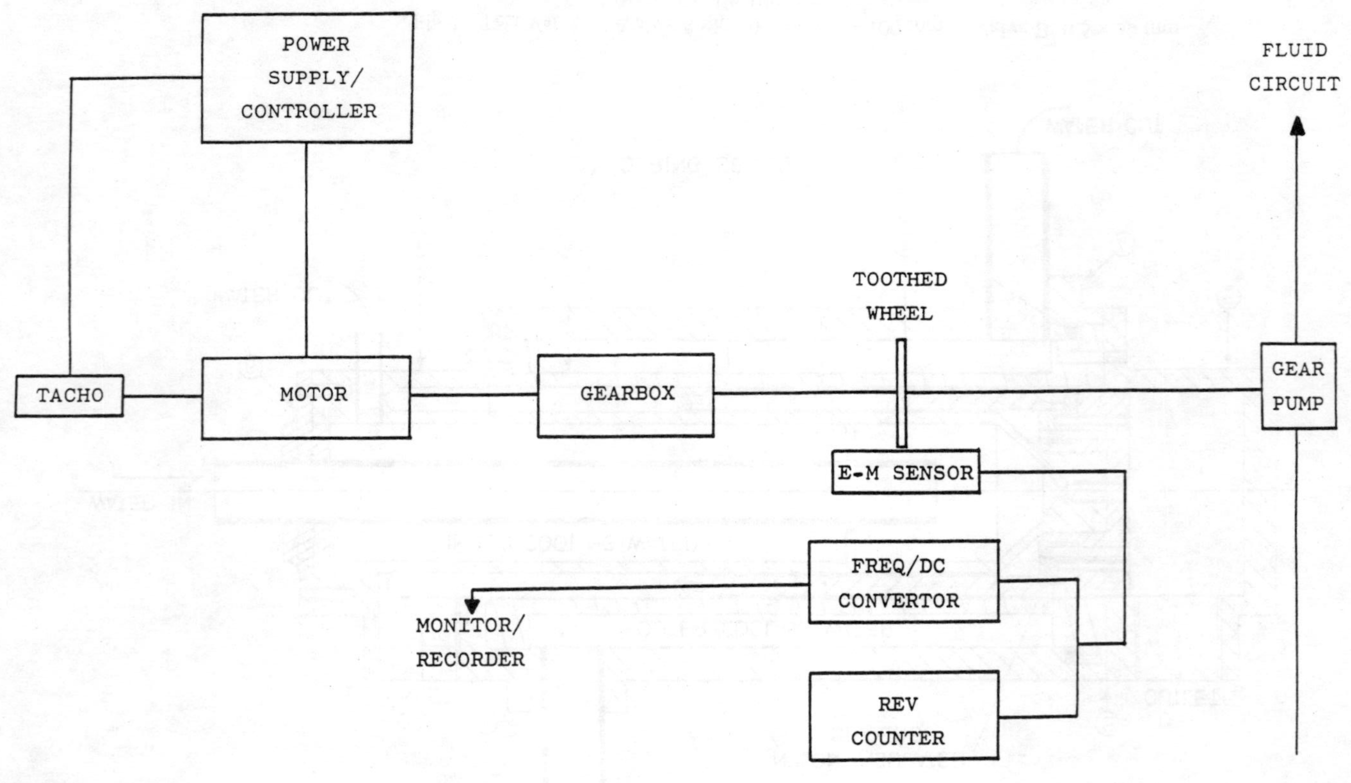

Fig 3. Speed control system.

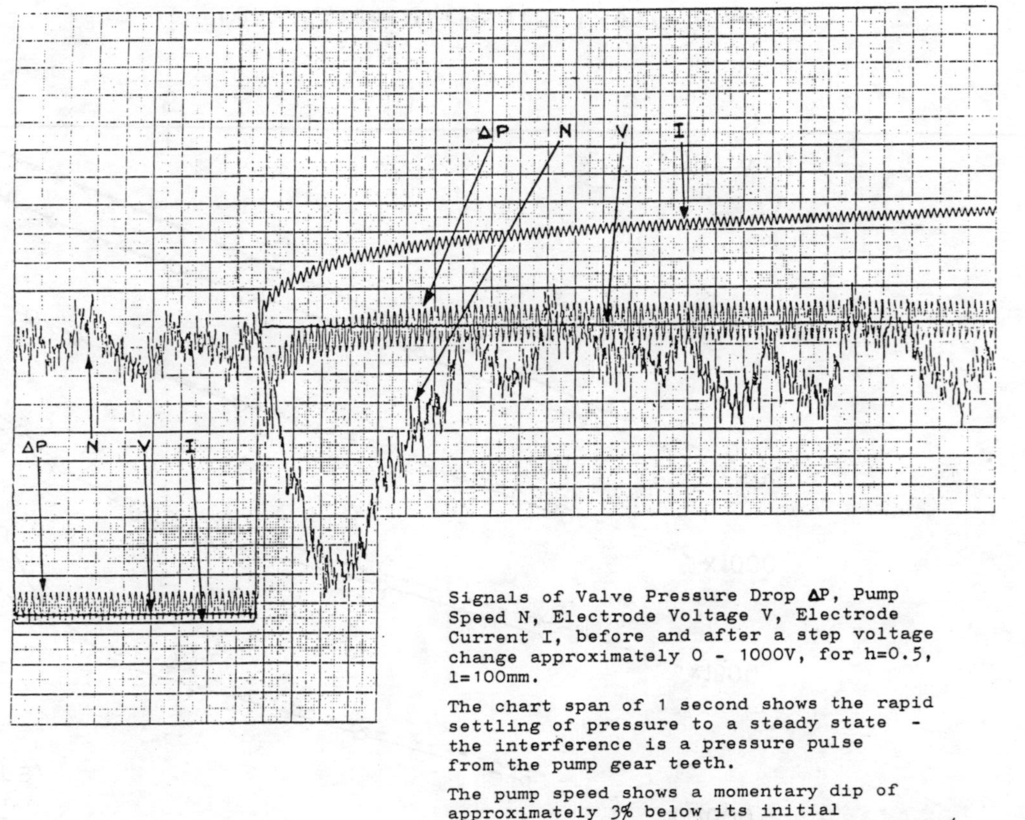

Fig 4. Typical Pressure recording - Definition Steady State.

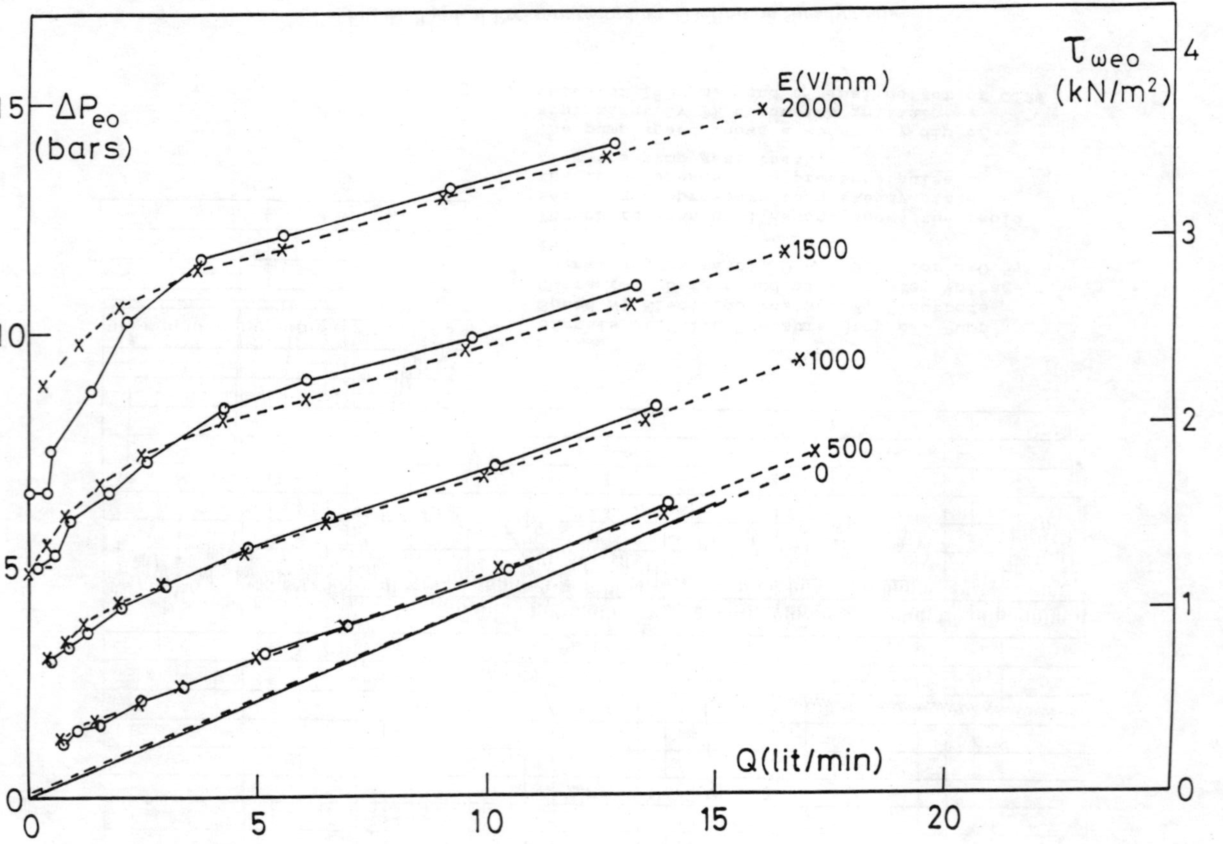

Fig 5. ΔP_{eo} versus Q for values of E. h = 0.5 ℓ = 100 mm

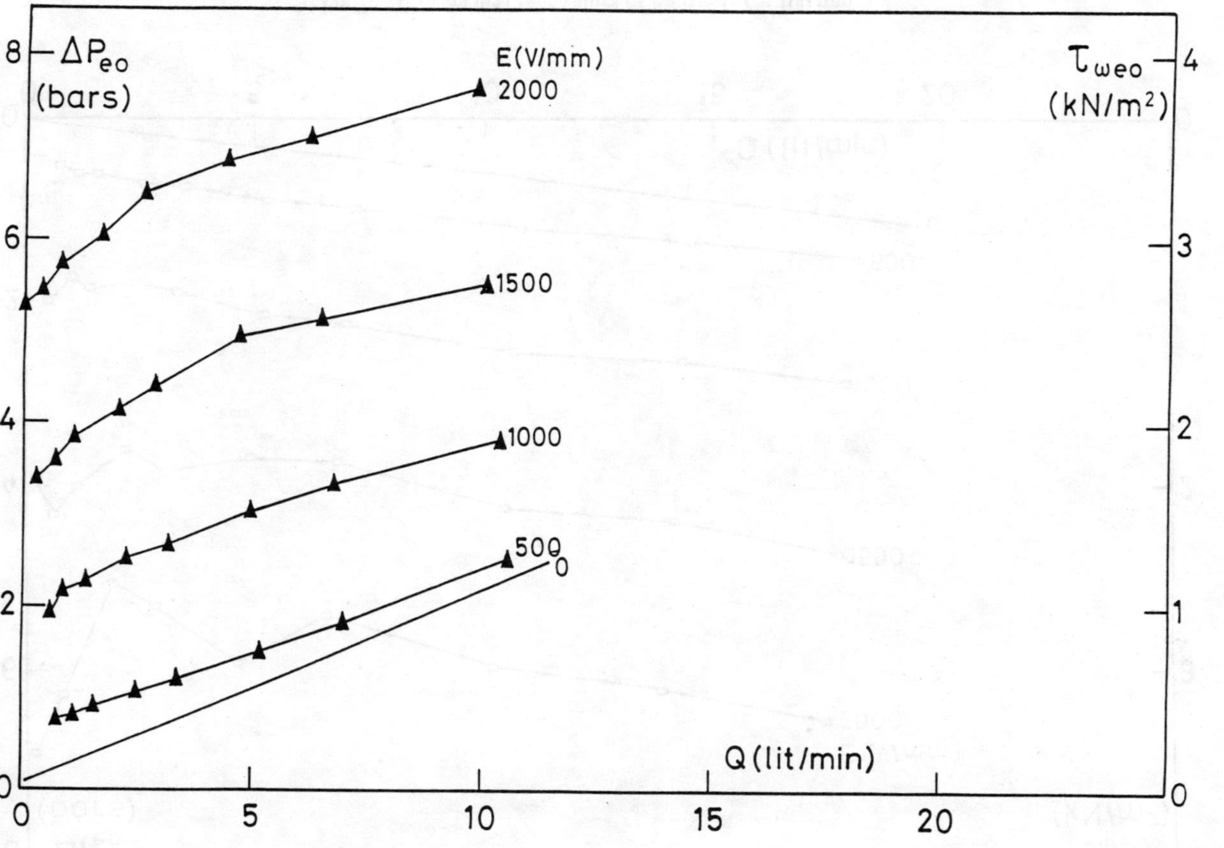

Fig 6. ΔP_{eo} versus Q for values of E. $h = 0.5$ $\ell = 50$ mm

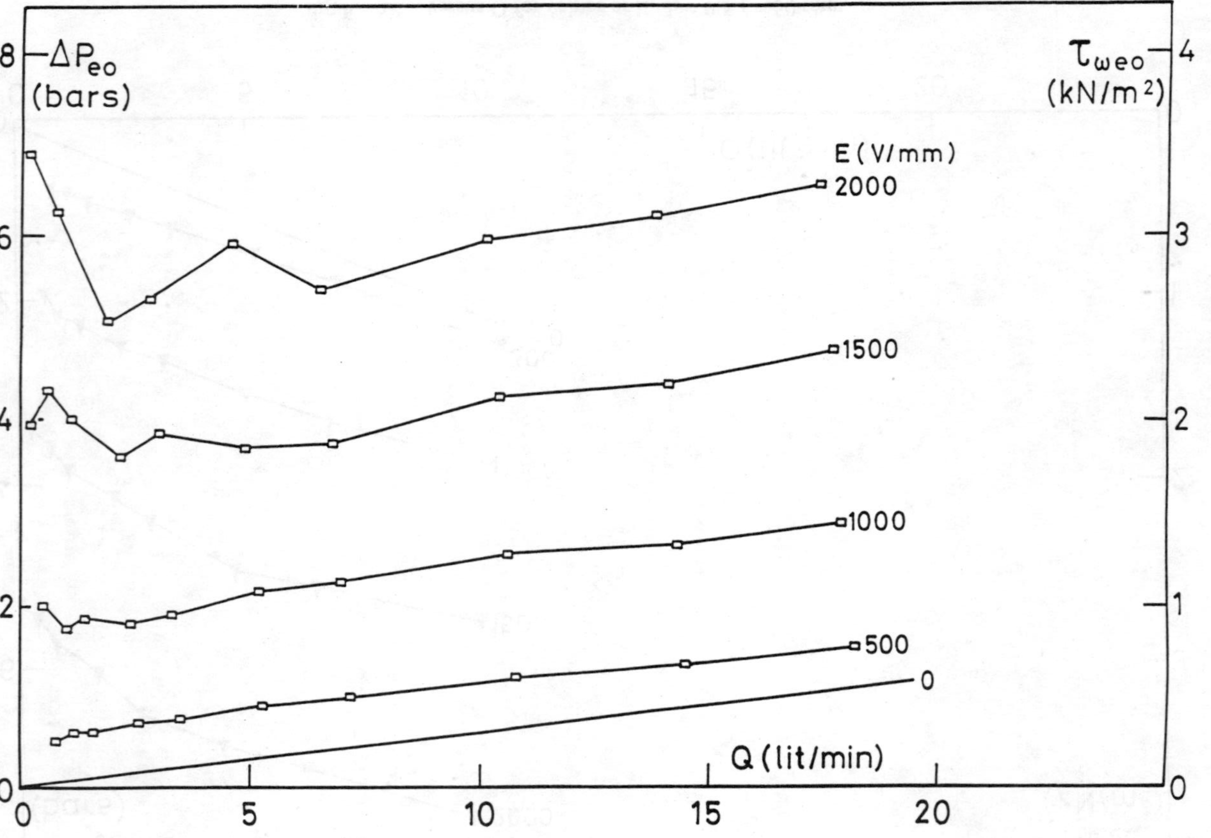

Fig 7. ΔP_{eo} versus Q for values of E. h = 1 ℓ = 100 mm

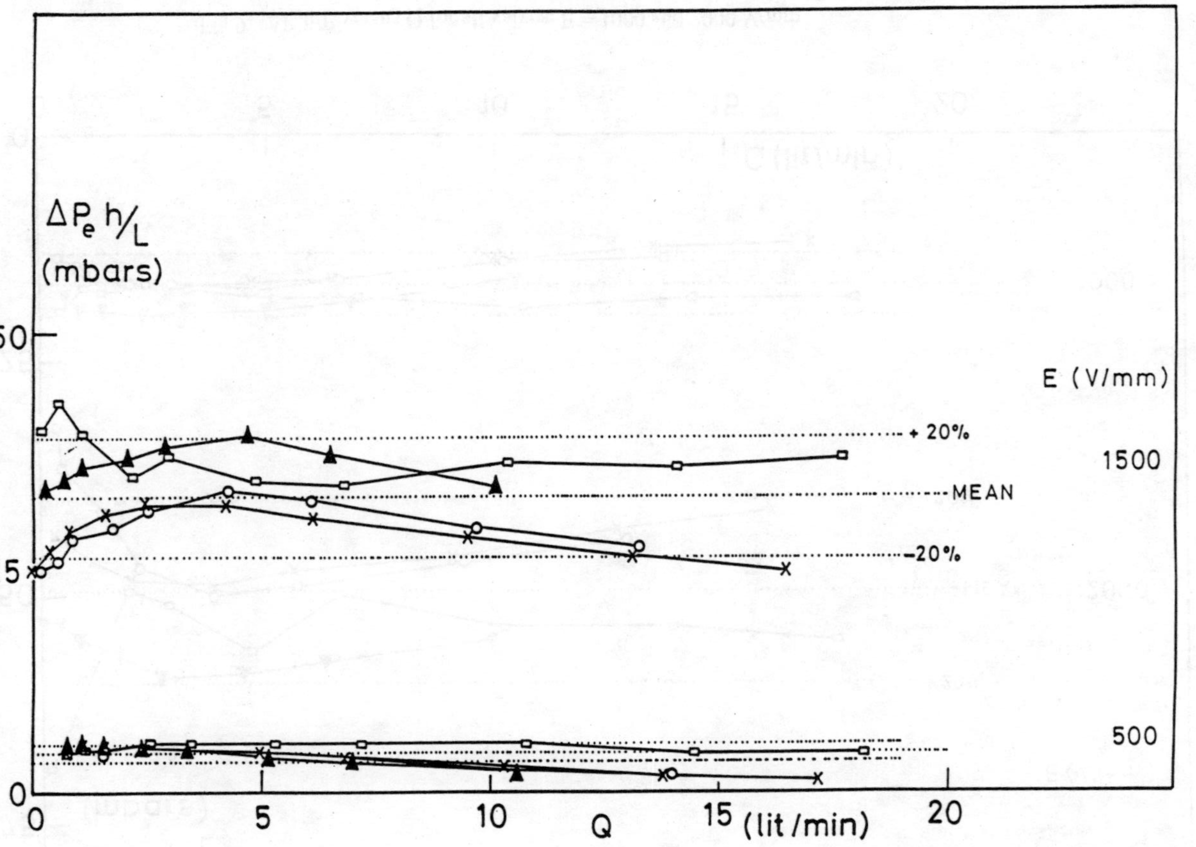

Fig 8. $\Delta P_e h/E$ versus Q for all valves, E = 500 and 1500 V/mm

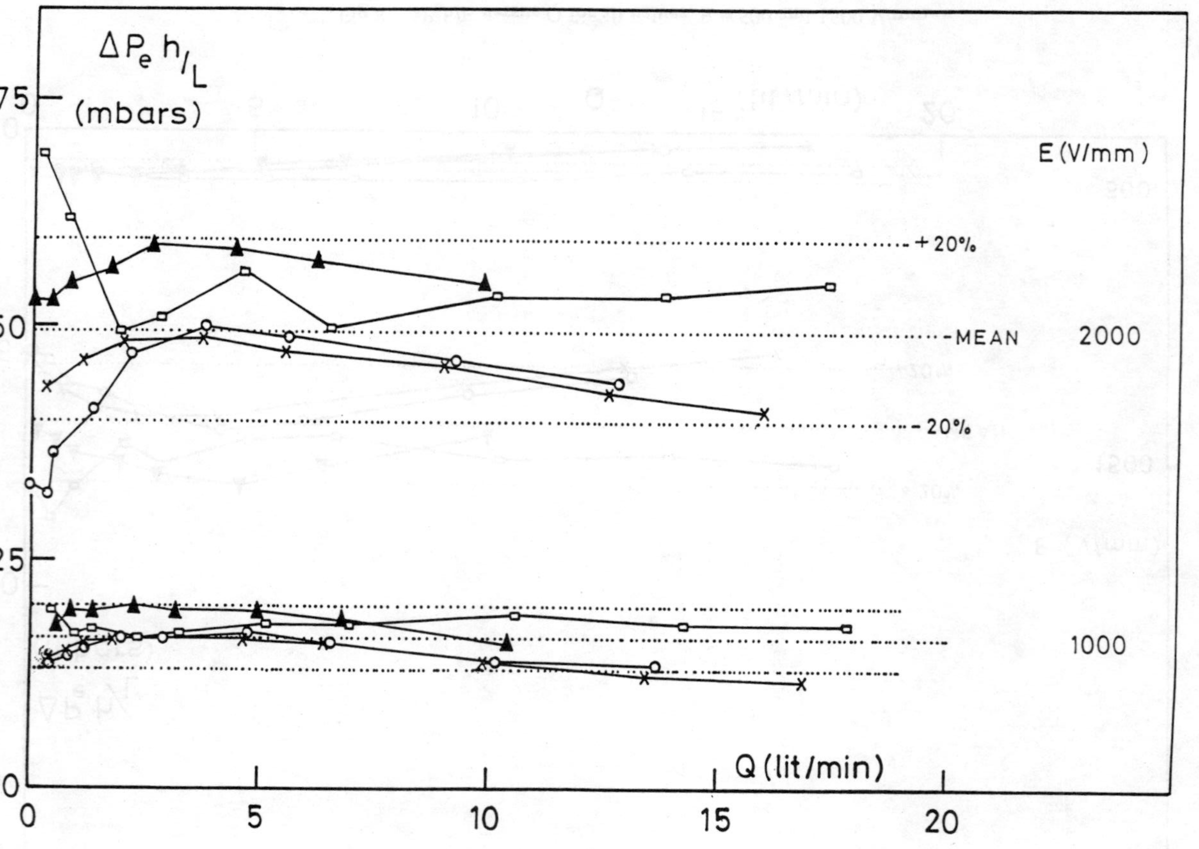

Fig 9. $\Delta P_e h/E$ versus Q for all valves, E = 1000 and 2000 V/mm

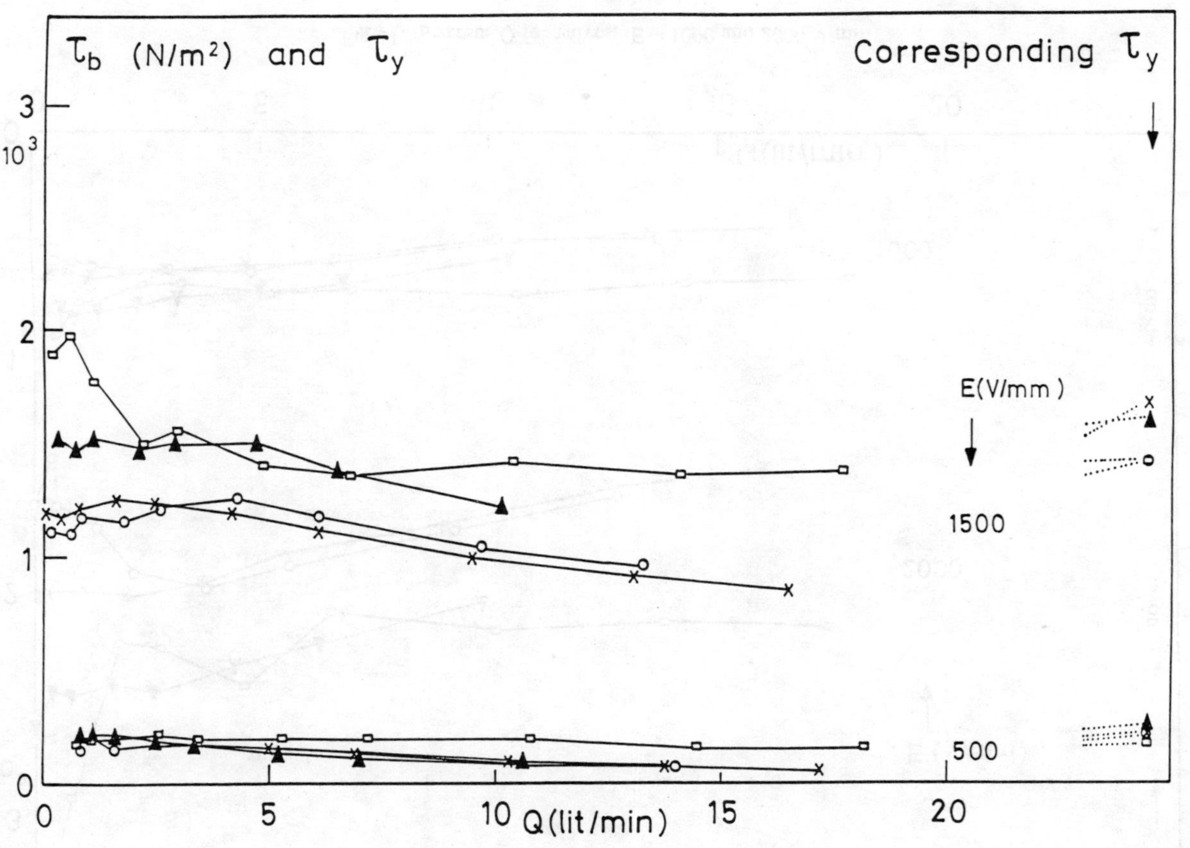

Fig 10. τ_b versus Q for all valves, E = 500 and 1500 V/mm

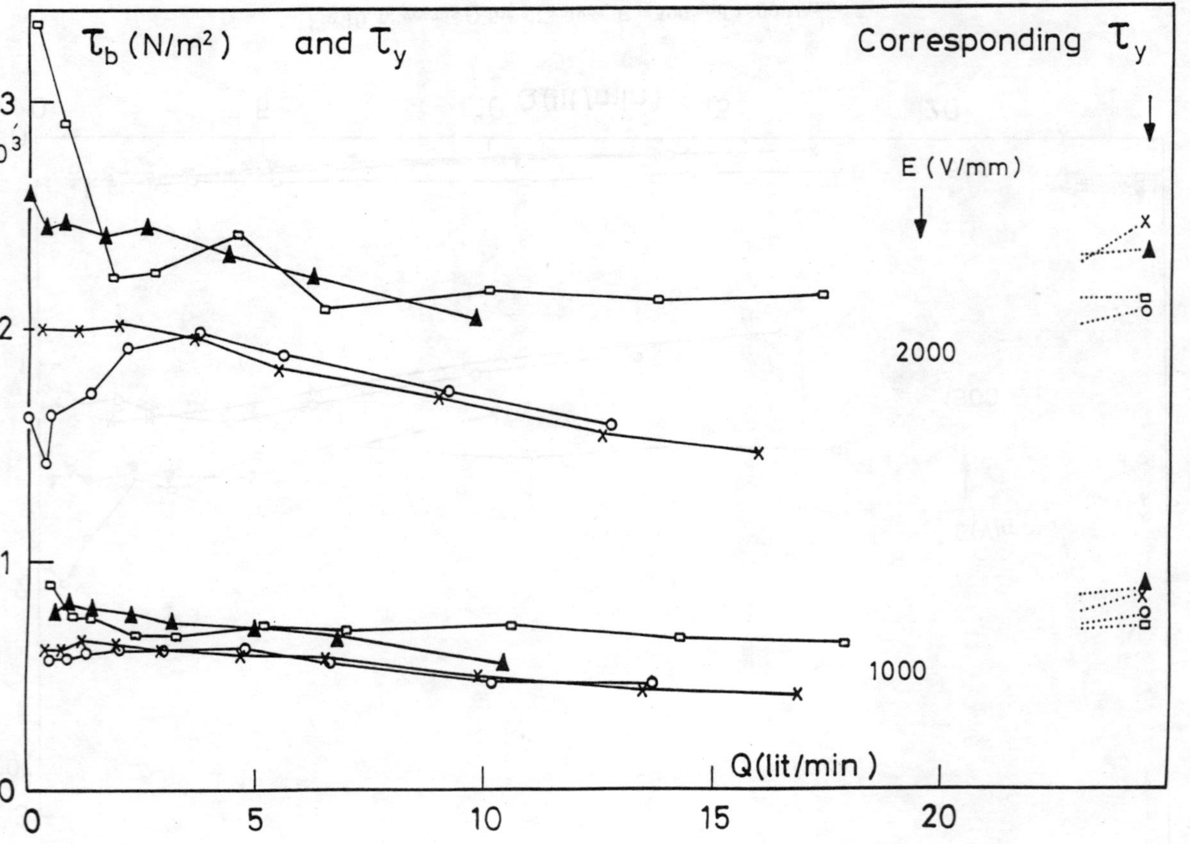

Fig 11. τ_b versus Q for valves, E = 1000 and 2000 V/mm

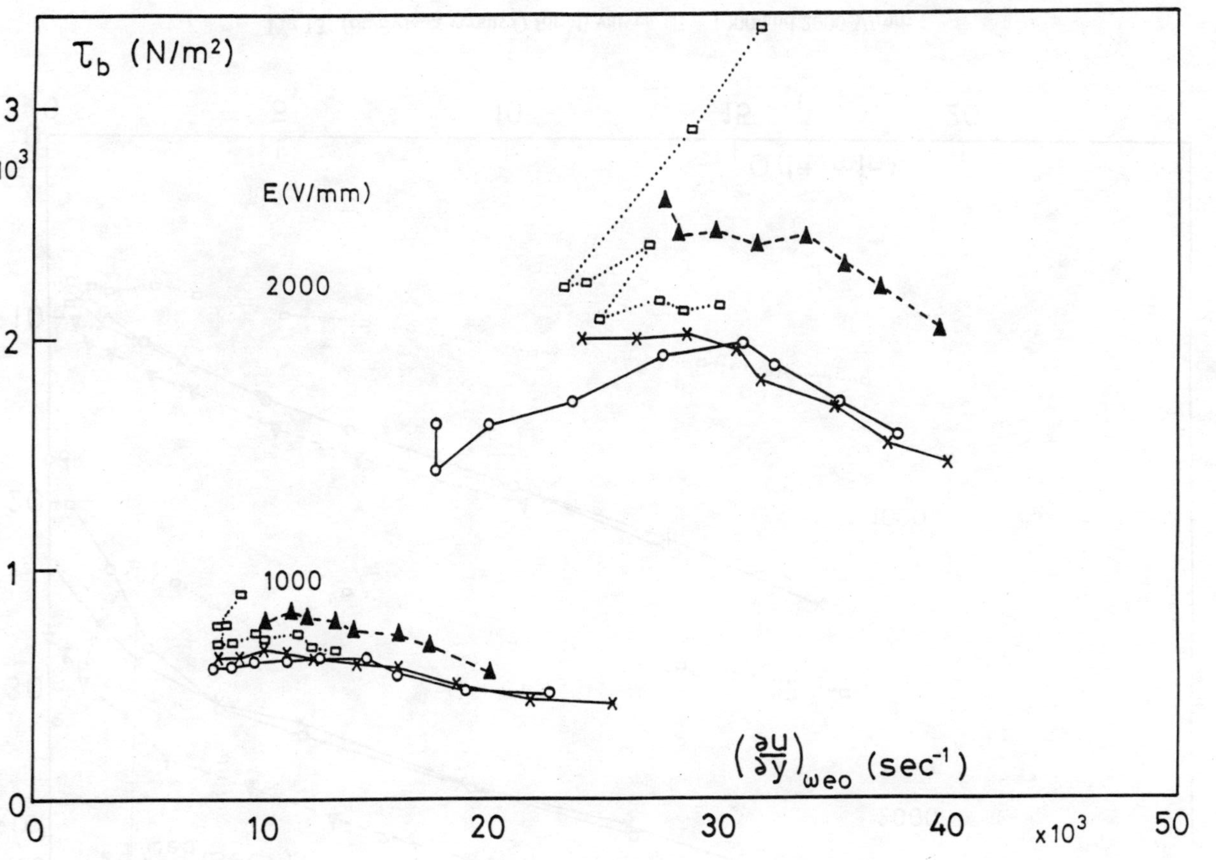

Fig 12. τ_b versus $(\partial u/\partial y)_{weo}$ for all valves, E = 1000 and 2000 V/mm

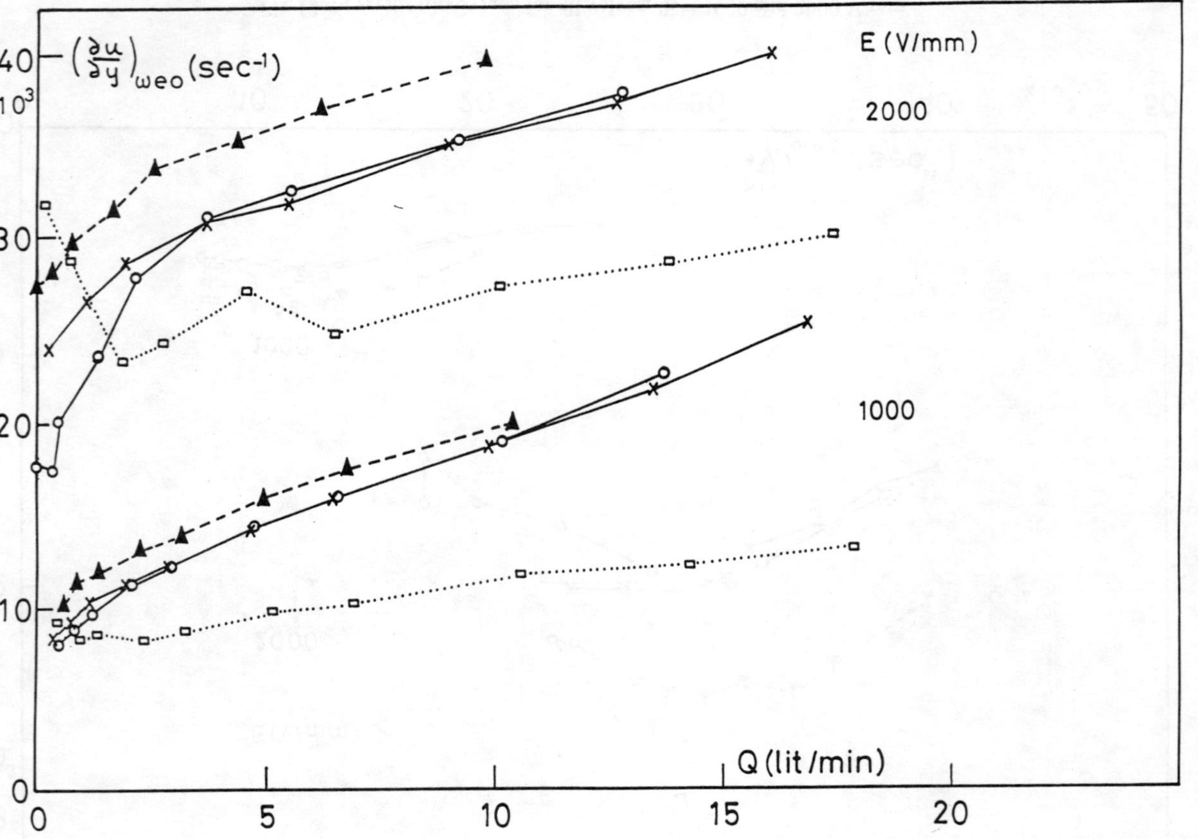

Fig 13. $(\partial u/\partial y)_{weo}$ versus Q for all valves, E = 1000 and 2000 V/mm

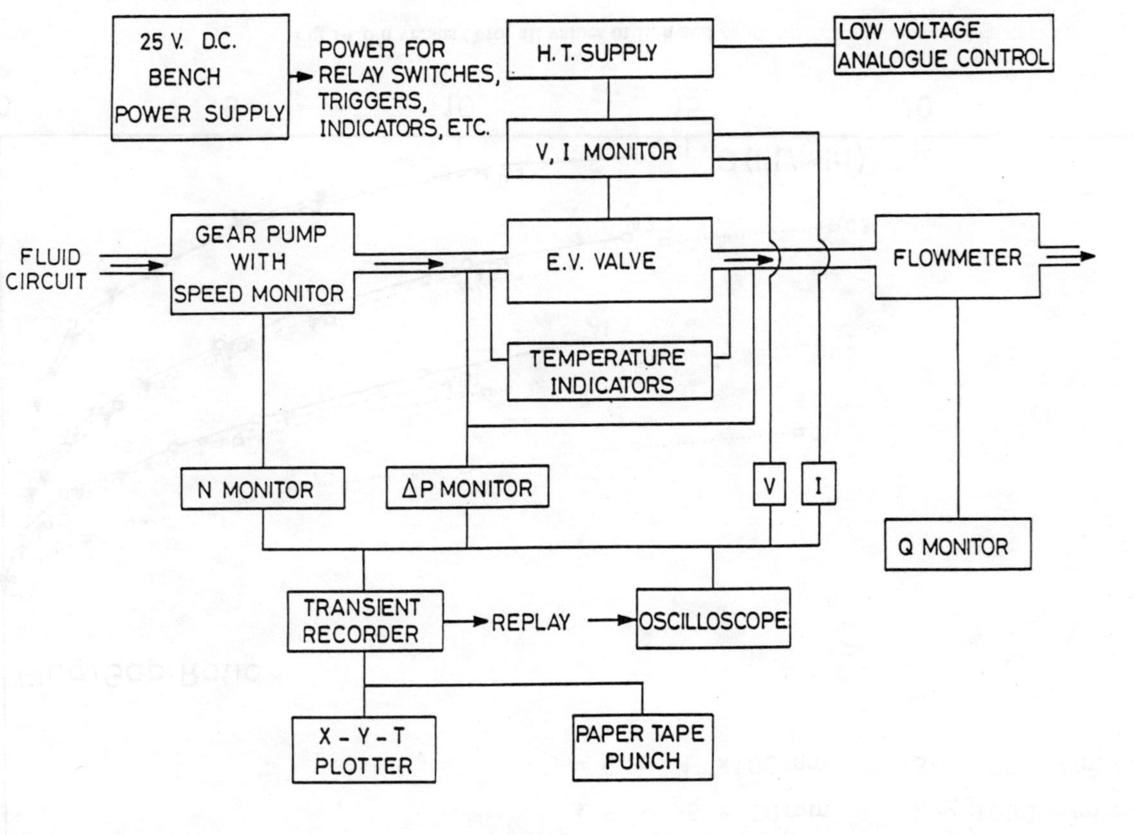

Fig 1. Test Rig

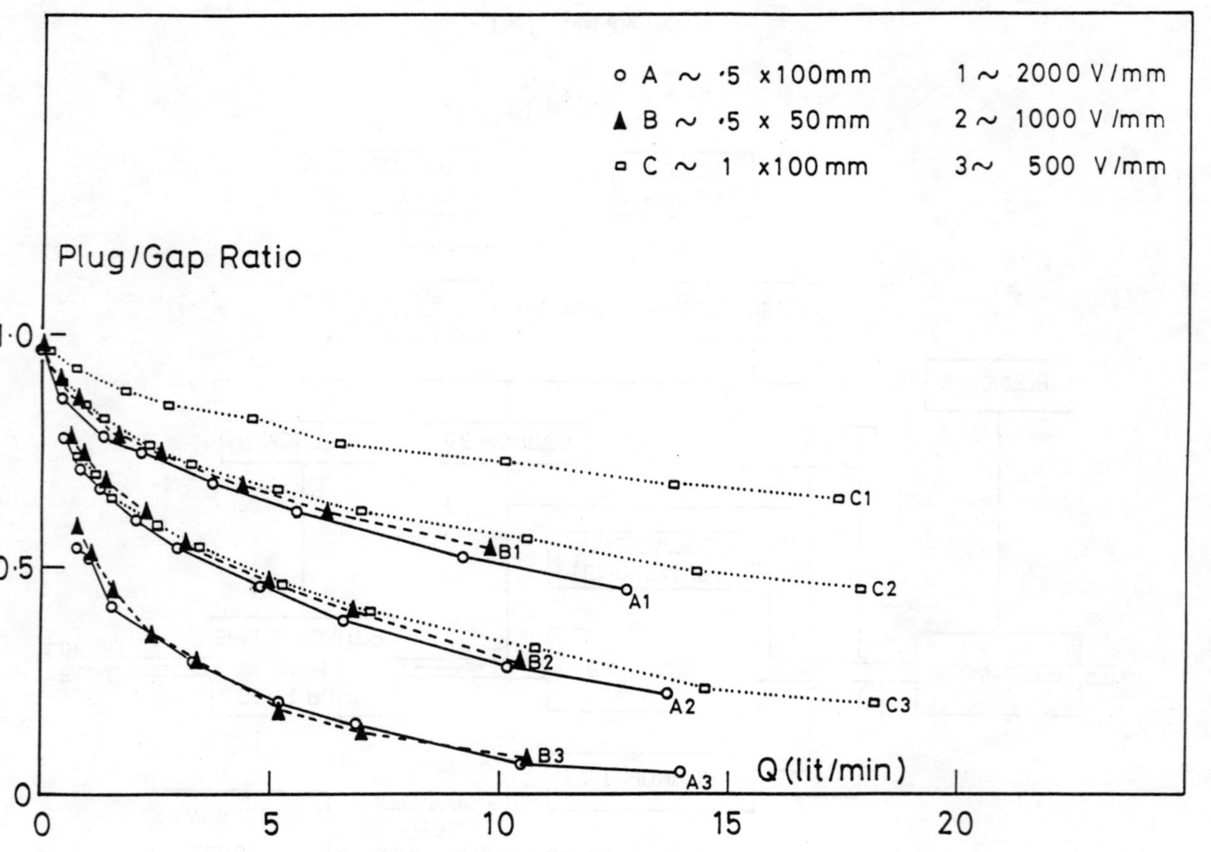

Fig 14. δ/h versus Q for all values of E, h and ℓ

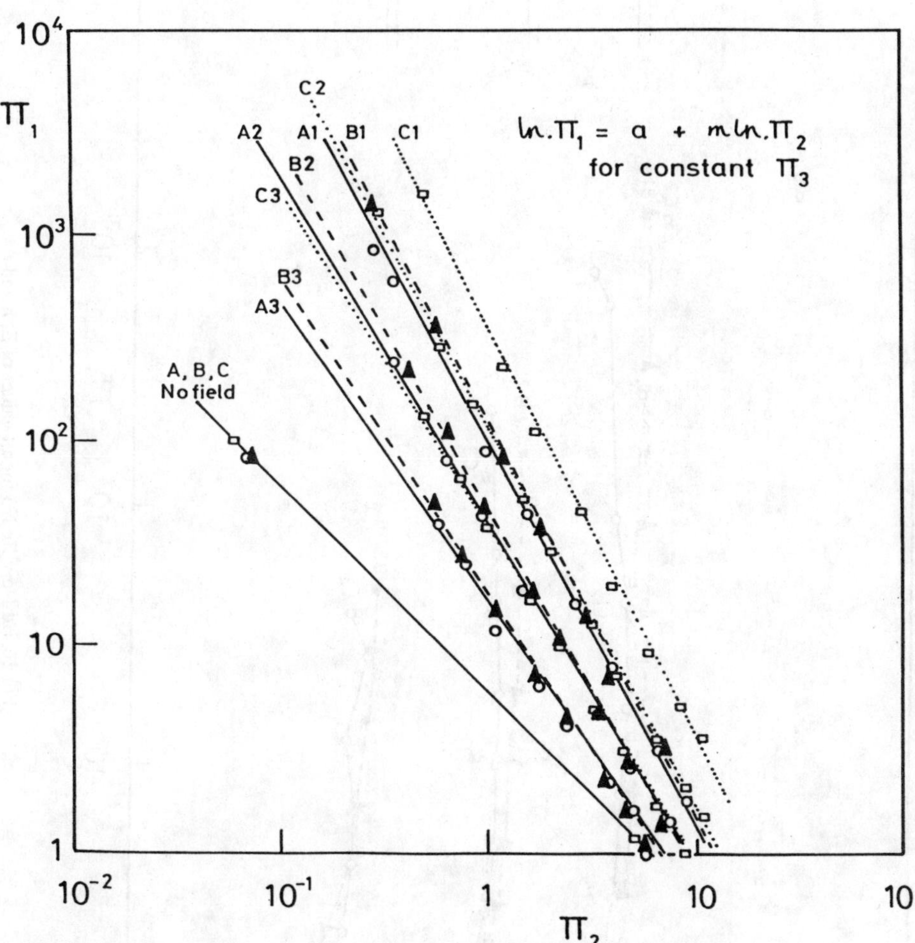

Fig 15. Cf versus Re No for lines of constant Hd No

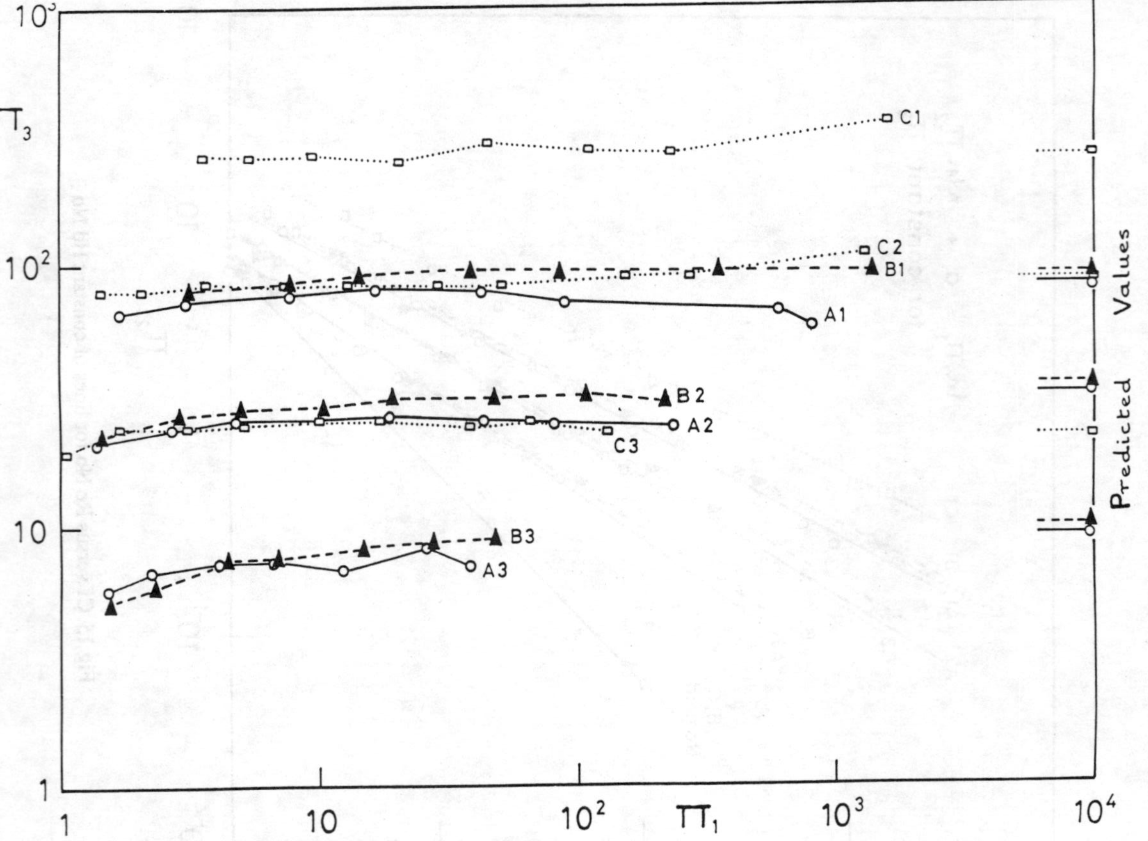

Fig 16. Hd No v Re No for all values of E, h and ℓ

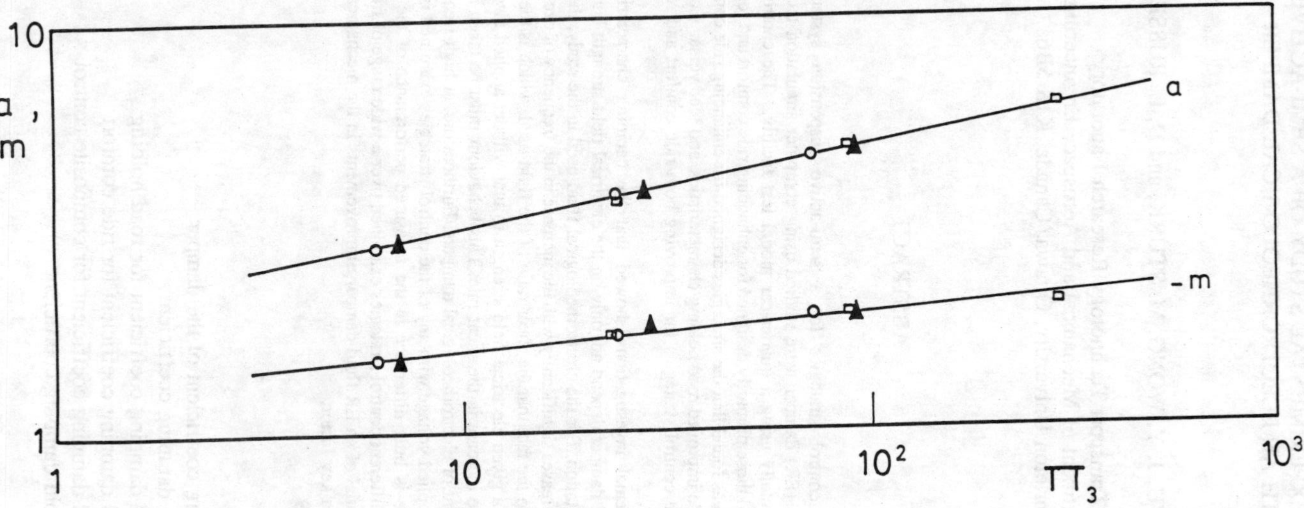

Fig 17. Correlation of slopes of Cf/Re No of Fig 15

SIMULATION AND EXPERIMENTAL STUDY OF A SEMI-ACTIVE SUSPENSION WITH AN ELECTRORHEOLOGICAL DAMPER

X. M. WU, J. Y. WONG, M. STURK, and D. L. RUSSELL

Transport Technology Research Laboratory
Department of Mechanical and Aerospace Engineering
Carleton University, Ottawa, Canada, K1S 5B6

ABSTRACT

Various control strategies for a semi-active suspension system with an electrorheological (ER) damper were studied using computer simulation techniques, as well as experimentally using a quarter-car model test facility. The control strategies examined included those primarily designed for enhancing ride comfort and for improving road holding. It was found that the strategies designed for enhancing ride comfort do not necessarily provide improved road holding characteristics, and vice versa. Consequently, various composite control strategies for improving both ride comfort and road holding were investigated.

Experimental investigations showed that the damping characteristics of an electrorheological damper depend not only on the electrical field strength but also on the frequency of excitation. For the electrorheological fluid used in the study, the equivalent damping ratio decreases significantly with the increase in the frequency of excitation. This is primarily due to the fact that the shear ratio of the fluid used, which is the ratio of the shear strength at a given electrical field strength to that without applied electrical field, decreases with the increase in the shear rate. This behaviour must be taken into account in the development of electrorheological dampers. Furthermore, at high frequencies, the duration of the applied voltage with any of the control strategies examined is very short. As a result, there is little difference in the measured performance of the semi-active suspension with different control strategies examined over a wide range of frequency. To achieve the potential of an ER fluid damper, improvements in the mechanical behaviour of ER fluids are a key factor.

Nomenclature

C_2	damping coefficient of the damper
C_c	critical damping coefficient
C_d	desired damping coefficient for road holding
C_r	desired damping coefficient for ride comfort
C_{rd}	desired damping coefficient for composite control
C_{th}	threshold damping coefficient

k_1	equivalent spring rate of the tire
k_2	spring rate of the suspension
m_1	unsprung mass
m_2	sprung mass
$x, \dot{x}, \ddot{x}$	vertical displacement, velocity and acceleration of the tire contact point with the road surface, respectively
$y, \dot{y}, \ddot{y}$	vertical displacement, velocity and acceleration of unsprung mass, respectively
$z, \dot{z}, \ddot{z}$	vertical displacement, velocity and acceleration of sprung mass, respectively
α	ratio of dynamic tire deflection to static tire deflection
α_{th}	threshold ratio of dynamic tire deflection to static tire deflection

1. Introduction

To further improve ride comfort and road holding of ground vehicles, active and semi-active suspension systems have attracted considerable interest in recent years. While an active suspension can provide improved ride and handling, as well as the control of the height, roll, dive and squat of the vehicle body, it is complex and requires considerable external power to operate. In comparison, a semi-active suspension is less complex and requires much less power to operate. In this type of suspension, the conventional suspension spring is retained, while the damping force of the damper can be modulated in accordance with operating conditions. The regulating of the damping force can be achieved by adjusting the orifice area in the shock absorber, thus changing the resistance to fluid flow. More recently, the possible application of the electrorheological (ER) fluid to the development of controllable dampers has attracted considerable interest. Its resistance to flow is related to the electrical voltage applied across it. The process is continuous and reversible, and the response is almost instantaneous. By regulating the voltage applied across the flow of the ER fluid in a damper, the damping force can be varied in a convenient way[1,2]. It has been shown that when properly designed, the performance of a semi-active system may approach that of an active system under a variety of operating conditions[3].

In this paper, various control strategies for the semi-active suspension to achieve improved ride comfort and road holding are evaluated and compared, using computer simulation techniques. Based on the simulation results, the performance of a semi-active suspension with an ER damper having various control strategies was experimentally measured and evaluated, using a quarter-car model test facility in the laboratory.

2. Control Strategies for Semi-Active Suspensions

A semi-active suspension system with appropriate control strategy should provide improved ride comfort and road holding for a given suspension travel, in comparison with the conventional passive suspension. Ride comfort may be evaluated by the response of the sprung mass to the excitation from the ground. Usually, the transmissibility ratio (or transfer function) is used as a basis for assessing the vibration isolation characteristics. Road holding is related to the variation of the normal force between the tire and the road during vibration. It can be represented by the dynamic tire deflection or the displacement of the tire centre relative to the road profile[4]. To examine the vibration isolation and road holding characteristics of a semi-active suspension with different control strategies, a quarter-car model, as shown in Fig. 1, is used in the analysis.

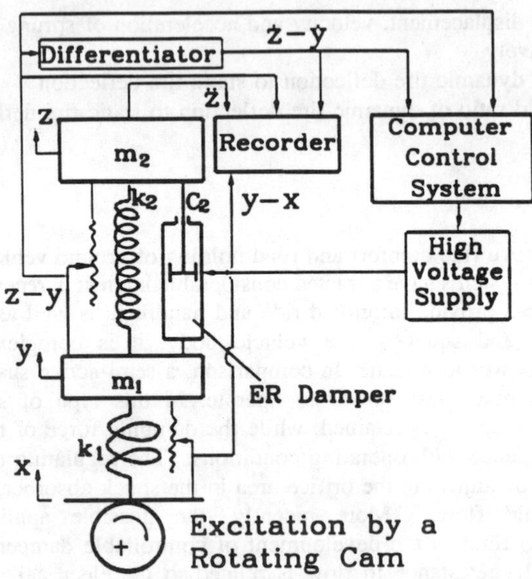

Figure 1 A quarter-car model

If the damping is assumed to be viscous, the equations of motion for the two degrees of freedom system are as follows:

$$m_1\ddot{y}=k_2(z-y)+C_2(\dot{z}-\dot{y})-k_1(y-x) \quad (1)$$

$$m_2\ddot{z}=-k_2(z-y)-C_2(\dot{z}-\dot{y}) \quad (2)$$

where m_1 is the unsprung mass, m_2 is the sprung mass, k_1 is the equivalent spring stiffness of the tire, k_2 is the suspension spring stiffness, and C_2 is the damping coefficient of the damper.

It should be noted that the static equilibrium positions of the sprung and unsprung masses are taken as the origins for the co-ordinates.

2.1 *Control Strategies for Improving Ride Comfort*

In the past, investigations into the control strategies for semi-active suspensions have been primarily concentrated on improving ride comfort (or vibration isolation). Three representative control strategies for improving ride comfort are outlined below. The objective is to reduce the sprung mass acceleration by modulating the damping force in accordance with operating conditions.

Strategy A. An on-off control strategy proposed by Krasnicki[5] and Margolis and Goshtasbpour[3]. This control strategy can be described as follows:

If $\dot{z}(\dot{z}-\dot{y})>0$, then the maximum damping is required,

and if $\dot{z}(\dot{z}-\dot{y})<0$, then the minimum damping is required.

where $\dot{z}$ is the velocity of the sprung mass; $\dot{z}-\dot{y}$ is the relative velocity across the damper. This strategy indicates that if the relative velocity of the sprung mass with respect to the unsprung mass is in the same direction as that of the sprung mass velocity, then a maximum damping force should be applied to reduce the sprung mass acceleration. On the other hand, if the two velocities are in the opposite direction, the damping force should be at a minimum to minimize the acceleration of the sprung mass. This control strategy is based on the "sky-hook" concept. It should be noted that this control strategy requires the measurement of the absolute velocity of the sprung mass. The accurate measurement of the absolute vibration velocity of the sprung mass on a moving vehicle is, however, very difficult to achieve. Thus, this control strategy is difficult to implement in practice.

Strategy B. A control strategy intended to minimize the sprung mass acceleration by continuously adjusting the damping force. From Eq. (2), it can be seen that to reduce the sprung mass acceleration to zero, the desired damping coefficient C_r is given by

$$C_r=-k_2(z-y)/(\dot{z}-\dot{y}) \tag{3}$$

where z-y and $\dot{z}-\dot{y}$ are the relative displacement and relative velocity across the damper, respectively.

It should be noted that this control strategy only requires the measurements of relative displacement and velocity between the sprung and unsprung mass, which can easily be made in practice. It should be pointed out that to satisfy Eq. (3), the desired damping coefficient may be negative under certain circumstances. Since negative damping can not be realized in practice without active power input, the minimum possible damping should be applied in this case. On the other hand, if the desired damping coefficient, as determined by Eq. (3), is beyond the range that the active damper can supply, the maximum possible damping should then be applied. This control strategy is similar to that proposed by Alanoly and Sankar[6] and Jolly and Miller[7].

Strategy C. A control strategy similar to strategy B, but instead of continuously adjusting the damping force, it will be set at either the maximum or the minimum value, dependent upon the operating conditions.

From Eq. 3, the desired damping coefficient C_r is

$C_r=-k_2(z-y)/(\dot{z}-\dot{y})$

If $C_r \le C_{th}$, then damping is set at the minimum level

if $C_r > C_{th}$, then damping is set at the maximum level
where C_{th} is referred to as the threshold damping coefficient. Based on the results of simulation and experimental study, the suitable value of C_{th} may be taken as 0.3 C_c, where C_c is the critical damping coefficient for the sprung-mass system.

The three control strategies described above are referred to as ride comfort control strategies A, B and C, respectively.

2.2 *Control Strategies for Improving Road Holding*

It should be understood that the strategies described in the previous section are designed for improving ride comfort. It will be shown later that these strategies do not necessarily offer improved road holding. To improve road holding, the following two strategies are studied. The objective is to maintain the normal load between the tire and the road at a certain level by modulating the damping force in accordance with operating conditions.

Strategy A. An on-off control strategy for modulating the damping force based on the direction of the relative velocity between the sprung and unsprung mass.

If $\dot{z}-\dot{y} \geq 0$, then the damping is set at the minimum level,
and if $\dot{z}-\dot{y} < 0$, then the damping is set at the maximum level.

This strategy indicates that if the relative velocity is positive, that is, the damping force exerted on the tire is upward, then the value of the damping coefficient should be set at the minimum level to help maintain the normal load between the tire and the road. On the other hand, if the damping force exerted on the tire is downward, the damping is set at the maximum level to enhance road holding.

Strategy B. A control strategy intended to minimize the dynamic tire deflection by continuously adjusting the damping force. From Eq. (1), it can be seen that to reduce the dynamic tire deflection (y-x) to zero, the desired damping coefficient C_d is given by

$$C_d=-[k_2(z-y)-m_1\ddot{y}]/(\dot{z}-\dot{y}) \tag{4}$$

It should be pointed out that to satisfy Eq. (4), the desired damping coefficient may be negative under certain circumstances. Since negative damping cannot be realized in practice without active power input, the minimum possible damping should be applied in this case. On the other hand, if the desired damping coefficient, as determined by Eq. (4), is beyond the range that the active damper can supply, the maximum possible damping should then be applied.

The two control strategies described above are referred to as road holding control strategies A and B, respectively.

2.3 *Control Strategies for Enhancing Both Ride Comfort and Road Holding*

The control strategies described in section 2.2 are intended for improving road holding. It will be shown later that these strategies do not necessarily offer improved ride comfort. In an attempt to provide a proper balance between ride comfort and road holding, the following three control strategies, which will be referred to as composite control strategies A, B and C, are examined.

Strategy A. A composite control strategy based on the ratio of the dynamic tire deflection to the static tire deflection. The damping coefficient C_{rd} for this control strategy is set in accordance with the following condition

$$C_{rd}=\alpha C_d+(1-\alpha)C_r \tag{5}$$

where α is the ratio of the dynamic tire deflection to the static tire deflection, that is, $\alpha=|y-x|/[(m_1+m_2)g/k_1]$, and C_d and C_r are the desired damping coefficients for road holding and for ride comfort, respectively, as defined previously.

It should be noted that when α=0, the damping coefficient $C_{rd}=C_r$, which means that the control is focused on improving ride comfort. On the other hand, if α=1, the damping coefficient $C_{rd}=C_d$, which means that the control is concentrated on improving road holding.

It should be mentioned that the dynamic tire deflection (y-x), which is used as a control parameter, can be derived from the measured accelerations of the sprung and unsprung masses as follows:

$$y-x=-(m_1\ddot{y}+m_2\ddot{z})/k_1 \tag{6}$$

Strategy B. The damping coefficient C_{rd} is set to the value of either C_r or C_d defined previously, dependent upon the value of α, that is

$C_{rd}=C_r$, if $\alpha< \alpha_{th}$,

and $C_{rd}=C_d$, if $\alpha> \alpha_{th}$.

where α_{th} is the threshold value for the ratio of the dynamic tire deflection to the static tire deflection. Based on the results of simulation and experimental study, α_{th} may be taken as 0.5.

This strategy indicates that if the dynamic tire deflection is low, the control is focused on improving ride comfort. On the other hand, when the dynamic tire deflection is higher than a predetermined value, the control is concentrated on improving road holding.

Strategy C. Similar to strategy B described above, the damping coefficient C_{rd} depends upon the value of α. This control strategy is described as follows:

C_{rd} is set at the maximum damping value, if $\alpha> \alpha_{th}$,

and $C_{rd}=C_r$, if $\alpha< \alpha_{th}$.

3. Simulation Study of Various Control Strategies

3.1 *Vibration Isolation*

The various control strategies described above are evaluated and compared using computer simulations based on the quarter-car model shown in Fig 1. The parameters of the model are: m_1=6.2 kg, m_2=44.2 kg, k_1=24.5 kN/m, k_2=2.846 kN/m. The excitation from the ground is assumed to be sinusoidal with a amplitude of road profile up to 10 mm. For the semi-active suspension with the continuous control strategies, the damping ratio can vary continuously from 0.11 to 1.0, and for the suspension with an on-off control, the damping ratio is taken to be 0.11 (minimum) when the control is off and 1.0 (maximum) when it is on.

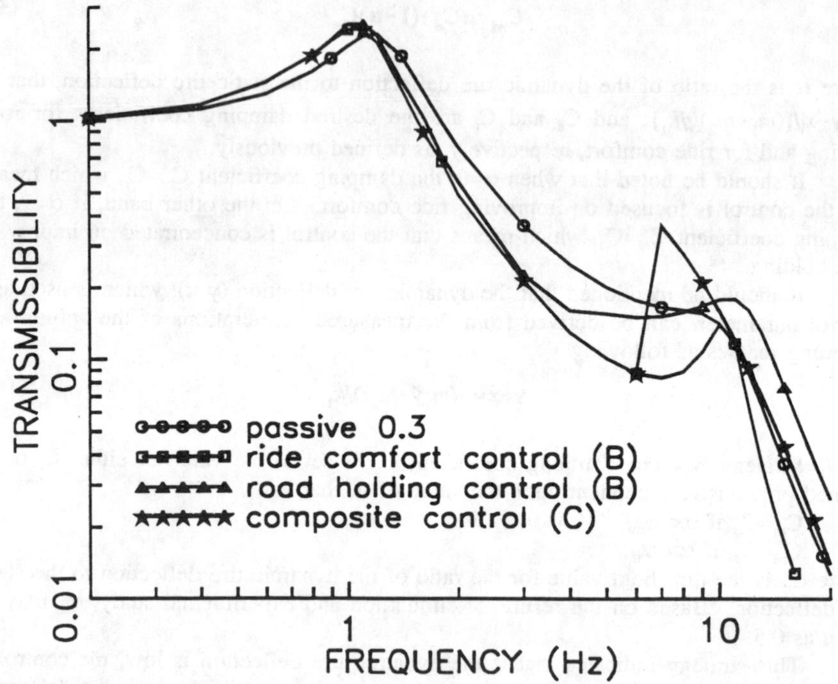

Figure 2 A comparison of the transmissibilities for different control strategies based on simulation results

Figure 2 shows the transmissibility, which is the ratio of the root mean square (rms) value of the acceleration of the sprung mass to that of the tire-ground contact point, for semi-active suspensions with the various control strategies described previously. For comparison, the transmissibility of a passive suspension with a damping ratio of 0.3 is also shown in Fig. 2.

The control strategy A for ride comfort underperforms the control strategy B, and the control strategy C yields similar results to those for the control strategy B. Therefore only the simulation results for the control strategy B for ride comfort are shown in Fig. 2.

The transmissibilities for the semi-active suspensions with the control strategy B for road holding and with the composite control strategy C are also shown in Fig. 2. It should be mentioned that the control strategy A for road holding underperforms the control strategy B and therefore is not included in the figure.

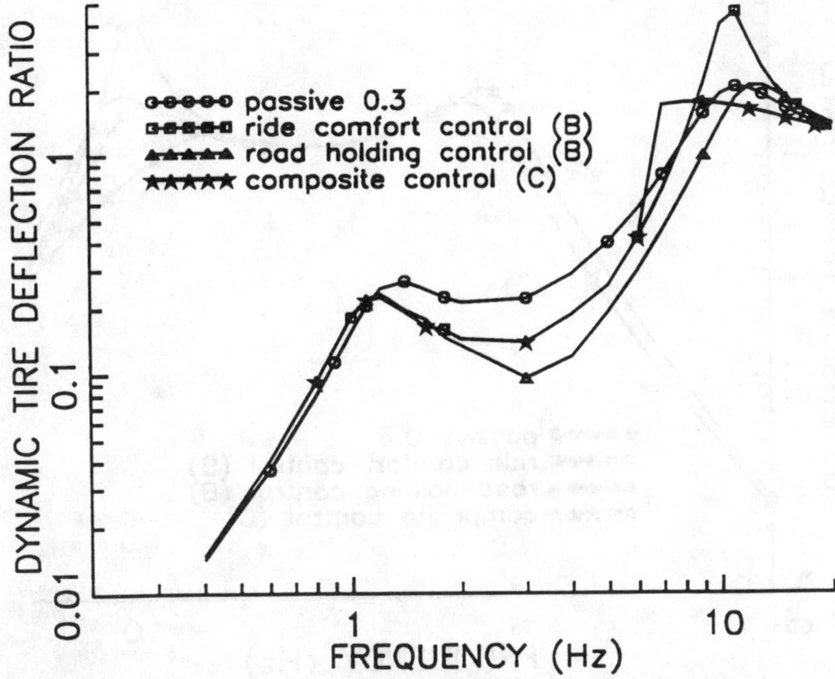

Figure 3 A comparison of the dynamic tire deflection ratios for different control strategies based on simulation results

From Fig. 2, it can be seen that the transmissibility of the suspension with the control strategy B for ride comfort is comparable to that of the conventional passive system around the natural frequency of the sprung mass and is lower at other frequencies. The transmissibility of the suspension with the control strategy B for road holding is comparable to that with the control strategy B for ride comfort at frequencies below 3 Hz, beyond that it is higher. The transmissibilities of the suspension with the composite control strategies A and B are similar and are comparable to, or lower than, that of the passive system at frequencies below the natural frequency of the unsprung mass, but higher beyond that. The transmissibility of the suspension with the composite control strategy C is similar to those of the composite control strategies A and B at frequencies below 6 Hz, where the maximum damping is applied. Beyond that it increases and reaches a peak at approximately 7 Hz. At frequencies higher than 7 Hz, the transmissibility decreases and subsequently is comparable to that of the passive system. For simplicity, only the transmissibility for the composite control strategy C is shown in Fig. 2.

Based on the results shown in Fig. 2, it can be said that from the vibration isolation point of view, the control strategy B for ride comfort is superior to all others.

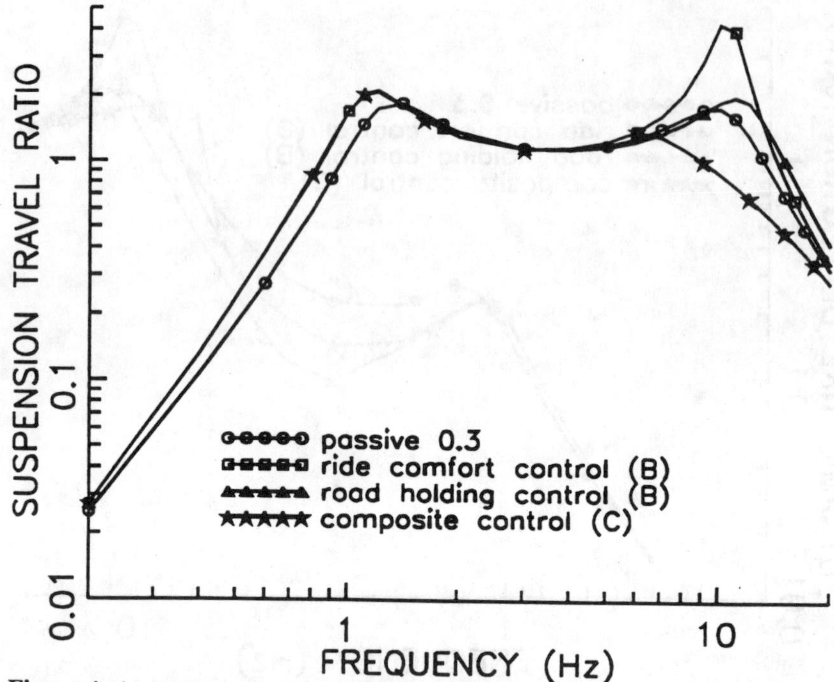

Figure 4 A comparison of the suspension travel ratios for different control strategies based on simulation results

The strategies designed for improving road holding, as well as the composite control strategies studied, are unable to provide the same level of vibration isolation as the strategy specifically designed for improving ride comfort, such as the control strategy B for ride comfort. This is due to the fact that for the suspension to exert force (i.e., the adjustable damping force in a semi-active system) to control the normal force between the tire and the ground, this control force must be reacted against the sprung mass, hence increasing the force applied to the sprung mass (i.e., vehicle body) and causing a deterioration in the vibration isolation characteristics.

3.2 *Road Holding*

Figure 3 shows the dynamic tire deflection ratio, which is the ratio of the rms value of dynamic tire deflection to that of the road profile, for semi-active suspensions with the various control strategies described previously. For comparison, the dynamic tire deflection ratio for the conventional passive system with damping ratio of 0.3 is also shown in Fig. 3.

From Fig. 3, it can be seen that the semi-active suspension with the control strategy B for road holding outperforms all others over a wide range of frequency, although it has slightly higher dynamic tire deflection ratio than the conventional passive system and the suspension with the composite control strategy C at frequencies higher than the natural frequency of the unsprung mass. It should be noted that with the composite control strategy C, the dynamic tire deflection ratio rises rapidly at a frequency of 6 Hz, where the maximum damping ratio of 1.0 is in effect. The dynamic tire deflection ratio reaches a peak at approximately 7 Hz, and beyond that it decreases. This is similar to the variation of the transmissibility with frequency for the composite control strategy C shown in Fig. 2. It should be pointed out that with the composite control strategy C, the dynamic tire deflection ratio is the lowest at frequencies higher than the natural frequency of the unsprung mass.

Figure 4 shows the suspension travel ratios, which is the ratio of the rms value of the relative displacement between the sprung and unsprung mass to that of the ground profile, for various control strategies. It can be seen that over a wide range of frequency, all control strategies perform in a similar way. However, from a frequency just below the natural frequency of the unsprung mass (i.e., approximately 6 Hz) onward, the composite control strategy C outperforms all others examined.

A comparison of the performances of various control strategies with sinusoidal excitations at various frequencies, obtained using computer simulation techniques, is summarized in Table 1.

Table 1: Comparison of Simulated Performances with Various Control Strategies at Four Discrete Frequencies

Performance parameter	Control strategy	Frequency, Hz			
		1.2	7	10	15
Transmissibility	passive, $\zeta=0.3$	2.37	0.17	0.14	0.04
	ride comfort (B)	2.13	0.08	0.14	0.02
	road holding (B)	2.29	0.15	0.18	0.08
	composite (C)	2.13	0.36	0.16	0.04
Dynamic tire deflection ratio	passive, $\zeta=0.3$	0.25	0.80	1.90	1.63
	ride comfort (B)	0.22	0.69	3.96	1.84
	road holding (B)	0.23	0.45	1.40	1.81
	composite (C)	0.22	1.68	1.73	1.46
Suspension travel ratio	passive, $\zeta=0.3$	1.76	1.44	1.80	0.71
	ride comfort (B)	2.13	1.65	4.36	0.86
	road holding (B)	2.10	1.37	1.92	1.04
	composite (C)	2.13	1.32	0.9	0.48

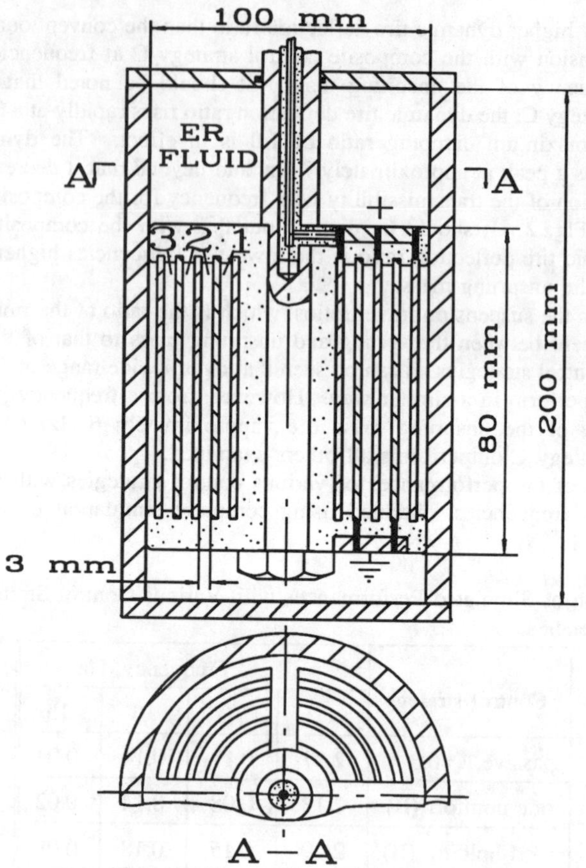

Figure 5 Schematic of the ER fluid damper used in the experiments

4. Experimental Study of Various Control Strategies

The various control strategies for the semi-active suspension were studied experimentally using a quarter-car model test facility with an electrorheological (ER) damper. The model has a sprung mass of 44.2 kg, an unsprung mass of 6.2 kg, the stiffness of the spring representing the suspension is 2.846 kN/m, and the stiffness of the spring representing the tire is 24.5 kN/m. The ER damper used in the experimental study is shown in Fig. 5. The excitation is from an eccentric cam with an amplitude of 7.5 mm, which is driven by a variable speed motor. The ER fluid used is a mixture of silicone base oil and starch "AnalaR", supplied by BDH Inc. and with a mass ratio of 5

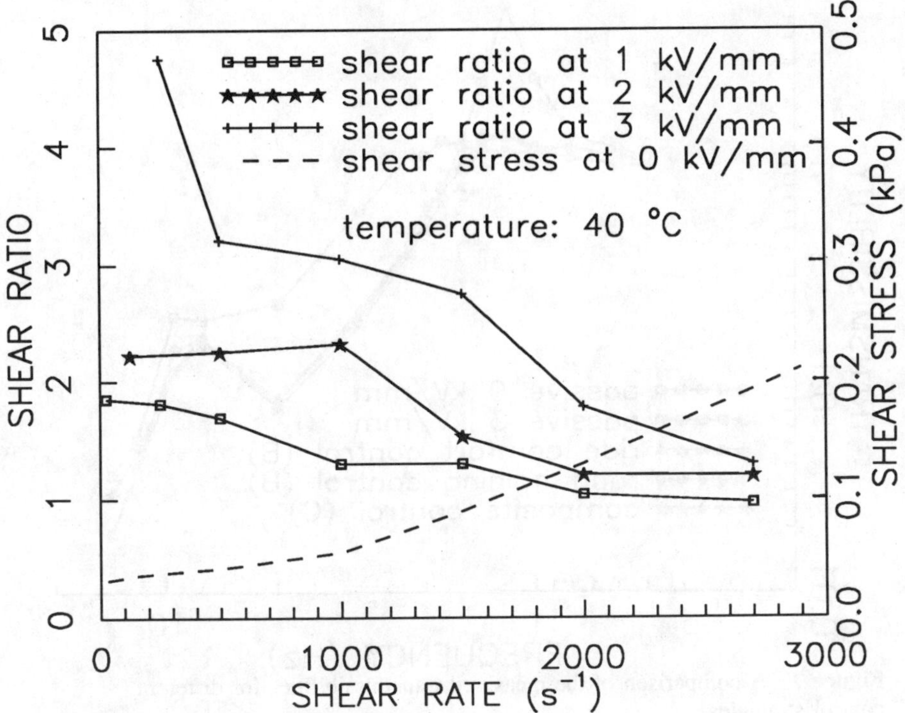

Figure 6 Variation of the shear ratio with the shear rate for the ER fluid used in the experiments

to 1. The rheological properties of the ER fluid were measured using a Couette rheometer (i.e., concentric cylinders), with an inside diameter of 0.145 m for the stator and an outside diameter of 0.139 m for the rotor. The rotor is driven by a variable speed motor and the stator is restrained by a torque cell. The fluid shear stress is derived from the readings of the torque cell. Figure 6 shows the variation of shear stress at zero electrical field strength, as well as that of the shear ratio for the ER fluid at three electrical field strengths of 1, 2 ad 3 kV/mm, as a function of shear rate at a temperature of 40°C. Shear ratio is the ratio of the shear stress developed at a given electrical field strength across the fluid to that at zero electrical field strength. It can be seen that as the shear rate increases the shear ratio decreases. This will have significant influence on the characteristics of the ER damper, particularly at high frequency of excitation, which will be discussed further later.

Since the relative displacement between the sprung and unsprung mass and that between the unsprung mass and ground input were used as control parameters in some of the control strategies examined, linear potentiometers were installed to monitor these displacements. In some control strategies, the relative velocity between the sprung and unsprung mass was also used as a control parameter. In this case, an analog differentiator

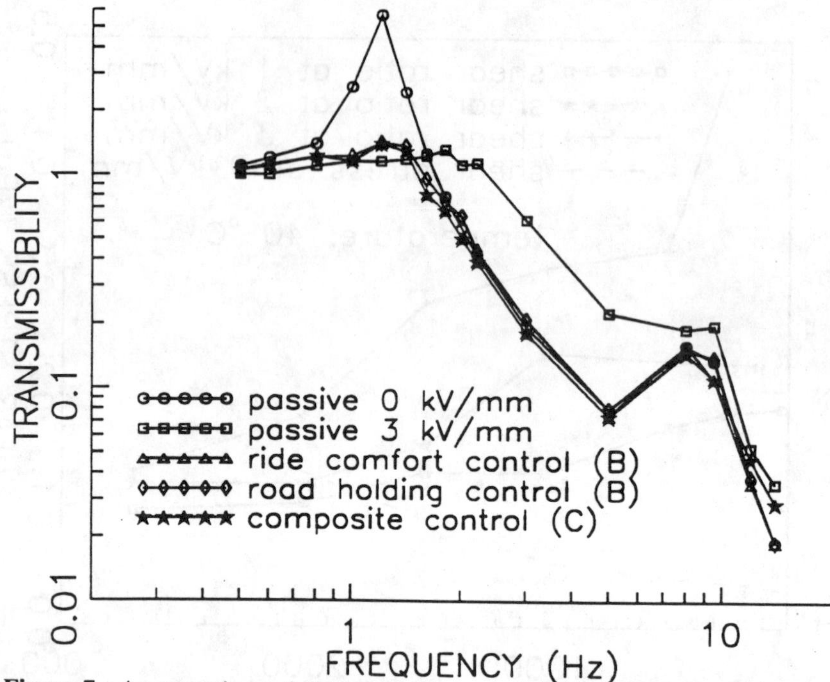

Figure 7 A comparison of the measured transmissibilities for different control strategies

was used to differentiate the relative displacement and to obtain the relative velocity.

Figure 7 shows the transmissibilities for various control strategies, measured using the quarter-car model test rig with the ER damper described previously. For comparison, the transmissibilities of the system with the ER damper at fixed electrical field strengths of 0 and 3 kV/mm are also shown. As noted previously, the shear stress developed by the ER fluid at a given electrical field strength is a function of shear rate, which is related to the frequency of excitation. As a result, for a fixed electrical field strength, the equivalent viscous damping ratio for the ER damper varies with frequency. Figure 8 shows the variation of the equivalent damping ratio with frequency of excitation at electrical field strengths of 0 and 3 kV/mm. This is obtained by comparing the curves shown in Fig. 7 at 0 and 3 kV/mm with those for a system with a viscous damper having various damping ratios. It can be seen from Fig. 8 that at 0 kV/mm without any ER effect, the damping ratio varies in a narrow range over a wide range of frequency. However, at 3 kV/mm, the damping ratio varies significantly with frequency. At frequencies up to 2 Hz, the equivalent damping ratio for the ER damper is a constant of approximately 1.5. Beyond that the equivalent damping ratio decreases rapidly. At frequencies higher than 10 Hz, the equivalent damping ratio decreases to approximately

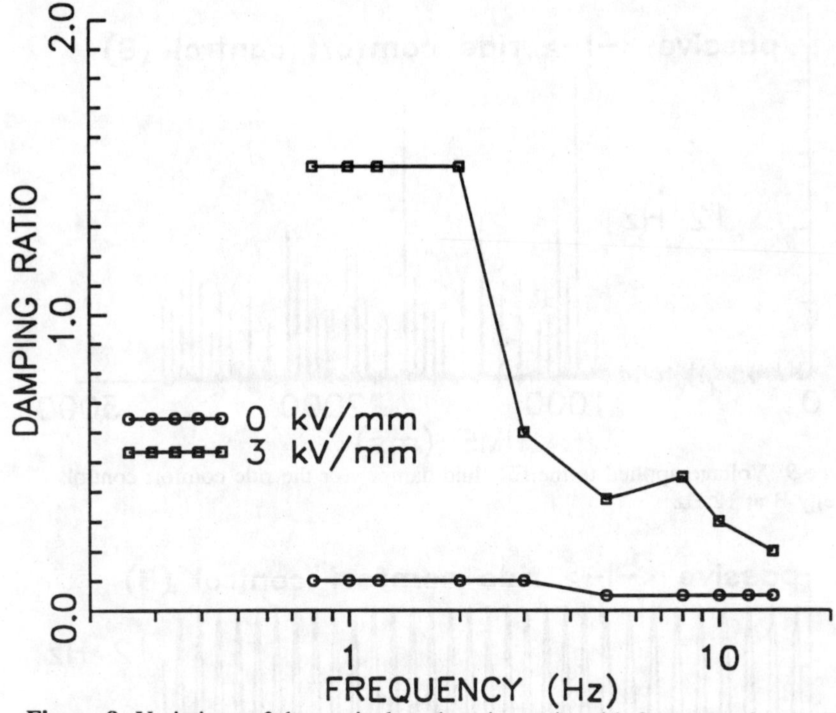

Figure 8 Variations of the equivalent damping ratio with frequency for the ER fluid damper tested

0.2. This means that at higher frequencies, even though a high electrical field is applied to the ER fluid in the damper, the damping force generated is close to that at zero electrical field strength. This is primarily due to the phenomenon shown in Figs. 6 and 8, which indicates that the ER effect diminishes with the increase in shear rate. Based on the data shown in Fig. 8 and the sprung mass and suspension spring stiffness of the quarter-car model test rig, the damping coefficient of the ER damper at frequency below 2 Hz is estimated to 1135 N·s/m and that at frequency of 10 Hz is approximately 213 N·s/m. For this reason, the transmissibilities for the passive system with electrical field strengths of 0 and 3 kV/mm and for the systems with various control strategies are very close at frequencies higher than 10 Hz. Furthermore, at high frequencies (such as 12 Hz), the duration that the control voltage is applied to the fluid in the ER damper is very short, as shown in Fig. 9. Consequently, it is unable to change the mechanical properties of the fluid as intended, and the behaviour of the semi-active system is essentially the same as that of a passive system as shown in Fig. 10. On the other hand, at low frequencies, the duration that the control voltage is applied to the fluid in the ER damper is sufficiently long, as shown in Fig. 11, so that the intended change in the mechanical properties of the ER fluid can be realized. As a result, the potential benefits of the semi-active system can

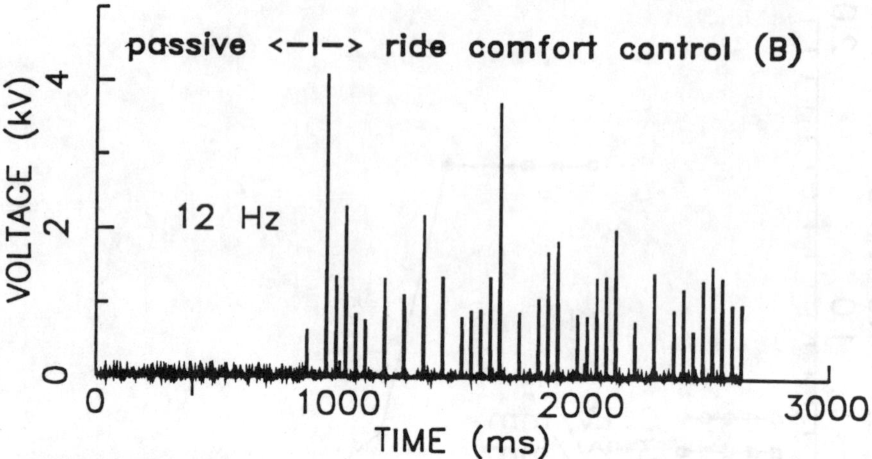

Figure 9 Voltage applied to the ER fluid damper for the ride comfort control strategy B at 12 Hz

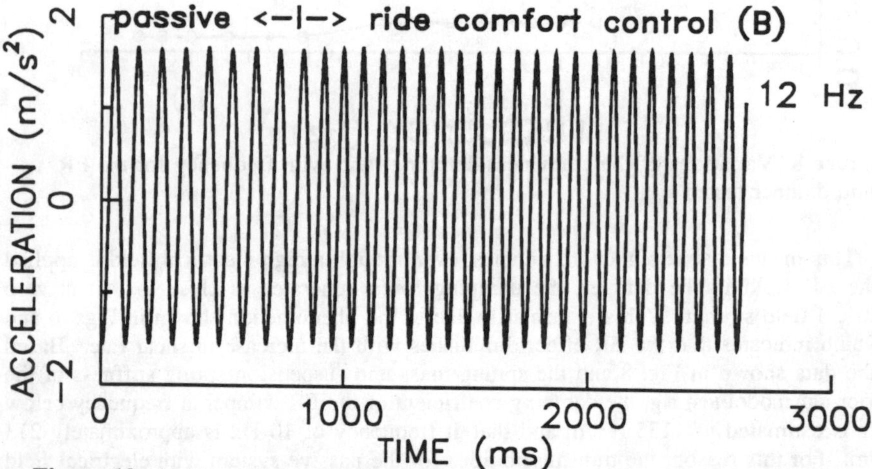

Figure 10 A comparison of the measured sprung mass acceleration before and after the activation of the ride comfort control strategy B at 12 Hz

be achieved, as shown in Fig. 12. It should be pointed out that the voltage applied to the ER damper and the acceleration response of the sprung mass shown in Figs. 9-12 were measured simultaneously. Figure 13 shows the measured dynamic tire deflection ratio as a function of frequency of excitation for various control strategies. It can be seen that at frequencies around the sprung mass natural frequency, the dynamic tire deflection ratio

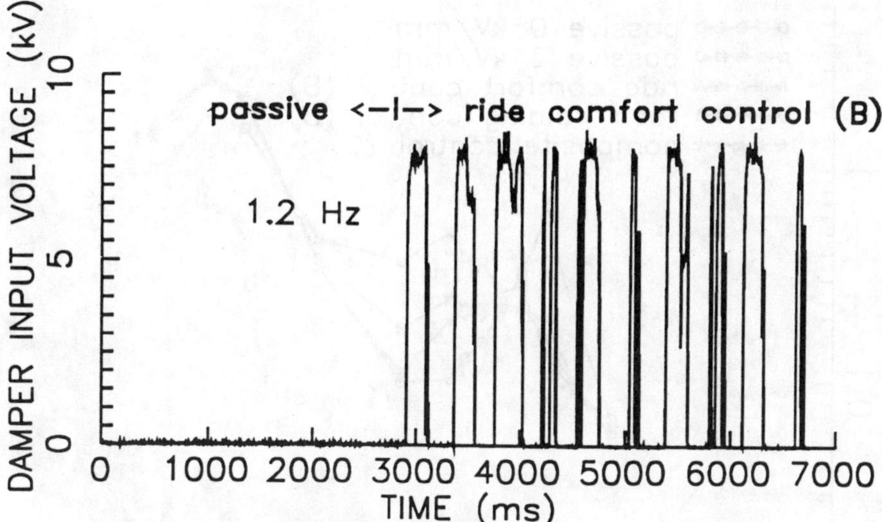

Figure 11 Voltage applied to the ER fluid damper for the ride comfort control strategy B at 1.2 Hz

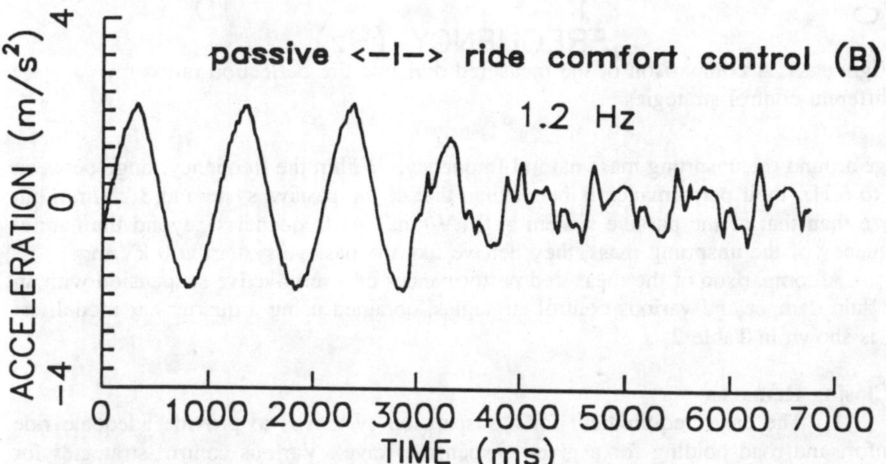

Figure 12 A comparison of the measured sprung mass acceleration before and after the activation of the ride comfort control strategy B at 1.2 Hz

of the passive system with no voltage applied to the damper is the highest. At frequencies around the unsprung mass natural frequency, the passive system with an applied voltage of 3 kV/mm has the lowest dynamic tire deflection ratio. All other control strategies perform in a similar way. They perform better than the passive system at 0 kV/mm at low frequencies but are worse than the passive system with 3 kV/mm at the high frequency

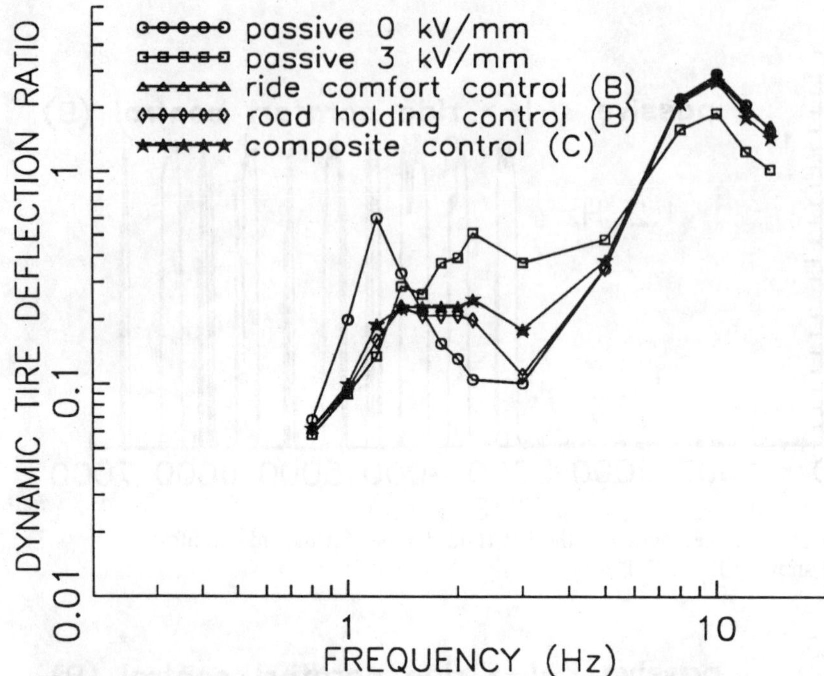

Figure 13 A comparison of the measured dynamic tire deflection ratios for different control strategies

range around the unsprung mass natural frequency. Within the frequency range between 1.5 to 6 Hz, their performance is better than that of the passive system at 3 kV/mm but worse than that of the passive system at 0 kV/mm. At frequencies beyond the natural frequency of the unsprung mass, they behave like the passive system at 0 kV/mm.

A comparison of the measured performances of a semi-active suspension with an ER fluid damper and various control strategies, obtained using a quarter-car model test rig, is shown in Table 2.

5. Closing Remarks

1. The basic requirement for a suspension system is to provide adequate ride comfort and road holding for a given suspension travel. Various control strategies for semi-active suspension to achieve improved ride comfort and/or road holding were investigated.

Simulation results show that the strategies designed for enhancing ride comfort do not necessarily provide improved road holding characteristics, and vice versa. This is primarily due to the fact that for the suspension to exert force (i.e., the adjustable damping force in a semi-active system) to control the normal force between the tire and the ground, this control force must react against the sprung mass. This increases the force applied to the sprung mass (i.e., the vehicle body) and causes a deterioration in ride comfort. Consequently, a suitable composite control strategy to accommodate the requirement of improving both the ride comfort and the road holding is required.

2. The performance of a semi-active suspension system with an electrorheological fluid damper and having various control strategies was measured using a quarter-car model test rig.

It is found that the damping characteristics of the ER fluid damper used in the study depend not only on the electrical field strength but also on the frequency of excitation. For the ER fluid used, the equivalent damping ratio decreases significantly with the increase in the frequency of excitation. This is primarily due to the fact that the

Table 2: Comparison of Measured Performance with Various Control Strategies at Four Discrete Frequencies

Performance parameter	Control strategy	Frequency, Hz			
		1.2	5	10	14
Transmissibility	passive, 0 kV/mm	5.68	0.08	0.14	0.02
	passive, 3 kV/mm	1.16	0.23	0.20	0.04
	ride comfort (B)	1.41	0.07	0.11	0.02
	road holding (B)	1.42	0.08	0.14	0.02
	composite (C)	1.41	0.07	0.14	0.03
Dynamic tire deflection ratio	passive, 0 kV/mm	0.60	0.35	2.92	1.57
	passive, 3 kV/mm	0.14	0.48	1.90	1.03
	ride comfort (B)	0.19	0.38	2.91	1.60
	road holding (B)	0.16	0.35	2.70	1.62
	composite (C)	0.19	0.38	2.87	1.46
Suspension travel ratio	passive, 0 kV/mm	4.76	1.27	2.02	0.48
	passive, 3 kV/mm	0.32	1.32	1.66	1.20
	ride comfort (B)	1.11	1.29	2.00	0.49
	road holding (B)	1.00	1.28	1.92	0.45
	composite (C)	1.11	1.28	1.92	0.45

shear ratio of the ER fluid used decreases with the increase in the shear rate. Furthermore, at high frequencies, the duration of the applied voltage with any of the control strategies examined is very short. Consequently, there is little difference in the measured performance of the semi-active suspension with different control strategies examined over a wide frequency range. To achieve the potential of an ER fluid damper, improvements in the mechanical behaviour of ER fluids are a key factor.

Acknowledgements

The work described in this paper was supported in part by a research grant awarded to Professor J. Y. Wong by the Natural Sciences and Engineering Research Council of Canada.

The assistance provided by Mr. Jamie Catania in the preparation of the paper is appreciated.

References

1. Wu, X. M., Stanway, R. and Sproston J. L., 1990, "Electrorheological Fluids and Their Applications in Power Transmission and Active Suspension", Proc. of 22nd Int. Symposium on Automotive Technology and Automation, Florence, Italy, 14-18, May, 1990, pp. 823-830.

2. Wong, J. Y., Wu, X. M., Sturk, M. and Bortolotto, C. M., 1992, "On the Applications of Electro-Rheological Fluids to the Development of Semi-Active Suspension Systems for Ground Vehicles", Proceedings of the Canadian Society of Mechanical Engineering Forum SCGM 1992, Transport 1992, Montreal, Canada, Vol. 1, pp. 186-191.

3. Margolis, D. L. and Goshtasbpour, W., 1984, "The Chatter of Semi-Active On-Off Suspension and Its Cure", Vehicle System Dynamics, V. 13, pp. 129-144.

4. Wong, J. Y., 1993, "Theory of Ground Vehicles", Second Edition, John Wiley.

5. Krasnicki, E. J., 1981, "The Experimental Performance of an "on-off" Active Damper", Shock and Vibration Bulletin, V. 51, part 1, pp.125-131.

6. Alanoly, J. and Sankar, S., 1987, 'A New Concept in Semi-Active Vibration Isolation', Transactions of the American Society of Mechanical Engineers, Journal of Mechanisms, Transmissions and Automation in Design, Vol. 109, pp. 242-247.

7. Jolly, M. R. and Miller, L. R., 1989, "The Control of Semi-Active Damper Using Relative Feedback Signals", SAE Paper No. 892483.

Applications of Electrorheological Fluid in Shock Absorbers

Wei Chenguan

Department of Vehicular Engineering, Beijing Institute of Technology, Baishiqiao Road, Beijing, China

and

Fu Zhao

Department of Vehicular Engineering, Beijing Institute of Technology, Baishiqiao Road, Beijing, China

ABSTRACT

This paper deals with the applications of electrorheological fluid (ERF) in shock absorbers. This kind of shock absorber (ERF shock absorber) whose damping force can be controlled continuously and quickly with electric signals will be used in many kinds of mechanical equipment for vibration control. The typical structures of the ERF shock absorber are mentioned. The requirements of the ERF employed in shock absorbers are discussed. A new kind of shock absorber and its control system are developed in this paper. The testing results of the ERF shock absorber are offered.

1. The ERF (Electrorheological Fluid) Shock Absorber

ERF and its engineering applications have gained more and more attention in recent years. This is because the viscosity of ERF cannot only be varied by the change in voltage of an electric field, but also be continuously and reversibly controlled. Furthermore it is highly responsive. Thus ERF can be used to solve some difficult engineering applications.

So far, ERF usage has been proposed in shock absorbers, clutches, engine mounts, liquid fuel, automatic control devices and brakes as well as for various applications in the aviation industry, etc. [1]. Among those the most prospective applications are in shock absorbers and clutches. This paper deals with the application of ERF in shock absorbers, that is the ERF shock absorber.

The traditional shock absorber whose damping force is generated by throttle pressure is passive. Being varied by the change in area of the orifice, the damping force can not be controlled smoothly and precisely over a wide range of velocities.

So only when the road conditions change remarkably, may the shock absorber offer different forces to meet the demands of comfort and safety. While the damping force of the ERF shock absorber is actively controlled by a control system. Through exact sensing of the road conditions and the continuous change in the viscosity of the fluid the damping force of the ERF shock absorber can be quickly and smoothly controlled to realize the optimal cooperation between shock absorber and elastic element.

Compared with the traditional hydraulic shock absorber, the ERF shock absorber can more properly meet the demands of comfort and safety. Therefore, the ERF shock absorber is an important development in the evolution of shock absorber design.

2. Typical Structures of the ERF Shock Absorber

The history of the ERF shock absorber is very short, so the research work is still in the exploratory stage. In general, ERF shock absorbers can be divided into the following typical structures: a structure with fixed electrode plates and a structure with sliding electrode plates[2]. See Fig. 1.

Different structures have different advantages. Usually a by—pass line is installed inside the ERF shock absorber with fixed plates and the electrode plates are located in the by—pass line. When the piston moves reciprocally in the chamber, ERF flows between the upper and the lower chambers through the electrode plates. The viscosity of the ERF can be varied by changing the differential voltage between two electrode plates to adapt the damping force to road conditions. With large electrode plates and a long flow passage, this kind of structure has a high reliability. However, because of the by—pass line, the dimension of the shock absorber is increased and with many electrode plates heat dissipation is decreased.

In the structure with sliding plates, generally, one of the plates is installed in the piston. Thus the plate and the piston move together. ERF flows between the upper and the lower chambers through the orifices in the piston. This kind of structure is compact and effectively dissipates heat. But, due to the restriction of stroke length, the axial dimension of the piston is limited, thus the area of the electrode plates can not be large enough to meet the demands of high damping.

Practically, the ERF shock absorber should be designed on the basis of the traditional structure of the hydraulic shock absorber with high working reliability. Thus the ERF shock absorber should have both high control sensitivity and high working

[illegible]

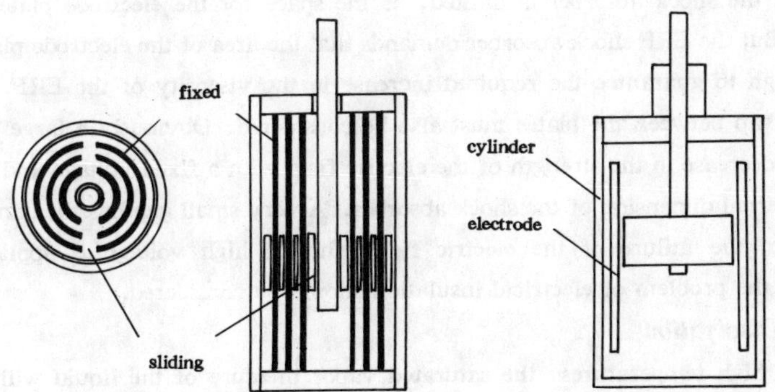

Figure 1 Typical structures of the ERF shock absorber

[illegible]

reliability to provide an adaptable damping force according to road conditions so as to meet the requirements of rider comfort and driving safety.

3. Problems in the Structural Design of the ERF Shock Absorber

As mentioned before, the study of the ERF shock absorber still remains in a period of exploration. So more efforts should be put forth to practically utilize the ERF shock absorber in vehicles to gain a commercial profit. The main problems are as follows:

3. 1. *Electric Field*

The ERF shock absorber needs a very high voltage differential to make the viscosity of the ERF change dramatically in a very short period of time. The external dimensions of the shock absorber is limited, so the space for the electrode plates is also limited. But the ERF shock absorber demands that the area of the electrode plates be large enough to guarantee the required increase in the viscosity of the ERF. In addition, the gap between the plates must also be considered. Obviously a large gap will lead to a decrease in the strength of the electric field with a fixed voltage and increase the external dimension of the shock absorber. A very small gap has the danger of resulting in the failure of the electric field when a high voltage is applied. Additionally, the problem of electrical insulation should be considered.

3. 2. *Heat Dissipation*

First, at high temperatures, the saturated vapor pressure of the liquid will be decreased, which makes it easy for emulsification to occur. Especially in the rebound cycle of shock absorber, high temperatures and negative pressures result in the formation of vapor bubbles in the liquid. Though the damping force of the ERF shock absorber is adapted by changing the viscosity of the ERF, naturally, it is still an energy consuming shock absorber whose damping force is generated by the pressure drop across the orifice. This kind of shock absorber requires the liquid to be incompressible. When there is air in the liquid, the liquid becomes elastic and its performance is not stable. This will negatively influence the generation of the damping force. Serious emulsification even leads to the failure of shock absorber.

Secondly, the damping force of the shock absorber is adapted to the road conditions by changing the viscosity of the ERF. High temperatures make the viscosity of the ERF decrease. Thus in spite of a high voltage application, it would not be easy for the shock absorber to provide a high damping force. In addition, vapor bubbles generated by the emulsification of the liquid at high temperatures would lead to the electric field failing and make the damping force decrease unexpectedly. Thus, a continuous damping force can not be guaranteed.

Therefore, heat dissipation in the structural design of the ERF shock absorber in order to protect the shock absorber from failing to work in a high temperature environment is a very important problem for the researcher to solve.

3. 3. *Requirements of ERF*

ERF characteristics are important factors in its application in shock absorbers. Without a high quality ERF, the ERF shock absorber can never provide an immediate and suitable damping force. For the ERF shock absorber application, a high quality ERF should meet the following requirements:

A. Viscosity:

Viscosity is an important performance characteristic of ERF. Generally, the original viscosity of the ERF, that is, the viscosity of the ERF without an electric field being applied should be low. When the electric field is applied, the viscosity of the ERF should dramatically increase with the increase in field strength so that the shock absorber can provide the required damping force. When the electric field is reduced or turned off, the viscosity of the ERF should decrease simultaneously. Thus good repeatability and high response sensitivity of the ERF are required to guarantee that the shock absorber will provide a suitable damping force in its practical working process.

B. Thermal Stability

As mentioned in the above discussion of heat dissipation, the ERF should have good thermal stability, that is, the viscosity of the liquid should change little even at high temperatures. In reality a shock absorber often works in a high temperature environment. Therefore, besides considering heat dissipation in the structural design of the ERF shock absorber, the high thermal stability of the ERF should also be taken into account in order that the shock absorber will provide enough of a damping force at high temperatures.

C. Critical Voltage

The viscosity of the ERF does not always increase with a strengthening of the electric field. Actually, there is a critical voltage. When the voltage of the electric field is higher than the critical voltage, the viscosity of the ERF will no longer increase with a strengthening of the electric field. It is even possible for the viscosity to decrease with an increase of the electric field strength. Therefore, the critical voltage of the ERF should be much higher than the maximum voltage of the electric field so that the ERF always works in the safe voltage range. This will guarantee that the ERF shock absorber provides enough of a damping force.

D. Dielectric Performance

Generally, the ERF is required to have high dielectric strength. But for the

ERF employed in shock absorbers, that requirement is not essential. As mentioned before, when the shock absorber works the fluid flows very fast between the upper and lower chambers, which will probably leads to the occurrence of vapor bubbles. With the bubbles which can not bear high voltage the failure of electric field which makes the damping force decrease remarkably is easy to occur. However, if there are electric particles in the ERF, the current can flow through the electric particles. Then high voltage can be applied on the ERF without leading to the failure of electric field. Therefore, the shock absorber can provide a continuous damping force. That is the special requirement of the ERF for shock absorbers.

E. Other Requirements

Besides all of the above mentioned requirements, the ERF is required to have high flowability , wearing capacity and to be absent of electrophoresis.

3. 4. *Control System of the ERF Shock Absorber*

Compared with the traditional shock absorber, the ERF shock absorber has the advantage of being able to adapt its damping force according to different road conditions to meet the demands of comfort and safety. Road conditions change fast, so the damping force provided by the shock absorber should also change fast. Thus, the ERF shock absorber should have high control sensitivity. The control circuit of the shock absorber should consist of the following parts: (See figure 2.)

A. Sensors

Sensors collect the information of rider comfort and driving safety in different road conditions (such as velocity, displacement or acceleration) and transfer the information to the control unit.

B. Control Unit

In accordance with sensor information, the control unit determines the magnitude of the voltage to be applied to the electrodes and transmits a signal to the variable voltage power supply.

C. Variable Voltage Power Supply

According to the signal from the control unit, the power supply applies the required voltage to the electrode plates.

D. The ERF Shock Absorber

The ERF shock absorber provides a damping force corresponding to the strength of the electric field.

Generally, the ER effect of ERF is very fast, so the response sensitivity of the whole system mainly depends on the sensitivity of other parts in the control scheme. Therefore, the control system should be as simple as possible to reduce response hysteresis such that the overall sensitivity of the system can be improved to guarantee that the shock absorber can provide optimal damping [illegible] road conditions.

[illegible] and Flexibility

[illegible] road [illegible] a shock absorber has to work in completely different automobiles, so the structure of the ERF shock absorber should be designed on the basis of the structure of the traditional hydraulic shock absorber which is reliable and durable. Thus the ERF shock absorber can have not only high control sensitivity but also as high a reliability as the hydraulic shock absorber and be durable.

[illegible] Cost Design

[illegible] cost design is also an important problem which must be taken into account in the structural design of the ERF shock absorber.

4. [illegible] the ERF Shock Absorber

[illegible] of structure based on the traditional hydraulic shock absorber has been developed in our case. We tested the damping performance of the ERF shock absorber with [illegible]. The tests show that damping force provided by the ERF shock absorber can [illegible] the requirement of shock [illegible].

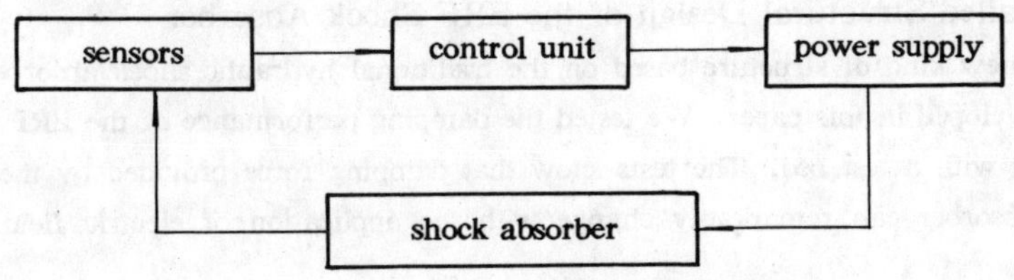

Figure 2 Control system of the ERF shock absorber

5. Conclusions

1) With a [illegible] shock absorber [illegible] an important development in shock absorber design.

The ERF shock absorber realizes its optimal damping control by the method of changing the viscosity of the ERF through varying the strength of the electric field so that the requirements of [illegible] are met.

2) The following problems should be taken into account in the research of the ERF shock absorber:

a. In structural design, the external dimensions should be as small as possible. Less ERF should be needed. The structure should be compact to improve the ability of dissipating heat and maximize [illegible] reliability.

b. An ERF with good performance characteristics is an important factor for the ERF shock absorber to provide suitable damping. Therefore, the development of a new kind of ERF which [illegible] the demands of the ERF shock absorber and has high reliability is very important.

c. The system for controlling the ERF shock absorbers should be capable of high response speed so that the ERF shock absorber can provide suitable damping that the shock absorber can provide optimal damping force under any road conditions.

3.5. *Reliability and Durability*

Due to different road conditions, a shock absorber has to work in obviously different circumstances. So the structure of the ERF shock absorber should be designed on the basis of the structure of the traditional hydraulic shock absorber which is reliable and durable. Thus the ERF shock absorber can have not only high control sensitivity but also as high a reliability as the hydraulic shock absorber and be durable.

3.6. *Seal Design*

Seal design is also an important problem which must be taken into account in the structural design of the ERF shock absorber.

4. Detailed Structural Design of the ERF Shock Absorber

A new kind of structure based on the traditional hydraulic shock absorber has been developed in this paper. We tested the damping performance of the ERF shock absorber with a test bed. The tests show that damping force provided by the ERF shock absorber can remarkably change with the application of electric field. See figure 3.

5. Conclusions

1) With a controllable damping force, the ERF shock absorber has become an important development in shock absorber design.

2) The ERF shock absorber realizes its optimal damping control by the method of changing the viscosity of the ERF through varying the strength of the electric field to meet the requirements of safety and comfort.

3) The following problems should be taken into account in the research of the ERF shock absorber:

A. In structural design, the external dimension should be as small as possible. Less ERF should be required. The structure should be compact to improve the ability of dissipating heat and maximize seal reliability.

B. An ERF with good performance characteristic is an important factor for the ERF shock absorber to provide suitable damping. Therefore, the development of a new kind of ERF which can meet the demands of the ERF shock absorber and has high reliability is very important.

C. The system for controlling the ERF shock absorber should be simple to improve the response speed so that the ERF shock absorber can provide a suitable damping

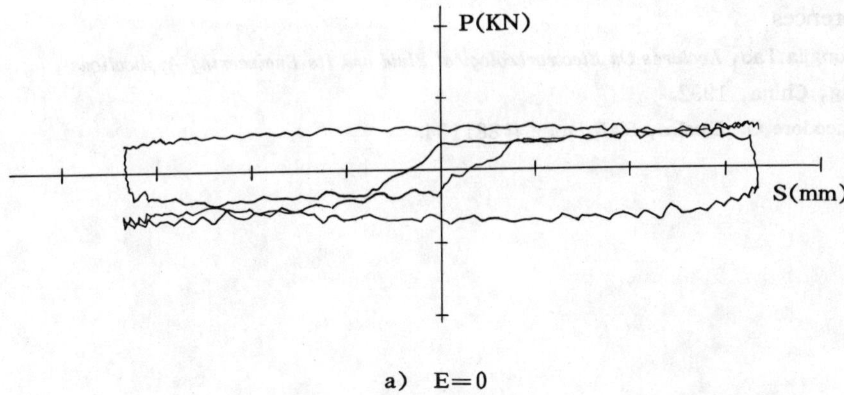

a) E=0

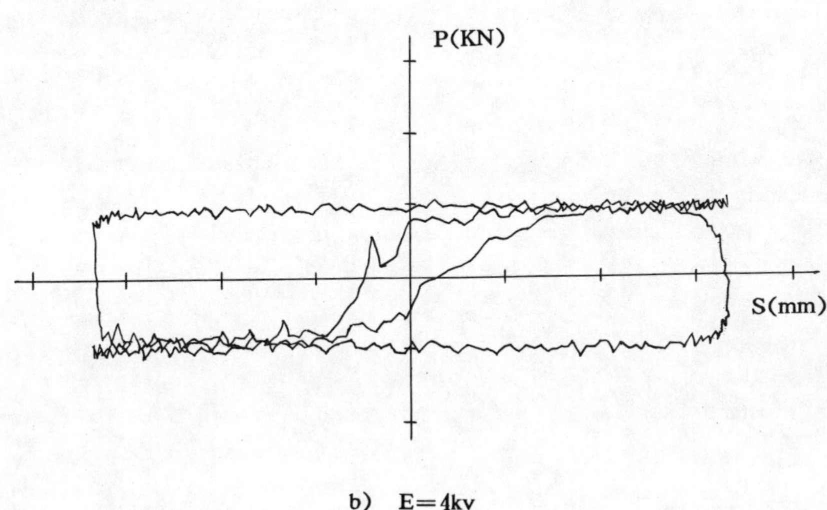

b) E=4kv

Figure 3 Damping performance of the ERF shock absorber

force immediately according to road conditions.

D. The software for controlling the ERF shock absorber should be developed.

References

1. Rongjia Tao, *Lectures On Electrorheological Fluid and Its Engineering Applications*, Beijing, China, 1992.

2. Theodore G. Duclos, *SAE Paper* #881134.

ON-OFF EXCITATION SWITCH FOR ER DEVICES

R. TOZER
The University, Mappin Street,
Sheffield, S1 3JD, UK

C. T. ORRELL
The University, Mappin Street,
Sheffield, S1 3JD, UK

and

W A. BULLOUGH
The University, Mappin Street,
Sheffield, S1 3JD, UK

ABSTRACT

A very important part of any ER fluids based system is the provision of excitation; this is particularly so since the electrical load is conductive and highly capacitive. In this environment the surge current at switch on is large and there are problems in dumping the charge to effect a rapid reduction of the voltage level.

This paper describes a voltage switch which when linked to a steady state supply can be used to effect on-off control. The arrangement is somewhat an analogous to the ER catch concept insomuch as it switches one of the ER controller electrodes either to high potential or to earth rails, the other electrode being permanently connected to earth (just as the catch attaches to a steady input motion or a brake). If the supply has a low output impedance then the excitation can be rapidly slewed in both ongoing and offgoing directions.

The circuit and its operation is described in detail and brief performance characteristics are shown for the switch when connected to a typical, contemporary ER device.

1. Introduction

This paper describes an on/off switch to control the supply of HT to an ER device. The switch offers a low impedance to HT when on and a a low impedance short circuit across the ER device when off, thus rapidly discharging the capacitance associated with the ER device.

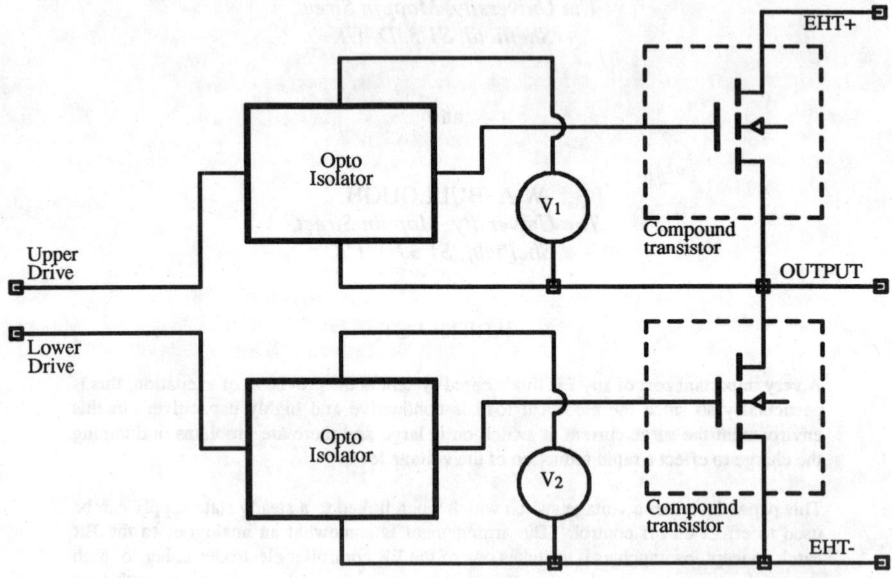

Figure 1. A block diagram of the switch. The EHT supply is connected between the terminals indicated and the load is connected between the output node and either EHT+ or EHT- as required. Any one of the three terminals EHT+, EHT- and OUTPUT may be connected to earth providing that the EHT supply is capable of operating with whatever earth scheme is chosen.

The lack of availability of suitably controllable high tension supplies for ER application research has inhibited the exploitation of the fast response time of ER fluids. There exist a number of applications where a switch is an attractive drive possibility. Typical of these applications is the high speed clutch [1] where rapid switching overcomes hysteretic problems. In a valve application the speed of response from zero to maximum voltage is equally important, in this case to activate a fail safe, stiffening damper. A multi ER head machine such as a coil winder bank, would benefit from the provision of common HT rail with individual control switches.

The present work was directed towards the provision of a laboratory standard switch i.e. one capable of working flexibly over a range of programmed duties. The voltage supply is provided by the back up steady state HT. unit. Duration of excitation is determined by the input from a signal generator or digital control circuit.

Since the electron-hydraulic time constant of an ER device is short (less than 1 msec) and the ER device is being used mainly on account of this factor, the HT. rise and fall times need to be a maximum of 50 to 100 μsec. This requires the switch to have a low output impedance so that the switch does not add substantially to the time constant of the ER equivalent circuit [2].

The switch described here was designed to operate with HT supply voltages of up to 4 kV and the circuit was designed to limit the switching current at around 600 mA (although this figure could easily be increased if required). The high voltage capability of the switch frees the ER engineer (at least in applications adequately controlled by on-off switches) from the need to keep the inter-electrode spacing narrow. Wider interelectrode gaps facilitate a reduction in the zero volts shear stress and thereby reduce heating problems.

2. The Switch Circuit

2.1 General Details

The circuit described here is based on compound transistors consisting of a number of high voltage power FETs in series. The switch is formed by using two such compound transistors, one connected as a common source switch and one as a source follower that behaves effectively as an active load. The general arrangement of the circuit is outlined in figure 1.

The two compound transistors in figure 1 are identical. Since the drive for the top device must be applied with respect to the circuit output which will on occasions be at a potential close to that of the main high tension supply, the top device must be driven via an opto-isolator capable of withstanding steady potential difference of up to 4kV.

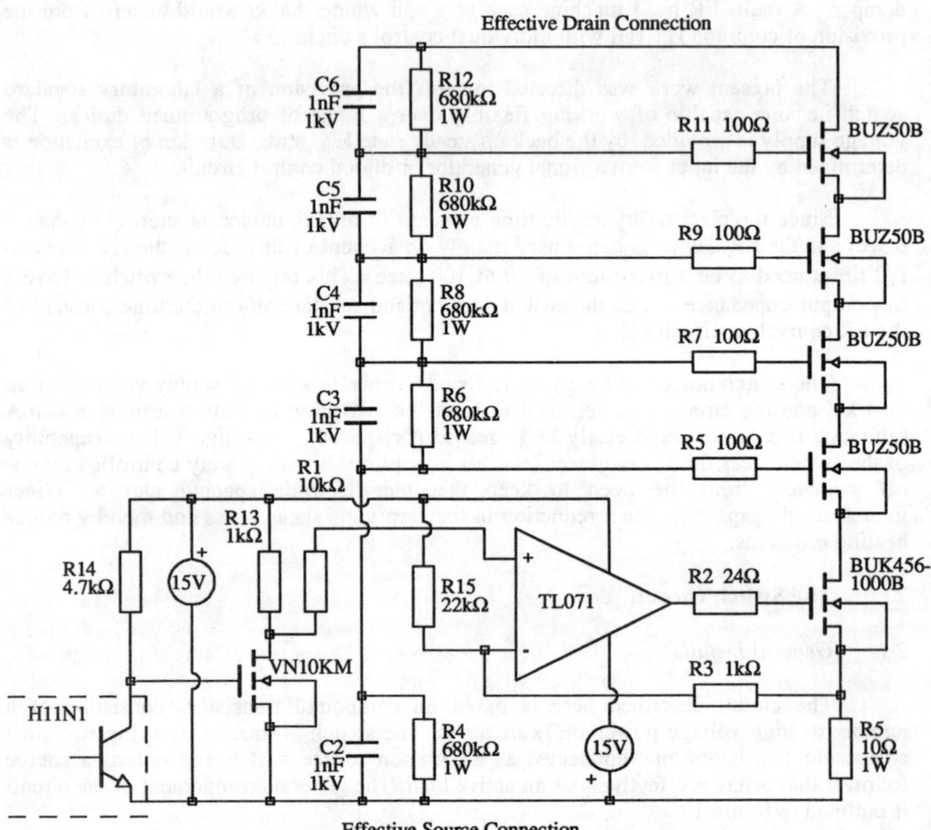

Figure 2. The circuit diagram of one of the compound transistors. Each power MOSFET is protected by an 810V zener diode between drain and gate and a 15V zener diode between gate and source.

In order to achieve maximum application flexibility and maintain complete isolation between the control input and the HT circuit, the bottom switch drive, which must be with respect to the negative end of the HT supply is also driven by an opto isolator. The switch earth may thus be used with either positive or negative HT systems. The supplies V_1 and V_2, which provide the isolated side of the opto isolators and the input circuits of the top and bottom compound transistors are derived from a small d.c. to d.c. converter which is also designed to offer d. c. isolation at potential differences of up to 4kV between any two of its three windings.

In the following sections, the elements of figure 1 are described in detail and results are presented which demonstrate the effectiveness of the circuit as an on - off drive for an electro-rheological device.

2.2 *The Compound Transistors*

Each compound transistor consists of a chain of five 1000V power MOSFETs, as shown in figure 2. A critical requirement in this application is the maintenance of equitable voltage sharing across each device. This sharing must be effective during both long term (d.c.) and short term (transient) events. The resistors R_4, R_6, R_8, R_{10} and R_{12} maintain long term equity of sharing while the capacitors C_2, C_3, C_4, C_5 and C_6 maintain sharing equity during transient events. The time constant of each RC pair (for example, R_4C_2) in this case is 680μs and this figure effectively marks the boundary between what the circuit regards as d. c. and what it regards as transient.

The tolerance of the resistors and capacitors used in the sharing network has a direct bearing on the accuracy of the sharing process. The tolerance must be sufficiently tight to ensure that no FET in the chain can be subjected to either transient or d. c. voltage levels that exceed its maximum drain source voltage specification. In this case the maximum drain-source voltage that each FET could tolerate was 1000V and with perfect sharing there would be a maximum of 800V across each device so the 5% tolerance of the resistors and capacitors used offered a comfortable margin of safety. The values of the resistors and capacitors in the sharing network were chosen so that leakage currents at the FET gate nodes and inter-electrode capacitance spreads from FET to FET had a negligible effect on both the long term and the transient sharing behaviour. With an applied voltage of 4kV across the chain the power loss in the sharing resistors is 4.7W. The capacitors in the sharing chain do not in themselves dissipate energy but in rapid switching applications the power supply must provide the energy necessary to charge the capacitors. Since the expected load capacitance may be as large as 10nF, and since the capacitance of the five sharing capacitors in series is 200pF, most of the energy required for capacitive charging is used by the load. As an example, for a lkHz square wave output of 0V to 4kV, the power lost by the power supply because of the sharing capacitors is 1. 6W and that lost because of the load capacitance is 80W. It is worth noting that the 80W lost to the power supply is ultimately dissipated in the FETs so adequate heat sinking must be arranged.

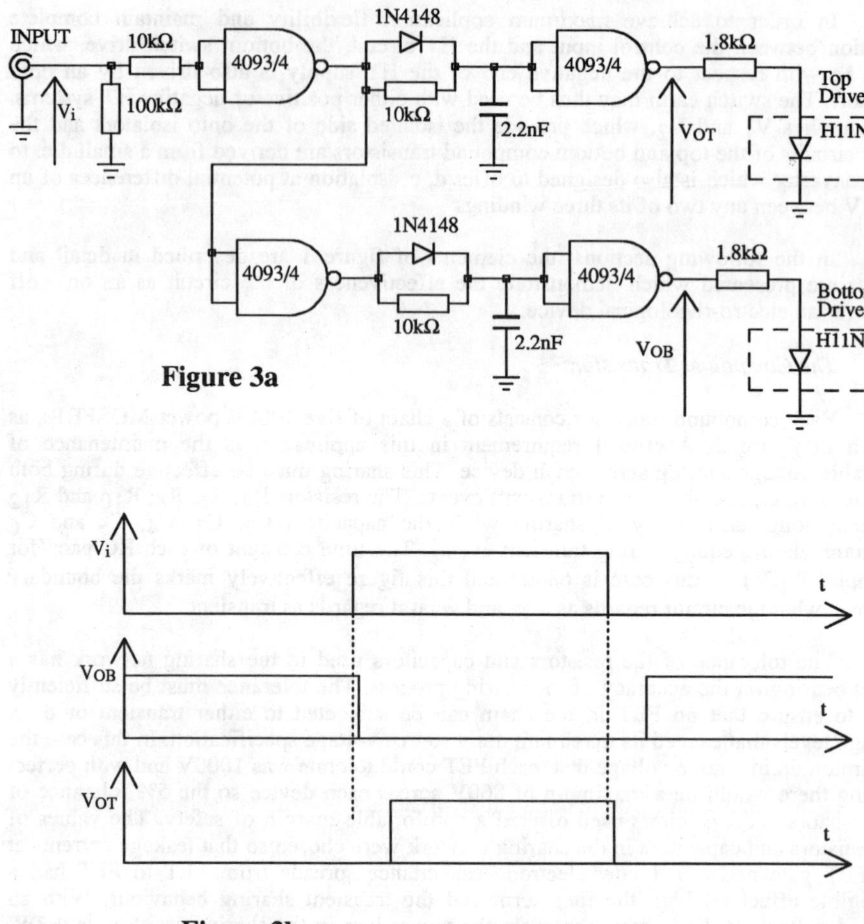

Figure 3a

Figure 3b

Figure 3. The drive circuit of figure 3a ensures that the switches are never on together. For each drive figure 3b shows that the turn off delay is much shorter than the turn on delay thus preventing simultaneous conduction in both top and bottom switches

The bottom FET in the chain controls the current that flows through the chain. The arrangement used here gives the chain a well defined transconductance, defined by R_5. With $R_5 = 10\Omega$ as shown, the transconductance of the compound transistor is 0.1S and the maximum current that the chain can support is approximately 600mA, a factor of five below the current carrying capacity of the FETs. Although the maximum current could be increased if necessary by reducing R_5, limiting its value also limits the severity of the damage to the system that would occur in the event of a transient short circuit at the load node. Limiting the current carrying capacity of the chain affects the available slew rate when the load is capacitive. For a worst case load capacitance of 10nF and a maximum output current of 600mA, the output slew rate is $60MVs^{-1}$ leading to a time to charge the load to 4kV of 67μs. This figure was considered adequate for the applications of interest.

2.3 *Drive Circuit*

A possibility that must be avoided in circuits such as that of figure 1 is both compound transistors being simultaneously in an "on" state. In this case the possibility of simultaneous conduction has been avoided by introducing a small time delay between the turning off of one device and the turning on of the other. The drive circuit is shown in figure 3a and the relative timings of the input, V_i, and outputs, V_{OT} and V_{OB} are shown in figure 3b. The delay is generated by the resistor-capacitor-diode combinations which ensure virtually no turn off delay but introduce a turn on delay of approximately 0. 7RC, ≅15μs in this case. Although the resulting modification of drive pulse width may appear inconvenient from the point of view of pulse width modulated control applications, the delay is constant and could be compensated for with ease in a digitally based control system.

2.4 *Drive Isolation for the Compound Transistors*

In order to control the compound transistors a control voltage must be applied between their gates and their sources. To maintain isolation from the HT supply, the drive signal is electrically isolated from all parts of the system connected to the HT. As the purpose of this design was on-off control, digital opto-isolators were used. The choice of opto-isolator is not critical providing that:

a) it does not introduce delays that are significant compared to those deliberately introduced by the drive circuit described in the previous section,

b) it can isolate effectively at the maximum intended operating voltage of the circuit and

c) its operation is not affected by the large rates of change of voltage across the isolating medium that occur in high voltage switching applications.

The devices used here could withstand a steady input-output voltage difference of 4kV, could tolerate a maximum rate of change of input-output voltage difference of $2 \times 10^9 Vs^{-1}$ and could operate at data rates of up to 5MHz.

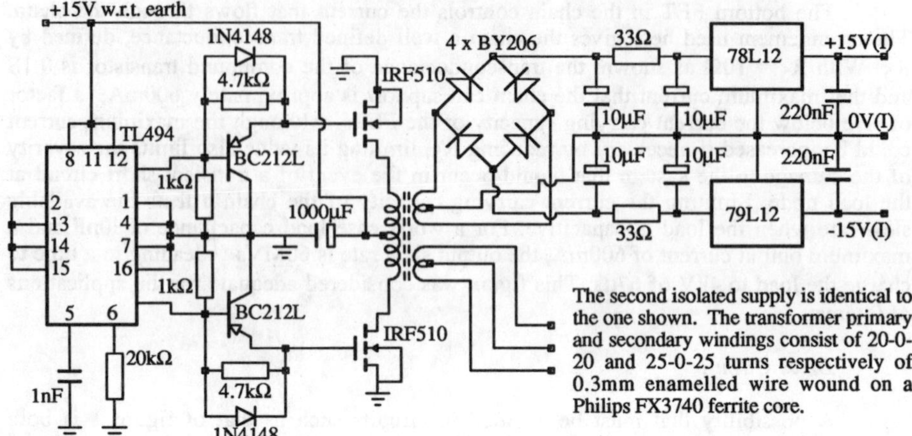

Figure 4. The circuit diagram of the isolated power supply for the compound transistors and their associated opto isolators. Only one secondary circuit is shown

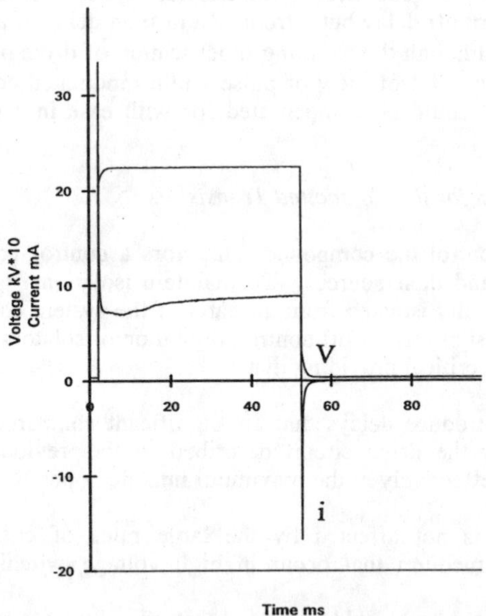

Figure 5. Full Switching Cycle

It was necessary to provide two isolated low voltage power supplies to power both the isolated side of the opto-isolator and the operational amplifier in the compound transistors. The isolated supplies of ±15V were derived using a simple forward converter with a transformer wound in such a way as to offer the required isolation. The isolated power supply circuit is shown in figure 4.

3. Specimen Test Results

Figure 5 shows a full switching cycle test result. The ER load was a concentric cylinder catch [1]. (60 mm dia x 60 mm long, 0.75 mm gap speed 100 rpm, fluid Lipol 30/W).

Details of the initial charging spike are not truly representative of the event due to the finite sampling rate set on the data acquisition unit. Current decay duration times (after switching) depend on the properties of the ER fluid, are seen to be roughly the same in both directions and will impose a limiting cycling rate. Whilst the current continues to change the voltage supplied will carry on settling.

The result was measured using a unidirectional drive polarity. This can be changed say to combat any dielectrophoretic tendencies in a fluid. In this event polarity can be reversed on alternate excitation applied periods, by using two switch modules in an H bridge configuration. This requires only one voltage supply.

4. Discussions

The device described in this paper is primarily for experimental or prototype use. In a production mode the design would be influenced by the specific load (suffix L) size and the availability and price economics of components.

In optimising an ER device the inter electrode gap size h and surface area A may be altered. For example, non linearities aside, the load resistance and capacitance vary as $R_L \alpha h/A$ and $C_L \alpha A/h$ respectively. Hence the larger the gap size, the greater V (to maintain V/h and the electro-stress) and the smaller C_L. The effect of such a manoeuvre would be to steepen the voltage slew rate but, also to cause more power to be dissipated in the switch since the stored energy in the load capacitance would increase linearly with applied voltage (for constant E).

The exact reduction in current i drawn will depend on the linearity of the electric load. This will change from fluid to fluid. Nevertheless the current to be switched will fall as h is increased, without detriment to the performance on the speed connection of the switch.

In very general and comparative terms the switch off performance is vastly superior to that obtained when the previous switch was used on the same clutch, in that instance employing an open circuit method of removing drive - Fig 6.

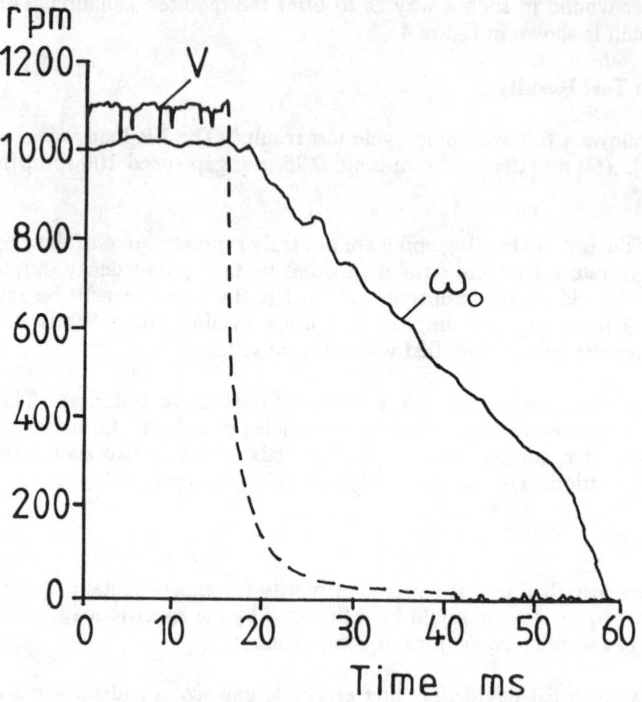

Figure 6. Comparative figure for open circuit 'switch off [1].
Decline of ω_2 is fixed by inertia of rotating parts.

5. Conclusions

A circuit capable of rapidly switching an ER device both on and off has been described. The slew rate with a 10nF load capacitance is $60MVs^{-1}$ in both the turn on and the turn off directions and this figure could be increased if necessary by the simple expedient of changing two resistors. The circuit has been shown to operate successfully with an ER device as a load and is presently being used to evaluate the effectiveness of pulse width modulated control strategies for ER devices. It is expected that further developments will lead to a circuit offering a linear input-output relationship, thus enabling a direct comparison of pulse width modulated and proportional control strategies.

6. References

1. A R Johnson et al. Proc, Japan Soc. Mech. Engs, Conf. on Motion and Power Transmission, Hiroshima, p 1016-1021 'Electro-Rheological Clutch Under Inertial Loading'.
2. M Whittle et al. Electrical Characterisation of an ER Valve.Proc. 3rd Int. ERF Conf. Carbondale, Ill, p. 343-366.

Acknowledgements

Thanks are due to Messrs B Stobbs, for assembling the excitation circuit and J Makin for the use of the test figure from his provisional PhD thesis.

HYDRODYNAMIC PRESSURE GENERATION WITH AN ELECTRO-RHEOLOGICAL FLUID. PART I - UNEXCITED FLUID.

T. H. LEEK
S. LINGARD
The University of Hong Kong, Dept. of Mechanical Engineering, Pokfulam Road, Hong Kong.

W. A. BULLOUGH
R. J. ATKIN
The University of Sheffield, Smart Machines Unit, P.O. Box 600, Mappin Street, Sheffield, S1 4DU, UK.

ABSTRACT

The work reported, in conjunction with part II, forms an early step in the development of controllable and fast responding electro-rheological (ER) fluid bearings. The practical behaviour of an unexcited ER fluid under conditions of combined Couette and Poiseuille flow, a combination of shear mode flow and valve mode flow in effect, is investigated. By comparison between theoretical predictions and experimental measurements of the steady state hydrodynamic pressure generated using ER fluid it is shown that, for the conditions examined, a bearing flow can be considered on the basis of a continuum analysis.

1. Nomenclature

B	Bingham number
C_n	constants of integration
h	step height of bearing
l_t	length to pressure tapping
l_{tc}	arc length to pressure tapping
p	pressure
P	dimensionless pressure gradient
q	volume rate of flow per unit width
r	radius
R_0, R_h	inner and outer radius of bearing
u	linear velocity of fluid
U	linear velocity of moving surface
w	bearing width
$\dot{\gamma}$	shear rate
η	Newtonian viscosity
η_p	plastic viscosity
Θ	absolute temperature
τ	shear stress

τ_e	yield shear stress
ω	angular velocity of fluid
Ω	angular velocity of moving surface

B and P are defined by

$$B = \frac{\tau_e h}{\eta_p U} \qquad P = \frac{dp}{dx}\frac{h^2}{\eta_p}\frac{1}{U} \tag{1}$$

for flow between flat plates and

$$B = \frac{\tau_e}{\eta_p}\frac{h}{\Omega R_0} \qquad P = \frac{1}{R_0}\frac{dp}{d\theta}\frac{h^2}{\eta_p}\frac{1}{\Omega R_0} \tag{2}$$

for flow between concentric cylinders.

2. Introduction

The growing number of potential applications for electro-rheological (ER) fluids[1] establishes the need to investigate the fundamental properties of such a fluid. Of particular importance is how to design with an ER fluid. Physically the fluid is a two-phase mixture, however, the use of a continuum assumption for the fluid to provide at least approximations to its in-service behaviour would be useful in engineering design. The present paper is the forerunner, perhaps, of a conclusive step in establishing the continuum principle for excited ER fluids in shear flows with pressure gradients.

Previous experimental work employing ER fluids has focused principally on circumstances of either purely Couette flow or purely Poiseuille flow. Herein the condition of combined Couette and Poiseuille flow is examined. Simply, the question of whether it is possible to use continuum theory together with ER fluid data, obtained from straightforward viscometric tests, to predict hydrodynamic pressures generated using the fluid, is posed. This question cannot be answered with any authority at present, the answer is of use to those involved in developing ER fluid engineering devices. In particular, the principal driving force behind this work is the potential contribution that ER fluid controlled bearings could make towards a new generation of flexible and durable high speed machines. The concept of ER fluid bearings has seen some theoretical analysis[2,3], the work herein, together with associated work[4,5], would appear to be the first efforts towards practical assessment. Previously reported[4] experimental data is included and discussed to illustrate the need for the improved experimental procedure used to obtain new data.

The work comprises four elements.

1. Measurement of fluid properties using conventional viscometry.
2. Development of the continuum theory.
3. Experimental measurement of hydrodynamic pressure generation.
4. Comparison of theory and practice.

Interest often centres solely around the behaviour of excited ER fluid. This work however, requires the consideration of unexcited fluid, performed herein, prior to the investigation of excited fluid which is reported in part II[5].

3. Fluid description and properties.

The ER fluid used in the tests comprised a dielectric liquid with lithium polymethacrylate particles averaging about 5 μm in diameter and occupying about 30% by volume of the mixture. The water content of the particles was 15-18% by weight. This is a typical 'wet' fluid[6] and it has been used in other ER fluid experimental investigations[7,8].

Unexcited ER fluid was tested in a conventional cone-on-plate viscometer across a shear rate range of 200 to 6300 s^{-1}. The cone used in the viscometer had a diameter of 50 mm. and a cone angle of 0.3°. This means that within a radius of approximately 1 mm from the cone tip, where the cone / plate separation is less than the average particle size, the fluid had a low particle concentration. Any effect of this is however considered negligible when compared to the total cone radius of 50 mm. It is possible that particles could become trapped between the cone and the plate and adversely affect the test results, if this occurred to any significant extent it would be evident in the comparison of repeated tests.

Prior to the extraction of a fluid sample for testing in the viscometer the ER fluid container was briefly shaken to ensure homogeneity of the fluid. Three nominal test temperatures were used, 22.5, 30.0, and 39.0 °C, to cover the temperature range of the step bearing tests, and the fluid temperature was maintained constant during each test to within ± 0.4 °C.

A complete test involved the following procedure performed after thermal equilibrium had been achieved:

1. Measure shear stress for various shear rates three times. The fluid so tested is referred to as unworked.
2. Work the fluid for 36 minutes at $\dot{\gamma} = 3152$ s^{-1}.
3. Reduce shear rate to zero briefly then record shear stresses three more times, the results are referred to as applying to worked fluid.

Each test was repeated using a new fluid sample. Data was collected by repeatedly setting the shear rate at a required value and recording the corresponding shear stress. Measurements were taken whilst the shear rate was both sequentially increased and sequentially decreased, and were taken rapidly to minimise the effect of any time dependence of the fluid. The data collected is shown in figures 1 and 2 where straight lines have been fitted through the average value of τ at each $\dot{\gamma}$. The variation in the repeated data from both the same test and from different tests is in general small, being significant only at low shear rates for the worked fluid. This is possibly due to the fluid relaxing after being worked and so being somewhat unstable at low shear rates. There is no obvious problem with particle size and cone / plate separation.

Figures 1 and 2 indicate that the fluid can be closely approximated as a Bingham plastic, defined by

$$\begin{aligned} \tau &= \eta_p \dot{\gamma} + \tau_e \operatorname{sgn}(\dot{\gamma}) \qquad && |\tau| \geq |\tau_e| \\ \dot{\gamma} &= 0 && |\tau| \leq |\tau_e| \end{aligned} \tag{3}$$

The values for η_p and τ_e from the straight lines in figures 1 and 2 are given in table 1 and show that τ_e for unworked fluid is approximately constant with temperature. When the fluid is worked τ_e tends to increase and appears to become dependent upon temperature. The plastic viscosity, η_p, of both unworked and worked fluid decreases as temperature increases, as would normally be expected, and does not appear to change by any significant amount when the fluid is worked.

unworked fluid			worked fluid		
Temp./°C	τ_e / Nm^{-2}	η_p / Nsm^{-2}	Temp./°C	τ_e / Nm^{-2}	η_p / Nsm^{-2}
22.5	26.7	0.188	22.7	148	0.187
30.0	32.5	0.107	30.0	79.8	0.113
39.2	22.1	0.0611	39.0	51.1	0.0585

Table 1 : Yield shear stress and plastic viscosity of ER fluid at different temperatures from cone-on-plate viscometer.

The possibility of an unexcited ER fluid exhibiting a yield shear stress has been noted previously[9] and can result from stabilising agents in the fluids[10]. Although the data suggests a yield shear stress for the fluid used here it is not proven to exist to its exact definition (that is at $\dot{\gamma} = 0$). However, the Bingham model is a better fit to the data than the Newtonian model. Figure 1 suggests that the difference between the two models is small, it is however application dependent and cannot necessarily be ignored.

Bearing	l_{tc} / mm	h / mm	w / mm
1	10.72	0.25	5.10
2	14.88	0.244	6.86
3	10.16	0.28	6.91
4	10.16	0.464	6.91

Table 2 : **Bearing dimensions, (all bearings are part circular with R_0 = 29.25 mm.).**

5. Theory

The flow solutions employed are given in outline only, being detailed elsewhere[3,11]. The following assumptions are made :

i) the flow in the bearing is fully developed, one dimensional, laminar, and isothermal,
ii) the fluid is a Bingham plastic, and is incompressible.

Figure 5 shows the form of the velocity profiles for a Newtonian fluid and a Bingham plastic flowing in a plane step bearing at a net zero volumetric flowrate. In the Bingham plastic a core of unyielded fluid occurs in region 2, where $|\tau| \leq |\tau_e|$, while in regions 1 and 3 $|\tau| \geq |\tau_e|$ and the fluid is being sheared. For a Bingham plastic flowing between parallel walls there are four possible forms of velocity profile[3,11] :

i) A core existing within the flow as in figure 5.
ii) A core attached to the moving surface.
iii) A core attached to the stationary surface.
iv) No core, the fluid is yielded all across the gap.

For the experimental conditions and theoretical assumptions made in this work it has been proved[11] for both plane and part-circular bearing geometries that only profiles of type i) in this list, that is of the form shown in figure 5, are possible. Such a conclusion can be drawn without recourse to analysis by considering the requirements:

a) For zero flow rate there must be some fluid flowing 'forwards' and some flowing 'backwards'.
b) There is no slip at the boundaries.
c) The shear rate cannot be equal to zero without a core region being present, so fluid velocity does not change from being positive to negative or *vice versa* other than across a core region.

Considering the plane bearing shown in figure 3(i) with h_2 set to zero. Using the assumptions given above the equation of motion for the fluid in the bearing is

$$\frac{dp}{dx} = \frac{d\tau_{yx}}{dy} \tag{5}$$

Combining this with the Bingham plastic constitutive equation (3) results in

$$\eta_p u = \frac{1}{2}\frac{dp}{dx}y^2 - \tau^* y + C_1 y + C_2 \qquad |\tau_{yx}| \geq |\tau_e| \tag{6}$$

where $\tau^* = \pm\, \tau_e$ depending upon whether $\tau_{y\dot{x}}$ is positive or negative; shear rate and therefore shear stress changes sign across a core so the sign of τ_e must be changed accordingly.

Equation (6) is applied separately to the regions of sheared flow, above and below the core, to determine the velocity profile of the flow with the constants, C_n, being obtained using the appropriate boundary conditions from

a,b) No slip at the walls.
c) At the boundaries of a core $\tau_{yx} = \tau^*$, *i.e.* $\dot{\gamma} = 0$.
d) From equilibrium of the forces on a core under steady flow conditions the core has a thickness given by

$$y_2 - y_1 = \frac{2\tau^*}{\frac{dp}{dx}} . \tag{7}$$

The condition of no slip of the walls has been identified[12] as questionable with ERF as 'wall slip' is not unusual in concentrated suspensions. Unknowns such as this form the reason for the present investigation, being a test of whether continuum assumptions and conditions can be used as close approximations. Having correctly determined the velocity profile the volumetric flowrate is straightforwardly obtained from

$$q = \int_0^h u\,dy = \int_0^{y_1} u\,dy + \int_{y_1}^{y_2} u\,dy + \int_{y_2}^{h} u\,dy \tag{8}$$

Using the dimensionless parameters B and P, as defined in equation (1), and imposing the required condition of volumetric flowrate equal to zero gives

$$\frac{q}{Uh} = \frac{P}{6}\left[\left(\frac{y_1}{h}\right)^3 - 2 - \left(\frac{y_2}{h}\right)^3 + 3\frac{y_2}{h}\right] = 0 \tag{9}$$

$$\frac{y_2}{h} = \frac{y_1}{h} + 2\frac{B}{P} = \frac{1}{2} + \frac{B}{P} + \frac{1}{P - 2B}$$

This is only valid when $0 \leq y_1 \leq y_2 \leq h$, which can be expressed as the general condition

$$0 \leq 2\frac{B}{P} \leq 1 - \sqrt{\frac{2}{|P|}} \tag{10}$$

An explicit solution to equation (9) is not available, discrete values of B versus P can however be obtained using for example the Newton-Raphson method for determining the roots of an equation. Only one of the three roots so found satisfies the validity condition. Curve fitting to discrete results provides the following expressions giving P from values of B to an accuracy of $\pm 0.5\%$.

$$\begin{aligned} P &= 6 + 2.69B - 2.53\times10^{-2}B^2 \qquad && 0 \leq B \leq 6 \\ P &= 6.86 + 2.43B - 5.03\times10^{-3}B^2 \qquad && 6 \leq B \leq 25 \end{aligned} \tag{11}$$

Theoretical pressures in the step bearing are calculated using these expressions. When τ_e and therefore B is set to zero equation (11) reduces to the solution for the Newtonian model of

$$P\,(\text{Newtonian fluid}) = \frac{dp}{dx}\frac{h^2}{\eta U} = 6 \tag{12}$$

A part-circular bearing as shown in figure 3(ii), rather than a plane bearing, is used in the experiments. Solution for the part circular geometry follows a similar procedure to that above. The equation of motion of the fluid is now

$$2\tau_{r\theta}r + r^2\frac{d\tau_{r\theta}}{dr} = r\frac{dp}{d\theta} \tag{13}$$

which when combined with the Bingham plastic constitutive equations (expressed appropriate to circumferential flow) gives the following general flow equation equivalent to equation (6) above.

$$\eta_p \omega = (\frac{1}{2}\frac{dp}{d\theta} - \tau^*) \ln r - \frac{C_1}{2r^2} + C_2 \quad (14)$$

This equation is applied to find the angular velocity profile by the application of boundary conditions equivalent to those listed above for plane bearings. The flow rate is then obtained and is set equal to zero. The solution is given in appendix A, using *P* and *B* as defined by equation (2).

The effect of the curvature of the bearing upon the *P* and *B* relationship is illustrated in figure 6. Note that *P* and *B* for the plane bearing, equation (1), and for the part-circular bearing, equation (2), are equivalent if x is taken to be equal to $R_0\theta$, *i.e.* if the arc length, l_{tc}, rather than the plane length, l_t, of the bearing is taken. For the bearings used in this work (dimensions given in table 2) h/R_0 is of the order of 0.01. As might be intuitively expected and as shown on figure 6 such a value of h/R_0 has negligible effect upon the *P* vs. *B* relationship as long as the arc length of the bearing is used, this is particularly true for small values of τ_e and therefore *B*. As τ_e and therefore *B* increases so does the significance of bearing curvature.

Pressures for the ER fluid tests are therefore predicted using equation (11), with *P* and *B* defined by equation (1). Note that the pressure rises linearly along the bearing length (since the pressure gradient is independent of x), so the pressure at the tapping, p_t, being the required pressure, is obtained from

$$\frac{dp}{dx} = \frac{p_t}{l_{tc}} \qquad P = \frac{p_t}{l_{tc}}\frac{h^2}{\eta_p U} \quad (15)$$

The largest value of *B* used in this work is 0.217. This gives from equation (11) P = 6.58, that is the Newtonian value plus 10%. Whilst the shear yield stress of 27 $\mathrm{Nm^{-2}}$ is small it cannot be as easily ignored as the curvature effect.

6. Results

The agreement between theory and experiment for control experiments[4], using a mineral oil, has shown the experimental arrangement and theoretical basis to be fundamentally sound. The experiment is therefore a valid test of the continuum behaviour, or otherwise, of the ER fluid.

To ensure the required limiting pressure was measured, that is zero or near zero net flowrate is achieved, preliminary tests were conducted with a varying load on the bearing. Pressures levelled off to a steady value as expected[4], and the constant load used for further data collection was established.

Tests with ER fluid were conducted within a temperature range of 24 to 38 °C, and were kept short to justify the use of unworked fluid data from the viscometry tests. The ER fluid container was shaken before a test sample was extracted to ensure homogeneous fluid. Experimental and theoretical limiting pressures are compared through the ratio $p_{exp.}/p_{theory}$, the theoretical pressure, p_{theory}, being given by equations (11) and (15). The properties determined in the viscometry tests are employed in the calculations. This comparison of experimental and theoretical results is shown in figure 7 for bearings one, two and three, in which $p_{exp.}/p_{theory}$ varies between 0.81 and 1.01 for ER fluid.

As noted previously[4] the agreement between theory and experiment for ER fluid is somewhat less than that for mineral oil. It has been proposed[4] that the additional errors obtained when using ER fluid, and the variation of this error with temperature as shown in figure 7, are due to significant deviation of the test away from the assumptions of isothermal flow and a test temperature being given by the thermocouple at the bearing entrance. This deviation being fluid dependent. Theoretical explanations for temperature discrepancies have been discussed more fully elsewhere[4], suffice to say here that the combined effects of working the fluid, heat generated by Coulomb friction, and heat transfer between the fluid, the bearing and the atmosphere, would be expected to influence the test temperature and it is whether this effect is significant that is important.

The crucial temperature is that which occurs inside the bearing. For this reason bearings were manufactured to include a thermocouple inside the bearing at the position of the step, all other features of the bearing design remaining the same. Tests using these bearings, under the same conditions as before, indicated that a significant difference in fluid temperature can occur between that at the bearing entrance and that at the step, the latter tending to be greater than the former. The differences could be as much as 2 °C for ER fluid at test temperatures close to atmospheric, about 23 °C, and decreased as test temperature increased. The occurrence of the larger errors at the lower test temperatures corresponds to where the fluid viscosity is most sensitive to temperature. This evidence matches the observations of the data shown in figure 7 being that temperature errors and reduce as the test temperature increases. As an example a 2 °C change in temperature of ER fluid at a bulk temperature of 23 °C corresponds to a change in plastic viscosity of approximately 13 %.

To minimise temperature errors further tests were performed using a thermally insulating chamber around the experimental apparatus. The chamber temperature was adjusted using a hot air blower and temperatures recorded using three thermocouples placed in the fluid reservoir, at the bearing entrance, and at the step within the bearing respectively. Results were recorded only when thermal equilibrium was achieved such that the three temperatures were within 0.25 °C. In this way data was obtained at close to isothermal conditions. The data for testing with both mineral oil and with ER fluid are shown in figure 8 and reflect, for both fluids but particularly for ER fluid, the improved temperature stability. The results are significant not so much in the fact that they are so close to 1.0, although that is not undesirable, but more so in not showing the large temperature dependency of $p_{exp.}/p_{theory}$ for ER fluid that was previously encountered.

7. Discussion

The properties of an unexcited ER fluid have been determined from conventional viscometry tests, showing the fluid to be well approximated as a Bingham plastic with a small yield shear stress. The ER fluid was then tested in a part circular Rayleigh step bearing and the experimental results compared with theoretical predictions made using continuum theory and the Bingham plastic properties already established.

Results from tests using ER fluid and bearings 1, 2 and 3 in atmosphere showed some unexpected temperature dependence, and as a result interpretation is somewhat complicated. Further tests have however shown that significant temperature errors are likely to have occurred in the original ER fluid tests, enough to account for the discrepancies in the theoretical and experimental comparisons.

The possibility of any important deviation away from continuum behaviour of the ER fluid being masked by temperature considerations is disproved by the final tests conducted under controlled temperature conditions. The results of these tests are shown in figure 8 and confirm the continuum behaviour of the ER fluid.

8. Conclusions

The ER fluid tested has the properties of a Bingham plastic as shown by figures 1 and 2 and given in table 1, and is somewhat time dependent with τ_e increasing if the fluid is worked for 36 minutes.

The results of tests using ER fluid, taking into account temperature effects where necessary, show that shear flow behaviour of the ER fluid can be modelled on a continuum basis using fluid properties determined from standard viscometry tests. This conclusion is restricted at this time to the experimental conditions encountered where the minimum dimension of the flow channel is approximately 50 times the average particle size.

Of further interest is the behaviour of excited fluid in shear flows, reported in part II[4] of this work, and the behaviour of both unexcited and excited ER fluid in thinner films. Both of these investigations require the sound experimental and theoretical basis provided by the work performed herein.

9. Acknowledgements

The authors wish to acknowledge the financial support of The Croucher Foundation and The British Council.

10. References

1. Section IV. Applications of ER fluids. Electrorheological fluids, Proceedings of the International Conference 1991, (World Scientific, Singapore, 1992).
2. A.D.Dimarogonas and A.Kollias, *'Electrorheological Fluid-Controlled "Smart" Journal Bearings'*, J. of Tribology Transactions, June 1992.
3. J.A.Tichy, *'Hydrodynamic lubrication theory for the Bingham plastic flow model'*, J. Rheol 35(4), May 1991.
4. T.H.Leek *et al.*, *'An experimental investigation of the flow of an electro-rheological fluid in a Rayleigh step bearing'*, accepted for publication in J. Phys. D: Appl. Phys. (1993).
5. T.H.Leek *et al.*, *'Hydrodynamic pressure generation with an electro-rheological fluid. Part II - excited fluid.*
6. H.Block and J.P.Kelly, *'Electro-rheology'*, J. Phys. D: Appl. Phys. 21 (1988) 1661-1677.
7. W.A.Bullough *et al.*, *'Fast pick-up and drop load performance of a low electrical and mechanical time constant, electro-rheologically based clutch'*, Mechatronic systems engineering 1 (1992) 315-327.
8. A.Hosseini-Sianaki *et al.*, *'Experimental measurements of the dynamic torque response of an electro-rheological fluid in the shear mode'*, Electrorheological fluids, Proceedings of the International Conference 1991, (World Scientific, Singapore, 1992) 219-235.
9. D.L.Klass and T.W.Martinek, *'Electroviscous fluids I. Rheological properties'*, J. Appl. Phys. 38 (1967) 67-74.
10. D.Brooks, *'Applicability of simplified expressions for design with electrorheological fluids'*, Proc. Smart Materials Conf. Blacksburg Va., (Technomic Press, Lancaster Pa., 1991).
11. T.H.Leek *et al.*, *'Solutions for one dimensional combined Couette and Poiseuille flow of an ER fluid as a Bingham plastic'*, report number HKTHL023 of Hong Kong / Sheffield ERF initiative, (to be published).
12. J.E.Stangroom, *The Bingham plastic model of ER fluids and its implications'*, Electrorheological fluids Proc. 2nd Intl. Conf., Raleigh North Carolina 1989 (Technomic, Lancaster Pa., 1990) 199-206.

11. Appendix A

Presented in this appendix is an outline of the solution[11] for idealised flow of a Bingham plastic in a part circular bearing, as shown in figure 3(ii), at a condition of zero flow rate such that a core of unyielded fluid exists in the flow. The radii of the internal and external surfaces of the bearing are R_0 and R_h respectively, such that the bearing height h equals R_h - R_0. The core exists in the region $R_1 \leq r \leq R_2$. Certain common factors appear in the solutions, the following identities are therefore used for brevity.

$$F^+ = \frac{1}{2}P + \frac{h}{R_0}B \qquad F^- = \frac{1}{2}P - \frac{h}{R_0}B \tag{A.1}$$

where B and P are defined in equation (2).

The following boundary conditions apply.

1. At $r = R_0$, $\omega = \Omega$.
2. At $r = R_1$, $\dot{\gamma}_r = 0$.
3. At $r = R_2$, $\dot{\gamma}_r = 0$.
4. At $r = R_h$, $\omega = 0$.
5. Core thickness satisfies steady state equilibrium requirements.
6. $\omega_{r\,=\,R1} = \omega_{r\,=\,R2}$.

These conditions in conjunction with equation (14) provide the angular velocity profile as three expressions covering respectively the three regions of below the core, the core itself, and above the core, care must be taken in analysis to follow the change in sign of shear rate and therefore of shear stress across the core region.

$$R_0 \le r \le R_1 \qquad \frac{\omega}{\Omega} = \frac{1}{2}(F^+)\left(\frac{R_0}{h}\right)^2\left[\ln\left(\frac{r}{R_0}\right)^2 + \left(\frac{R_1}{r}\right)^2 - \left(\frac{R_1}{R_0}\right)^2\right] + 1$$

$$R_1 \le r \le R_2 \qquad \frac{\omega}{\Omega} = \frac{1}{2}(F^-)\left(\frac{R_0}{h}\right)^2\left[\ln\left(\frac{R_2}{R_h}\right)^2 + 1 - \left(\frac{R_2}{R_h}\right)^2\right] \tag{A.2}$$

$$R_2 \le r \le R_h \qquad \frac{\omega}{\Omega} = \frac{1}{2}(F^-)\left(\frac{R_0}{h}\right)^2\left[\ln\left(\frac{r}{R_h}\right)^2 + \left(\frac{R_2}{r}\right)^2 - \left(\frac{R_2}{R_h}\right)^2\right]$$

together with

$$\frac{2}{F^-}\left(\frac{h}{R_0}\right)^2 = \left[\ln\left(\frac{R_2}{R_h}\right)^2 + 1 - \left(\frac{R_2}{R_h}\right)^2\right] - \left(\frac{R_2}{R_1}\right)^2\left[\ln\left(\frac{R_1}{R_0}\right)^2 + 1 - \left(\frac{R_1}{R_0}\right)^2\right] \tag{A.3}$$

$$\left(\frac{R_2}{R_1}\right)^2 = \frac{F^+}{F^-}$$

The obvious conditions for this flow profile to occur are $R_0 \le R_1 \le R_2 \le R_h$. These conditions are in fact derivatives of the more fundamental requirements

$$|\tau_{walls}| \ge \tau_e , \qquad \tau_{R_2} = \tau^* , \qquad \tau_{R_1} = -\tau^* \tag{A.4}$$

which if required can be used to determine more explicit (but complex) conditions.

Volumetric flowrate per unit width, q, is given by appropriate evaluation of

$$q = \int_{R_0}^{R_h} \omega r \, dr$$
$$= \left[\frac{\omega r^2}{2}\right]_{R_0}^{R_h} - \frac{1}{2}\left\{\int_{R_0}^{R_1} r^2 \frac{d\omega}{dr} dr + \int_{R_1}^{R_2} r^2 \frac{d\omega}{dr} dr + \int_{R_2}^{R_h} r^2 \frac{d\omega}{dr} dr\right\} \tag{A.5}$$

which gives

$$\frac{q}{\Omega R_0^2} = -\frac{1}{2} - \frac{1}{4}\left(\frac{R_1}{h}\right)^2 (F^+)\left[1 - \left(\frac{R_0}{R_1}\right)^2 - \ln\left(\frac{R_1}{R_0}\right)^2\right]$$
$$- \frac{1}{4}\left(\frac{R_2}{h}\right)^2 (F^-)\left[\left(\frac{R_h}{R_2}\right)^2 - 1 + \ln\left(\frac{R_2}{R_h}\right)^2\right] \tag{A.6}$$

The required solution of zero flowrate is obtained by setting equation (A.6) equal to zero and solving simultaneously with equations (A.3). The problem is then

$$f(B, P, \frac{h}{R_0}) = 0 \tag{A.7}$$

An explicit expression for P as a function of $(B, h/R_0)$ cannot be obtained, discrete solutions must be obtained by numerical analysis. For a particular value of h/R_0 there is only one value of P for each value of B that satisfies both A.7 and the validity requirements of A.4.

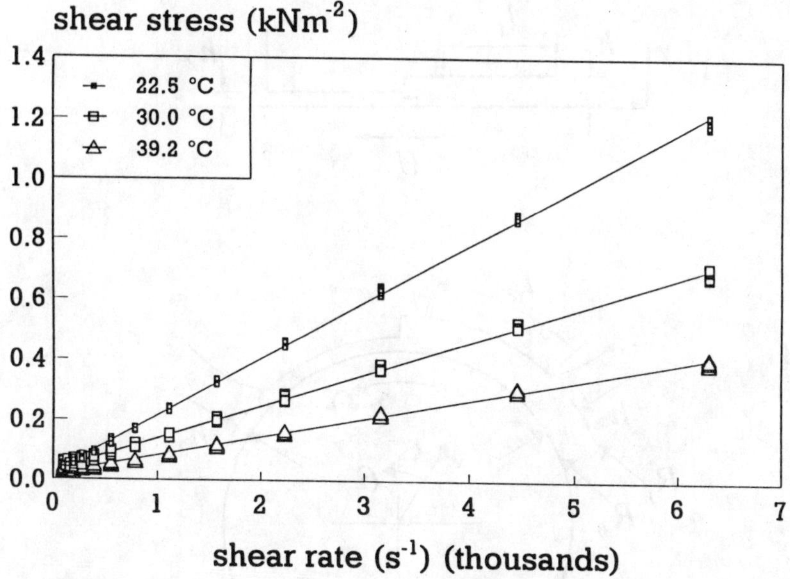

Figure 1 : Shear stress / shear rate data from cone-on-plate tests, unworked ER fluid.

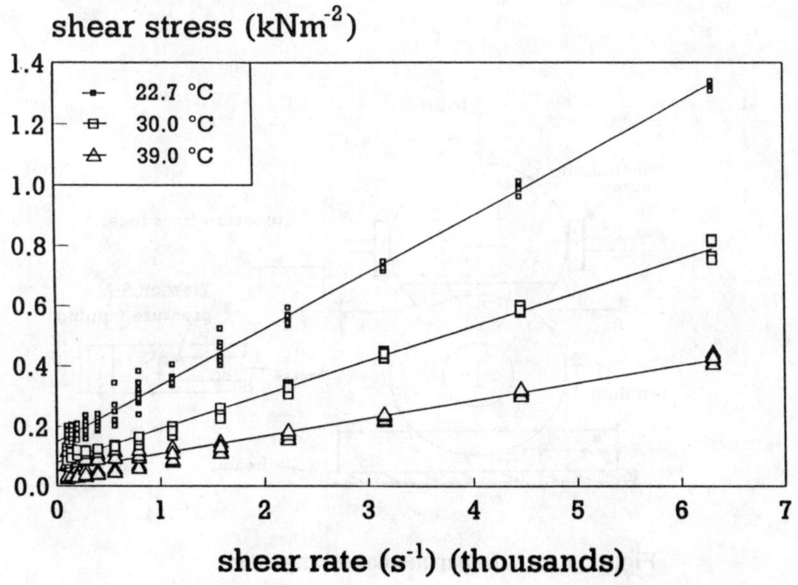

Figure 2 : Shear stress / shear rate data from cone-on-plate tests, worked ER fluid.

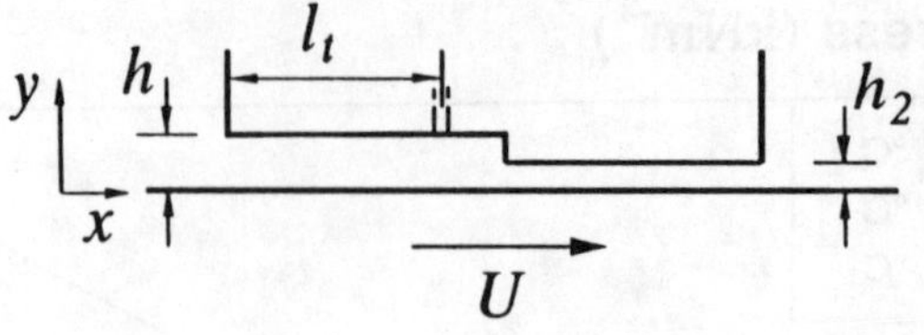

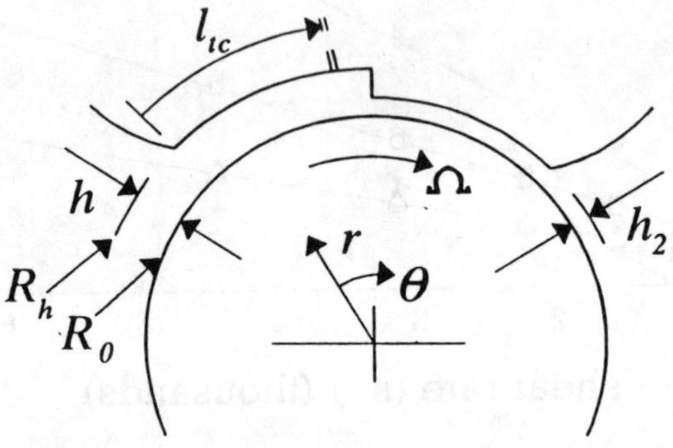

Figure 3 : Plane and part circular step bearings.

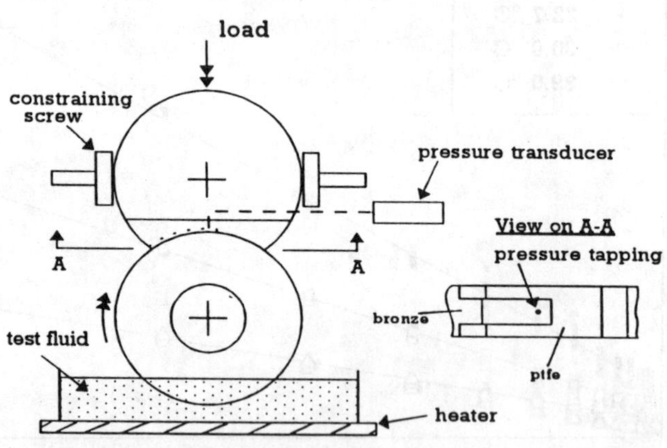

Figure 4 : Experimental arrangement.

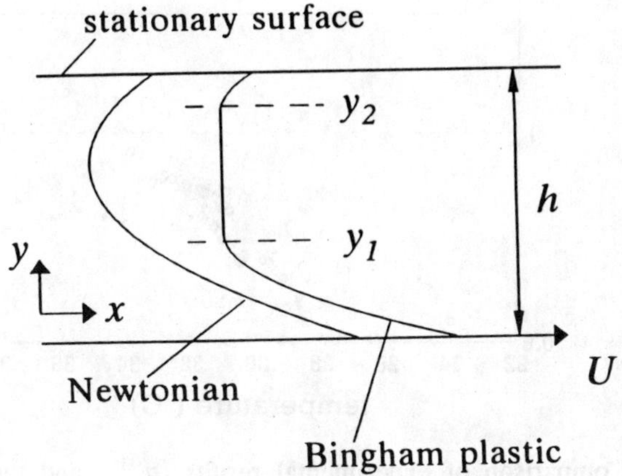

Figure 5 : Velocity profiles of a Newtonian fluid and a Bingham plastic at a zero flow rate.

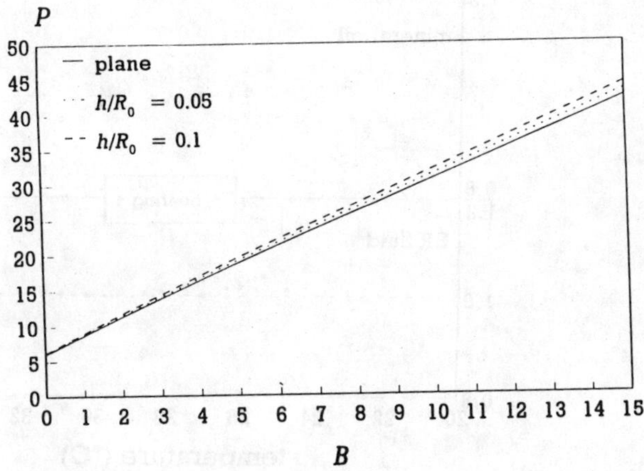

Figure 6 : The effect of curvature upon the P versus B relationship.

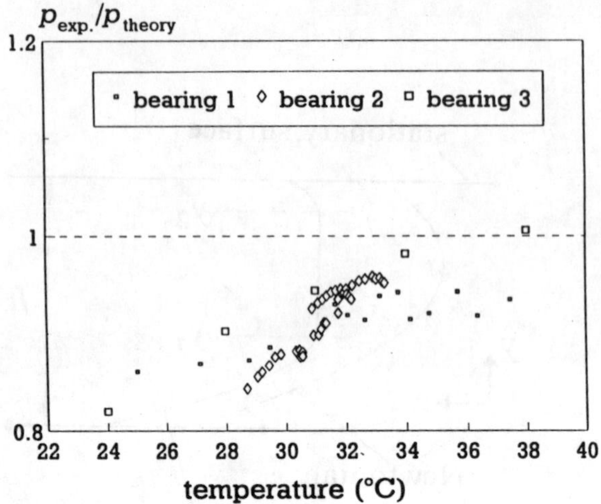

Figure 7 : Comparison of experimental results, $p_{exp.}$, and theoretical predictions, p_{theory}, for ER fluid, bearings 1, 2, and 3.

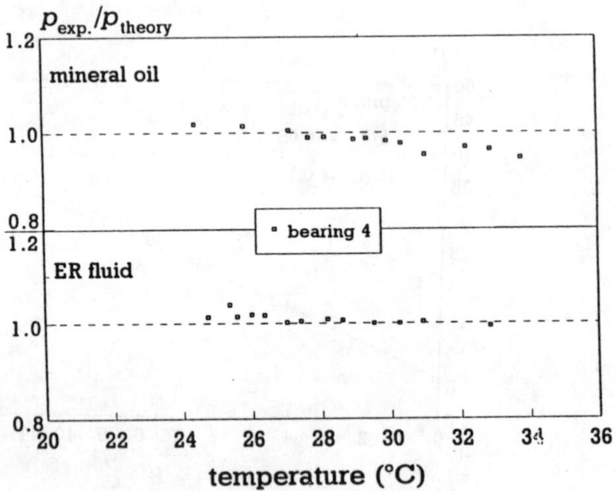

Figure 8 : Comparison of experimental results, $p_{exp.}$, and theoretical predictions, p_{theory}, for mineral oil and ER fluid, bearing 4.

HYDRODYNAMIC PRESSURE GENERATION WITH AN ELECTRO-RHEOLOGICAL FLUID. PART II - EXCITED FLUID.

T. H. LEEK
S. LINGARD
The University of Hong Kong, Dept. of Mechanical Engineering,
Pokfulam Road, Hong Kong.

W. A. BULLOUGH
R. J. ATKIN
The University of Sheffield, Smart Machines Unit,
P.O. Box 600, Mappin Street, Sheffield, S1 4DU, UK.

ABSTRACT

The behaviour of an excited electro-rheological (ER) fluid under conditions of steady state hydrodynamic pressure generation is investigated experimentally and compared to theoretical analyses developed by assuming the fluid to be a Bingham material. This is a direct continuation of the investigation of unexcited ER fluid reported in Part I. Apparent properties of the excited ER fluid determined from viscometer tests are used in theoretical predictions of fluid behaviour. Data for the same fluid under different conditions, reported in the literature, are considered alongside the new data. The comparisons made indicate that, at least for the conditions examined, the Bingham plastic model is acceptable for engineering design calculations.

1. Nomenclature

B	Bingham number
E	electric field
h	step height of bearing, fluid gap in viscometry
l_t	linear length of bearing to pressure tapping
l_{tc}	arc length of bearing to pressure tapping
L	height of fluid in viscometer
p	pressure
dp/dx	pressure gradient
P	non-dimensional pressure gradient
r	radius
R_0, R_h	radius of inner and outer surface
T	torque
U	linear velocity of moving surface
V	voltage
w	width of bearing
x,y,z	coordinate axes
W,X	geometric parameters of the viscometry equipment

$\mathring{\gamma}$	shear rate / s^{-1}
$\mathring{\gamma}_n$	nominal shear rate
η_p	plastic viscosity
Θ	absolute temperature
τ	shear stress
τ_m	measured shear stress
τ_e	yield shear stress
Ω	angular velocity

2. Introduction

Alongside research into the fundamental physics of ER fluids and the development of improved fluids[1] there is an important need to study practical ER fluid behaviour. It has been suggested[2] that the properties of an excited ER fluid depend upon shear rate, temperature, and flow condition and geometry in addition to electric field; at the same time the fluid is a suspension of solid particles in a liquid carrier. Combined, this makes effective and innovative design with an ER fluid an almost impossible task without the use of simplifying approximations. Adopting a necessarily straightforward approach therefore, this work examines whether some of the problems and unknowns of ER fluid behaviour could, in the circumstances of steady state hydrodynamic pressure generation, be sidestepped by designers. Until such time as theories are developed which successfully explain the physical phenomena of ER fluid flow practical investigations of this type are useful in attempting to progress towards commercial ER fluid applications.

The uncertainties of designing with an ER fluid are

i. the continuum problem,

ii. the consistency problem.

These are, more explicitly, whether the two phase solid - liquid mixture can be considered as a simple continuous liquid phase with no special consideration for the presence of solid particles, and whether self similar behaviour occurs between devices and instruments of different sizes, designs, and flow conditions when a consistent theoretical approach is taken. These problems are examined herein for steady state hydrodynamic film flow. Data is obtained from a Couette flow viscometer for excited ER fluid and this is analyzed by assuming a continuum theory to apply, giving approximate continuum properties for the ER fluid. These properties and a consistent theoretical analysis are then used to predict the behaviour of excited ER fluid in the hitherto uninvestigated flow condition of combined Couette and Poiseuille flow, that is in circumstances of hydrodynamic pressure generation in a part-circular Rayleigh step bearing. Comparison between the theory and practical measurements enables a critical analysis of the behaviour of the ER fluid. This work improves upon a prior investigation of excited ER fluid flow[3], for which fluid properties were not available, and is based upon the scientific and theoretical approach developed in part I[4] where the behaviour of unexcited fluid was examined.

3. Fluid description and model

The ER fluid used comprised a dielectric liquid containing lithium polymethacrylate particles averaging about 5 μm in diameter and occupying about 30% by volume of the mixture. The water content of the particles is 15-18% by weight. This is the same type of ER fluid that was used in part I[4] of this work, under unexcited conditions, and has been used under excited conditions in circumstances of Poiseuille flow[5].

From cone-on-plate viscometry tests[4] the unexcited ER fluid has been shown to be best approximated as a Bingham plastic, defined by

$$\begin{aligned} \tau &= \eta_p \dot{\gamma} + \tau_e \operatorname{sgn}(\dot{\gamma}) && |\tau| \geq |\tau_e| \\ \dot{\gamma} &= 0 && |\tau| \leq |\tau_e| \end{aligned} \qquad (1)$$

with a yield shear stress of 27 Nm^{-2} and a plastic viscosity versus temperature relationship of

$$\eta_p = 1.43 \times 10^{-10} e^{6203/\Theta} \qquad (2)$$

where η_p (Nsm^{-2}) and Θ (K). This is for fluid that has not been worked for any length of time, when worked the yield shear stress of the fluid appears to increase.

A Couette type concentric cylinder with the facility to apply an electric field across the fluid gap was used to investigate the properties of the excited fluid. The details of the equipment are as follows :

Radius of inner cylinder, R_0 = 13.49 mm.
Radius of outer cylinder, R_h = 14.005 mm.
Size of fluid gap, $h = R_h - R_0$ = 0.515 mm.
Length in contact with fluid, L = 40 mm.

Data was recorded at two temperatures, 30 °C and 35 °C, by measuring the torque on the outer cylinder for a range of speeds of rotation, Ω, of the inner cylinder and at a potential difference V across the fluid gap; this was performed for 25 different voltages at each of the two temperatures. Examples of the results are shown on figure 1 where the nominal shear rate and measured shear stress were obtained from the recorded parameters by

$$\dot{\gamma}_n = \frac{\Omega R_h}{h} \qquad \tau_m = \frac{T}{2\pi R_h^2 L} \tag{3}$$

and the electric field strength is given by

$$E = \frac{V}{h} \tag{4}$$

The same trend appeared in the data at both temperatures. At low electric fields the τ_m versus $\dot{\gamma}_n$ relationship is close to linear across the whole range of shear rates but at higher electric fields the approximate linear relationship breaks down at the lower levels of nominal shear rate. Similar deviation from any smooth curve or line has been noted previously[6] and is not unusual. A second trend, again common to the data at both temperatures and not unexpected[7] was the reduction in slope of the data with increase in electric field.

For the purpose of analysis in the present work the fluid is assumed to be a Bingham plastic, as defined in equation (1), with a yield shear stress that depends upon electric field intensity. The nominal shear rate versus measured shear stress data is therefore analyzed to provide the η_p and τ_e values at each value of applied electric field. Analysis of the flow of a Bingham plastic in the viscometer[8] provides

$$\tau_m = \eta_p \dot{\gamma}_n X + \tau_e W$$

$$W = \frac{\ln\left(\frac{R_h}{R_0}\right)^2}{\left(\frac{R_h}{R_0}\right)^2 - 1} \qquad X = \frac{2\left(\frac{R_0}{R_h}\right)}{\left(\frac{R_h}{R_0}\right) + 1} \tag{5}$$

valid only when

$$\frac{\eta_p \dot{\gamma}_n}{\tau_e} \geq \frac{1}{2}\frac{R_h}{h}\left[\left(\frac{R_h}{R_0}\right)^2 - \ln\left(\frac{R_h}{R_0}\right)^2 - 1\right] \tag{6}$$

Best straight lines of the form

$$\tau_m = s\dot{\gamma}_n + t \tag{7}$$

are fitted to the linear portion of each set of τ_m vs. $\mathring{\gamma}_n$ test data by linear regression. The required Bingham plastic parameters are then given by comparing equations (5) and (7) thus

$$\tau_e = \frac{t}{W} \qquad \eta_p = \frac{s}{X} \tag{8}$$

The τ_e and η_p values so obtained are shown in figures 2 and 3 which include fitted lines. The values of η_p given by equation (2) are also included on figure 3. Figure 2 shows that the calculated yield shear stress values increase smoothly with electric field, and that although test temperature might be expected to have an effect[6] it is negligible in this instance. τ_e can be expressed

$$\begin{aligned} 0 \le E \le 0.39 \qquad & \tau_e = 187.6E \\ 0.39 < E \le 2.7 \qquad & \tau_e = 205E^2 + 362E - 102 \end{aligned} \tag{9}$$

$$(E\ (\mathrm{kV/mm}),\ \tau_e\ (\mathrm{Pa}))$$

In general, as shown on figure 3, the calculated values of η_p are subject to considerable variation. At low electric field and at 35 °C the plastic viscosity calculated from the concentric cylinder viscometry agrees closely with that determined at zero field in cone-on-plate viscometry given by equation (2). However, at 30 °C the discrepancy between the two values is approximately 20 %. The trend discernible from figure 3 is for η_p to decrease as electric field increases. The following expression is used to represent the data at 35 °C.

$$\begin{aligned} 0 \le E \le 3.0 \qquad & \eta_p = 0.074 - 0.0213E \\ E \ge 3.0 \qquad & \eta_p = 0.01 \end{aligned} \tag{10}$$

$$(E\ (\mathrm{kV/mm}),\ \eta_p\ (\mathrm{Pas}))$$

The second expression is necessary because a negative plastic viscosity is not possible within the Bingham plastic model.

The determination of a value from the slope of a line that is an approximate fit to data points which are themselves subject to error is a somewhat inaccurate procedure, particularly when the slope is small. This is precisely the situation encountered here in obtaining values for η_p. As can be inferred from figure 1, at high values of yield shear stress the slope of any fitted straight line can vary as much as five times (0.01 - 0.05) for errors in measured shear stress of less than 10 %. This goes some way towards explaining the variability in η_p shown in figure 3.

To remain straightforward the above analysis of the viscometry data makes no attempt to account for deviations away from the ideal Bingham plastic model, in particular those variations at lower shear rate illustrated on figure 1. The validity condition, equation (6), indicates whether, as assumed in the derivation of equation (5), the modulus of the shear stress in the Bingham plastic is greater than yield across the whole of the fluid gap. Since the torque is constant with radius the shear stress reduces with radius, such that when the condition of equation (6) is not satisfied there exists (theoretically) a region or core of unyielded fluid attached to the outer cylinder. As a result the complete theoretical τ_m vs. $\dot{\gamma}_n$ relationship for a Bingham plastic is, beginning at zero shear rate, a smooth curve which as $\dot{\gamma}_n$ increases has decreasing gradient, becoming a straight line at the point defined by equation (6). Although this expected deviation from linearity does not account for that obtained, the point at which equation (5) becomes invalid (by condition (6)) does tend to correspond with the point at which the measured shear stress begins to adopt unexpected values *i.e.* theoretical and actual behaviour tends towards enhanced disagreement when a core is predicted to exist in the flow. It is possible that the applicability of the Bingham plastic model reduces when a core is predicted to occur in the flow; this is potentially significant in the application of the model.

A second aspect of the analysis of the viscometry data is that the electric field strength as calculated by equation (4) is a convenient approximation only. More correctly

$$E = \frac{V}{r \ln \frac{R_h}{R_0}} \tag{11}$$

and since τ_e is a function of E then it is in turn a function of r. Assuming

$$\tau_e = aE^2 + bE + c \tag{12}$$

then flow in the viscometer of a Bingham plastic with this dependence of yield shear stress upon radius gives[9,10]

$$\tau_m = \dot{\gamma}_n \eta_p X + \left[a(E_{r=R_h})^2 + b \frac{2(E_{r=R_h})}{\frac{R_h}{R_0} + 1} + cW \right] \tag{13}$$

as the equivalent to equation (5), with the condition, as before, that this applies only when there is flow across the whole gap. The constants a, b, and c are then determined by minimising the sum of the differences between the predicted intercepts, given by the term in square brackets in equation (13), and the actual intercepts of the straight lines fitted to the data as in equation (7). Proceeding in this manner gives a maximum difference in τ_e from the expression in equation (9) of 0.6 %. A similarly negligible influence upon the results is obtained if the η_p variation with E and therefore with r is incorporated into the analysis.

4. Test apparatus

The test apparatus is designed for the measurement of hydrodynamic pressures generated using an ER fluid under an electric field in a part circular Rayleigh step bearing. The same experimental equipment as for unexcited fluid[4] is used, the essentials of which being a step bearing loaded against a rotating disc, which is partly immersed in the ER fluid, and an appropriately positioned pressure tapping, as shown in figure 4. Fluid is drawn into the bearing from the reservoir by the rotating disc, the latter having a diameter of 58.5 mm. and a constant speed for each test of 207 r.p.m. Pressures were measured using semiconductor strain gauge pressure transducers. The only modification made to the equipment for this work is the addition of a D.C. high voltage supply and satisfaction of the necessary insulation requirements.

Pressures within the bearing are measured at the repeatable limiting condition of zero flow rate, that is $h_2 = 0$ in the schematic diagram of the step bearing shown in figure 5. The thickness of the fluid film is then given by h. Tests were conducted at steady state conditions in atmosphere, approximately 23 °C, and temperature was measured by a thermocouple positioned in the ER fluid at the bearing entrance. Such an arrangement for temperature measurement was identified as being error prone in tests using unexcited ER fluid[4], however the data on figures 2 and 3 suggests that temperature sensitivity is not a major factor in the current tests. Other than these temperature considerations the test arrangement and concepts, being the practical achievement of near zero flow rate and the use of a zero flow rate theory, have been shown[4] to be an effective test of fluid behaviour.

Bearing	l_{tc} / mm.	h / mm.	w / mm.
1	14.88	0.244	6.86
2	10.16	0.28	6.91

Table 1 : Step bearing dimensions.

Bearing design is illustrated in figure 4, two bearings were tested having the dimensions given in table 1. The ptfe insert on the bearings provides the required electrical insulation and forms the bearing step and side walls with the latter employed to provide a one-dimensional flow condition.

As shown on figure 1 the viscometry tests using the ER fluid cover a range of nominal shear rates up to 2600 s^{-1}. In comparison the bearing dimensions and experimental apparatus used give a nominal shear rate in the bearings, as defined in equation (3), of approximately 2536 s^{-1}. However, at the condition of zero flow rate in the bearing the shear rates will be considerably higher than this nominal value. The viscometry data does not therefore fully cover the experimental conditions within the bearing. This is part of the test of the Bingham plastic model since the model includes the implicit assumption that fluid parameters do not vary with shear rate and do not therefore have to be measured across wide ranges of shear rates.

5. Theory

The theory for the flow of a Bingham plastic in the step bearing at a zero flow rate was developed in part I[4]. Experimental and fluid parameters are linked through the expressions

$$\begin{aligned} P &= 6 + 2.69B - 2.53\times10^{-2}B^2 \qquad 0 \le B \le 6 \\ P &= 6.86 + 2.43B - 5.03\times10^{-3}B^2 \qquad 6 \le B \le 25 \end{aligned} \tag{14}$$

with P and B defined by

$$B = \frac{\tau_e}{\eta_p}\frac{h}{U}$$
$$P = \frac{dp}{dx}\frac{h^2}{\eta_p}\frac{1}{U} \qquad \frac{dp}{dx} = \frac{p}{l_{tc}} \tag{15}$$

Equation (14) is a close approximation to discrete solutions of an exact implicit P versus B relationship for flow in a plane bearing at zero flow rate, as developed in part I of this work. It has been shown[4,8] that the use of l_{tc}, the arc length to the pressure tapping of the part circular bearing, and the small ratio of film thickness to disc radius in the tests, means that the errors introduced by application of this plane theory to the part circular case are negligible. Since the curvature of the fluid film in the bearing, as measured by the ratio h/R_0, is similar to that in the viscometer discussed in section 2 the effect upon electric field and hence upon fluid properties is reasonably assumed to be similarly negligible. Electric field is therefore as defined by equation (4).

Theoretical pressures in the step bearing, denoted by p_{theory}, can thus be predicted from equations (14) and (15) and the fluid properties and test geometry. The inverse of equation (15), for example to calculate the apparent yield shear stress from a measured pressure is approximately

$$
\begin{aligned}
B &= -2.17 + 3.51\times10^{-1}P + 1.56\times10^{-3}P^2 \qquad (0 \le P \le 22) \\
B &= -2.76 + 4.03\times10^{-1}P + 4.19\times10^{-4}P^2 \qquad (22 \le P \le 60)
\end{aligned}
\tag{16}
$$

6. The effect of particle size.

The theory which results in equation (14), gives the velocity profile of the flow as shown in figure 6. Three regions can be identified, region 2 being a core of unyielded fluid existing within the flow, and regions and 3 of yielded fluid either side of the core. Denoting the size of these regions by f_1, f_2 and f_3 then

$$
\begin{aligned}
\frac{f_1}{h} &= \frac{1}{2} - \frac{B}{P} + \frac{1}{P-2B} \\
\frac{f_2}{h} &= 2\frac{B}{P} \\
\frac{f_3}{h} &= \frac{1}{2} - \frac{B}{P} - \frac{1}{P-2B}
\end{aligned}
\tag{17}
$$

The heights, h, of the step bearings used in this work were chosen to be approximately 50 times the average particle diameter so that the movement of particles within the bearing is not restricted. If particles are to exist in the 3 regions in the flow and if a velocity profile is to occur that is similar to the smooth continuum profile assumed in the theory, then the dimensions of the regions are important. In the limit a region must be at least one particle diameter, that is greater than approximately $0.02h$ for the average particle size of 5μm and film thickness of $\sim$ 0.25 mm. used in this work.

Figure 7 shows the variation in the size of the three regions against B as predicted by continuum theory. It can be seen that the effect of there being a likely minimum core size of one particle, that is a restriction of $f_2/h \ge 0.02$, is negligible since the curve for core size is so steep for small cores.

In fact the enforcement of a minimum core size in the theoretical analysis indicates that minimum core sizes of up to $0.2h$ have an effect of less than 1% upon the values of P obtained for particular values of B.

Of most importance is, as shown in figure 7, the size of region 3. For particle diameter = $0.02h$ region three reduces to 3, 2, and 1 particle diameters at $B \approx 16, 29$ and 80 respectively. For larger ratios of particle diameter to film thickness region three approaches 1 particle diameter at significantly lower values of B.

7. Experimental results.

At a particular applied electric field the pressure in the bearing rises with applied load to a steady value at which additional load had no significant effect. These limiting (zero flow rate pressures) are shown in figure 8 against electric field. Note that there is not expected to be any correlation between the data from the two bearings due to their differing dimensions. If a test was left running for any length of time, particularly with an electric field applied, the measured temperature of the ER fluid would increase significantly. In order to achieve results at reasonably constant temperatures data was recorded during only brief periods of operation of the equipment, of the order of 10 seconds. For the two bearings the measured temperature varied between 34 and 36 °C during the tests and, since it is not considered critical, is taken to be 35 °C.

Generated pressure clearly increases with E, as is expected if τ_e is increasing. Theoretical and experimental pressures are compared using the parameter $p_{exp.}/p_{theory}$ on figure 9. In this figure the fluid properties used to determine p_{theory} are those given by equations (9) and (10).

The data can also be assessed in comparison with independently obtained data for the same fluid at a similar temperature. Equation (16) enables the apparent yield shear stress exhibited by the fluid in the step bearings to be calculated from the experimental pressures. This data is compared in figure 10 with apparent yield shear stress values exhibited by the fluid in tests of an ER fluid valve[5], that is pressure driven axial flow between two concentric cylinders. The data for the valve is determined from pressure versus flowrate measurements made on the valve, together with a Bingham plastic analysis of the flow, and the valve geometry. In the derivation of the data shown on figure 10 it is necessary to assume values for the plastic viscosity to apply; those given by equation (10) are used.

8. Discussion

The agreement between theory and experiment from tests with two different step bearings as shown on figure 9 is, for all but one point, remarkably good and is strong evidence that the experimental pressures can be predicted to acceptable accuracy using the continuum approach adopted. There is no known explanation for the apparently errant point on figure 9, it is possibly due to incorrect measurement of experimental parameters. Additional data points may be necessary to confirm no unexpected behaviour. The consistency, or otherwise, of the ER fluid behaviour is further examined using the available data in figure 10. This figure is essentially comparing the behaviour of the ER fluid in the three flow conditions of Couette flow (viscometry), Poiseuille flow (valve tests[5]), and combined Couette and Poiseuille flow, and also across a wide range of shear rates - the theory predicting different ranges to have occurred in each of the flow conditions. Considering the potential errors present, in particular those associated with pump/flowrate measurement in the valve tests[5], figure 10 shows good general agreement between behaviour in the different flow conditions.

The largest value of B that is calculated to occur in the bearing tests is approximately 120 and is for bearing 1 at 3.2 kV/mm. By equation (10) the low value of 0.01 is then used for plastic viscosity, a value of 0.015 may be just as applicable making B around 80. For bearing 2 a value of B of around 80 occurs at E = 2.8 kV/mm. It would appear that the tests are operating just at the boundary of where continuum theory might be expected to break down according to the analysis of section 5 and figure 7, any effect of this could however not be discerned. The possibility of the region sizes as discussed being significant in the applicability or otherwise of continuum analysis requires examination. This would be best achieved by investigation of thinner fluid films. For the potential use of ER fluids in controllable bearings films of perhaps an order of magnitude thinner than those employed in this work are required. Particle size may have to be reduced by a similar magnitude.

The theory predicts a core of unyielded fluid to occur in the bearing flow whenever flow rate is zero. Although it was proposed from the viscometry analysis that the predicted presence of a core in the fluid flow may be a point at which continuum analysis becomes unreliable this does not appear to have been significant in the bearing tests.

The significance of the plastic viscosity to the data analysis should not be neglected. As indicated the expression for η_p given in equation (10) has been used in the calculation of data on figures 9 and 10. An example of the effect of neglecting the variation of plastic viscosity with electric field is shown in figure 11 where calculations to produce values of p_{theory} have used the plastic viscosity at 35 °C given by equation (2), constant with electric field strength. Discrepancies between theory and experiment become on average 20 % larger. It is clear that the plastic viscosity values and variation are significant.

9. Conclusions

There is strong evidence to suggest that the excited ER fluid under the conditions of hydrodynamic pressure generation imposed conforms to the continuum Bingham plastic model used. For situations where the size of the particles in the fluid are of the order of one fiftieth of the gap height the hydrodynamic pressure can be controlled and, with prior knowledge of fluid properties from standard laboratory tests, be predicted. Furthermore the ER fluid tested behaves in a consistent manner in the three geometries and flow conditions encountered.

Of further interest is the behaviour of the fluid in thinner films and the response time of the ER fluid in the step bearing to the application of the electric field. One of the potential advantages of using ER fluid in engineering machinery is improved controllability by virtue of a rapid response. Such investigations form part of a continuing research program.

10. Acknowledgements

The authors would like to acknowledge the financial support of The Croucher Foundation and The British Council.

11. References

1. For a recent review see *Electrorheological fluids proceedings of the international conference 1991*, ed. R.Tao (World Scientific, Singapore, 1992).
2. F.P.Boyle, *'Performance characterization of ER fluids: durability'*, Electrorheological fluids, Proceedings of the International Conference 1991, (World Scientific, Singapore, 1992) 236-245.
3. T.H.Leek *et al.*, *'An experimental investigation of the flow of an electrorheological fluid in a Rayleigh step bearing'*, accepted for publication in J.Phys. D: Appl. Phys. (1993).
4. T.H.Leek *et al.*, *'Hydrodynamic pressure generation with an electro-rheological fluid. Part I - unexcited fluid'*.
5. W.A.Bullough, D.J.Peel, *'The field controlled liquid state; a conceptual means for the implementation of advanced mechatronic systems'*, Proc. I. Mech. E. Mechatronics Conf. 1990, Cambridge, 171-177.
6. H.Block and J.P.Kelly, *'Electro-rheology'*, J. Phys. D: Appl. Phys. 21

(1988) 1661-1677.

7. J.E.Stangroom, *'The Bingham plastic model of ER fluids and its implications'*, Electrorheological fluids Proc. 2nd Intl. Conf., Raleigh North Carolina 1989 (Technomic, Lancaster Pa., 1990) 199-206.
8. K.J.Korane, *'Putting ER fluids to work'*, Machine Design, May 9, 1991, 52-60.
9. T.H.Leek *et al.*, *'Solutions for one dimensional combined Couette and Poiseuille flow of an ER fluid as a Bingham plastic'*, report no. HKTHL023, Hong Kong / Sheffield ERF initiative, (to be published).
10. T.H.Leek, *'Report on the properties of an ER fluid'*, report no. HKTHL022, Hong Kong / Sheffield ERF, (internal document).

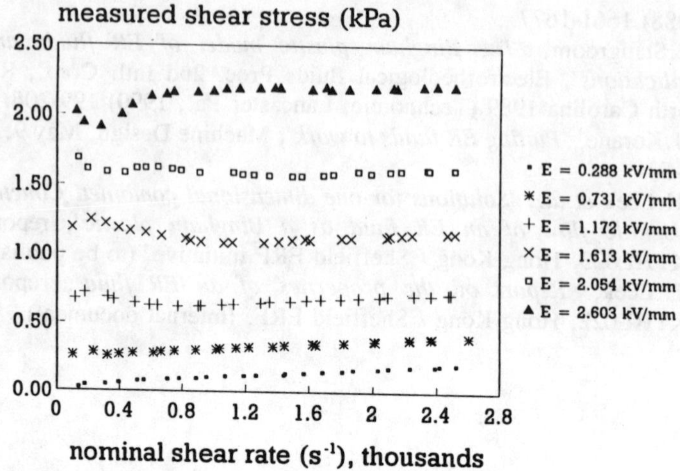

1. Examples of data obtained from concentric cylinder viscometer at 30 °C.

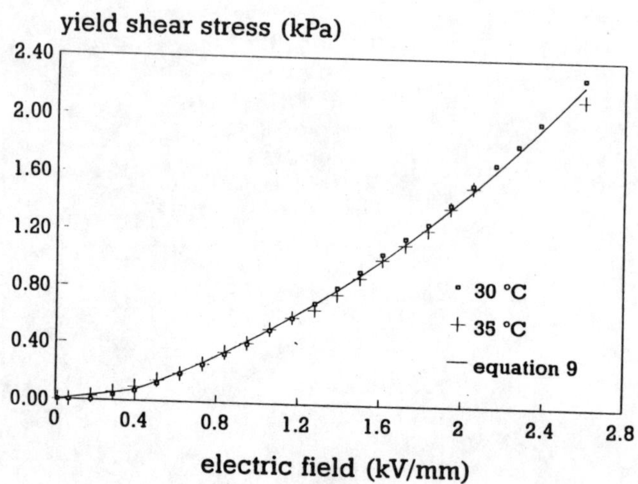

2. Yield shear stress obtained from Bingham plastic analysis of viscometer data.

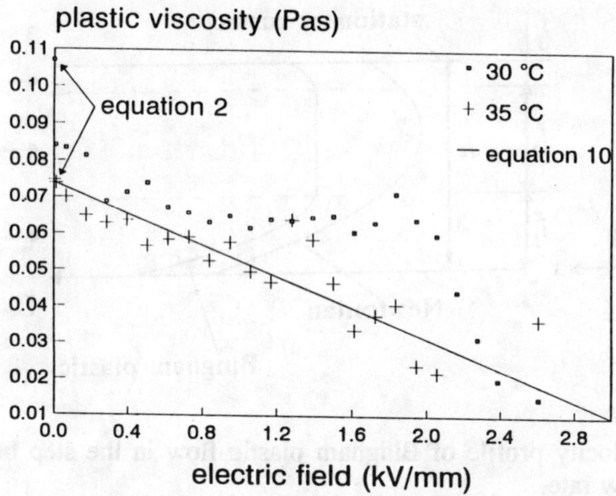

3. Plastic viscosity obtained from Bingham plastic analysis of viscometer data.

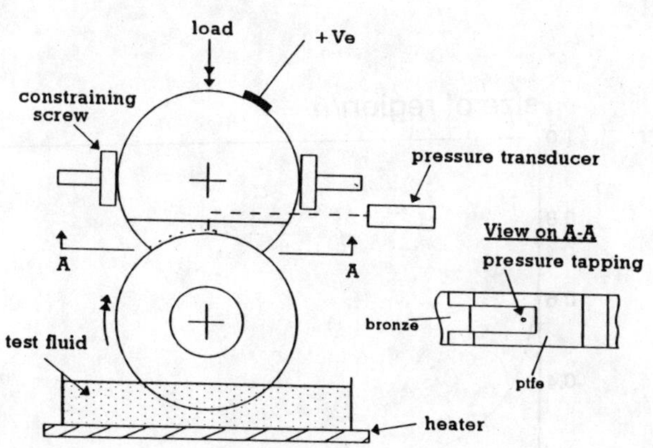

4. Experimental arrangement.

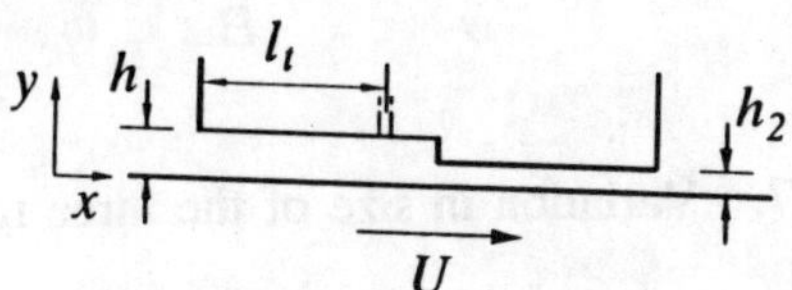

5. Rayleigh step bearing.

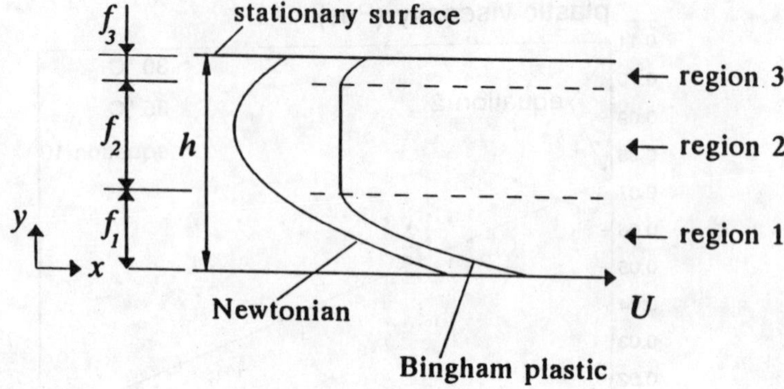

6. Velocity profile of Bingham plastic flow in the step bearing at a zero flow rate.

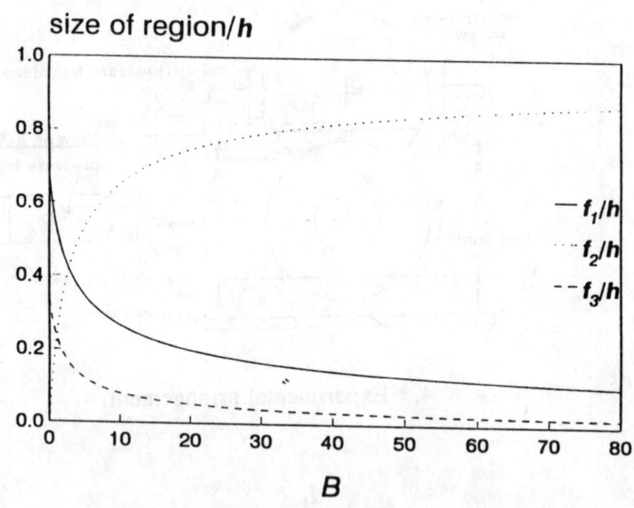

7. Variation in size of the three regions of the flow.

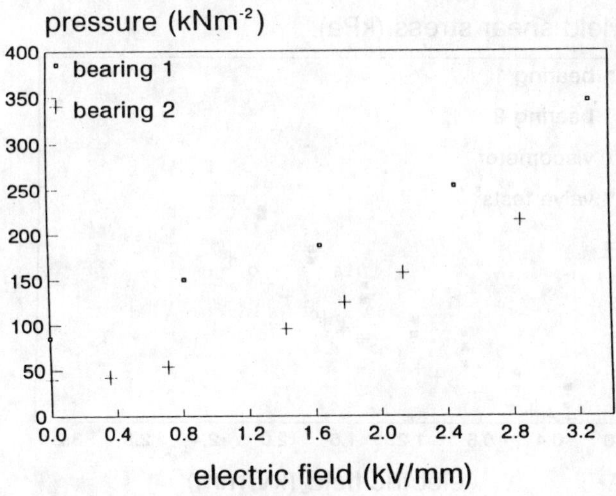

8. Limiting pressures measured in the step bearings versus electric field.

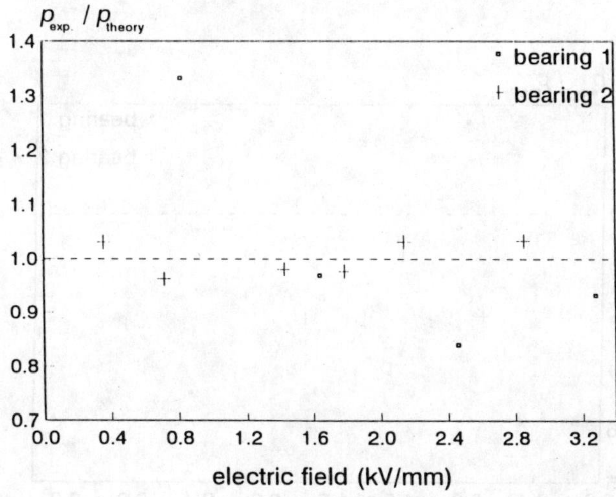

9. Comparison of experimental results and theoretical predictions of the pressures generated in the step bearings.

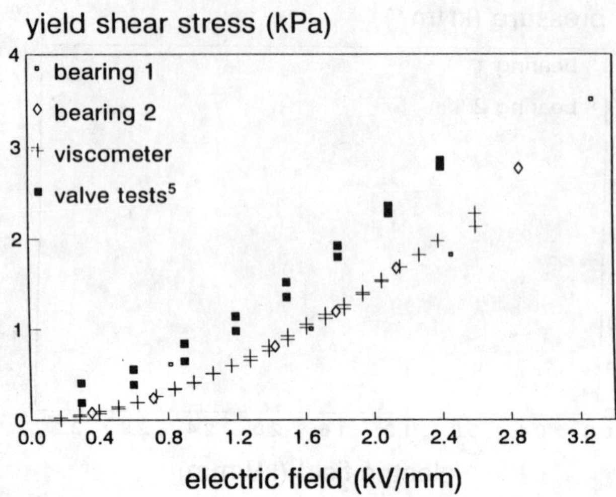

10. Comparison of yield shear stress values exhibited by the ER fluid in different flow conditions.

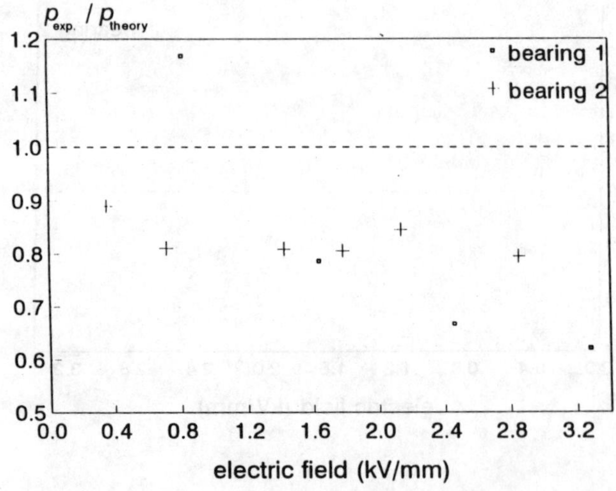

11. The effect of plastic viscosity.

SELECTION OF COMMERCIAL

ELECTRO-RHEOLOGICAL DEVICES

DOUGLAS ALAN BROOKS
Advanced Fluid Systems Limited
10/14 Pensbury Industrial Estate
Pensbury Street
London SW8 4TJ
England

ABSTRACT

The generation of ideas for devices using electro-rheological (ER) fluids is a relatively straightforward process. However, the development of commercially viable ER products is not. From an engineering perspective the science is of the utmost interest, but from the commercial viewpoint the market is fundamental. What are the problems encountered with existing ER fluids and how do we select an idea that minimises technical risk yet confers a competitive edge?

It is suggested that current fluids are appropriate to a range of industrial applications but market penetration is inhibited by a fixed approach to the design of devices. A suggested way forward is the use of ER fluids in compression as opposed to the conventional arrangement of fluids in shear. A promising application is considered which has evolved from the consideration of valve designs using the compressive stress capability of ER fluids.

HISTORY

The early days of my association with ER fluids was directed towards the more basic understanding of ER fluids in flow. This work generated a number of interesting devices as well as a number of novel ideas for controlling either the pressure loss or flow rate through a valve. However, none of the devices were commercially orientated. During the mid 80s American Cyanamid became involved with ER fluids and so began the current renaissance of the technology. Why did their involvement begin the renaissance of a virtually dormant technology? Simply because they began to aggressively market the technology, this made others take notice, and so sparked off an ever increasing interest and awareness of the technology. It is almost a decade since their involvement and you could argue that not a lot has happened since then, but I think you would be profoundly mistaken. From the initial push in the mid 80s the foundations have been steadily laid, new materials developed, new concepts elaborated, new ideas put forward, and new designs evaluated. Yet still the elusive first commercial product has to be realised, why? Is it a lack of will, of investment, of fluid performance, or a lack of ideas? Contrary to popular belief, the generation of ideas for ER devices is a relatively straightforward process. I am confident that, together with four of five participants from this or any other conference, we could in some 30 - 45 minutes generate maybe 20 - 30 ideas. The problem in any new and wide ranging technology, such as ER fluids, in not the generation of ideas but the selection from these ideas viable commercial ones.

It is perhaps this difficulty that divides industrialists and academics. Academics see the broad range of possible products and cannot understand why there are not at least some on the market. The industrialist however, know all to well, that the road from concept to product is difficult and expensive. It is not only the technical development but also the productionisation of the product, the marketing and selling of the product, the servicing of the market as well as the after sales servicing of the product, and of course the disposal of the product at the end of its life that are important. It is not surprising therefore, that before an industrialist sets out to launch a product it is planned in some detail so as to minimise the risk of failure.

So how do we go about selecting devices and what are the risks associated with selection?

SELECTION & RISK

Those who may have been expecting, and hoping, for a list of opportunities and a set of rules on how to select the best ones are in for a disappointment. Whilst I do have my own list, as well as a set of selection criteria, these are clearly very sensitive material and I will not disclose them. However, there is one well defined principle - reduce risk wherever possible.

Broadly there are two risk factors to be considered in any development process, the technical and the commercial. These factors are interlinked. In the hard world of commercial reality the ideal technical solution is rarely, if ever, the ideal commercial solution! So how do we define both the ideal technical and the ideal commercial solution? Perhaps we could define the ideal technical solution as one that would meet all the technical requirements over the required lifetime of the product with zero, or negligible, risk of failure. Whereas the ideal commercial solution might be a product that has the lowest manufacturing, selling, and distribution costs whilst commanding the highest possible selling cost. That is the largest gross profit margin.

Clearly then within these two definitions there is ample scope for conflict. The cost to produce the ideal technical solution may be far in excess of the price that any customer is prepared to pay for it. Alternatively the ideal commercial solution may lead to a product of such low quality that only the foolish would purchase it. What we require is a compromise, a compromise between maximising both the commercial returns and the excellence of the technical solution. The compromise may well be along the lines of a product that the customer, you or I either directly or indirectly, will pay for and be wholly satisfied with. That is, the perceived value to the customer is commensurate with the financial investment required. I have dwelt on definitions to illustrate the conflict between a good idea and a good commercial product. It is one of the features of ER technology that the good idea is often not a good commercial product. The commercialisation of ER technology requires that the good idea and the good product merge.

Should we be surprised then that a technology discovered in the late 30s is still in its commercial infancy? Upon serious reflection the answer is an unambiguous no. Not only were the supporting technologies, colloid science, high voltage amplifiers, control theory, and microelectronics lacking then, many of them did not even exist. It was simply impossible for the technology, at that time, to develop into commercial products. Now that we have the supporting technology there are attempts to redress the issue by overstating the technology and making unreal claims over unspecified fluids. Why does this occur? Because we have to deal with the scepticism that expresses itself in phrases like "if its any good why has not anyone exploited it before?" or "it a solution looking for a problem" or "its only a laboratory curiosity" or "its been around for 40 years with nothing happening so it will never happen". There are many rebuttals to these derisive comments, but perhaps the most damming is to simply point to the fact that commercial products are beginning to emerge.

Whilst we may see the appearance of some embryonic commercial products in well defined environments this should not delude us into thinking that all is solved and it is only commercial inertia and investment policy that stops the dawn of a new ER age. There still remain a significant number of technical barriers as well as a significant number of commercial barriers to be breached before the technology will have wide spread applicability.

What are the key issues that currently bedevil ER technology? Are their ways
we can circumvent them or turn them to our advantage?

KEY ISSUES

The issues can be broken down technically and commercially as follows:

Technical Issues

1 The yield stress, however defined, is too low.

2 The fluid requires to much power for most applications.

3 The temperature range is to narrow.

Market Issues

1 electronic control is required, why not maximising the benefits and advantages obtainable?

2 ER fluid devices are reactive, why not exploit the opportunity offered?

You will observe that while the technical issues have a negative flavour, the market issues can always be expressed positively. The difference, although cosmetic, carries an important message. It rather like saying to someone, "You're looking more beautiful than ever" as opposed to "You're not as ugly as you used to be" the same statement, but weighted more positively.

However, to address the issues we need to try and define the problem a little more precisely. If the yield stress is to low, what does it need to be? What power level is acceptable? What temperature range is required? Can we define our problem clearly or are we always chasing the wind.

The Magnitude of the Yield Stress

It is clear that even the term yield stress is problematic. It has been pointed out that we may well be dealing with a number of so called yield stresses, both static and dynamic. Not only do the magnitudes vary, the dependence on flow rate, in the case of valve or duct flow, and shear rate, in Couette flow, vary. The static situation usually provides the largest value and is often quoted by fluid developers wishing to impress or acquire funds. Couette flow is usually the most severe test and is usually quoted by the large multinational fluid developers as it is the most cautious value. Value flow usually gives values somewhere between the two.

Not only do ER fluids withstand shear forces they also exhibit compressive and tensile features. Clearly this behaviour cannot exist in a liquid but is predictable for a solid. It is surprising therefore that until now little attention has been focused towards this behaviour in ER fluids as it can lead to some significant advantages. One of the curious aspects that adds to the confusion is the fact that because pressure, shear stress, yield stress, and compressive stress, are all measured in N/m^2 they are often taken as the same parameter. I know of a least one situation where claims are made for large compressive and tensile stresses that are implied to be yield stresses. Such claims are just plain confusing to the uninitiated and leave the prospective user frustrated.

Electrical Power

When we turn to power requirements the question really is how much power is required by the device and how much power is available during operation. If the ratio of power required to power available is less than one then were OK. At one it becomes marginal, above one its simply no good. Clearly lower power demand is better as the power supply is smaller and cheaper. In addition to this power demand is also temperature dependent, which leads to the third question, that of limited temperature range.

Temperature Range

Similar to the statement that the yield stress it to low, saying the temperature range is to small is unhelpful. Similarly saying we need a temperature range of -55°C to +200°C is equally unhelpful. There are temperature windows where devices operate, some are well regulated, the weaving industry for example is tightly controlled, some areas less so, for example offices, factories, warehouses. Others are more wide ranging, with some of the severest being the automotive and aerospace requirements. It is unfortunate that one of the largest potential markets, that of automotive, has one of the severest temperature requirements, as well as some of the most demanding performance versus cost conditions attached. This hinders market penetration

LINKAGE

I now want to turn to a basic design issue that I term linkage. Duclos[1] first publicly raised this issue in an SAE paper, in simplified from it says that for given requirement of energised force transmission to un-energised force transmission the volume of fluid required depends on the ratio of the viscosity to the yield stress squared. That is changing the geometry, length, width, gap of, say a valve will not give additional advantage. The performance is dictated by the ratio of viscosity to yield stress squared.

Similarly the power requirement is a function of the geometry, which we have seen above is effectively fixed, and the product of the field and the current density.

So, to improve mechanical performance we need to increase yield stress or reduce viscosity, ie improve the fluid's stress transfer capability. To improve electrical performance we need to reduce current density, again therefore, we need to improve the fluid.

Is this the only way forward? If it is we can only wait until the fluid developers produce and market the next generation ER fluid. However, given that many researchers are still mixing their own first generation ER fluids, starch and sunflower oil, despite the fact that second generation ER fluids are readily available, we could have to wait some time to see any significant progress. Can we approach the problem from another direction and ask ourselves can we break the linkage between viscosity and yield stress and if so how?

BREAKING THE LINK

Breaking the link calls for one thing, and one thing only:

imaginative design

or creativity in the use of ER fluids. It is my belief that to much time and effort has been spent either re-inventing starch and silicone fluid, building simple parallel plate clutches, or constructing basic dashpot dampers. Apart from a few isolated examples there seems to be no effort directed towards novel ways of using existing fluids to enhance device performance. To show that it can be done I will illustrate one route that breaks the link.

USING ER FLUIDS IN COMPRESSION

ER dashpot dampers and cylindrical clutches use ER fluids in shear. However, ER fluids also have a compressive stress capability. For the last nine months we have been studying this feature, (funded by the European Commission Brite-EuRam Feasibility Scheme), and designed and developed both a compressive ER valve and uni-directional shock absorber. So what do I mean by using ER fluids in compression? Basically, rather than load the fluid in shear we subject it to a compressive or tensile stress. If an ER fluid is placed between two unbounded electrodes and a field applied the material will resist both a compressive and tensile load. Clearly, if a tensile stress results in a tensile strain, the effective field applied will reduce and rapid fracture may result. However, a compressive strain increases the field which enhances load bearing capability. We can use such a feature in a valve as shown in the slide. Two valves are shown in figure 1, one using the fluid in compression the other using shear forces. The valve consists of a body with a spring supported poppet valve assembly. The poppet valve assembly is extended to provide an upper floating electrode. The lower electrode is insulated from the body. The poppet valve is permanently biased to an open position with the spring. The valve can be designed so that full closure is not possible and flow through the device can always occur. Alternatively it can be allowed to fully close thus operating as a selectable, and controllable non return valve. The design on the right shows a similar arrangement but using static shear forces as opposed to compressive forces. Alternative arrangements can be easily conceived to use valve flow forces.

In all cases the closing of the poppet valve is controlled by the voltage applied to the electrode which acts on the fluid and uses the compressive stress behaviour. The final height being determined by the flow forces on the poppet valve and the compressive stress induced in the ER fluid. The design is essentially uni-directional but the principle may be used as bi-directional with a pair of poppet valves. As long as the device receives a voltage it allows flow in either direction. Reducing the applied voltage reduces the poppet valve gap and increases the flow loss through the valve. With zero voltage the device can be shut fully.

The main advantage is to reduce the linkage between viscosity and yield stress. The compressive stress in this arrangement typically operates as a static yield stress, that is a fluid with a 3.0 kPa excess stress in Couette flow exhibits a yield stress in the order 10-12 kPa under compression. There is a simple formula that relates yield stress to force via the radius of the electrode. It is then simple a matter of balancing the poppet valve closure forces against this resisting force. The link between viscosity and yield stress is not completely broken however. The pressure loss through the poppet valve is a radial flow loss and is viscosity dependent, however the force required to hold open the poppet valve is significantly less than a comparable shear requirement. Furthermore the it is quite possible to totally separate the ER fluid and the operating fluid. The ER fluid can be isolated by means of a diaphragm seal arrangement from the fluid that flows through the poppet valve. This enables a hybrid of conventional hydraulics and ER fluid control which completely severs the link.

Shock Absorber

Shock absorbers are not only used on cars. Every moving object possesses kinetic energy. If the object is required to change direction or stop, its kinetic energy must be dissipated. Unless the energy dissipation is controlled, shock forces will cause damage to the structure of the object or to associated equipment. The simplest form of energy dissipator is a spring or rubber bumper. However, these devices only store energy which has to be dissipated elsewhere in the system. A dashpot is slightly better since it provides a means of energy conversion. Unfortunately, this energy conversion occurs chiefly at the start of the stroke and imposes a significant initial shock load.

Linear decelerators or industrial shock absorbers are an attempt to overcome the problem. They are designed to dissipate energy linearly over the entire stroke. Being industrial they are located in factories are have a less severe operating temperature window than automotive requirements.

An elementary ER dissipation device based on a dashpot is mechanically simple. However, when detailed calculations are conducted the energy absorbtion of an ER decelerator according to this principle is limited and consequently has no advantage over an existing conventional device unless the fluid can be improved by a least on order of magnitude, 20-30 kPa being required. This limitation can be overcome using a compressive ER valve. A valve member is constructed so that it can be retained in the open position during flow by the introduction of a force which opposes the flow forces tending to close the valve. The opposing force is generated directly from an electrical input by using either the compressive characteristic of the electrically stressed fluid between the electrodes.

Figure 2 shows a schematic of a possible shock absorber using a compressive stress valve. By applying a voltage between the two electrodes the ER fluid solidifies and the valve can be held open. The pressure drop generated this valve can be calculated using the expressions for the laminar flow of a fluid through a radial gap. The pressure drop through the valve is a function of its geometry and the opening of the valve. Consequently the kinetic energy of the impacting load can be dissipated by controlling the pressure drop through the valve and will result in the load being arrested. As the load is decelerated the flow rate through valve decreases and hence the pressure drop through the valve decreases. This can be compensated by reducing the valve opening. This is achieved by reducing the voltage applied between electrodes. Hence by controlling the compressive stress characteristics of the ER fluid the opening of valve can be controlled and controlled deceleration of the load can be achieved. The fluid inertial forces due to the impact are significant and the valve needs to be shielded from them. This is achieved by diverting the fluid away from the valve by a diverter. Naturally a general arrangement of the device appears more complex than the simplified schematic as shown here, but the essential features are all included. The additional complications come only from the particular mechanical engineering constraints.

In this particular configuration the pressure is monitored and used as the control signal. A schematic of the control system is shown. For linear deceleration constant internal pressure is required and a typical control strategy is outlined in figure 3. Upon impact of the load the piston is moved causing an increase in internal pressure. This pressure characteristic can be compared with a characteristic either stored in the memory of a microcomputer or generated by the microcomputer from measurements of pressure against time. The error signal is computed and processed to give the required output voltage to the high voltage power supply. The signal processing is such that the non-linear closure characteristics of the valve are accounted for. The nature of either the stored characteristic or the generated one can be modified as required for specific purposes. Hence deceleration profile of a non-linear nature can be envisaged.

Whilst this works effectively, it is not particularly cost effective and a non-contact magnetic displacement sensor is used in practice.

MARKET ISSUES

In the above description of an ER industrial shock absorber we have barely touched on market issues. The essential feature is to make a virtue out of the need for control and electronics. The electronic "intelligence" is a requirement so it should be maximised. The device can sense both the external and internal environments and adjust continuously. Consequent advantages are:

Reactive features are added, the damper reacts to the load and adjust itself to optimise the deceleration or damping.

The measurement of speed and displacement gives an estimate of the mass and propelling force, hence we now have data on the object. Is it full, half full, or even empty? The shock absorber becomes part of the quality control procedure for the production line.

A running count of the number of operations can be implemented.

The controller has a communication link and can either download or upload data. Hence it can be remotely reprogrammed for different conditions or transfer data to other areas using an industrial standard protocol.

These then are some of the inherent advantages of involving electronics which can be expressed plainly as "providing information and data".

CONCLUSIONS

To recap the ground covered in this presentation I hope I have stressed the need to align possible technical excellence with market opportunities. In this way some of the less favourable questions posed of ER technology can be answered.

I also addressed the three key issues of yield stress, power requirement, and temperature range. Consideration of these led to what I term "linkage", the apparent link between two basic fluid properties, viscosity and yield stress, and the output of an ER device. It was intimated that unless this link could be broken early penetration of ER devices into larger markets would have to wait for fluid development. Fortunately, this was not the only route and the use of imaginative design solutions would help. One possible solution was put forward and some simple geometries indicated as well as the application of this into an otherwise unpenetrable market, industrial shock absorbers.

As a company we have moved on a considerably distance in the design of "link breaking" design and have developed what are effectively ER amplifiers. These are far simpler, more reliable, and more effective solutions than those detailed here. For example, the most promising design creates the illusion of a 10 kPa fluid from a 1 kPa fluid and at the same time reduce the electrode area by a factor of seven. However, lack of time and commercial prudence does not permit me to disclose more.

So before I close let me leave you with two closing points vital to the selection process:

1 *REDUCE RISK*

2 *IMAGINATIVE DESIGNS CAN OVERCOME FLUID LIMITATIONS*

REFERENCES

1 T. G. Duclos *SAE Paper* 881134 1988
Design of Devices Using Electrorheological Fluids

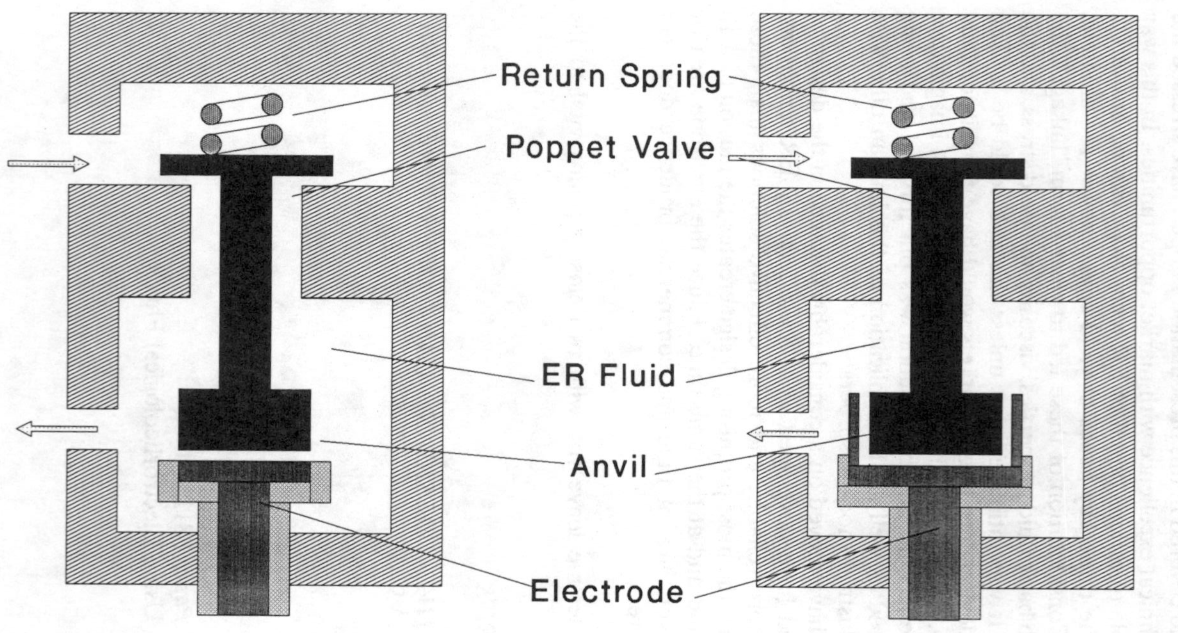

POPPET VALVE USING ER FLUID IN COMPRESSION

POPPET VALVE USING ER IN SHEAR

Figure 1 Alternative Valve Arrangements

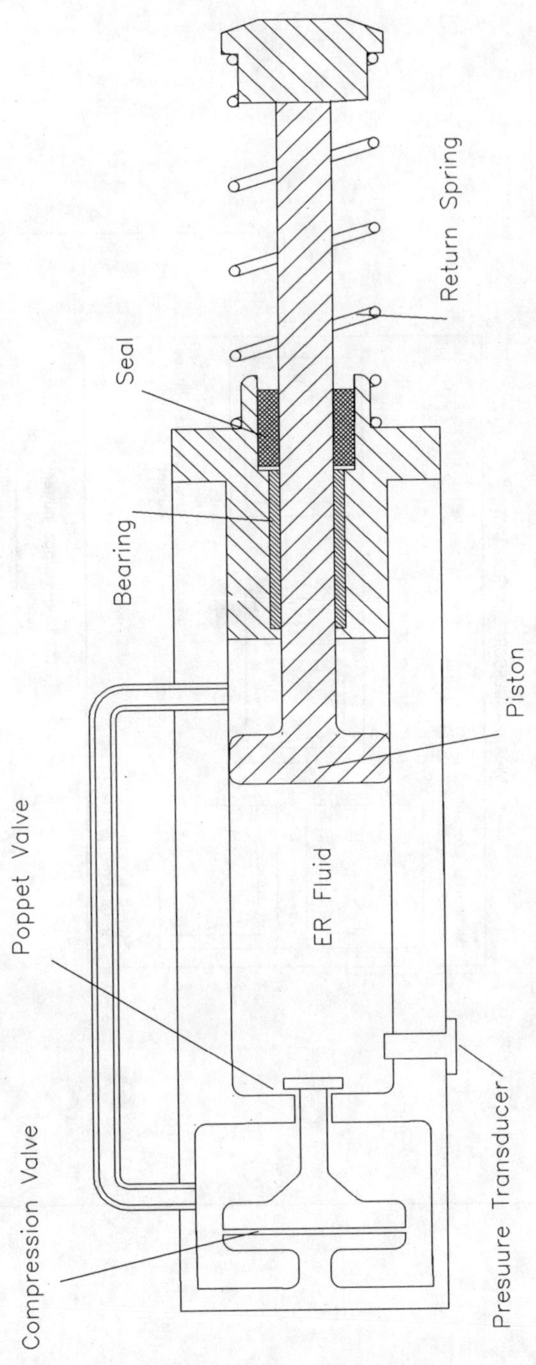

FIGURE 2 Basic Shock Absorber Design

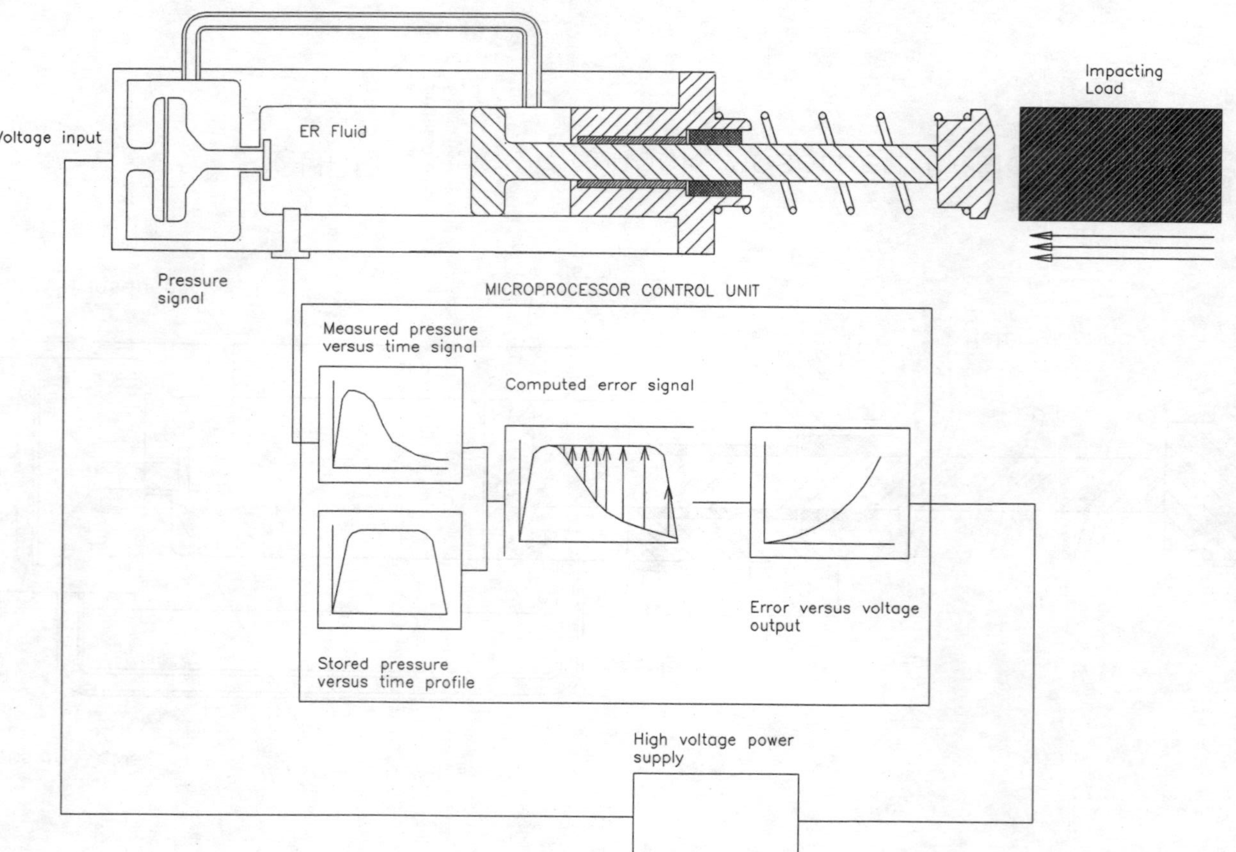

FIGURE 3 Basic Control Procedure

HEAT TRANSFER MODELLING OF A CYLINDRICAL ER CATCH

R Smyth, K H Tan and W A Bullough
Department of Mechanical and Process Engineering
The University, PO Box 600, Sheffield, S1 4DU, United Kingdom

ABSTRACT

In electro-rheological (ER) devices the control of temperature is often of paramount importance. For a device through which the ER fluid is not able to flow continuously to and from a reservoir, where it may be cooled, this can be a problem. Such a situation occurs in the ER catch. The heat generated there will predominantly be dissipated from the outer surface of the drive/input. The rate of heat transfer is thus principally a function of the speed and area of that surface and the temperatures of the ER fluid and atmosphere. Since these factors reflect on the levels of electro-stress, current requirement and viscosity of the ER fluid, an opportunity exists for optimisation of catch performance.

The paper shows the results of an investigation into the effects on the cooling/heating problem of varying the radii of the clutch rotors, their relative rotational speed and inter-electrode spacing. Equilibrium fluid temperatures are confirmed by experimental evidence. The effects of heat generation in run up to speed and clutch locked periods, (through electro viscous drag and dielectric loading respectively) are quantified and compared with the case of a contemporary ER fluid in a cylindrical catch on zero volts idling. At any given operating condition, uniform temperature, viscosity and constant electro stress are assumed throughout the fluid.

1. Introduction

A fundamental limitation on a machine lies in its capacity to dissipate the heat it generates. ER devices are particularly susceptible to overheating at the present time. This is due to the relatively high shear rates that are encountered in practice, the high contemporary zero volts viscosity of a fluid of useful electro-stress and limitations on the amount of fluid that can be employed. One method of cooling an ER device on which fluid power engineering is based, would be to circulate pre-cooled fluid through the zone of heating. This is effective in ER flow mode machines but is not an easy option in the shear mode type of device where "natural" cooling is consequently most important.

The total heat transfer problem in an ER clutch is very complex. Constantly excited, controlled slippage devices may produce temperature gradients large enough to significantly alter the local viscosity and electro stress with consequential disruption of the velocity profile and subsequently the controllability of the slip rate. Digitally excited (volts on-off) devices are therefore worthy of examination in respect of clutch control.

In order to make a start on the temperature control problem of such a system it is necessary to separate the operational conditions before modelling them. If this can be done on a quantitative basis, and the cumulative model tested to substantiate the approach, then the various contributions to the heating rate can be assessed from a comparison of the product of the heat generation rate and residence time patterns to be encountered in service. This is then to be balanced with the cooling characteristics and a further step forward can be planned. The conditions in an unexcited and slipping phase, a switching operation and locked plates are thus considered individually for the simplest construction of clutch, a cylindrical device. The problem is further simplified by assuming that the temperature in the ER fluid is that which exists on the inner surface of a surrounding driver cylinder which rotates at a set speed - Fig. 1. The variables are then reduced for one fluid to (i) effective radius, (ii) the conditions existing at its outer surface, and (iii) the inner (driven rotor) radius.

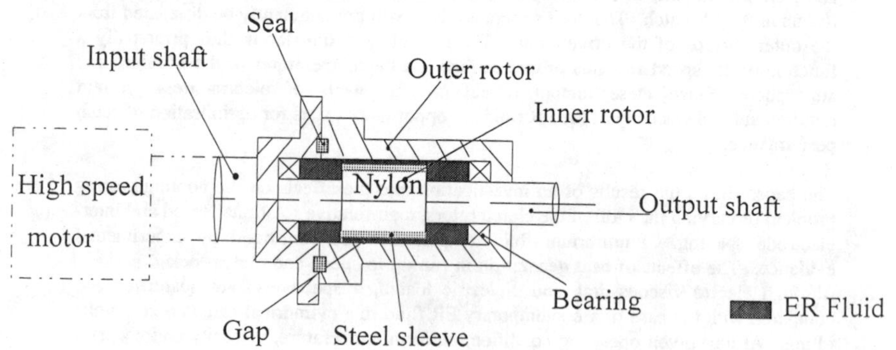

Fig. 1 ER Clutch

All analyses should be made as easy to apply as is possible. Thus the present effort is conducted on a torque per unit length of rotor basis (since the heat generated and its dissipation rate both depend on the length of a cylindrical clutch.). For a given fluid the input power of a slipping clutch at zero volts is taken to be due to $\mu_o \dot{\gamma}$ only (the product of viscosity and shear rate in the inter electrode gap) and the specific resistive heating effect of a given fluid, being relatively small, is reckoned for a given fluid to depend only on the applied electric field strength (E) which is alone assumed to fix the electro-stress τ_e. Thus, the results are easily translated to situations where one or two driven/driving rotor surfaces (double acting) and/or where there are one or two heat transfer surfaces are involved and/or for different durations of volts on/off. The present analysis is based on the Bingham plastic form of clutch-continuum model with steady heat transfer ensuing. However, the value of μ_o (the zero volts and plastic viscosity though $\neq f(\dot{\gamma})$) does vary with fluid temperature θ_{ev} and in the present case the significance of this factor is included by correlating base fluid viscosity and θ whilst taking the effect of volume fraction on mixture viscosity to be constant over a range of temperature. This fundamental assumption is verified to some extent by previous experience [1] and the fluid manufacturers (AFS Ltd, London) data.

2. Energy Generation and Heat Transfer at $\omega_2 = 0$ and Unexcited

2.1 Underlying Assumptions

In this analysis only the shear stresses on the cylindrical surfaces of the clutch are considered important. All heat transferred is assumed to be dissipated by the external cylindrical surface. Uniform conditions are taken to apply throughout the fluid film i.e. the fluid is an isothermal continuum which endures uniform shear stresses and velocity gradients yet, the effect of fluid temperature is considered not to change the level of electro-stress τ_e. A further assumption is that $\tau_e \neq f(\dot{\gamma})$. The flow is taken to be both laminar and fully developed in the circumferential plane; radial and end effects are neglected. The flow of the fluid at zero volts is essentially Newtonian.

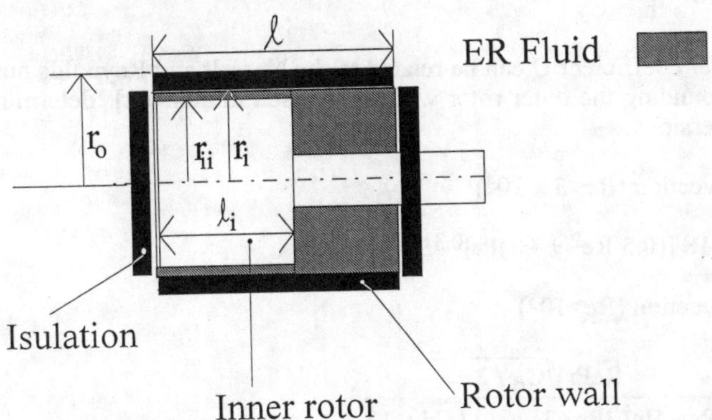

Fig 2. Schematic diagram of Clutch Model

With reference to the convention given in Fig 2, a simple combined radial conduction and convection analysis with the equilibrium heat transfer rate Q linked to the outer rotor surface and fluid temperatures θ_o and θ_{ev} respectively gives:

$$(\theta_{ev} - 20) = Q\left(\frac{1}{2\pi k_s \ell}\ln\left(\frac{r_o}{r_i}\right) + \frac{1}{2\pi U r_o \ell}\right) \quad (1)$$

$$(\theta_{ev} - \theta_o) = Q\left(\ln\left(\frac{r_o}{r_i}\right)\frac{1}{2\pi k_s \ell}\right) \quad (2)$$

$$(\theta_o - 20) = Q\left(\frac{1}{2\pi U r_o \ell}\right) \quad (3)$$

$$U = \frac{Nu(k_a)}{2r_o} \quad (4)$$

The rate of energy generated by viscous shear is given by ωT where $T = \tau.2\pi r^2{}_i \ell_i$ or,

$$Q_g = \mu_o \frac{\omega^2 r_i^3}{h}(2\pi\ell_i) \quad (5)$$

The heat transfer coefficient U can be related to the Nusselt and Reynolds numbers of the air stream surrounding the outer rotor with the Nusselt number [2] determined from the following expressions:

For mixed convection ($Re<5 \times 10^4$)

$$Nu = 0.18\,[(0.5\,Re^2 + Gr)Pr]^{0.315} \quad (6)$$

For forced convection ($Re>10^5$)

$$Nu = \frac{RePr\sqrt{C_d/2}}{5Pr + 5\ln(3Pr+1) + \sqrt{2/Cd} - 12} \quad (7)$$

where C_d is obtained from either

$$1\Big/\sqrt{C_d} = -1.828 + 1.77\ln\left(Re\sqrt{C_d}\right) \qquad \left(Re\sqrt{C_d} > 950\right) \quad (7a)$$

or

$$1\Big/\sqrt{C_d} = -3.68 + 2.04\ln\left(Re\sqrt{C_d}\right) \qquad \left(Re\sqrt{C_d} < 950\right) \quad (7b)$$

For a stagnant ambient air condition well removed from the rotor, there are 3 modes of convection: *free* convection, *combined free and forced* convection and *forced* convection. These modes are well justified for a horizontal rotating cylinder with a smooth surface finish in stagnant air. Any variation of rotor inclination angle to the horizontal or surface finish would invalidate the relevant equations.

A solution for the temperature θ_{ev} was achieved with the systematic approach illustrated in Fig. 3. The Reynolds number was determined in order to decide the appropriate convection equation to be used, (equation 6 or 7). Thereafter simultaneous equations are obtained for θ_{ev}. A computer programme was written for the solution of these.

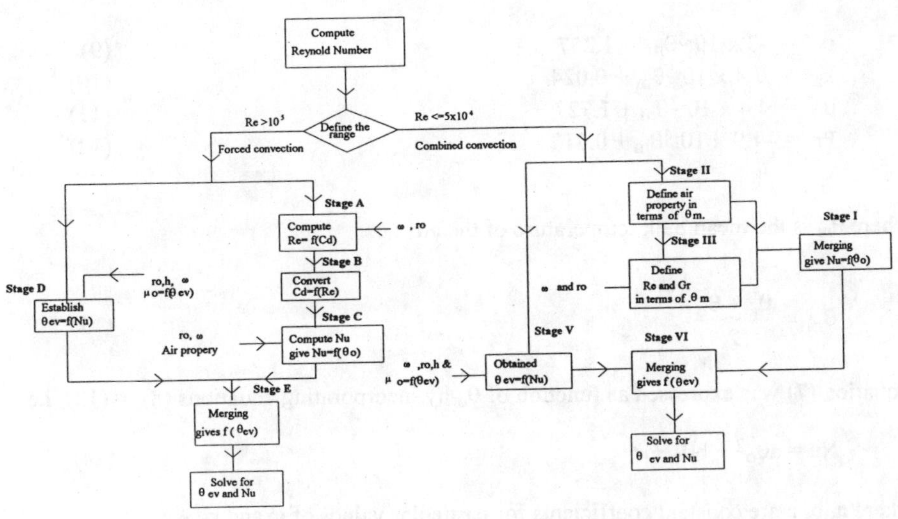

Fig 3. The solution procedure for the mathematical model of the ER Clutch heat transfer behaviour

2.2. *Calculation Procedure*

2.2.1 *Effect of Reynolds number on the convection mode*

The Reynolds number is defined as $$Re = 4\omega r_o^2/\upsilon \qquad (8)$$

The air kinematic viscosity υ was taken at the bulk air temperature θ_m which is the average value of the surface temperature and the ambient temperature. By linearlising υ

over the temperature range 20°C to 100°C, υ was obtained as a function of mean bulk temperature. Using equations 9 and 11 below, the Reynolds number from equation (8) gives the mode of convection.

There is an undefined region [2] sandwiched between the *mixed* and *forced* convection regions which was classified as *mixed* convection, which gives a somewhat higher temperature θ_o than would be the case if forced convection had been assumed.

2.2.2 *Combined Free and Forced Convection Mode (Re≤5 x 10⁵)*

Referring to stage I, II and III in Fig. 3, in the calculation of the Nusselt number in equation (6) for this mode of convection, the properties of air were linearised in the range of 20°C to 100°C as follows:

$$\rho = -3 \times 10^{-3}\theta_m + 1.257 \quad (9)$$
$$k_a = 7.4 \times 10^{-5}\theta_m + 0.024 \quad (10)$$
$$\mu = 4.4 \times 10^{-3}\theta_m + 1.727 \quad (11)$$
$$Pr = -1.9 \times 10^{-4}\theta_m + 0.712 \quad (12)$$

where θ_m is the mean bulk temperature of the air, i.e.

$$\theta_m = \frac{\theta_a + \theta_o}{2} \quad (13)$$

Equation (7) was expressed as function of θ_o by incorporating equations (8) to (13), i.e.

$$Nu = a\theta_o^2 + b\theta_o + c \quad (14)$$

where a, b, c are constant coefficients for particular values of ω and r_o.
A curve fitting technique was used to calculate the coefficients a, b and c.

Referring to stage C (Fig 3) the kinematic viscosity of the base oil of one ER fluid (Lipol 30/W) ν_b is represented as (see Fig. 4).

$$\nu_b = 0.0045\ \theta_{ev}^2 - 0.841\ \theta_{ev} + 40.9 \quad (15)$$

The mixture viscosity is increased from this by a factor determined from manufacturers figures for mixture and base oils at 30°C.

Substituting equation (15), (4) and (5) into (1) gives an equation for θ_{ev}, i.e.

$$\theta_{ev} = f(Nu, h, r_o, r_i, \ell_i, \ell, \omega, k_s) \quad (16)$$

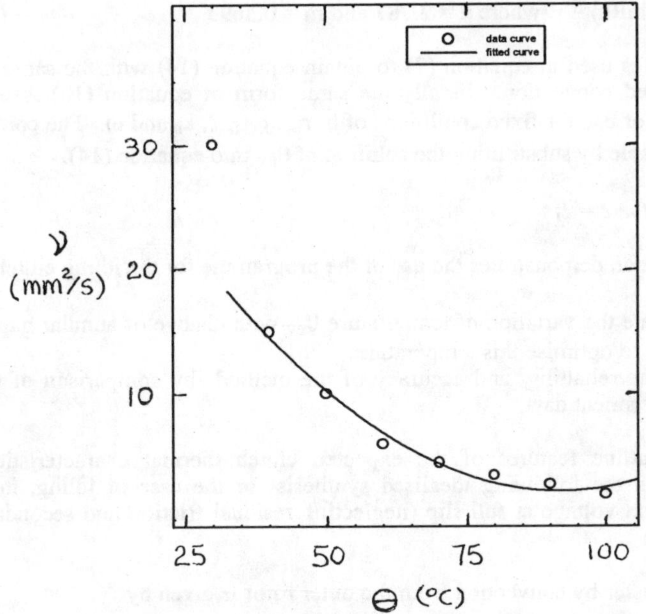

Fig 4. Curve fitting for base oil used in Lipol/30W

The Nusselt number in equation (16) can be determined from equation (14) after obtaining θ_o from (2) to give the following relationship:

$$Nu = f(\theta_{ev}, h, r_o, r_i, \ell_i, \ell, \omega, k_s) \tag{17}$$

Substituting (17) into (16) gives a final equation for θ_{ev} in the form of

$$A\theta_{ev}^4 + B\theta_{ev}^3 + C\theta_{ev}^2 + D\theta_{ev} + F = 0 \tag{18}$$

where A, B, C, D and F are constants for fixed values of h, r_o, r_i, ℓ_i, ℓ, k_s and ω. The Newton Raphson method was employed to solve equation (18) for θ_{ev} at a given geometric configuration and fixed angular speed.

2.2.3 Forced Convection Mode (Re>10⁵)

This mode of heat transfer (stage A, B and C in Fig. 3), is accounted for in equation (7), with the parameter C_d obtained from equation (7a) or (7b). An exponential curve fit of C_d was used as follows:

$$C_d = (m\ln Re)^{-n} \quad \text{where } n = 2.787 \text{ and } m = 0.5693 \tag{19}$$

This expression is used in equation (7) to obtain equation (14) with the same procedure as that for mixed convection. Finally the same form of equation (18) is obtained to provide a value of θ_{ev} for fixed conditions of h, r_i, r_o, ℓ_i, ℓ, k_s and ω. The corresponding Nu is also available by substituting the solution of θ_{ev} into equation (14).

2.3 *Results for $\ell = \ell_i$*

This section demonstrates the use of the programme for the idling clutch to:

(a) Investigate the variation of temperature θ_{ev} with change of annular gap thickness and outer radius to optimise this temperature.
(b) Verify the reliability and accuracy of the method by comparison of theoretical results and experiment data.

Some outline features of the expected clutch thermal characteristics can be appreciated from the following idealised synthesis for the case of idling, i.e. running without excitation voltage at full slip (neglecting residual friction and secondary modes of heat transfer):

Rate of heat transfer by convection from the outer rotor is given by

$$Q = UA(\theta_o - 20) \qquad \text{where } A = 2\pi r_o \ell_i$$

But at steady state, a temperature ratio k can be expressed as

$$k = \frac{\theta_o - 20}{\theta_{ev} - 20} = \frac{\theta_o - 20}{\Delta\theta} \qquad \text{where k is a constant}$$

Hence for steady state thermal conditions ($Q = Q_g$), from equation 5 we obtain

$$\Delta\theta = \frac{\mu_o \omega^2 r_{ii}^3}{kUr_o h} \tag{20}$$

Thus it seems that:

(a) The temperature difference is independent of the length of the rotor when $\ell_i = \ell$.
(b) The temperature difference is inversely proportional to the heat transfer coefficient, outer radius r_o and the gap thickness h. However, the geometrical variables h and r_o are subjected to certain constraints with respect to device size, power, torque and efficiency whereas U depends on the ambient temperature and the air flow condition.
(c) Any increase in viscosity, and angular speed and rotor radius in particular, increases the temperature difference. The choice of viscosity and of the rotational speed of the rotor are however, determined by the available fluid and duty of the clutch, often leaving the change of rotor radius as the only flexible and significant parameter to

influence the variation of temperature θ_{ev}. It can also be seen that a reduction of the radius r_{ii} causes the temperature to decrease significantly. Thus for a given torque requirement it appears prudent to design a long slender clutch rather than a short one if the torsional stiffness and size remain acceptable. However, the problem is one of quantification of the effect of conditions on U. In this respect, for the values of physical parameters:

$r_i = 30.5$ mm, $r_{ii} = 30$ mm, $r_o = 40$ mm, $\ell = 60$ mm, $\ell_i = 60$ mm, h = 0.5 mm

a set of calculations were made to produce the equilibrium temperatures θ_{ev} at rotational speeds from 250 rpm to 3000 rpm and for $r_o = 60$ and 80 mm (see Fig. 5).
Similar calculations were repeated for r_o fixed at 40 mm and h = 0.5, 1 and 2 mm. Also, $r_{ii} = 10$, 20 and 30 mm are varied for fixed r_o and h. (See Figs. 6 and 7)

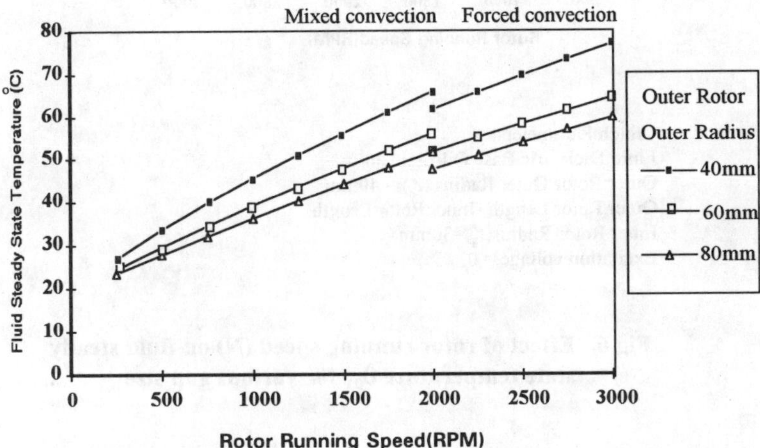

Clutch Parameters:
Fluid:Dielectric Base Oil/Lipol 30/w
Inner Rotor Radius (r_{ii}) =30mm
Outer Rotor Length=Inner Rotor Length
Gap Size (h) = 0.5mm
Excitation voltage = 0

Fig 5. Effect of rotor running speed (N) on fluid steady state temperature θ_{ev} for various outer rotor radii (r_o)

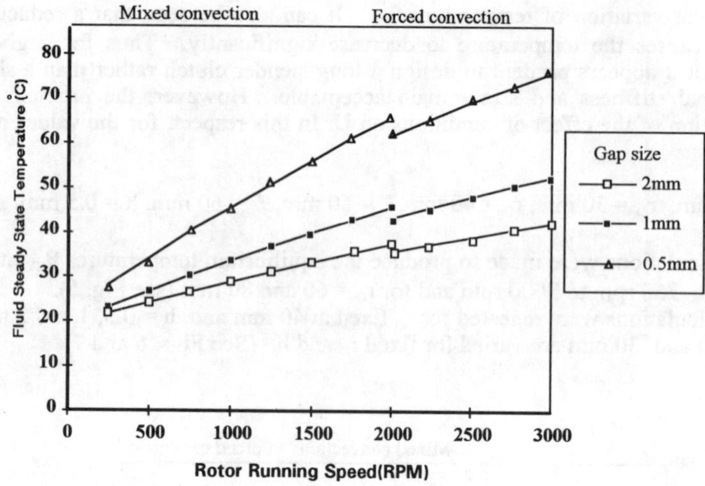

Clutch Parameters:
Fluid:Dielectric Base Oil/Lipol 30/w
Outer Rotor Outer Radius (r_o) = 40mm
Outer Rotor Length=Inner Rotor Length
Inner Rotor Radius(r_{ii})=30mm
Excitation voltage = 0

Fig 6. Effect of rotor running speed (N) on fluid steady state temperature θ_{ev} for various gap size

In Fig. 8 a typical θ_{ev} versus time test result is shown. The slow continuation of temperature rise is thought due to an increase in the test enclosure and running gear temperatures. Assumed equilibrium test conditions are compared with theory in Figs. 9a and 9b.

3. Other Operational Modes

The relationship between the likely dominant zero volts slip mode and the switching modes in normal on-off control depends on the specific operational cycle to which the clutch is subjected. Both heat transfer and electrical power dissipation are process/load form, fluid and geometry dependant parameters and are dealt with only in a general and comparative fashion, viz:-

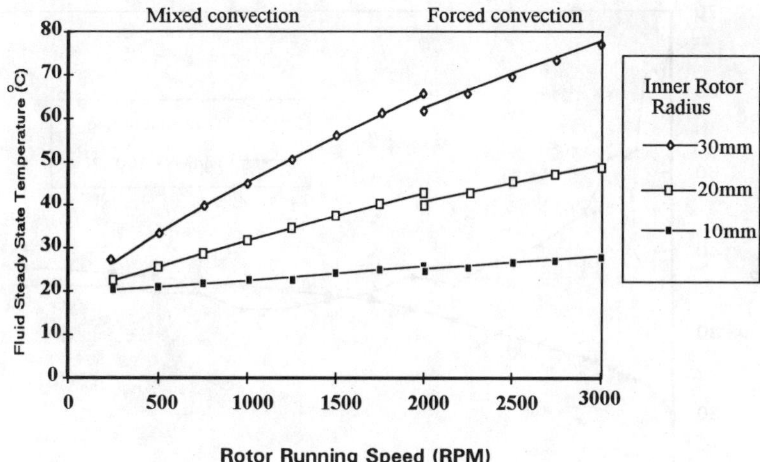

Clutch Parameters:
Fluid:Dielectric Base Oil/Lipol 30/w
Outer Rotor Outer Radius (r_o) = 40mm
Outer Rotor Length=Inner Rotor Length
Gap Size (h) = 0.5mm
Excitation voltage = 0

Fig. 7 Effect of rotor running speed (N) on fluid steady state temperature θ_{ev} for various inner rotor radii (r_{ii})

3.1 The Switching Phase

Consider the motion switching cycle to start by the application of excitation at 1 and terminate at 3 in Fig. 10a. The torque is assumed to be activated instantaneously and be fixed by E alone with a subsequent smooth run up to speed of the driven rotor (ω_2) from zero to the constant driver speed ω_1.

Slip work = $\int_o^{t'}(T_e + T_0)(\omega_1 - \omega_2)\,dt$ where t' is the run up time

It will be noted that both T_0 and $\omega_1 - \omega_2$ fall as ω_2 increases. Thus the real (full) line of Q_g due to slip in Fig. 10b always lies below a projected linear progression (shown dotted) from Q_g max to $Q_g = 0$. Hence the area under the dotted line is always an overestimate of the area under the full line. For a limiting boundary condition of $T_{0\,max} \le T_e$ the overestimated value of this integral is: $T_e\ \omega_1\ t'$. When $T_0 > T_e$ the control ratio T_e/T_0 falls to less than unity and the clutch becomes relatively uncontrollable. Typically, for the 101 mPas Lipol 30/W fluid used in one test, $\mu_o\dot{\gamma}$ = 1kPa at 1500 rpm, compared with τ_{emax} = 2 kPa at the same shear rate.

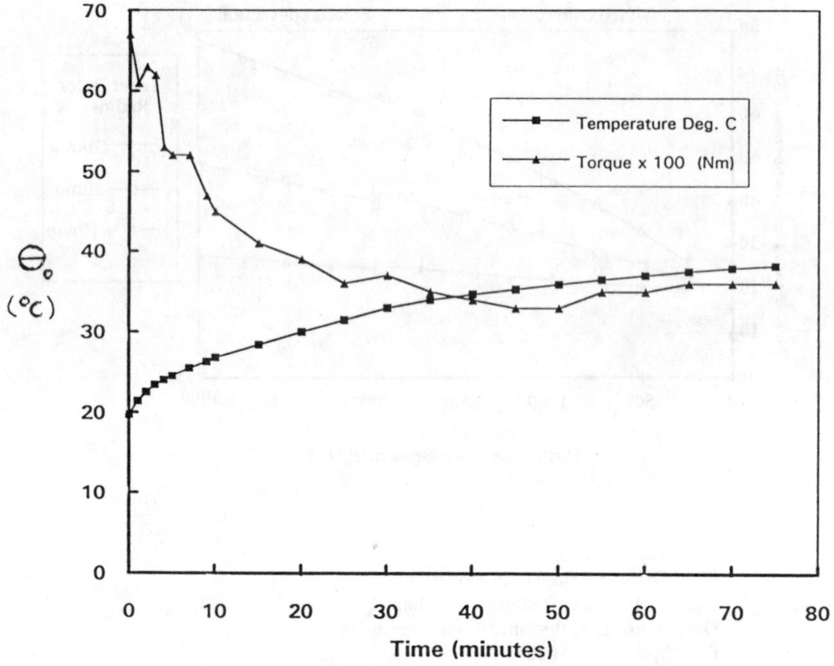

Fig. 8 Clutch Rig Temperature Trial at 500 rpm

The real process shows good matching between load and drive since maximum torque is generated at switch on and the least torque when approaching the required locked state - this is contrary to say the action of a solenoid. The precise shape of the actual Q_g versus t will depend on the specific process, and in particular, the inertias of the driven rotor and load. To avoid such complications and yet give some quantification to the various heat generation phases the over estimate is used. Thus, the amount of heat generated per switching process is $T_e\omega_1 t'$ whilst that generated during a period t" of full slip is $T_0\omega_1 t'' = T_e\omega_1 t''$. Hence, the ratio of the heating effect (though only an estimate) depends on the relative durations of t' and t". For example a 50 milliseconds full slip duration and a 5 millisec switching duration would give only an average 10% max heating effect occuring in the run up period. In such a process the over estimate is justified. A similar argument can be made for a braking process given equality of energisation and de energisation phenomena [3].

3.2 *Plates Locked Phase*

When the plates are locked together the only loading is electrical/resistive: iV. Although the surge current can be many times the value of the steady (resistive) current its duration is but of the order hundreds of microseconds in the test clutch and is of little consequence from a heating point of view.

3.3 *Practical Cycle Calculations*

Data taken from [4] is used to show (in conjunction with the above assumptions) typical amounts of power and energy consumed in the various parts of a clutch cycle. The figures are for a clutch of r_i = 30 mm, $\ell = \ell_i$ = 60 mm and a fluid of approximate τ_e

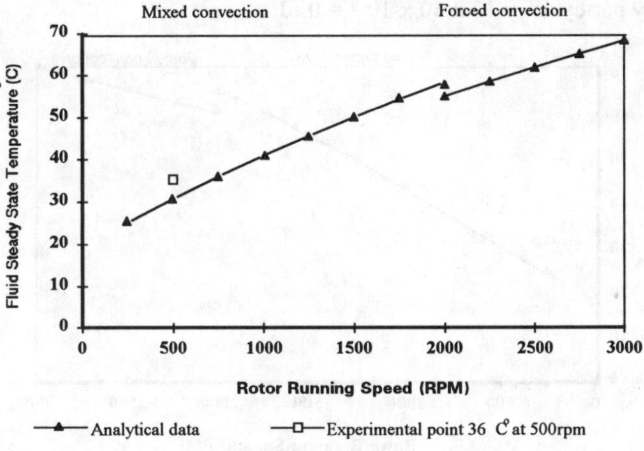

Clutch Parameters:
Fluid:Dielectric Base Oil/Lipol 30/w
Inner Rotor Radius (r_{ii}) = 30mm
Outer Rotor Length= 80mm
Inner Rotor Length= 60mm
Outer Rotor Outer Radius (r_o)=40mm
Gap Size (h) = 0.5mm
Excitation voltage = 0

Fig 9a. Comparison of analytical data with experimental results

= 3 kPa. At an excitation step of 3 kVolts/mm an overshoot current of 40 x 10^{-3} amps decays to 'steady state' current of 8 x 10^{-3} amps over 0.5 x 10^{-3}s, figures being estimated for a rotational speed of 1500 rpm, an electrode separation of 0.5 mm and running temperature of 30°C. The overall cycle is assumed to comprise equal run up and run down periods t' of 2.5 x 10^{-3}s separated by a 'locked plates' period t''' of 50 x 10^{-3}s . The 5 m/s uniform speed equates to four pole motor speed and a $\dot{\gamma}$ of 9,500s^{-1} on a 0.5mm gap. Based on these premises:

(i) Q_{gmax} at zero volts slip = 150W. Energy per cycle = 150 x 50 x 10^{-3}J = 7.5J
(ii) Run up/down periods: power based on $T_e = T_{0max}$ = 150W.
Energy per cycle = 150 x 5 x 10^{-3} = 0.75J
(iii) Electrical Power on overshoot estimated as 0.5 i V = 40 x 10^{-3} x 1500/2 = 30W
Energy per cycle (2 x switch on basis) = 30 x 0.5 x 10^{-3} x 2 = 0.03J
(iv) Electrical Power consumed with plates locked = 1500 x 8 x 10^{-3} = 12W.
Energy per cycle = 12 x 50 x 10^{-3} = 0.6J

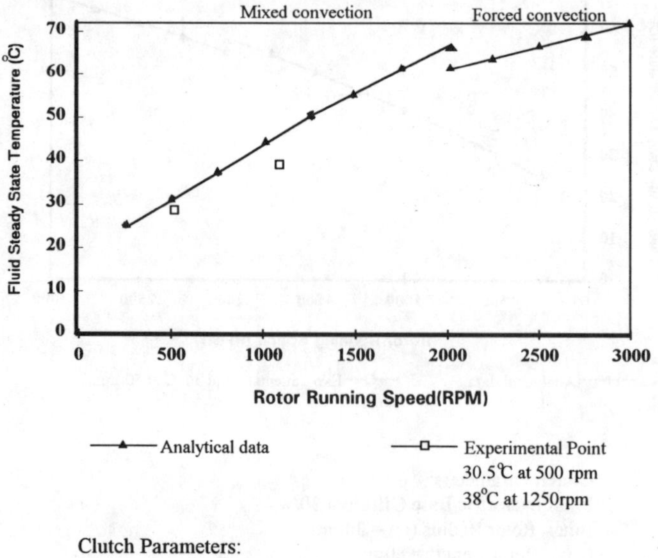

Clutch Parameters:
Fluid:Silicone S30/w
Inner Rotor Radius (r_{ii}) = 30mm
Outer Rotor Length= 80mm
Inner Rotor Length= 60mm
Outer Rotor Outer Radius (r_o)=40mm
Gap Size (h) = 0.5mm
Excitation voltage = 0

Fig 9b. Comparison of analytical data with experimental results

The assumption of taking the zero volts slip period to be the dominant source of heat generation is generally proved for the case of t'/t" ≤ 1 i.e. for relative low inertia mechanisms operated at high speed and where the run up period is relatively short compared to the idling or rotors locked periods.

4. Discussion of Results

The analytical results show that changing the different parameters has various influences on the steady state fluid temperature θ_{ev} as seen in Figs. 5 to 7. There are several common features in the graphs which are worth noting. All are compatible with the (simplified) approach equation (20).

(a) All the system temperature curves diverge as the speed increases.
(b) The curves are made up of two regions governed by the Reynolds Number, i.e. the mixed flow and forced flow regions. The boundary usually occurs at 2000 rpm for outer radius less than 60 mm and at 1000 rpm for a radius greater than 80 mm.
(c) The temperature θ_{ev} increases almost linearly with ω_1. The rate is greatest from 250 rpm to 1000 rpm, thereafter having a declining rate of change.
(d) For most cases in the forced convection mode (Fig. 5 to 7), a doubling in the size of the operating parameters r_{ii}, h and γ_o, results in a very significant change of equilibrium temperature.

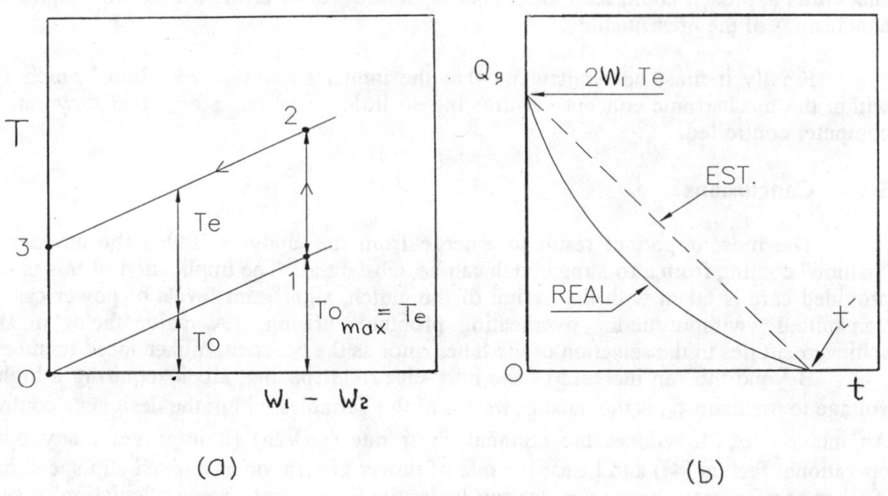

Fig 10. Switched Duty Cycle (a) torque v slip rate (b) Q_g v time

In summary, for any given speed these curves show quantitatively how θ_{ev} can be reduced by the use of (in ascending order of importance) a large outer radius r_o, a large gap size h, and a small driven rotor radius r_{ii}. The main drawback in reducing the rotor radius is the consequent need to extend its length to retain the torque. This may cause problems in installation and loss of torsional stiffness.

The tests done to date to confirm the analysis, have taken place on two fluids (Fig. 9a and Fig. 9b). Both have relatively low base fluid viscosities (circa 20 mPas at 30 °C) and therefore are not very sensitive to temperature. An existing test rig [4, 7] was used for the experiments.

Radiation from the equipment has been neglected in calculations of heat loss. Also, the resistance to heat transfer which occurs at the inner surface of the outer rotor has been assumed small in comparison to that at the outer surface. Liquid to solid surface coefficients are normally much more conductive to heat transfer than at solid to gas. This is fortunate as no transport property data is available for the ER fluid. In the case of rotors locked in the 'yield held' mode, the temperature gradient in the fluid is likely to be very small as it is across the thickness of the outer rotor. The closeness of experimental data to calculated predictions tends to support the simple approach used here.

It should be understood that the viscosity of the overall fluid mixture has been estimated from that of the base oil, as given in the manufacturers data for temperature dependance, and an approximate model for the added effect of solids content. Since the accuracy of these procedures will depend on the specific constituents used, because of the original and general nature of the present work, and to ease computation, the base oil viscosity/temperature relationship has been estimated by a quadratic. If adopted without check this approach could lead to an unacceptable level of error; it does not impair the conclusions of the present study.

Finally it must be pointed out that the model applies to ER fluids which fall within the mechatronic concept i.e. drawing so little electrical power that they can be computer controlled.

5. Conclusions

The most important result to emerge from the analyses is that the amount of "natural" cooling from a rotating clutch can be substantial. The implication of this is that provided care is taken with the design of the clutch, significant levels of power can be transmitted without undue overheating problems arising. A major factor in this achievement lies in the selection of the outer rotor as the constant, higher speed member.

Beyond this, an increase of the inter-electrode spacing, albeit requiring a higher voltage to maintain τ_e, is the most powerful of the variables within the designer's control. An increase of h reduces the nominal shear rate ($\omega D/2h$) (it improves many other operational factors [4]) and hence the rate of power generation at a given slip speed but, if taken too far, may impair the electron-hydraulic time delay. Some alleviation of such a consequence may be had from higher temperature operation when τ_e often declines less (or sometimes not at all) with increased shear rate [7]. Though the resistance and capacitance of the electrical load will thereby be affected, the change that this causes to the heating load has not been calculated.

Some reduction in fluid working temperature may be had from a reduction of the diameter of the inner rotor (for the same torque) at the expense of an increase in it's length. This will probably increase the surge current (via capacitance)and increase the resistive load (reduced resistance)on account of the increased electrode area[5]. Speed reduction seems to be a major factor in reducing temperature but, this, like the other changes will affect the dynamics of the system and electric heating rates. Total mechatronic steady and unsteady design is called for if and when specific application and fluid characteristic data becomes available.

In the case of the application of clutches in any mechanism in which the time spent in the switching is relatively short, the technique presented seems suitable for the assessment of fluid temperature. In terms of steady speed governance of say a rotary clutch the position depends on the relative duration at the three operational conditions, the properties of the ER fluid and what extreme outer radii can be tolerated, this being the least influential of the designers control parameters.

6. References

1. D. Brooks, *Applications of Simple Expressions for Design with ER Fluids. Proc Conf. Recent Advances in Adaptive and Sensory Materials and their Applications*, Blacksburg Va. 1992 p. 524-536, Technomic Press.

2. H. Y. Wong, *Heat Transfer For Engineers*, Longmans, 1977.

3. D. J. Peel and W. A. Bullough, *Effect of Flow Rate, Excitation Level and Solids Content on the Time Response in an Electro-rheological Valve*, Jnl. Intel. Matl. Systems and Structures. **Vol. 4** No. 1, Jan 1993, p. 54-64.

4. W A Bullough et al. *Power Transmission via the ER Catch: Continuum Basis, Performance Predictions, Infrastructure Requirements*. Proc. IMechE. Conf. Machines, Actuators and Controls. p.37-44. London HQ. March 31 1993

5. M. Whittle et al *A Model for the Electrical Characteristics of an ER Valve*, Jnl. Mod. Phys. B. **Vol. 15-16**. p.93-108, 1992.

6. A R Johnson et al, *Electro-rheological Catch/Clutch Simulations* - see these proceedings.

7. J Makin et al, *The Effect of Solid Fraction Concentration on the Performance of ER Fluid in the Flow Mode.* - see these proceedings.

Nomenclature

A	surface area of outer rotor (m^2)
C_d	drag coefficient
c_p	constant pressure specific heat of air at 20°C (J/kg°C)
D	diameter (m)

E	electric field strength (volts/h)
g	gravitational acceleration = 9.81 m/s^2
h	the annular gap size, $r_0 - r_i$ (m)
k	temperature difference ratio $(\theta_0 - \theta_a)/\Delta\theta$
k_a	thermal conductivity of air at 20°C (W/m°C)
k_s	thermal conductivity of rotor = 15 W/m°C
ℓ	the effective length of the outer rotor(m)
ℓ_i	the effective length of the inner rotor(m)
N	angular speed (rpm)
Q	rate of heat transfer (W)
Q_g	heat generated in the fluid (W)
r	radius (m)
r_i	the internal radius of the outer rotor(m)
r_o	the external radius of the outer rotor(m)
r_{ii}	the radius of the driven rotor(m)
θ	temperature (°C)
θ_{ev}	steady state temperature of the ER fluid (°C)
θ_o	outer surface temperature of the outer rotor in (°C)
θ_a	ambient temperature of air =20°C
θ_m	mean bulk air temperature $(\theta_0 + \theta_a)/2$ (°C)
t	time
t'	run up time from $\omega_2 = 0$ to $\omega_2 = \omega_1$
t''	duration of full slip, $\omega_2 = 0$
t'''	duration of plates locked, $\omega_2 = \omega_1$
T	torque(Nm)
U	surface heat transfer coefficient (W/m^2°C)
$\dot{\gamma}$	shear rate, $\omega r/h$ (s^{-1})
β	coefficient of volumetric thermal expansion of air at 20°C (K^{-1})
ρ	density of air (kg/m^3)
μ_o	dynamic viscosity of ER fluid at zero voltage (Pas)
μ	dynamic viscosity of air at 20°C (Pas)
ν	kinematic viscosity of air at 20°C (m^2/s)
ν_b	kinematic viscosity of base oil of ER fluid (m^2/s)
τ	shear stress in ER fluid
$\Delta\theta$	temperature difference = $\theta_o - \theta_a$ (°C)
ω	running speed of a rotor (rad/sec)

subscripts

o	viscous (zero volts) components - except when defined above
1	driving (outer rotor)
2	driven (inner) rotor
e	electro stress

Gr Grashof number $= \dfrac{\beta g \Delta\theta 8 r_o^3 \rho^2}{\mu^2}$

Re Reynolds number $= \dfrac{\rho \omega 4 r_o^2}{\mu}$

Pr Prandtl number $= \dfrac{c_p \mu}{k_a}$

Nu Nusselt number $= \dfrac{U 2 r_o}{k_a}$

Index

Absorber 60, 84
ac fields 38, 39, 46
Aggregation of chains 522
Attraction between spheres 141-142, 146-149

Base liquid 3, 37
Bearing, ER 608 625
Bingham fluid 19, 61, 76, 328
Bingham law 328
Bingham material 294
Bingham model 45, 56, 623, 625
Bingham plastic 45, 56, 478, 526, 538, 543, 617
Body-centered tetragonal (bct) lattice 314, 320
Brownian diffusion 5, 45
Brownian force 131

Catastrophic failure 389
Cellular structure 25
Chain 25, 81
Chain deformation 280
Chain Formation 117
Characteristic length 119
Characteristic time 19, 119
Charge transfer 23, 26
Clutch 60, 84, 493, 526, 543, 649, 658
Coarseness of the chain 522
Coarsening power law of ferrofluids 190-191
Coarsening process 119, 190-191
Coarsening time 121
Cole-Cole model 5, 10
Column formation 8, 39
Composite fabrication 520
Conductance of Liquid and Spheres
  Low Field 142-143
  High Field 144
Conduction model 40-144
Conductivity of fluid 23, 39, 50
conductivity of Surface 149-150
Constitutive equation 538
Control strategy 569
Controlled strained rheometer 270
Controller indications 384
Couette flow 5, 23, 76, 100, 477, 609, 647
Coupling constant of ferrofluids 174, 192
Cox-Merz Rule 246
Cryogenic ER fluids 37-42
Curing process 516
Current density 43, 46

Damper 46, 569, 649
dc fields 3, 5, 38, 100, 141

Dielectric loss 4, 6, 67, 79
Dielectric mismatch 278, 649
Dielectric relaxation 78, 81
Dielectropheretic assembly 516
Dimensional analysis 207-208, 545
Dipolar interaction 27, 283, 518
Dipole model 268
Dispersion 4, 10, 67, 217, 218
Drag coefficient 285
Dynamic Response Model 203-206

Effective charge method 393, 395
Electric field induced second harmonic generation (EFISH) 463, 464
Electric field 23, 24, 590
Electric model 427
Electrode morphology 256-261
  and shear stress 256-258
  and particle velocity 259-260
Electropheresis 350, 516
Ellipsoids model 334
ER fluid materials,
  Excited 625
  Gel-based 67, 68
  Unexcited 611
  Zeolite-based 68

Faraday cage 454
Ferroelectric 60
Ferrofluids, confined 173-174, 191-192 175-180
Ferromagnetic dispersed phase 22, 25

Fibril formation 3, 45, 61, 66, 516
Finite element analysis 272, 280
Flow, Couette 5, 23, 76, 100, 609, 647
Flow, laminar 526, 545
Flow in ER valve 290
Frequency response 521, 582

Heat dissipation 590
Heat transfer 657, 661, 672
Hexagonal arrays 340, 400, 408, 410
Hobson relation 157
Hydraulic machines 421
Hydrodynamic pressure generation 610
Hydrous ER fluid 453
Hysteresis 39

Independent droplet model 122-124
Intelligent fluid 440
Inter-column impedances 12
Interaction force 19
Interfacial polarization 3, 11, 82
Intrinsic time scale 132
Ion mobility 69
Ionic conduction 69, 70
Ionic polarization 71

Johnson-Rabeck effect 149

Leakage current 148
Light scattering in ER Fluids 117-119
Light scattering in Ferrofluids 180, 193-198
Light transmissions in Ferrofluids 182
Linear optical properties 468
Linear oscillating shear 206

Magneto-rheological suspensions 22, 24, 43, 45
Magnetic permittivity 26
Mason number 26, 123, 234, 248-249, 294
Maxwell-Wagner equation 307
Maxwell-Wagner relaxation time 422
Monodisperse suspensions 208, 211-215
Morphology 60, 66
Mount 60, 67, 84
Multi-component ER fluid 441
Multipolar interactions 155-159

Negative stress 237-238, 242-243
Nelder-Mead routine 429
No-slip boundary condition 253-254
Newtonian fluid 76
Noise 524
Nonlinear conductance 427
Nonlinear deformation 216-218
Nonlinear optical properties 470

On-off excitation switch 597
One-sided Fourier transform 424
Onsager effect 144
Optimal microscopy 8
Order parameters 133,135-136
Oscillating plate rheometer 330
Oscillatory Mason number 234

Particle size 623
Particle-wall dry condition 254
Particulate 8, 41, 60, 64, 66
Peierls-Landau fluctuation 37, 315, 324
Periodic deformation 440, 441
Permittivity mismatch parameter 308
Phase separation model 224-226
  at equilibrium 164-167
  Coulomb energy 225-226
  Correlation length 230
  Critical fields 227-231
  High and low density phases 223-224
Phase Transition 38
Piezoelectric hydrophone 523
Platen stress 300
Poiseuille-flow 477, 527, 629
Polarization 3,10,27,39,72
Poly-dispersion effect 215, 309
Polyrethane 67, 71
Power consumption 46, 60, 84
Pressure coupling 412, 427
Pressure generation 610
Pulse rectangular 3, 5, 18

Quiescent ER fluids 5

Rectangular arrays 403, 410
Relative permittivity 298, 299, 329
Restoring force 159-160
Reynolds number 284, 545, 661
Reciprocating mechanism 494

Relaxation 211-215
Relaxation frequency 3, 10
Rheometer 5, 17
Rotating shaft seal 526
Rotational viscometer 344, 345, 352
Rough electrodes 253-254

Scaling behavior of ferrofluids 198-199
Scanning electron microscope 65
Sediment 68
Sedimentation 68, 72
Semi-active suspension system 569
Semiconductor polymer based ER Fluids 234-236
Shear flow 100
Shear modulus 317
Shear rate 18, 24, 73, 86, 527, 538, 611
Shear resistance 294
Shear strain 45, 100, 101, 108
Shear stress 23, 24, 43, 45, 61, 64, 68, 74, 86, 511, 527, 611, 626
Sheared lattices 161-163
Sheared particle arrays 400
Shock absorber 84, 569, 587, 592, 649, 650
Silicon oil 4, 68, 85, 100
Solid fraction concentration 372
Spanned strand model 234
Static particle array 399
Sticking condition 205
Strain stimulus 103, 104
Stress 102, 103, 104
    Induced 4, 43, 511
    Yield 5, 10, 43, 545, 611, 626, 647
Structure 5, 10
Structure deformation 328
Super-paramagnetic behavior of ferrofluids 174, 192
Surface tension in ferrofluids 166
Surfactant 69, 84

Temporal evolution model 129-133
Tension-like internal force 286
Thermal fluctuations 121-122
Thermal motion 130-131
Thermal stability 591
Thermoset polymer 519
Thixotropy 337
Torque 18, 23, 43, 84, 498
Transient pressure responses 485, 486, 487

Valve 38, 76, 477, 538, 647
Valve design 538
Vibration isolation 573
Viscosity 4, 6, 14, 26, 37, 40, 43, 101, 512, 528, 591, 611, 650, 662
Viscoelastic medium 24
Viscoelasticity 105, 111, 267, 330, 340
Viscometer 23, 76
Volume fraction 3, 50, 68, 73, 522

Water free ER fluids 82
Worm-like structure 178,179

Yield stress 43, 60, 62, 85, 109, 252, 253, 254, 545, 611, 626, 647